STUDENT SOLUTIONS MANUAL

Robert A. Adams
University of British Columbia

Calculus: A Complete Course

Fifth Edition

Robert A. Adams

Toronto

0-201-79803-4

Acquisitions Editor: Leslie Carson
Developmental Editor: Pamela Voves
Production Editor: Gillian Scobie
Production Coordinator: Andrea Falkenberg

2 3 4 5   07 06 05 04 03

Printed and bound in Canada.

FOREWORD

These solutions are provided for the benefit of students using the textbook *Calculus: A Complete Course* *(Fifth Edition)* by R. A. Adams, published by Pearson Education Canada. For the most part, the solutions are detailed, especially in exercises on core material and techniques. Occasionally some details are omitted — for example, in exercises on applications of integration, the evaluation of the integrals encountered is not always given with the same degree of detail as the evaluation of integrals found in those exercises dealing specifically with techniques of integration. With a few exceptions, only the even-numbered exercises in the text are solved in this Manual.

As a student of Calculus, you should use this Manual with caution. It is always more beneficial to attempt exercises and problems on your own, before you look at solutions prepared by others. If you use these solutions as "study material" prior to attempting the exercises, you can lose much of the benefit that can follow from diligent attempts to develop your own analytical powers. When you have tried unsuccessfully to solve a problem, then look at these solutions to try to get a "hint" for a second attempt.

The author is grateful to Joanna Kwan and Valerie Adams who assisted with the preparation and typesetting the solutions for the second edition, to Winnie Poon who helped with some of the solutions for the third edition, and also to Ken MacKenzie of McGill University who did a very thorough checking of the solutions, thereby exposing for correction several errors that would otherwise have made it into this printed edition. Of course, the author accepts full responsibility for any errors that remain and will be grateful to readers who call them to his attention.

April, 2002.

R. A. Adams
Department of Mathematics
The University of British Columbia
Vancouver, B.C., Canada. V6T 1Z2
adms@math.ubc.ca

CONTENTS

CHAPTER P. PRELIMINARIES

Section P.1 Real Numbers and the Real Line (page 11)

2. $\dfrac{1}{11} = 0.09090909\cdots = 0.\overline{09}$

4. If $x = 3.277777\cdots$, then $10x - 32 = 0.77777\cdots$ and $100x - 320 = 7 + (10x - 32)$, or $90x = 295$. Thus $x = 295/90 = 59/18$.

6. Two different decimal expansions can represent the same number. For instance, both $0.999999\cdots = 0.\overline{9}$ and $1.000000\cdots = 1.\overline{0}$ represent the number 1.

8. $x < 2$ and $x \geq -3$ define the interval $[-3, 2[$.

10. $x \leq -1$ defines the interval $]-\infty, -1]$.

12. $x < 4$ or $x \geq 2$ defines the interval $]-\infty, \infty[$, that is, the whole real line.

14. If $3x + 5 \leq 8$, then $3x \leq 8 - 5 - 3$ and $x \leq 1$. Solution: $]-\infty, 1]$

16. If $\dfrac{6-x}{4} \geq \dfrac{3x-4}{2}$, then $6 - x \geq 6x - 8$. Thus $14 \geq 7x$ and $x \leq 2$. Solution: $]-\infty, 2]$

18. If $x^2 < 9$, then $|x| < 3$ and $-3 < x < 3$. Solution: $]-3, 3[$

20. Given: $(x+1)/x \geq 2$.
CASE I. If $x > 0$, then $x + 1 \geq 2x$, so $x \leq 1$.
CASE II. If $x < 0$, then $x + 1 \leq 2x$, so $x \geq 1$. (not possible)
Solution: $]0, 1]$.

22. Given $6x^2 - 5x \leq -1$, then $(2x-1)(3x-1) \leq 0$, so either $x \leq 1/2$ and $x \geq 1/3$, or $x \leq 1/3$ and $x \geq 1/2$. The latter combination is not possible. The solution set is $[1/3, 1/2]$.

24. Given $x^2 - x \leq 2$, then $x^2 - x - 2 \leq 0$ so $(x-2)(x+1) \leq 0$. This is possible if $x \leq 2$ and $x \geq -1$ or if $x \geq 2$ and $x \leq -1$. The latter situation is not possible. The solution set is $[-1, 2]$.

26. Given: $\dfrac{3}{x-1} < \dfrac{2}{x+1}$.
CASE I. If $x > 1$ then $(x-1)(x+1) > 0$, so that $3(x+1) < 2(x-1)$. Thus $x < -5$. There are no solutions in this case.
CASE II. If $-1 < x < 1$, then $(x-1)(x+1) < 0$, so $3(x+1) > 2(x-1)$. Thus $x > -5$. In this case all numbers in $]-1, 1[$ are solutions.
CASE III. If $x < -1$, then $(x-1)(x+1) > 0$, so that $3(x+1) < 2(x-1)$. Thus $x < -5$. All numbers $x < -5$ are solutions.
Solutions: $]-\infty, -5[\cup]-1, 1[$.

28. If $|x - 3| = 7$, then $x - 3 = \pm 7$, so $x = -4$ or $x = 10$.

30. If $|1 - t| = 1$, then $1 - t = \pm 1$, so $t = 0$ or $t = 2$.

32. If $\left|\dfrac{s}{2} - 1\right| = 1$, then $\dfrac{s}{2} - 1 = \pm 1$, so $s = 0$ or $s = 4$.

34. If $|x| \leq 2$, then x is in $[-2, 2]$.

36. If $|t + 2| < 1$, then $-2 - 1 < t < -2 + 1$, so t is in $]-3, -1[$.

38. If $|2x + 5| < 1$, then $-5 - 1 < 2x < -5 + 1$, so x is in $]-3, -2[$.

40. If $\left|2 - \dfrac{x}{2}\right| < \dfrac{1}{2}$, then $x/2$ lies between $2 - (1/2)$ and $2 + (1/2)$. Thus x is in $]3, 5[$.

42. $|x - 3| < 2|x| \Leftrightarrow x^2 - 6x + 9 = (x-3)^2 < 4x^2$ $\Leftrightarrow 3x^2 + 6x - 9 > 0 \Leftrightarrow 3(x+3)(x-1) > 0$. This inequality holds if $x < -3$ or $x > 1$.

44. The equation $|x - 1| = 1 - x$ holds if $|x - 1| = -(x-1)$, that is, if $x - 1 < 0$, or, equivalently, if $x < 1$.

Section P.2 Cartesian Coordinates in the Plane (page 18)

2. From $A(-1, 2)$ to $B(4, -10)$, $\Delta x = 4 - (-1) = 5$ and $\Delta y = -10 - 2 = -12$. $|AB| = \sqrt{5^2 + (-12)^2} = 13$.

4. From $A(0.5, 3)$ to $B(2, 3)$, $\Delta x = 2 - 0.5 = 1.5$ and $\Delta y = 3 - 3 = 0$. $|AB| = 1.5$.

6. Arrival point: $(-2, -2)$. Increments $\Delta x = -5$, $\Delta y = 1$. Starting point was $(-2 - (-5), -2 - 1)$, that is, $(3, -3)$.

8. $x^2 + y^2 = 2$ represents a circle of radius $\sqrt{2}$ centred at the origin.

10. $x^2 + y^2 = 0$ represents the origin.

12. $y < x^2$ represents all points lying below the parabola $y = x^2$.

14. The vertical line through $(\sqrt{2}, -1.3)$ is $x = \sqrt{2}$; the horizontal line through that point is $y = -1.3$.

16. Line through $(-2, 2)$ with slope $m = 1/2$ is $y = 2 + (1/2)(x + 2)$, or $x - 2y = -6$.

18. Line through $(a, 0)$ with slope $m = -2$ is $y = 0 - 2(x - a)$, or $y = 2a - 2x$.

20. At $x = 3$, the height of the line $x - 4y = 7$ is $y = (3 - 7)/4 = -1$. Thus $(3, -1)$ lies on the line.

22. The line through $(-2, 1)$ and $(2, -2)$ has slope $m = (-2 - 1)/(2 + 2) = -3/4$ and equation $y = 1 - (3/4)(x + 2)$ or $3x + 4y = -2$.

24. The line through $(-2, 0)$ and $(0, 2)$ has slope $m = (2 - 0)/(0 + 2) = 1$ and equation $y = 2 + x$.

26. If $m = -1/2$ and $b = -3$, then the line has equation $y = -(1/2)x - 3$, or $x + 2y = -6$.

28. $x + 2y = -4$ has x-intercept $a = -4$ and y-intercept $b = -4/2 = -2$. Its slope is $-b/a = 2/(-4) = -1/2$.

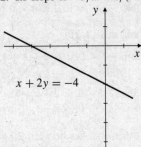

Fig. P.2.28

30. $1.5x - 2y = -3$ has x-intercept $a = -3/1.5 = -2$ and y-intercept $b = -3/(-2) = 3/2$. Its slope is $-b/a = 3/4$.

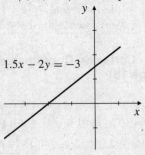

Fig. P.2.30

32. line through $(-2, 2)$ parallel to $2x + y = 4$ is $2x + y = -2$; line perpendicular to $2x + y = 4$ is $x - 2y = -6$.

34. We have

$$2x + y = 8 \implies 14x + 7y = 56$$
$$5x - 7y = 1 \qquad\quad 5x - 7y = 1.$$

Adding these equations gives $19x = 57$, so $x = 3$ and $y = 8 - 2x = 2$. The intersection point is $(3, 2)$.

36. The line $(x/2) - (y/3) = 1$ has x-intercept $a = 2$, and y-intercept $b = -3$.

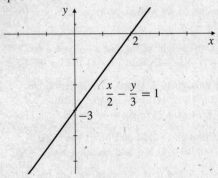

Fig. P.2.36

38. The line through $(-2, 5)$ and $(k, 1)$ has x-intercept 3, so also passes through $(3, 0)$. Its slope m satisfies

$$\frac{1 - 0}{k - 3} = m = \frac{0 - 5}{3 + 2} = -1.$$

Thus $k - 3 = -1$, and so $k = 2$.

40. $-40°$ and $-40°$ is the same temperature on both the Fahrenheit and Celsius scales.

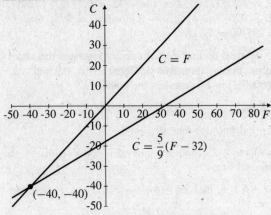

Fig. P.2.40

42. $A = (0, 0)$, $B = (1, \sqrt{3})$, $C = (2, 0)$

$$|AB| = \sqrt{(1 - 0)^2 + (\sqrt{3} - 0)^2} = \sqrt{4} = 2$$
$$|AC| = \sqrt{(2 - 0)^2 + (0 - 0)^2} = \sqrt{4} = 2$$
$$|BC| = \sqrt{(2 - 1)^2 + (0 - \sqrt{3})^2} = \sqrt{4} = 2.$$

Since $|AB| = |AC| = |BC|$, triangle ABC is equilateral.

44. If $M = (x_m, y_m)$ is the midpoint of $P_1 P_2$, then the displacement of M from P_1 equals the displacement of P_2 from M:

$$x_m - x_1 = x_2 - x_m, \quad y_m - y_1 = y_2 - y_m.$$

Thus $x_m = (x_1 + x_2)/2$ and $y_m = (y_1 + y_2)/2$.

46. Let the coordinates of P be $(x, 0)$ and those of Q be $(X, -2X)$. If the midpoint of PQ is $(2, 1)$, then

$$(x + X)/2 = 2, \quad (0 - 2X)/2 = 1.$$

The second equation implies that $X = -1$, and the second then implies that $x = 5$. Thus P is $(5, 0)$.

48. $\sqrt{(x - 2)^2 + y^2} = \sqrt{x^2 + (y - 2)^2}$ says that (x, y) is equidistant from $(2, 0)$ and $(0, 2)$. Thus (x, y) must lie on the line that is the right bisector of the line from $(2, 0)$ to $(0, 2)$. A simpler equation for this line is $x = y$.

50. For any value of k, the coordinates of the point of intersection of $x + 2y = 3$ and $2x - 3y = -1$ will also satisfy the equation

$$(x + 2y - 3) + k(2x - 3y + 1) = 0$$

because they cause both expressions in parentheses to be 0. The equation above is linear in x and y, and so represents a straight line for any choice of k. This line will pass through $(1, 2)$ provided $1 + 4 - 3 + k(2 - 6 + 1) = 0$, that is, if $k = 2/3$. Therefore, the line through the point of intersection of the two given lines and through the point $(1, 2)$ has equation

$$x + 2y - 3 + \frac{2}{3}(2x - 3y + 1) = 0,$$

or, on simplification, $x = 1$.

Section P.3   Graphs of Quadratic Equations (page 25)

2. $x^2 + (y - 2)^2 = 4$, or $x^2 + y^2 - 4y = 0$

4. $(x - 3)^2 + (y + 4)^2 = 25$, or $x^2 + y^2 - 6x + 8y = 0$.

6. $x^2 + y^2 + 4y = 0$
$x^2 + y^2 + 4y + 4 = 4$
$x^2 + (y + 2)^2 = 4$
centre: $(0, -2)$; radius 2.

8. $x^2 + y^2 - 2x - y + 1 = 0$
$x^2 - 2x + 1 + y^2 - y + \frac{1}{4} = \frac{1}{4}$
$(x - 1)^2 + \left(y - \frac{1}{2}\right)^2 = \frac{1}{4}$
centre: $(1, 1/2)$; radius $1/2$.

10. $x^2 + y^2 < 4$ represents the open disk consisting of all points lying inside the circle of radius 2 centred at the origin.

12. $x^2 + (y - 2)^2 \leq 4$ represents the closed disk consisting of all points lying inside or on the circle of radius 2 centred at the point $(0, 2)$.

14. Together, $x^2 + y^2 \leq 4$ and $(x + 2)^2 + y^2 \leq 4$ represent the region consisting of all points that are inside or on both the circle of radius 2 centred at the origin and the circle of radius 2 centred at $(-2, 0)$.

16. $x^2 + y^2 - 4x + 2y > 4$ can be rewritten $(x - 2)^2 + (y + 1)^2 > 9$. This equation, taken together with $x + y > 1$, represents all points that lie both outside the circle of radius 3 centred at $(2, -1)$ and above the line $x + y = 1$.

18. The exterior of the circle with centre $(2, -3)$ and radius 4 is given by $(x - 2)^2 + (y + 3)^2 > 16$, or $x^2 + y^2 - 4x + 6y > 3$.

20. $x^2 + y^2 > 4$, $(x - 1)^2 + (y - 3)^2 < 10$

22. The parabola with focus $(0, -1/2)$ and directrix $y = 1/2$ has equation $x^2 = -2y$.

24. The parabola with focus $(-1, 0)$ and directrix $x = 1$ has equation $y^2 = -4x$.

26. $y = -x^2$ has focus $(0, -1/4)$ and directrix $y = 1/4$.

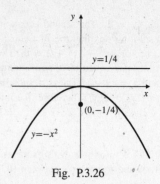

Fig. P.3.26

28. $x = y^2/16$ has focus $(4, 0)$ and directrix $x = -4$.

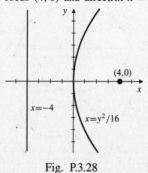

Fig. P.3.28

30. a) If $y = mx$ is shifted to the right by amount x_1, the equation $y = m(x - x_1)$ results. If (a, b) satisfies this equation, then $b = m(a - x_1)$, and so $x_1 = a - (b/m)$. Thus the shifted equation is
$y = m(x - a + (b/m)) = m(x - a) + b$.

b) If $y = mx$ is shifted vertically by amount y_1, the equation $y = mx + y_1$ results. If (a, b) satisfies this equation, then $b = ma + y_1$, and so $y_1 = b - ma$. Thus the shifted equation is $y = mx + b - ma = m(x - a) + b$, the same equation obtained in part (a).

32. $4y = \sqrt{x + 1}$

34. $(y/2) = \sqrt{4x + 1}$

36. $x^2 + y^2 = 5$ shifted up 2, left 4 gives $(x + 4)^2 + (y - 2)^2 = 5$.

38. $y = \sqrt{x}$ shifted down 2, left 4 gives $y = \sqrt{x + 4} - 2$.

40. $y = x^2 - 6$, $y = 4x - x^2$. Subtracting these equations gives
$2x^2 - 4x - 6 = 0$, or $2(x - 3)(x + 1) = 0$. Thus $x = 3$ or $x = -1$. The corresponding values of y are 3 and -5. The intersection points are $(3, 3)$ and $(-1, -5)$.

42. $2x^2 + 2y^2 = 5$, $xy = 1$. The second equation says that $y = 1/x$. Substituting this into the first equation gives $2x^2 + (2/x^2) = 5$, or $2x^4 - 5x^2 + 2 = 0$. This equation factors to $(2x^2 - 1)(x^2 - 2) = 0$, so its solutions are $x = \pm 1/\sqrt{2}$ and $x = \pm\sqrt{2}$. The corresponding values of y are given by $y = 1/x$. Therefore, the intersection points are $(1/\sqrt{2}, \sqrt{2})$, $(-1/\sqrt{2}, -\sqrt{2})$, $(\sqrt{2}, 1/\sqrt{2})$, and $(-\sqrt{2}, -1/\sqrt{2})$.

44. $9x^2 + 16y^2 = 144$ is an ellipse with major axis between $(-4, 0)$ and $(4, 0)$ and minor axis between $(0, -3)$ and $(0, 3)$.

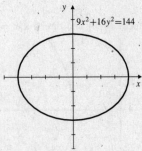

Fig. P.3.44

46. $(x - 1)^2 + \dfrac{(y + 1)^2}{4} = 4$ is an ellipse with centre at $(1, -1)$, major axis between $(1, -5)$ and $(1, 3)$ and minor axis between $(-1, -1)$ and $(3, -1)$.

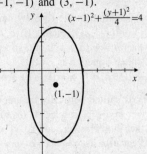

Fig. P.3.46

48. $x^2 - y^2 = -1$ is a rectangular hyperbola with centre at the origin and passing through $(0, \pm 1)$. Its asymptotes are $y = \pm x$.

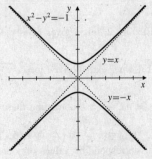

Fig. P.3.48

50. $(x - 1)(y + 2) = 1$ is a rectangular hyperbola with centre at $(1, -2)$ and passing through $(2, -1)$ and $(0, -3)$. Its asymptotes are $x = 1$ and $y = -2$.

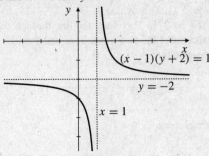

Fig. P.3.50

52. Replacing x with $-x$ and y with $-y$ reflects the graph in both axes. This is equivalent to rotating the graph $180°$ about the origin.

Section P.4 Functions and Their Graphs (page 35)

2. $f(x) = 1 - \sqrt{x}$; domain $[0, \infty)$, range $(-\infty, 1]$

4. $F(x) = 1/(x - 1)$; domain $(-\infty, 1) \cup (1, \infty)$, range $(-\infty, 0) \cup (0, \infty)$

6. $g(x) = \dfrac{1}{1 - \sqrt{x - 2}}$; domain $(2, 3) \cup (3, \infty)$, range $(-\infty, 0) \cup (0, \infty)$. The equation $y = g(x)$ can be solved for $x = 2 - (1 - (1/y))^2$ so has a real solution provided $y \neq 0$.

8.

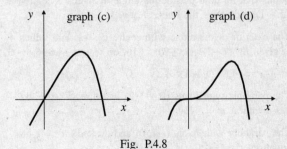
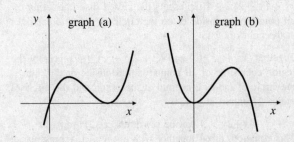

Fig. P.4.8

a) is the graph of $x(1 - x)^2$, which is positive for $x > 0$.

b) is the graph of $x^2 - x^3 = x^2(1 - x)$, which is positive if $x < 1$.

c) is the graph of $x - x^4$, which is positive if $0 < x < 1$ and behaves like x near 0.

d) is the graph of $x^3 - x^4$, which is positive if $0 < x < 1$ and behaves like x^3 near 0.

10.

x	$f(x) = x^{2/3}$
0	0
± 0.5	0.62996
± 1	1
± 1.5	1.3104
± 2	1.5874

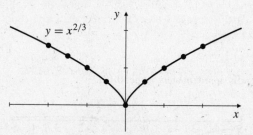

Fig. P.4.10

12. $f(x) = x^3 + x$ is odd: $f(-x) = -f(x)$

14. $f(x) = \dfrac{1}{x^2 - 1}$ is even: $f(-x) = f(x)$

16. $f(x) = \dfrac{1}{x + 4}$ is odd about $(-4, 0)$:
$f(-4 - x) = -f(-4 + x)$

18. $f(x) = x^3 - 2$ is odd about $(0, -2)$:
$f(-x) + 2 = -(f(x) + 2)$

20. $f(x) = |x + 1|$ is even about $x = -1$:
$f(-1 - x) = f(-1 + x)$

22. $f(x) = \sqrt{(x - 1)^2}$ is even about $x = 1$:
$f(1 - x) = f(1 + x)$

24.

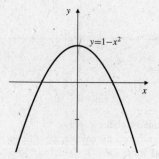

26.

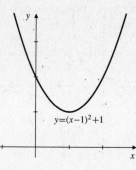

28.

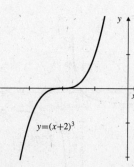

30.

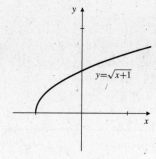

32.

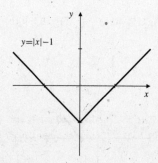

34.

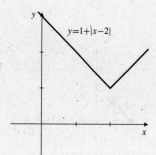

36.

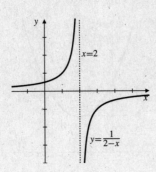

38.

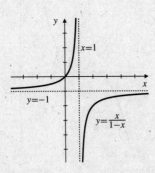

40.

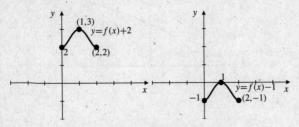

Fig. P.4.40(a)   Fig. P.4.40(b)

42.

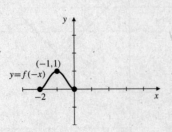

44.

46.

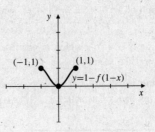

48. Range is approximately $(-\infty, 0.17]$.

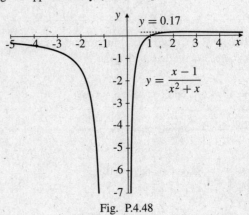

Fig. P.4.48

50.

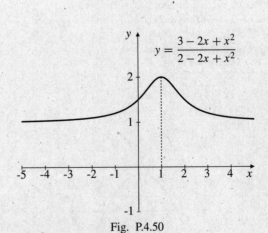

Fig. P.4.50

Apparent symmetry about $x = 1$.
This can be confirmed by calculating $f(2-x)$, which turns out to be equal to $f(x)$.

52.

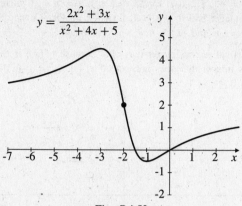

$$y = \frac{2x^2 + 3x}{x^2 + 4x + 5}$$

Fig. P.4.52

Apparent symmetry about $(-2, 2)$.
This can be confirmed by calculating shifting the graph right by 2 (replace x with $x-2$) and then down 2 (subtract 2). The result is $-5x/(1 + x^2)$, which is odd.

Section P.5 Combining Functions to Make New Functions (page 41)

2. $f(x) = \sqrt{1 - x}$, $g(x) = \sqrt{1 + x}$.
$Df =] -\infty, 1]$, $D(g) = [-1, \infty[$.
$D(f + g) = D(f - g) = D(fg) = [-1, 1]$,
$D(f/g) =] -1, 1]$, $D(g/f) = [-1, 1[$.
$(f + g)(x) = \sqrt{1 - x} + \sqrt{1 + x}$
$(f - g)(x) = \sqrt{1 - x} - \sqrt{1 + x}$
$(fg)(x) = \sqrt{1 - x^2}$
$(f/g)(x) = \sqrt{(1 - x)/(1 + x)}$
$(g/f)(x) = \sqrt{(1 + x)/(1 - x)}$

4.

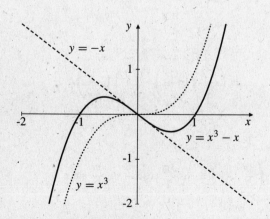

$y = -x$

$y = x^3 - x$

$y = x^3$

6.

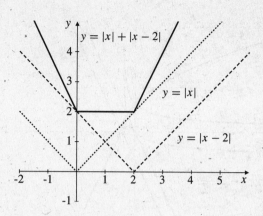

$y = |x| + |x - 2|$

$y = |x|$

$y = |x - 2|$

8. $f(x) = 2/x$, $g(x) = x/(1 - x)$.
$f \circ f(x) = 2/(2/x) = x$; $D(f \circ f) = \{x : x \neq 0\}$
$f \circ g(x) = 2/(x/(1 - x)) = 2(1 - x)/x$;
   $D(f \circ g) = \{x : x \neq 0, 1\}$
$g \circ f(x) = (2/x)/(1 - (2/x)) = 2/(x - 2)$;
   $D(g \circ f) = \{x : x \neq 0, 2\}$
$g \circ g(x) = (x/(1 - x))/(1 - (x/(1 - x))) = x/(1 - 2x)$;
   $D(g \circ g) = \{x : x \neq 1/2, 1\}$

10. $f(x) = (x + 1)/(x - 1) = 1 + 2/(x - 1)$, $g(x) = \text{sgn}(x)$.
$f \circ f(x) = 1 + 2/(1 + (2/(x - 1) - 1)) = x$;
$D(f \circ f) = \{x : x \neq 1\}$
$f \circ g(x) = \dfrac{\text{sgn}\, x + 1}{\text{sgn}\, x - 1} = 0$; $D(f \circ g) =] -\infty, 0[$
$g \circ f(x) = \text{sgn}\left(\dfrac{x + 1}{x - 1}\right) = \begin{cases} 1 & \text{if } x < -1 \text{ or } x > 1 \\ -1 & \text{if } -1 < x < 1 \end{cases}$;
   $D(g \circ f) = \{x : x \neq -1, \ 1\}$
$g \circ g(x) = \text{sgn}(\text{sgn}(x)) = \text{sgn}(x)$; $D(g \circ g) = \{x : x \neq 0\}$

	$f(x)$	$g(x)$	$f \circ g(x)$
11.	x^2	$x + 1$	$(x + 1)^2$
12.	$x - 4$	$x + 4$	x
13.	$\sqrt{x}$	x^2	$|x|$
14.	$2x^3 + 3$	$x^{1/3}$	$2x + 3$
15.	$(x + 1)/x$	$1/(x - 1)$	x
16.	$1/(x + 1)^2$	$x - 1$	$1/x^2$

18.

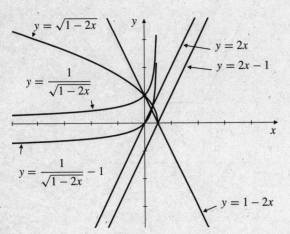

Fig. P.5.18

20.

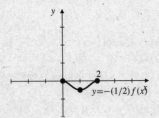

22.

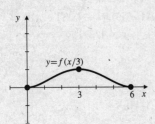

24.

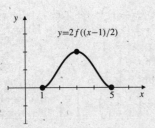

26.

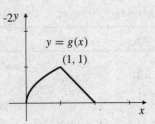

28. $\lfloor x \rfloor = 0$ for $0 \leq x < 1$; $\lceil x \rceil = 0$ for $-1 \leq x < 0$.

30. $\lceil -x \rceil = -\lfloor x \rfloor$ is true for all real x; if $x = n + y$ where n is an integer and $0 \leq y < 1$, then $-x = -n - y$, so that $\lceil -x \rceil = -n$ and $\lfloor x \rfloor = n$.

32. $f(x)$ is called the integer part of x because $|f(x)|$ is the largest integer that does not exceed x; i.e. $|x| = |f(x)| + y$, where $0 \leq y < 1$.

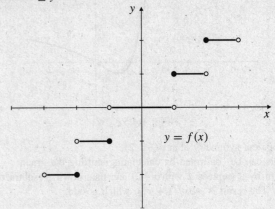

Fig. P.5.32

34. f even $\Leftrightarrow f(-x) = f(x)$
f odd $\Leftrightarrow f(-x) = -f(x)$
f even and odd $\Rightarrow f(x) = -f(x) \Rightarrow 2f(x) = 0$
$\Rightarrow f(x) = 0$

Section P.6   The Trigonometric Functions (page 55)

2. $\tan \dfrac{-3\pi}{4} = -\tan \dfrac{3\pi}{4} = -1$

4. $\sin\left(\dfrac{7\pi}{12}\right) = \sin\left(\dfrac{\pi}{4} + \dfrac{\pi}{3}\right)$

$= \sin\dfrac{\pi}{4}\cos\dfrac{\pi}{3} + \cos\dfrac{\pi}{4}\sin\dfrac{\pi}{3}$

$= \dfrac{1}{\sqrt{2}}\dfrac{1}{2} + \dfrac{1}{\sqrt{2}}\dfrac{\sqrt{3}}{2} = \dfrac{1+\sqrt{3}}{2\sqrt{2}}$

6. $\sin\dfrac{11\pi}{12} = \sin\dfrac{\pi}{12}$

$= \sin\left(\dfrac{\pi}{3} - \dfrac{\pi}{4}\right)$

$= \sin\dfrac{\pi}{3}\cos\dfrac{\pi}{4} - \cos\dfrac{\pi}{3}\sin\dfrac{\pi}{4}$

$= \left(\dfrac{\sqrt{3}}{2}\right)\left(\dfrac{1}{\sqrt{2}}\right) - \left(\dfrac{1}{2}\right)\left(\dfrac{1}{\sqrt{2}}\right)$

$= \dfrac{\sqrt{3}-1}{2\sqrt{2}}$

8. $\sin(2\pi - x) = -\sin x$

10. $\cos\left(\dfrac{3\pi}{2} + x\right) = \cos\dfrac{3\pi}{2}\cos x - \sin\dfrac{3\pi}{2}\sin x$

$= (-1)(-\sin x) = \sin x$

12. $\dfrac{\tan x - \cot x}{\tan x + \cot x} = \dfrac{\left(\dfrac{\sin x}{\cos x} - \dfrac{\cos x}{\sin x}\right)}{\left(\dfrac{\sin x}{\cos x} + \dfrac{\cos x}{\sin x}\right)}$

$\qquad = \dfrac{\left(\dfrac{\sin^2 x - \cos^2 x}{\cos x \sin x}\right)}{\left(\dfrac{\sin^2 x + \cos^2 x}{\cos x \sin x}\right)}$

$\qquad = \sin^2 x - \cos^2 x$

14. $(1 - \cos x)(1 + \cos x) = 1 - \cos^2 x = \sin^2 x$ implies $\dfrac{1 - \cos x}{\sin x} = \dfrac{\sin x}{1 + \cos x}$. Now

$\dfrac{1 - \cos x}{\sin x} = \dfrac{1 - \cos 2\left(\dfrac{x}{2}\right)}{\sin 2\left(\dfrac{x}{2}\right)}$

$\qquad = \dfrac{1 - \left(1 - 2\sin^2\left(\dfrac{x}{2}\right)\right)}{2 \sin \dfrac{x}{2} \cos \dfrac{x}{2}}$

$\qquad = \dfrac{\sin \dfrac{x}{2}}{\cos \dfrac{x}{2}} = \tan \dfrac{x}{2}$

16. $\dfrac{\cos x - \sin x}{\cos x + \sin x} = \dfrac{(\cos x - \sin x)^2}{(\cos x + \sin x)(\cos x - \sin x)}$

$\qquad = \dfrac{\cos^2 x - 2\sin x \cos x + \sin^2 x}{\cos^2 x - \sin^2 x}$

$\qquad = \dfrac{1 - \sin(2x)}{\cos(2x)}$

$\qquad = \sec(2x) - \tan(2x)$

18. $\cos 3x = \cos(2x + x)$

$\qquad = \cos 2x \cos x - \sin 2x \sin x$

$\qquad = (2\cos^2 x - 1)\cos x - 2\sin^2 x \cos x$

$\qquad = 2\cos^3 x - \cos x - 2(1 - \cos^2 x)\cos x$

$\qquad = 4\cos^3 x - 3\cos x$

20. $\sin \dfrac{x}{2}$ has period 4π.

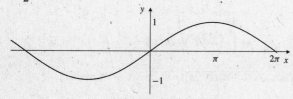

Fig. P.6.20

22. $\cos \dfrac{\pi x}{2}$ has period 4.

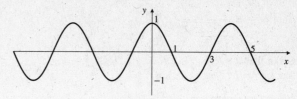

Fig. P.6.22

24.

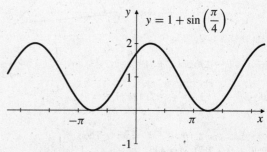

$y = 1 + \sin\left(\dfrac{\pi}{4}\right)$

26. $\tan x = 2$ where x is in $[0, \dfrac{\pi}{2}]$. Then
$\sec^2 x = 1 + \tan^2 x = 1 + 4 = 5$. Hence,
$\sec x = \sqrt{5}$ and $\cos x = \dfrac{1}{\sec x} = \dfrac{1}{\sqrt{5}}$,
$\sin x = \tan x \cos x = \dfrac{2}{\sqrt{5}}$.

28. $\cos x = -\dfrac{5}{13}$ where x is in $\left[\dfrac{\pi}{2}, \pi\right]$. Hence,
$\sin x = \sqrt{1 - \cos^2 x} = \sqrt{1 - \dfrac{25}{169}} = \dfrac{12}{13}$,
$\tan x = -\dfrac{12}{5}$.

30. $\tan x = \dfrac{1}{2}$ where x is in $[\pi, \dfrac{3\pi}{2}]$. Then,
$\sec^2 x = 1 + \dfrac{1}{4} = \dfrac{5}{4}$. Hence,
$\sec x = -\dfrac{\sqrt{5}}{2}, \quad \cos x = -\dfrac{2}{\sqrt{5}}$,
$\sin x = \tan x \cos x = -\dfrac{1}{\sqrt{5}}$.

32. $b = 2, \quad B = \dfrac{\pi}{3}$

$\dfrac{2}{a} = \tan B = \sqrt{3} \Rightarrow a = \dfrac{2}{\sqrt{3}}$

$\dfrac{2}{c} = \sin B = \dfrac{\sqrt{3}}{2} \Rightarrow c = \dfrac{4}{\sqrt{3}}$

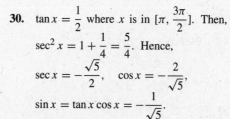

34. $\sin A = \dfrac{a}{c} \Rightarrow a = c \sin A$

36. $\cos B = \dfrac{a}{c} \Rightarrow a = c \cos B$

38. $\sin A = \dfrac{a}{c} \Rightarrow c = \dfrac{a}{\sin A}$

40. $\sin A = \dfrac{a}{c}$

42. $\sin A = \dfrac{a}{c} = \dfrac{a}{\sqrt{a^2 + b^2}}$

44. Given that $a = 2$, $b = 2$, $c = 3$.
Since $a^2 = b^2 + c^2 - 2bc \cos A$,

$$\cos A = \frac{a^2 - b^2 - c^2}{-2bc}$$

$$= \frac{4 - 4 - 9}{-2(2)(3)} = \frac{3}{4}.$$

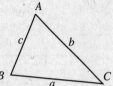

46. Given that $a = 2$, $b = 3$, $C = \dfrac{\pi}{4}$:

$$c^2 = a^2 + b^2 - 2ab \cos C = 4 + 9 - 2(2)(3) \cos \frac{\pi}{4} = 13 - \frac{12}{\sqrt{2}}.$$

Hence, $c = \sqrt{13 - \dfrac{12}{\sqrt{2}}} \approx 2.12479.$

48. Given that $a = 2$, $b = 3$, $C = 35°$. Then
$c^2 = 4 + 9 - 2(2)(3) \cos 35°$, hence $c \approx 1.78050$.

50. If $a = 1, b = \sqrt{2}, A = 30°$, then $\dfrac{\sin B}{b} = \dfrac{\sin A}{a} = \dfrac{1}{2}$.

Thus $\sin B = \dfrac{\sqrt{2}}{2} = \dfrac{1}{\sqrt{2}}$, $B = \dfrac{\pi}{4}$ or $\dfrac{3\pi}{4}$, and

$$C = \pi - \left(\frac{\pi}{4} + \frac{\pi}{6}\right) = \frac{7\pi}{12} \text{ or } C = \pi - \left(\frac{3\pi}{4} + \frac{\pi}{6}\right) = \frac{\pi}{12}.$$

Thus, $\cos C = \cos \dfrac{7\pi}{12} = \cos \left(\dfrac{\pi}{4} + \dfrac{\pi}{3}\right) = \dfrac{1 - \sqrt{3}}{2\sqrt{2}}$ or

$$\cos C = \cos \frac{\pi}{12} = \cos \left(\frac{\pi}{3} - \frac{\pi}{4}\right) = \frac{1 + \sqrt{3}}{2\sqrt{2}}.$$

Hence,

$$\begin{aligned}
c^2 &= a^2 + b^2 - 2ab \cos C \\
&= 1 + 2 - 2\sqrt{2} \cos C \\
&= 3 - (1 - \sqrt{3}) \text{ or } 3 - (1 + \sqrt{3}) \\
&= 2 + \sqrt{3} \text{ or } 2 - \sqrt{3}.
\end{aligned}$$

Hence, $c = \sqrt{2 + \sqrt{3}}$ or $\sqrt{2 - \sqrt{3}}$.

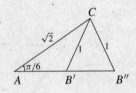

Fig. P.6.50

52. See the following diagram. Since $\tan 40° = h/a$, therefore $a = h/\tan 40°$. Similarly, $b = h/\tan 70°$.
Since $a + b = 2$ km, therefore,

$$\frac{h}{\tan 40°} + \frac{h}{\tan 70°} = 2$$

$$h = \frac{2(\tan 40° \tan 70°)}{\tan 70° + \tan 40°} \approx 1.286 \text{ km.}$$

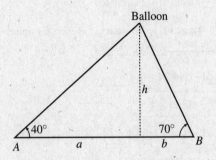

Fig. P.6.52

54. From Exercise 53, area $= \frac{1}{2} ac \sin B$. By Cosine Law,
$\cos B = \dfrac{a^2 + c^2 - b^2}{2ac}$. Thus,

$$\begin{aligned}
\sin B &= \sqrt{1 - \left(\frac{a^2 + c^2 - b^2}{2ac}\right)^2} \\
&= \frac{\sqrt{-a^4 - b^4 - c^4 + 2a^2b^2 + 2b^2c^2 + 2a^2c^2}}{2ac}.
\end{aligned}$$

Hence, Area $= \dfrac{\sqrt{-a^4 - b^4 - c^4 + 2a^2b^2 + 2b^2c^2 + 2a^2c^2}}{4}$
square units. Since,

$$\begin{aligned}
s(s &- a)(s - b)(s - c) \\
&= \frac{b + c + a}{2} \frac{b + c - a}{2} \frac{a - b + c}{2} \frac{a + b - c}{2} \\
&= \frac{1}{16}\left((b + c)^2 - a^2\right)\left(a^2 - (b - c)^2\right) \\
&= \frac{1}{16}\left(a^2\left((b + c)^2 + (b - c)^2\right) - a^4 - (b^2 - c^2)^2\right) \\
&= \frac{1}{16}\left(2a^2b^2 + 2a^2c^2 - a^4 - b^4 - c^4 + 2b^2c^2\right)
\end{aligned}$$

Thus $\sqrt{s(s - a)(s - b)(s - c)} =$ Area of triangle.

CHAPTER 1. LIMITS AND CONTINUITY

Section 1.1 Examples of Velocity, Growth Rate, and Area   (page 61)

2.

h	Avg. vel. over $[2, 2+h]$
1	5.0000
0.1	4.1000
0.01	4.0100
0.001	4.0010
0.0001	4.0001

4. Average volocity on $[2, 2+h]$ is

$$\frac{(2+h)^2 - 4}{(2+h) - 2} = \frac{4 + 4h + h^2 - 4}{h} = \frac{4h + h^2}{h} = 4 + h.$$

As h approaches 0 this average velocity approaches 4 m/s

6. Average velocity over $[t, t+h]$ is

$$\frac{3(t+h)^2 - 12(t+h) + 1 - (3t^2 - 12t + 1)}{(t+h) - t}$$

$$= \frac{6th + 3h^2 - 12h}{h} = 6t + 3h - 12 \text{ m/s}.$$

This average velocity approaches $6t - 12$ m/s as h approaches 0.
At $t = 1$ the velocity is $6 \times 1 - 12 = -6$ m/s.
At $t = 2$ the velocity is $6 \times 2 - 12 = 0$ m/s.
At $t = 3$ the velocity is $6 \times 3 - 12 = 6$ m/s.

8. Average velocity over $[t - k, t + k]$ is

$$\frac{3(t+k)^2 - 12(t+k) + 1 - [3(t-k)^2 - 12(t-k) + 1]}{(t+k) - (t-k)}$$

$$= \frac{1}{2k}\big(3t^2 + 6tk + 3k^2 - 12t - 12k + 1 - 3t^2 + 6tk - 3k^2$$

$$+ 12t - 12k + 1\big)$$

$$= \frac{12tk - 24k}{2k} = 6t - 12 \text{ m/s},$$

which is the velocity at time t from Exercise 7.

10. Average velocity over $[1, 1+h]$ is

$$\frac{2 + \dfrac{1}{\pi}\sin \pi(1+h) - \left(2 + \dfrac{1}{\pi}\sin \pi\right)}{h}$$

$$= \frac{\sin(\pi + \pi h)}{\pi h} = \frac{\sin \pi \cos(\pi h) + \cos \pi \sin(\pi h)}{\pi h}$$

$$= -\frac{\sin(\pi h)}{\pi h}.$$

h	Avg. vel. on $[1, 1+h]$
1.0000	0
0.1000	-0.983631643
0.0100	-0.999835515
0.0010	-0.999998355

12. We sketched a tangent line to the graph on page 55 in the text at $t = 20$. The line appeared to pass through the points $(10, 0)$ and $(50, 1)$. On day 20 the biomass is growing at about $(1 - 0)/(50 - 10) = 0.025$ mm^2/d.

14. a)

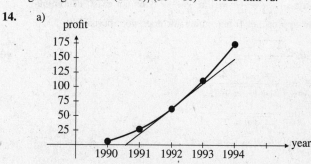

Fig. 1.1.14

b) Average rate of increase in profits between 1992 and 1994 is

$$\frac{174 - 62}{1994 - 1992} = \frac{112}{2} = 56 \text{ (thousand\$/yr)}.$$

c) Drawing a tangent line to the graph in (a) at $t = 1992$ and measuring its slope, we find that the rate of increase of profits in 1992 is about 43 thousand\$/year.

Section 1.2 Limits of Functions   (page 70)

2. From inspecting the graph

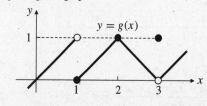

Fig. 1.2.2

we see that

$$\lim_{x \to 1} g(x) \text{ does not exist}$$

(left limit is 1, right limit is 0)

$$\lim_{x \to 2} g(x) = 1, \qquad \lim_{x \to 3} g(x) = 0.$$

4. $\lim_{x \to 1+} g(x) = 0$

6. $\lim_{x \to 3-} g(x) = 0$

8. $\displaystyle\lim_{x\to 2} 3(1-x)(2-x) = 3(-1)(2-2) = 0$

10. $\displaystyle\lim_{t\to -4}\frac{t^2}{4-t} = \frac{(-4)^2}{4+4} = 2$

12. $\displaystyle\lim_{x\to -1}\frac{x^2-1}{x+1} = \lim_{x\to -1}(x-1) = -2$

14. $\displaystyle\lim_{x\to -2}\frac{x^2+2x}{x^2-4} = \lim_{x\to -2}\frac{x}{x-2} = \frac{-2}{-4} = \frac{1}{2}$

16. $\displaystyle\lim_{h\to 0}\frac{3h+4h^2}{h^2-h^3} = \lim_{h\to 0}\frac{3+4h}{h-h^2}$ does not exist; denominator approaches 0 but numerator does not approach 0.

18. $\displaystyle\lim_{h\to 0}\frac{\sqrt{4+h}-2}{h}$
$\displaystyle = \lim_{h\to 0}\frac{4+h-4}{h(\sqrt{4+h}+2)}$
$\displaystyle = \lim_{h\to 0}\frac{1}{\sqrt{4+h}+2} = \frac{1}{4}$

20. $\displaystyle\lim_{x\to -2}|x-2| = |-4| = 4$

22. $\displaystyle\lim_{x\to 2}\frac{|x-2|}{x-2} = \lim_{x\to 2}\begin{cases} 1, & \text{if } x>2 \\ -1, & \text{if } x<2. \end{cases}$
Hence, $\displaystyle\lim_{x\to 2}\frac{|x-2|}{x-2}$ does not exist.

24. $\displaystyle\lim_{x\to 2}\frac{\sqrt{4-4x+x^2}}{x-2}$
$\displaystyle = \lim_{x\to 2}\frac{|x-2|}{x-2}$ does not exist.

26. $\displaystyle\lim_{x\to 1}\frac{x^2-1}{\sqrt{x+3}-2} = \lim_{x\to 1}\frac{(x-1)(x+1)(\sqrt{x+3}+2)}{(x+3)-4}$
$\displaystyle = \lim_{x\to 1}(x+1)(\sqrt{x+3}+2) = (2)(\sqrt{4}+2) = 8$

28. $\displaystyle\lim_{s\to 0}\frac{(s+1)^2-(s-1)^2}{s} = \lim_{s\to 0}\frac{4s}{s} = 4$

30. $\displaystyle\lim_{x\to -1}\frac{x^3+1}{x+1}$
$\displaystyle = \lim_{x\to -1}\frac{(x+1)(x^2-x+1)}{x+1} = 3$

32. $\displaystyle\lim_{x\to 8}\frac{x^{2/3}-4}{x^{1/3}-2}$
$\displaystyle = \lim_{x\to 8}\frac{(x^{1/3}-2)(x^{1/3}+2)}{(x^{1/3}-2)}$
$\displaystyle = \lim_{x\to 8}(x^{1/3}+2) = 4$

34. $\displaystyle\lim_{x\to 2}\left(\frac{1}{x-2}-\frac{1}{x^2-4}\right)$
$\displaystyle = \lim_{x\to 2}\frac{x+2-1}{(x-2)(x+2)}$
$\displaystyle = \lim_{x\to 2}\frac{x+1}{(x-2)(x+2)}$ does not exist.

36. $\displaystyle\lim_{x\to 0}\frac{|3x-1|-|3x+1|}{x}$
$\displaystyle = \lim_{x\to 0}\frac{(3x-1)^2-(3x+1)^2}{x\,(|3x-1|+|3x+1|)}$
$\displaystyle = \lim_{x\to 0}\frac{-12x}{x\,(|3x-1|+|3x+1|)} = \frac{-12}{1+1} = -6$

38. $f(x) = x^3$
$\displaystyle\lim_{h\to 0}\frac{f(x+h)-f(x)}{h} = \lim_{h\to 0}\frac{(x+h)^3-x^3}{h}$
$\displaystyle = \lim_{h\to 0}\frac{3x^2h+3xh^2+h^3}{h}$
$\displaystyle = \lim_{h\to 0}3x^2+3xh+h^2 = 3x^2$

40. $f(x) = 1/x^2$
$\displaystyle\lim_{h\to 0}\frac{f(x+h)-f(x)}{h} = \lim_{h\to 0}\frac{\dfrac{1}{(x+h)^2}-\dfrac{1}{x^2}}{h}$
$\displaystyle = \lim_{h\to 0}\frac{x^2-(x^2+2xh+h^2)}{h(x+h)^2x^2}$
$\displaystyle = \lim_{h\to 0}-\frac{2x+h}{(x+h)^2x^2} = -\frac{2x}{x^4} = -\frac{2}{x^3}$

42. $f(x) = 1/\sqrt{x}$
$\displaystyle\lim_{h\to 0}\frac{f(x+h)-f(x)}{h} = \lim_{h\to 0}\frac{\dfrac{1}{\sqrt{x+h}}-\dfrac{1}{\sqrt{x}}}{h}$
$\displaystyle = \lim_{h\to 0}\frac{\sqrt{x}-\sqrt{x+h}}{h\sqrt{x}\sqrt{x+h}}$
$\displaystyle = \lim_{h\to 0}\frac{x-(x+h)}{h\sqrt{x}\sqrt{x+h}(\sqrt{x}+\sqrt{x+h})}$
$\displaystyle = \lim_{h\to 0}\frac{-1}{\sqrt{x}\sqrt{x+h}(\sqrt{x}+\sqrt{x+h})}$
$\displaystyle = \frac{-1}{2x^{3/2}}$

44. $\displaystyle\lim_{x\to \pi/4}\cos x = \cos \pi/4 = 1/\sqrt{2}$

46. $\displaystyle\lim_{x\to 2\pi/3}\sin x = \sin 2\pi/3 = \sqrt{3}/2$

48.

x	$(1-\cos x)/x^2$
± 1.0	0.45969769
± 0.1	0.49958347
± 0.01	0.49999583
± 0.001	0.49999996
0.0001	0.50000000

It appears that $\displaystyle\lim_{x\to 0}\frac{1-\cos x}{x^2} = \frac{1}{2}$.

50. $\displaystyle\lim_{x\to 2+}\sqrt{2-x}$ does not exist.

52. $\displaystyle\lim_{x\to -2+}\sqrt{2-x} = 2$

54. $\lim\limits_{x\to 0-} \sqrt{x^3 - x} = 0$

56. $\lim\limits_{x\to 0+} \sqrt{x^2 - x^4} = 0$

58. $\lim\limits_{x\to a+} \dfrac{|x-a|}{x^2 - a^2} = \lim\limits_{x\to a+} \dfrac{x-a}{x^2 - a^2} = \dfrac{1}{2a}$

60. $\lim\limits_{x\to 2+} \dfrac{x^2 - 4}{|x+2|} = \dfrac{0}{4} = 0$

62. $\lim\limits_{x\to -1+} f(x) = \lim\limits_{x\to -1+} x^2 + 1 = 1 + 1 = 2$

64. $\lim\limits_{x\to 0-} f(x) = \lim\limits_{x\to 0-} x^2 + 1 = 1$

66. If $\lim x \to a\, f(x) = 4$ and $\lim\limits_{x\to a} g(x) = -2$, then

   a) $\lim\limits_{x\to a}\big(f(x) + g(x)\big) = 4 + (-2) = 2$

   b) $\lim\limits_{x\to a} f(x)\cdot g(x) = 4 \times (-2) = -8$

   c) $\lim\limits_{x\to a} 4g(x) = 4(-2) = -8$

   d) $\lim\limits_{x\to a} \dfrac{f(x)}{g(x)} = \dfrac{4}{-2} = -2$

68. If $\lim\limits_{x\to 0} \dfrac{f(x)}{x^2} = -2$ then

   $\lim_{x\to 0} f(x) = \lim_{x\to 0} x^2\,\dfrac{f(x)}{x^2} = 0 \times (-2) = 0$,

   and similarly, $\lim_{x\to 0} \dfrac{f(x)}{x} = \lim\limits_{x\to 0} x\,\dfrac{f(x)}{x^2} = 0 \times (-2) = 0.$

70.

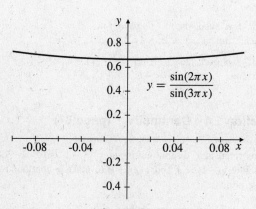

Fig. 1.2.70

$\lim_{x\to 0} \sin(2\pi x)/\sin(3\pi x) = 2/3$

72.

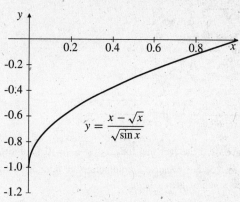

Fig. 1.2.72

$\lim\limits_{x\to 0+} \dfrac{x - \sqrt{x}}{\sqrt{\sin x}} = -1$

74. Since $\sqrt{5 - 2x^2} \le f(x) \le \sqrt{5 - x^2}$ for $-1 \le x \le 1$, and $\lim_{x\to 0} \sqrt{5 - 2x^2} = \lim_{x\to 0} \sqrt{5 - x^2} = \sqrt{5}$, we have $\lim_{x\to 0} f(x) = \sqrt{5}$ by the squeeze theorem.

76. a)

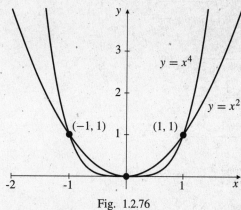

Fig. 1.2.76

   b) Since the graph of f lies between those of x^2 and x^4, and since these latter graphs come together at $(\pm 1, 1)$ and at $(0, 0)$, we have $\lim_{x\to \pm 1} f(x) = 1$ and $\lim_{x\to 0} f(x) = 0$ by the squeeze theorem.

78. $f(x) = s\sin\dfrac{1}{x}$ is defined for all $x \ne 0$; its domain is $(-\infty, 0) \cup (0, \infty)$. Since $|\sin t| \le 1$ for all t, we have $|f(x)| \le |x|$ and $-|x| \le f(x) \le |x|$ for all $x \ne 0$. Since $\lim_{x\to 0} (-|x|) = 0 = \lim_{x\to 0} |x|$, we have $\lim_{x\to 0} f(x) = 0$ by the squeeze theorem.

Section 1.3 Limits at Infinity and Infinite Limits (page 77)

2. $\lim\limits_{x\to \infty} \dfrac{x}{x^2 - 4} = \lim\limits_{x\to \infty} \dfrac{1/x}{1 - (4/x^2)} = \dfrac{0}{1} = 0$

4. $\lim\limits_{x \to -\infty} \dfrac{x^2 - 2}{x - x^2}$

$= \lim\limits_{x \to -\infty} \dfrac{1 - \dfrac{2}{x^2}}{\dfrac{1}{x} - 1} = \dfrac{1}{-1} = -1$

6. $\lim\limits_{x \to \infty} \dfrac{x^2 + \sin x}{x^2 + \cos x} = \lim\limits_{x \to \infty} \dfrac{1 + \dfrac{\sin x}{x^2}}{1 + \dfrac{\cos x}{x^2}} = \dfrac{1}{1} = 1$

We have used the fact that $\lim_{x \to \infty} \dfrac{\sin x}{x^2} = 0$ (and similarly for cosine) because the numerator is bounded while the denominator grows large.

8. $\lim\limits_{x \to \infty} \dfrac{2x - 1}{\sqrt{3x^2 + x + 1}}$

$= \lim\limits_{x \to \infty} \dfrac{x\left(2 - \dfrac{1}{x}\right)}{|x|\sqrt{3 + \dfrac{1}{x} + \dfrac{1}{x^2}}}$ (but $|x| = x$ as $x \to \infty$)

$= \lim\limits_{x \to \infty} \dfrac{2 - \dfrac{1}{x}}{\sqrt{3 + \dfrac{1}{x} + \dfrac{1}{x^2}}} = \dfrac{2}{\sqrt{3}}$

10. $\lim\limits_{x \to -\infty} \dfrac{2x - 5}{|3x + 2|} = \lim\limits_{x \to -\infty} \dfrac{2x - 5}{-(3x + 2)} = -\dfrac{2}{3}$

12. $\lim\limits_{x \to 3} \dfrac{1}{(3 - x)^2} = \infty$

14. $\lim\limits_{x \to 3+} \dfrac{1}{3 - x} = -\infty$

16. $\lim\limits_{x \to -2/5} \dfrac{2x + 5}{5x + 2}$ does not exist.

18. $\lim\limits_{x \to -2/5+} \dfrac{2x + 5}{5x + 2} = \infty$

20. $\lim\limits_{x \to 1-} \dfrac{x}{\sqrt{1 - x^2}} = \infty$

22. $\lim\limits_{x \to 1-} \dfrac{1}{|x - 1|} = \infty$

24. $\lim\limits_{x \to 1+} \dfrac{\sqrt{x^2 - x}}{x - x^2} = \lim\limits_{x \to 1+} \dfrac{-1}{\sqrt{x^2 - x}} = -\infty$

26. $\lim\limits_{x \to \infty} \dfrac{x^3 + 3}{x^2 + 2} = \lim\limits_{x \to \infty} \dfrac{x + \dfrac{3}{x^2}}{1 + \dfrac{2}{x^2}} = \infty$

28. $\lim\limits_{x \to \infty} \left(\dfrac{x^2}{x + 1} - \dfrac{x^2}{x - 1} \right) = \lim\limits_{x \to \infty} \dfrac{-2x^2}{x^2 - 1} = -2$

30. $\lim\limits_{x \to \infty} \left(\sqrt{x^2 + 2x} - \sqrt{x^2 - 2x} \right)$

$= \lim\limits_{x \to \infty} \dfrac{x^2 + 2x - x^2 + 2x}{\sqrt{x^2 + 2x} + \sqrt{x^2 - 2x}}$

$= \lim\limits_{x \to \infty} \dfrac{4x}{x\sqrt{1 + \dfrac{2}{x}} + x\sqrt{1 - \dfrac{2}{x}}}$

$= \lim\limits_{x \to \infty} \dfrac{4}{\sqrt{1 + \dfrac{2}{x}} + \sqrt{1 - \dfrac{2}{x}}} = \dfrac{4}{2} = 2$

32. $\lim\limits_{x \to -\infty} \dfrac{1}{\sqrt{x^2 + 2x} - x} = \lim\limits_{x \to -\infty} \dfrac{1}{|x|(\sqrt{1 + (2/x)} + 1)} = 0$

34. Since $\lim\limits_{x \to \infty} \dfrac{2x - 5}{|3x + 2|} = \dfrac{2}{3}$ and $\lim\limits_{x \to -\infty} \dfrac{2x - 5}{|3x + 2|} = -\dfrac{2}{3}$, $y = \pm(2/3)$ are horizontal asymptotes of $y = (2x - 5)/|3x + 2|$. The only vertical asymptote is $x = -2/3$, which makes the denominator zero.

36. $\lim\limits_{x \to 1} f(x) = \infty$

38. $\lim\limits_{x \to 2-} f(x) = 2$

40. $\lim\limits_{x \to 3+} f(x) = \infty$

42. $\lim\limits_{x \to 4-} f(x) = 0$

44. $\lim\limits_{x \to 5+} f(x) = 0$

46. horizontal: $y = 1$; vertical: $x = 1$, $x = 3$.

48. $\lim\limits_{x \to 3-} \lfloor x \rfloor = 2$

50. $\lim\limits_{x \to 2.5} \lfloor x \rfloor = 2$

52. $\lim\limits_{x \to -3-} \lfloor x \rfloor = -4$

54. $\lim\limits_{x \to 0+} f(x) = L$
    (a) If f is even, then $f(-x) = f(x)$.
    Hence, $\lim\limits_{x \to 0-} f(x) = L$.
    (b) If f is odd, then $f(-x) = -f(x)$.
    Therefore, $\lim\limits_{x \to 0-} f(x) = -L$.

Section 1.4 Continuity (page 87)

2. g has removable discontinuities at $x = -1$ and $x = 2$. Redefine $g(-1) = 1$ and $g(2) = 0$ to make g continuous at those points.

4. Function f is discontinuous at $x = 1$, 2, 3, 4, and 5. f is left continuous at $x = 4$ and right continuous at $x = 2$ and $x = 5$.

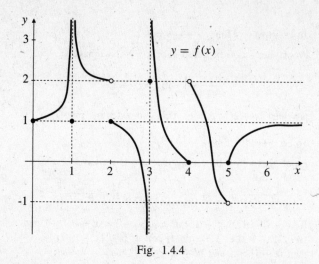

Fig. 1.4.4

6. sgn x is not defined at $x = 0$, so cannot be either continuous or discontinuous there. (Functions can be continuous or discontinuous only at points in their domains!)

8. $f(x) = \begin{cases} x & \text{if } x < -1 \\ x^2 & \text{if } x \geq -1 \end{cases}$ is continuous everywhere on the real line except at $x = -1$ where it is right continuous, but not left continuous.

$$\lim_{x \to -1-} f(x) = \lim_{x \to -1-} x = -1 \neq 1$$
$$= f(-1) = \lim_{x \to -1+} x^2 = \lim_{x \to -1+} f(x).$$

10. $f(x) = \begin{cases} x^2 & \text{if } x \leq 1 \\ 0.987 & \text{if } x > 1 \end{cases}$ is continuous everywhere except at $x = 1$, where it is left continuous but not right continuous because $0.987 \neq 1$. Close, as they say, but no cigar.

12. $C(t)$ is discontinuous only at the integers. It is continuous on the left at the integers, but not on the right.

14. Since $\dfrac{1+t^3}{1-t^2} = \dfrac{(1+t)(1-t+t^2)}{(1+t)(1-t)} = \dfrac{1-t+t^2}{1-t}$ for $t \neq -1$, we can define the function to be $3/2$ at $t = -1$ to make it continuous there. The continuous extension is $\dfrac{1-t+t^2}{1-t}$.

16. Since
$$\frac{x^2 - 2}{x^4 - 4} = \frac{(x - \sqrt{2})(x + \sqrt{2})}{(x - \sqrt{2})(x + \sqrt{2})(x^2 + 2)} = \frac{x + \sqrt{2}}{(x + \sqrt{2})(x^2 + 2)}$$
for $x \neq \sqrt{2}$, we can define the function to be $1/4$ at $x = \sqrt{2}$ to make it continuous there. The continuous extension is $\dfrac{x + \sqrt{2}}{(x + \sqrt{2})(x^2 + 2)}$. (Note: cancelling the $x + \sqrt{2}$ factors provides a further continuous extension to $x = -\sqrt{2}$.

18. $\lim_{x \to 3-} g(x) = 3 - m$ and $\lim_{x \to 3+} g(x) = 1 - 3m = g(3)$. Thus g will be continuous at $x = 3$ if $3 - m = 1 - 3m$, that is, if $m = -1$.

20. The Max-Min Theorem says that a continuous function defined on a closed, finite interval must have maximum and minimum values. It does not say that other functions cannot have such values. The Heaviside function is not continuous on $[-1, 1]$ (because it is discontinuous at $x = 0$), but it still has maximum and minimum values. Do not confuse a theorem with its converse.

22. Let the numbers be x and y, where $x \geq 0$, $y \geq 0$, and $x + y = 8$. If S is the sum of their squares then

$$S = x^2 + y^2 = x^2 + (8 - x)^2$$
$$= 2x^2 - 16x + 64 = 2(x - 4)^2 + 32.$$

Since $0 \leq x \leq 8$, the maximum value of S occurs at $x = 0$ or $x = 8$, and is 64. The minimum value occurs at $x = 4$ and is 32.

24. If x desks are shipped, the shipping cost per desk is

$$C = \frac{245x - 30x^2 + x^3}{x} = x^2 - 30x + 245$$
$$= (x - 15)^2 + 20.$$

This cost is minimized if $x = 15$. The manufacturer should send 15 desks in each shipment, and the shipping cost will then be \$20 per desk.

26. $f(x) = x^2 + 4x + 3 = (x + 1)(x + 3)$
$f(x) > 0$ on $]-\infty, -3[$ and $]-1, \infty[$
$f(x) < 0$ on $]-3, -1[$.

28. $f(x) = \dfrac{x^2 - 1}{x^2 - 4} = \dfrac{(x - 1)(x + 1)}{(x - 2)(x + 2)}$
$f = 0$ at $x = \pm 1$.
f is not defined at $x = \pm 2$.
$f(x) > 0$ on $]-\infty, -2[$, $]-1, 1[$, and $]2, \infty[$.
$f(x) < 0$ on $]-2, -1[$ and $]1, 2[$.

30. $f(x) = x^3 + x - 1$, $f(0) = -1$, $f(1) = 1$.
Since f is continuous and changes sign between 0 and 1, it must be zero at some point between 0 and 1 by IVT.

32. $F(x) = (x - a)^2(x - b)^2 + x$. Without loss of generality, we can assume that $a < b$. Being a polynomial, F is continuous on $[a, b]$. Also $F(a) = a$ and $F(b) = b$. Since $a < \frac{1}{2}(a + b) < b$, the Intermediate-Value Theorem guarantees that there is an x in (a, b) such that $F(x) = (a + b)/2$.

34. The domain of an even function is symmetric about the y-axis. Since f is continuous on the right at $x = 0$, therefore it must be defined on an interval $[0, h]$ for some $h > 0$. Being even, f must therefore be defined on $[-h, h]$. If $x = -y$, then

$$\lim_{x \to 0-} f(x) = \lim_{y \to 0+} f(-y) = \lim_{y \to 0+} f(y) = f(0).$$

Thus, f is continuous on the left at $x = 0$. Being continuous on both sides, it is therefore continuous.

36. max 0.133 at $x = 1.437$; min -0.232 at $x = -1.805$

38. max 1.510 at $x = 0.465$; min 0 at $x = 0$ and $x = 1$

40. root $x = 0.739$

42. roots $x = -0.7244919590$ and $x = 1.220744085$

Section 1.5 The Formal Definition of Limit (page 94)

2. Since 1.2% of 8,000 is 96, we require the edge length x of the cube to satisfy $7904 \le x^3 \le 8096$. It is sufficient that $19.920 \le x \le 20.079$. The edge of the cube must be within 0.079 cm of 20 cm.

4. $4 - 0.1 \le x^2 \le 4 + 0.1$
$1.9749 \le x \le 2.0024$

6. $-2 - 0.01 \le \dfrac{1}{x} \le -2 + 0.01$
$-\dfrac{1}{2.01} \ge x \ge -\dfrac{1}{1.99}$
$-0.5025 \le x \le -0.4975$

8. We need $-0.01 \le \sqrt{2x + 3} - 3 \le 0.01$. Thus

$$2.99 \le \sqrt{2x + 3} \le 3.01$$
$$8.9401 \le 2x + 3 \le 9.0601$$
$$2.97005 \le x \le 3.03005$$
$$3 - 0.02995 \le x - 3 \le 0.03005.$$

Here $\delta = 0.02995$ will do.

10. We need $1 - 0.05 \le 1/(x + 1) \le 1 + 0.05$, or $1.0526 \ge x + 1 \ge 0.9524$. This will occur if $-0.0476 \le x \le 0.0526$. In this case we can take $\delta = 0.0476$.

12. To be proved: $\lim\limits_{x \to 2}(5 - 2x) = 1$.
Proof: Let $\epsilon > 0$ be given. Then $|(5 - 2x) - 1| < \epsilon$ holds if $|2x - 4| < \epsilon$, and so if $|x - 2| < \delta = \epsilon/2$. This confirms the limit.

14. To be proved: $\lim\limits_{x \to 2}\dfrac{x - 2}{1 + x^2} = 0$.
Proof: Let $\epsilon > 0$ be given. Then

$$\left|\frac{x - 2}{1 + x^2} - 0\right| = \frac{|x - 2|}{1 + x^2} \le |x - 2| < \epsilon$$

provided $|x - 2| < \delta = \epsilon$.

16. To be proved: $\lim\limits_{x \to -2}\dfrac{x^2 + 2x}{x + 2} = -2$.
Proof: Let $\epsilon > 0$ be given. For $x \ne -2$ we have

$$\left|\frac{x^2 + 2x}{x + 2} - (-2)\right| = |x + 2| < \epsilon$$

provided $|x + 2| < \delta = \epsilon$. This completes the proof.

18. To be proved: $\lim\limits_{x \to -1}\dfrac{x + 1}{x^2 - 1} = -\dfrac{1}{2}$.
Proof: Let $\epsilon > 0$ be given. If $x \ne -1$, we have

$$\left|\frac{x + 1}{x^2 - 1} - \frac{1}{2}\right| = \left|\frac{1}{x - 1} - \left(-\frac{1}{2}\right)\right| = \frac{|x + 1|}{2|x - 1|}.$$

If $|x + 1| < 1$, then $-2 < x < 0$, so $-3 < x - 1 < -1$ and $|x - 1| > 1$. Ler $\delta = \min(1, 2\epsilon)$. If $0 < |x - (-1)| < \delta$ then $|x - 1| > 1$ and $|x + 1| < 2\epsilon$. Thus

$$\left|\frac{x + 1}{x^2 - 1} - \frac{1}{2}\right| = \frac{|x + 1|}{2|x - 1|} < \frac{2\epsilon}{2} = \epsilon.$$

This completes the required proof.

20. To be proved: $\lim\limits_{x \to 2} x^3 = 8$.
Proof: Let $\epsilon > 0$ be given. We have $|x^3 - 8| = |x - 2||x^2 + 2x + 4|$. If $|x - 2| < 1$, then $1 < x < 3$ and $x^2 < 9$. Therefore $|x^2 + 2x + 4| \le 9 + 2 \times 3 + 4 = 19$. If $|x - 2| < \delta = \min(1, \epsilon/19)$, then

$$|x^3 - 8| = |x - 2||x^2 + 2x + 4| < \frac{\epsilon}{19} \times 19 = \epsilon.$$

This completes the proof.

22. We say that $\lim_{x \to -\infty} f(x) = L$ if the following condition holds: for every number $\epsilon > 0$ there exists a number $R > 0$, depending on ϵ, such that

$$x < -R \quad \text{implies} \quad |f(x) - L| < \epsilon.$$

24. We say that $\lim_{x \to \infty} f(x) = \infty$ if the following condition holds: for every number $B > 0$ there exists a number $R > 0$, depending on B, such that

$$x > R \quad \text{implies} \quad f(x) > B.$$

26. We say that $\lim_{x \to a-} f(x) = \infty$ if the following condition holds: for every number $B > 0$ there exists a number $\delta > 0$, depending on B, such that

$$a - \delta < x < a \quad \text{implies} \quad f(x) > B.$$

28. To be proved: $\lim_{x \to 1-} \dfrac{1}{x - 1} = -\infty$. Proof: Let $B > 0$ be given. We have $\dfrac{1}{x - 1} < -B$ if $0 > x - 1 > -1/B$, that is, if $1 - \delta < x < 1$, where $\delta = 1/B$. This completes the proof.

30. To be proved: $\lim_{x \to \infty} \sqrt{x} = \infty$. Proof: Let $B > 0$ be given. We have $\sqrt{x} > B$ if $x > R$ where $R = B^2$. This completes the proof.

32. To be proved: if $\lim\limits_{x\to a} g(x) = M$, then there exists $\delta > 0$ such that if $0 < |x - a| < \delta$, then $|g(x)| < 1 + |M|$.
Proof: Taking $\epsilon = 1$ in the definition of limit, we obtain a number $\delta > 0$ such that if $0 < |x - a| < \delta$, then $|g(x) - M| < 1$. It follows from this latter inequality that

$$|g(x)| = |(g(x) - M) + M| \le |G(x) - M| + |M| < 1 + |M|.$$

34. To be proved: if $\lim\limits_{x\to a} g(x) = M$ where $M \ne 0$, then there exists $\delta > 0$ such that if $0 < |x - a| < \delta$, then $|g(x)| > |M|/2$.
Proof: By the definition of limit, there exists $\delta > 0$ such that if $0 < |x - a| < \delta$, then $|g(x) - M| < |M|/2$ (since $|M|/2$ is a positive number). This latter inequality implies that

$$|M| = |g(x) + (M - g(x))| \le |g(x)| + |g(x) - M| < |g(x)| + \frac{|M|}{2}.$$

It follows that $|g(x)| > |M| - (|M|/2) = |M|/2$, as required.

36. To be proved: if $\lim\limits_{x\to a} f(x) = L$ and $\lim\limits_{x\to a} f(x) = M \ne 0$, then $\lim\limits_{x\to a} \dfrac{f(x)}{g(x)} = \dfrac{L}{M}$.
Proof: By Exercises 33 and 35 we have

$$\lim_{x\to a} \frac{f(x)}{g(x)} = \lim_{x\to a} f(x) \times \frac{1}{g(x)} = L \times \frac{1}{M} = \frac{L}{M}.$$

38. To be proved: if $f(x) \le g(x) \le h(x)$ in an open interval containing $x = a$ (say, for $a - \delta_1 < x < a + \delta_1$, where $\delta_1 > 0$), and if $\lim_{x\to a} f(x) = \lim_{x\to a} h(x) = L$, then also $\lim_{x\to a} g(x) = L$.
Proof: Let $\epsilon > 0$ be given. Since $\lim_{x\to a} f(x) = L$, there exists $\delta_2 > 0$ such that if $0 < |x - a| < \delta_2$, then $|f(x) - L| < \epsilon/3$. Since $\lim_{x\to a} h(x) = L$, there exists $\delta_3 > 0$ such that if $0 < |x - a| < \delta_3$, then $|h(x) - L| < \epsilon/3$. Let $\delta = \min(\delta_1, \delta_2, \delta_3)$. If $0 < |x - a| < \delta$, then

$$\begin{aligned}
|g(x) - L| &= |g(x) - f(x) + f(x) - L| \\
&\le |g(x) - f(x)| + |f(x) - L| \\
&\le |h(x) - f(x)| + |f(x) - L| \\
&= |h(x) - L + L - f(x)| + |f(x) - L| \\
&\le |h(x) - L| + |f(x) - L| + |f(x) - L| \\
&< \frac{\epsilon}{3} + \frac{\epsilon}{3} + \frac{\epsilon}{3} = \epsilon.
\end{aligned}$$

Thus $\lim_{x\to a} g(x) = L$.

Review Exercises 1 (page 95)

2. The average rate of change of $1/x$ over $[-2, -1]$ is

$$\frac{(1/(-1)) - (1/(-2))}{-1 - (-2)} = \frac{-1/2}{1} = -\frac{1}{2}.$$

4. The rate of change of $1/x$ at $x = -3/2$ is

$$\begin{aligned}
\lim_{h\to 0} \frac{\dfrac{1}{-(3/2) + h} - \left(\dfrac{1}{-3/2}\right)}{h} &= \lim_{h\to 0} \frac{\dfrac{2}{2h - 3} + \dfrac{2}{3}}{h} \\
&= \lim_{h\to 0} \frac{2(3 + 2h - 3)}{3(2h - 3)h} \\
&= \lim_{h\to 0} \frac{4}{3(2h - 3)} = -\frac{4}{9}.
\end{aligned}$$

6. $\lim\limits_{x\to 2} \dfrac{x^2}{1 - x^2} = \dfrac{2^2}{1 - 2^2} = -\dfrac{4}{3}$

8. $\lim\limits_{x\to 2} \dfrac{x^2 - 4}{x^2 - 5x + 6} = \lim\limits_{x\to 2} \dfrac{(x - 2)(x + 2)}{(x - 2)(x - 3)} = \lim\limits_{x\to 2} \dfrac{x + 2}{x - 3} = -4$

10. $\lim\limits_{x\to 2-} \dfrac{x^2 - 4}{x^2 - 4x + 4} = \lim\limits_{x\to 2-} \dfrac{x + 2}{x - 2} = -\infty$

12. $\lim\limits_{x\to 4} \dfrac{2 - \sqrt{x}}{x - 4} = \lim\limits_{x\to 4} \dfrac{4 - x}{(2 + \sqrt{x})(x - 4)} = -\dfrac{1}{4}$

14. $\lim\limits_{h\to 0} \dfrac{h}{\sqrt{x + 3h} - \sqrt{x}} = \lim\limits_{h\to 0} \dfrac{h(\sqrt{x + 3h} + \sqrt{x})}{(x + 3h) - x}$

$$= \lim_{h\to 0} \frac{\sqrt{x + 3h} + \sqrt{x}}{3} = \frac{2\sqrt{x}}{3}$$

16. $\lim\limits_{x\to 0} \sqrt{x - x^2}$ does not exist because $\sqrt{x - x^2}$ is not defined for $x < 0$.

18. $\lim\limits_{x\to 1-} \sqrt{x - x^2} = 0$

20. $\lim\limits_{x\to -\infty} \dfrac{2x + 100}{x^2 + 3} = \lim\limits_{x\to -\infty} \dfrac{(2/x) + (100/x^2)}{1 + (3/x^2)} = 0$

22. $\lim\limits_{x\to\infty} \dfrac{x^4}{x^2 - 4} = \lim\limits_{x\to\infty} \dfrac{x^2}{1 - (4/x^2)} = \infty$

24. $\lim\limits_{x\to 1/2} \dfrac{1}{\sqrt{x - x^2}} = \dfrac{1}{\sqrt{1/4}} = 2$

26. $\lim\limits_{x\to\infty} \dfrac{\cos x}{x} = 0$ by the squeeze theorem, since

$$-\frac{1}{x} \le \frac{\cos x}{x} \le \frac{1}{x} \quad \text{for all } x > 0$$

and $\lim_{x\to\infty} (-1/x) = \lim_{x\to\infty} (1/x) = 0$.

28. $\lim\limits_{x\to 0}\sin\dfrac{1}{x^2}$ does not exist; $\sin(1/x^2)$ takes the values -1 and 1 in any interval $(-\delta, \delta)$, where $\delta > 0$, and limits, if they exist, must be unique.

30. $\lim\limits_{x\to\infty}[x + \sqrt{x^2 - 4x + 1}] = \infty + \infty = \infty$

32. $f(x) = \dfrac{x}{x+1}$ is continuous everywhere on its domain, which consists of all real numbers except $x = -1$. It is discontinuous nowhere.

34. $f(x) = \begin{cases} x^2 & \text{if } x > 1 \\ x & \text{if } x \le 1 \end{cases}$ is defined and continuous everywhere, and so discontinuous nowhere. Observe that $\lim_{x\to 1-} f(x) = 1 = \lim_{x\to 1+} f(x)$.

36. $f(x) = H(9 - x^2) = \begin{cases} 1 & \text{if } -3 \le x \le 3 \\ 0 & \text{if } x < -3 \text{ or } x > 3 \end{cases}$ is defined everywhere and discontinuous at $x = \pm 3$. It is right continuous at -3 and left continuous at 3.

38. $f(x) = \begin{cases} |x|/|x+1| & \text{if } x \ne -1 \\ 1 & \text{if } x = -1 \end{cases}$ is defined everywhere and discontinuous at $x = -1$ where it is neither left nor right continuous since $\lim_{x\to -1} f(x) = \infty$, while $f(-1) = 1$.

Challenging Problems 1 (page 96)

2. For x near 0 we have $|x - 1| = 1 - x$ and $|x + 1| = x + 1$. Thus

$$\lim_{x\to 0}\frac{x}{|x-1| - |x+1|} = \lim_{x\to 0}\frac{x}{(1-x)-(x+1)} = -\frac{1}{2}.$$

4. Let $y = x^{1/6}$. Then we have

$$\lim_{x\to 64}\frac{x^{1/3} - 4}{x^{1/2} - 8} = \lim_{y\to 2}\frac{y^2 - 4}{y^3 - 8}$$

$$= \lim_{y\to 2}\frac{(y-2)(y+2)}{(y-2)(y^2+2y+4)}$$

$$= \lim_{y\to 2}\frac{y+2}{y^2+2y+4} = \frac{4}{12} = \frac{1}{3}.$$

6. $r_+(a) = \dfrac{-1 + \sqrt{1+a}}{a}, \; r_-(a) = \dfrac{-1 - \sqrt{1+a}}{a}.$

a) $\lim_{a\to 0} r_-(a)$ does not exist. Observe that the right limit is $-\infty$ and the left limit is ∞.

b) From the following table it appears that $\lim_{a\to 0} r_+(a) = 1/2$, the solution of the linear equation $2x - 1 = 0$ which results from setting $a = 0$ in the quadratic equation $ax^2 + 2x - 1 = 0$.

a	$r_+(a)$
1	0.41421
0.1	0.48810
−0.1	0.51317
0.01	0.49876
−0.01	0.50126
0.001	0.49988
−0.001	0.50013

c) $\lim\limits_{a\to 0} r_+(a) = \lim\limits_{a\to 0}\dfrac{\sqrt{1+a} - 1}{a}$

$$= \lim_{a\to 0}\frac{(1+a) - 1}{a(\sqrt{1+a} + 1)}$$

$$= \lim_{a\to 0}\frac{1}{\sqrt{1+a} + 1} = \frac{1}{2}.$$

8. a) To be proved: if f is a continuous function defined on a closed interval $[a, b]$, then the range of f is a closed interval.
Proof: By the Max-Min Theorem there exist numbers u and v in $[a, b]$ such that $f(u) \le f(x) \le f(v)$ for all x in $[a, b]$. By the Intermediate-Value Theorem, $f(x)$ takes on all values between $f(u)$ and $f(v)$ at values of x between u and v, and hence at points of $[a, b]$. Thus the range of f is $[f(u), f(v)]$, a closed interval.

b) If the domain of the continuous function f is an open interval, the range of f can be any interval (open, closed, half open, finite, or infinite).

10. $f(x) = \dfrac{1}{x - x^2} = \dfrac{1}{\frac{1}{4} - \left(\frac{1}{4} - x + x^2\right)} = \dfrac{1}{\frac{1}{4} - \left(x - \frac{1}{2}\right)^2}.$
Observe that $f(x) \ge f(1/2) = 4$ for all x in $(0, 1)$.

CHAPTER 2. DIFFERENTIATION

Section 2.1 Tangent Lines and Their Slopes (page 102)

2. Since $y = x/2$ is a straight line, its tangent at any point $(a, a/2)$ on it is the same line $y = x/2$.

4. The slope of $y = 6 - x - x^2$ at $x = -2$ is

$$m = \lim_{h \to 0} \frac{6 - (-2 + h) - (-2 + h)^2 - 4}{h}$$

$$= \lim_{h \to 0} \frac{3h - h^2}{h} = \lim_{h \to 0}(3 - h) = 3.$$

The tangent line at $(-2, 4)$ is $y = 3x + 10$.

6. The slope of $y = \dfrac{1}{x^2 + 1}$ at $(0, 1)$ is

$$m = \lim_{h \to 0} \frac{1}{h}\left(\frac{1}{h^2 + 1} - 1\right) = \lim_{h \to 0} \frac{-h}{h^2 + 1} = 0.$$

The tangent line at $(0, 1)$ is $y = 1$.

8. The slope of $y = \dfrac{1}{\sqrt{x}}$ at $x = 9$ is

$$m = \lim_{h \to 0} \frac{1}{h}\left(\frac{1}{\sqrt{9 + h}} - \frac{1}{3}\right)$$

$$= \lim_{h \to 0} \frac{3 - \sqrt{9 + h}}{3h\sqrt{9 + h}} \cdot \frac{3 + \sqrt{9 + h}}{3 + \sqrt{9 + h}}$$

$$= \lim_{h \to 0} \frac{9 - 9 - h}{3h\sqrt{9 + h}(3 + \sqrt{9 + h})}$$

$$= -\frac{1}{3(3)(6)} = -\frac{1}{54}.$$

The tangent line at $(9, \frac{1}{3})$ is $y = \frac{1}{3} - \frac{1}{54}(x - 9)$, or $y = \frac{1}{2} - \frac{1}{54}x$.

10. The slope of $y = \sqrt{5 - x^2}$ at $x = 1$ is

$$m = \lim_{h \to 0} \frac{\sqrt{5 - (1 + h)^2} - 2}{h}$$

$$= \lim_{h \to 0} \frac{5 - (1 + h)^2 - 4}{h\left(\sqrt{5 - (1 + h)^2} + 2\right)}$$

$$= \lim_{h \to 0} \frac{-2 - h}{\sqrt{5 - (1 + h)^2} + 2} = -\frac{1}{2}$$

The tangent line at $(1, 2)$ is $y = 2 - \frac{1}{2}(x - 1)$, or $y = \frac{5}{2} - \frac{1}{2}x$.

12. The slope of $y = \dfrac{1}{x}$ at $(a, \frac{1}{a})$ is

$$m = \lim_{h \to 0} \frac{1}{h}\left(\frac{1}{a + h} + \frac{1}{a}\right) = \lim_{h \to 0} \frac{a - a - h}{h(a + h)(a)} = -\frac{1}{a^2}.$$

The tangent line at $(a, \frac{1}{a})$ is $y = \frac{1}{a} - \frac{1}{a^2}(x - a)$, or $y = \frac{2}{a} - \frac{x}{a^2}$.

14. The slope of $f(x) = (x - 1)^{4/3}$ at $x = 1$ is

$$m = \lim_{h \to 0} \frac{(1 + h - 1)^{4/3} - 0}{h} = \lim_{h \to 0} h^{1/3} = 0.$$

The graph of f has a tangent line with slope 0 at $x = 1$. Since $f(1) = 0$, the tangent has equation $y = 0$

16. The slope of $f(x) = |x^2 - 1|$ at $x = 1$ is

$$m = \lim_{h \to 0} \frac{|(1 + h)^2 - 1| - |1 - 1|}{h} = \lim_{h \to 0} \frac{|2h + h^2|}{h},$$

which does not exist, and is not $-\infty$ or ∞. The graph of f has no tangent at $x = 1$.

18. The slope of $y = x^2 - 1$ at $x = x_0$ is

$$m = \lim_{h \to 0} \frac{[(x_0 + h)^2 - 1] - (x_0^2 - 1)}{h}$$

$$= \lim_{h \to 0} \frac{2x_0 h + h^2}{h} = 2x_0.$$

If $m = -3$, then $x_0 = -\frac{3}{2}$. The tangent line with slope $m = -3$ at $(-\frac{3}{2}, \frac{5}{4})$ is $y = \frac{5}{4} - 3(x + \frac{3}{2})$, that is, $y = -3x - \frac{13}{4}$.

20. The slope of $y = x^3 - 3x$ at $x = a$ is

$$m = \lim_{h \to 0} \frac{1}{h}\left[(a + h)^3 - 3(a + h) - (a^3 - 3a)\right]$$

$$= \lim_{h \to 0} \frac{1}{h}\left[a^3 + 3a^2 h + 3ah^2 + h^3 - 3a - 3h - a^3 + 3a\right]$$

$$= \lim_{h \to 0}[3a^2 + 3ah + h^2 - 3] = 3a^2 - 3.$$

At points where the tangent line is parallel to the x-axis, the slope is zero, so such points must satisfy $3a^2 - 3 = 0$. Thus, $a = \pm 1$. Hence, the tangent line is parallel to the x-axis at the points $(1, -2)$ and $(-1, 2)$.

22. The slope of the curve $y = 1/x$ at $x = a$ is

$$m = \lim_{h \to 0} \frac{\frac{1}{a + h} - \frac{1}{a}}{h} = \lim_{h \to 0} \frac{a - (a + h)}{ah(a + h)} = -\frac{1}{a^2}.$$

The tangent at $x = a$ is perpendicular to the line $y = 4x - 3$ if $-1/a^2 = -1/4$, that is, if $a = \pm 2$. The corresponding points on the curve are $(-2, -1/2)$ and $(2, 1/2)$.

24. The curves $y = kx^2$ and $y = k(x - 2)^2$ intersect at $(1, k)$. The slope of $y = kx^2$ at $x = 1$ is

$$m_1 = \lim_{h \to 0} \frac{k(1 + h)^2 - k}{h} = \lim_{h \to 0} (2 + h)k = 2k.$$

The slope of $y = k(x - 2)^2$ at $x = 1$ is

$$m_2 = \lim_{h \to 0} \frac{k(2 - (1 + h))^2 - k}{h} = \lim_{h \to 0} (-2 + h)k = -2k.$$

The two curves intersect at right angles if $2k = -1/(-2k)$, that is, if $4k^2 = 1$, which is satisfied if $k = \pm 1/2$.

26. Horizontal tangent at $(-1, 8)$ and $(2, -19)$.

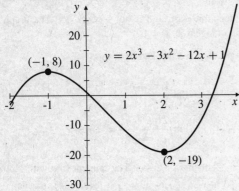

Fig. 2.1.26

28. Horizontal tangent at $(a, 2)$ and $(-a, -2)$ for all $a > 1$. No tangents at $(1, 2)$ and $(-1, -2)$.

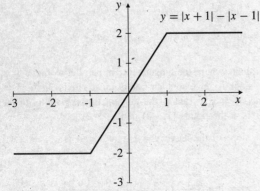

Fig. 2.1.28

30. Horizontal tangent at $(0, 1)$. No tangents at $(-1, 0)$ and $(1, 0)$.

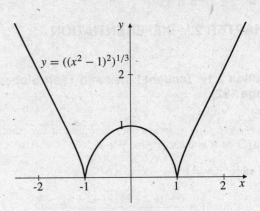

Fig. 2.1.30

32. The slope of $P(x)$ at $x = a$ is

$$m = \lim_{h \to 0} \frac{P(a + h) - P(a)}{h}.$$

Since $P(a + h) = a_0 + a_1 h + a_2 h^2 + \cdots + a_n h^n$ and $P(a) = a_0$, the slope is

$$m = \lim_{h \to 0} \frac{a_0 + a_1 h + a_2 h^2 + \cdots + a_n h^n - a_0}{h}$$

$$= \lim_{h \to 0} a_1 + a_2 h + \cdots + a_n h^{n-1} = a_1.$$

Thus the line $y = \ell(x) = m(x - a) + b$ is tangent to $y = P(x)$ at $x = a$ if and only if $m = a_1$ and $b = a_0$, that is, if and only if

$$P(x) - \ell(x) = a_2(x - a)^2 + a_3(x - a)^3 + \cdots + a_n(x - a)^n$$

$$= (x - a)^2 \left[a_2 + a_3(x - a) + \cdots + a_n(x - a)^{n-2} \right]$$

$$= (x - a)^2 Q(x)$$

where Q is a polynomial.

Section 2.2   The Derivative   (page 110)

2.

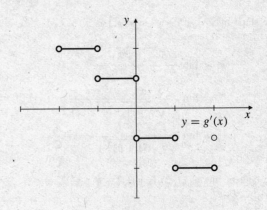

4.

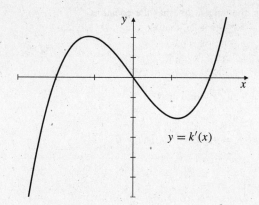

$y = k'(x)$

$y = f(x) = |x^2 - 1| - |x^2 - 4|$

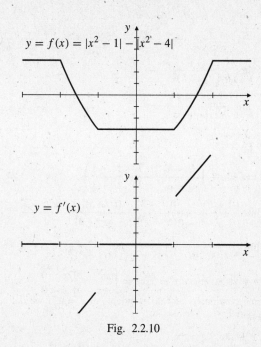

$y = f'(x)$

Fig. 2.2.10

6. Assuming the tick marks are spaced 1 unit apart, the function g is differentiable on the intervals $]-2, -1[$, $]-1, 0[$, $]0, 1[$, and $]1, 2[$.

8. $y = f(x)$ has horizontal tangents at the points near $1/2$ and $3/2$ where $f'(x) = 0$

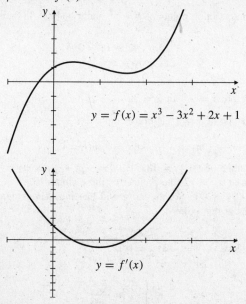

$y = f(x) = x^3 - 3x^2 + 2x + 1$

$y = f'(x)$

Fig. 2.2.8

10. $y = f(x)$ is constant on the intervals $(-\infty, -2)$, $(-1, 1)$, and $(2, \infty)$. It is not differentiable at $x = \pm 2$ and $x = \pm 1$.

12. $f(x) = 1 + 4x - 5x^2$

$$f'(x) = \lim_{h \to 0} \frac{1 + 4(x + h) - 5(x + h)^2 - (1 + 4x - 5x^2)}{h}$$

$$= \lim_{h \to 0} \frac{4h - 10xh - 5h^2}{h} = 4 - 10x$$

14. $s = \dfrac{1}{3 + 4t}$

$$\frac{ds}{dt} = \lim_{h \to 0} \frac{1}{h}\left[\frac{1}{3 + 4(t + h)} - \frac{1}{3 + 4t}\right]$$

$$= \lim_{h \to 0} \frac{3 + 4t - 3 - 4t - 4h}{h(3 + 4t)[3 + (4t + h)]} = -\frac{4}{(3 + 4t)^2}$$

16. $f(x) = \frac{3}{4}\sqrt{2 - x}$

$$f'(x) = \lim_{h \to 0} \frac{\frac{3}{4}\sqrt{2 - (x + h)} - \frac{3}{4}\sqrt{2 - x}}{h}$$

$$= \lim_{h \to 0} \frac{3}{4}\left[\frac{2 - x - h - 2 + x}{h(\sqrt{2 - (x + h)} + \sqrt{2 - x})}\right]$$

$$= -\frac{3}{8\sqrt{2 - x}}$$

18. $z = \dfrac{s}{1 + s}$

$$\frac{dz}{ds} = \lim_{h \to 0} \frac{1}{h}\left[\frac{s + h}{1 + s + h} - \frac{s}{1 + s}\right]$$

$$= \lim_{h \to 0} \frac{(s + h)(1 + s) - s(1 + s + h)}{h(1 + s)(1 + s + h)} = \frac{1}{(1 + s)^2}$$

20. $y = \dfrac{1}{x^2}$

$$y' = \lim_{h \to 0} \frac{1}{h} \left[\frac{1}{(x+h)^2} - \frac{1}{x^2} \right]$$

$$= \lim_{h \to 0} \frac{x^2 - (x+h)^2}{hx^2(x+h)^2} = -\frac{2}{x^3}$$

22. $f(t) = \dfrac{t^2 - 3}{t^2 + 3}$

$$f'(t) = \lim_{h \to 0} \frac{1}{h} \left(\frac{(t+h)^2 - 3}{(t+h)^2 + 3} - \frac{t^2 - 3}{t^2 + 3} \right)$$

$$= \lim_{h \to 0} \frac{[(t+h)^2 - 3](t^2 + 3) - (t^2 - 3)[(t+h)^2 + 3]}{h(t^2 + 3)[(t+h)^2 + 3]}$$

$$= \lim_{h \to 0} \frac{12th + 6h^2}{h(t^2 + 3)[(t+h)^2 + 3]} = \frac{12t}{(t^2 + 3)^2}$$

24. Since $g(x) = x^2 \operatorname{sgn} x = x|x| = \begin{cases} x^2 & \text{if } x > 0 \\ -x^2 & \text{if } x < 0 \end{cases}$, g will become continuous and differentiable at $x = 0$ if we define $g(0) = 0$.

26. $y = x^3 - 2x$

x	$\dfrac{f(x) - f(1)}{x - 1}$	x	$\dfrac{f(x) - f(1)}{x - 1}$
0.9	0.71000	1.1	1.31000
0.99	0.97010	1.01	1.03010
0.999	0.99700	1.001	1.00300
0.9999	0.99970	1.0001	1.00030

$$\frac{d}{dx}(x^3 - 2x)\Big|_{x=1} = \lim_{h \to 0} \frac{(1+h)^3 - 2(1+h) - (-1)}{h}$$

$$= \lim_{h \to 0} \frac{h + 3h^2 + h^3}{h}$$

$$= \lim_{h \to 0} 1 + 3h + h^2 = 1$$

28. The slope of $y = 5 + 4x - x^2$ at $x = 2$ is

$$\frac{dy}{dx}\Big|_{x=2} = \lim_{h \to 0} \frac{5 + 4(2+h) - (2+h)^2 - 9}{h}$$

$$= \lim_{h \to 0} \frac{-h^2}{h} = 0.$$

Thus, the tangent line at $x = 2$ has the equation $y = 9$.

30. The slope of $y = \dfrac{t}{t^2 - 2}$ at $t = -2$ and $y = -1$ is

$$\frac{dy}{dt}\Big|_{t=-2} = \lim_{h \to 0} \frac{1}{h} \left[\frac{-2+h}{(-2+h)^2 - 2} - (-1) \right]$$

$$= \lim_{h \to 0} \frac{-2 + h + [(-2+h)^2 - 2]}{h[(-2+h)^2 - 2]} = -\frac{3}{2}.$$

Thus, the tangent line has the equation
$y = -1 - \frac{3}{2}(t + 2)$, that is, $y = -\frac{3}{2}t - 4$.

32. $f'(x) = -17x^{-18}$ for $x \neq 0$

34. $\dfrac{dy}{dx} = \dfrac{1}{3} x^{-2/3}$ for $x \neq 0$

36. $\dfrac{d}{dt} t^{-2.25} = -2.25 t^{-3.25}$ for $t > 0$

38. $\dfrac{d}{ds} \sqrt{s}\Big|_{s=9} = \dfrac{1}{2\sqrt{s}}\Big|_{s=9} = \dfrac{1}{6}.$

40. $f'(8) = -\dfrac{2}{3} x^{-5/3}\Big|_{x=8} = -\dfrac{1}{48}$

42. The slope of $y = \sqrt{x}$ at $x = x_0$ is

$$\frac{dy}{dx}\Big|_{x=x_0} = \frac{1}{2\sqrt{x_0}}.$$

Thus, the equation of the tangent line is

$$y = \sqrt{x_0} + \frac{1}{2\sqrt{x_0}}(x - x_0), \text{ that is, } y = \frac{x + x_0}{2\sqrt{x_0}}.$$

44. The intersection points of $y = x^2$ and $x + 4y = 18$ satisfy

$$4x^2 + x - 18 = 0$$
$$(4x + 9)(x - 2) = 0.$$

Therefore $x = -\frac{9}{4}$ or $x = 2$.
The slope of $y = x^2$ is $m_1 = \dfrac{dy}{dx} = 2x$.
At $x = -\dfrac{9}{4}$, $m_1 = -\dfrac{9}{2}$. At $x = 2$, $m_1 = 4$.
The slope of $x + 4y = 18$, i.e. $y = -\frac{1}{4}x + \frac{18}{4}$, is $m_2 = -\frac{1}{4}$.
Thus, at $x = 2$, the product of these slopes is
$(4)(-\frac{1}{4}) = -1$. So, the curve and line intersect at right angles at that point.

46. The slope of $y = \dfrac{1}{x}$ at $x = a$ is

$$\frac{dy}{dx}\Big|_{x=a} = -\frac{1}{a^2}.$$

If the slope is -2, then $-\dfrac{1}{a^2} = -2$, or $a = \pm\dfrac{1}{\sqrt{2}}$. Therefore, the equations of the two straight lines are
$y = \sqrt{2} - 2\left(x - \dfrac{1}{\sqrt{2}}\right)$ and $y = -\sqrt{2} - 2\left(x + \dfrac{1}{\sqrt{2}}\right)$,
or $y = -2x \pm 2\sqrt{2}$.

48. If a line is tangent to $y = x^2$ at (t, t^2), then its slope is $\left.\dfrac{dy}{dx}\right|_{x=t} = 2t$. If this line also passes through (a, b), then its slope satisfies

$$\frac{t^2 - b}{t - a} = 2t, \quad \text{that is } t^2 - 2at + b = 0.$$

Hence $t = \dfrac{2a \pm \sqrt{4a^2 - 4b}}{2} = a \pm \sqrt{a^2 - b}$.
If $b < a^2$, i.e. $a^2 - b > 0$, then $t = a \pm \sqrt{a^2 - b}$ has two real solutions. Therefore, there will be two distinct tangent lines passing through (a, b) with equations $y = b + 2\left(a \pm \sqrt{a^2 - b}\right)(x - a)$. If $b = a^2$, then $t = a$. There will be only one tangent line with slope $2a$ and equation $y = b + 2a(x - a)$.
If $b > a^2$, then $a^2 - b < 0$. There will be no real solution for t. Thus, there will be no tangent line.

50. Let $f(x) = x^{-n}$. Then

$$
\begin{aligned}
f'(x) &= \lim_{h \to 0} \frac{(x+h)^{-n} - x^{-n}}{h} \\
&= \lim_{h \to 0} \frac{1}{h}\left(\frac{1}{(x+h)^n} - \frac{1}{x^n}\right) \\
&= \lim_{h \to 0} \frac{x^n - (x+h)^n}{hx^n(x+h)^n} \\
&= \lim_{h \to 0} \frac{x - (x+h)}{hx^n((x+h))^n} \times \\
&\qquad \left(x^{n-1} + x^{n-2}(x+h) + \cdots + (x+h)^{n-1}\right) \\
&= -\frac{1}{x^{2n}} \times nx^{n-1} = -nx^{-(n+1)}.
\end{aligned}
$$

52. Let $f(x) = x^{1/n}$. Then

$$
\begin{aligned}
f'(x) &= \lim_{h \to 0} \frac{(x+h)^{1/n} - x^{1/n}}{h} \quad (\text{let } x+h = a^n, \ x = b^n) \\
&= \lim_{a \to b} \frac{a - b}{a^n - b^n} \\
&= \lim_{a \to b} \frac{1}{a^{n-1} + a^{n-2}b + a^{n-3}b^2 + \cdots + b^{n-1}} \\
&= \frac{1}{nb^{n-1}} = \frac{1}{n} x^{(1/n)-1}.
\end{aligned}
$$

54. Let

$$f'(a+) = \lim_{h \to 0+} \frac{f(a+h) - f(a)}{h}$$

$$f'(a-) = \lim_{h \to 0-} \frac{f(a+h) - f(a)}{h}$$

If $f'(a+)$ is finite, call the half-line with equation $y = f(a) + f'(a+)(x - a)$, $(x \geq a)$, the *right tangent line* to the graph of f at $x = a$. Similarly, if $f'(a-)$ is finite, call the half-line $y = f(a) + f'(a-)(x - a)$, $(x \leq a)$, the *left tangent line*. If $f'(a+) = \infty$ (or $-\infty$), the right tangent line is the half-line $x = a$, $y \geq f(a)$ (or $x = a$, $y \leq f(a)$). If $f'(a-) = \infty$ (or $-\infty$), the right tangent line is the half-line $x = a$, $y \leq f(a)$ (or $x = a$, $y \geq f(a)$). The graph has a tangent line at $x = a$ if and only if $f'(a+) = f'(a-)$. (This includes the possibility that both quantities may be $+\infty$ or both may be $-\infty$.) In this case the right and left tangents are two opposite halves of the same straight line. For $f(x) = x^{2/3}$, $f'(x) = \frac{2}{3}x^{-1/3}$. At $(0, 0)$, we have $f'(0+) = +\infty$ and $f'(0-) = -\infty$. In this case both left and right tangents are the *positive* y-axis, and the curve does not have a tangent line at the origin. For $f(x) = |x|$, we have

$$f'(x) = \operatorname{sgn}(x) = \begin{cases} 1 & \text{if } x > 0 \\ -1 & \text{if } x < 0. \end{cases}$$

At $(0, 0)$, $f'(0+) = 1$, and $f'(0-) = -1$. In this case the right tangent is $y = x$, $(x \geq 0)$, and the left tangent is $y = -x$, $(x \leq 0)$. There is no tangent line.

Section 2.3 Differentiation Rules (page 119)

2. $y = 4x^{1/2} - \dfrac{5}{x}, \quad y' = 2x^{-1/2} + 5x^{-2}$

4. $f(x) = \dfrac{6}{x^3} + \dfrac{2}{x^2} - 2, \quad f'(x) = -\dfrac{18}{x^4} - \dfrac{4}{x^3}$

6. $y = x^{45} - x^{-45} \quad y' = 45x^{44} + 45x^{-46}$

8. $y = 3\sqrt[3]{t^2} - \dfrac{2}{\sqrt{t^3}} = 3t^{2/3} - 2t^{-3/2}$

$\dfrac{dy}{dt} = 2t^{-1/3} + 3t^{-5/2}$

10. $F(x) = (3x - 2)(1 - 5x)$
$F'(x) = 3(1 - 5x) + (3x - 2)(-5) = 13 - 30x$

12. $g(t) = \dfrac{1}{2t - 3}, \quad g'(t) = -\dfrac{2}{(2t - 3)^2}$

14. $y = \dfrac{4}{3 - x}, \quad y' = \dfrac{4}{(3 - x)^2}$

16. $g(y) = \dfrac{2}{1 - y^2}, \quad g'(y) = \dfrac{4y}{(1 - y^2)^2}$

18. $g(u) = \dfrac{u\sqrt{u} - 3}{u^2} = u^{-1/2} - 3u^{-2}$

$g'(u) = -\dfrac{1}{2}u^{-3/2} + 6u^{-3} = \dfrac{12 - u\sqrt{u}}{2u^3}$

20. $z = \dfrac{x - 1}{x^{2/3}} = x^{1/3} - x^{-2/3}$

$\dfrac{dz}{dx} = \dfrac{1}{3}x^{-2/3} + \dfrac{2}{3}x^{-5/3} = \dfrac{x + 2}{3x^{5/3}}$

22. $z = \dfrac{t^2 + 2t}{t^2 - 1}$

$z' = \dfrac{(t^2 - 1)(2t + 2) - (t^2 + 2t)(2t)}{(t^2 - 1)^2}$

$= -\dfrac{2(t^2 + t + 1)}{(t^2 - 1)^2}$

24. $f(x) = \dfrac{x^3 - 4}{x + 1}$

$f'(x) = \dfrac{(x + 1)(3x^2) - (x^3 - 4)(1)}{(x + 1)^2}$

$= \dfrac{2x^3 + 3x^2 + 4}{(x + 1)^2}$

26. $F(t) = \dfrac{t^2 + 7t - 8}{t^2 - t + 1}$

$F'(t) = \dfrac{(t^2 - t + 1)(2t + 7) - (t^2 + 7t - 8)(2t - 1)}{(t^2 - t + 1)^2}$

$= \dfrac{-8t^2 + 18t - 1}{(t^2 - t + 1)^2}$

28. $f(r) = (r^{-2} + r^{-3} - 4)(r^2 + r^3 + 1)$

$f'(r) = (-2r^{-3} - 3r^{-4})(r^2 + r^3 + 1)$
$\qquad + (r^{-2} + r^{-3} - 4)(2r + 3r^2)$

or

$f(r) = -2 + r^{-1} + r^{-2} + r^{-3} + r - 4r^2 - 4r^3$

$f'(r) = -r^{-2} - 2r^{-3} - 3r^{-4} + 1 - 8r - 12r^2$

30. $y = \dfrac{(x^2 + 1)(x^3 + 2)}{(x^2 + 2)(x^3 + 1)}$

$= \dfrac{x^5 + x^3 + 2x^2 + 2}{x^5 + 2x^3 + x^2 + 2}$

$y' = \dfrac{(x^5 + 2x^3 + x^2 + 2)(5x^4 + 3x^2 + 4x)}{(x^5 + 2x^3 + x^2 + 2)^2}$

$\qquad - \dfrac{(x^5 + x^3 + 2x^2 + 2)(5x^4 + 6x^2 + 2x)}{(x^5 + 2x^3 + x^2 + 2)^2}$

$= \dfrac{2x^7 - 3x^6 - 3x^4 - 6x^2 + 4x}{(x^5 + 2x^3 + x^2 + 2)^2}$

$= \dfrac{2x^7 - 3x^6 - 3x^4 - 6x^2 + 4x}{(x^2 + 2)^2(x^3 + 1)^2}$

32. $f(x) = \dfrac{(\sqrt{x} - 1)(2 - x)(1 - x^2)}{\sqrt{x}(3 + 2x)}$

$= \left(1 - \dfrac{1}{\sqrt{x}}\right) \cdot \dfrac{2 - x - 2x^2 + x^3}{3 + 2x}$

$f'(x) = \left(\dfrac{1}{2}x^{-3/2}\right)\dfrac{2 - x - 2x^2 + x^3}{3 + 2x} + \left(1 - \dfrac{1}{\sqrt{x}}\right)$

$\qquad \times \dfrac{(3 + 2x)(-1 - 4x + 3x^2) - (2 - x - 2x^2 + x^3)(2)}{(3 + 2x)^2}$

$= \dfrac{(2 - x)(1 - x^2)}{2x^{3/2}(3 + 2x)}$

$\qquad + \left(1 - \dfrac{1}{\sqrt{x}}\right)\dfrac{4x^3 + 5x^2 - 12x - 7}{(3 + 2x)^2}$

34. $\dfrac{d}{dx}\left(\dfrac{f(x)}{x^2}\right)\bigg|_{x=2} = \dfrac{x^2 f'(x) - 2x f(x)}{x^4}\bigg|_{x=2}$

$= \dfrac{4 f'(2) - 4 f(2)}{16} = \dfrac{4}{16} = \dfrac{1}{4}$

36. $\dfrac{d}{dx}\left(\dfrac{f(x)}{x^2 + f(x)}\right)\bigg|_{x=2}$

$= \dfrac{(x^2 + f(x)) f'(x) - f(x)(2x + f'(x))}{(x^2 + f(x))^2}\bigg|_{x=2}$

$= \dfrac{(4 + f(2)) f'(2) - f(2)(4 + f'(2))}{(4 + f(2))^2} = \dfrac{18 - 14}{6^2} = \dfrac{1}{9}$

38. $\dfrac{d}{dt}\left[\dfrac{t(1 + \sqrt{t})}{5 - t}\right]\bigg|_{t=4}$

$= \dfrac{d}{dt}\left[\dfrac{t + t^{3/2}}{5 - t}\right]\bigg|_{t=4}$

$= \dfrac{(5 - t)(1 + \frac{3}{2}t^{1/2}) - (t + t^{3/2})(-1)}{(5 - t)^2}\bigg|_{t=4}$

$= \dfrac{(1)(4) - (12)(-1)}{(1)^2} = 16$

40. $\dfrac{d}{dt}[(1 + t)(1 + 2t)(1 + 3t)(1 + 4t)]\bigg|_{t=0}$

$= (1)(1 + 2t)(1 + 3t)(1 + 4t) + (1 + t)(2)(1 + 3t)(1 + 4t) +$

$(1 + t)(1 + 2t)(3)(1 + 4t) + (1 + t)(1 + 2t)(1 + 3t)(4)\bigg|_{t=0}$

$= 1 + 2 + 3 + 4 = 10$

42. For $y = \dfrac{x + 1}{x - 1}$ we calculate

$$y' = \dfrac{(x - 1)(1) - (x + 1)(1)}{(x - 1)^2} = -\dfrac{2}{(x - 1)^2}.$$

At $x = 2$ we have $y = 3$ and $y' = -2$. Thus, the equation of the tangent line is $y = 3 - 2(x - 2)$, or $y = -2x + 7$. The normal line is $y = 3 + \frac{1}{2}(x - 2)$, or $y = \frac{1}{2}x + 2$.

44. If $y = x^2(4 - x^2)$, then

$$y' = 2x(4 - x^2) + x^2(-2x) = 8x - 4x^3 = 4x(2 - x^2).$$

24

The slope of a horizontal line must be zero, so $4x(2 - x^2) = 0$, which implies that $x = 0$ or $x = \pm\sqrt{2}$. At $x = 0$, $y = 0$ and at $x = \pm\sqrt{2}$, $y = 4$.
Hence, there are two horizontal lines that are tangent to the curve. Their equations are $y = 0$ and $y = 4$.

46. If $y = \dfrac{x + 1}{x + 2}$, then

$$y' = \frac{(x + 2)(1) - (x + 1)(1)}{(x + 2)^2} = \frac{1}{(x + 2)^2}.$$

In order to be parallel to $y = 4x$, the tangent line must have slope equal to 4, i.e.,

$$\frac{1}{(x + 2)^2} = 4, \qquad \text{or } (x + 2)^2 = \tfrac{1}{4}.$$

Hence $x + 2 = \pm\tfrac{1}{2}$, and $x = -\tfrac{3}{2}$ or $-\tfrac{5}{2}$. At $x = -\tfrac{3}{2}$, $y = -1$, and at $x = -\tfrac{5}{2}$, $y = 3$.
Hence, the tangent is parallel to $y = 4x$ at the points $\left(-\tfrac{3}{2}, -1\right)$ and $\left(-\tfrac{5}{2}, 3\right)$.

48. Since $\dfrac{1}{\sqrt{x}} = y = x^2 \Rightarrow x^{5/2} = 1$, therefore $x = 1$ at the intersection point. The slope of $y = x^2$ at $x = 1$ is $2x\Big|_{x=1} = 2$. The slope of $y = \dfrac{1}{\sqrt{x}}$ at $x = 1$ is

$$\frac{dy}{dx}\bigg|_{x=1} = -\frac{1}{2}x^{-3/2}\bigg|_{x=1} = -\frac{1}{2}.$$

The product of the slopes is $(2)\left(-\tfrac{1}{2}\right) = -1$. Hence, the two curves intersect at right angles.

50. The tangent to $y = x^2/(x - 1)$ at $(a, a^2/(a - 1))$ has slope

$$m = \frac{(x - 1)2x - x^2(1)}{(x - 1)^2}\bigg|_{x=a} = \frac{a^2 - 2a}{(a - 1)^2}.$$

The equation of the tangent is

$$y - \frac{a^2}{a - 1} = \frac{a^2 - 2a}{(a - 1)^2}(x - a).$$

This line passes through $(2, 0)$ provided

$$0 - \frac{a^2}{a - 1} = \frac{a^2 - 2a}{(a - 1)^2}(2 - a),$$

or, upon simplification, $3a^2 - 4a = 0$. Thus we can have either $a = 0$ or $a = 4/3$. There are two tangents through $(2, 0)$. Their equations are $y = 0$ and $y = -8x + 16$.

52. $f(x) = |x^3| = \begin{cases} x^3 & \text{if } x \geq 0 \\ -x^3 & \text{if } x < 0 \end{cases}$. Therefore f is differentiable everywhere except *possibly* at $x = 0$, However,

$$\lim_{h \to 0+} \frac{f(0 + h) - f(0)}{h} = \lim_{h \to 0+} h^2 = 0$$

$$\lim_{h \to 0-} \frac{f(0 + h) - f(0)}{h} = \lim_{h \to 0-} (-h^2) = 0.$$

Thus $f'(0)$ exists and equals 0. We have

$$f'(x) = \begin{cases} 3x^2 & \text{if } x \geq 0 \\ -3x^2 & \text{if } x < 0. \end{cases}$$

54. To be proved:

$$(f_1 f_2 \cdots f_n)'$$
$$= f_1' f_2 \cdots f_n + f_1 f_2' \cdots f_n + \cdots + f_1 f_2 \cdots f_n'$$

Proof: The case $n = 2$ is just the Product Rule. Assume the formula holds for $n = k$ for some integer $k > 2$. Using the Product Rule and this hypothesis we calculate

$$(f_1 f_2 \cdots f_k f_{k+1})'$$
$$= [(f_1 f_2 \cdots f_k) f_{k+1}]'$$
$$= (f_1 f_2 \cdots f_k)' f_{k+1} + (f_1 f_2 \cdots f_k) f_{k+1}'$$
$$= (f_1' f_2 \cdots f_k + f_1 f_2' \cdots f_k + \cdots + f_1 f_2 \cdots f_k') f_{k+1}$$
$$\quad + (f_1 f_2 \cdots f_k) f_{k+1}'$$
$$= f_1' f_2 \cdots f_k f_{k+1} + f_1 f_2' \cdots f_k f_{k+1} + \cdots$$
$$\quad + f_1 f_2 \cdots f_k' f_{k+1} + f_1 f_2 \cdots f_k f_{k+1}'$$

so the formula is also true for $n = k + 1$. The formula is therefore for all integers $n \geq 2$ by induction.

Section 2.4 The Chain Rule (page 125)

2. $y = \left(1 - \dfrac{x}{3}\right)^{99}$

$$y' = 99\left(1 - \frac{x}{3}\right)^{98}\left(-\frac{1}{3}\right) = -33\left(1 - \frac{x}{3}\right)^{98}$$

4. $\dfrac{dy}{dx} = \dfrac{d}{dx}\sqrt{1 - 3x^2} = \dfrac{-6x}{2\sqrt{1 - 3x^2}} = -\dfrac{3x}{\sqrt{1 - 3x^2}}$

6. $z = (1 + x^{2/3})^{3/2}$

$z' = \tfrac{3}{2}(1 + x^{2/3})^{1/2}(\tfrac{2}{3}x^{-1/3}) = x^{-1/3}(1 + x^{2/3})^{1/2}$

8. $y = (1 - 2t^2)^{-3/2}$

$y' = -\tfrac{3}{2}(1 - 2t^2)^{-5/2}(-4t) = 6t(1 - 2t^2)^{-5/2}$

10. $f(t) = |2 + t^3|$

$f'(t) = [\text{sgn}\,(2 + t^3)](3t^2) = \dfrac{3t^2(2 + t^3)}{|2 + t^3|}$

12. $y = (2 + |x|^3)^{1/3}$

$y' = \tfrac{1}{3}(2 + |x|^3)^{-2/3}(3|x|^2)\text{sgn}\,(x)$

$\quad = |x|^2(2 + |x|^3)^{-2/3}\left(\dfrac{x}{|x|}\right) = x|x|(2 + |x|^3)^{-2/3}$

14. $f(x) = \left(1 + \sqrt{\dfrac{x-2}{3}}\,\right)^4$

$f'(x) = 4\left(1 + \sqrt{\dfrac{x-2}{3}}\,\right)^3 \left(\dfrac{1}{2}\sqrt{\dfrac{3}{x-2}}\right)\left(\dfrac{1}{3}\right)$

$ = \dfrac{2}{3}\sqrt{\dfrac{3}{x-2}}\left(1 + \sqrt{\dfrac{x-2}{3}}\,\right)^3$

16. $y = \dfrac{x^5\sqrt{3+x^6}}{(4+x^2)^3}$

$y' = \dfrac{1}{(4+x^2)^6}\left((4+x^2)^3\left[5x^4\sqrt{3+x^6} + x^5\left(\dfrac{3x^5}{\sqrt{3+x^6}}\right)\right]\right.$

$\left. - x^5\sqrt{3+x^6}\big[3(4+x^2)^2(2x)\big]\right)$

$ = \dfrac{(4+x^2)\big[5x^4(3+x^6)+3x^{10}\big] - x^5(3+x^6)(6x)}{(4+x^2)^4\sqrt{3+x^6}}$

$ = \dfrac{60x^4 - 3x^6 + 32x^{10} + 2x^{12}}{(4+x^2)^4\sqrt{3+x^6}}$

18.

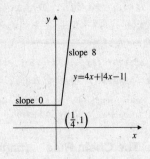

slope 8

$y = 4x + |4x - 1|$

slope 0

$\left(\dfrac{1}{4}, 1\right)$

20. $\dfrac{d}{dx}x^{3/4} = \dfrac{d}{dx}\sqrt{x\sqrt{x}} = \dfrac{1}{2\sqrt{x\sqrt{x}}}\left(\sqrt{x} + \dfrac{x}{2\sqrt{x}}\right) = \dfrac{3}{4}x^{-1/4}$

22. $\dfrac{d}{dt}f(2t+3) = 2f'(2t+3)$

24. $\dfrac{d}{dx}\left[f\left(\dfrac{2}{x}\right)\right]^3 = 3\left[f\left(\dfrac{2}{x}\right)\right]^2 f'\left(\dfrac{2}{x}\right)\left(\dfrac{-2}{x^2}\right)$

$\phantom{\dfrac{d}{dx}\left[f\left(\dfrac{2}{x}\right)\right]^3} = -\dfrac{2}{x^2}f'\left(\dfrac{2}{x}\right)\left[f\left(\dfrac{2}{x}\right)\right]^2$

26. $\dfrac{d}{dt}f(\sqrt{3+2t}) = f'(\sqrt{3+2t})\dfrac{2}{2\sqrt{3+2t}}$

$\phantom{\dfrac{d}{dt}f(\sqrt{3+2t})} = \dfrac{1}{\sqrt{3+2t}}f'(\sqrt{3+2t})$

28. $\dfrac{d}{dt}f\big(2f(3f(x))\big)$

$= f'\big(2f(3f(x))\big)\cdot 2f'(3f(x))\cdot 3f'(x)$

$= 6f'(x)f'(3f(x))f'\big(2f(3f(x))\big)$

30. $\dfrac{d}{dx}\left(\dfrac{\sqrt{x^2-1}}{x^2+1}\right)\bigg|_{x=-2}$

$= \dfrac{(x^2+1)\dfrac{x}{\sqrt{x^2-1}} - \sqrt{x^2-1}(2x)}{(x^2+1)^2}\bigg|_{x=-2}$

$= \dfrac{(5)\left(-\dfrac{2}{\sqrt{3}}\right) - \sqrt{3}(-4)}{25} = \dfrac{2}{25\sqrt{3}}$

32. $f(x) = \dfrac{1}{\sqrt{2x+1}}$

$f'(4) = -\dfrac{1}{(2x+1)^{3/2}}\bigg|_{x=4} = -\dfrac{1}{27}$

34. $F(x) = (1+x)(2+x)^2(3+x)^3(4+x)^4$

$F'(x) = (2+x)^2(3+x)^3(4+x)^4 +$

$\qquad 2(1+x)(2+x)(3+x)^3(4+x)^4 +$

$\qquad 3(1+x)(2+x)^2(3+x)^2(4+x)^4 +$

$\qquad 4(1+x)(2+x)^2(3+x)^3(4+x)^3$

$F'(0) = (2^2)(3^3)(4^4) + 2(1)(2)(3^3)(4^4) +$

$\qquad 3(1)(2^2)(3^2)(4^4) + 4(1)(2^2)(3^3)(4^3)$

$\qquad = 4(2^2 \cdot 3^3 \cdot 4^4) = 110,592$

36. The slope of $y = \sqrt{1+2x^2}$ at $x=2$ is

$$\dfrac{dy}{dx}\bigg|_{x=2} = \dfrac{4x}{2\sqrt{1+2x^2}}\bigg|_{x=2} = \dfrac{4}{3}.$$

Thus, the equation of the tangent line at $(2,3)$ is $y = 3 + \dfrac{4}{3}(x-2)$, or $y = \dfrac{4}{3}x + \dfrac{1}{3}$.

38. The slope of $y = (ax+b)^8$ at $x = \dfrac{b}{a}$ is

$$\dfrac{dy}{dx}\bigg|_{x=b/a} = 8a(ax+b)^7\bigg|_{x=b/a} = 1024ab^7.$$

The equation of the tangent line at $x = \dfrac{b}{a}$ and $y = (2b)^8 = 256b^8$ is

$y = 256b^8 + 1024ab^7\left(x - \dfrac{b}{a}\right)$, or $y = 2^{10}ab^7 x - 3\times 2^8 b^8$.

40. Given that $f(x) = (x-a)^m(x-b)^n$ then

$f'(x) = m(x-a)^{m-1}(x-b)^n + n(x-a)^m(x-b)^{n-1}$

$\qquad = (x-a)^{m-1}(x-b)^{n-1}(mx - mb + nx - na)$.

If $x \neq a$ and $x \neq b$, then $f'(x) = 0$ if and only if

$$mx - mb + nx - na = 0,$$

which is equivalent to

$$x = \dfrac{n}{m+n}a + \dfrac{m}{m+n}b.$$

This point lies lies between a and b.

42. $4(7x^4 - 49x^2 + 54)/x^7$

44. $5/8$

46. It may happen that $k = g(x + h) - g(x) = 0$ for values of h arbitrarily close to 0 so that the division by k in the "proof" is not justified.

Section 2.5   Derivatives of Trigonometric Functions   (page 131)

2. $\dfrac{d}{dx} \cot x = \dfrac{d}{dx} \dfrac{\cos x}{\sin x} = \dfrac{-\cos^2 x - \sin^2 x}{\sin^2 x} = -\csc^2 x$

4. $y = \sin \dfrac{x}{5}, \quad y' = \dfrac{1}{5} \cos \dfrac{x}{5}.$

6. $y = \sec ax, \quad y' = a \sec ax \tan ax.$

8. $\dfrac{d}{dx} \sin \dfrac{\pi - x}{3} = -\dfrac{1}{3} \cos \dfrac{\pi - x}{3}$

10. $y = \sin(Ax + B), \quad y' = A \cos(Ax + B)$

12. $\dfrac{d}{dx} \cos(\sqrt{x}) = -\dfrac{1}{2\sqrt{x}} \sin(\sqrt{x})$

14. $\dfrac{d}{dx} \sin(2 \cos x) = \cos(2 \cos x)(-2 \sin x)$
$= -2 \sin x \cos(2 \cos x)$

16. $g(\theta) = \tan(\theta \sin \theta)$
$g'(\theta) = (\sin \theta + \theta \cos \theta) \sec^2(\theta \sin \theta)$

18. $y = \sec(1/x), \quad y' = -(1/x^2) \sec(1/x) \tan(1/x)$

20. $G(\theta) = \dfrac{\sin a\theta}{\cos b\theta}$
$G'(\theta) = \dfrac{a \cos b\theta \cos a\theta + b \sin a\theta \sin b\theta}{\cos^2 b\theta}.$

22. $\dfrac{d}{dx}(\cos^2 x - \sin^2 x) = \dfrac{d}{dx} \cos(2x)$
$= -2 \sin(2x) = -4 \sin x \cos x$

24. $\dfrac{d}{dx}(\sec x - \csc x) = \sec x \tan x + \csc x \cot x$

26. $\dfrac{d}{dx} \tan(3x) \cot(3x) = \dfrac{d}{dx}(1) = 0$

28. $\dfrac{d}{dt}(t \sin t + \cos t) = \sin t + t \cos t - \sin t = t \cos t$

30. $\dfrac{d}{dx} \dfrac{\cos x}{1 + \sin x} = \dfrac{(1 + \sin x)(-\sin x) - \cos(x)(\cos x)}{(1 + \sin x)^2}$
$= \dfrac{-\sin x - 1}{(1 + \sin x)^2} = \dfrac{-1}{1 + \sin x}$

32. $g(t) = \sqrt{(\sin t)/t}$
$g'(t) = \dfrac{1}{2\sqrt{(\sin t)/t}} \times \dfrac{t \cos t - \sin t}{t^2}$
$= \dfrac{t \cos t - \sin t}{2t^{3/2} \sqrt{\sin t}}$

34. $z = \dfrac{\sin \sqrt{x}}{1 + \cos \sqrt{x}}$
$z' = \dfrac{(1 + \cos \sqrt{x})(\cos \sqrt{x}/2\sqrt{x}) - (\sin \sqrt{x})(-\sin \sqrt{x}/2\sqrt{x})}{(1 + \cos \sqrt{x})^2}$
$= \dfrac{1 + \cos \sqrt{x}}{2\sqrt{x}(1 + \cos \sqrt{x})^2} = \dfrac{1}{2\sqrt{x}(1 + \cos \sqrt{x})}$

36. $f(s) = \cos(s + \cos(s + \cos s))$
$f'(s) = -[\sin(s + \cos(s + \cos s))]$
$\times [1 - (\sin(s + \cos s))(1 - \sin s)]$

38. Differentiate both sides of $\cos(2x) = \cos^2 x - \sin^2 x$ and divide by -2 to get $\sin(2x) = 2 \sin x \cos x$.

40. The slope of $y = \tan(2x)$ at $(0, 0)$ is $2 \sec^2(0) = 2$. Therefore the tangent and normal lines to $y = \tan(2x)$ at $(0, 0)$ have equations $y = 2x$ and $y = -x/2$, respectively.

42. The slope of $y = \cos^2 x$ at $(\pi/3, 1/4)$ is $-\sin(2\pi/3) = -\sqrt{3}/2$. Therefore the tangent and normal lines to $y = \tan(2x)$ at $(0, 0)$ have equations $y = (1/4) - (\sqrt{3}/2)(x - (\pi/3))$ and $y = (1/4) + (2/\sqrt{3})(x - (\pi/3))$, respectively.

44. For $y = \sec(x°) = \sec\left(\dfrac{x\pi}{180}\right)$ we have

$$\dfrac{dy}{dx} = \dfrac{\pi}{180} \sec\left(\dfrac{x\pi}{180}\right) \tan\left(\dfrac{x\pi}{180}\right).$$

At $x = 60$ the slope is $\dfrac{\pi}{180}(2\sqrt{3}) = \dfrac{\pi\sqrt{3}}{90}$.

Thus, the normal line has slope $-\dfrac{90}{\pi\sqrt{3}}$ and has equation
$y = 2 - \dfrac{90}{\pi\sqrt{3}}(x - 60)$.

46. The slope of $y = \tan(2x)$ at $x = a$ is $2 \sec^2(2a)$. The tangent there is normal to $y = -x/8$ if $2 \sec^2(2a) = 8$, or $\cos(2a) = \pm 1/2$. The only solutions in $(-\pi/4, \pi/4)$ are $a = \pm\pi/6$. The corresponding points on the graph are $(\pi/6, \sqrt{3})$ and $(-\pi/6, -\sqrt{3})$.

48. $\dfrac{d}{dx} \tan x = \sec^2 x = 0$ nowhere.
$\dfrac{d}{dx} \cot x = -\csc^2 x = 0$ nowhere.
Thus neither of these functions has a horizontal tangent.

50. $y = 2x + \sin x$ has no horizontal tangents because $dy/dx = 2 + \cos x \geq 1$ everywhere.

52. $y = x + 2 \cos x$ has horizontal tangents at $x = \pi/6$ and $x = 5\pi/6$ because $dy/dx = 1 - 2 \sin x = 0$ at those points.

54. $\lim\limits_{x \to \pi} \sec(1 + \cos x) = \sec(1 - 1) = \sec 0 = 1$

56. $\lim\limits_{x \to 0} \cos\left(\dfrac{\pi - \pi \cos^2 x}{x^2}\right) = \lim\limits_{x \to 0} \cos \pi \left(\dfrac{\sin x}{x}\right)^2 = \cos \pi = -1$

58. f will be differentiable at $x = 0$ if

$$2 \sin 0 + 3 \cos 0 = b, \quad \text{and}$$

$$\frac{d}{dx}(2 \sin x + 3 \cos x)\bigg|_{x=0} = a.$$

Thus we need $b = 3$ and $a = 2$.

60. 1

62. a) As suggested by the figure in the problem, the square of the length of chord AP is $(1 - \cos \theta)^2 + (0 - \sin \theta)^2$, and the square of the length of arc AP is θ^2. Hence

$$(1 + \cos \theta)^2 + \sin^2 \theta < \theta^2,$$

and, since squares cannot be negative, each term in the sum on the left is less than θ^2. Therefore

$$0 \leq |1 - \cos \theta| < |\theta|, \quad 0 \leq |\sin \theta| < |\theta|.$$

Since $\lim_{\theta \to 0} |\theta| = 0$, the squeeze theorem implies that

$$\lim_{\theta \to 0} 1 - \cos \theta = 0, \quad \lim_{\theta \to 0} \sin \theta = 0.$$

From the first of these, $\lim_{\theta \to 0} \cos \theta = 1$.

b) Using the result of (a) and the addition formulas for cosine and sine we obtain

$$\lim_{h \to 0} \cos(\theta_0 + h) = \lim_{h \to 0}(\cos \theta_0 \cos h - \sin \theta_0 \sin h) = \cos \theta_0$$

$$\lim_{h \to 0} \sin(\theta_0 + h) = \lim_{h \to 0}(\sin \theta_0 \cos h + \cos \theta_0 \sin h) = \sin \theta_0.$$

This says that cosine and sine are continuous at any point θ_0.

Section 2.6 The Mean-Value Theorem (page 139)

2. If $f(x) = \dfrac{1}{x}$, and $f'(x) = -\dfrac{1}{x^2}$ then

$$\frac{f(2) - f(1)}{2 - 1} = \frac{1}{2} - 1 = -\frac{1}{2} = -\frac{1}{c^2} = f'(c)$$

where $c = \sqrt{2}$ lies between 1 and 2.

4. If $f(x) = \cos x + (x^2/2)$, then $f'(x) = x - \sin x > 0$ for $x > 0$. By the MVT, if $x > 0$, then $f(x) - f(0) = f'(c)(x - 0)$ for some $c > 0$, so $f(x) > f(0) = 1$. Thus $\cos x + (x^2/2) > 1$ and $\cos x > 1 - (x^2/2)$ for $x > 0$. Since both sides of the inequality are even functions, it must hold for $x < 0$ as well.

6. Let $f(x) = (1 + x)^r - 1 - rx$ where $r > 1$. Then $f'(x) = r(1 + x)^{r-1} - r$. If $-1 \leq x < 0$ then $f'(x) < 0$; if $x > 0$, then $f'(x) > 0$. Thus $f(x) > f(0) = 0$ if $-1 \leq x < 0$ or $x > 0$. Thus $(1 + x)^r > 1 + rx$ if $-1 \leq x < 0$ or $x > 0$.

8. If $f(x) = x^2 + 2x + 2$ then $f'(x) = 2x + 2 = 2(x + 1)$. Evidently, $f'(x) > 0$ if $x > -1$ and $f'(x) < 0$ if $x < -1$. Therefore, f is increasing on $(-1, \infty)$ and decreasing on $]-\infty, -1[$.

10. If $f(x) = x^3 + 4x + 1$, then $f'(x) = 3x^2 + 4$. Since $f'(x) > 0$ for all real x, hence $f(x)$ is increasing on the whole real line, i.e., on $]-\infty, \infty[$.

12. If $f(x) = \dfrac{1}{x^2 + 1}$ then $f'(x) = \dfrac{-2x}{(x^2 + 1)^2}$. Evidently, $f'(x) > 0$ if $x < 0$ and $f'(x) < 0$ if $x > 0$. Therefore, f is increasing on $]-\infty, 0[$ and decreasing on $]0, \infty[$.

14. If $f(x) = x - 2\sin x$, then $f'(x) = 1 - 2\cos x = 0$ at $x = \pm \pi/3 + 2n\pi$ for $n = 0, \pm 1, \pm 2, \ldots$. f is decreasing on $]-\pi/3 + 2n\pi, \pi + 2n\pi[$. f is increasing on $]\pi/3 + 2n\pi, -\pi/3 + 2(n + 1)\pi[$ for integers n.

16. If $x_1 < x_2 < \ldots < x_n$ belong to I, and $f(x_i) = 0$, $(1 \leq i \leq n)$, then there exists y_i in (x_i, x_{i+1}) such that $f'(y_i) = 0$, $(1 \leq i \leq n - 1)$ by MVT.

18. $f(x) = \begin{cases} x + 2x^2 \sin(1/x) & \text{if } x \neq 0 \\ 0 & \text{if } x = 0. \end{cases}$

a) $f'(0) = \displaystyle\lim_{h \to 0} \frac{f(0 + h) - f(0)}{h}$

$= \displaystyle\lim_{h \to 0} \frac{h + 2h^2 \sin(1/h)}{h}$

$= \displaystyle\lim_{h \to 0}(1 + 2h \sin(1/h) = 1,$

because $|2h \sin(1/h)| \leq 2|h| \to 0$ as $h \to 0$.

b) For $x \neq 0$, we have

$$f'(x) = 1 + 4x \sin(1/x) - 2 \cos(1/x).$$

There are numbers x arbitrarily close to 0 where $f'(x) = -1$; namely, the numbers $x = \pm 1/(2n\pi)$, where $n = 1, 2, 3, \ldots$. Since $f'(x)$ is continuous at every $x \neq 0$, it is negative in a small interval about every such number. Thus f cannot be increasing on any interval containing $x = 0$.

Section 2.7 Using Derivatives (page 145)

2. If $y = 1/x$, then $\Delta y \approx (-1/x^2)\Delta x$. If $\Delta x = (2/100)x$, then $\Delta y \approx (-2/100)/x = (-2/100)y$, so y decreases by about 2%.

4. If $y = x^3$, then $\Delta y \approx 3x^2 \Delta x$. If $\Delta x = (2/100)x$, then $\Delta y \approx (6/100)x^3 = (6/100)y$, so y increases by about 6%.

6. If $y = x^{-2/3}$, then $\Delta y \approx (-2/3)x^{-5/3}\,\Delta x$. If $\Delta x = (2/100)x$, then $\Delta y \approx (-4/300)x^{2/3} = (-4/300)y$, so y decreases by about 1.33%.

8. If V is the volume and x is the edge length of the cube then $V = x^3$. Thus $\Delta V \approx 3x^2\,\Delta x$. $\Delta V = -(6/100)V$, then $-6x^3/100 = 3x^2\,\Delta x$, so $\Delta x \approx -(2/100)x$. The edge of the cube decreases by about 2%.

10. If $A = s^2$, then $s = \sqrt{A}$ and $ds/dA = 1/(2\sqrt{A})$. If $A = 16$ m^2, then the side is changing at rate $ds/dA = 1/8$ m/m^2.

12. Since $A = \pi D^2/4$, the rate of change of area with respect to diameter is $dA/dD = \pi D/2$ square units per unit.

14. Let A be the area of a square, s be its side length and L be its diagonal. Then, $L^2 = s^2 + s^2 = 2s^2$ and $A = s^2 = \frac{1}{2}L^2$, so $\dfrac{dA}{dL} = L$. Thus, the rate of change of the area of a square with respect to its diagonal L is L.

16. Let s be the side length and V be the volume of a cube. Then $V = s^3 \Rightarrow s = V^{1/3}$ and $\dfrac{ds}{dV} = \frac{1}{3}V^{-2/3}$. Hence, the rate of change of the side length of a cube with respect to its volume V is $\frac{1}{3}V^{-2/3}$.

18. If $f(x) = x^3 - 12x + 1$, then $f'(x) = 3(x^2 - 4)$. The critical points of f are $x = \pm 2$. f is increasing on $(-\infty, -2)$ and $(2, \infty)$ where $f'(x) > 0$, and is decreasing on $(-2, 2)$ where $f'(x) < 0$.

20. If $y = 1 - x - x^5$, then $y' = -1 - 5x^4 < 0$ for all x. Thus y has no critical points and is decreasing on the whole real line.

22. If $f(x) = x + 2\sin x$, then $f'(x) = 1 + 2\cos x > 0$ if $\cos x > -1/2$. Thus f is increasing on the intervals $]-(4\pi/3) + 2n\pi, (4\pi/3) + 2n\pi[$ where n is any integer.

24. CPs $x = -1.366025$ and $x = 0.366025$

26. CP $x = 0.521350$

28. Flow rate $F = kr^4$, so $\Delta F \approx 4kr^3\,\Delta r$. If $\Delta F = F/10$, then
$$\Delta r \approx \frac{F}{40kr^3} = \frac{kr^4}{40kr^3} = 0.025r.$$
The flow rate will increase by 10% if the radius is increased by about 2.5%.

30. If price = p, then revenue is $R = 4{,}000p - 10p^2$.

   a) Sensitivity of R to p is $dR/dp = 4{,}000 - 20p$. If $p = 100, 200$, and 300, this sensitivity is 2,000 $/\$$, 0 $/\$$, and $-2{,}000$ $/\$$ respectively.

   b) The distributor should charge \$200. This maximizes the revenue.

32. Daily profit if production is x sheets per day is $P(x)$ where
$$P(x) = 8x - 0.005x^2 - 1{,}000.$$

   a) Marginal profit $P'(x) = 8 - 0.01x$. This is positive if $x < 800$ and negative if $x > 800$.

   b) To maximize daily profit, production should be 800 sheets/day.

34. Daily profit $P = 13x - Cx = 13x - 10x - 20 - \dfrac{x^2}{1000}$
$$= 3x - 20 - \frac{x^2}{1000}$$
Graph of P is a parabola opening downward. P will be maximum where the slope is zero:
$$0 = \frac{dP}{dx} = 3 - \frac{2x}{1000} \quad \text{so } x = 1500$$
Should extract 1500 tonnes of ore per day to maximize profit.

36. If $y = Cp^{-r}$, then the elasticity of y is
$$-\frac{p}{y}\frac{dy}{dp} = -\frac{p}{Cp^{-r}}(-r)Cp^{-r-1} = r.$$

Section 2.8 Higher-Order Derivatives (page 150)

2. $y = x^2 - \dfrac{1}{x}$ $y'' = 2 - \dfrac{2}{x^3}$

$y' = 2x + \dfrac{1}{x^2}$ $y''' = \dfrac{6}{x^4}$

4. $y = \sqrt{ax+b}$ $y'' = -\dfrac{a^2}{4(ax+b)^{3/2}}$

$y' = \dfrac{a}{2\sqrt{ax+b}}$ $y''' = \dfrac{3a^3}{8(ax+b)^{5/2}}$

6. $y = x^{10} + 2x^8$ $y'' = 90x^8 + 112x^6$

$y' = 10x^9 + 16x^7$ $y''' = 720x^7 + 672x^5$

8. $y = \dfrac{x-1}{x+1}$ $y'' = -\dfrac{4}{(x+1)^3}$

$y' = \dfrac{2}{(x+1)^2}$ $y''' = \dfrac{12}{(x+1)^4}$

10. $y = \sec x$ $y'' = \sec x\tan^2 x + \sec^3 x$

$y' = \sec x\tan x$ $y''' = \sec x\tan^3 x + 5\sec^3 x\tan x$

12. $y = \dfrac{\sin x}{x}$

$y' = \dfrac{\cos x}{x} - \dfrac{\sin x}{x^2}$

$y'' = \dfrac{(2-x^2)\sin x}{x^3} - \dfrac{2\cos x}{x^2}$

$y''' = \dfrac{(6-x^2)\cos x}{x^3} + \dfrac{3(x^2-2)\sin x}{x^4}$

14. $f(x) = \dfrac{1}{x^2} = x^{-2}$

$f'(x) = -2x^{-3}$

$f''(x) = -2(-3)x^{-4} = 3!x^{-4}$

$f^{(3)}(x) = -2(-3)(-4)x^{-5} = -4!x^{-5}$

Conjecture:

$$f^{(n)}(x) = (-1)^n(n+1)!x^{-(n+2)} \qquad \text{for } n = 1, 2, 3, \ldots$$

Proof: Evidently, the above formula holds for $n = 1, 2$ and 3. Assume it holds for $n = k$,

i.e., $f^{(k)}(x) = (-1)^k(k+1)!x^{-(k+2)}$. Then

$$\begin{aligned} f^{(k+1)}(x) &= \frac{d}{dx}f^{(k)}(x) \\ &= (-1)^k(k+1)![(-1)(k+2)]x^{-(k+2)-1} \\ &= (-1)^{k+1}(k+2)!x^{-[(k+1)+2]}. \end{aligned}$$

Thus, the formula is also true for $n = k + 1$. Hence it is true for $n = 1, 2, 3, \ldots$ by induction.

16. $f(x) = \sqrt{x} = x^{1/2}$

$f'(x) = \frac{1}{2}x^{-1/2}$

$f''(x) = \frac{1}{2}(-\frac{1}{2})x^{-3/2}$

$f'''(x) = \frac{1}{2}(-\frac{1}{2})(-\frac{3}{2})x^{-5/2}$

$f^{(4)}(x) = \frac{1}{2}(-\frac{1}{2})(-\frac{3}{2})(-\frac{5}{2})x^{-7/2}$

Conjecture:

$$f^{(n)}(x) = (-1)^{n-1}\frac{1 \cdot 3 \cdot 5 \cdots (2n-3)}{2^n}x^{-(2n-1)/2} \quad (n \ge 2).$$

Proof: Evidently, the above formula holds for $n = 2, 3$ and 4. Assume that it holds for $n = k$, i.e.

$$f^{(k)}(x) = (-1)^{k-1}\frac{1 \cdot 3 \cdot 5 \cdots (2k-3)}{2^k}x^{-(2k-1)/2}.$$

Then

$$\begin{aligned} f^{(k+1)}(x) &= \frac{d}{dx}f^{(k)}(x) \\ &= (-1)^{k-1}\frac{1 \cdot 3 \cdot 5 \cdots (2k-3)}{2^k} \cdot \left[\frac{-(2k-1)}{2}\right]x^{-[(2k-1)/2]-1} \\ &= (-1)^{(k+1)-1}\frac{1 \cdot 3 \cdot 5 \cdots (2k-3)[2(k+1)-3]}{2^{k+1}}x^{-[2(k+1)-1]/2}. \end{aligned}$$

Thus, the formula is also true for $n = k + 1$. Hence, it is true for $n \ge 2$ by induction.

18. $f(x) = x^{2/3}$

$f'(x) = \frac{2}{3}x^{-1/3}$

$f''(x) = \frac{2}{3}(-\frac{1}{3})x^{-4/3}$

$f'''(x) = \frac{2}{3}(-\frac{1}{3})(-\frac{4}{3})x^{-7/3}$

Conjecture:

$$f^{(n)}(x) = 2(-1)^{n-1}\frac{1 \cdot 4 \cdot 7 \cdots (3n-5)}{3^n}x^{-(3n-2)/3} \text{ for } n \ge 2.$$

Proof: Evidently, the above formula holds for $n = 2$ and 3. Assume that it holds for $n = k$, i.e.

$$f^{(k)}(x) = 2(-1)^{k-1}\frac{1 \cdot 4 \cdot 7 \cdots (3k-5)}{3^k}x^{-(3k-2)/3}.$$

Then,

$$\begin{aligned} f^{(k+1)}(x) &= \frac{d}{dx}f^{(k)}(x) \\ &= 2(-1)^{k-1}\frac{1 \cdot 4 \cdot 7 \cdots (3k-5)}{3^k} \cdot \left[\frac{-(3k-2)}{3}\right]x^{-[(3k-2)/3]-1} \\ &= 2(-1)^{(k+1)-1}\frac{1 \cdot 4 \cdot 7 \cdots (3k-5)[3(k+1)-5]}{3(k+1)}x^{-[3(k+1)-2]/3}. \end{aligned}$$

Thus, the formula is also true for $n = k + 1$. Hence, it is true for $n \ge 2$ by induction.

20. $f(x) = x\cos x$

$f'(x) = \cos x - x\sin x$

$f''(x) = -2\sin x - x\cos x$

$f'''(x) = -3\cos x + x\sin x$

$f^{(4)}(x) = 4\sin x + x\cos x$

This suggests the formula (for $k = 0, 1, 2, \ldots$)

$$f^{(n)}(x) = \begin{cases} n\sin x + x\cos x & \text{if } n = 4k \\ n\cos x - x\sin x & \text{if } n = 4k+1 \\ -n\sin x - x\cos x & \text{if } n = 4k+2 \\ -n\cos x + x\sin x & \text{if } n = 4k+3 \end{cases}$$

Differentiating any of these four formulas produces the one for the next higher value of n, so induction confirms the overall formula.

22. $f(x) = \dfrac{1}{|x|} = |x|^{-1}$. Recall that $\dfrac{d}{dx}|x| = \operatorname{sgn} x$, so

$$f'(x) = -|x|^{-2}\operatorname{sgn} x.$$

If $x \ne 0$ we have

$$\frac{d}{dx}\operatorname{sgn} x = 0 \quad \text{and} \quad (\operatorname{sgn} x)^2 = 1.$$

Thus we can calculate successive derivatives of f using the product rule where necessary, but will get only one nonzero term in each case:

$$\begin{aligned} f''(x) &= 2|x|^{-3}(\operatorname{sgn} x)^2 = 2|x|^{-3} \\ f^{(3)}(x) &= -3!|x|^{-4}\operatorname{sgn} x \\ f^{(4)}(x) &= 4!|x|^{-5}. \end{aligned}$$

The pattern suggests that

$$f^{(n)}(x) = \begin{cases} -n!|x|^{-(n+1)}\operatorname{sgn} x & \text{if } n \text{ is odd} \\ n!|x|^{-(n+1)} & \text{if } n \text{ is even} \end{cases}$$

Differentiating this formula leads to the same formula with n replaced by $n+1$ so the formula is valid for all $n \geq 1$ by induction.

24. If $y = \tan(kx)$, then $y' = k\sec^2(kx)$ and

$$y'' = 2k^2 \sec^2(kx)\tan(kx)$$
$$= 2k^2(1 + \tan^2(kx))\tan(kx) = 2k^2 y(1 + y^2).$$

26. To be proved: if $f(x) = \sin(ax + b)$, then

$$f^{(n)}(x) = \begin{cases} (-1)^k a^n \sin(ax+b) & \text{if } n = 2k \\ (-1)^k a^n \cos(ax+b) & \text{if } n = 2k+1 \end{cases}$$

for $k = 0, 1, 2, \ldots$ Proof: The formula works for $k = 0$ ($n = 2 \times 0 = 0$ and $n = 2 \times 0 + 1 = 1$):

$$\begin{cases} f^{(0)}(x) = f(x) = (-1)^0 a^0 \sin(ax+b) = \sin(ax+b) \\ f^{(1)}(x) = f'(x) = (-1)^0 a^1 \cos(ax+b) = a\cos(ax+b) \end{cases}$$

Now assume the formula holds for some $k \geq 0$. If $n = 2(k+1)$, then

$$f^{(n)}(x) = \frac{d}{dx} f^{(n-1)}(x) = \frac{d}{dx} f^{(2k+1)}(x)$$
$$= \frac{d}{dx}\left((-1)^k a^{2k+1} \cos(ax+b) \right)$$
$$= (-1)^{k+1} a^{2k+2} \sin(ax+b)$$

and if $n = 2(k+1) + 1 = 2k+3$, then

$$f^{(n)}(x) = \frac{d}{dx}\left((-1)^{k+1} a^{2k+2} \sin(ax+b) \right)$$
$$= (-1)^{k+1} a^{2k+3} \cos(ax+b).$$

Thus the formula also holds for $k+1$. Therefore it holds for all positive integers k by induction.

28. $(fg)'' = (f'g + fg')' = f''g + f'g' + f'g' + fg''$
$$= f''g + 2f'g' + fg''$$

30. Let a, b, and c be three points in I where f vanishes; that is, $f(a) = f(b) = f(c) = 0$. Suppose $a < b < c$. By the Mean-Value Theorem, there exist points r in (a, b) and s in (b, c) such that $f'(r) = f'(s) = 0$. By the Mean-Value Theorem applied to f' on $[r, s]$, there is some point t in (r, s) (and therefore in I) such that $f''(t) = 0$.

32. Given that $f(0) = f(1) = 0$ and $f(2) = 1$:

a) By MVT,

$$f'(a) = \frac{f(2) - f(0)}{2 - 0} = \frac{1 - 0}{2 - 0} = \frac{1}{2}$$

for some a in $(0, 2)$.

b) By MVT, for some r in $(0, 1)$,

$$f'(r) = \frac{f(1) - f(0)}{1 - 0} = \frac{0 - 0}{1 - 0} = 0.$$

Also, for some s in $(1, 2)$,

$$f'(s) = \frac{f(2) - f(1)}{2 - 1} = \frac{1 - 0}{2 - 1} = 1.$$

Then, by MVT applied to f' on the interval $[r, s]$, for some b in (r, s),

$$f''(b) = \frac{f'(s) - f'(r)}{s - r} = \frac{1 - 0}{s - r}$$
$$= \frac{1}{s - r} > \frac{1}{2}$$

since $s - r < 2$.

c) Since $f''(x)$ exists on $[0, 2]$, therefore $f'(x)$ is continuous there. Since $f'(r) = 0$ and $f'(s) = 1$, and since $0 < \frac{1}{7} < 1$, the Intermediate-Value Theorem assures us that $f'(c) = \frac{1}{7}$ for some c between r and s.

Section 2.9 Implicit Differentiation (page 156)

2. $x^3 + y^3 = 1$
$3x^2 + 3y^2 y' = 0$, so $y' = -\dfrac{x^2}{y^2}$.

4. $x^3 y + xy^5 = 2$
$3x^2 y + x^3 y' + y^5 + 5xy^4 y' = 0$
$y' = \dfrac{-3x^2 y - y^5}{x^3 + 5xy^4}$

6. $x^2 + 4(y-1)^2 = 4$
$2x + 8(y-1)y' = 0$, so $y' = \dfrac{2x}{4(1-y)}$

8. $x\sqrt{x+y} = 8 - xy$
$\sqrt{x+y} + x\dfrac{1}{2\sqrt{x+y}}(1 + y') = -y - xy'$
$2(x+y) + x(1 + y') = -2\sqrt{x+y}(y + xy')$
$y' = -\dfrac{3x + 2y + 2y\sqrt{x+y}}{x + 2x\sqrt{x+y}}$

10. $x^2 y^3 - x^3 y^2 = 12$
$2xy^3 + 3x^2 y^2 y' - 3x^2 y^2 - 2x^3 yy' = 0$
At $(-1, 2)$: $-16 + 12y' - 12 + 4y' = 0$, so the slope is
$y' = \dfrac{12 + 16}{12 + 4} = \dfrac{28}{16} = \dfrac{7}{4}$.
Thus, the equation of the tangent line is
$y = 2 + \frac{7}{4}(x+1)$, or $7x - 4y + 15 = 0$.

12. $x + 2y + 1 = \dfrac{y^2}{x-1}$

$1 + 2y' = \dfrac{(x-1)2yy' - y^2(1)}{(x-1)^2}$

At $(2, -1)$ we have $1 + 2y' = -2y' - 1$ so $y' = -\frac{1}{2}$.
Thus, the equation of the tangent is
$y = -1 - \frac{1}{2}(x - 2)$, or $x + 2y = 0$.

14. $\tan(xy^2) = (2/\pi)xy$
$(\sec^2(xy^2))(y^2 + 2xyy') = (2/\pi)(y + xy')$.
At $(-\pi, 1/2)$: $2((1/4) - \pi y') = (1/\pi) - 2y'$, so
$y' = (\pi - 2)/(4\pi(\pi - 1))$. The tangent has equation

$$y = \frac{1}{2} + \frac{\pi - 2}{4\pi(\pi - 1)}(x + \pi).$$

16. $\cos\left(\dfrac{\pi y}{x}\right) = \dfrac{x^2}{y} - \dfrac{17}{2}$

$\left[-\sin\left(\dfrac{\pi y}{x}\right)\right]\dfrac{\pi(xy' - y)}{x^2} = \dfrac{2xy - x^2 y'}{y^2}$.

At $(3, 1)$: $-\dfrac{\sqrt{3}}{2}\dfrac{\pi(3y' - 1)}{9} = 6 - 9y'$,

so $y' = (108 - \sqrt{3}\pi)/(162 - 3\sqrt{3}\pi)$. The tangent has equation

$$y = 1 + \frac{108 - \sqrt{3}\pi}{162 - 3\sqrt{3}\pi}(x - 3).$$

18. $x^2 + 4y^2 = 4$, $\quad 2x + 8yy' = 0$, $\quad 2 + 8(y')^2 + 8yy'' = 0$.
Thus, $y' = \dfrac{-x}{4y}$ and

$y'' = \dfrac{-2 - 8(y')^2}{8y} = -\dfrac{1}{4y} - \dfrac{x^2}{16y^3} = \dfrac{-4y^2 - x^2}{16y^3} = -\dfrac{1}{4y^3}$.

20. $x^3 - 3xy + y^3 = 1$
$3x^2 - 3y - 3xy' + 3y^2 y' = 0$
$6x - 3y' - 3y' - 3xy'' + 6y(y')^2 + 3y^2 y'' = 0$
Thus

$$y' = \frac{y - x^2}{y^2 - x}$$

$$y'' = \frac{-2x + 2y' - 2y(y')^2}{y^2 - x}$$

$$= \frac{2}{y^2 - x}\left[-x + \left(\frac{y - x^2}{y^2 - x}\right) - y\left(\frac{y - x^2}{y^2 - x}\right)^2\right]$$

$$= \frac{2}{y^2 - x}\left[\frac{-2xy}{(y^2 - x)^2}\right] = \frac{4xy}{(x - y^2)^3}.$$

22. $Ax^2 + By^2 = C$
$2Ax + 2Byy' = 0 \Rightarrow y' = -\dfrac{Ax}{By}$
$2A + 2B(y')^2 + 2Byy'' = 0$.
Thus,

$$y'' = \frac{-A - B(y')^2}{By} = \frac{-A - B\left(\dfrac{Ax}{By}\right)^2}{By}$$

$$= \frac{-A(By^2 + Ax^2)}{B^2 y^3} = -\frac{AC}{B^2 y^3}.$$

24. Maple gives the slope as $\dfrac{206}{55}$.

26. Maple gives the value $-\dfrac{855,000}{371,293}$.

28. The slope of the ellipse $\dfrac{x^2}{a^2} + \dfrac{y^2}{b^2} = 1$ is found from

$$\frac{2x}{a^2} + \frac{2y}{b^2}y' = 0, \quad \text{i.e. } y' = -\frac{b^2 x}{a^2 y}.$$

Similarly, the slope of the hyperbola $\dfrac{x^2}{A^2} - \dfrac{y^2}{B^2} = 1$ at (x, y) satisfies

$$\frac{2x}{A^2} - \frac{2y}{B^2}y' = 0, \quad \text{or } y' = \frac{B^2 x}{A^2 y}.$$

If the point (x, y) is an intersection of the two curves, then

$$\frac{x^2}{a^2} + \frac{y^2}{b^2} = \frac{x^2}{A^2} - \frac{y^2}{B^2}$$

$$x^2\left(\frac{1}{A^2} - \frac{1}{a^2}\right) = y^2\left(\frac{1}{B^2} + \frac{1}{b^2}\right).$$

Thus, $\dfrac{x^2}{y^2} = \dfrac{b^2 + B^2}{B^2 b^2} \cdot \dfrac{A^2 a^2}{a^2 - A^2}$.

Since $a^2 - b^2 = A^2 + B^2$, therefore $B^2 + b^2 = a^2 - A^2$,
and $\dfrac{x^2}{y^2} = \dfrac{A^2 a^2}{B^2 b^2}$. Thus, the product of the slope of the two curves at (x, y) is

$$-\frac{b^2 x}{a^2 y} \cdot \frac{B^2 x}{A^2 y} = -\frac{b^2 B^2}{a^2 A^2} \cdot \frac{A^2 a^2}{B^2 b^2} = -1.$$

Therefore, the curves intersect at right angles.

30. $\dfrac{x - y}{x + y} = \dfrac{x}{y} + 1 \Leftrightarrow xy - y^2 = x^2 + xy + xy + y^2$

$$\Leftrightarrow x^2 + 2y^2 + xy = 0$$

Differentiate with respect to x:

$$2x + 4yy' + y + xy' = 0 \quad \Rightarrow \quad y' = -\frac{2x + y}{4y + x}.$$

However, since $x^2 + 2y^2 + xy = 0$ can be written

$$x + xy + \frac{1}{4}y^2 + \frac{7}{4}y^2 = 0, \text{ or } (x + \frac{y}{2})^2 + \frac{7}{4}y^2 = 0,$$

the only solution is $x = 0$, $y = 0$, and these values do not satisfy the original equation. There are no points on the given curve.

Section 2.10   Antiderivatives and Initial-Value Problems   (page 162)

2. $\int x^2\,dx = \frac{1}{3}x^3 + C$

4. $\int x^{12}\,dx = \frac{1}{13}x^{13} + C$

6. $\int (x + \cos x)\,dx = \frac{x^2}{2} + \sin x + C$

8. $\int \frac{1 + \cos^3 x}{\cos^2 x}\,dx = \int (\sec^2 x + \cos x)\,dx = \tan x + \sin x + C$

10. $\int (A + Bx + Cx^2)\,dx = Ax + \frac{B}{2}x^2 + \frac{C}{3}x^3 + K$

12. $\int \frac{6(x - 1)}{x^{4/3}}\,dx = \int (6x^{-1/3} - 6x^{-4/3})\,dx$
$$= 9x^{2/3} + 18x^{-1/3} + C$$

14. $105 \int (1 + t^2 + t^4 + t^6)\,dt$
$$= 105(t + \frac{1}{3}t^3 + \frac{1}{5}t^5 + \frac{1}{7}t^7) + C$$
$$= 105t + 35t^3 + 21t^5 + 15t^7 + C$$

16. $\int \sin\left(\frac{x}{2}\right)\,dx = -2\cos\left(\frac{x}{2}\right) + C$

18. $\int \sec(1 - x)\tan(1 - x)\,dx = -\sec(1 - x) + C$

20. Since $\frac{d}{dx}\sqrt{x + 1} = \frac{1}{2\sqrt{x + 1}}$, therefore
$$\int \frac{4}{\sqrt{x + 1}}\,dx = 8\sqrt{x + 1} + C.$$

22. Since $\frac{d}{dx}\sqrt{x^2 + 1} = \frac{x}{\sqrt{x^2 + 1}}$, therefore
$$\int \frac{2x}{\sqrt{x^2 + 1}}\,dx = 2\sqrt{x^2 + 1} + C.$$

24. $\int \sin x \cos x\,dx = \int \frac{1}{2}\sin(2x)\,dx = -\frac{1}{4}\cos(2x) + C$

26. $\int \sin^2 x\,dx = \int \frac{1 - \cos(2x)}{2}\,dx = \frac{x}{2} - \frac{\sin(2x)}{4} + C$

28. Given that
$$\begin{cases} y' = x^{-2} - x^{-3} \\ y(-1) = 0, \end{cases}$$
then $y = \int (x^{-2} - x^{-3})\,dx = -x^{-1} + \frac{1}{2}x^{-2} + C$
and $0 = y(-1) = -(-1)^{-1} + \frac{1}{2}(-1)^{-2} + C$ so $C = -\frac{3}{2}$.
Hence, $y(x) = -\frac{1}{x} + \frac{1}{2x^2} - \frac{3}{2}$ which is valid on the interval $(-\infty, 0)$.

30. Given that
$$\begin{cases} y' = x^{1/3} \\ y(0) = 5, \end{cases}$$
then $y = \int x^{1/3}\,dx = \frac{3}{4}x^{4/3} + C$ and $5 = y(0) = C$.
Hence, $y(x) = \frac{3}{4}x^{4/3} + 5$ which is valid on the whole real line.

32. Given that
$$\begin{cases} y' = x^{-9/7} \\ y(1) = -4, \end{cases}$$
then $y = \int x^{-9/7}\,dx = -\frac{7}{2}x^{-2/7} + C.$
Also, $-4 = y(1) = -\frac{7}{2} + C$, so $C = -\frac{1}{2}$. Hence, $y = -\frac{7}{2}x^{-2/7} - \frac{1}{2}$, which is valid in the interval $(0, \infty)$.

34. For $\begin{cases} y' = \sin(2x) \\ y(\pi/2) = 1 \end{cases}$, we have
$$y = \int \sin(2x)\,dx = -\frac{1}{2}\cos(2x) + C$$
$$1 = -\frac{1}{2}\cos\pi + C = \frac{1}{2} + C \implies C = \frac{1}{2}$$
$$y = \frac{1}{2}\big(1 - \cos(2x)\big) \quad \text{(for all } x\text{)}.$$

36. For $\begin{cases} y' = \sec^2 x \\ y(\pi) = 1 \end{cases}$, we have
$$y = \int \sec^2 x\,dx = \tan x + C$$
$$1 = \tan\pi + C = C \implies C = 1$$
$$y = \tan x + 1 \quad \text{(for } \pi/2 < x < 3\pi/2\text{)}.$$

38. Given that
$$\begin{cases} y'' = x^{-4} \\ y'(1) = 2 \\ y(1) = 1, \end{cases}$$
then $y' = \int x^{-4}\,dx = -\frac{1}{3}x^{-3} + C.$
Since $2 = y'(1) = -\frac{1}{3} + C$, therefore $C = \frac{7}{3}$, and $y' = -\frac{1}{3}x^{-3} + \frac{7}{3}$. Thus
$$y = \int \left(-\frac{1}{3}x^{-3} + \frac{7}{3}\right)\,dx = \frac{1}{6}x^{-2} + \frac{7}{3}x + D,$$

and $1 = y(1) = \frac{1}{6} + \frac{7}{3} + D$, so that $D = -\frac{3}{2}$. Hence, $y(x) = \frac{1}{6}x^{-2} + \frac{7}{3}x - \frac{3}{2}$, which is valid in the interval $(0, \infty)$.

40. Given that
$$\begin{cases} y'' = 5x^2 - 3x^{-1/2} \\ y'(1) = 2 \\ y(1) = 0, \end{cases}$$

we have $y' = \int 5x^2 - 3x^{-1/2}\,dx = \frac{5}{3}x^3 - 6x^{1/2} + C.$

Also, $2 = y'(1) = \frac{5}{3} - 6 + C$ so that $C = \dfrac{19}{3}$. Thus,
$y' = \frac{5}{3}x^3 - 6x^{1/2} + \frac{19}{3}$, and

$$y = \int \left(\frac{5}{3}x^3 - 6x^{1/2} + \frac{19}{3}\right) dx = \frac{5}{12}x^4 - 4x^{3/2} + \frac{19}{3}x + D.$$

Finally, $0 = y(1) = \frac{5}{12} - 4 + \frac{19}{3} + D$ so that $D = -\frac{11}{4}$.
Hence, $y(x) = \frac{5}{12}x^4 - 4x^{3/2} + \frac{19}{3}x - \frac{11}{4}$.

42. For $\begin{cases} y'' = x + \sin x \\ y(0) = 2 \\ y'(0) = 0 \end{cases}$ we have

$$y' = \int (x + \sin x)\,dx = \frac{x^2}{2} - \cos x + C_1$$

$$0 = 0 - \cos 0 + C_1 \implies C_1 = 1$$

$$y = \int \left(\frac{x^2}{2} - \cos x + 1\right) dx = \frac{x^3}{6} - \sin x + x + C_2$$

$$2 = 0 - \sin 0 + 0 + C_2 \implies C_2 = 2$$

$$y = \frac{x^3}{6} - \sin x + x + 2.$$

44. Let r_1 and r_2 be distinct rational roots of the equation
$ar(r - 1) + br + c = 0$
Let $y = Ax^{r_1} + Bx^{r_2}$ $(x > 0)$
Then $y' = Ar_1 x^{r_1 - 1} + Br_2 x^{r_2 - 1}$,
and $y'' = Ar_1(r_1 - 1)x^{r_1 - 2} + Br_2(r_2 - 1)x^{r_2 - 2}$. Thus
$ax^2 y'' + bxy' + cy$
$= ax^2(Ar_1(r_1 - 1)x^{r_1 - 2} + Br_2(r_2 - 1)x^{r_2 - 2}$
$\quad + bx(Ar_1 x^{r_1 - 1} + Br_2 x^{r_2 - 1}) + c(Ax^{r_1} + Bx^{r_2})$
$= A\big(ar_1(r_1 - 1) + br_1 + c\big)x^{r_1}$
$\quad + B\big(ar_2(r_2 - 1) + br_2 + c\big)x^{r_2}$
$= 0x^{r_1} + 0x^{r_2} \equiv 0 \quad (x > 0)$

46. Consider
$$\begin{cases} x^2 y'' - 6y = 0 \\ y(1) = 1 \\ y'(1) = 1. \end{cases}$$

Let $y = x^r$, $y' = rx^{r-1}$, $y'' = r(r-1)x^{r-2}$. Substituting these expressions into the differential equation we obtain
$$x^2[r(r-1)x^{r-2}] - 6x^r = 0$$
$$[r(r-1) - 6]x^r = 0.$$
Since this equation must hold for all $x > 0$, we must have
$$r(r-1) - 6 = 0$$
$$r^2 - r - 6 = 0$$
$$(r - 3)(r + 2) = 0.$$
There are two roots: $r_1 = -2$, and $r_2 = 3$. Thus the differential equation has solutions of the form $y = Ax^{-2} + Bx^3$. Then $y' = -2Ax^{-3} + 3Bx^2$. Since $1 = y(1) = A + B$ and $1 = y'(1) = -2A + 3B$, therefore $A = \frac{2}{5}$ and $B = \frac{3}{5}$. Hence, $y = \frac{2}{5}x^{-2} + \frac{3}{5}x^3$.

Section 2.11 Velocity and Acceleration (page 169)

2. $x = 4 + 5t - t^2$, $v = 5 - 2t$, $a = -2$.

a) The point is moving to the right if $v > 0$, i.e., when $t < \frac{5}{2}$.

b) The point is moving to the left if $v < 0$, i.e., when $t > \frac{5}{2}$.

c) The point is accelerating to the right if $a > 0$, but $a = -2$ at all t; hence, the point never accelerates to the right.

d) The point is accelerating to the left if $a < 0$, i.e., for all t.

e) The particle is speeding up if v and a have the same sign, i.e., for $t > \frac{5}{2}$.

f) The particle is slowing down if v and a have opposite sign, i.e., for $t < \frac{5}{2}$.

g) Since $a = -2$ at all t, $a = -2$ at $t = \frac{5}{2}$ when $v = 0$.

h) The average velocity over $[0, 4]$ is
$$\frac{x(4) - x(0)}{4} = \frac{8 - 4}{4} = 1.$$

4. $x = \dfrac{t}{t^2 + 1}$, $v = \dfrac{(t^2 + 1)(1) - (t)(2t)}{(t^2 + 1)^2} = \dfrac{1 - t^2}{(t^2 + 1)^2}$,
$a = \dfrac{(t^2 + 1)^2(-2t) - (1 - t^2)(2)(t^2 + 1)(2t)}{(t^2 + 1)^4} = \dfrac{2t(t^2 - 3)}{(t^2 + 1)^3}.$

a) The point is moving to the right if $v > 0$, i.e., when $1 - t^2 > 0$, or $-1 < t < 1$.

b) The point is moving to the left if $v < 0$, i.e., when $t < -1$ or $t > 1$.

c) The point is accelerating to the right if $a > 0$, i.e., when $2t(t^2 - 3) > 0$, that is, when $t > \sqrt{3}$ or $-\sqrt{3} < t < 0$.

d) The point is accelerating to the left if $a < 0$, i.e., for $t < -\sqrt{3}$ or $0 < t < \sqrt{3}$.

e) The particle is speeding up if v and a have the same sign, i.e., for $t < -\sqrt{3}$, or $-1 < t < 0$ or $1 < t < \sqrt{3}$.

f) The particle is slowing down if v and a have opposite sign, i.e., for $-\sqrt{3} < t < -1$, or $0 < t < 1$ or $t > \sqrt{3}$.

g) $v = 0$ at $t = \pm 1$. At $t = -1$, $a = \dfrac{-2(-2)}{(2)^3} = \dfrac{1}{2}$.

At $t = 1$, $a = \dfrac{2(-2)}{(2)^3} = -\dfrac{1}{2}$.

h) The average velocity over $[0, 4]$ is
$$\frac{x(4) - x(0)}{4} = \frac{\frac{4}{17} - 0}{4} = \frac{1}{17}.$$

6. Given that $y = 100 - 2t - 4.9t^2$, the time t at which the ball reaches the ground is the positive root of the equation $y = 0$, i.e., $100 - 2t - 4.9t^2 = 0$, namely,

$$t = \frac{-2 + \sqrt{4 + 4(4.9)(100)}}{9.8} \approx 4.318 \text{ s}.$$

The average velocity of the ball is $\dfrac{-100}{4.318} = -23.16$ m/s. Since $-23.159 = v = -2 - 9.8t$, then $t \simeq 2.159$ s.

8. Let $y(t)$ be the height of the projectile t seconds after it is fired upward from ground level with initial speed v_0. Then

$$y''(t) = -9.8, \quad y'(0) = v_0, \quad y(0) = 0.$$

Two antidifferentiations give

$$y = -4.9t^2 + v_0 t = t(v_0 - 4.9t).$$

Since the projectile returns to the ground at $t = 10$ s, we have $y(10) = 0$, so $v_0 = 49$ m/s. On Mars, the acceleration of gravity is 3.72 m/s^2 rather than 9.8 m/s^2, so the height of the projectile would be

$$y = -1.86t^2 + v_0 t = t(49 - 1.86t).$$

The time taken to fall back to ground level on Mars would be $t = 49/1.86 \approx 26.3$ s.

10. To get to $3h$ metres above Mars, the ball would have to be thrown upward with speed

$$v_M = \sqrt{6g_M h} = \sqrt{6g_M v_0^2/(2g)} = v_0\sqrt{3g_M/g}.$$

Since $g_M = 3.72$ and $g = 9.80$, we have $v_M \approx 1.067 v_0$ m/s.

12. If the cliff is h ft high, then the height of the rock t seconds after it is thrown down is $y = h - 32t - 16t^2$ ft. The rock hits the ground ($y = 0$) at time

$$t = \frac{-32 + \sqrt{32^2 + 64h}}{32} = -1 + \frac{1}{4}\sqrt{16 + h} \text{ s}.$$

Its speed at that time is

$$v = -32 - 32t = -8\sqrt{16 + h} = -160 \text{ ft/s}.$$

Solving this equation for h gives the height of the cliff as 384 ft.

14. $x = At^2 + Bt + C$, $v = 2At + B$.
The average velocity over $[t_1, t_2]$ is
$$\frac{x(t_2) - x(t_1)}{t_2 - t_1}$$
$$= \frac{At_2^2 + Bt_1 + C - At_1^2 - Bt_1 - C}{t_2 - t_1}$$
$$= \frac{A(t_2^2 - t_1^2) + B(t_2 - t_1)}{(t_2 - t_1)}$$
$$= \frac{A(t_2 + t_1)(t_2 - t_1) + B(t_2 - t_1)}{(t_2 - t_1)}$$
$$= A(t_2 + t_1) + B.$$
The instantaneous velocity at the midpoint of $[t_1, t_2]$ is
$$v\left(\frac{t_2 + t_1}{2}\right) = 2A\left(\frac{t_2 + t_1}{2}\right) + B = A(t_2 + t_1) + B.$$
Hence, the average velocity over the interval is equal to the instantaneous velocity at the midpoint.

16. This exercise and the next three refer to the following figure depicting the velocity of a rocket fired from a tower as a function of time since firing.

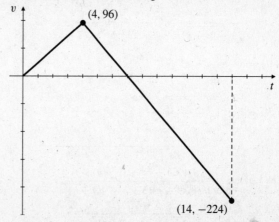

Fig. 2.11.16

The rocket's acceleration while its fuel lasted is the slope of the first part of the graph, namely $96/4 = 24$ ft/s.

18. As suggested in Example 1 on page 154 of the text, the distance travelled by the rocket while it was falling from its maximum height to the ground is the area between the velocity graph and the part of the t-axis where $v < 0$. The area of this triangle is $(1/2)(14 - 7)(224) = 784$ ft. This is the maximum height the rocket achieved.

20. Let $s(t)$ be the distance the car travels in the t seconds after the brakes are applied. Then $s''(t) = -t$ and the velocity at time t is given by

$$s'(t) = \int (-t)\, dt = -\frac{t^2}{2} + C_1,$$

where $C_1 = 20$ m/s (that is, 72km/h) as determined in Example 6. Thus

$$s(t) = \int \left(20 - \frac{t^2}{2}\right) dt = 20t - \frac{t^3}{6} + C_2,$$

where $C_2 = 0$ because $s(0) = 0$. The time taken to come to a stop is given by $s'(t) = 0$, so it is $t = \sqrt{40}$ s. The distance travelled is

$$s = 20\sqrt{40} - \frac{1}{6}40^{3/2} \approx 84.3 \text{ m}.$$

Review Exercises 2   (page 171)

2.
$$\frac{d}{dx}\sqrt{1-x^2} = \lim_{h\to 0}\frac{\sqrt{1-(x+h)^2} - \sqrt{1-x^2}}{h}$$
$$= \lim_{h\to 0}\frac{1-(x+h)^2 - (1-x^2)}{h(\sqrt{1-(x+h)^2} + \sqrt{1-x^2})}$$
$$= \lim_{h\to 0}\frac{-2x-h}{\sqrt{1-(x+h)^2} + \sqrt{1-x^2}} = -\frac{x}{\sqrt{1-x^2}}$$

4. $g(t) = \dfrac{t-5}{1+\sqrt{t}}$

$$g'(9) = \lim_{h\to 0}\frac{\dfrac{4+h}{1+\sqrt{9+h}} - 1}{h}$$
$$= \lim_{h\to 0}\frac{(3+h-\sqrt{9+h})(3+h+\sqrt{9+h})}{h(1+\sqrt{9+h})(3+h+\sqrt{9+h})}$$
$$= \lim_{h\to 0}\frac{9+6h+h^2 - (9+h)}{h(1+\sqrt{9+h})(3+h+\sqrt{9+h})}$$
$$= \lim_{h\to 0}\frac{5+h}{(1+\sqrt{9+h})(3+h+\sqrt{9+h})}$$
$$= \frac{5}{24}$$

6. At $x = \pi$ the curve $y = \tan(x/4)$ has slope $(\sec^2(\pi/4))/4 = 1/2$. The normal to the curve there has equation $y = 1 - 2(x - \pi)$.

8.
$$\frac{d}{dx}\frac{1+x+x^2+x^3}{x^4} = \frac{d}{dx}(x^{-4} + x^{-3} + x^{-2} + x^{-1})$$
$$= -4x^{-5} - 3x^{-4} - 2x^{-3} - x^{-2}$$
$$= -\frac{4+3x+2x^2+x^3}{x^5}$$

10.
$$\frac{d}{dx}\sqrt{2+\cos^2 x} = \frac{-2\cos x \sin x}{2\sqrt{2+\cos^2 x}} = \frac{-\sin x \cos x}{\sqrt{2+\cos^2 x}}$$

12.
$$\frac{d}{dt}\frac{\sqrt{1+t^2}-1}{\sqrt{1+t^2}+1}$$
$$= \frac{(\sqrt{1+t^2}+1)\dfrac{t}{\sqrt{1+t^2}} - (\sqrt{1+t^2}-1)\dfrac{t}{\sqrt{1+t^2}}}{(\sqrt{1+t^2}+1)^2}$$
$$= \frac{2t}{\sqrt{1+t^2}(\sqrt{1+t^2}+1)^2}$$

14.
$$\lim_{x\to 2}\frac{\sqrt{4x+1}-3}{x-2} = \lim_{h\to 0}4\frac{\sqrt{9+4h}-3}{4h}$$
$$= \frac{d}{dx}4\sqrt{x}\Big|_{x=9} = \frac{4}{2\sqrt{9}} = \frac{2}{3}$$

16.
$$\lim_{x\to -a}\frac{(1/x^2) - (1/a^2)}{x+a} = \lim_{h\to 0}\frac{\dfrac{1}{(-a+h)^2} - \dfrac{1}{(-a)^2}}{h}$$
$$= \frac{d}{dx}\frac{1}{x^2}\Big|_{x=-a} = \frac{2}{a^3}$$

18.
$$\frac{d}{dx}[f(\sqrt{x})]^2 = 2f(\sqrt{x})f'(\sqrt{x})\frac{1}{2\sqrt{x}} = \frac{f(\sqrt{x})f'(\sqrt{x})}{\sqrt{x}}$$

20.
$$\frac{d}{dx}\frac{f(x)-g(x)}{f(x)+g(x)}$$
$$= \frac{1}{(f(x)+g(x))^2}\Big[f(x)+g(x))(f'(x)-g'(x))$$
$$- (f(x)-g(x))(f'(x)+g'(x))\Big]$$
$$= \frac{2(f'(x)g(x) - f(x)g'(x))}{(f(x)+g(x))^2}$$

22.
$$\frac{d}{dx}f\left(\frac{g(x^2)}{x}\right) = \frac{2x^2 g'(x^2) - g(x^2)}{x^2}f'\left(\frac{g(x^2)}{x}\right)$$

24.
$$\frac{d}{dx}\sqrt{\frac{\cos f(x)}{\sin g(x)}}$$
$$= \frac{1}{2}\sqrt{\frac{\sin g(x)}{\cos f(x)}}$$
$$\times \frac{-f'(x)\sin f(x)\sin g(x) - g'(x)\cos f(x)\cos g(x)}{(\sin g(x))^2}$$

26. $3\sqrt{2x}\sin(\pi y) + 8y\cos(\pi x) = 2$
$3\sqrt{2}\sin(\pi y) + 3\pi\sqrt{2x}\cos(\pi y)y' + 8y'\cos(\pi x)$
$-8\pi y\sin(\pi x) = 0$
At $(1/3, 1/4)$: $3 + \pi y' + 4y' - \pi\sqrt{3} = 0$, so the slope there is $y' = \dfrac{\pi\sqrt{3}-3}{\pi+4}$.

28. $\displaystyle\int\frac{1+x}{\sqrt{x}}\,dx = \int(x^{-1/2} + x^{1/2})\,dx = 2\sqrt{x} + \frac{2}{3}x^{3/2} + C$

30. $\displaystyle\int (2x+1)^4\,dx = \int (16x^4 + 32x^3 + 24x^2 + 8x + 1)\,dx$

$$= \frac{16x^5}{5} + 8x^4 + 8x^3 + 4x^2 + x + C$$

or, equivalently,

$$\int (2x+1)^4\,dx = \frac{(2x+1)^5}{10} + C$$

32. If $g'(x) = \sin(x/3) + \cos(x/6)$, then

$$g(x) = -3\cos(x/3) + 6\sin(x/6) + C.$$

If $(\pi, 2)$ lies on $y = g(x)$, then $-(3/2) + 3 + C = 2$, so $C = 1/2$ and $g(x) = -3\cos(x/3) + 6\sin(x/6) + (1/2)$.

34. If $f'(x) = f(x)$ and $g(x) = x\,f(x)$, then

$$g'(x) = f(x) + xf'(x) = (1+x)f(x)$$
$$g''(x) = f(x) + (1+x)f'(x) = (2+x)f(x)$$
$$g'''(x) = f(x) + (2+x)f'(x) = (3+x)f(x)$$

Conjecture: $g^{(n)}(x) = (n+x)f(x)$ for $n = 1, 2, 3, \ldots$
Proof: The formula is true for $n = 1, 2,$ and 3 as shown above. Suppose it is true for $n = k$; that is, suppose $g^{(k)}(x) = (k+x)f(x)$. Then

$$g^{(k+1)}(x) = \frac{d}{dx}\big((k+x)f(x)\big)$$
$$= f(x) + (k+x)f'(x) = ((k+1)+x)f(x).$$

Thus the formula is also true for $n = k+1$. It is therefore true for all positive integers n by induction.

36. The tangent to $y = \sqrt{2+x^2}$ at $x = a$ has slope $a/\sqrt{2+a^2}$ and equation

$$y = \sqrt{2+a^2} + \frac{a}{\sqrt{2+a^2}}(x-a).$$

This line passes through $(0, 1)$ provided

$$1 = \sqrt{2+a^2} - \frac{a^2}{\sqrt{2+a^2}}$$
$$\sqrt{2+a^2} = 2 + a^2 - a^2 = 2$$
$$2 + a^2 = 4$$

The possibilities are $a = \pm\sqrt{2}$, and the equations of the corrresponding tangent lines are $y = 1 \pm (x/\sqrt{2})$.

38. $\dfrac{d}{dx}\big(\sin^n x \cos(nx)\big)$

$$= n\sin^{n-1} x \cos x \cos(nx) - n\sin^n x \sin(nx)$$
$$= n\sin^{n-1} x[\cos x \cos(nx) - \sin x \sin(nx)]$$
$$= n\sin^{n-1} x \cos((n+1)x)$$

$\dfrac{d}{dx}\big(\cos^n x \sin(nx)\big)$

$$= -n\cos^{n-1} x \sin x \sin(nx) + n\cos^n x \cos(nx)$$
$$= n\cos^{n-1} x[\cos x \cos(nx) - \sin x \sin(nx)]$$
$$= n\cos^{n-1} x \cos((n+1)x)$$

$\dfrac{d}{dx}\big(\cos^n x \cos(nx)\big)$

$$= -n\cos^{n-1} x \sin x \cos(nx) - n\cos^n x \sin(nx)$$
$$= -n\cos^{n-1} x[\sin x \cos(nx) + \cos x \sin(nx)]$$
$$= -n\cos^{n-1} x \sin((n+1)x)$$

40. The average profit per tonne if x tonnes are exported is $P(x)/x$, that is the slope of the line joining $(x, P(x))$ to the origin. This slope is maximum if the line is tangent to the graph of $P(x)$. In this case the slope of the line is $P'(x)$, the marginal profit.

42. $PV = kT$. Differentiate with respect to P holding T constant to get

$$V + P\frac{dV}{dP} = 0$$

Thus the isothermal compressibility of the gas is

$$\frac{1}{V}\frac{dV}{dP} = \frac{1}{V}\left(-\frac{V}{P}\right) = -\frac{1}{P}.$$

44. The first ball has initial height 60 m and initial velocity 0, so its height at time t is

$$y_1 = 60 - 4.9t^2 \text{ m}.$$

The second ball has initial height 0 and initial velocity v_0, so its height at time t is

$$y_2 = v_0 t - 4.9t^2 \text{ m}.$$

The two balls collide at a height of 30 m (at time T, say). Thus

$$30 = 60 - 4.9T^2$$
$$30 = v_0 T - 4.9T^2.$$

Thus $v_0 T = 60$ and $T^2 = 30/4.9$. The initial upward speed of the second ball is

$$v_0 = \frac{60}{T} = 60\sqrt{\frac{4.9}{30}} \approx 24.25 \text{ m/s}.$$

At time T, the velocity of the first ball is

$$\left.\frac{dy_1}{dt}\right|_{t=T} = -9.8T \approx -24.25 \text{ m/s}.$$

At time T, the velocity of the second ball is

$$\left.\frac{dy_2}{dt}\right|_{t=T} = v_0 - 9.8T = 0 \text{ m/s}.$$

46. $P = 2\pi\sqrt{L/g} = 2\pi L^{1/2}g^{-1/2}$.

a) If L remains constant, then

$$\Delta P \approx \frac{dP}{dg}\,\Delta g = -\pi L^{1/2}g^{-3/2}\,\Delta g$$

$$\frac{\Delta P}{P} \approx \frac{-\pi L^{1/2}g^{-3/2}}{2\pi L^{1/2}g^{-1/2}}\,\Delta g = -\frac{1}{2}\frac{\Delta g}{g}.$$

If g increases by 1%, then $\Delta g/g = 1/100$, and $\Delta P/P = -1/200$. Thus P decreases by 0.5%.

b) If g remains constant, then

$$\Delta P \approx \frac{dP}{dL}\,\Delta L = \pi L^{-1/2}g^{-1/2}\,\Delta L$$

$$\frac{\Delta P}{P} \approx \frac{\pi L^{-1/2}g^{-1/2}}{2\pi L^{1/2}g^{-1/2}}\,\Delta L = \frac{1}{2}\frac{\Delta L}{L}.$$

If L increases by 2%, then $\Delta L/L = 2/100$, and $\Delta P/P = 1/100$. Thus P increases by 1%.

Challenging Problems 2   (page 172)

2. $f'(x) = 1/x$, $f(2) = 9$.

a)
$$\lim_{x\to 2}\frac{f(x^2+5)-f(9)}{x-2} = \lim_{h\to 0}\frac{f(9+4h+h^2)-f(9)}{h}$$

$$= \lim_{h\to 0}\frac{f(9+4h+h^2)-f(9)}{4h+h^2}\times\frac{4h+h^2}{h}$$

$$= \lim_{k\to 0}\frac{f(9+k)-f(9)}{k}\times\lim_{h\to 0}(4+h)$$

$$= f'(9)\times 4 = \frac{4}{9}$$

b)
$$\lim_{x\to 2}\frac{\sqrt{f(x)}-3}{x-2} = \lim_{h\to 0}\frac{\sqrt{f(2+h)}-3}{h}$$

$$= \lim_{h\to 0}\frac{f(2+h)-9}{h}\times\frac{1}{\sqrt{f(2+h)}+3}$$

$$= f'(2)\times\frac{1}{6} = \frac{1}{12}.$$

4. $f(x) = \begin{cases} x, & \text{if } x = 1,\ 1/2,\ 1/3,\ \dots \\ x^2 & \text{otherwise} \end{cases}$

a) f is continuous except at $1/2$, $1/3$, $1/4$, ... It is continuous at $x = 1$ and $x = 0$ (and everywhere else). Note that

$$\lim_{x\to 1}x^2 = 1 = f(1),$$

$$\lim_{x\to 0}x^2 = \lim_{x\to 0}x = 0 = f(0)$$

b) If $a = 1/2$ and $b = 1/3$, then

$$\frac{f(a)+f(b)}{2} = \frac{1}{2}\left(\frac{1}{2}+\frac{1}{3}\right) = \frac{5}{12}.$$

If $1/3 < x < 1/2$, then $f(x) = x^2 < 1/4 < 5/12$. Thus the statement is FALSE.

c) By (a) f cannot be differentiable at $x = 1/2$, $1/2$, ... It is not differentiable at $x = 0$ either, since

$$\lim_{h\to 0}h - 0h = 1 \neq 0 = \lim_{h\to 0}\frac{h^2-0}{h}.$$

f is differentiable elsewhere, including at $x = 1$ where its derivative is 2.

6.   Given that $f'(0) = k$, $f(0) \neq 0$, and $f(x+y) = f(x)f(y)$, we have

$$f(0) = f(0+0) = f(0)f(0) \quad\Longrightarrow\quad f(0) = 0 \text{ or } f(0) = 1.$$

Thus $f(0) = 1$.

$$f'(x) = \lim_{h\to 0}\frac{f(x+h)-f(x)}{h}$$

$$= \lim_{h\to 0}\frac{f(x)f(h)-f(x)}{h} = f(x)f'(0) = kf(x).$$

8.   a)
$$f'(x) = \lim_{k\to 0}\frac{f(x+k)-f(x)}{k}\quad(\text{let } k = -h)$$

$$= \lim_{h\to 0}\frac{f(x-h)-f(x)}{-h} = \lim_{h\to 0}\frac{f(x)-f(x-h)}{h}.$$

$$f'(x) = \frac{1}{2}\big(f'(x)+f'(x)\big)$$

$$= \frac{1}{2}\left(\lim_{h\to 0}\frac{f(x+h)-f(x)}{h}\right.$$

$$\left.+\lim_{h\to 0}\frac{f(x)-f(x-h)}{h}\right)$$

$$= \lim_{h\to 0}\frac{f(x+h)-f(x-h)}{2h}.$$

b) The change of variables used in the first part of (a) shows that

$$\lim_{h\to 0}\frac{f(x+h)-f(x)}{h}\quad and\quad \lim_{h\to 0}\frac{f(x)-f(x-h)}{h}$$

are always equal if either exists.

c) If $f(x) = |x|$, then $f'(0)$ does not exist, but

$$\lim_{h\to 0}\frac{f(0+h)-f(0-h)}{2h} = \lim_{h\to 0}\frac{|h|-|h|}{h} = \lim_{h\to 0}\frac{0}{h} = 0.$$

10.   By symmetry, any line tangent to both curves must pass through the origin.

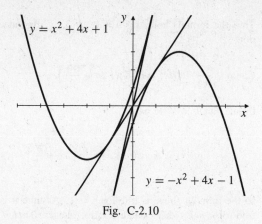

Fig. C-2.10

The tangent to $y = x^2 + 4x + 1$ at $x = a$ has equation

$$y = a^2 + 4a + 1 + (2a + 4)(x - a)$$
$$= (2a + 4)x - (a^2 - 1),$$

which passes through the origin if $a = \pm 1$. The two common tangents are $y = 6x$ and $y = 2x$.

12. The point $Q = (a, a^2)$ on $y = x^2$ that is closest to $P = (3, 0)$ is such that PQ is normal to $y = x^2$ at Q. Since PQ has slope $a^2/(a - 3)$ and $y = x^2$ has slope $2a$ at Q, we require

$$\frac{a^2}{a - 3} = -\frac{1}{2a},$$

which simplifies to $2a^3 + a - 3 = 0$. Observe that $a = 1$ is a solution of this cubic equation. Since the slope of $y = 2x^3 + x - 3$ is $6x^2 + 1$, which is always positive, the cubic equation can have only one real solution. Thus $Q = (1, 1)$ is the point on $y = x^2$ that is closest to P. The distance from P to the curve is $|PQ| = \sqrt{5}$ units.

14. Parabola $y = x^2$ has tangent $y = 2ax - a^2$ at (a, a^2). Parabola $y = Ax^2 + Bx + C$ has tangent

$$y = (2Ab + B)x - Ab^2 + C$$

at $(b, Ab^2 + Bb + C)$. These two tangents coincide if

$$2Ab + B = 2a \qquad\qquad (*)$$
$$Ab^2 - C = a^2.$$

The two curves have one (or more) common tangents if $(*)$ has real solutions for a and b. Eliminating a between the two equations leads to

$$(2Ab + B)^2 = 4Ab^2 - 4C,$$

or, on simplification,

$$4A(A - 1)b^2 + 4ABb + (B^2 + 4C) = 0.$$

This quadratic equation in b has discriminant

$$D = 16A^2B^2 - 16A(A-1)(B^2+4C) = 16A(B^2-4(A-1)C).$$

There are five cases to consider:

CASE I. If $A = 1$, $B \neq 0$, then $(*)$ gives

$$b = -\frac{B^2 + 4C}{4B}, \quad a = \frac{B^2 - 4C}{4B}.$$

There is a single common tangent in this case.

CASE II. If $A = 1$, $B = 0$, then $(*)$ forces $C = 0$, which is not allowed. There is no common tangent in this case.

CASE III. If $A \neq 1$ but $B^2 = 4(A - 1)C$, then

$$b = \frac{-B}{2(A - 1)} = a.$$

There is a single common tangent, and since the points of tangency on the two curves coincide, the two curves are tangent to each other.

CASE IV. If $A \neq 1$ and $B^2 - 4(A - 1)C < 0$, there are no real solutions for b, so there can be no common tangents.

CASE V. If $A \neq 1$ and $B^2 - 4(A - 1)C > 0$, there are two distinct real solutions for b, and hence two common tangent lines.

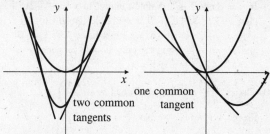

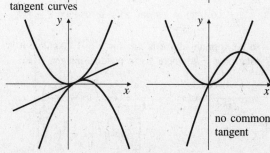

Fig. C-2.14

16.　a) $y = x^4 - 2x^2$ has horizontal tangents at points x satisfying $4x^3 - 4x = 0$, that is, at $x = 0$ and $x = \pm 1$. The horizontal tangents are $y = 0$ and $y = -1$. Note that $y = -1$ is a double tangent; it is tangent at the two points $(\pm 1, -1)$.

b) The tangent to $y = x^4 - 2x^2$ at $x = a$ has equation

$$y = a^4 - 2a^2 + (4a^3 - 4a)(x - a)$$
$$= 4a(a^2 - 1)x - 3a^4 + 2a^2.$$

Similarly, the tangent at $x = b$ has equation

$$y = 4b(b^2 - 1)x - 3b^4 + 2b^2.$$

These tangents are the same line (and hence a double tangent) if

$$4a(a^2 - 1) = 4b(b^2 - 1)$$
$$- 3a^4 + 2a^2 = -3b^4 + 2b^2.$$

The second equation says that either $a^2 = b^2$ or $3(a^2 + b^2) = 2$; the first equation says that $a^3 - b^3 = a - b$, or, equivalently, $a^2 + ab + b^2 = 1$. If $a^2 = b^2$, then $a = -b$ ($a = b$ is not allowed). Thus $a^2 = b^2 = 1$ and the two points are $(\pm 1, -1)$ as discovered in part (a). If $a^2 + b^2 = 2/3$, then $ab = 1/3$. This is not possible since it implies that

$$0 = a^2 + b^2 - 2ab = (a - b)^2 > 0.$$

Thus $y = -1$ is the only double tangent to $y = x^4 - 2x^2$.

c) If $y = Ax + B$ is a double tangent to $y = x^4 - 2x^2 + x$, then $y = (A - 1)x + B$ is a double tangent to $y = x^4 - 2x^2$. By (b) we must have $A - 1 = 0$ and $B = -1$. Thus the only double tangent to $y = x^4 - 2x^2 + x$ is $y = x - 1$.

18. a) Claim: $\dfrac{d^n}{dx^n} \cos(ax) = a^n \cos\left(ax + \dfrac{n\pi}{2}\right)$.

Proof: For $n = 1$ we have

$$\frac{d}{dx} \cos(ax) = -a \sin(ax) = a \cos\left(ax + \frac{\pi}{2}\right),$$

so the formula above is true for $n = 1$. Assume it is true for $n = k$, where k is a positive integer. Then

$$\frac{d^{k+1}}{dx^{k+1}} \cos(ax) = \frac{d}{dx}\left[a^k \cos\left(ax + \frac{k\pi}{2}\right)\right]$$
$$= a^k\left[-a \sin\left(ax + \frac{k\pi}{2}\right)\right]$$
$$= a^{k+1} \cos\left(ax + \frac{(k+1)\pi}{2}\right).$$

Thus the formula holds for $n = 1, 2, 3, \ldots$ by induction.

b) Claim: $\dfrac{d^n}{dx^n} \sin(ax) = a^n \sin\left(ax + \dfrac{n\pi}{2}\right)$.

Proof: For $n = 1$ we have

$$\frac{d}{dx} \sin(ax) = a \cos(ax) = a \sin\left(ax + \frac{\pi}{2}\right),$$

so the formula above is true for $n = 1$. Assume it is true for $n = k$, where k is a positive integer. Then

$$\frac{d^{k+1}}{dx^{k+1}} \sin(ax) = \frac{d}{dx}\left[a^k \sin\left(ax + \frac{k\pi}{2}\right)\right]$$
$$= a^k\left[a \cos\left(ax + \frac{k\pi}{2}\right)\right]$$
$$= a^{k+1} \sin\left(ax + \frac{(k+1)\pi}{2}\right).$$

Thus the formula holds for $n = 1, 2, 3, \ldots$ by induction.

c) Note that

$$\frac{d}{dx}(\cos^4 x + \sin^4 x) = -4\cos^3 x \sin x + 4\sin^3 x \cos x$$
$$= -4 \sin x \cos x(\cos^2 - \sin^2 x)$$
$$= -2 \sin(2x) \cos(2x)$$
$$= -\sin(4x) = \cos\left(4x + \frac{\pi}{2}\right).$$

It now follows from part (a) that

$$\frac{d^n}{dx^n}(\cos^4 x + \sin^4 x) = 4^{n-1} \cos\left(4x + \frac{n\pi}{2}\right).$$

CHAPTER 3. TRANSCENDENTAL FUNCTIONS

Section 3.1 Inverse Functions (page 181)

2. $f(x) = 2x - 1$. If $f(x_1) = f(x_2)$, then $2x_1 - 1 = 2x_2 - 1$. Thus $2(x_1 - x_2) = 0$ and $x_1 = x_2$. Hence, f is one-to-one.
Let $y = f^{-1}(x)$. Thus $x = f(y) = 2y - 1$, so $y = \frac{1}{2}(x + 1)$. Thus $f^{-1}(x) = \frac{1}{2}(x + 1)$.
$D(f) = R(f^{-1}) =]-\infty, \infty[$.
$R(f) = D(f^{-1}) =]-\infty, \infty[$.

4. $f(x) = -\sqrt{x - 1}$ for $x \geq 1$.
If $f(x_1) = f(x_2)$, then $-\sqrt{x_1 - 1} = -\sqrt{x_2 - 1}$ and $x_1 - 1 = x_2 - 1$. Thus $x_1 = x_2$ and f is one-to-one.
Let $y = f^{-1}(x)$. Then $x = f(y) = -\sqrt{y - 1}$ so $x^2 = y - 1$ and $y = x^2 + 1$. Thus, $f^{-1}(x) = x^2 + 1$.
$D(f) = R(f^{-1}) = [1, \infty[. \ R(f) = D(f^{-1}) =]-\infty, 0]$.

6. $f(x) = 1 + \sqrt[3]{x}$. If $f(x_1) = f(x_2)$, then $1 + \sqrt[3]{x_1} = 1 + \sqrt[3]{x_2}$ so $x_1 = x_2$. Thus, f is one-to-one.
Let $y = f^{-1}(x)$ so that $x = f(y) = 1 + \sqrt[3]{y}$. Thus $y = (x - 1)^3$ and $f^{-1}(x) = (x - 1)^3$.
$D(f) = R(f^{-1}) =]-\infty, \infty[$.
$R(f) = D(f^{-1}) =]-\infty, \infty[$.

8. $f(x) = (1 - 2x)^3$. If $f(x_1) = f(x_2)$, then $(1 - 2x_1)^3 = (1 - 2x_2)^3$ and $x_1 = x_2$. Thus, f is one-to-one.
Let $y = f^{-1}(x)$. Then $x = f(y) = (1 - 2y)^3$ so $y = \frac{1}{2}(1 - \sqrt[3]{x})$. Thus, $f^{-1}(x) = \frac{1}{2}(1 - \sqrt[3]{x})$.
$D(f) = R(f^{-1}) =]-\infty, \infty[$.
$R(f) = D(f^{-1}) =]-\infty, \infty[$.

10. $f(x) = \dfrac{x}{1 + x}$. If $f(x_1) = f(x_2)$, then $\dfrac{x_1}{1 + x_1} = \dfrac{x_2}{1 + x_2}$. Hence $x_1(1 + x_2) = x_2(1 + x_1)$ and, on simplification, $x_1 = x_2$. Thus, f is one-to-one.
Let $y = f^{-1}(x)$. Then $x = f(y) = \dfrac{y}{1 + y}$ and $x(1 + y) = y$. Thus $y = \dfrac{x}{1 - x} = f^{-1}(x)$.
$D(f) = R(f^{-1}) =]-\infty, -1[\cup]-1, \infty[$.
$R(f) = D(f^{-1}) =]-\infty, 1[\cup]1, \infty[$.

12. $f(x) = \dfrac{x}{\sqrt{x^2 + 1}}$. If $f(x_1) = f(x_2)$, then
$$\frac{x_1}{\sqrt{x_1^2 + 1}} = \frac{x_2}{\sqrt{x_2^2 + 1}}. \qquad (*)$$
Thus $x_1^2(x_2^2 + 1) = x_2^2(x_1^2 + 1)$ and $x_1^2 = x_2^2$.
From (*), x_1 and x_2 must have the same sign. Hence, $x_1 = x_2$ and f is one-to-one.
Let $y = f^{-1}(x)$. Then $x = f(y) = \dfrac{y}{\sqrt{y^2 + 1}}$, and
$x^2(y^2 + 1) = y^2$. Hence $y^2 = \dfrac{x^2}{1 - x^2}$. Since $f(y)$ and y have the same sign, we must have $y = \dfrac{x}{\sqrt{1 - x^2}}$, so
$$f^{-1}(x) = \frac{x}{\sqrt{1 - x^2}}.$$
$D(f) = R(f^{-1}) =]-\infty, \infty[$.
$R(f) = D(f^{-1}) =]-1, 1[$.

14. $h(x) = f(2x)$. Let $y = h^{-1}(x)$. Then $x = h(y) = f(2y)$ and $2y = f^{-1}(x)$. Thus $h^{-1}(x) = y = \frac{1}{2}f^{-1}(x)$.

16. $m(x) = f(x - 2)$. Let $y = m^{-1}(x)$. Then $x = m(y) = f(y - 2)$, and $y - 2 = f^{-1}(x)$. Hence $m^{-1}(x) = y = f^{-1}(x) + 2$.

18. $q(x) = \dfrac{f(x) - 3}{2}$ Let $y = q^{-1}(x)$. Then $x = q(y) = \dfrac{f(y) - 3}{2}$ and $f(y) = 2x + 3$. Hence $q^{-1}(x) = y = f^{-1}(2x + 3)$.

20. $s(x) = \dfrac{1 + f(x)}{1 - f(x)}$. Let $y = s^{-1}(x)$.
Then $x = s(y) = \dfrac{1 + f(y)}{1 - f(y)}$. Solving for $f(y)$ we obtain
$f(y) = \dfrac{x - 1}{x + 1}$. Hence $s^{-1}(x) = y = f^{-1}\left(\dfrac{x - 1}{x + 1}\right)$.

22. $g(x) = x^3$ if $x \geq 0$, and $g(x) = x^{1/3}$ if $x < 0$.
Suppose $f(x_1) = f(x_2)$. If $x_1 \geq 0$ and $x_2 \geq 0$ then $x_1^3 = x_2^3$ so $x_1 = x_2$.
Similarly, $x_1 = x_2$ if both are negative. If x_1 and x_2 have opposite sign, then so do $g(x_1)$ and $g(x_2)$.
Therefore g is one-to-one. Let $y = g^{-1}(x)$. Then
$$x = g(y) = \begin{cases} y^3 & \text{if } y \geq 0 \\ y^{1/3} & \text{if } y < 0. \end{cases}$$
Thus $g^{-1}(x) = y = \begin{cases} x^{1/3} & \text{if } x \geq 0 \\ x^3 & \text{if } x < 0. \end{cases}$

24. $y = f^{-1}(x) \Leftrightarrow x = f(y) = y^3 + y$. To find $y = f^{-1}(2)$ we solve $y^3 + y = 2$ for y. Evidently $y = 1$ is the only solution, so $f^{-1}(2) = 1$.

26. $h(x) = -3$ if $x|x| = -4$, that is, if $x = -2$. Thus $h^{-1}(-3) = -2$.

28. If $f(x) = \dfrac{4x^3}{x^2 + 1}$, then

$$f'(x) = \frac{(x^2+1)(12x^2) - 4x^3(2x)}{(x^2+1)^2} = \frac{4x^2(x^2+3)}{(x^2+1)^2}.$$

Since $f'(x) > 0$ for all x, except $x = 0$, f must be one-to-one and so it has an inverse.
If $y = f^{-1}(x)$, then $x = f(y) = \dfrac{4y^3}{y^2+1}$, and

$$1 = f'(y) = \frac{(y^2+1)(12y^2 y') - 4y^3(2yy')}{(y^2+1)^2}.$$

Thus $y' = \dfrac{(y^2+1)^2}{4y^4 + 12y^2}$. Since $f(1) = 2$, therefore $f^{-1}(2) = 1$ and

$$\left(f^{-1}\right)'(2) = \frac{(y^2+1)^2}{4y^4+12y^2}\bigg|_{y=1} = \frac{1}{4}.$$

30. If $f(x) = x\sqrt{3+x^2}$ and $y = f^{-1}(x)$, then $x = f(y) = y\sqrt{3+y^2}$, so,

$$1 = y'\sqrt{3+y^2} + y\frac{2yy'}{2\sqrt{3+y^2}} \quad\Rightarrow\quad y' = \frac{\sqrt{3+y^2}}{3+2y^2}.$$

Since $f(-1) = -2$ implies that $f^{-1}(-2) = -1$, we have

$$\left(f^{-1}\right)'(-2) = \frac{\sqrt{3+y^2}}{3+2y^2}\bigg|_{y=-1} = \frac{2}{5}.$$

Note: $f(x) = x\sqrt{3+x^2} = -2 \Rightarrow x^2(3+x^2) = 4$
$\Rightarrow x^4 + 3x^2 - 4 = 0 \Rightarrow (x^2+4)(x^2-1) = 0$.
Since $(x^2+4) = 0$ has no real solution, therefore $x^2 - 1 = 0$ and $x = 1$ or -1. Since it is given that $f(x) = -2$, therefore x must be -1.

32. $g(x) = 2x + \sin x \Rightarrow g'(x) = 2 + \cos x \geq 1$ for all x.
Therefore g is increasing, and so one-to-one and invertible on the whole real line.

$y = g^{-1}(x) \Leftrightarrow x = g(y) = 2y + \sin y$. For $y = g^{-1}(2)$, we need to solve $2y + \sin y - 2 = 0$. The root is between 0 and 1; to five decimal places $g^{-1}(2) = y \approx 0.68404$. Also

$$1 = \frac{dx}{dx} = (2 + \cos y)\frac{dy}{dx}$$

$$(g^{-1})'(2) = \frac{dy}{dx}\bigg|_{x=2} = \frac{1}{2+\cos y} \approx 0.36036.$$

34. If $y = (f \circ g)^{-1}(x)$, then $x = f \circ g(y) = f(g(y))$. Thus $g(y) = f^{-1}(x)$ and $y = g^{-1}(f^{-1}(x)) = g^{-1} \circ f^{-1}(x)$. That is, $(f \circ g)^{-1} = g^{-1} \circ f^{-1}$.

36. Let $f(x)$ be an even function. Then $f(x) = f(-x)$. Hence, f is not one-to-one and it is not invertible. Therefore, it cannot be self-inverse.
An odd function $g(x)$ may be self-inverse if its graph is symmetric about the line $x = y$. Examples are $g(x) = x$ and $g(x) = 1/x$.

38. First we consider the case where the domain of f is a closed interval. Suppose that f is one-to-one and continuous on $[a, b]$, and that $f(a) < f(b)$. We show that f must be increasing on $[a, b]$. Suppose not. Then there are numbers x_1 and x_2 with $a \leq x_1 < x_2 \leq b$ and $f(x_1) > f(x_2)$. If $f(x_1) > f(a)$, let u be a number such that $u < f(x_1)$, $f(x_2) < u$, and $f(a) < u$. By the Intermediate-Value Theorem there exist numbers c_1 in $]a, x_1[$ and c_2 in $]x_1, x_2[$ such that $f(c_1) = u = f(c_2)$, contradicting the one-to-oneness of f. A similar contradiction arises if $f(x_1) \leq f(a)$ because, in this case, $f(x_2) < f(b)$ and we can find c_1 in $]x_1, x_2[$ and c_2 in $]x_2, b[$ such that $f(c_1) = f(c_2)$. Thus f must be increasing on $[a, b]$.

A similar argument shows that if $f(a) > f(b)$, then f must be decreasing on $[a, b]$.

Finally, if the interval I where f is defined is not necessarily closed, the same argument shows that if $[a, b]$ is a subinterval of I on which f is increasing (or decreasing), then f must also be increasing (or decreasing) on any intervals of either of the forms $[x_1, b]$ or $[a, x_2]$, where x_1 and x_2 are in I and $x_1 \leq a < b \leq x_2$. So f must be increasing (or decreasing) on the whole of I.

Section 3.2 Exponential and Logarithmic Functions (page 185)

2. $2^{1/2}8^{1/2} = 2^{1/2}2^{3/2} = 2^2 = 4$

4. $(\frac{1}{2})^x 4^{x/2} = \dfrac{2^x}{2^x} = 1$

6. If $\log_4(\frac{1}{8}) = y$ then $4^y = \frac{1}{8}$, or $2^{2y} = 2^{-3}$. Thus $2y = -3$ and $\log_4(\frac{1}{8}) = y = -\frac{3}{2}$.

8. $4^{3/2} = 8 \quad\Rightarrow\quad \log_4 8 = \frac{3}{2} \quad\Rightarrow\quad 2^{\log_4 8} = 2^{3/2} = 2\sqrt{2}$

10. Since $\log_a\left(x^{1/(\log_a x)}\right) = \dfrac{1}{\log_a x}\log_a x = 1$, therefore $x^{1/(\log_a x)} = a^1 = a$.

12. $\log_x\left(x(\log_y y^2)\right) = \log_x(2x) = \log_x x + \log_x 2$
$$= 1 + \log_x 2 = 1 + \frac{1}{\log_2 x}$$

14. $\log_{15} 75 + \log_{15} 3 = \log_{15} 225 = 2$
$$\text{(since } 15^2 = 225)$$

16. $2\log_3 12 - 4\log_3 6 = \log_3\left(\dfrac{4^2 \cdot 3^2}{2^4 \cdot 3^4}\right)$

$\qquad\qquad = \log_3(3^{-2}) = -2$

18. $\log_\pi(1 - \cos x) + \log_\pi(1 + \cos x) - 2\log_\pi \sin x$

$\qquad = \log_\pi\left[\dfrac{(1-\cos x)(1+\cos x)}{\sin^2 x}\right] = \log_\pi \dfrac{\sin^2 x}{\sin^2 x}$

$\qquad = \log_\pi 1 = 0$

20. $\log_3 5 = (\log_{10} 5)/(\log_{10} 3 \approx 1.46497$

22. $x^{\sqrt{2}} = 3,\ \sqrt{2}\log_{10} x = \log_{10} 3,$
$x = 10^{(\log_{10} 3)/\sqrt{2}} \approx 2.17458$

24. $\log_3 x = 5,\ (\log_{10} x)/(\log_{10} 3) = 5,$
$\log_{10} x = 5\log_{10} 3,\ x = 10^{5\log_{10} 3} = 3^5 = 243$

26. Let $\log_a x = u,\ \log_a y = v$.
Then $x = a^u,\ y = a^v$.
Thus $\dfrac{x}{y} = \dfrac{a^u}{a^v} = a^{u-v}$

and $\log_a\left(\dfrac{x}{y}\right) = u - v = \log_a x - \log_a y$.

28. Let $\log_b x = u,\ \log_b a = v$.
Thus $b^u = x$ and $b^v = a$.
Therefore $x = b^u = b^{v(u/v)} = a^{u/v}$
and $\log_a x = \dfrac{u}{v} = \dfrac{\log_b x}{\log_b a}$.

30. First observe that $\log_9 x = \log_3 x / \log_3 9 = \frac{1}{2}\log_3 x$. Now
$2\log_3 x + \log_9 x = 10$
$\log_3 x^2 + \log_3 x^{1/2} = 10$
$\log_3 x^{5/2} = 10$
$x^{5/2} = 3^{10}$, so $x = (3^{10})^{2/5} = 3^4 = 81$

32. Note that $\log_x(1/2) = -\log_x 2 = -1/\log_2 x$.
Since $\lim_{x\to 0+}\log_2 x = -\infty$, therefore
$\lim_{x\to 0+}\log_x(1/2) = 0$.

34. Note that $\log_x 2 = 1/\log_2 x$.
Since $\lim_{x\to 1-}\log_2 x = 0-$, therefore
$\lim_{x\to 1-}\log_x 2 = -\infty$.

36. $y = f^{-1}(x) \Rightarrow x = f(y) = a^y$

$\Rightarrow 1 = \dfrac{dx}{dx} = ka^y\dfrac{dy}{dx}$

$\Rightarrow \dfrac{dy}{dx} = \dfrac{1}{ka^y} = \dfrac{1}{kx}$.

Thus $(f^{-1})'(x) = 1/(kx)$.

Section 3.3 The Natural Logarithm and Exponential (page 195)

2. $\ln(e^{1/2}e^{2/3}) = \frac{1}{2} + \frac{2}{3} = \frac{7}{6}$

4. $e^{(3\ln 9)/2} = 9^{3/2} = 27$

6. $e^{2\ln\cos x} + \left(\ln e^{\sin x}\right)^2 = \cos^2 x + \sin^2 x = 1$

8. $4\ln\sqrt{x} + 6\ln(x^{1/3}) = 2\ln x + 2\ln x = 4\ln x$

10. $\ln(x^2 + 6x + 9) = \ln[(x+3)^2] = 2\ln(x+3)$

12. $3^x = 9^{1-x} \Rightarrow 3^x = 3^{2(1-x)}$
$\Rightarrow\ x = 2(1-x)\ \Rightarrow\ x = \frac{2}{3}$

14. $2^{x^2-3} = 4^x = 2^{2x} \Rightarrow x^2 - 3 = 2x$
$x^2 - 2x - 3 = 0 \Rightarrow (x-3)(x+1) = 0$
Hence, $x = -1$ or 3.

16. $\ln(x^2 - x - 2) = \ln[(x-2)(x+1)]$ is defined if $(x-2)(x+1) > 0$, that is, if $x < -1$ or $x > 2$. The domain is the union $(-\infty, -1) \cup (2, \infty)$.

18. $\ln(x^2 - 2) \le \ln x$ holds if $x^2 > 2$, $x > 0$, and $x^2 - 2 \le x$. Thus we need $x > \sqrt{2}$ and $x^2 - x - 2 \le 0$. This latter inequality says that $(x-2)(x+1) \le 0$, so it holds for $-1 \le x \le 2$. The solution set of the given inequality is $(\sqrt{2}, 2]$.

20. $y = xe^x - x, \qquad y' = e^x + xe^x - 1$

22. $y = x^2 e^{x/2}, \qquad y' = 2xe^{x/2} + \frac{1}{2}x^2 e^{x/2}$

24. $y = \ln|3x - 2|, \qquad y' = \dfrac{3}{3x-2}$

26. $y = 2\ln\sqrt{x^2+2} = 2(\frac{1}{2})\ln(x^2+2) = \ln(x^2+2)$
$y' = \dfrac{2x}{x^2+2}$

28. $f(x) = e^{x^2}, \qquad f'(x) = (2x)e^{x^2}$

30. $x = e^{3t}\ln t, \qquad \dfrac{dx}{dt} = 3e^{3t}\ln t + \dfrac{1}{t}e^{3t}$

32. $f(x) = \dfrac{e^x - e^{-x}}{e^x + e^{-x}}$
$f'(x) = \dfrac{(e^x + e^{-x})(e^x + e^{-x}) - (e^x - e^{-x})(e^x - e^{-x})}{(e^x + e^{-x})^2}$
$\qquad = \dfrac{4}{(e^x + e^{-x})^2}$

34. $y = e^{-x}\cos x, \qquad y' = -e^{-x}\cos x - e^{-x}\sin x$

36. $y = x\ln x - x$
$y' = \ln x + x\left(\dfrac{1}{x}\right) - 1 = \ln x$

38. $y = \ln|\sin x|, \qquad y' = \dfrac{\cos x}{\sin x} = \cot x$

40. $y = 2^{(x^2-3x+8)}, \qquad y' = (2x-3)(\ln 2)2^{(x^2-3x+8)}$

42. $h(t) = t^x - x^t, \qquad h'(t) = xt^{x-1} - x^t\ln x$

44. $g(x) = \log_x(2x+3) = \dfrac{\ln(2x+3)}{\ln x}$

$$g'(x) = \dfrac{\ln x\left(\dfrac{2}{2x+3}\right) - [\ln(2x+3)]\left(\dfrac{1}{x}\right)}{(\ln x)^2}$$

$$= \dfrac{2x\ln x - (2x+3)\ln(2x+3)}{x(2x+3)(\ln x)^2}$$

46. Given that $y = \left(\dfrac{1}{x}\right)^{\ln x}$, let $u = \ln x$. Then $x = e^u$ and

$y = \left(\dfrac{1}{e^u}\right)^u = (e^{-u})^u = e^{-u^2}$. Hence,

$$\dfrac{dy}{dx} = \dfrac{dy}{du}\cdot\dfrac{du}{dx} = (-2ue^{-u^2})\left(\dfrac{1}{x}\right) = -\dfrac{2\ln x}{x}\left(\dfrac{1}{x}\right)^{\ln x}.$$

48. $y = \ln|x + \sqrt{x^2 - a^2}|$

$$y' = \dfrac{1 + \dfrac{2x}{2\sqrt{x^2-a^2}}}{x + \sqrt{x^2-a^2}} = \dfrac{1}{\sqrt{x^2-a^2}}$$

50. $y = (\cos x)^x - x^{\cos x} = e^{x\ln\cos x} - e^{(\cos x)(\ln x)}$

$$y' = e^{x\ln\cos x}\left[\ln\cos x + x\left(\dfrac{1}{\cos x}\right)(-\sin x)\right]$$

$$- e^{(\cos x)(\ln x)}\left[-\sin x\ln x + \dfrac{1}{x}\cos x\right]$$

$$= (\cos x)^x(\ln\cos x - x\tan x)$$

$$- x^{\cos x}\left(-\sin x\ln x + \dfrac{1}{x}\cos x\right)$$

52. Since

$$\dfrac{d}{dx}(ax^2 + bx + c)e^x = (2ax + b)e^x + (ax^2 + bx + c)e^x$$

$$= [ax^2 + (2a+b)x + (b+c)]e^x$$

$$= [Ax^2 + Bx + C]e^x.$$

Thus, differentiating $(ax^2 + bx + c)e^x$ produces another function of the same type with different constants. Any number of differentiations will do likewise.

54. $f(x) = \ln(2x+1)$　　　　$f'(x) = 2(2x+1)^{-1}$

$f''(x) = (-1)2^2(2x+1)^{-2}$　　$f'''(x) = (2)2^3(2x+1)^{-3}$

$f^{(4)}(x) = -(3!)2^4(2x+1)^{-4}$

Thus, if $n = 1, 2, 3, \ldots$ we have
$f^{(n)}(x) = (-1)^{n-1}(n-1)!2^n(2x+1)^{-n}$.

56. Given that $x^{x^{x^{\cdot^{\cdot^{\cdot}}}}} = a$ where $a > 0$, then

$$\ln a = x^{x^{x^{\cdot^{\cdot^{\cdot}}}}}\ln x = a\ln x.$$

Thus $\ln x = \dfrac{1}{a}\ln a = \ln a^{1/a}$, so $x = a^{1/a}$.

58. $F(x) = \dfrac{\sqrt{1+x}(1-x)^{1/3}}{(1+5x)^{4/5}}$

$\ln F(x) = \frac{1}{2}\ln(1+x) + \frac{1}{3}\ln(1-x) - \frac{4}{5}\ln(1+5x)$

$\dfrac{F'(x)}{F(x)} = \dfrac{1}{2(1+x)} - \dfrac{1}{3(1-x)} - \dfrac{4}{(1+5x)}$

$F'(0) = F(0)\left[\dfrac{1}{2} - \dfrac{1}{3} - \dfrac{4}{1}\right] = (1)\left[\dfrac{1}{2} - \dfrac{1}{3} - 4\right] = -\dfrac{23}{6}$

60. Since $y = x^2 e^{-x^2}$, then

$$y' = 2xe^{-x^2} - 2x^3 e^{-x^2} = 2x(1-x)(1+x)e^{-x^2}.$$

The tangent is horizontal at $(0,0)$ and $\left(\pm 1, \dfrac{1}{e}\right)$.

62. Since $y = \ln x$ and $y' = \dfrac{1}{x} = 4$ then $x = \frac{1}{4}$ and
$y = \ln\frac{1}{4} = -\ln 4$. The tangent line of slope 4 is
$y = -\ln 4 + 4(x - \frac{1}{4})$, i.e., $y = 4x - 1 - \ln 4$.

64. The slope of $y = \ln x$ at $x = a$ is $y' = \dfrac{1}{x}\Big|_{x=a} = \dfrac{1}{a}$. The
line from $(0,0)$ to $(a, \ln a)$ is tangent to $y = \ln x$ if

$$\dfrac{\ln a - 0}{a - 0} = \dfrac{1}{a}$$

i.e., if $\ln a = 1$, or $a = e$. Thus, the line is $y = \dfrac{x}{e}$.

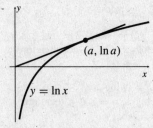

Fig. 3.3.64

66. The tangent line to $y = a^x$ which passes through the origin is tangent at the point (b, a^b) where

$$\dfrac{a^b - 0}{b - 0} = \dfrac{d}{dx}a^x\Big|_{x=b} = a^b\ln a.$$

Thus $\dfrac{1}{b} = \ln a$, so $a^b = a^{1/\ln a} = e$. The line $y = x$
will intersect $y = a^x$ provided the slope of this tangent
line does not exceed 1, i.e., provided $\dfrac{e}{b} \le 1$, or $e\ln a \le 1$.
Thus we need $a \le e^{1/e}$.

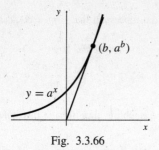

Fig. 3.3.66

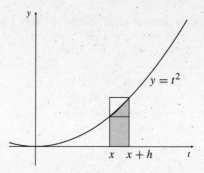

Fig. 3.3.76

Comparing this area with that of the two rectangles, we see that

$$hx^2 < F(x + h) - F(x) < h(x + h)^2.$$

Hence, the Newton quotient for $F(x)$ satisfies

$$x^2 < \frac{F(x + h) - F(x)}{h} < (x + h)^2.$$

Letting h approach 0 from the right (by the Squeeze Theorem applied to one-sided limits)

$$\lim_{h \to 0+} \frac{F(x + h) - F(x)}{h} = x^2.$$

If $h < 0$ and $0 < x + h < x$, then

$$(x + h)^2 < \frac{F(x + h) - F(x)}{h} < x^2,$$

so similarly,

$$\lim_{h \to 0-} \frac{F(x + h) - F(x)}{h} = x^2.$$

Combining these two limits, we obtain

$$\frac{d}{dx} F(x) = \lim_{h \to 0} \frac{F(x + h) - F(x)}{h} = x^2.$$

Therefore $F(x) = \int x^2 \, dx = \frac{1}{3}x^3 + C$. Since $F(0) = C = 0$, therefore $F(x) = \frac{1}{3}x^3$. For $x = 2$, the area of the region is $F(2) = \frac{8}{3}$ square units.

Section 3.4   Growth and Decay   (page 203)

68. $xe^y + y - 2x = \ln 2 \Rightarrow e^y + xe^y y' + y' - 2 = 0.$
At $(1, \ln 2)$, $2 + 2y' + y' - 2 = 0 \Rightarrow y' = 0.$
Therefore, the tangent line is $y = \ln 2.$

70. $F_{A,B}(x) = Ae^x \cos x + Be^x \sin x$

$\dfrac{d}{dx} F_{A,B}(x)$

$= Ae^x \cos x - Ae^x \sin x + Be^x \sin x + Be^x \cos x$

$= (A + B)e^x \cos x + (B - A)e^x \sin x = F_{A+B,B-A}(x)$

72. $\dfrac{d}{dx}(Ae^{ax} \cos bx + Be^{ax} \sin bx)$

$= Aae^{ax} \cos bx - Abe^{ax} \sin bx + Bae^{ax} \sin bx$
$\quad + Bbe^{ax} \cos bx$

$= (Aa + Bb)e^{ax} \cos bx + (Ba - Ab)e^{ax} \sin bx.$

(a) If $Aa + Bb = 1$ and $Ba - Ab = 0$, then $A = \dfrac{a}{a^2 + b^2}$

and $B = \dfrac{b}{a^2 + b^2}$. Thus

$$\int e^{ax} \cos bx \, dx$$

$$= \frac{1}{a^2 + b^2}\left(ae^{ax} \cos bx + be^{ax} \sin bx\right) + C.$$

(b) If $Aa + Bb = 0$ and $Ba - Ab = 1$, then $A = \dfrac{-b}{a^2 + b^2}$

and $B = \dfrac{a}{a^2 + b^2}$. Thus

$$\int e^{ax} \sin bx \, dx$$

$$= \frac{1}{a^2 + b^2}\left(ae^{ax} \sin bx - be^{ax} \cos bx\right) + C.$$

74. $\ln \dfrac{x}{y} = \ln\left(x\,\dfrac{1}{y}\right) = \ln x + \ln \dfrac{1}{y} = \ln x - \ln y.$

76. Let $x > 0$, and $F(x)$ be the area bounded by $y = t^2$, the t-axis, $t = 0$ and $t = x$. For $h > 0$, $F(x+h) - F(x)$ is the shaded area in the following figure.

2. $\displaystyle\lim_{x \to \infty} x^{-3}e^x = \lim_{x \to \infty} \frac{e^x}{x^3} = \infty$

4. $\displaystyle\lim_{x \to \infty} \frac{x - 2e^{-x}}{x + 3e^{-x}} = \lim_{x \to \infty} \frac{1 - 2/(xe^x)}{1 + 3/(xe^x)} = \frac{1 - 0}{1 + 0} = 1$

6. $\displaystyle\lim_{x \to 0+} \frac{\ln x}{x} = -\infty$

8. $\displaystyle\lim_{x \to \infty} \frac{(\ln x)^3}{\sqrt{x}} = 0$ (power wins)

45

10. Let $y(t)$ be the number of kg undissolved after t hours. Thus, $y(0) = 50$ and $y(5) = 20$. Since $y'(t) = ky(t)$, therefore $y(t) = y(0)e^{kt} = 50e^{kt}$. Then

$$20 = y(5) = 50e^{5k} \Rightarrow k = \tfrac{1}{5} \ln \tfrac{2}{5}.$$

If 90% of the sugar is dissolved at time T then $5 = y(T) = 50e^{kT}$, so

$$T = \frac{1}{k} \ln \frac{1}{10} = \frac{5\ln(0.1)}{\ln(0.4)} \approx 12.56.$$

Hence, 90% of the sugar will dissolved in about 12.56 hours.

12. Let $P(t)$ be the percentage remaining after t years. Thus $P'(t) = kP(t)$ and $P(t) = P(0)e^{kt} = 100e^{kt}$. Then,

$$50 = P(1690) = 100e^{1690k} \Rightarrow k = \frac{1}{1690} \ln \frac{1}{2} \approx 0.0004101.$$

a) $P(100) = 100e^{100k} \approx 95.98$, i.e., about 95.98% remains after 100 years.

b) $P(1000) = 100e^{1000k} \approx 66.36$, i.e., about 66.36% remains after 1000 years.

14. Let $N(t)$ be the number of bacteria in the culture t days after the culture was set up. Thus $N(3) = 3N(0)$ and $N(7) = 10 \times 10^6$. Since $N(t) = N(0)e^{kt}$, we have

$$3N(0) = N(3) = N(0)e^{3k} \Rightarrow k = \tfrac{1}{3} \ln 3.$$
$$10^7 = N(7) = N(0)e^{7k} \Rightarrow N(0) = 10^7 e^{-(7/3)\ln 3} \approx 770400.$$

There were approximately 770,000 bacteria in the culture initially. (Note that we are approximating a discrete quantity (number of bacteria) by a continuous quantity $N(t)$ in this exercise.)

16. Since

$$I'(t) = kI(t) \Rightarrow I(t) = I(0)e^{kt} = 40e^{kt},$$
$$15 = I(0.01) = 40e^{0.01k} \Rightarrow k = \frac{1}{0.01} \ln \frac{15}{40} = 100 \ln \frac{3}{8},$$

thus,

$$I(t) = 40\exp\left(100t \ln \frac{3}{8}\right) = 40\left(\frac{3}{8}\right)^{100t}.$$

18. Let $y(t)$ be the value of the investment after t years. Thus $y(0) = 1000$ and $y(5) = 1500$. Since $y(t) = 1000e^{kt}$ and $1500 = y(5) = 1000e^{5k}$, therefore, $k = \tfrac{1}{5} \ln \tfrac{3}{2}$.

a) Let t be the time such that $y(t) = 2000$, i.e.,

$$1000e^{kt} = 2000$$
$$\Rightarrow t = \frac{1}{k} \ln 2 = \frac{5\ln 2}{\ln(\frac{3}{2})} = 8.55.$$

Hence, the doubling time for the investment is about 8.55 years.

b) Let $r\%$ be the effective annual rate of interest; then

$$1000(1 + \frac{r}{100}) = y(1) = 1000e^k$$
$$\Rightarrow r = 100(e^k - 1) = 100[\exp\left(\tfrac{1}{5} \ln \tfrac{3}{2}\right) - 1]$$
$$= 8.447.$$

The effective annual rate of interest is about 8.45%.

20. Let $i\%$ be the effective rate, then an original investment of $\$A$ will grow to $\$A\left(1 + \dfrac{i}{100}\right)$ in one year. Let $r\%$ be the nominal rate per annum compounded n times per year, then an original investment of $\$A$ will grow to

$$\$A\left(1 + \frac{r}{100n}\right)^n$$

in one year, if compounding is performed n times per year. For $i = 9.5$ and $n = 12$, we have

$$\$A\left(1 + \frac{9.5}{100}\right) = \$A\left(1 + \frac{r}{1200}\right)^{12}$$
$$\Rightarrow r = 1200\left(\sqrt[12]{1.095} - 1\right) = 9.1098.$$

The nominal rate of interest is about 9.1098%.

22. a) The concentration $x(t)$ satisfies $\dfrac{dx}{dt} = a - bx(t)$. This says that $x(t)$ is increasing if it is less than a/b and decreasing if it is greater than a/b. Thus, the limiting concentration is a/b.

b) The differential equation for $x(t)$ resembles that of Exercise 21(b), except that $y(x)$ is replaced by $x(t)$, and b is replaced by $-b$. Using the result of Exercise 21(b), we obtain, since $x(0) = 0$,

$$x(t) = \left(x(0) - \frac{a}{b}\right)e^{-bt} + \frac{a}{b}$$
$$= \frac{a}{b}\left(1 - e^{-bt}\right).$$

c) We will have $x(t) = \tfrac{1}{2}(a/b)$ if $1 - e^{-bt} = \tfrac{1}{2}$, that is, if $e^{-bt} = \tfrac{1}{2}$, or $-bt = \ln(1/2) = -\ln 2$. The time required to attain half the limiting concentration is $t = (\ln 2)/b$.

24. Let $T(t)$ be the temperature of the object t minutes after its temperature was $45°$ C. Thus $T(0) = 45$ and $T(40) = 20$. Also $\dfrac{dT}{dt} = k(T+5)$. Let $u(t) = T(t)+5$, so $u(0) = 50$, $u(40) = 25$, and $\dfrac{du}{dt} = \dfrac{dT}{dt} = k(T+5) = ku$. Thus,

$$u(t) = 50e^{kt},$$
$$25 = u(40) = 50e^{40k},$$
$$\Rightarrow k = \frac{1}{40}\ln\frac{25}{50} = \frac{1}{40}\ln\frac{1}{2}.$$

We wish to know t such that $T(t) = 0$, i.e., $u(t) = 5$, hence

$$5 = u(t) = 50e^{kt}$$

$$t = \frac{40\ln\left(\dfrac{5}{50}\right)}{\ln\left(\dfrac{1}{2}\right)} = 132.88 \text{ min.}$$

Hence, it will take about $(132.88 - 40) = 92.88$ minutes more to cool to $0°$ C.

26. By the solution given for the logistic equation, we have

$$y_1 = \frac{Ly_0}{y_0 + (L - y_0)e^{-k}}, \qquad y_2 = \frac{Ly_0}{y_0 + (L - y_0)e^{-2k}}$$

Thus $y_1(L - y_0)e^{-k} = (L - y_1)y_0$, and $y_2(L - y_0)e^{-2k} = (L - y_2)y_0$.
Square the first equation and thus eliminate e^{-k}:

$$\left(\frac{(L - y_1)y_0}{y_1(L - y_0)}\right)^2 = \frac{(L - y_2)y_0}{y_2(L - y_0)}$$

Now simplify: $y_0y_2(L - y_1)^2 = y_1^2(L - y_0)(L - y_2)$
$y_0y_2L^2 - 2y_1y_0y_2L + y_0y_1^2y_2 = y_1^2L^2 - y_1^2(y_0 + y_2)L + y_0y_1^2y_2$

Assuming $L \neq 0$, $L = \dfrac{y_1^2(y_0 + y_2) - 2y_0y_1y_2}{y_1^2 - y_0y_2}$.

If $y_0 = 3$, $y_1 = 5$, $y_2 = 6$, then
$L = \dfrac{25(9) - 180}{25 - 18} = \dfrac{45}{7} \approx 6.429$.

28. The solution $y = \dfrac{Ly_0}{y_0 + (L - y_0)e^{-kt}}$ is valid on the largest interval containing $t = 0$ on which the denominator does not vanish.
If $y_0 > L$ then $y_0 + (L - y_0)e^{-kt} = 0$ if
$t = t^* = -\dfrac{1}{k}\ln\dfrac{y_0}{y_0 - L}$.
Then the solution is valid on (t^*, ∞). $\lim_{t\to t^{*}+} y(t) = \infty$.

30.
$$y(t) = \frac{L}{1 + Me^{-kt}}$$
$$200 = y(0) = \frac{L}{1 + M}$$
$$1{,}000 = y(1) = \frac{L}{1 + Me^{-k}}$$
$$10{,}000 = \lim_{t\to\infty} y(t) = L$$

Thus $200(1 + M) = L = 10{,}000$, so $M = 49$. Also $1{,}000(1 + 49e^{-k}) = L = 10{,}000$, so $e^{-k} = 9/49$ and $k = \ln(49/9) \approx 1.695$.

Section 3.5 The Inverse Trigonometric Functions (page 212)

2. $\cos^{-1}\left(-\dfrac{1}{2}\right) = \dfrac{2\pi}{3}$

4. $\sec^{-1}\sqrt{2} = \dfrac{\pi}{4}$

6. $\cos(\sin^{-1} 0.7) = \sqrt{1 - \sin^2(\arcsin 0.7)}$
$\qquad = \sqrt{1 - 0.49} = \sqrt{0.51}$

8. $\sin^{-1}(\cos 40°) = 90° - \cos^{-1}(\cos 40°) = 50°$

10. $\sin\left(\cos^{-1}(-\tfrac{1}{3})\right) = \sqrt{1 - \cos^2(\arccos(-\tfrac{1}{3}))}$
$\qquad = \sqrt{1 - \tfrac{1}{9}} = \dfrac{\sqrt{8}}{3} = \dfrac{2\sqrt{2}}{3}$

12. $\tan(\tan^{-1} 200) = 200$

14. $\cos(\sin^{-1} x) = \sqrt{1 - \sin^2\left(\sin^{-1} x\right)} = \sqrt{1 - x^2}$

16. $\tan(\arctan x) = x \Rightarrow \sec(\arctan x) = \sqrt{1 + x^2}$
$\qquad \Rightarrow \cos(\arctan x) = \dfrac{1}{\sqrt{1 + x^2}}$
$\qquad \Rightarrow \sin(\arctan x) = \dfrac{x}{\sqrt{1 + x^2}}$

18. $\cos(\sec^{-1}x) = \dfrac{1}{x} \Rightarrow \sin(\sec^{-1}x) = \sqrt{1 - \dfrac{1}{x^2}} = \dfrac{\sqrt{x^2 - 1}}{|x|}$
$\qquad \Rightarrow \tan(\sec^{-1}x) = \sqrt{x^2 - 1}\,\text{sgn}\,x$
$\qquad = \begin{cases} \sqrt{x^2 - 1} & \text{if } x \geq 1 \\ -\sqrt{x^2 - 1} & \text{if } x \leq -1 \end{cases}$

20. $y = \tan^{-1}(ax + b)$, $\qquad y' = \dfrac{a}{1 + (ax + b)^2}$.

22. $f(x) = x\sin^{-1} x$
$f'(x) = \sin^{-1} x + \dfrac{x}{\sqrt{1 - x^2}}$.

24. $u = z^2\sec^{-1}(1 + z^2)$
$\dfrac{du}{dz} = 2z\sec^{-1}(1 + z^2) + \dfrac{z^2(2z)}{(1 + z^2)\sqrt{(1 + z^2)^2 - 1}}$
$\qquad = 2z\sec^{-1}(1 + z^2) + \dfrac{2z^2\text{sgn}(z)}{(1 + z^2)\sqrt{z^2 + 2}}$

26. $y = \sin^{-1}\left(\dfrac{a}{x}\right)$ $(|x| > |a|)$

$$y' = \frac{1}{\sqrt{1 - \left(\dfrac{a}{x}\right)^2}}\left[-\frac{a}{x^2}\right] = -\frac{a}{|x|\sqrt{x^2 - a^2}}$$

28. $H(t) = \dfrac{\sin^{-1} t}{\sin t}$

$$H'(t) = \frac{\sin t\left(\dfrac{1}{\sqrt{1 - t^2}}\right) - \sin^{-1} t \cos t}{\sin^2 t}$$

$$= \frac{1}{(\sin t)\sqrt{1 - t^2}} - \csc t \cot t \sin^{-1} t$$

30. $y = \cos^{-1}\left(\dfrac{a}{\sqrt{a^2 + x^2}}\right)$

$$y' = -\left(1 - \frac{a^2}{a^2 + x^2}\right)^{-1/2}\left[-\frac{a}{2}(a^2 + x^2)^{-3/2}(2x)\right]$$

$$= \frac{a\,\mathrm{sgn}\,(x)}{a^2 + x^2}$$

32. $y = a\cos^{-1}\left(1 - \dfrac{x}{a}\right) - \sqrt{2ax - x^2}$ $(a > 0)$

$$y' = -a\left[1 - \left(1 - \frac{x}{a}\right)^2\right]^{-1/2}\left(-\frac{1}{a}\right) - \frac{2a - 2x}{2\sqrt{2ax - x^2}}$$

$$= \frac{x}{\sqrt{2ax - x^2}}$$

34. If $y = \sin^{-1} x$, then $y' = \dfrac{1}{\sqrt{1 - x^2}}$. If the slope is 2 then

$\dfrac{1}{\sqrt{1 - x^2}} = 2$ so that $x = \pm\dfrac{\sqrt{3}}{2}$. Thus the equations of the two tangent lines are

$$y = \frac{\pi}{3} + 2\left(x - \frac{\sqrt{3}}{2}\right) \text{ and } y = -\frac{\pi}{3} + 2\left(x + \frac{\sqrt{3}}{2}\right).$$

36. Since the domain of $\sec^{-1}$ consists of two disjoint intervals $(-\infty, -1]$ and $[1, \infty)$, the fact that the derivative of $\sec^{-1}$ is positive wherever defined does not imply that $\sec^{-1}$ is increasing over its whole domain, only that it is increasing on each of those intervals taken independently. In fact, $\sec^{-1}(-1) = \pi > 0 = \sec^{-1}(1)$ even though $-1 < 1$.

38. $\cot^{-1} x = \arctan(1/x)$;

$$\frac{d}{dx}\cot^{-1} x = \frac{1}{1 + \dfrac{1}{x^2}}\frac{-1}{x^2} = -\frac{1}{1 + x^2}$$

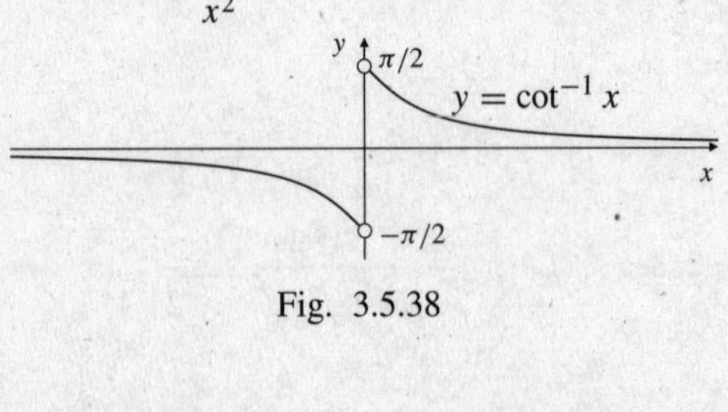

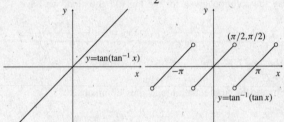

Fig. 3.5.38

Remark: the domain of $\cot^{-1}$ can be extended to include 0 by defining, say, $\cot^{-1} 0 = \pi/2$. This will make $\cot^{-1}$ right-continuous (but not continuous) at $x = 0$. It is also possible to define $\cot^{-1}$ in such a way that it is continuous on the whole real line, but we would then lose the identity $\cot^{-1} x = \tan^{-1}(1/x)$, which we prefer to maintain for calculation purposes.

40. If $g(x) = \tan(\tan^{-1} x)$ then

$$g'(x) = \frac{\sec^2(\tan^{-1} x)}{1 + x^2}$$

$$= \frac{1 + [\tan(\tan^{-1} x)]^2}{1 + x^2} = \frac{1 + x^2}{1 + x^2} = 1.$$

If $h(x) = \tan^{-1}(\tan x)$ then h is periodic with period π, and

$$h'(x) = \frac{\sec^2 x}{1 + \tan^2 x} = 1$$

provided that $x \neq (k + \frac{1}{2})\pi$ where k is an integer. $h(x)$ is not defined at odd multiples of $\dfrac{\pi}{2}$.

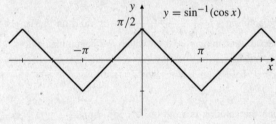

Fig. 3.5.40(a)   Fig. 3.5.40(b)

42. $\dfrac{d}{dx}\sin^{-1}(\cos x) = \dfrac{1}{\sqrt{1 - \cos^2 x}}(-\sin x)$

$$= \begin{cases} -1 & \text{if } \sin x > 0 \\ 1 & \text{if } \sin x < 0 \end{cases}$$

$\sin^{-1}(\cos x)$ is continuous everywhere and differentiable everywhere except at $x = n\pi$ for integers n.

Fig. 3.5.42

44. $\dfrac{d}{dx}\tan^{-1}(\cot x) = \dfrac{1}{1 + \cot^2 x}(-\csc^2 x) = -1$ except at integer multiples of π.

$\tan^{-1}(\cot x)$ is continuous and differentiable everywhere except at $x = n\pi$ for integers n. It is not defined at those points.

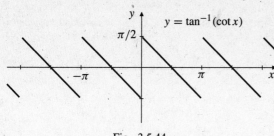

Fig. 3.5.44

46. If $x \geq 1$ and $y = \tan^{-1} \sqrt{x^2 - 1}$, then $\tan y = \sqrt{x^2 - 1}$ and $\sec y = x$, so that $y = \sec^{-1} x$.

If $x \leq -1$ and $y = \pi - \tan^{-1} \sqrt{x^2 - 1}$, then $\frac{\pi}{2} < y < \frac{3\pi}{2}$, so $\sec y < 0$. Therefore

$$\tan y = \tan(\pi - \tan^{-1} \sqrt{x^2 - 1}) = -\sqrt{x^2 - 1}$$
$$\sec^2 y = 1 + (x^2 - 1) = x^2$$
$$\sec y = x,$$

because both x and $\sec y$ are negative. Thus $y = \sec^{-1} x$ in this case also.

48. If $x \geq 1$ and $y = \sin^{-1} \dfrac{\sqrt{x^2 - 1}}{x}$, then $0 \leq y < \frac{\pi}{2}$ and

$$\sin y = \frac{\sqrt{x^2 - 1}}{x}$$
$$\cos^2 y = 1 - \frac{x^2 - 1}{x^2} = \frac{1}{x^2}$$
$$\sec^2 y = x^2.$$

Thus $\sec y = x$ and $y = \sec^{-1} x$.

If $x \leq -1$ and $y = \pi - \sin^{-1} \dfrac{\sqrt{x^2 - 1}}{x}$, then $\frac{\pi}{2} \leq y < \frac{3\pi}{2}$ and $\sec y < 0$. Therefore

$$\sin y = \sin\left(\pi - \sin^{-1} \frac{\sqrt{x^2 - 1}}{x}\right) = \frac{\sqrt{x^2 - 1}}{x}$$
$$\cos^2 y = 1 - \frac{x^2 - 1}{x^2} = \frac{1}{x^2}$$
$$\sec^2 y = x^2$$
$$\sec y = x,$$

because both x and $\sec y$ are negative. Thus $y = \sec^{-1} x$ in this case also.

50. Since $f(x) = x - \tan^{-1}(\tan x)$ then

$$f'(x) = 1 - \frac{\sec^2 x}{1 + \tan^2 x} = 1 - 1 = 0$$

if $x \neq -(k + \frac{1}{2})\pi$ where k is an integer. Thus, f is constant on intervals not containing odd multiples of $\frac{\pi}{2}$. $f(0) = 0$ but $f(\pi) = \pi - 0 = \pi$. There is no contradiction here because $f'\left(\dfrac{\pi}{2}\right)$ is not defined, so f is not constant on the interval containing 0 and π.

52. $y' = \dfrac{1}{1 + x^2} \Rightarrow y = \tan^{-1} x + C$

$y(0) = C = 1$

Thus, $y = \tan^{-1} x + 1$.

54. $y' = \dfrac{1}{\sqrt{1 - x^2}} \Rightarrow y = \sin^{-1} x + C$

$y(\frac{1}{2}) = \sin^{-1}\left(\frac{1}{2}\right) + C = 1$

$\Rightarrow \dfrac{\pi}{6} + C = 1 \Rightarrow C = 1 - \dfrac{\pi}{6}.$

Thus, $y = \sin^{-1} x + 1 - \dfrac{\pi}{6}.$

Section 3.6 Hyperbolic Functions (page 218)

2. $\cosh x \cosh y + \sinh x \sinh y$
$= \frac{1}{4}[(e^x + e^{-x})(e^y + e^{-y}) + (e^x - e^{-x})(e^y - e^{-y})]$
$= \frac{1}{4}(2e^{x+y} + 2e^{-x-y}) = \frac{1}{2}(e^{x+y} + e^{-(x+y)})$
$= \cosh(x + y)$.
$\sinh x \cosh y + \cosh x \sinh y$
$= \frac{1}{4}[(e^x - e^{-x})(e^y + e^{-y}) + (e^x + e^{-x})(e^y - e^{-y})]$
$= \frac{1}{2}(e^{x+y} - e^{-(x+y)}) = \sinh(x + y)$.
$\cosh(x - y) = \cosh[x + (-y)]$
$= \cosh x \cosh(-y) + \sinh x \sinh(-y)$
$= \cosh x \cosh y - \sinh x \sinh y$.
$\sinh(x - y) = \sinh[x + (-y)]$
$= \sinh x \cosh(-y) + \cosh x \sinh(-y)$
$= \sinh x \cosh y - \cosh x \sinh y$.

4. $y = \coth x = \dfrac{e^x + e^{-x}}{e^x - e^{-x}}$ $\qquad y = \operatorname{sech} x = \dfrac{2}{e^x + e^{-x}}$

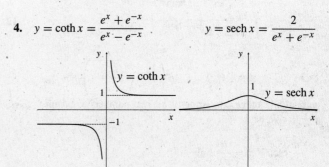

Fig. 3.6.4(a) Fig. 3.6.4(b)

$y = \operatorname{csch} x = \dfrac{2}{e^x - e^{-x}}$

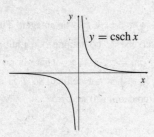

Fig. 3.6.4

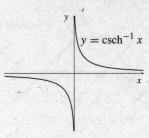

Fig. 3.6.8

6. Let $y = \sinh^{-1}\left(\dfrac{x}{a}\right) \Leftrightarrow x = a \sinh y \Rightarrow 1 = a(\cosh y)\dfrac{dy}{dx}$. Thus,

$$\frac{d}{dx}\sinh^{-1}\left(\frac{x}{a}\right) = \frac{1}{a\cosh y}$$
$$= \frac{1}{a\sqrt{1+\sinh^2 y}} = \frac{1}{\sqrt{a^2+x^2}}$$
$$\int \frac{dx}{\sqrt{a^2+x^2}} = \sinh^{-1}\frac{x}{a} + C. \qquad (a > 0)$$

Let $y = \cosh^{-1}\dfrac{x}{a} \Leftrightarrow x = a\operatorname{Cosh} y = a\cosh y$

for $y \geq 0$, $x \geq a$. We have $1 = a(\sinh y)\dfrac{dy}{dx}$. Thus,

$$\frac{d}{dx}\cosh^{-1}\frac{x}{a} = \frac{1}{a\sinh y}$$
$$= \frac{1}{a\sqrt{\cosh^2 y - 1}} = \frac{1}{\sqrt{x^2 - a^2}}$$
$$\int \frac{dx}{\sqrt{x^2 - a^2}} = \cosh^{-1}\frac{x}{a} + C. \qquad (a > 0,\ x \geq a)$$

Let $y = \tanh^{-1}\dfrac{x}{a} \Leftrightarrow x = a\tanh y \Rightarrow 1 = a(\operatorname{sech}^2 y)\dfrac{dy}{dx}$. Thus,

$$\frac{d}{dx}\tanh^{-1}\frac{x}{a} = \frac{1}{a\operatorname{sech}^2 y}$$
$$= \frac{a}{a^2 - a^2\tanh^2 x} = \frac{a}{a^2 - x^2}$$
$$\int \frac{dx}{a^2 - x^2} = \frac{1}{a}\tanh^{-1}\frac{x}{a} + C.$$

8. $\operatorname{csch}^{-1}x = \sinh^{-1}(1/x) = \ln\left(\dfrac{1}{x} + \sqrt{\dfrac{1}{x^2} + 1}\right)$ has

domain and range consisting of all real numbers x except $x = 0$. We have

$$\frac{d}{dx}\operatorname{csch}^{-1}x = \frac{d}{dx}\sinh^{-1}\frac{1}{x}$$
$$= \frac{1}{\sqrt{1 + \left(\dfrac{1}{x}\right)^2}}\left(\frac{-1}{x^2}\right) = \frac{-1}{|x|\sqrt{x^2 + 1}}.$$

10. Let $y = \operatorname{Sech}^{-1}x$ where $\operatorname{Sech} x = \operatorname{sech} x$ for $x \geq 0$. Hence, for $y \geq 0$,

$$x = \operatorname{sech} y \Leftrightarrow \frac{1}{x} = \cosh y$$
$$\Leftrightarrow \frac{1}{x} = \operatorname{Cosh} y \Leftrightarrow y = \operatorname{Cosh}^{-1}\frac{1}{x}.$$

Thus,

$$\operatorname{Sech}^{-1}x = \operatorname{Cosh}^{-1}\frac{1}{x}$$
$$D(\operatorname{Sech}^{-1}) = R(\operatorname{sech}) = (0, 1]$$
$$R(\operatorname{Sech}^{-1}) = D(\operatorname{sech}) = [0, \infty).$$

Also,

$$\frac{d}{dx}\operatorname{Sech}^{-1}x = \frac{d}{dx}\operatorname{Cosh}^{-1}\frac{1}{x}$$
$$= \frac{1}{\sqrt{\left(\dfrac{1}{x}\right)^2 - 1}}\left(\frac{-1}{x^2}\right) = \frac{-1}{x\sqrt{1 - x^2}}.$$

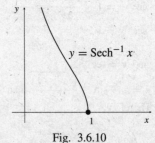

Fig. 3.6.10

12. Since

$$h_{L,M}(x) = L\cosh k(x - a) + M\sinh k(x - a)$$
$$h_{L,M}''(x) = Lk^2\cosh k(x - a) + Mk^2\sinh k(x - a)$$
$$= k^2 h_{L,M}(x)$$

hence, $h_{L,M}(x)$ is a solution of $y'' - k^2 y = 0$ and

$$h_{L,M}(x)$$
$$= \frac{L}{2}\left(e^{kx-ka} + e^{-kx+ka}\right) + \frac{M}{2}\left(e^{kx-ka} - e^{-kx+ka}\right)$$
$$= \left(\frac{L}{2}e^{-ka} + \frac{M}{2}e^{-ka}\right)e^{kx} + \left(\frac{L}{2}e^{ka} - \frac{M}{2}e^{ka}\right)e^{-kx}$$
$$= Ae^{kx} + Be^{-kx} = f_{A,B}(x)$$

where $A = \frac{1}{2}e^{-ka}(L + M)$ and $B = \frac{1}{2}e^{ka}(L - M)$.

Section 3.7 Second-Order Linear DEs with Constant Coefficients (page 228)

2.
$$y'' - 2y' - 3y = 0$$
auxiliary eqn $r^2 - 2r - 3 = 0 \implies r = -1, \ r = 3$
$$y = Ae^{-t} + Be^{3t}$$

4. $4y'' - 4y' - 3y = 0$
$4r^2 - 4r - 3 = 0 \implies (2r + 1)(2r - 3) = 0$
Thus, $r_1 = -\frac{1}{2}$, $r_2 = \frac{3}{2}$, and $y = Ae^{-(1/2)t} + Be^{(3/2)t}$.

6. $y'' - 2y' + y = 0$
$r^2 - 2r + 1 = 0 \implies (r - 1)^2 = 0$
Thus, $r = 1, \ 1$, and $y = Ae^t + Bte^t$.

8. $9y'' + 6y' + y = 0$
$9r^2 + 6r + 1 = 0 \implies (3r + 1)^2 = 0$
Thus, $r = -\frac{1}{3}, \ -\frac{1}{3}$, and $y = Ae^{-(1/3)t} + Bte^{-(1/3)t}$.

10. For $y'' - 4y' + 5y = 0$ the auxiliary equation is $r^2 - 4r + 5 = 0$, which has roots $r = 2 \pm i$. Thus, the general solution of the DE is $y = Ae^{2t}\cos t + Be^{2t}\sin t$.

12. Given that $y'' + y' + y = 0$, hence $r^2 + r + 1 = 0$. Since $a = 1$, $b = 1$ and $c = 1$, the discriminant is $D = b^2 - 4ac = -3 < 0$ and $-(b/2a) = -\frac{1}{2}$ and $\omega = \sqrt{3}/2$. Thus, the general solution is
$$y = Ae^{-(1/2)t}\cos\left(\frac{\sqrt{3}}{2}t\right) + Be^{-(1/2)t}\sin\left(\frac{\sqrt{3}}{2}t\right).$$

14. Given that $y'' + 10y' + 25y = 0$, hence $r^2 + 10r + 25 = 0 \implies (r + 5)^2 = 0 \implies r = -5$. Thus,
$$y = Ae^{-5t} + Bte^{-5t}$$
$$y' = -5e^{-5t}(A + Bt) + Be^{-5t}.$$
Since
$$0 = y(1) = Ae^{-5} + Be^{-5}$$
$$2 = y'(1) = -5e^{-5}(A + B) + Be^{-5},$$
we have $A = -2e^5$ and $B = 2e^5$.
Thus, $y = -2e^5 e^{-5t} + 2te^5 e^{-5t} = 2(t - 1)e^{-5(t-1)}$.

16. The auxiliary equation $r^2 - (2 + \epsilon)r + (1 + \epsilon)$ factors to $(r - 1 - \epsilon)(r - 1) = 0$ and so has roots $r = 1 + \epsilon$ and $r = 1$. Thus the DE $y'' - (2 + \epsilon)y' + (1 + \epsilon)y = 0$ has general solution $y = Ae^{(1+\epsilon)t} + Be^t$. The function $y_\epsilon(t) = \dfrac{e^{(1+\epsilon)t} - e^t}{\epsilon}$ is of this form with $A = -B = 1/\epsilon$. We have, substituting $\epsilon = h/t$,

$$\lim_{\epsilon \to 0} y_\epsilon(t) = \lim_{\epsilon \to 0} \frac{e^{(1+\epsilon)t} - e^t}{\epsilon}$$
$$= t \lim_{h \to 0} \frac{e^{t+h} - e^t}{h}$$
$$= t\left(\frac{d}{dt}e^t\right) = t\,e^t$$

which is, along with e^t, a solution of the CASE II DE $y'' - 2y' + y = 0$.

18. The auxiliary equation $ar^2 + br + c = 0$ has roots
$$r_1 = \frac{-b - \sqrt{D}}{2a}, \quad r_2 = \frac{-b + \sqrt{D}}{2a},$$
where $D = b^2 - 4ac$. Note that $a(r_2 - r_1) = \sqrt{D} = -(2ar_1 + b)$. If $y = e^{r_1 t}u$, then $y' = e^{r_1 t}(u' + r_1 u)$, and $y'' = e^{r_1 t}(u'' + 2r_1 u' + r_1^2 u)$. Substituting these expressions into the DE $ay'' + by' + cy = 0$, and simplifying, we obtain
$$e^{r_1 t}(au'' + 2ar_1 u' + bu') = 0,$$
or, more simply, $u'' - (r_2 - r_1)u' = 0$. Putting $v = u'$ reduces this equation to first order:
$$v' = (r_2 - r_1)v,$$
which has general solution $v = Ce^{(r_2-r_1)t}$. Hence
$$u = \int Ce^{(r_2-r_1)t}\,dt = Be^{(r_2-r_1)t} + A,$$
and $y = e^{r_1 t}u = Ae^{r_1 t} + Be^{r_2 t}$.

19. If $y = A\cos\omega t + B\sin\omega t$ then
$$y'' + \omega^2 y = -A\omega^2\cos\omega t - B\omega^2\sin\omega t$$
$$+ \omega^2(A\cos\omega t + B\sin\omega t) = 0$$
for all t. So y is a solution of (†).

20. If $f(t)$ is any solution of (†) then $f''(t) = -\omega^2 f(t)$ for all t. Thus,
$$\frac{d}{dt}\left[\omega^2\big(f(t)\big)^2 + \big(f'(t)\big)^2\right]$$
$$= 2\omega^2 f(t)f'(t) + 2f'(t)f''(t)$$
$$= 2\omega^2 f(t)f'(t) - 2\omega^2 f(t)f'(t) = 0$$

for all t. Thus, $\omega^2\big(f(t)\big)^2 + \big(f'(t)\big)^2$ is constant. (This can be interpreted as a conservation of energy statement.)

21. If $g(t)$ satisfies (†) and also $g(0) = g'(0) = 0$, then by Exercise 20,

$$\omega^2\big(g(t)\big)^2 + \big(g'(t)\big)^2$$
$$= \omega^2\big(g(0)\big)^2 + \big(g'(0)\big)^2 = 0.$$

Since a sum of squares cannot vanish unless each term vanishes, $g(t) = 0$ for all t.

22. If $f(t)$ is any solution of (†),
let $g(t) = f(t) - A\cos\omega t - B\sin\omega t$ where $A = f(0)$ and $B\omega = f'(0)$. Then g is also solution of (†). Also $g(0) = f(0) - A = 0$ and $g'(0) = f'(0) - B\omega = 0$. Thus, $g(t) = 0$ for all t by Exercise 24, and therefore $f(x) = A\cos\omega t + B\sin\omega t$. Thus, it is proved that every solution of (†) is of this form.

24. Because $y'' + 4y = 0$, therefore $y = A\cos 2t + B\sin 2t$. Now

$$y(0) = 2 \Rightarrow A = 2,$$
$$y'(0) = -5 \Rightarrow B = -\tfrac{5}{2}.$$

Thus, $y = 2\cos 2t - \tfrac{5}{2}\sin 2t$.

circular frequency $= \omega = 2$, frequency $= \dfrac{\omega}{2\pi} = \dfrac{1}{\pi} \approx 0.318$

period $= \dfrac{2\pi}{\omega} = \pi \approx 3.14$

amplitude $= \sqrt{(2)^2 + (-\tfrac{5}{2})^2} \simeq 3.20$

26. $y = \mathcal{A}\cos\big(\omega(t-c)\big) + \mathcal{B}\sin\big(\omega(t-c)\big)$
(easy to calculate $y'' + \omega^2 y = 0$)
$y = \mathcal{A}\big(\cos(\omega t)\cos(\omega c) + \sin(\omega t)\sin(\omega c)\big)$
$\quad + \mathcal{B}\big(\sin(\omega t)\cos(\omega c) - \cos(\omega t)\sin(\omega c)\big)$
$= \big(\mathcal{A}\cos(\omega c) - \mathcal{B}\sin(\omega c)\big)\cos\omega t$
$\quad + \big(\mathcal{A}\sin(\omega c) + \mathcal{B}\cos(\omega c)\big)\sin\omega t$
$= A\cos\omega t + B\sin\omega t$
where $A = \mathcal{A}\cos(\omega c) - \mathcal{B}\sin(\omega c)$ and $B = \mathcal{A}\sin(\omega c) + \mathcal{B}\cos(\omega c)$

28. $\begin{cases} y'' + \omega^2 y = 0 \\ y(a) = A \\ y'(a) = B \end{cases}$
$y = A\cos\big(\omega(t-a)\big) + \dfrac{B}{\omega}\sin\big(\omega(t-a)\big)$

30. Frequency $= \dfrac{\omega}{2\pi}$, $\omega^2 = \dfrac{k}{m}$ (k = spring const, m = mass)
Since the spring does not change, $\omega^2 m = k$ (constant)
For $m = 400$ gm, $\omega = 2\pi(24)$ (frequency = 24 Hz)
If $m = 900$ gm, then $\omega^2 = \dfrac{4\pi^2(24)^2(400)}{900}$

so $\omega = \dfrac{2\pi \times 24 \times 2}{3} = 32\pi$.

Thus frequency $= \dfrac{32\pi}{2\pi} = 16$ Hz

For $m = 100$ gm, $\omega = \dfrac{4\pi^2(24)^2 400}{100}$

so $\omega = 96\pi$ and frequency $= \dfrac{\omega}{2\pi} = 48$ Hz.

32. Expanding the hyperbolic functions in terms of exponentials,

$$y = e^{kt}\big[A\cosh\omega(t-t_0)B\sinh\omega(t-t_0)\big]$$
$$= e^{kt}\left[\frac{A}{2}e^{\omega(t-t_0)} + \frac{A}{2}e^{-\omega(t-t_0)}\right.$$
$$\left.+\frac{B}{2}e^{\omega(t-t_0)} - \frac{B}{2}e^{-\omega(t-t_0)}\right]$$
$$= A_1 e^{(k+\omega)t} + B_1 e^{(k-\omega)t}$$

where $A_1 = (A/2)e^{-\omega t_0} + (B/2)e^{-\omega t_0}$ and $B_1 = (A/2)e^{\omega t_0} - (B/2)e^{\omega t_0}$. Under the conditions of this problem we know that $Rr = k \pm \omega$ are the two real roots of the auxiliary equation $ar^2 + br + c = 0$, so $e^{(k\pm\omega)t}$ are independent solutions of $ay'' + by' + cy = 0$, and our function y must also be a solution. Since it involves two arbitrary constants, it is a general solution.

34. $\begin{cases} y'' + 4y' + 3y = 0 \\ y(3) = 1 \\ y'(3) = 0 \end{cases}$

The DE has auxiliary equation $r^2 + 4r + 3 = 0$ with roots $r = -2 + 1 = -1$ and $r = -2 - 1 = -3$ (i.e. $k \pm \omega$, where $k = -2$ and $\omega = 1$). By the second previous problem, a general solution can be expressed in the form $y = e^{-2t}[A\cosh(t-3) + B\sinh(t-3)]$ for which

$$y' = -2e^{-2t}[A\cosh(t-3) + B\sinh(t-3)]$$
$$+ e^{-2t}[A\sinh(t-3) + B\cosh(t-3)].$$

The initial conditions give

$$1 = y(3) = e^{-6}A$$
$$0 = y'(3) = -e^{-6}(-2A + B)$$

Thus $A = e^6$ and $B = 2A = 2e^6$. The IVP has solution

$$y = e^{6-2t}[\cosh(t-3) + 2\sinh(t-3)].$$

36. $y'' + y' - 2y = t$.
The auxiliary equation for $y'' + y' - 2y = 0$ is $r^2 + r - 2 = 0$, which has roots $r = -2$ and $r = 1$. Thus the complementary function is

$$y_h = C_1 e^{-2t} + C_2 e^t.$$

For a particular solution try
$y = At + B$. Then $y' = A$ and $y'' = 0$, so y satisfies the given equation if

$$t = A - 2(At + B) = A - 2B - 2At.$$

We require $A - 2B = 0$ and $-2A = 1$, so $A = -1/2$ and $B = -1/4$. The general solution of the given equation is

$$y = -\frac{2t + 1}{4} + C_1 e^{-2t} + C_2 e^t.$$

38. $y'' + y' - 2y = 20\cos(2t)$.
The auxiliary equation for $y'' + y' - 2y = 0$ is $r^2 + r - 2 = 0$, which has roots $r = -2$ and $r = 1$. Thus the complementary function is

$$y_h = C_1 e^{-2t} + C_2 e^t.$$

For a particular solution try
$$\begin{aligned} y &= A\cos(2t) + B\sin(2t).\ \text{Then} \\ y' &= -2A\sin(2t) + 2B\cos(2t)\ \text{and} \\ y'' &= -4A\cos(2t) - 4B\sin(2t), \end{aligned}$$
so y satisfies the given equation if $-6A + 2B = 20$ and $-2A - 6B = 0$. Thus we require $B = 1$ and $A = -3$. The general solution of the given equation is

$$y = \sin(2t) - 3\cos(2t) + C_1 e^{-2t} + C_2 e^t.$$

40. $y'' + y' - 2y = e^t$.
The auxiliary equation for $y'' + y' - 2y = 0$ is $r^2 + r - 2 = 0$, which has roots $r = -2$ and $r = 1$. Thus the complementary function is

$$y_h = C_1 e^{-2t} + C_2 e^t.$$

For a particular solution try
$y = Ate^t$. Then

$$y' = Ae^t(1 + t), \qquad y'' = Ae^t(2 + t),$$

so y satisfies the given equation if

$$e^t = Ae^t(2 + t + 1 + t - 2t) = 3Ae^t.$$

We require $A = 1/3$. The general solution of the given equation is

$$y = \frac{1}{3}te^t + C_1 e^{-2t} + C_3 e^t.$$

42. As in the solution of the previous problem, $y_1 = t^r$ satisfies the DE $at^2 y'' + bty' + cy = 0$ on $t > 0$ if r satisfies the auxiliary equation $ar(r - 1) + br + c = 0$, that is, $ar^2 + (b - a)4 + c = 0$. This equation has two equal real roots if $(b - a)^2 = 4ac$, in which case $r = -(b - a)/(2a)$, and $2ar + b - a = 0$. If $y_2 = t^r \ln t$, then

$$\begin{aligned} y_2' &= rt^{r-1}\ln t + t^{r-1} \\ y_2'' &= r(r-1)t^{r-2}\ln t + (2r-1)t^{r-2} \\ at^2 y_2'' &+ bty_2' + cy_2 \\ &= \big[ar(r-1) + br + c\big]t^r \ln t + \big[2ar + b - a\big]t^r = 0. \end{aligned}$$

In this case the DE has general solution

$$y = At^r + Bt^r \ln t \qquad (t > 0).$$

44. $2t^2 y'' - ty' - 2y = 0$
The auxiliary equation is $2r(r - 1) - r - 2 = 0$, or $2r^2 - 3r - 2 = 0$, which has roots $r = 2$ and $r = -1/2$. The general solution is $y = C_1 t^2 + C_2 t^{-1/2}$ for $t > 0$.

46. $t^2 y'' - 3ty' + 13y = 0$
The auxiliary equation is $r(r - 1) - 3r + 13 = 0$, or $r^2 - 4r + 13 = 0$, which has complex conjugate roots $r = 2 \pm 3i$. The general solution is $y = C_1 t^2 \cos(3\ln t) + C_2 t^2 \sin(3\ln t)$ for $t > 0$.

Review Exercises 3   (page 230)

2. $f(x) = \sec^2 x \tan x \Rightarrow f'(x) = 2\sec^2 x \tan^2 x + \sec^4 x > 0$ for x in $(-\pi/2, \pi/2)$, so f is increasing and therefore one-to-one and invertible there. The domain of f^{-1} is $(-\infty, \infty)$, the range of f. Since $f(\pi/4) = 2$, therefore $f^{-1}(2) = \pi/4$, and

$$(f^{-1})'(2) = \frac{1}{f'(f^{-1}(2))} = \frac{1}{f'(\pi/4)} = \frac{1}{8}.$$

4. Observe $f'(x) = e^{-x^2}(1 - 2x^2)$ is positive if $x^2 < 1/2$ and is negative if $x^2 > 1/2$. Thus f is increasing on $]-1/\sqrt{2}, 1/\sqrt{2}[$ and is decreasing on $]-\infty, -1/\sqrt{2}[$ and on $]1/\sqrt{2}, \infty[$.

6. $y = e^{-x}\sin x$, $(0 \le x \le 2\pi)$ has a horizontal tangent where

$$0 = \frac{dy}{dx} = e^{-x}(\cos x - \sin x).$$

This occurs if $\tan x = 1$, so $x = \pi/4$ or $x = 5\pi/4$. The points are $(\pi/4, e^{-\pi/4}/\sqrt{2})$ and $(5\pi/4, -e^{-5\pi/4}/\sqrt{2})$.

8. Let the length, radius, and volume of the clay cylinder at time t be ℓ, r, and V, respectively. Then $V = \pi r^2 \ell$, and

$$\frac{dV}{dt} = 2\pi r \ell \frac{dr}{dt} + \pi r^2 \frac{d\ell}{dt}.$$

Since $dV/dt = 0$ and $d\ell/dt = k\ell$ for some constant $k > 0$, we have

$$2\pi r \ell \frac{dr}{dt} = -k\pi r^2 \ell, \quad \Rightarrow \quad \frac{dr}{dt} = -\frac{kr}{2}.$$

That is, r is decreasing at a rate proportional to itself.

10. a) $\lim\limits_{h \to 0} \dfrac{a^h - 1}{h} = \lim\limits_{h \to 0} \dfrac{a^{0+h} - a^0}{h} = \dfrac{d}{dx} a^x \Big|_{x=0} = \ln a.$

Putting $h = 1/n$, we get $\lim\limits_{n \to \infty} n \left(a^{1/n} - 1 \right) = \ln a.$

b) Using the technique described in the exercise, we calculate

$$2^{10} \left(2^{1/2^{10}} - 1 \right) \approx 0.69338183$$

$$2^{11} \left(2^{1/2^{11}} - 1 \right) \approx 0.69326449$$

Thus $\ln 2 \approx 0.693.$

12. If $f(x) = (\ln x)/x$, then $f'(x) = (1 - \ln x)/x^2$. Thus $f'(x) > 0$ if $\ln x < 1$ (i.e., $x < e$) and $f'(x) < 0$ if $\ln x > 1$ (i.e., $x > e$). Since f is increasing to the left of e and decreasing to the right, it has a maximum value $f(e) = 1/e$ at $x = e$. Thus, if $x > 0$ and $x \neq e$, then

$$\frac{\ln x}{x} < \frac{1}{e}.$$

Putting $x = \pi$ we obtain $(\ln \pi)/\pi < 1/e$. Thus

$$\ln(\pi^e) = e \ln \pi < \pi = \pi \ln e = \ln e^{\pi},$$

and $\pi^e < e^{\pi}$ follows because $\ln$ is increasing.

14. a) $\dfrac{\ln x}{x} = \dfrac{\ln 2}{2}$ is satisfied if $x = 2$ or $x = 4$ (because $\ln 4 = 2 \ln 2$).

b) The line $y = mx$ through the origin intersects the curve $y = \ln x$ at $(b, \ln b)$ if $m = (\ln b)/b$. The same line intersects $y = \ln x$ at a different point $(x, \ln x)$ if $(\ln x)/x = m = (\ln b)/b$. This equation will have only one solution $x = b$ if the line $y = mx$ intersects the curve $y = \ln x$ only once, at $x = b$, that is, if the line is tangent to the curve at $x = b$. In this case m is the slope of $y = \ln x$ at $x = b$, so

$$\frac{1}{b} = m = \frac{\ln b}{b}.$$

Thus $\ln b = 1$, and $b = e$.

16. If $y = \cos^{-1} x$, then $x = \cos y$ and $0 \leq y \leq \pi$. Thus

$$\tan y = \operatorname{sgn} x \sqrt{\sec^2 y - 1} = \operatorname{sgn} x \sqrt{\frac{1}{x^2} - 1} = \frac{\sqrt{1 - x^2}}{x}.$$

Thus $\cos^{-1} x = \tan^{-1}((\sqrt{1 - x^2})/x).$

Since $\cot x = 1/\tan x$, $\cot^{-1} x = \tan^{-1}(1/x).$

$$\csc^{-1} x = \sin^{-1} \frac{1}{x} = \frac{\pi}{2} - \cos^{-1} \frac{1}{x}$$

$$= \frac{\pi}{2} - \tan^{-1} \frac{\sqrt{1 - (1/x)^2}}{1/x}$$

$$= \frac{\pi}{2} - \operatorname{sgn} x \tan^{-1} \sqrt{x^2 - 1}.$$

18. Let $T(t)$ be the temperature of the milk t minutes after it is removed from the refrigerator. Let $U(t) = T(t) - 20$. By Newton's law,

$$U'(t) = kU(t) \quad \Rightarrow \quad U(t) = U(0)e^{kt}.$$

Now $T(0) = 5 \Rightarrow U(0) = -15$ and $T(12) = 12 \Rightarrow U(12) = -8$. Thus

$$-8 = U(12) = U(0)e^{12k} = -15e^{12k}$$

$$e^{12k} = 8/15, \qquad k = \frac{1}{12} \ln(8/15).$$

If $T(s) = 18$, then $U(s) = -2$, so $-2 = -15e^{sk}$. Thus $sk = \ln(2/15)$, and

$$s = \frac{\ln(2/15)}{k} = 12\frac{\ln(2/15)}{\ln(8/15)} \approx 38.46.$$

It will take another $38.46 - 12 = 26.46$ min for the milk to warm up to $18°$.

20. Let $f(x) = e^x - 1 - x$. Then $f(0) = 0$ and by the MVT,

$$\frac{f(x)}{x} = \frac{f(x) - f(0)}{x - 0} = f'(c) = e^c - 1$$

for some c between 0 and x. If $x > 0$, then $c > 0$, and $f'(c) > 0$. If $x < 0$, then $c < 0$, and $f'(c) < 0$. In either case $f(x) = xf'(c) > 0$, which is what we were asked to show.

Challenging Problems 3 (page 231)

2. $\dfrac{dv}{dt} = -g - kv.$

a) Let $u(t) = -g - kv(t)$. Then $\dfrac{du}{dt} = -k\dfrac{dv}{dt} = -ku$, and

$$u(t) = u(0)e^{-kt} = -(g + kv_0)e^{-kt}$$

$$v(t) = -\frac{1}{k}\big(g + u(t)\big) = -\frac{1}{k}\big(g - (g + kv_0)e^{-kt}\big).$$

b) $\lim\limits_{t \to \infty} v(t) = -g/k$

c) $\dfrac{dy}{dt} = v(t) = -\dfrac{g}{k} + \dfrac{g + kv_0}{k}e^{-kt}, \quad y(0) = y_0$

$$y(t) = -\frac{gt}{k} - \frac{g + kv_0}{k^2}e^{-kt} + C$$

$$y_0 = -0 - \frac{g + kv_0}{k^2} + C \Rightarrow C = y_0 + \frac{g + kv_0}{k^2}$$

$$y(t) = y_0 - \frac{gt}{k} + \frac{g + kv_0}{k^2}\left(1 - e^{-kt}\right)$$

4. If $p = e^{-bt} y$, then $\dfrac{dp}{dt} = e^{-bt} \left(\dfrac{dy}{dt} - by \right)$.

The DE $\dfrac{dp}{dt} = kp \left(1 - \dfrac{p}{e^{-bt} M} \right)$ therefore transforms to

$$\frac{dy}{dt} = by + kpe^{bt} \left(1 - \frac{p}{e^{-bt} M} \right)$$
$$= (b+k)y - \frac{ky^2}{M} = Ky \left(1 - \frac{y}{L} \right),$$

where $K = b + k$ and $L = \dfrac{b+k}{k} M$. This is a standard Logistic equation with solution (as obtained in Section 3.4) given by

$$y = \frac{Ly_0}{y_0 + (L - y_0)e^{-Kt}},$$

where $y_0 = y(0) = p(0) = p_0$. Converting this solution back in terms of the function $p(t)$, we obtain

$$p(t) = \frac{Lp_0 e^{-bt}}{p_0 + (L - p_0)e^{-(b+k)t}}$$
$$= \frac{(b+k)Mp_0}{p_0 k e^{bt} + \big((b+k)M - kp_0\big)e^{-kt}}.$$

Since p represents a percentage, we must have $(b+k)M/k < 100$.

If $k = 10$, $b = 1$, $M = 90$, and $p_0 = 1$, then $\dfrac{b+k}{k} M = 99 < 100$. The numerator of the final expression for $p(t)$ given above is a constant. Therefore $p(t)$ will be largest when the derivative of the denominator,

$$f(t) = p_0 k e^{bt} + \big((b+k)M - kp_0\big)e^{-kt} = 10e^t + 980e^{-10t}$$

is zero. Since $f'(t) = 10e^t - 9,800e^{-10t}$, this will happen at $t = \ln(980)/11$. The value of p at this t is approximately 48.1. Thus the maximum percentage of potential clients who will adopt the technology is about 48.1%.

CHAPTER 4. SOME APPLICATIONS OF DERIVATIVES

Section 4.1 Related Rates (page 237)

2. As in Exercise 1, $dA/dt = 2x \, dx/dt$. If $dA/dt = -2$ ft^2/s and $x = 8$ ft, then $dx/dt = -2/(16)$. The side length is decreasing at 1/8 ft/s.

4. Let A and r denote the area and radius of the circle. Then

$$A = \pi r^2 \Rightarrow r = \sqrt{\frac{A}{\pi}}$$

$$\Rightarrow \frac{dr}{dt} = \left(\frac{1}{2\sqrt{A\pi}}\right) \frac{dA}{dt}.$$

When $\dfrac{dA}{dt} = -2$, and $A = 100$, $\dfrac{dr}{dt} = -\dfrac{1}{10\sqrt{\pi}}$. The radius is decreasing at the rate $\dfrac{1}{10\sqrt{\pi}}$ cm/min when the area is $100\,\text{cm}^2$.

6. Let the length, width, and area be l, w, and A at time t. Thus $A = lw$.

$$\frac{dA}{dt} = l\frac{dw}{dt} + w\frac{dl}{dt}$$

When $l = 16$, $w = 12$, $\dfrac{dw}{dt} = 3$, $\dfrac{dA}{dt} = 0$, we have

$$0 = 16 \times 3 + 12\frac{dl}{dt} \Rightarrow \frac{dl}{dt} = -\frac{48}{12} = -4$$

The length is decreasing at 4 m/s.

8. The volume V of the ball is given by

$$V = \frac{4}{3}\pi r^3 = \frac{4\pi}{3}\left(\frac{D}{2}\right)^3 = \frac{\pi}{6}D^3,$$

where $D = 2r$ is the diameter of the ball. We have

$$\frac{dV}{dt} = \frac{\pi}{2}D^2\frac{dD}{dt}.$$

When $D = 6$ cm, $dD/dt = -.5$ cm/h. At that time

$$\frac{dV}{dt} = \frac{\pi}{2}(36)(-0.5) = -9\pi \approx -28.3.$$

The volume is decreasing at about 28.3 cm^3/h.

10. Let V, r and h denote the volume, radius and height of the cylinder at time t. Thus, $V = \pi r^2 h$ and

$$\frac{dV}{dt} = 2\pi rh\frac{dr}{dt} + \pi r^2\frac{dh}{dt}.$$

If $V = 60$, $\dfrac{dV}{dt} = 2$, $r = 5$, $\dfrac{dr}{dt} = 1$, then

$$h = \frac{V}{\pi r^2} = \frac{60}{25\pi} = \frac{12}{5\pi}$$

$$\frac{dh}{dt} = \frac{1}{\pi r^2}\left(\frac{dV}{dt} - 2\pi rh\frac{dr}{dt}\right)$$

$$= \frac{1}{25\pi}\left(2 - 10\pi\frac{12}{5\pi}\right) = -\frac{22}{25\pi}.$$

The height is decreasing at the rate $\dfrac{22}{25\pi}$ cm/min.

12. Let the length, width and area at time t be x, y and A respectively. Thus $A = xy$ and

$$\frac{dA}{dt} = x\frac{dy}{dt} + y\frac{dx}{dt}.$$

If $\dfrac{dA}{dt} = 5$, $\dfrac{dx}{dt} = 10$, $x = 20$, $y = 16$, then

$$5 = 20\frac{dy}{dt} + 16(10) \Rightarrow \frac{dy}{dt} = -\frac{31}{4}.$$

Thus, the width is decreasing at $\dfrac{31}{4}$ m/s.

14. Since $x^2 y^3 = 72$, then

$$2xy^3\frac{dx}{dt} + 3x^2 y^2\frac{dy}{dt} = 0 \Rightarrow \frac{dy}{dt} = -\frac{2y}{3x}\frac{dx}{dt}.$$

If $x = 3$, $y = 2$, $\dfrac{dx}{dt} = 2$, then $\dfrac{dy}{dt} = -\dfrac{8}{9}$. Hence, the vertical velocity is $-\dfrac{8}{9}$ units/s.

16. From the figure, $x^2 + k^2 = s^2$. Thus

$$x\frac{dx}{dt} = s\frac{ds}{dt}.$$

When angle $PCA = 45°$, $x = k$ and $s = \sqrt{2}k$. The radar gun indicates that $ds/dt = 100$ km/h. Thus $dx/dt = 100\sqrt{2}k/k \approx 141$. The car is travelling at about 141 km/h.

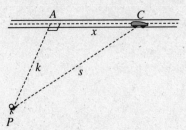

Fig. 4.1.16

18. Let the distances x and y be as shown at time t. Thus $x^2 + y^2 = 25$ and $2x\dfrac{dx}{dt} + 2y\dfrac{dy}{dt} = 0$.

If $\dfrac{dx}{dt} = \dfrac{1}{3}$ and $y = 3$, then $x = 4$ and $\dfrac{4}{3} + 3\dfrac{dy}{dt} = 0$ so $\dfrac{dy}{dt} = -\dfrac{4}{9}$.

The top of the ladder is slipping down at a rate of $\dfrac{4}{9}$ m/s.

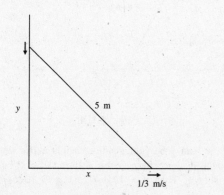

5 m

y

x

1/3 m/s

Fig. 4.1.18

20.

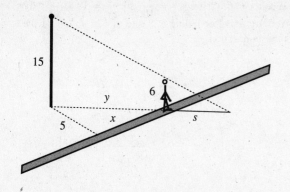

15

6

y

x

s

5

Fig. 4.1.20

Refer to the figure. s, y, and x are, respectively, the length of the woman's shadow, the distances from the woman to the lamppost, and the distances from the woman to the point on the path nearest the lamppost. From one of triangles in the figure we have

$$y^2 = x^2 + 25.$$

If $x = 12$, then $y = 13$. Moreover,

$$2y\frac{dy}{dt} = 2x\frac{dx}{dt}.$$

We are given that $dx/dt = 2$ ft/s, so $dy/dt = 24/13$ ft/s when $x = 12$ ft. Now the similar triangles in the figure show that

$$\frac{s}{6} = \frac{s+y}{15},$$

so that $s = 2y/3$. Hence $ds/dt = 48/39$. The woman's shadow is changing at rate 48/39 ft/s when she is 12 ft from the point on the path nearest the lamppost.

22. Let x, y be distances travelled by A and B from their positions at 1:00 pm in t hours.
Thus $\dfrac{dx}{dt} = 16$ km/h, $\dfrac{dy}{dt} = 20$ km/h.
Let s be the distance between A and B at time t.
Thus $s^2 = x^2 + (25+y)^2$

$$2s\frac{ds}{dt} = 2x\frac{dx}{dt} + 2(25+y)\frac{dy}{dt}$$

At 1:30 $\left(t = \tfrac{1}{2}\right)$ we have $x = 8$, $y = 10$, $s = \sqrt{8^2 + 35^2} = \sqrt{1289}$ so

$$\sqrt{1289}\frac{ds}{dt} = 8 \times 16 + 35 \times 20 = 828$$

and $\dfrac{ds}{dt} = \dfrac{828}{\sqrt{1289}} \approx 23.06$. At 1:30, the ships are separating at about 23.06 km/h.

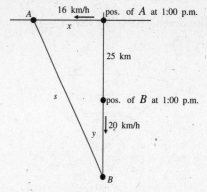

A 16 km/h   pos. of A at 1:00 p.m.

x

25 km

s pos. of B at 1:00 p.m.

y 20 km/h

B

Fig. 4.1.22

24. Let y be the height of balloon t seconds after release. Then $y = 5t$ m.
Let θ be angle of elevation at B of balloon at time t.
Then $\tan\theta = y/100$. Thus

$$\sec^2\theta\frac{d\theta}{dt} = \frac{1}{100}\frac{dy}{dt} = \frac{5}{100} = \frac{1}{20}$$

$$\left(1 + \tan^2\theta\right)\frac{d\theta}{dt} = \frac{1}{20}$$

$$\left[1 + \left(\frac{y}{100}\right)^2\right]\frac{d\theta}{dt} = \frac{1}{20}.$$

When $y = 200$ we have $5\dfrac{d\theta}{dt} = \dfrac{1}{20}$ so $\dfrac{d\theta}{dt} = \dfrac{1}{100}$.

The angle of elevation of balloon at B is increasing at a rate of $\dfrac{1}{100}$ rad/s.

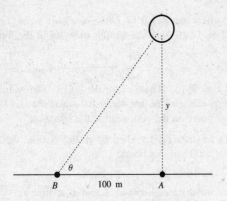

Fig. 4.1.24

26. Let r, h, and V be the top radius, depth, and volume of the water in the tank at time t. Then $\dfrac{r}{h} = \dfrac{10}{8}$ and

$V = \dfrac{1}{3}\pi r^2 h = \dfrac{\pi}{3}\dfrac{25}{16}h^3$. We have

$$\frac{1}{10} = \frac{\pi}{3}\frac{25}{16}3h^2\frac{dh}{dt} \Rightarrow \frac{dh}{dt} = \frac{16}{250\pi h^2}.$$

When $h = 4$ m, we have $\dfrac{dh}{dt} = \dfrac{1}{250\pi}$.

The water level is rising at a rate of $\dfrac{1}{250\pi}$ m/min when depth is 4 m.

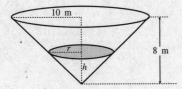

Fig. 4.1.26

28. Let r, h, and V be the top radius, depth, and volume of the water in the tank at time t. Then

$$\frac{r}{h} = \frac{3}{9} = \frac{1}{3}$$
$$V = \frac{1}{3}\pi r^2 h = \frac{\pi}{27}h^3$$
$$\frac{dV}{dt} = \frac{\pi}{9}h^2\frac{dh}{dt}.$$

If $\dfrac{dh}{dt} = 20$ cm/h $= \dfrac{2}{10}$ m/h when $h = 6$ m, then

$$\frac{dV}{dt} = \frac{\pi}{9} \times 36 \times \frac{2}{10} = \frac{4\pi}{5} \approx 2.51 \text{ m}^3/\text{h}.$$

Since water is coming in at a rate of 10 m³/h, it must be leaking out at a rate of $10 - 2.51 \approx 7.49$ m³/h.

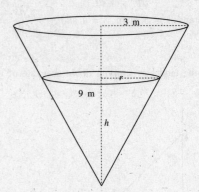

Fig. 4.1.28

30. Let P, x, and y be your position, height above centre, and horizontal distance from centre at time t. Let θ be the angle shown. Then $y = 10\sin\theta$, and $x = 10\cos\theta$. We have

$$\frac{dy}{dt} = 10\cos\theta\frac{d\theta}{dt}, \qquad \frac{d\theta}{dt} = 1 \text{ rpm} = 2\pi \text{ rad/min}.$$

When $x = 6$, then $\cos\theta = \dfrac{6}{10}$, so $\dfrac{dy}{dt} = 10 \times \dfrac{6}{10} \times 12\pi$. You are rising or falling at a rate of 12π m/min at the time in question.

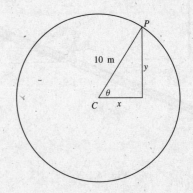

Fig. 4.1.30

32.
$$P = \frac{1}{3}x^{0.6}y^{0.4}$$
$$\frac{dP}{dt} = \frac{0.6}{3}x^{-0.4}y^{0.4}\frac{dx}{dt} + \frac{0.4}{3}x^{0.6}y^{-0.6}\frac{dy}{dt}.$$

If $dP/dt = 0$, $x = 40$, $dx/dt = 1$, and $y = 10,000$, then

$$\frac{dy}{dt} = -\frac{6y^{0.4}}{x^{0.4}}\frac{y^{0.6}}{4x^{0.6}}\frac{dx}{dt} = -\frac{6y}{4x}\frac{dx}{dt} = -375.$$

The daily expenses are decreasing at $375 per day.

34. Let x and y be the distances travelled from the intersection point by the boat and car respectively in t minutes. Then

$$\frac{dx}{dt} = 20 \times \frac{1000}{60} = \frac{1000}{3}\,\text{m/min}$$

$$\frac{dy}{dt} = 80 \times \frac{1000}{60} = \frac{4000}{3}\,\text{m/min}$$

The distance s between the boat and car satisfy

$$s^2 = x^2 + y^2 + 20^2, \qquad s\frac{ds}{dt} = x\frac{dx}{dt} + y\frac{dy}{dt}.$$

After one minute, $x = \dfrac{1000}{3}$, $y = \dfrac{4000}{3}$ so $s \approx 1374$. m. Thus

$$1374.5\frac{ds}{dt} = \frac{1000}{3}\frac{1000}{3} + \frac{4000}{3}\frac{4000}{3} \approx 1,888,889.$$

Hence $\dfrac{ds}{dt} \approx 1374.2$ m/min ≈ 82.45 km/h after 1 minute.

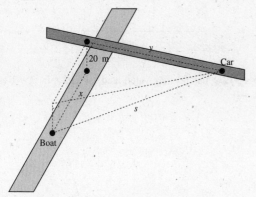

Fig. 4.1.34

36. Let V and h be the volume and depth of water in the pool at time t. If $h \le 2$, then

$$\frac{x}{h} = \frac{20}{2} = 10, \quad \text{so } V = \frac{1}{2}xh8 = 40h^2.$$

If $2 \le h \le 3$, then $V = 160 + 160(h - 2)$.

a) If $h = 2.5$m, then $-1 = \dfrac{dV}{dt} = 160\dfrac{dh}{dt}$.

So surface of water is dropping at a rate of $\dfrac{1}{160}$ m/min.

b) If $h = 1$m, then $-1 = \dfrac{dV}{dt} = 80h\dfrac{dh}{dt} = 80\dfrac{dh}{dt}$.

So surface of water is dropping at a rate of $\dfrac{1}{80}$ m/min.

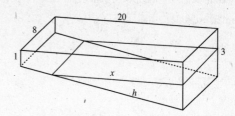

Fig. 4.1.36

38. Let x, y, and s be distances shown at time t. Then

$$s^2 = x^2 + 16, \qquad (15 - s)^2 = y^2 + 16$$

$$s\frac{ds}{dt} = x\frac{dx}{dt}, \qquad -(15 - s)\frac{ds}{dt} = y\frac{dy}{dt}.$$

When $x = 3$ and $\dfrac{dx}{dt} = \dfrac{1}{2}$, then $s = 5$ and $y = \sqrt{10^2 - 4^2} = \sqrt{84}$.

Also $\dfrac{ds}{dt} = \dfrac{3}{5}\left(\dfrac{1}{2}\right) = \dfrac{3}{10}$ so

$$\frac{dy}{dt} = -\frac{10}{\sqrt{84}}\frac{3}{10} = -\frac{3}{\sqrt{84}} \approx 0.327.$$

Crate B is moving toward Q at a rate of 0.327 m/s.

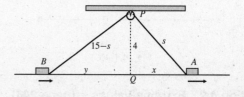

Fig. 4.1.38

40. Let y be height of ball t seconds after it drops. Thus $\dfrac{d^2y}{dt^2} = -9.8$, $\dfrac{dy}{dt}\big|_{t=0} = 0$, $y|_{t=0} = 20$, and

$$y = -4.9t^2 + 20, \qquad \frac{dy}{dt} = -9.8t.$$

Let s be distance of shadow of ball from base of pole. By similar triangles, $\dfrac{s - 10}{y} = \dfrac{s}{20}$.

$20s - 200 = sy$, $s = \dfrac{200}{20 - y}$

$20\dfrac{ds}{dt} = y\dfrac{ds}{dt} + s\dfrac{dy}{dt}$.

a) At $t = 1$, we have $\dfrac{dy}{dt} = -9.8$, $y = 15.1$,

$4.9\dfrac{ds}{dt} = \dfrac{200}{4.9}(-9.8)$.

The shadow is moving at a rate of 81.63 m/s after one second.

b) As the ball hits the ground, $y = 0$, $s = 10$, $t = \sqrt{\dfrac{20}{4.9}}$,

and $\dfrac{dy}{dt} = -9.8\sqrt{\dfrac{20}{4.9}}$, so $20\dfrac{ds}{dt} = 0 + 10\dfrac{dy}{dt}$.

Now $y = 0$ implies that $t = \sqrt{\dfrac{20}{4.9}}$. Thus

$$\dfrac{ds}{dt} = -\dfrac{1}{2}(9.8)\sqrt{\dfrac{20}{4.9}} \approx -9.90.$$

The shadow is moving at about 9.90 m/s when the ball hits the ground.

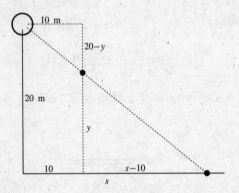

Fig. 4.1.40

Section 4.2 Extreme Values (page 246)

2. $f(x) = x + 2$ on $]-\infty, 0]$
abs max 2 at $x = 0$, no min.

4. $f(x) = x^2 - 1$
no max, abs min -1 at $x = 0$.

6. $f(x) = x^2 - 1$ on $]2, 3[$
no max or min values.

8. $f(x) = x^3 + x - 4$ on $]a, b[$
Since $f'(x) = 3x^2 + 1 > 0$ for all x, therefore f is increasing. Since $]a, b[$ is open, f has no max or min values.

10. $f(x) = \dfrac{1}{x - 1}$. Since $f'(x) = \dfrac{-1}{(x - 1)^2} < 0$ for all x in the domain of f, therefore f has no max or min values.

12. $f(x) = \dfrac{1}{x - 1}$ on $[2, 3]$
abs min $\frac{1}{2}$ at $x = 3$, abs max 1 at $x = 2$.

14. Let $f(x) = |x^2 - x - 2| = |(x - 2)(x + 1)|$ on $[-3, 3]$:
$f(-3) = 10$, $f(3) = 4$.
$f'(x) = (2x - 1)\mathrm{sgn}\,(x^2 - x - 2)$.
CP $x = 1/2$; SP $x = -1$, and $x = 2$. $f(1/2) = 9/4$,
$f(-1) = 0$, $f(2) = 0$.
Max value of f is 10 at $x = -3$; min value is 0 at $x = -1$ or $x = 2$.

16. $f(x) = (x + 2)^{(2/3)}$
no max, abs min 0 at $x = -2$.

18. $f(x) = x^2 + 2x$, $f'(x) = 2x + 2 = 2(x + 1)$
Critical point: $x = -1$.
$f(x) \to \infty$ as $x \to \pm\infty$.

		CP	
f'	$-$	-1	$+$
			$\longrightarrow x$
f	$\searrow$	abs min	$\nearrow$

Hence, $f(x)$ has no max value, and the abs min is -1 at $x = -1$.

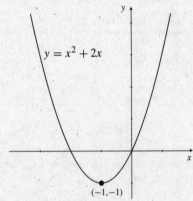

$y = x^2 + 2x$

$(-1, -1)$

Fig. 4.2.18

20. $f(x) = (x^2 - 4)^2$, $f'(x) = 4x(x^2 - 4) = 4x(x + 2)(x - 2)$
Critical points: $x = 0, \pm 2$.
$f(x) \to \infty$ as $x \to \pm\infty$.

		CP		CP		CP	
f'	$-$	-2	$+$	0	$-$	$+2$	$+$
							$\longrightarrow x$
f	$\searrow$	abs min	$\nearrow$	loc max	$\searrow$	abs min	$\nearrow$

Hence, $f(x)$ has abs min 0 at $x = \pm 2$ and loc max 16 at $x = 0$.

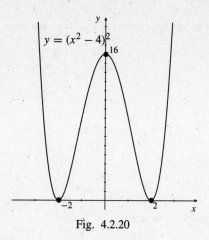

Fig. 4.2.20

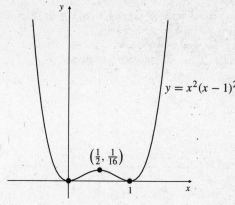

$y = x^2(x-1)^2$

$(\frac{1}{2}, \frac{1}{16})$

Fig. 4.2.24

22. $f(x) = x^4 + 4x$, $f'(x) = 4x^3 + 4 = 4(x^3 + 1)$
Critical point: $x = -1$.
$f(x) \to \infty$ as $x \to \pm\infty$.

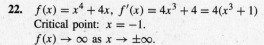

Hence, $f(x)$ has no max value but has abs min -3 at $x = -1$.

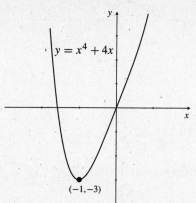

$y = x^4 + 4x$

$(-1, -3)$

Fig. 4.2.22

24. $f(x) = x^2(x-1)^2$,
$f'(x) = 2x(x-1)^2 + 2x^2(x-1) = 2x(2x-1)(x-1)$
Critical points: $x = 0$, $\frac{1}{2}$ and 1.
$f(x) \to \infty$ as $x \to \pm\infty$.

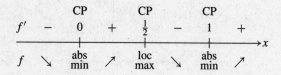

Hence, $f(x)$ has loc max $\frac{1}{16}$ at $x = \frac{1}{2}$ and abs min 0 at $x = 0$ and $x = 1$.

26. $f(x) = \dfrac{x}{x^2+1}$, $f'(x) = \dfrac{1-x^2}{(x^2+1)^2}$
Critical point: $x = \pm 1$.
$f(x) \to 0$ as $x \to \pm\infty$.

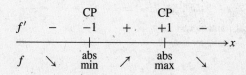

Hence, f has abs max $\frac{1}{2}$ at $x = 1$ and abs min $-\frac{1}{2}$ at $x = -1$.

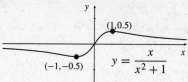

$(1, 0.5)$

$(-1, -0.5)$

$y = \dfrac{x}{x^2+1}$

Fig. 4.2.26

28. $f(x) = \dfrac{x}{\sqrt{x^4+1}}$, $f'(x) = \dfrac{1-x^4}{(x^4+1)^{3/2}}$
Critical points: $x = \pm 1$.
$f(x) \to 0$ as $x \to \pm\infty$.

Hence, f has abs max $\frac{1}{\sqrt{2}}$ at $x = 1$ and abs min $-\frac{1}{\sqrt{2}}$ at $x = -1$.

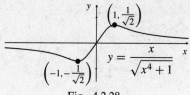

$\left(1, \frac{1}{\sqrt{2}}\right)$

$\left(-1, -\frac{1}{\sqrt{2}}\right)$

$y = \dfrac{x}{\sqrt{x^4+1}}$

Fig. 4.2.28

61

30. $f(x) = x + \sin x$, $f'(x) = 1 + \cos x \geq 0$
$f'(x) = 0$ at $x = \pm\pi$, $\pm 3\pi$, ...
$f(x) \to \pm\infty$ as $x \to \pm\infty$.
Hence, f has no max or min values.

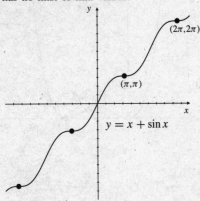

Fig. 4.2.30

32. $f(x) = x - 2\tan^{-1} x$, $f'(x) = 1 - \dfrac{2}{1+x^2} = \dfrac{x^2-1}{x^2+1}$
Critical points: $x = \pm 1$.
$f(x) \to \pm\infty$ as $x \to \pm\infty$.

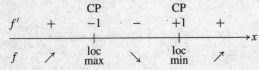

Hence, f has loc max $-1 + \dfrac{\pi}{2}$ at $x = -1$ and loc min
$1 - \dfrac{\pi}{2}$ at $x = 1$.

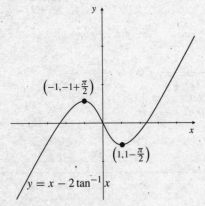

Fig. 4.2.32

34. $f(x) = e^{-x^2/2}$, $f'(x) = -x e^{-x^2/2}$
Critical point: $x = 0$.
$f(x) \to 0$ as $x \to \pm\infty$.

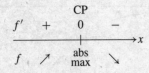

Hence, f has abs max 1 at $x = 0$ and no min value.

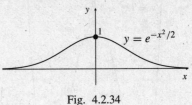

Fig. 4.2.34

36. $f(x) = x^2 e^{-x^2}$, $f'(x) = 2x e^{-x^2}(1 - x^2)$
Critical points: $x = 0, \pm 1$.
$f(x) \to 0$ as $x \to \pm\infty$.

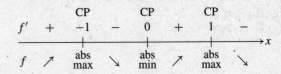

Hence, f has abs max $1/e$ at $x = \pm 1$ and abs min 0 at
$x = 0$.

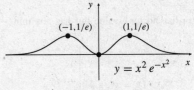

Fig. 4.2.36

38. Since $f(x) = |x + 1|$,

$$f'(x) = \operatorname{sgn}(x+1) = \begin{cases} 1, & \text{if } x > -1; \\ -1, & \text{if } x < -1. \end{cases}$$

-1 is a singular point; f has no max but has abs min 0 at
$x = -1$.
$f(x) \to \infty$ as $x \to \pm\infty$.

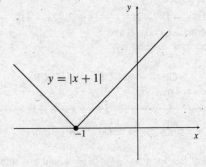

Fig. 4.2.38

40. $f(x) = \sin|x|$

$f'(x) = \operatorname{sgn}(x)\cos|x| = 0$ at $x = \pm\dfrac{\pi}{2},\ \pm\dfrac{3\pi}{2},\ \pm\dfrac{5\pi}{2},\ \dots$

0 is a singular point. Since $f(x)$ is an even function, its graph is symmetric about the origin.

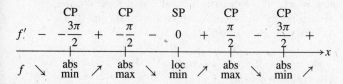

Hence, f has abs max 1 at $x = \pm(4k+1)\dfrac{\pi}{2}$ and abs min -1 at $x = \pm(4k+3)\dfrac{\pi}{2}$ where $k = 0, 1, 2, \dots$ and loc min 0 at $x = 0$.

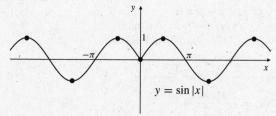

Fig. 4.2.40

42. $f(x) = (x-1)^{2/3} - (x+1)^{2/3}$

$f'(x) = \dfrac{2}{3}(x-1)^{-1/3} - \dfrac{2}{3}(x+1)^{-1/3}$

Singular point at $x = \pm 1$. For critical points:

$(x-1)^{-1/3} = (x+1)^{-1/3} \Rightarrow x-1 = x+1 \Rightarrow 2 = 0$, so there are no critical points.

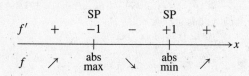

Hence, f has abs max $2^{2/3}$ at $x = -1$ and abs min $-2^{2/3}$ at $x = 1$.

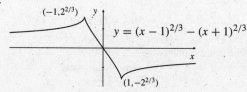

Fig. 4.2.42

44. $f(x) = x/\sqrt{x^4+1}$. f is continuous on $\mathbb{R}$, and $\lim_{x\to\pm\infty} f(x) = 0$. Since $f(1) > 0$ and $f(-1) < 0$, f must have both maximum and minimum values.

$$f'(x) = \frac{\sqrt{x^4+1} - x\dfrac{4x^3}{2\sqrt{x^4+1}}}{x^4+1} = \frac{1-x^4}{(x^4+1)^{3/2}}.$$

CP $x = \pm 1$. $f(\pm 1) = \pm 1/\sqrt{2}$. f has max value $1/\sqrt{2}$ and min value $-1/\sqrt{2}$.

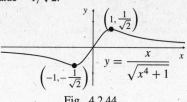

Fig. 4.2.44

46. $f(x) = x^2/\sqrt{4-x^2}$ is continuous on $(-2, 2)$, and $\lim_{x\to-2+} f(x) = \lim_{x\to 2-} f(x) = \infty$. Thus f can have no maximum value, but will have a minimum value.

$$f'(x) = \frac{2x\sqrt{4-x^2} - x^2\dfrac{-2x}{2\sqrt{4-x^2}}}{4-x^2} = \frac{8x - x^3}{(4-x^2)^{3/2}}.$$

CP $x = 0$, $x = \pm\sqrt{8}$. $f(0) = 0$, and $\pm\sqrt{8}$ is not in the domain of f. f has minimum value 0 at $x = 0$.

48. $f(x) = (\sin x)/x$ is continuous and differentiable on $\mathbb{R}$ except at $x = 0$ where it is undefined.

Since $\lim_{x\to 0} f(x) = 1$, and $|f(x)| < 1$ for all $x \neq 0$ (because $|\sin x| < |x|$), f cannot have a maximum value. Since $\lim_{x\to\pm\infty} f(x) = 0$ and since $f(x) < 0$ at some points, f must have a minimum value occurring at a critical point. In fact, since $|f(x)| \leq 1/|x|$ for $x \neq 0$ and f is even, the minimum value will occur at the two critical points closest to $x = 0$. (See Figure 2.20 on page 124 of the text.)

50. No. $f(x) = -x^2$ has abs max value 0, but $g(x) = |f(x)| = x^2$ has no abs max value.

Section 4.3 Concavity and Inflections (page 251)

2. $f(x) = 2x - x^2$, $f'(x) = 2 - 2x$, $f''(x) = -2 < 0$. Thus, f is concave down on $(-\infty, \infty)$.

4. $f(x) = x - x^3$, $f'(x) = 1 - 3x^2$, $f''(x) = -6x$.

f''	$+$	0	$-$
f	$\smile$	infl	$\frown$

6. $f(x) = 10x^3 + 3x^5$, $f'(x) = 30x^2 + 15x^4$, $f''(x) = 60x + 60x^3 = 60x(1 + x^2)$.

f''	$-$	0	$+$
f	$\frown$	infl	$\smile$

63

8. $f(x) = (2 + 2x - x^2)^2$, $f'(x) = 2(2 + 2x - x^2)(2 - 2x)$,
$f''(x) = 2(2 - 2x)^2 + 2(2 + 2x - x^2)(-2)$
$\qquad = 12x(x - 2)$.

$$
\begin{array}{ccccccc}
f'' & + & 0 & - & 2 & + & \\
\hline
f & \smile & \text{infl} & \frown & \text{infl} & \smile & \!\!\!\to x
\end{array}
$$

10. $f(x) = \dfrac{x}{x^2 + 3}$, $f'(x) = \dfrac{3 - x^2}{(x^2 + 3)^2}$,
$f''(x) = \dfrac{2x(x^2 - 9)}{(x^2 + 3)^3}$.

$$
\begin{array}{cccccccc}
f'' & - & -3 & + & 0 & - & 3 & + \\
\hline
f & \frown & \text{infl} & \smile & \text{infl} & \frown & \text{infl} & \smile
\end{array}
$$

12. $f(x) = \cos 3x$, $f'(x) = -3\sin 3x$, $f''(x) = -9\cos 3x$.
Inflection points: $x = \left(n + \frac{1}{2}\right)\dfrac{\pi}{3}$ for $n = 0$, ± 1, ± 2, ...
f is concave up on $\left(\dfrac{4n + 1}{6}\pi, \dfrac{4n + 3}{6}\pi\right)$ and concave
down on $\left(\dfrac{4n + 3}{6}\pi, \dfrac{4n + 5}{6}\pi\right)$.

14. $f(x) = x - 2\sin x$, $f'(x) = 1 - 2\cos x$, $f''(x) = 2\sin x$.
Inflection points: $x = n\pi$ for $n = 0$, ± 1, ± 2, ...
f is concave down on $\big((2n + 1)\pi, (2n + 2)\pi\big)$ and concave
up on $\big((2n)\pi, (2n + 1)\pi\big)$.

16. $f(x) = xe^x$, $f'(x) = e^x(1 + x)$,
$f''(x) = e^x(2 + x)$.

$$
\begin{array}{cccc}
f'' & - & -2 & + \\
\hline
f & \frown & \text{infl} & \smile
\end{array}
$$

18. $f(x) = \dfrac{\ln(x^2)}{x}$, $f'(x) = \dfrac{2 - \ln(x^2)}{x^2}$,
$f''(x) = \dfrac{-6 + 2\ln(x^2)}{x^3}$.
f has inflection point at $x = \pm e^{3/2}$ and f is undefined at
$x = 0$. f is concave up on $(-e^{3/2}, 0)$ and $(e^{3/2}, \infty)$; and
concave down on $(-\infty, -e^{3/2})$ and $(0, e^{3/2})$.

20. $f(x) = (\ln x)^2$, $f'(x) = \dfrac{2}{x}\ln x$,
$f''(x) = \dfrac{2(1 - \ln x)}{x^2}$ for all $x > 0$.

$$
\begin{array}{ccccc}
f'' & 0 & + & e & - \\
\hline
f & & \smile & \text{infl} & \frown
\end{array}
$$

22. $f(x) = (x - 1)^{1/3} + (x + 1)^{1/3}$,
$f'(x) = \frac{1}{3}[(x - 1)^{-2/3} + (x + 1)^{-2/3}]$,
$f''(x) = -\frac{2}{9}[(x - 1)^{-5/3} + (x + 1)^{-5/3}]$.
$f(x) = 0 \Leftrightarrow x - 1 = -(x + 1) \Leftrightarrow x = 0$.
Thus, f has inflection point at $x = 0$. $f''(x)$ is undefined
at $x = \pm 1$. f is defined at ± 1 and $x = \pm 1$ are also inflection points. f is concave up on $(-\infty, -1)$ and $(0, 1)$;
and down on $(-1, 0)$ and $(1, \infty)$.

24. $f(x) = 3x^3 - 36x - 3$, $f'(x) = 9(x^2 - 4)$, $f''(x) = 18x$.
The critical points are
$x = 2$, $f''(2) > 0 \Rightarrow$ local min;
$x = -2$, $f''(-2) < 0 \Rightarrow$ local max.

26. $f(x) = x + \dfrac{4}{x}$, $f'(x) = 1 - \dfrac{4}{x^2}$, $f''(x) = 8x^{-3}$.
The critical points are
$x = 2$, $f''(2) > 0 \Rightarrow$ local min;
$x = -2$, $f''(-2) < 0 \Rightarrow$ local max.

28. $f(x) = \dfrac{x}{2^x}$, $f'(x) = \dfrac{1 - x\ln 2}{2^x}$,
$f''(x) = \dfrac{\ln 2(x\ln 2 - 2)}{2^x}$.
The critical point is
$x = \dfrac{1}{\ln 2}$, $f''\left(\dfrac{1}{\ln 2}\right) < 0 \Rightarrow$ local max.

30. $f(x) = xe^x$, $f'(x) = e^x(1 + x)$, $f''(x) = e^x(2 + x)$.
The critical point is $x = -1$.
$f''(-1) > 0, \Rightarrow$ local min.

32. $f(x) = (x^2 - 4)^2$, $f'(x) = 4x^3 - 16x$, $f''(x) = 12x^2 - 16$.
The critical points are
$x = 0$, $f''(0) < 0 \Rightarrow$ local max;
$x = 2$, $f''(2) > 0 \Rightarrow$ local min;
$x = -2$, $f''(-2) > 0 \Rightarrow$ local min.

34. $f(x) = (x^2 - 3)e^x$,
$f'(x) = (x^2 + 2x - 3)e^x = (x + 3)(x - 1)e^x$,
$f''(x) = (x^2 + 4x - 1)e^x$.
The critical points are
$x = -3$, $f''(-3) < 0 \Rightarrow$ local max;
$x = 1$, $f''(1) > 0 \Rightarrow$ local min.

36. Since
$$
f(x) = \begin{cases} x^2 & \text{if } x \geq 0 \\ -x^2 & \text{if } x < 0, \end{cases}
$$
we have
$$
f'(x) = \begin{cases} 2x & \text{if } x \geq 0 \\ -2x & \text{if } x < 0 \end{cases} = 2|x|
$$
$$
f''(x) = \begin{cases} 2 & \text{if } x > 0 \\ -2 & \text{if } x < 0 \end{cases} = 2\operatorname{sgn} x.
$$

$f'(x) = 0$ if $x = 0$. Thus, $x = 0$ is a critical point of f. It is also an inflection point since the conditions of Definition 4.3.3 are satisfied. $f''(0)$ does not exist. If a the graph of a function has a tangent line, vertical or not, at x_0, and has opposite concavity on opposite sides of x_0, the x_0 is an inflection point of f, whether or not $f''(x_0)$ even exists.

38. Suppose that f has an inflection point at x_0. To be specific, suppose that $f''(x) < 0$ on (a, x_0) and $f''(x) > 0$ on (x_0, b) for some numbers a and b satisfying $a < x_0 < b$. If the graph of f has a non-vertical tangent line at x_0, then $f'(x_0)$ exists. Let

$$F(x) = f(x) - f(x_0) - f'(x_0)(x - x_0).$$

$F(x)$ represents the signed vertical distance between the graph of f and its tangent line at x_0. To show that the graph of f crosses its tangent line at x_0, it is sufficient to show that $F(x)$ has opposite signs on opposite sides of x_0. Observe that $F(x_0) = 0$, and $F'(x) = f'(x) - f'(x_0)$, so that $F'(x_0) = 0$ also. Since $F''(x) = f''(x)$, the assumptions above show that F' has a local minimum value at x_0 (by the First Derivative Test). Hence $F(x) > 0$ if $a < x < x_0$ or $x_0 < x < b$. It follows (by Theorem 6) that $F(x) < 0$ if $a < x < x_0$, and $F(x) > 0$ if $x_0 < x < b$. This completes the proof for the case of a nonvertical tangent.

If f has a vertical tangent at x_0, then its graph necessarily crosses the tangent (the line $x = x_0$) at x_0, since the graph of a function must cross any vertical line through a point of its domain that is not an endpoint.

40. Let there be a function f such that

$$f'(x_0) = f''(x_0) = \ldots = f^{(k-1)}(x_0) = 0,$$
$$f^{(k)}(x_0) \neq 0 \qquad \text{for some } k \geq 2.$$

If k is even, then f has a local min value at $x = x_0$ when $f^{(k)}(x_0) > 0$, and f has a local max value at $x = x_0$ when $f^{(k)}(x_0) < 0$.
If k is odd, then f has an inflection point at $x = x_0$.

42. We are given that

$$f(x) = \begin{cases} x^2 \sin \dfrac{1}{x}, & \text{if } x \neq 0; \\ 0, & \text{if } x = 0. \end{cases}$$

If $x \neq 0$, then

$$f'(x) = 2x \sin \frac{1}{x} - \cos \frac{1}{x}$$
$$f''(x) = 2 \sin \frac{1}{x} - \frac{2}{x} \cos \frac{1}{x} - \frac{1}{x^2} \sin \frac{1}{x}.$$

If $x = 0$, then

$$f'(x) = \lim_{h \to 0} \frac{h^2 \sin \dfrac{1}{h} - 0}{h} = 0.$$

Thus 0 is a critical point of f. There are points x arbitrarily close to 0 where $f(x) > 0$, for example $x = \dfrac{2}{(4n+1)\pi}$, and other such points where $f(x) < 0$, for example $x = \dfrac{2}{(4n+3)\pi}$. Therefore f does not have a local max or min at $x = 0$. Also, there are points arbitrarily close to 0 where $f''(x) > 0$, for example $x = \dfrac{1}{(2n+1)\pi}$, and other such points where $f''(x) < 0$, for instance $x = \dfrac{1}{2n\pi}$. Therefore f does not have constant concavity on any interval $(0, a)$ where $a > 0$, so 0 is not an inflection point of f either.

Section 4.4   Sketching the Graph of a Function   (page 261)

2.

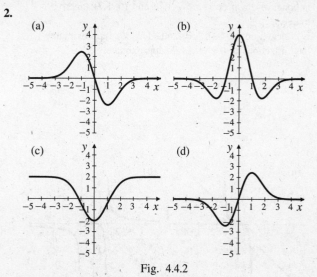

Fig. 4.4.2

The function graphed in Fig. 4.2(a):
is odd, is asymptotic to $y = 0$ at $\pm\infty$,
is increasing on $(-\infty, -1)$ and $(1, \infty)$,
is decreasing on $(-1, 1)$,
has CPs at $x = -1$ (max) and 1 (min),
is concave up on $(-\infty, -2)$ and $(0, 2)$ (approximately),
is concave down on $(-2, 0)$ and $(2, \infty)$ (approximately),
has inflections at $x = \pm 2$ (approximately).

The function graphed in Fig. 4.2(b):
is even, is asymptotic to $y = 0$ at $\pm\infty$,
is increasing on $(-1.7, 0)$ and $(1.7, \infty)$ (approximately),
is decreasing on $(-\infty, -1.7)$ and $(0, 1.7)$ (approximately),
has CPs at $x = 0$ (max) and ± 1.7 (min) (approximately),
is concave up on $(-2.5, -1)$ and $(1, 2.5)$ (approximately),
is concave down on $(-\infty, -2.5)$, $(-1, 1)$, and $(2.5, \infty)$ (approximately),
has inflections at ± 2.5 and ± 1 (approximately).

The function graphed in Fig. 4.2(c):
is even, is asymptotic to $y = 2$ at $\pm\infty$,
is increasing on $(0, \infty)$,
is decreasing on $(-\infty, 0)$,
has a CP at $x = 0$ (min),
is concave up on $(-1, 1)$ (approximately),
is concave down on $(-\infty, -1)$ and $(1, \infty)$ (approximately),
has inflections at $x = \pm 1$ (approximately).

The function graphed in Fig. 4.2(d):
is odd, is asymptotic to $y = 0$ at $\pm\infty$,
is increasing on $(-1, 1)$,
is decreasing on $(-\infty, -1)$ and $(1, \infty)$,
has CPs at $x = -1$ (min) and 1 (max),
is concave down on $(-\infty, -1.7)$ and $(0, 1.7)$ (approximately),
is concave up on $(-1.7, 0)$ and $(1.7, \infty)$ (approximately),
has inflections at 0 and ± 1.7 (approximately).

The function graphed in Fig. 4.4(a):
is odd, is asymptotic to $x = \pm 1$ and $y = x$,
is increasing on $(-\infty, -1.5)$, $(-1, 1)$, and $(1.5, \infty)$ (approximately),
is decreasing on $(-1.5, -1)$ and $(1, 1.5)$ (approximately),
has CPs at $x = -1.5$, $x = 0$, and $x = 1.5$,
is concave up on $(0, 1)$ and $(1, \infty)$,
is concave down on $(-\infty, -1)$ and $(-1, 0)$,
has an inflection at $x = 0$.

The function graphed in Fig. 4.4(b):
is odd, is asymptotic to $x = \pm 1$ and $y = 0$,
is increasing on $(-\infty, -1)$, $(-1, 1)$, and $(1, \infty)$,
has a CP at $x = 0$,
is concave up on $(-\infty, -1)$ and $(0, 1)$,
is concave down on $(-1, 0)$ and $(1, \infty)$,
has an inflection at $x = 0$.

The function graphed in Fig. 4.4(c):
is odd, is asymptotic to $x = \pm 1$ and $y = 0$,
is increasing on $(-\infty, -1)$, $(-1, 1)$, and $(1, \infty)$,
has no CP,
is concave up on $(-\infty, -1)$ and $(0, 1)$,
is concave down on $(-1, 0)$ and $(1, \infty)$,
has an inflection at $x = 0$.

The function graphed in Fig. 4.4(d):
is odd, is asymptotic to $y = \pm 2$,
is increasing on $(-\infty, -0.7)$ and $(0.7, \infty)$ (approximately),
is decreasing on $(-0.7, 0.7)$ (approximately),
has CPs at $x = \pm 0.7$ (approximately),
is concave up on $(-\infty, -1)$ and $(0, 1)$ (approximately),
is concave down on $(-1, 0)$ and $(1, \infty)$ (approximately),
has an inflection at $x = 0$ and $x = \pm 1$ (approximately).

4.

(a)

(b)

(c)

(d)

Fig. 4.4.4

6. According to the given properties:
Oblique asymptote: $y = x - 1$.
Critical points: $x = 0$, 2. Singular point: $x = -1$.
Local max 2 at $x = 0$; local min 0 at $x = 2$.

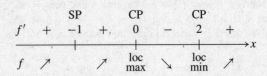

Inflection points: $x = -1$, 1, 3.

Since $\lim\limits_{x \to \pm\infty} \big(f(x) + 1 - x\big) = 0$, the line $y = x - 1$ is an oblique asymptote.

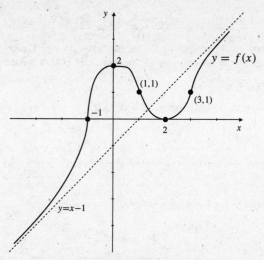

Fig. 4.4.6

8. $y = x(x^2 - 1)^2$, $y' = (x^2 - 1)(5x^2 - 1)$, $y'' = 4x(5x^2 - 3)$.
From y: Intercepts: $(0, 0)$, $(1, 0)$. Symmetry: odd (i.e., about the origin).

From y': Critical point: $x = \pm 1$, $\pm \dfrac{1}{\sqrt{5}}$.

		CP		CP		CP		CP	
y'	$+$	-1	$-$	$-\dfrac{1}{\sqrt{5}}$	$+$	$\dfrac{1}{\sqrt{5}}$	$-$	1	$+$
y	$\nearrow$	loc max	$\searrow$	loc min	$\nearrow$	loc max	$\searrow$	loc min	$\nearrow$

From y'': Inflection points at $x = 0$, $\pm\sqrt{\dfrac{3}{5}}$.

y''	$-$	$-\sqrt{\dfrac{3}{5}}$	$+$	0	$-$	$\sqrt{\dfrac{3}{5}}$	$+$
y	$\frown$	infl	$\smile$	infl	$\frown$	infl	$\smile$

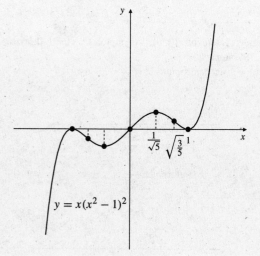

$y = x(x^2 - 1)^2$

Fig. 4.4.8

10. $y = \dfrac{x - 1}{x + 1} = 1 - \dfrac{2}{x + 1}$, $y' = \dfrac{2}{(x + 1)^2}$, $y'' = \dfrac{-4}{(x + 1)^3}$.
From y: Intercepts: $(0, -1)$, $(1, 0)$. Asymptotes: $y = 1$ (horizontal), $x = -1$ (vertical). No obvious symmetry. Other points: $(-2, 3)$.
From y': No critical point.

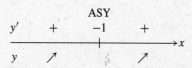

		ASY	
y'	$+$	-1	$+$
y	$\nearrow$		$\nearrow$

From y'': No inflection point.

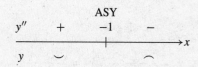

		ASY	
y''	$+$	-1	$-$
y	$\smile$		$\frown$

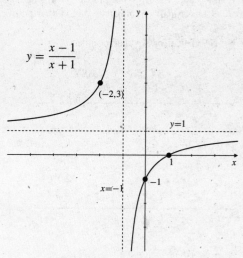

$y = \dfrac{x - 1}{x + 1}$

$(-2, 3)$

$y = 1$

$x = -1$

Fig. 4.4.10

12. $y = \dfrac{1}{4 + x^2}$, $y' = \dfrac{-2x}{(4 + x^2)^2}$, $y'' = \dfrac{6x^2 - 8}{(4 + x^2)^3}$.
From y: Intercept: $(0, \frac{1}{4})$. Asymptotes: $y = 0$ (horizontal). Symmetry: even (about y-axis).
From y': Critical point: $x = 0$.

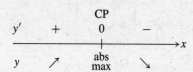

		CP	
y'	$+$	0	$-$
y	$\nearrow$	abs max	$\searrow$

From y'': $y'' = 0$ at $x = \pm\dfrac{2}{\sqrt{3}}$.

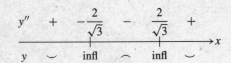

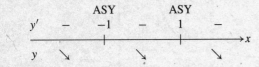

Fig. 4.4.12

14. $y = \dfrac{x}{x^2 - 1}$, $y' = -\dfrac{x^2 + 1}{(x^2 - 1)^2}$, $y'' = \dfrac{2x(x^2 + 3)}{(x^2 - 1)^3}$.
From y: Intercept: $(0, 0)$. Asymptotes: $y = 0$ (horizontal), $x = \pm 1$ (vertical). Symmetry: odd. Other points: $(2, \frac{2}{3})$, $(-2, -\frac{2}{3})$.
From y': No critical or singular points.

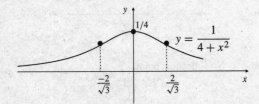

From y'': $y'' = 0$ at $x = 0$.

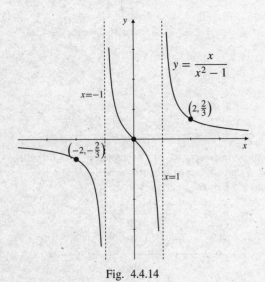

Fig. 4.4.14

16. $y = \dfrac{x^3}{x^2 - 1}$, $y' = \dfrac{x^2(x^2 - 3)}{(x^2 - 1)^2}$, $y'' = \dfrac{2x(x^2 + 3)}{(x^2 - 1)^3}$.
From y: Intercept: $(0, 0)$. Asymptotes: $x = \pm 1$ (vertical), $y = x$ (oblique). Symmetry: odd. Other points: $\left(\pm\sqrt{3}, \pm\dfrac{3\sqrt{3}}{2}\right)$.
From y': Critical point: $x = 0$, $\pm\sqrt{3}$.

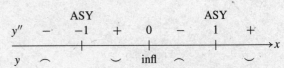

From y'': $y'' = 0$ at $x = 0$.

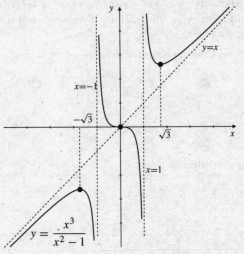

Fig. 4.4.16

18. $y = \dfrac{x^2}{x^2 + 1}$, $y' = \dfrac{2x}{(x^2 + 1)^2}$, $y'' = \dfrac{2(1 - 3x^2)}{(x^2 + 1)^3}$.
From y: Intercept: $(0, 0)$. Asymptotes: $y = 1$ (horizontal). Symmetry: even.
From y': Critical point: $x = 0$.

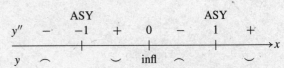

From y'': $y'' = 0$ at $x = \pm\dfrac{1}{\sqrt{3}}$.

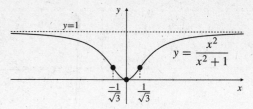

Fig. 4.4.18

From y'': y'' is negative for all x.

20. $y = \dfrac{x^2 - 2}{x^2 - 1}$, $y' = \dfrac{2x}{(x^2 - 1)^2}$, $y'' = \dfrac{-2(3x^2 + 1)}{(x^2 - 1)^3}$.
From y: Intercept: $(0, 2)$, $(\pm\sqrt{2}, 0)$. Asymptotes: $y = 1$ (horizontal), $x = \pm 1$ (vertical). Symmetry: even.
From y': Critical point: $x = 0$.

		ASY		CP		ASY	
f'	$-$	-1	$-$	0	$+$	1	$+$
f	$\searrow$		$\searrow$	loc min	$\nearrow$		$\nearrow$

From y'': $y'' = 0$ nowhere.

		ASY		ASY	
y'	$-$	-1	$+$	1	$-$
y	$\frown$		$\smile$		$\frown$

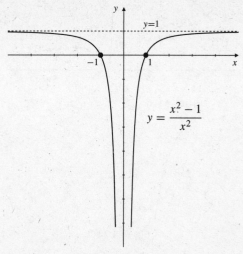

Fig. 4.4.22

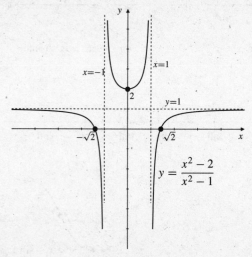

Fig. 4.4.20

22. $y = \dfrac{x^2 - 1}{x^2} = 1 - \dfrac{1}{x^2}$, $y' = \dfrac{2}{x^3}$, $y'' = -\dfrac{6}{x^4}$.
From y: Intercepts: $(\pm 1, 0)$. Asymptotes: $y = 1$ (horizontal), $x = 0$ (vertical). Symmetry: even.
From y': No critical points.

		ASY	
y'	$-$	0	$+$
y	$\searrow$		$\nearrow$

24. $y = \dfrac{(2 - x)^2}{x^3}$, $y' = -\dfrac{(x - 2)(x - 6)}{x^4}$,
$y'' = \dfrac{2(x^2 - 12x + 24)}{x^5} = \dfrac{2(x - 6 + 2\sqrt{3})(x - 6 - 2\sqrt{3})}{x^5}$.
From y: Intercept: $(2, 0)$. Asymptotes: $y = 0$ (horizontal), $x = 0$ (vertical). Symmetry: none obvious. Other points: $(-2, -2)$, $(-10, -0.144)$.
From y': Critical points: $x = 2, 6$.

		ASY		CP		CP	
y'	$-$	0	$-$	2	$+$	6	$-$
y	$\searrow$		$\searrow$	loc min	$\nearrow$	loc max	$\searrow$

From y'': $y'' = 0$ at $x = 6 \pm 2\sqrt{3}$.

				$6 + 2\sqrt{3}$		$6 - 2\sqrt{3}$	
y''	$-$	0	$+$		$-$		$+$
y	$\frown$		$\smile$	infl	$\frown$	infl	$\smile$

69

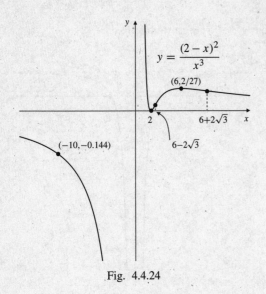

$$y = \frac{(2-x)^2}{x^3}$$

(6, 2/27)

2

$6+2\sqrt{3}$

$6-2\sqrt{3}$

(−10, −0.144)

Fig. 4.4.24

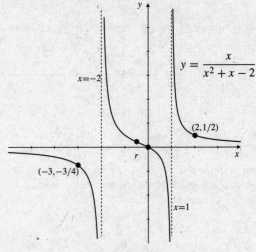

$x = -2$

$$y = \frac{x}{x^2 + x - 2}$$

(2, 1/2)

r

(−3, −3/4)

$x = 1$

Fig. 4.4.26

28. $y = x + \sin x,\ y' = 1 + \cos x,\ y'' = -\sin x.$
From y: Intercept: $(0,0)$. Other points: $(k\pi, k\pi)$, where k is an integer. Symmetry: odd.
From y': Critical point: $x = (2k+1)\pi$, where k is an integer.

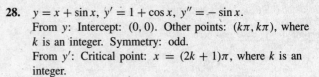

		CP		CP		CP	
f'	+	$-\pi$	+	π	−	3π	+
f	↗		↗		↗		↗

From y'': $y'' = 0$ at $x = k\pi$, where k is an integer.

y''	+	-2π	−	$-\pi$	+	0	−	π	+	2π	−
y	⌣	infl	⌢	infl	⌣	infl	⌢	infl	⌣	infl	⌢

26. $y = \dfrac{x}{x^2 + x - 2} = \dfrac{x}{(2+x)(x-1)},$
$y' = \dfrac{-(x^2 + 2)}{(x+2)^2(x-1)^2},\ y'' = \dfrac{2(x^3 + 6x + 2)}{(x+2)^3(x-1)^3}.$
From y: Intercepts: $(0,0)$. Asymptotes: $y = 0$ (horizontal), $x = 1$, $x = -2$ (vertical). Other points: $(-3, -\frac{3}{4})$, $(2, \frac{1}{2})$.
From y': No critical point.

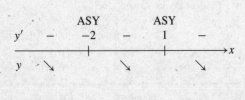

		ASY		ASY	
y'	−	-2	−	1	−
y	↘		↘		↘

From y'': $y'' = 0$ if $f(x) = x^3 + 6x + 2 = 0$. Since $f'(x) = 3x^2 + 6 \ge 6$, f is increasing and can only have one root. Since $f(0) = 2$ and $f(-1) = -5$, that root must be between -1 and 0. Let the root be r.

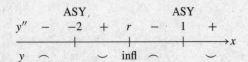

		ASY				ASY	
y''	−	-2	+	r	−	1	+
y	⌢		⌣	infl	⌢		⌣

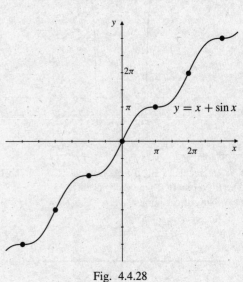

2π

π

$y = x + \sin x$

π

2π

Fig. 4.4.28

30. $y = e^{-x^2}$, $y' = -2xe^{-x^2}$, $y'' = (4x^2 - 2)e^{-x^2}$.
From y: Intercept: $(0, 1)$. Asymptotes: $y = 0$ (horizontal). Symmetry: even.
From y': Critical point: $x = 0$.

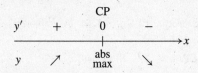

From y'': $y'' = 0$ at $x = \pm \dfrac{1}{\sqrt{2}}$.

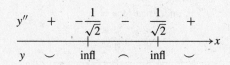

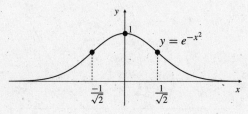

Fig. 4.4.30

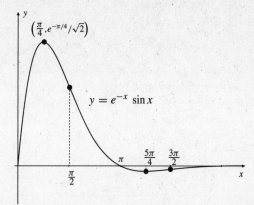

Fig. 4.4.32

34. $y = x^2 e^x$, $y' = (2x + x^2)e^x = x(2 + x)e^x$,
$y'' = (x^2 + 4x + 2)e^x = (x + 2 - \sqrt{2})(x + 2 + \sqrt{2})e^x$.
From y: Intercept: $(0, 0)$.
Asymptotes: $y = 0$ as $x \to -\infty$.
From y': Critical point: $x = 0$, $x = -2$.

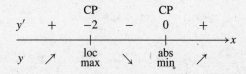

From y'': $y'' = 0$ at $x = -2 \pm \sqrt{2}$.

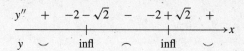

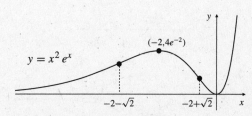

Fig. 4.4.34

32. $y = e^{-x} \sin x \qquad (x \geq 0)$,
$y' = e^{-x}(\cos x - \sin x)$, $y'' = -2e^{-x}\cos x$.
From y: Intercept: $(k\pi, 0)$, where k is an integer. Asymptotes: $y = 0$ as $x \to \infty$.
From y': Critical points: $x = \dfrac{\pi}{4} + k\pi$, where k is an integer.

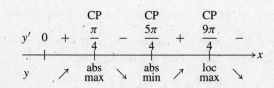

From y'': $y'' = 0$ at $x = (k + \frac{1}{2})\pi$, where k is an integer.

36. $y = \dfrac{\ln x}{x^2} \qquad (x > 0)$,
$y' = \dfrac{1 - 2\ln x}{x^3}$, $y'' = \dfrac{6\ln x - 5}{x^4}$.
From y: Intercepts: $(1, 0)$. Asymptotes: $y = 0$, since $\lim\limits_{x \to \infty} \dfrac{\ln x}{x^2} = 0$, and $x = 0$, since $\lim\limits_{x \to 0+} \dfrac{\ln x}{x^2} = -\infty$.
From y': Critical point: $x = e^{1/2}$.

From y'': $y'' = 0$ at $x = e^{5/6}$.

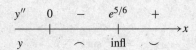

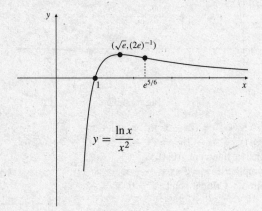

Fig. 4.4.36

38. $y = \dfrac{x}{\sqrt{x^2 + 1}}$, $y' = (x^2 + 1)^{-3/2}$, $y'' = -3x(x^2 + 1)^{-5/2}$.

From y: Intercept: $(0, 0)$. Asymptotes: $y = 1$ as $x \to \infty$, and $y = -1$ as $x \to -\infty$. Symmetry: odd.

From y': No critical point. $y' > 0$ and y is increasing for all x.

From y'': $y'' = 0$ at $x = 0$.

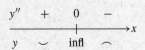

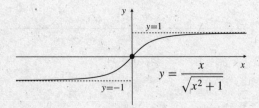

Fig. 4.4.38

40. According to Theorem 5 of Section 4.4,

$$\lim_{x \to 0+} x \ln x = 0.$$

Thus,

$$\lim_{x \to 0} x \ln |x| = \lim_{x \to 0+} x \ln x = 0.$$

If $f(x) = x \ln |x|$ for $x \neq 0$, we may define $f(0)$ such that $f(0) = \lim\limits_{x \to 0} x \ln |x| = 0$. Then f is continuous on the whole real line and

$$f'(x) = \ln |x| + 1, \qquad f''(x) = \frac{1}{|x|} \text{sgn}\,(x).$$

From f: Intercept: $(0, 0)$, $(\pm 1, 0)$. Asymptotes: none. Symmetry: odd.

From f': CP: $x = \pm \dfrac{1}{e}$. SP: $x = 0$.

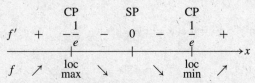

From f'': f'' is undefined at $x = 0$.

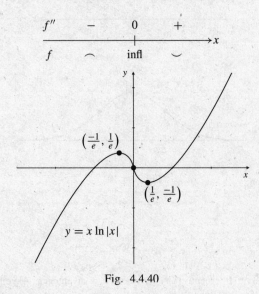

Fig. 4.4.40

Section 4.5 Extreme-Value Problems (page 269)

2. Let the numbers be x and $\dfrac{8}{x}$ where $x > 0$. Their sum is $S = x + \dfrac{8}{x}$. Since $S \to \infty$ as $x \to \infty$ or $x \to 0+$, the minimum sum must occur at a critical point:

$$0 = \frac{dS}{dx} = 1 - \frac{8}{x^2} \Rightarrow x = 2\sqrt{2}.$$

Thus, the smallest possible sum is $2\sqrt{2} + \dfrac{8}{2\sqrt{2}} = 4\sqrt{2}$.

4. Let the numbers be x and $16 - x$. Let $P(x) = x^3(16 - x)^5$. Since $P(x) \to -\infty$ as $x \to \pm\infty$, so the maximum must occur at a critical point:

$$\begin{aligned} 0 = P'(x) &= 3x^2(16 - x)^5 - 5x^3(16 - x)^4 \\ &= x^2(16 - x)^4(48 - 8x). \end{aligned}$$

The critical points are 0, 6 and 16. Clearly, $P(0) = P(16) = 0$, and $P(6) = 216 \times 10^5$. Thus, $P(x)$ is maximum if the numbers are 6 and 10.

6. If the numbers are x and $n - x$, then $0 \le x \le n$ and the sum of their squares is

$$S(x) = x^2 + (n - x)^2.$$

Observe that $S(0) = S(n) = n^2$. For critical points:

$$0 = S'(x) = 2x - 2(n - x) = 2(2x - n) \Rightarrow x = n/2.$$

Since $S(n/2) = n^2/2$, this is the smallest value of the sum of squares.

8. Let the width and the length of a rectangle of given perimeter $2P$ be x and $P - x$. Then the area of the rectangle is

$$A(x) = x(P - x) = Px - x^2.$$

Since $A(x) \to -\infty$ as $x \to \pm\infty$ the maximum must occur at a critical point:

$$0 = \frac{dA}{dx} = P - 2x \Rightarrow x = \frac{P}{2}.$$

Hence, the width and the length are $\dfrac{P}{2}$ and $(P - \dfrac{P}{2}) = \dfrac{P}{2}$. Since the width equals the length, it is a square.

10. Let the various dimensions be as shown in the figure. Since $h = 10 \sin\theta$ and $b = 20 \cos\theta$, the area of the triangle is

$$A(\theta) = \tfrac{1}{2}bh = 100 \sin\theta \cos\theta$$
$$= 50 \sin 2\theta \quad \text{for } 0 < \theta < \frac{\pi}{2}.$$

Since $A(\theta) \to 0$ as $\theta \to 0$ and $\theta \to \dfrac{\pi}{2}$, the maximum must be at a critial point:

$$0 = A'(\theta) = 100 \cos 2\theta \Rightarrow 2\theta = \frac{\pi}{2} \Rightarrow \theta = \frac{\pi}{4}.$$

Hence, the largest possible area is

$$A(\pi/4) = 50 \sin\left[2\left(\frac{\pi}{4}\right)\right] = 50\,\text{m}^2.$$

(Remark: alternatively, we may simply observe that the largest value of $\sin 2\theta$ is 1; therefore the largest possible area is $50(1) = 50$ m^2.)

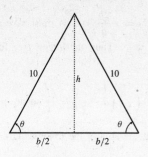

Fig. 4.5.10

12. Let x be as shown in the figure. The perimeter of the rectangle is

$$P(x) = 4x + 2\sqrt{R^2 - x^2} \qquad (0 \le x \le R).$$

For critical points:

$$0 = \frac{dP}{dx} = 4 + \frac{-2x}{\sqrt{R^2 - x^2}}$$
$$\Rightarrow 2\sqrt{R^2 - x^2} = x \Rightarrow x = \frac{2R}{\sqrt{5}}.$$

Since

$$\frac{d^2P}{dx^2} = \frac{-2R^2}{(R^2 - x^2)^{3/2}} < 0$$

therefore $P(x)$ is concave down on $[0, R]$, so it must have an absolute maximum value at $x = \dfrac{2R}{\sqrt{5}}$. The largest perimeter is therefore

$$P\left(\frac{2R}{\sqrt{5}}\right) = 4\left(\frac{2R}{\sqrt{5}}\right) + \sqrt{R^2 - \frac{4R^2}{5}} = \frac{10R}{\sqrt{5}} \text{ units.}$$

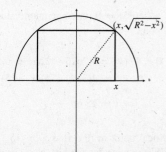

Fig. 4.5.12

14. See the diagrams below.

a) The area of the rectangle is $A = xy$. Since

$$\frac{y}{a - x} = \frac{b}{a} \Rightarrow y = \frac{b(a - x)}{a}.$$

Thus, the area is

$$A = A(x) = \frac{bx}{a}(a - x) \qquad (0 \le x \le a).$$

For critical points:

$$0 = A'(x) = \frac{b}{a}(a - 2x) \Rightarrow x = \frac{a}{2}.$$

Since $A''(x) = -\dfrac{2b}{a} < 0$, A must have a maximum

value at $x = \dfrac{a}{2}$. Thus, the largest area for the rectangle is

$$\frac{b}{a}\left(\frac{a}{2}\right)\left(a - \frac{a}{2}\right) = \frac{ab}{4} \text{ square units,}$$

that is, half the area of the triangle ABC.

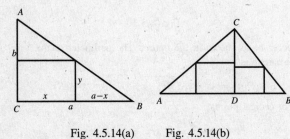

Fig. 4.5.14(a)　　　　Fig. 4.5.14(b)

(b) This part has the same answer as part (a). To see this, let $CD \perp AB$, and solve separate problems for the largest rectangles in triangles ACD and BCD as shown. By part (a), both maximizing rectangles have the same height, namely half the length of CD. Thus, their union is a rectangle of area half of that of triangle ABC.

16. Let x be the side of the cut-out squares. Then the volume of the box is

$$V(x) = x(70 - 2x)(150 - 2x) \qquad (0 \le x \le 35).$$

Since $V(0) = V(35) = 0$, the maximum value will occur at a critical point:

$$0 = V'(x) = 4(2625 - 220x + 3x^2)$$
$$= 4(3x - 175)(x - 15)$$
$$\Rightarrow x = 15 \text{ or } \frac{175}{3}.$$

The only critical point in $[0, 35]$ is $x = 15$. Thus, the largest possible volume for the box is

$$V(15) = 15(70 - 30)(150 - 30) = 72,000\,\text{cm}^3.$$

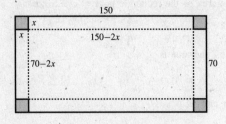

Fig. 4.5.16

18. If the manager charges $\$(40 + x)$ per room, then $(80 - 2x)$ rooms will be rented.
The total income will be $\$(80 - 2x)(40 + x)$ and the total cost will be $\$(80 - 2x)(10) + (2x)(2)$. Therefore, the profit is

$$P(x) = (80 - 2x)(40 + x) - [(80 - 2x)(10) + (2x)(2)]$$
$$= 2400 + 16x - 2x^2 \qquad \text{for } x > 0.$$

If $P'(x) = 16 - 4x = 0$, then $x = 4$. Since $P''(x) = -4 < 0$, P must have a maximum value at $x = 4$. Therefore, the manager should charge $\$44$ per room.

20. This problem is similar to the previous one except that the 10 in the numerator of the second fraction in the expression for T is replaced with a 4. This has no effect on the critical point of T, namely $x = 5$, which now lies outside the appropriate interval $0 \le x \le 4$. Minimum T must occur at an endpoint. Note that

$$T(0) = \frac{12}{15} + \frac{4}{39} = 0.9026$$
$$T(4) = \frac{1}{15}\sqrt{12^2 + 4^2} = 0.8433.$$

The minimum travel time corresponds to $x = 4$, that is, to driving in a straight line to B.

22. Let the dimensions of the rectangle be as shown in the figure. Clearly,

$$x = a\sin\theta + b\cos\theta,$$
$$y = a\cos\theta + b\sin\theta.$$

Therefore, the area is

$$A(\theta) = xy$$
$$= (a\sin\theta + b\cos\theta)(a\cos\theta + b\sin\theta)$$
$$= ab + (a^2 + b^2)\sin\theta\cos\theta$$
$$= ab + \frac{1}{2}(a^2 + b^2)\sin 2\theta \qquad \text{for } 0 \le \theta \le \frac{\pi}{2}.$$

If $A'(\theta) = (a^2 + b^2)\cos 2\theta = 0$, then $\theta = \dfrac{\pi}{4}$. Since $A''(\theta) = -2(a^2 + b^2)\sin 2\theta < 0$ when $0 \le \theta \le \dfrac{\pi}{2}$, therefore $A(\theta)$ must have a maximum value at $\theta = \dfrac{\pi}{4}$. Hence, the area of the largest rectangle is

$$A\left(\frac{\pi}{4}\right) = ab + \frac{1}{2}(a^2 + b^2)\sin\left(\frac{\pi}{2}\right)$$
$$= ab + \frac{1}{2}(a^2 + b^2) = \frac{1}{2}(a + b)^2 \quad \text{sq. units.}$$

(Note: $x = y = \dfrac{a}{\sqrt{2}} + \dfrac{b}{\sqrt{2}}$ indicates that the rectangle containing the given rectangle with sides a and b, has largest area when it is a square.)

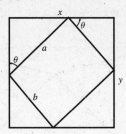

Fig. 4.5.22

24. The longest beam will have length equal to the minimum of $L = x + y$, where x and y are as shown in the figure below:

$$x = \frac{a}{\cos\theta}, \quad y = \frac{b}{\sin\theta}.$$

Thus,

$$L = L(\theta) = \frac{a}{\cos\theta} + \frac{b}{\sin\theta} \quad \left(0 < \theta < \frac{\pi}{2}\right).$$

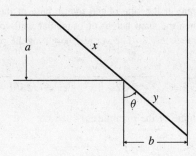

Fig. 4.5.24

If $L'(\theta) = 0$, then

$$\frac{a\sin\theta}{\cos^2\theta} - \frac{b\cos\theta}{\sin^2\theta} = 0$$

$$\Leftrightarrow \quad \frac{a\sin^3\theta - b\cos^3\theta}{\cos^2\theta\sin^2\theta} = 0$$

$$\Leftrightarrow \quad a\sin^3\theta - b\cos^3\theta = 0$$

$$\Leftrightarrow \quad \tan^3\theta = \frac{b}{a}$$

$$\Leftrightarrow \quad \tan\theta = \frac{b^{1/3}}{a^{1/3}}.$$

Clearly, $L(\theta) \to \infty$ as $\theta \to 0+$ or $\theta \to \frac{\pi}{2}-$. Thus, the minimum must occur at $\theta = \tan^{-1}\left(\frac{b^{1/3}}{a^{1/3}}\right)$. Using the triangle above for $\tan\theta = \frac{b^{1/3}}{a^{1/3}}$, it follows that

$$\cos\theta = \frac{a^{1/3}}{\sqrt{a^{2/3} + b^{2/3}}}, \quad \sin\theta = \frac{b^{1/3}}{\sqrt{a^{2/3} + b^{2/3}}}.$$

Hence, the minimum is

$$L(\theta) = \frac{a}{\left(\dfrac{a^{1/3}}{\sqrt{a^{2/3} + b^{2/3}}}\right)} + \frac{b}{\left(\dfrac{b^{1/3}}{\sqrt{a^{2/3} + b^{2/3}}}\right)}$$

$$= \left(a^{2/3} + b^{2/3}\right)^{3/2} \text{ units.}$$

26. Let θ be the angle of inclination of the ladder. The height of the fence is

$$h(\theta) = 6\sin\theta - 2\tan\theta \quad \left(0 < \theta < \frac{\pi}{2}\right).$$

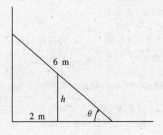

Fig. 4.5.26

For critical points:

$$0 = h'(\theta) = 6\cos\theta - 2\sec^2\theta$$

$$\Rightarrow 3\cos\theta = \sec^2\theta \Rightarrow 3\cos^3\theta = 1$$

$$\Rightarrow \cos\theta = \left(\tfrac{1}{3}\right)^{1/3}.$$

Since $h''(\theta) = -6\sin\theta - 4\sec^2\theta\tan\theta < 0$ for $0 < \theta < \frac{\pi}{2}$, therefore $h(\theta)$ must be maximum at $\theta = \cos^{-1}\left(\tfrac{1}{3}\right)^{1/3}$. Then

$$\sin\theta = \frac{\sqrt{3^{2/3} - 1}}{3^{1/3}}, \quad \tan\theta = \sqrt{3^{2/3} - 1}.$$

Thus, the maximum height of the fence is

$$h(\theta) = 6\left(\frac{\sqrt{3^{2/3} - 1}}{3^{1/3}}\right) - 2\sqrt{3^{2/3} - 1}$$

$$= 2(3^{2/3} - 1)^{3/2} \approx 2.24 \text{ m.}$$

28. The square of the distance from $(8, 1)$ to the curve $y = 1 + x^{3/2}$ is

$$
\begin{aligned}
S &= (x - 8)^2 + (y - 1)^2 \\
&= (x - 8)^2 + (1 + x^{3/2} - 1)^2 \\
&= x^3 + x^2 - 16x + 64.
\end{aligned}
$$

Note that y, and therefore also S, is only defined for $x \geq 0$. If $x = 0$ then $S = 64$. Also, $S \to \infty$ if $x \to \infty$. For critical points:

$$
0 = \frac{dS}{dx} = 3x^2 + 2x - 16 = (3x + 8)(x - 2)
$$
$$
\Rightarrow x = -\tfrac{8}{3} \text{ or } 2.
$$

Only $x = 2$ is feasible. At $x = 2$ we have $S = 44 < 64$. Therefore the minimum distance is $\sqrt{44} = 2\sqrt{11}$ units.

30. Let the radius and the height of the circular cylinder be r and h. By similar triangles,

$$
\frac{h}{R - r} = \frac{H}{R} \Rightarrow h = \frac{H(R - r)}{R}.
$$

Hence, the volume of the circular cylinder is

$$
\begin{aligned}
V(r) &= \pi r^2 h = \frac{\pi r^2 H (R - r)}{R} \\
&= \pi H \left(r^2 - \frac{r^3}{R} \right) \quad \text{for } 0 \leq r \leq R.
\end{aligned}
$$

Since $V(0) = V(R) = 0$, the maximum value of V must be at a critical point. If $\dfrac{dV}{dr} = \pi H \left(2r - \dfrac{3r^2}{R} \right) = 0$, then $r = \dfrac{2R}{3}$. Therefore the cylinder has maximum volume if its radius is $r = \dfrac{2R}{3}$ units, and its height is

$$
h = \frac{H \left(R - \dfrac{2R}{3} \right)}{R} = \frac{H}{3} \text{ units.}
$$

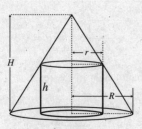

Fig. 4.5.30

32.

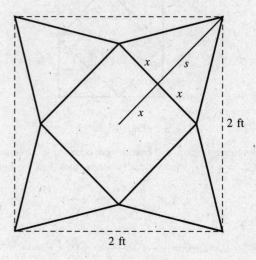

Fig. 4.5.32

From the figure, if the side of the square base of the pyramid is $2x$, then the slant height of triangular walls of the pyramid is $s = \sqrt{2} - x$. The vertical height of the pyramid is

$$
h = \sqrt{s^2 - x^2} = \sqrt{2 - 2\sqrt{2}x + x^2 - x^2} = \sqrt{2}\sqrt{1 - \sqrt{2}x}.
$$

Thus the volume of the pyramid is

$$
V = \frac{4\sqrt{2}}{3} x^2 \sqrt{1 - \sqrt{2}x},
$$

for $0 \leq x \leq 1/\sqrt{2}$. $V = 0$ at both endpoints, so the maximum will occur at an interior critical point. For CP:

$$
0 = \frac{dV}{dx} = \frac{4\sqrt{2}}{3} \left[2x\sqrt{1 - \sqrt{2}x} - \frac{\sqrt{2}x^2}{2\sqrt{1 - \sqrt{2}x}} \right]
$$
$$
4x(1 - \sqrt{2}x) = \sqrt{2}x^2
$$
$$
4x = 5\sqrt{2}x^2 \quad , x = 4/(5\sqrt{2}).
$$

$V(4/(5\sqrt{2})) = 32\sqrt{2}/(75\sqrt{5})$. The largest volume of such a pyramid is $32\sqrt{2}/(75\sqrt{5})$ ft^3.

34. Let h and r be the length and radius of the cylindrical part of the tank. The volume of the tank is

$$
V = \pi r^2 h + \tfrac{4}{3}\pi r^3.
$$

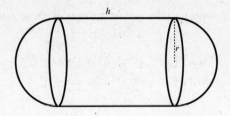

Fig. 4.5.34

If the cylindrical wall costs \$$k$ per unit area and the hemispherical wall \$$2k$ per unit area, then the total cost of the tank wall is

$$C = 2\pi r h k + 8\pi r^2 k$$
$$= 2\pi r k \frac{V - \frac{4}{3}\pi r^3}{\pi r^2} + 8\pi r^2 k$$
$$= \frac{2Vk}{r} + \frac{16}{3}\pi r^2 k \qquad (0 < r < \infty).$$

Since $C \to \infty$ as $r \to 0+$ or $r \to \infty$, the minimum cost must occur at a critical point. For critical points,

$$0 = \frac{dC}{dr} = -2Vkr^{-2} + \frac{32}{3}\pi r k \quad \Leftrightarrow \quad r = \left(\frac{3V}{16\pi}\right)^{1/3}.$$

Since $V = \pi r^2 h + \frac{4}{3}\pi r^3$,

$$r^3 = \frac{3}{16\pi}\left(\pi r^2 h + \frac{4}{3}\pi r^3\right) \Rightarrow r = \frac{1}{4}h$$
$$\Rightarrow h = 4r = 4\left(\frac{3V}{16\pi}\right)^{1/3}.$$

Hence, in order to minimize the cost, the radius and length of the cylindrical part of the tank should be $\left(\dfrac{3V}{16\pi}\right)^{1/3}$ and $4\left(\dfrac{3V}{16\pi}\right)^{1/3}$ units respectively.

36. If the path of the light ray is as shown in the figure then the time of travel from A to B is

$$T = T(x) = \frac{\sqrt{a^2 + x^2}}{v_1} + \frac{\sqrt{b^2 + (c - x)^2}}{v_2}.$$

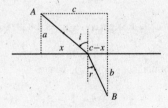

Fig. 4.5.36

To minimize T, we look for a critical point:

$$0 = \frac{dT}{dx} = \frac{1}{v_1}\frac{x}{\sqrt{a^2 + x^2}} - \frac{1}{v_2}\frac{c - x}{\sqrt{b^2 + (c - x)^2}}$$
$$= \frac{1}{v_1}\sin i - \frac{1}{v_2}\sin r.$$

Thus,
$$\frac{\sin i}{\sin r} = \frac{v_1}{v_2}.$$

38. The curve $y = 1 + 2x - x^3$ has slope $m = y' = 2 - 3x^2$. Evidently m is greatest for $x = 0$, in which case $y = 1$ and $m = 2$. Thus the tangent line with maximal slope has equation $y = 1 + 2x$.

40. Let h and r be the height and base radius of the cone and R be the radius of the sphere. From similar triangles,

$$\frac{r}{\sqrt{h^2 + r^2}} = \frac{R}{h - R}$$
$$\Rightarrow \quad h = \frac{2r^2 R}{r^2 - R^2} \qquad (r > R).$$

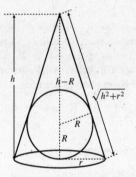

Fig. 4.5.40

Then the volume of the cone is

$$V = \frac{1}{3}\pi r^2 h = \frac{2}{3}\pi R \frac{r^4}{r^2 - R^2} \qquad (R < r < \infty).$$

Clearly $V \to \infty$ if $r \to \infty$ or $r \to R+$. Therefore to minimize V, we look for a critical point:

$$0 = \frac{dV}{dr} = \frac{2}{3}\pi R \left[\frac{(r^2 - R^2)(4r^3) - r^4(2r)}{(r^2 - R^2)^2}\right]$$
$$\Leftrightarrow \quad 4r^5 - 4r^3 R^2 - 2r^5 = 0$$
$$\Leftrightarrow \quad r = \sqrt{2}R.$$

Hence, the smallest possible volume of a right circular cone which can contain sphere of radius R is

$$V = \frac{2}{3}\pi R \left(\frac{4R^4}{2R^2 - R^2}\right) = \frac{8}{3}\pi R^3 \text{ cubic units.}$$

42. Let r be the radius of the circular arc and θ be the angle shown in the left diagram below. Thus,

$$2r\theta = 100 \quad \Rightarrow \quad r = \frac{50}{\theta}.$$

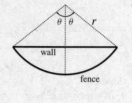

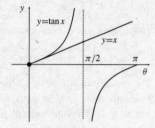

Fig. 4.5.42(a)   Fig. 4.5.42(b)

The area of the enclosure is

$$A = \frac{2\theta}{2\pi}\pi r^2 - (r\cos\theta)(r\sin\theta)$$

$$= \frac{50^2}{\theta} - \frac{50^2}{\theta^2}\frac{\sin 2\theta}{2}$$

$$= 50^2\left(\frac{1}{\theta} - \frac{\sin 2\theta}{2\theta^2}\right)$$

for $0 < \theta \le \pi$. Note that $A \to \infty$ as $\theta \to 0+$, and for $\theta = \pi$ we are surrounding the entire enclosure with fence (a circle) and not using the wall at all. Evidently this would not produce the greatest enclosure area, so the maximum area must correspond to a critical point of A:

$$0 = \frac{dA}{d\theta} = 50^2\left(-\frac{1}{\theta^2} - \frac{2\theta^2(2\cos 2\theta) - \sin 2\theta(4\theta)}{4\theta^4}\right)$$

$$\Leftrightarrow \quad \frac{1}{\theta^2} + \frac{\cos 2\theta}{\theta^2} = \frac{\sin 2\theta}{\theta^3}$$

$$\Leftrightarrow \quad 2\theta\cos^2\theta = 2\sin\theta\cos\theta$$

$$\Leftrightarrow \quad \cos\theta = 0 \quad \text{or} \quad \tan\theta = \theta.$$

Observe that $\tan\theta = \theta$ has no solutions in $(0, \pi]$. (The graphs of $y = \tan\theta$ and $y = \theta$ cross at $\theta = 0$ but nowhere else between 0 and π.) Thus, the greatest enclosure area must correspond to $\cos\theta = 0$, that is, to $\theta = \frac{\pi}{2}$. The largest enclosure is thus semicircular, and has area $\frac{2}{\pi}(50)^2 = \frac{5000}{\pi}$ m^2.

44. Let the various distances be as labelled in the diagram.

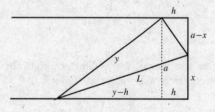

Fig. 4.5.44

From the geometry of the various triangles in the diagram we have

$$x^2 = h^2 + (a - x)^2 \Rightarrow h^2 = 2ax - a^2$$

$$y^2 = a^2 + (y - h)^2 \Rightarrow h^2 = 2hy - a^2$$

hence $hy = ax$. Then

$$L^2 = x^2 + y^2 = x^2 + \frac{a^2x^2}{h^2}$$

$$= x^2 + \frac{a^2x^2}{2ax - a^2} = \frac{2ax^3}{2ax - a^2}$$

for $\frac{a}{2} < x \le a$. Clearly, $L \to \infty$ as $x \to \frac{a}{2}+$, and $L(a) = \sqrt{2}a$. For critical points of L^2:

$$0 = \frac{d(L^2)}{dx} = \frac{(2ax - a^2)(6ax^2) - (2ax^3)(2a)}{(2ax - a^2)^2}$$

$$= \frac{2a^2x^2(4x - 3a)}{(2ax - a^2)^2}.$$

The only critical point in $\left(\frac{a}{2}, a\right]$ is $x = \frac{3a}{4}$. Since

$$L\left(\frac{3a}{4}\right) = \frac{3\sqrt{3}a}{4} < L(a),$$ therefore the least possible length for the fold is $\frac{3\sqrt{3}a}{4}$ cm.

Section 4.6   Finding Roots of Equations (page 278)

2. $f(x) = x^2 - 3$, $f'(x) = 2x$.
Newton's formula $x_{n+1} = g(x_n)$, where

$$g(x) = x - \frac{x^2 - 3}{2x} = \frac{x^2 + 3}{2x}.$$

Starting with $x_0 = 1.5$, get $x_4 = x_5 = 1.73205080757$.

4. $f(x) = x^3 + 2x^2 - 2$, $f'(x) = 3x^2 + 4x$.
Newton's formula $x_{n+1} = g(x_n)$, where

$$g(x) = x - \frac{x^3 + 2x^2 - 2}{3x^2 + 4x} = \frac{2x^3 + 2x^2 + 2}{3x^2 + 4x}.$$

Starting with $x_0 = 1.5$, get $x_5 = x_6 = 0.839286755214$.

6. $f(x) = x^3 + 3x^2 - 1$, $f'(x) = 3x^2 + 6x$.
Newton's formula $x_{n+1} = g(x_n)$, where

$$g(x) = x - \frac{x^3 + 3x^2 - 1}{3x^2 + 6x} = \frac{2x^3 + 3x^2 + 1}{3x^2 + 6x}.$$

Because $f(-3) = -1$, $f(-2) = 3$, $f(-1) = 1$, $f(0) = -1$, $f(1) = 3$, there are roots between -3 and -2, between -1 and 0, and between 0 and 1.
Starting with $x_0 = -2.5$, get $x_5 = x_6 = -2.87938524157$.
Starting with $x_0 = -0.5$, get $x_4 = x_5 = -0.652703644666$.
Starting with $x_0 = 0.5$, get $x_4 = x_5 = 0.532088886328$.

8. $f(x) = x^2 - \cos x$, $f'(x) = 2x + \sin x$.
Newton's formula is $x_{n+1} = g(x_n)$, where

$$g(x) = x - \frac{x^2 - \cos x}{2x + \sin x}.$$

The graphs of $\cos x$ and x^2, suggest a root near $x = \pm 0.8$. Starting with $x_0 = 0.8$, get $x_3 = x_4 = 0.824132312303$. The other root is the negative of this one, because $\cos x$ and x^2 are both even functions.

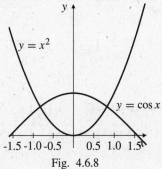

Fig. 4.6.8

10. A graphing calculator shows that the equation

$$(1 + x^2)\sqrt{x} - 1 = 0$$

has a root near $x = 0.6$. Use of a solve routine or Newton's method gives $x = 0.56984029099806$.

12. Let $f(x) = \dfrac{\sin x}{1 + x^2}$. Since $|f(x)| \le 1/(1 + x^2) \to 0$ as $x \to \pm\infty$ and $f(0) = 0$, the maximum and minimum values of f will occur at the two critical points of f that are closest to the origin on the right and left, respectively. For CP:

$$0 = f'(x) = \frac{(1 + x^2)\cos x - 2x \sin x}{(1 + x^2)^2}$$
$$0 = (1 + x^2)\cos x - 2x \sin x$$

with $0 < x < \pi$ for the maximum and $-\pi < x < 0$ for the minimum. Solving this equation using a solve routine or Newton's Method starting, say, with $x_0 = 1.5$, we get $x = \pm 0.79801699184239$. The corresponding max and min values of f are ± 0.437414158279.

14. For $x^2 = 0$ we have $x_{n+1} = x_n - (x_n^2/(2x_n)) = x_n/2$.
If $x_0 = 1$, then $x_1 = 1/2$, $x_2 = 1/4$, $x_3 = 1/8$.

a) $x_n = 1/2^n$, by induction.

b) x_n approximates the root $x = 0$ to within 0.0001 provided $2^n > 10,000$. We need $n \ge 14$ to ensure this.

c) To ensure that x_n^2 is within 0.0001 of 0 we need $(1/2^n)^2 < 0.0001$, that is, $2^{2n} > 10,000$. We need $n \ge 7$.

d) Convergence of Newton approximations to the root $x = 0$ of $x^2 = 0$ is slower than usual because the derivative $2x$ of x^2 is zero at the root.

16. Newton's Method formula for $f(x) = x^{1/3}$ is

$$x_{n+1} = x_n - \frac{x_n^{1/3}}{(1/3)x_n^{-2/3}} = x_n - 3x_n = -2x_n.$$

If $x_0 = 1$, then $x_1 = -2$, $x_2 = 4$, $x_3 = -8$, $x_4 = 16$, and, in general, $x_n = (-2)^n$. The successive "approximations" oscillate ever more widely, diverging from the root at $x = 0$.

18. To solve $1 + \frac{1}{4}\sin x = x$, start with $x_0 = 1$ and iterate $x_{n+1} = 1 + \frac{1}{4}\sin x_n$. x_5 and x_6 round to 1.23613.

20. To solve $(x + 9)^{1/3} = x$, start with $x_0 = 2$ and iterate $x_{n+1} = (x_n + 9)^{1/3}$. x_4 and x_5 round to 2.24004.

22. To solve $x^3 + 10x - 10 = 0$, start with $x_0 = 1$ and iterate $x_{n+1} = 1 - \frac{1}{10}x_n^3$. x_7 and x_8 round to 0.92170.

24. Let $g(x) = f(x) - x$ for $a \le x \le b$. g is continuous (because f is), and since $a \le f(x) \le b$ whenever $a \le x \le b$ (by condition (i)), we know that $g(a) \ge 0$ and $g(b) \le 0$. By the Intermediate-Value Theorem there exists r in $[a, b]$ such that $g(r) = 0$, that is, such that $f(r) = r$.

Section 4.7 Linear Approximations (page 284)

2. $f(x) = x^{-3}$, $f'(x) = -3x^{-4}$, $f(2) = 1/8$, $f'(2) = -3/16$.
Linearization at $x = 2$: $L(x) = \frac{1}{8} - \frac{3}{16}(x - 2)$.

4. $f(x) = \sqrt{3 + x^2}$, $f'(x) = x/\sqrt{3 + x^2}$, $f(1) = 2$, $f'(1) = 1/2$.
Linearization at $x = 1$: $L(x) = 2 + \frac{1}{2}(x - 1)$.

6. $f(x) = x^{-1/2}$, $f'(x) = (-1/2)x^{-3/2}$, $f(4) = 1/2$, $f'(4) = -1/16$.
Linearization at $x = 4$: $L(x) = \frac{1}{2} - \frac{1}{16}(x - 4)$.

8. $f(x) = \cos(2x)$, $f'(x) = -2\sin(2x)$, $f(\pi/3) = -1/2$, $f'(\pi/3) = -\sqrt{3}$.
Linearization at $x = \pi/3$: $L(x) = -\frac{1}{2} - \sqrt{3}\left(x - \frac{\pi}{3}\right)$.

10. $f(x) = \tan x$, $f'(x) = \sec^2 x$, $f(\pi/4) = 1$, $f'(\pi/4) = 2$.
Linearization at $x = \pi/4$: $L(x) = 1 + 2\left(x - \frac{\pi}{4}\right)$.

12. If V and x are the volume and side length of the cube, then $V = x^3$. If $x = 20$ cm and $\Delta V = -12$ cm^3, then

$$-12 = \Delta V \approx \frac{dV}{dx}\,\Delta x = 3x^2\,\Delta x = 1,200\,\Delta x,$$

so that $\Delta x = -1/100$. The edge length must decrease by about 0.01 cm in to decrease the volume by 12 cm^3.

14. $a = g[R/(R+h)]^2$ implies that

$$\Delta a \approx \frac{da}{dh}\,\Delta h = gR^2\,\frac{-2}{(R+h)^3}\,\Delta h.$$

If $h = 0$ and $\Delta h = 10$ mi, then

$$\Delta a \approx -\frac{20g}{R} = -\frac{20 \times 32}{3960} \approx 0.16 \text{ ft/s}^2.$$

16. Let $f(x) = \sqrt{x}$, then $f'(x) = \frac{1}{2}x^{-1/2}$ and $f''(x) = -\frac{1}{4}x^{-3/2}$. Hence,

$$\sqrt{47} = f(47) \approx f(49) + f'(49)(47-49)$$

$$= 7 + \left(\frac{1}{14}\right)(-2) = \frac{48}{7} \approx 6.8571429.$$

Clearly, if $x \geq 36$, then

$$|f''(x)| \leq \frac{1}{4 \times 6^3} = \frac{1}{864} = K.$$

Since $f''(x) < 0$, f is concave down. Therefore, the error $E = \sqrt{47} - \dfrac{48}{7} < 0$ and

$$|E| < \frac{K}{2}(47-49)^2 = \frac{1}{432}.$$

Thus,

$$\frac{48}{7} - \frac{1}{432} < \sqrt{47} < \frac{48}{7}$$

$$6.8548 < \sqrt{47} < 6.8572.$$

18. Let $f(x) = \dfrac{1}{x}$, then $f'(x) = -\dfrac{1}{x^2}$ and $f''(x) = \dfrac{2}{x^3}$. Hence,

$$\frac{1}{2.003} = f(2.003) \approx f(2) + f'(2)(0.003)$$

$$= \frac{1}{2} + \left(-\frac{1}{4}\right)(0.003) = 0.49925.$$

If $x \geq 2$, then $|f''(x)| \leq \frac{2}{8} = \frac{1}{4}$. Since $f''(x) > 0$ for $x > 0$, f is concave up. Therefore, the error

$$E = \frac{1}{2.003} - 0.49925 > 0$$

and

$$|E| < \frac{1}{8}(0.003)^2 = 0.000001125.$$

Thus,

$$0.49925 < \frac{1}{2.003} < 0.49925 + 0.000001125$$

$$0.49925 < \frac{1}{2.003} < 0.499251125.$$

20. Let $f(x) = \sin x$, then $f'(x) = \cos x$ and $f''(x) = -\sin x$. Hence,

$$\sin\left(\frac{\pi}{5}\right) = f\left(\frac{\pi}{6}+\frac{\pi}{30}\right) \approx f\left(\frac{\pi}{6}\right) + f'\left(\frac{\pi}{6}\right)\left(\frac{\pi}{30}\right)$$

$$= \frac{1}{2} + \frac{\sqrt{3}}{2}\left(\frac{\pi}{30}\right) \approx 0.5906900.$$

If $x \leq \dfrac{\pi}{4}$, then $|f''(x)| \leq \dfrac{1}{\sqrt{2}}$. Since $f''(x) < 0$ on $0 < x \leq 90°$, f is concave down. Therefore, the error E is negative and

$$|E| < \frac{1}{2\sqrt{2}}\left(\frac{\pi}{30}\right)^2 = 0.0038772.$$

Thus,

$$0.5906900 - 0.0038772 < \sin\left(\frac{\pi}{5}\right) < 0.5906900$$

$$0.5868128 < \sin\left(\frac{\pi}{5}\right) < 0.5906900.$$

22. Let $f(x) = \sin x$, then $f'(x) = \cos x$ and $f''(x) = -\sin x$. The linearization at $x = 30° = \pi/6$ gives

$$\sin(33°) = \sin\left(\frac{\pi}{6}+\frac{\pi}{60}\right)$$

$$\approx \sin\frac{\pi}{6} + \cos\frac{\pi}{6}\left(\frac{\pi}{60}\right)$$

$$= \frac{1}{2} + \frac{\sqrt{3}}{2}\left(\frac{\pi}{60}\right) \approx 0.545345.$$

Since $f''(x) < 0$ between 30° and 33°, the error E in the above approximation is negative: $\sin(33°) < 0.545345$. For $30° \leq t \leq 33°$, we have

$$|f''(t)| = \sin t \leq \sin(33°) < 0.545345.$$

Thus the error satisfies

$$|E| \leq \frac{0.545345}{2}\left(\frac{\pi}{60}\right)^2 < 0.000747.$$

Therefore

$$0.545345 - 0.000747 < \sin(33°) < 0.545345$$

$$0.544598 < \sin(33°) < 0.545345.$$

24. From the solution to Exercise 16, the linearization to $f(x) = x^{1/2}$ at $x = 49$ has value at $x = 47$ given by

$$L(47) = f(49) + f'(49)(47-49) \approx 6.8571429.$$

Also, $6.8548 \le \sqrt{47} \le 6.8572$, and, since $f''(x) = -1/(4(\sqrt{x})^3)$,

$$\frac{-1}{4(6.8548)^3} \le \frac{-1}{4(\sqrt{47})^3} \le f''(x) \le \frac{-1}{4(7)^3}$$

for $47 \le x \le 49$. Thus, on that interval, $M \le f''(x) \le N$, where $M = -0.000776$ and $N = -0.000729$. By Corollary C,

$$L(47) + \frac{M}{2}(47-49)^2 \le f(47) \le L(47) + \frac{N}{2}(47-49)^2$$
$$6.855591 \le \sqrt{47} \le 6.855685.$$

Using the midpoint of this interval as a new approximation for $\sqrt{47}$ ensures that the error is no greater than half the length of the interval:

$$\sqrt{47} \approx 6.855638, \quad |\text{error}| \le 0.000047.$$

26. From the solution to Exercise 22, the linearization to $f(x) = \sin x$ at $x = 30° = \pi/6$ has value at $x = 33° = \pi/6 + \pi/60$ given by

$$L(33°) = f(\pi/6) + f'(\pi/6)(\pi/60) \approx 0.545345.$$

Also, $0.544597 \le \sin(33°) \le 5.545345$, and, since $f''(x) = -\sin x$,

$$-\sin(33°) \le f''(x) \le -\sin(30°)$$

for $30° \le x \le 33°$. Thus, on that interval, $M \le f''(x) \le N$, where $M = -0.545345$ and $N = -0.5$. By Corollary C,

$$L(33°) + \frac{M}{2}(\pi/60)^2 \le \sin(33°) \le L(33°) + \frac{N}{2}(\pi/60)^2$$
$$0.544597 \le \sin(33°) \le 0.544660.$$

Using the midpoint of this interval as a new approximation for $\sin(33°)$ ensures that the error is no greater than half the length of the interval:

$$\sin(33°) \approx 0.544629, \quad |\text{error}| \le 0.000031.$$

28. The linearization of $f(x)$ about $x = 2$ is

$$L(x) = f(2) + f'(2)(x-2) = 4 - (x-2).$$

Thus $L(3) = 3$. Also, since $1/(2x) \le f''(x) \le 1/x$ for $x > 0$, we have for $2 \le x \le 3$, $(1/6) \le f''(x) \le (1/2)$. Thus

$$3 + \frac{1}{2}\left(\frac{1}{6}\right)(3-2)^2 \le f(3) \le 3 + \frac{1}{2}\left(\frac{1}{2}\right)(3-2)^2.$$

The best approximation for $f(3)$ is the midpoint of this interval: $f(3) \approx 3\frac{1}{6}$.

30. If $f(\theta) = \sin\theta$, then $f'(\theta) = \cos\theta$ and $f''(\theta) = -\sin\theta$. Since $f(0) = 0$ and $f'(0) = 1$, the linearization of f at $\theta = 0$ is $L(\theta) = 0 + 1(\theta - 0) = \theta$.
If $0 \le t \le \theta$, then $f''(t) \le 0$, so $0 \le \sin\theta \le \theta$.
If $0 \ge t \ge \theta$, then $f''(t) \ge 0$, so $0 \ge \sin\theta \ge \theta$.
In either case, $|\sin t| \le |\sin\theta| \le |\theta|$ if t is between 0 and θ. Thus the error $E(\theta)$ in the approximation $\sin\theta \approx \theta$ satisfies

$$|E(\theta)| \le \frac{|\theta|}{2}|\theta|^2 = \frac{|\theta|^3}{2}.$$

If $|\theta| \le 17° = 17\pi/180$, then

$$\frac{|E(\theta)|}{|\theta|} \le \frac{1}{2}\left(\frac{17\pi}{180}\right)^2 \approx 0.044.$$

Thus the percentage error is less than 5%.

Section 4.8 Taylor Polynomials (page 292)

2. If $f(x) = \cos x$, then $f'(x) = -\sin x$, $f''(x) = -\cos x$, and $f'''(x) = \sin x$. In particular, $f(\pi/4) = f'''(\pi/4) = 1/\sqrt{2}$ and $f'(\pi/4) = f''(\pi/4) = -1/\sqrt{2}$. Thus

$$P_3(x) = \frac{1}{\sqrt{2}}\left[1 - \left(x - \frac{\pi}{4}\right) - \frac{1}{2}\left(x - \frac{\pi}{4}\right)^2 + \frac{1}{6}\left(x - \frac{\pi}{4}\right)^3\right].$$

4.
$$f(x) = \sec x \qquad\qquad f(0) = 1$$
$$f'(x) = \sec x \tan x \qquad\qquad f'(0) = 0$$
$$f''(x) = 2\sec^3 x - \sec x \qquad f''(0) = 1$$
$$f'''(x) = (6\sec^2 x - 1)\sec x \tan x \qquad f'''(0) = 0$$

Thus $P_3(x) = 1 + (x^2/2)$.

6.
$$f(x) = \frac{1}{2+x} \qquad\qquad f(1) = \frac{1}{3}$$
$$f'(x) = \frac{-1}{(2+x)^2} \qquad\qquad f'(1) = \frac{-1}{9}$$
$$f''(x) = \frac{2!}{(2+x)^3} \qquad\qquad f''(1) = \frac{2!}{27}$$
$$f'''(x) = \frac{-3!}{(2+x)^4} \qquad\qquad f'''(1) = \frac{-3!}{3^4}$$
$$\vdots \qquad\qquad\qquad \vdots$$
$$f^{(n)}(x) = \frac{(-1)^n n!}{(2+x)^{n+1}} \qquad f^{(n)}(1) = \frac{(-1)^n n!}{3^{n+1}}$$

Thus

$$P_n(x) = \frac{1}{3} - \frac{1}{9}(x-1) + \frac{1}{27}(x-1)^2 - \cdots + \frac{(-1)^n}{3^{n+1}}(x-1)^n.$$

8. Since $f(x) = \sqrt{x}$, then $f'(x) = \frac{1}{2}x^{-1/2}$, $f''(x) = -\frac{1}{4}x^{-3/2}$ and $f'''(x) = \frac{3}{8}x^{-5/2}$. Hence,

$$\sqrt{61} \approx f(64) + f'(64)(61 - 64) + \frac{1}{2}f''(64)(61 - 64)^2$$

$$= 8 + \frac{1}{16}(-3) - \frac{1}{2}\left(\frac{1}{2048}\right)(-3)^2 \approx 7.8103027.$$

The error is $R_2 = R_2(f; 64, 61) = \dfrac{f'''(X)}{3!}(61 - 64)^3$ for some X between 61 and 64. Clearly $R_2 < 0$. If $t \geq 49$, and in particular $61 \leq t \leq 64$, then

$$|f'''(t)| \leq \tfrac{3}{8}(49)^{-5/2} = 0.0000223 = K.$$

Hence,

$$|R_2| \leq \frac{K}{3!}|61 - 64|^3 = 0.0001004.$$

Since $R_2 < 0$, therefore,

$$7.8103027 - 0.0001004 < \sqrt{61} < 7.8103027$$
$$7.8102023 < \sqrt{61} < 7.8103027.$$

10. Since $f(x) = \tan^{-1} x$, then

$$f'(x) = \frac{1}{1 + x^2}, \quad f''(x) = \frac{-2x}{(1 + x^2)^2}, \quad f'''(x) = \frac{-2 + 6x^2}{(1 + x^2)^3}.$$

Hence,

$$\tan^{-1}(0.97) \approx f(1) + f'(1)(0.97 - 1) + \tfrac{1}{2}f''(1)(0.97 - 1)^2$$

$$= \frac{\pi}{4} + \frac{1}{2}(-0.03) + \left(-\frac{1}{4}\right)(-0.03)^2$$

$$= 0.7701731.$$

The error is $R_2 = \dfrac{f'''(X)}{3!}(-0.03)^3$ for some X between 0.97 and 1. Note that $R_2 < 0$. If $0.97 \leq t \leq 1$, then

$$|f'''(t)| \leq f'''(1) = \frac{-2 + 6}{(1.97)^3} < 0.5232 = K.$$

Hence,

$$|R_2| \leq \frac{K}{3!}|0.97 - 1|^3 < 0.0000024.$$

Since $R_2 < 0$,

$$0.7701731 - 0.0000024 < \tan^{-1}(0.97) < 0.7701731$$
$$0.7701707 < \tan^{-1}(0.97) < 0.7701731.$$

12. Since $f(x) = \sin x$, then $f'(x) = \cos x$, $f''(x) = -\sin x$ and $f'''(x) = -\cos x$. Hence,

$$\sin(47°) = f\left(\frac{\pi}{4} + \frac{\pi}{90}\right)$$

$$\approx f\left(\frac{\pi}{4}\right) + f'\left(\frac{\pi}{4}\right)\left(\frac{\pi}{90}\right) + \frac{1}{2}f''\left(\frac{\pi}{4}\right)\left(\frac{\pi}{90}\right)^2$$

$$= \frac{1}{\sqrt{2}} + \frac{1}{\sqrt{2}}\left(\frac{\pi}{90}\right) - \frac{1}{2\sqrt{2}}\left(\frac{\pi}{90}\right)^2$$

$$\approx 0.7313587.$$

The error is $R_2 = \dfrac{f'''(X)}{3!}\left(\dfrac{\pi}{90}\right)^3$ for some X between $45°$ and $47°$. Observe that $R_2 < 0$. If $45° \leq t \leq 47°$, then

$$|f'''(t)| \leq |-\cos 45°| = \frac{1}{\sqrt{2}} = K.$$

Hence,

$$|R_2| \leq \frac{K}{3!}\left(\frac{\pi}{90}\right)^3 < 0.0000051.$$

Since $R_2 < 0$, therefore

$$0.7313587 - 0.0000051 < \sin(47°) < 0.7313587$$
$$0.7313536 < \sin(47°) < 0.7313587.$$

14. For $f(x) = \cos x$ we have

$$f'(x) = -\sin x \qquad f''(x) = -\cos x \qquad f'''(x) = \sin x$$
$$f^{(4)}(x) = \cos x \qquad f^{(5)}(x) = -\sin x \qquad f^{(6)}(x) = -\cos x.$$

The Taylor's Formula for f with $a = 0$ and $n = 6$ is

$$\cos x = 1 - \frac{x^2}{2!} + \frac{x^4}{4!} - \frac{x^6}{6!} + R_6(f; 0, x)$$

where the Lagrange remainder R_6 is given by

$$R_6 = R_6(f; 0, x) = \frac{f^{(7)}(X)}{7!}x^7 = \frac{\sin X}{7!}x^7,$$

for some X between 0 and x.

16. Given that $f(x) = \dfrac{1}{1 - x}$, then

$$f'(x) = \frac{1}{(1 - x)^2}, \quad f''(x) = \frac{2}{(1 - x)^3}.$$

In general,

$$f^{(n)}(x) = \frac{n!}{(1 - x)^{(n+1)}}.$$

Since $a = 0$, $f^{(n)}(0) = n!$. Hence, for $n = 6$, the Taylor's Formula is

$$\frac{1}{1 - x} = f(0) + \sum_{n=1}^{6}\frac{f^{(n)}(0)}{n!}x^n + R_6(f; 0, x)$$

$$= 1 + x + x^2 + x^3 + x^4 + x^5 + x^6 + R_6(f; 0, x).$$

The Langrange remainder is

$$R_6(f;0,x) = \frac{f^{(7)}(X)}{7!}x^7 = \frac{x^7}{(1-X)^8}$$

for some X between 0 and x.

18. Given that $f(x) = \tan x$, then

$$f'(x) = \sec^2 x$$
$$f''(x) = 2\sec^2 x \tan x$$
$$f^{(3)}(x) = 6\sec^4 x - 4\sec^2 x$$
$$f^{(4)}(x) = 8\tan x(3\sec^4 x - \sec^2 x).$$

Given that $a = 0$ and $n = 3$, the Taylor's Formula is

$$\tan x = f(0) + f'(0)x + \frac{f''(0)}{2!}x^2 + \frac{f'''(0)}{3!}x^3 + R_3(f;0,x)$$

$$= x + \frac{2}{3!}x^3 + R_3(f;0,x)$$

$$= x + \frac{1}{3}x^3 + \frac{2}{15}x^5.$$

The Lagrange remainder is

$$R_3(f;0,x) = \frac{f^{(4)}(X)}{4!}x^4 = \frac{\tan X(3\sec^4 X - \sec^2 C)}{3}x^4$$

for some X between 0 and x.

20. For e^u, $P_4(u) = 1 + u + \frac{u^2}{2!} + \frac{u^3}{3!} + \frac{u^4}{4!}$. Let $u = -x^2$. Then for e^{-x^2}:

$$P_8(x) = 1 - x^2 + \frac{x^4}{2!} - \frac{x^6}{3!} + \frac{x^8}{4!}.$$

22. $\sin x = \sin\left(\pi + (x - \pi)\right) = -\sin(x - \pi)$

$$P_5(x) = -(x - \pi) + \frac{(x - \pi)^3}{3!} - \frac{(x - \pi)^5}{5!}$$

24. $\cos(3x - \pi) = -\cos(3x)$

$$P_8(x) = -1 + \frac{3^2 x^2}{2!} - \frac{3^4 x^4}{4!} + \frac{3^6 x^6}{6!} - \frac{3^8 x^8}{8!}.$$

26. For $\ln(1 + x)$ about $x = 0$ we have

$$P_{2n+1}(x) = x - \frac{x^2}{2} + \frac{x^3}{3} - \cdots + \frac{x^{2n+1}}{2n+1}.$$

For $\ln(1 - x)$ about $x = 0$ we have

$$P_{2n+1}(x) = -x - \frac{x^2}{2} - \frac{x^3}{3} - \cdots - \frac{x^{2n+1}}{2n+1}.$$

For $\tanh^{-1} x = \frac{1}{2}\ln(1 + x) - \frac{1}{2}\ln(1 - x)$,

$$P_{2n+1}(x) = x + \frac{x^3}{3} + \frac{x^5}{5} + \cdots + \frac{x^{2n+1}}{2n+1}.$$

28. In Taylor's Formulas for $f(x) = \sin x$ with $a = 0$, only odd powers of x have nonzero coefficients. Accordingly we can take terms up to order x^{2n+1} and the remainder after the $0x^{2n+2}$. The formula is

$$\sin x = x - \frac{x^3}{3!} + \frac{x^5}{5!} - \cdots + (-1)^n \frac{x^{2n+1}}{(2n+1)!} + R_{2n+2},$$

where

$$R_{2n+2}(f;0,x) = (-1)^{n+1} \frac{\cos X}{(2n+3)!}x^{2n+3}$$

for some X between 0 and x.

In order to use the formula to approximate $\sin(1)$ correctly to 5 decimal places, we need $|R_{2n+2}(f;0,1)| < 0.000005$. Since $|\cos X| \leq 1$, it is sufficient to have $1/(2n+3)! < 0.000005$. $n = 3$ will do since $1/9! \approx 0.000003$. Thus

$$\sin(1) \approx 1 - \frac{1}{3!} + \frac{1}{5!} - \frac{1}{7!} \approx 0.84147$$

correct to five decimal places.

Section 4.9 Indeterminate Forms (page 298)

2. $\displaystyle\lim_{x \to 2} \frac{\ln(2x - 3)}{x^2 - 4} \quad \left[\frac{0}{0}\right]$

$$= \frac{\left(\dfrac{2}{2x - 3}\right)}{2x} = \frac{1}{2}.$$

4. $\displaystyle\lim_{x \to 0} \frac{1 - \cos ax}{1 - \cos bx} \quad \left[\frac{0}{0}\right]$

$$= \lim_{x \to 0} \frac{a \sin ax}{b \sin bx} \quad \left[\frac{0}{0}\right]$$

$$= \lim_{x \to 0} \frac{a^2 \cos ax}{b^2 \cos bx} = \frac{a^2}{b^2}.$$

6. $\displaystyle\lim_{x \to 1} \frac{x^{1/3} - 1}{x^{2/3} - 1} \quad \left[\frac{0}{0}\right]$

$$= \lim_{x \to 1} \frac{(\frac{1}{3})x^{-2/3}}{(\frac{2}{3})x^{-1/3}} = \frac{1}{2}.$$

8. $\displaystyle\lim_{x\to 0}\frac{1-\cos x}{\ln(1+x^2)} \quad \begin{bmatrix}0\\0\end{bmatrix}$

$\displaystyle = \lim_{x\to 0}\frac{\sin x}{\left(\dfrac{2x}{1+x^2}\right)}$

$\displaystyle = \lim_{x\to 0}(1+x^2)\lim_{x\to 0}\frac{\sin x}{2x}$

$\displaystyle = \lim_{x\to 0}\frac{\cos x}{2} = \frac{1}{2}.$

10. $\displaystyle\lim_{x\to 0}\frac{10^x-e^x}{x} \quad \begin{bmatrix}0\\0\end{bmatrix}$

$\displaystyle = \lim_{x\to 0}\frac{10^x\ln 10 - e^x}{1} = \ln 10 - 1.$

12. $\displaystyle\lim_{x\to 1}\frac{\ln(ex)-1}{\sin\pi x} \quad \begin{bmatrix}0\\0\end{bmatrix}$

$\displaystyle = \lim_{x\to 1}\frac{\dfrac{1}{x}}{\pi\cos(\pi x)} = -\frac{1}{\pi}.$

14. $\displaystyle\lim_{x\to 0}\frac{x-\sin x}{x^3} \quad \begin{bmatrix}0\\0\end{bmatrix}$

$\displaystyle = \lim_{x\to 0}\frac{1-\cos x}{3x^2} \quad \begin{bmatrix}0\\0\end{bmatrix}$

$\displaystyle = \lim_{x\to 0}\frac{\sin x}{6x} \quad \begin{bmatrix}0\\0\end{bmatrix}$

$\displaystyle = \lim_{x\to 0}\frac{\cos x}{6} = \frac{1}{6}.$

16. $\displaystyle\lim_{x\to 0}\frac{2-x^2-2\cos x}{x^4} \quad \begin{bmatrix}0\\0\end{bmatrix}$

$\displaystyle = \lim_{x\to 0}\frac{-2x+2\sin x}{4x^3} \quad \begin{bmatrix}0\\0\end{bmatrix}$

$\displaystyle = -\frac{1}{2}\lim_{x\to 0}\frac{x-\sin x}{x^3}$

$\displaystyle = -\frac{1}{2}\left(\frac{1}{6}\right) = -\frac{1}{12} \quad \text{(by Exercise 14)}.$

18. $\displaystyle\lim_{r\to\pi/2}\frac{\ln\sin r}{\cos r} \quad \begin{bmatrix}0\\0\end{bmatrix}$

$\displaystyle = \lim_{r\to\pi/2}\frac{\left(\dfrac{\cos r}{\sin r}\right)}{-\sin r} = 0.$

20. $\displaystyle\lim_{x\to 1-}\frac{\cos^{-1}x}{x-1} \quad \begin{bmatrix}0\\0\end{bmatrix}$

$\displaystyle = \lim_{x\to 1-}\frac{-\left(\dfrac{1}{\sqrt{1-x^2}}\right)}{1} = -\infty.$

22. $\displaystyle\lim_{t\to(\pi/2)-}(\sec t-\tan t) \quad [\infty-\infty]$

$\displaystyle = \lim_{t\to(\pi/2)-}\frac{1-\sin t}{\cos t} \quad \begin{bmatrix}0\\0\end{bmatrix}$

$\displaystyle = \lim_{t\to(\pi/2)-}\frac{-\cos t}{-\sin t} = 0.$

24. Since $\displaystyle\lim_{x\to 0+}\sqrt{x}\ln x = \lim_{x\to 0+}\frac{\ln x}{x^{-1/2}} \quad \begin{bmatrix}0\\0\end{bmatrix}$

$\displaystyle = \lim_{x\to 0+}\frac{\left(\dfrac{1}{x}\right)}{\left(-\dfrac{1}{2}\right)x^{-3/2}} = 0,$

hence $\displaystyle\lim_{x\to 0+}x^{\sqrt{x}}$

$\displaystyle = \lim_{x\to 0+}e^{\sqrt{x}\ln x} = e^0 = 1.$

26. $\displaystyle\lim_{x\to 1+}\left(\frac{x}{x-1}-\frac{1}{\ln x}\right) \quad [\infty-\infty]$

$\displaystyle = \lim_{x\to 1+}\frac{x\ln x-x+1}{(x-1)(\ln x)} \quad \begin{bmatrix}0\\0\end{bmatrix}$

$\displaystyle = \lim_{x\to 1+}\frac{\ln x}{\ln x+1-\dfrac{1}{x}} \quad \begin{bmatrix}0\\0\end{bmatrix}$

$\displaystyle = \lim_{x\to 1+}\frac{\dfrac{1}{x}}{\dfrac{1}{x}+\dfrac{1}{x^2}}$

$\displaystyle = \lim_{x\to 1+}\frac{x}{x+1} = \frac{1}{2}.$

28. Let $y=\left(\dfrac{\sin x}{x}\right)^{1/x^2}.$

$\displaystyle\lim_{x\to 0}\ln y = \lim_{x\to 0}\frac{\ln\left(\dfrac{\sin x}{x}\right)}{x^2} \quad \begin{bmatrix}0\\0\end{bmatrix}$

$\displaystyle = \lim_{x\to 0}\frac{\left(\dfrac{x}{\sin x}\right)\left(\dfrac{x\cos x-\sin x}{x^2}\right)}{2x}$

$\displaystyle = \lim_{x\to 0}\frac{x\cos x-\sin x}{2x^2\sin x} \quad \begin{bmatrix}0\\0\end{bmatrix}$

$\displaystyle = \lim_{x\to 0}\frac{-x\sin x}{4x\sin x+2x^2\cos x}$

$\displaystyle = \lim_{x\to 0}\frac{-\sin x}{4\sin x+2x\cos x} \quad \begin{bmatrix}0\\0\end{bmatrix}$

$\displaystyle = \lim_{x\to 0}\frac{-\cos x}{6\cos x-2x\sin x} = -\frac{1}{6}.$

Thus, $\displaystyle\lim_{x\to 0}\left(\frac{\sin x}{x}\right)^{1/x^2} = e^{-1/6}.$

30. $\displaystyle\lim_{x\to 0+}\frac{\csc x}{\ln x}\quad\left[-\frac{\infty}{\infty}\right]$

$\displaystyle =\lim_{x\to 0+}\frac{-\csc x\cot x}{\dfrac{1}{x}}\quad\left[-\frac{\infty}{\infty}\right]$

$\displaystyle =\lim_{x\to 0+}\frac{-x\cos x}{\sin^2 x}\quad\left[\frac{0}{0}\right]$

$\displaystyle =-\left(\lim_{x\to 0+}\cos x\right)\lim_{x\to 0+}\frac{1}{2\sin x\cos x}$

$=-\infty.$

32. Let $y=(1+\tan x)^{1/x}.$

$\displaystyle\lim_{x\to 0}\ln y=\lim_{x\to 0}\frac{\ln(1+\tan x)}{x}\quad\left[\frac{0}{0}\right]$

$\displaystyle =\lim_{x\to 0}\frac{\sec^2 x}{1+\tan x}=1.$

Thus, $\displaystyle\lim_{x\to 0}(1+\tan x)^{1/x}=e.$

34. $\displaystyle\lim_{h\to 0}\frac{f(x+3h)-3f(x+h)+3f(x-h)-f(x-3h)}{h^3}$

$\displaystyle =\lim_{h\to 0}\frac{3f'(x+3h)-3f'(x+h)-3f'(x-h)+3f'(x-3h)}{3h^2}$

$\displaystyle =\lim_{h\to 0}\frac{3f''(x+3h)-f''(x+h)+f''(x-h)-3f''(x-3h)}{2h}$

$\displaystyle =\lim_{h\to 0}\frac{9f'''(x+3h)-f'''(x+h)-f'''(x-h)+9f'''(x-3h)}{2}$

$=8f'''(x).$

Review Exercises 4   (page 299)

2.   a) Since F must be continuous at $r=R$, we have

$$\frac{mgR^2}{R^2}=mkR,\quad\text{or}\quad k=\frac{g}{R}.$$

b) The rate of change of F as r decreases from R is

$$\left(-\frac{d}{dr}(mkr)\right)\Bigg|_{r=R}=-mk=-\frac{mg}{R}.$$

The rate of change of F as r increases from R is

$$\left(-\frac{d}{dr}\frac{mgR^2}{r^2}\right)\Bigg|_{r=R}=-\frac{2mgR^2}{R^3}=-2\frac{mg}{R}.$$

Thus F decreases as r increases from R at twice the rate at which it decreases as r decreases from R.

4. If $pV=5.0T$, then

$$\frac{dp}{dt}V+p\frac{dV}{dt}=5.0\frac{dT}{dt}.$$

a) If $T=400$ K, $dT/dt=4$ K/min, and $V=2.0$ m^3, then $dV/dt=0$, so $dp/dt=5.0(4)/2.0=10$. The pressure is increasing at 10 kPa/min.

b) If $T=400$ K, $dT/dt=0$, $V=2$ m^3, and $dV/dt=0.05$ m^3/min, then $p=5.0(400)/2=1{,}000$ kPa, and $2\,dp/dt+1{,}000(0.05)=0$, so $dp/dt=-25$. The pressure is decreasing at 25 kPa/min.

6. If she charges $\$x$ per bicycle, her total profit is $\$P$, where

$$P=(x-75)N(x)=4.5\times 10^6\frac{x-75}{x^2}.$$

Evidently $P\le 0$ if $x\le 75$, and $P\to 0$ as $x\to\infty$. P will therefore have a maximum value at a critical point in $(75,\infty)$. For CP:

$$0=\frac{dP}{dx}=4.5\times 10^6\frac{x^2-(x-75)2x}{x^4},$$

from which we obtain $x=150$. She should charge $\$150$ per bicycle and order $N(150)=200$ of them from the manufacturer.

8.

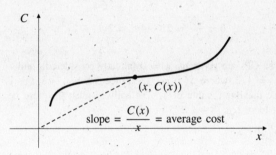

Fig. R-4.8

a) For minimum $C(x)/x$, we need

$$0=\frac{d}{dx}\frac{C(x)}{x}=\frac{xC'(x)-C(x)}{x^2},$$

so $C'(x)=C(x)/x$; the marginal cost equals the average cost.

b) The line from $(0,0)$ to $(x,C(x))$ has smallest slope at a value of x which makes it tangent to the graph of $C(x)$. Thus $C'(x)=C(x)/x$, the slope of the line.

c) The line from $(0,0)$ to $(x,C(x))$ can be tangent to the graph of $C(x)$ at more than one point. Not all such points will provide a minimum value for the average cost. (In the figure, one such line will make the average cost maximum.)

10. If x more trees are planted, the yield of apples will be

$$Y=(60+x)(800-10x)$$
$$=10(60+x)(80-x)$$
$$=10(4{,}800+20x-x^2).$$

This is a quadratic expression with graph opening downward; its maximum occurs at a CP:

$$0 = \frac{dY}{dx} = 10(20 - 2x) = 20(10 - x).$$

Thus 10 more trees should be planted to maximize the yield.

12. The narrowest hallway in which the table can be turned horizontally through $180°$ has width equal to twice the greatest distance from the origin (the centre of the table) to the curve $x^2 + y^4 = 1/8$ (the edge of the table). We maximize the square of this distance, which we express as a function of y:

$$S(y) = x^2 + y^2 = y^2 + \frac{1}{8} - y^4, \quad (0 \le y \le (1/8)^{1/4}).$$

Note that $S(0) = 1/8$ and $S((1/8)^{1/4}) = 1/\sqrt{8} > S(0)$. For CP:

$$0 = \frac{dS}{dy} = 2y - 4y^3 = 2y(1 - 2y^2).$$

The CPs are given by $y = 0$ (already considered), and $y^2 = 1/2$, where $S(y) = 3/8$. Since $3/8 > 1/\sqrt{8}$, this is the maximum value of S. The hallway must therefore be at least $2\sqrt{3/8} \approx 1.225$ m wide.

14.

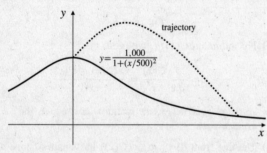

Fig. R-4.14

If the origin is at sea level under the launch point, and $x(t)$ and $y(t)$ are the horizontal and vertical coordinates of the cannon ball's position at time t s after it is fired, then

$$\frac{d^2x}{dt^2} = 0, \qquad \frac{d^2y}{dt^2} = -32.$$

At $t = 0$, we have $dx/dt = dy/dt = 200/\sqrt{2}$, so

$$\frac{dx}{dt} = \frac{200}{\sqrt{2}}, \qquad \frac{dy}{dt} = -32t + \frac{200}{\sqrt{2}}.$$

At $t = 0$, we have $x = 0$ and $y = 1{,}000$. Thus the position of the ball at time t is given by

$$x = \frac{200t}{\sqrt{2}}, \qquad y = -16t^2 + \frac{200t}{\sqrt{2}} + 1{,}000.$$

We can obtain the Cartesian equation for the path of the cannon ball by solving the first equation for t and substituting into the second equation:

$$y = -16\frac{2x^2}{200^2} + x + 1{,}000.$$

The cannon ball strikes the ground when

$$-16\frac{2x^2}{200^2} + x + 1{,}000 = \frac{1{,}000}{1 + (x/500)^2}.$$

Graphing both sides of this equation suggests a solution near $x = 1{,}900$. Newton's Method or a solve routine then gives $x \approx 1{,}873$. The horizontal range is about $1{,}873$ ft.

16.
$$\sin^2 x = \frac{1}{2}\left(1 - \cos(2x)\right)$$
$$= \frac{1}{2}\left[1 - \left(1 - \frac{2^2x^2}{2!} + \frac{2^4x^4}{4!} - \frac{2^6x^6}{6!} + O(x^8)\right)\right]$$
$$= x^2 - \frac{x^4}{3} + \frac{2x^6}{45} + O(x^8)$$
$$\lim_{x\to 0}\frac{3\sin^2 x - 3x^2 + x^4}{x^6}$$
$$= \lim_{x\to 0}\frac{3x^2 - x^4 + \frac{2}{15}x^6 + O(x^8) - 3x^2 - x^4}{x^6}$$
$$= \lim_{x\to 0}\frac{2}{15} + O(x^2) = \frac{2}{15}.$$

18. The second approximation x_1 is the x-intercept of the tangent to $y = f(x)$ at $x = x_0 = 2$; it is the x-intercept of the line $2y = 10x - 19$. Thus $x_1 = 19/10 = 1.9$.

20. The square of the distance from $(2, 0)$ to $(x, \ln x)$ is $S(x) = (x - 2)^2 + (\ln x)^2$, for $x > 0$. Since $S(x) \to \infty$ as $x \to \infty$ or $x \to 0+$, the minimum value of $S(x)$ will occur at a critical point. For CP:

$$0 = S'(x) = 2\left(x - 2 + \frac{\ln x}{x}\right).$$

We solve this equation using a TI-85 solve routine; $x \approx 1.6895797$. The minimum distance from the origin to $y = e^x$ is $\sqrt{S(x)} \approx 0.6094586$.

Challenging Problems 4 (page 301)

2. Let the speed of the tank be v where $v = \frac{dy}{dt} = ky$. Thus, $y = Ce^{kt}$. Given that at $t = 0$, $y = 4$, then $4 = y(0) = C$. Also given that at $t = 10$, $y = 2$, thus,

$$2 = y(10) = 4e^{10k} \Rightarrow k = -\frac{1}{10}\ln 2.$$

Hence, $y = 4e^{(-\frac{1}{10}\ln 2)t}$ and $v = \dfrac{dy}{dt} = (-\dfrac{1}{10}\ln 2)y$. The slope of the curve $xy = 1$ is $m = \dfrac{dy}{dx} = -\dfrac{1}{x^2}$. Thus, the equation of the tangent line at the point $\left(\dfrac{1}{y_0}, y_0\right)$ is

$$y = y_0 - \dfrac{1}{\left(\dfrac{1}{y_0}\right)^2}\left(x - \dfrac{1}{y_0}\right), \quad \text{i.e.,} \quad y = 2y_0 - xy_0^2.$$

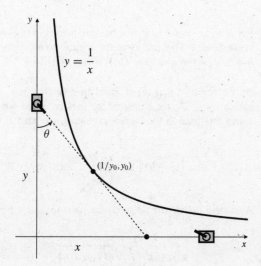

Fig. C-4.2

Hence, the x-intercept is $x = \dfrac{2}{y_0}$ and the y-intercept is $y = 2y_0$. Let θ be the angle between the gun and the y-axis. We have

$$\tan\theta = \dfrac{x}{y} = \dfrac{\left(\dfrac{2}{y_0}\right)}{2y_0} = \dfrac{1}{y_0^2} = \dfrac{4}{y^2}$$

$$\Rightarrow \quad \sec^2\theta \dfrac{d\theta}{dt} = \dfrac{-8}{y^3}\dfrac{dy}{dt}.$$

Now

$$\sec^2\theta = 1 + \tan^2\theta = 1 + \dfrac{16}{y^4} = \dfrac{y^4 + 16}{y^4},$$

so

$$\dfrac{d\theta}{dt} = -\dfrac{8y}{y^4 + 16}\dfrac{dy}{dt} = -\dfrac{8ky^2}{y^4 + 16}.$$

The maximum value of $\dfrac{y^2}{y^4 + 16}$ occurs at a critical point:

$$0 = \dfrac{(y^4 + 16)2y - y^2(4y^3)}{(y^4 + 16)^2}$$

$$\Leftrightarrow \quad 2y^5 = 32y,$$

or $y = 2$. Therefore the maximum rate of rotation of the gun turret must be

$$-8k\dfrac{2^2}{2^4 + 16} = -k = \dfrac{1}{10}\ln 2 \approx 0.0693 \text{ rad/m},$$

and occurs when your tank is 2 km from the origin.

4. $P = 2\pi\sqrt{L/g} = 2\pi L^{1/2}g^{-1/2}$.

a) If L remains constant, then

$$\Delta P \approx \dfrac{dP}{dg}\Delta g = -\pi L^{1/2}g^{-3/2}\Delta g$$

$$\dfrac{\Delta P}{P} \approx \dfrac{-\pi L^{1/2}g^{-3/2}}{2\pi L^{1/2}g^{-1/2}}\Delta g = -\dfrac{1}{2}\dfrac{\Delta g}{g}.$$

If g increases by 1%, then $\Delta g/g = 1/100$, and $\Delta P/P = -1/200$. Thus P decreases by 0.5%.

b) If g remains constant, then

$$\Delta P \approx \dfrac{dP}{dL}\Delta L = \pi L^{-1/2}g^{-1/2}\Delta L$$

$$\dfrac{\Delta P}{P} \approx \dfrac{\pi L^{-1/2}g^{-1/2}}{2\pi L^{1/2}g^{-1/2}}\Delta L = \dfrac{1}{2}\dfrac{\Delta L}{L}.$$

If L increases by 2%, then $\Delta L/L = 2/100$, and $\Delta P/P = 1/100$. Thus P increases by 1%.

6. If the depth of liquid in the tank at time t is $y(t)$, then the surface of the liquid has radius $r(t) = Ry(t)/H$, and the volume of liquid in the tank at that time is

$$V(t) = \dfrac{\pi}{3}\left(\dfrac{Ry(t)}{H}\right)^2 y(t) = \dfrac{\pi R^2}{3H^2}\left(y(t)\right)^3.$$

By Torricelli's law, $dV/dt = -k\sqrt{y}$. Thus

$$\dfrac{\pi R^2}{3H^2}3y^2\dfrac{dy}{dt} = \dfrac{dV}{dt} = -k\sqrt{y},$$

or, $dy/dt = -k_1 y^{-3/2}$, where $k_1 = kH^2/(\pi R^2)$. If $y(t) = y_0\left(1 - \dfrac{t}{T}\right)^{2/5}$, then $y(0) = y_0$, $y(T) = 0$, and

$$\dfrac{dy}{dt} = \dfrac{2}{5}y_0\left(1 - \dfrac{t}{T}\right)^{-3/5}\left(-\dfrac{1}{T}\right) = -k_1 y^{-3/2},$$

where $k_1 = 2y_0/(5T)$. Thus this function $y(t)$ satisfies the conditions of the problem.

8. The slope of $y = x^3 + ax^2 + bx + c$ is

$$y' = 3x^2 + 2ax + b,$$

which $\to \infty$ as $x \to \pm\infty$. The quadratic expression y' takes each of its values at two different points except its minimum value, which is achieved only at one point given by $y'' = 6x + 2a = 0$. Thus the tangent to the cubic at $x = -a/3$ is not parallel to any other tangent. This tangent has equation

$$
\begin{aligned}
y &= -\frac{a^3}{27} + \frac{a^3}{9} - \frac{ab}{3} + c \\
&\quad + \left(\frac{a^2}{3} - \frac{2a^2}{3} + b\right)\left(x + \frac{a}{3}\right) \\
&= -\frac{a^3}{27} + c + \left(b - \frac{a^2}{3}\right)x.
\end{aligned}
$$

10.

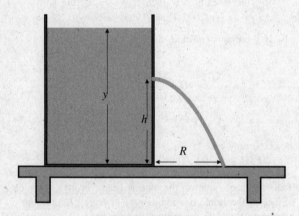

Fig. C-4.10

a) Let the origin be at the point on the table directly under the hole. If a water particle leaves the tank with horizontal velocity v, then its position $(X(t), Y(t))$, t seconds later, is given by

$$
\begin{array}{ll}
\dfrac{d^2X}{dt^2} = 0 & \dfrac{d^2Y}{dt^2} = -g \\[2mm]
\dfrac{dX}{dt} = v & \dfrac{dY}{dt} = -gt \\[2mm]
X = vt & Y = -\dfrac{1}{2}gt^2 + h.
\end{array}
$$

The range R of the particle (i.e., of the spurt) is the value of X when $Y = 0$, that is, at time $t = \sqrt{2h/g}$. Thus $R = v\sqrt{2h/g}$.

b) Since $v = k\sqrt{y - h}$, the range R is a function of y, the depth of water in the tank.

$$
R = k\sqrt{\frac{2}{g}}\sqrt{h(y - h)}.
$$

For a given depth y, R will be maximum if $h(y - h)$ is maximum. This occurs at the critical point $h = y/2$ of the quadratic $Q(h) = h(y - h)$.

c) By the result of part (c) of Problem 3 (with y replaced by $y - h$, the height of the surface of the water above the drain in the current problem), we have

$$
y(t) - h = (y_0 - h)\left(1 - \frac{t}{T}\right)^2, \quad \text{for } 0 \le t \le T.
$$

As shown above, the range of the spurt at time t is

$$
R(t) = k\sqrt{\frac{2}{g}}\sqrt{h\big(y(t) - h\big)}.
$$

Since $R = R_0$ when $y = y_0$, we have

$$
k = \frac{R_0}{\sqrt{\dfrac{2}{g}}\sqrt{h(y_0 - h)}}.
$$

Therefore $R(t) = R_0 \dfrac{\sqrt{h\big(y(t) - h\big)}}{\sqrt{h(y_0 - h)}} = R_0\left(1 - \dfrac{t}{T}\right).$

CHAPTER 5.   INTEGRATION

Section 5.1   Sums and Sigma Notation (page 307)

2. $\displaystyle\sum_{j=1}^{100}\frac{j}{j+1}=\frac{1}{2}+\frac{2}{3}+\frac{3}{4}+\cdots+\frac{100}{101}$

4. $\displaystyle\sum_{i=0}^{n-1}\frac{(-1)^i}{i+1}=1-\frac{1}{2}+\frac{1}{3}-\cdots+\frac{(-1)^{n-1}}{n}$

6. $\displaystyle\sum_{j=1}^{n}\frac{j^2}{n^3}=\frac{1}{n^3}+\frac{4}{n^3}+\frac{9}{n^3}+\cdots+\frac{n^2}{n^3}$

8. $\displaystyle\sum_{k=2}^{n}\frac{e^{-k}}{n+k}=\frac{e^{-2}}{n+2}+\frac{e^{-3}}{n+3}+\cdots+\frac{e^{-n}}{n+n}$

10. $2+2+2+\cdots+2$ (200 terms) equals $\displaystyle\sum_{i=1}^{200}2$

12. $1+2x+3x^2+4x^3+\cdots+100x^{99}=\displaystyle\sum_{i=1}^{100}ix^{i-1}$

14. $1-x+x^2-x^3+\cdots+x^{2n}=\displaystyle\sum_{i=0}^{2n}(-1)^ix^i$

16. $\dfrac{1}{2}+\dfrac{2}{4}+\dfrac{3}{8}+\dfrac{4}{16}+\cdots+\dfrac{n}{2^n}=\displaystyle\sum_{i=1}^{n}\frac{i}{2^i}$

18. $\displaystyle\sum_{k=-5}^{m}\frac{1}{k^2+1}=\sum_{i=1}^{m+6}\frac{1}{((i-6)^2+1)}$

20. $\displaystyle\sum_{j=1}^{1,000}(2j+3)=\frac{2(1,000)(1,001)}{2}+3,000=1,004,000$

22. $\displaystyle\sum_{k=1}^{n-1}(3k^2-4)=3\frac{(n-1)n(2n-1)}{6}-4(n-1)=n^3-\frac{3}{2}n^2-\frac{7}{2}n+4$

24. $\displaystyle\sum_{i=1}^{n}(2^i-i^2)=2^{n+1}-2-\frac{1}{6}n(n+1)(2n+1)$

26. $\displaystyle\sum_{i=0}^{n}e^{i/n}=\frac{e^{(n+1)/n}-1}{e^{1/n}-1}$

28. $1+x+x^2+\cdots+x^n=\begin{cases}\dfrac{1-x^{n+1}}{1-x} & \text{if } x\neq 1\\[2mm] n+1 & \text{if } x=1\end{cases}$

30. Let $f(x)=1+x+x^2+\cdots+x^{100}=\dfrac{x^{101}-1}{x-1}$ if $x\neq 1$. Then

$$f'(x)=1+2x+3x^2+\cdots+100x^{99}$$
$$=\frac{d}{dx}\frac{x^{101}-1}{x-1}=\frac{100x^{101}-101x^{100}+1}{(x-1)^2}.$$

32. Let $s=\dfrac{1}{2}+\dfrac{2}{4}+\dfrac{3}{8}+\cdots+\dfrac{n}{2^n}$. Then

$$\frac{s}{2}=\frac{1}{4}+\frac{2}{8}+\frac{3}{16}+\cdots+\frac{n}{2^{n+1}}.$$

Subtracting these two sums, we get

$$\frac{s}{2}=\frac{1}{2}+\frac{1}{4}+\frac{1}{8}+\cdots+\frac{1}{2^n}-\frac{n}{2^{n+1}}$$
$$=\frac{1}{2}\frac{1-(1/2^n)}{1-(1/2)}-\frac{n}{2^{n+1}}$$
$$=1-\frac{n+2}{2^{n+1}}.$$

Thus $s=2+(n+2)/2^n$.

34. $\displaystyle\sum_{n=1}^{10}(n^4-(n-1)^4)=10^4-0^4=10,000$

36. $\displaystyle\sum_{i=m}^{2m}\left(\frac{1}{i}-\frac{1}{i+1}\right)=\frac{1}{m}-\frac{1}{2m+1}=\frac{m+1}{m(2m+1)}$

38. The number of small shaded squares is $1+2+\cdots+n$. Since each has area 1, the total area shaded is $\sum_{i=1}^{n}i$. But this area consists of a large right-angled triangle of area $n^2/2$ (below the diagonal), and n small triangles (above the diagonal) each of area $1/2$. Equating these areas, we get

$$\sum_{i=1}^{n}i=\frac{n^2}{2}+n\frac{1}{2}=\frac{n(n+1)}{2}.$$

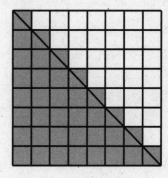

Fig. 5.1.38

40. The formula $\sum_{i=1}^{n} i = n(n+1)/2$ holds for $n = 1$, since it says $1 = 1$ in this case. Now assume that it holds for $n =$ some number $k \geq 1$; that is, $\sum_{i=1}^{k} i = k(k+1)/2$. Then for $n = k+1$, we have

$$\sum_{i=1}^{k+1} i = \sum_{i=1}^{k} i + (k+1) = \frac{k(k+1)}{2} + (k+1) = \frac{(k+1)(k+2)}{2}.$$

Thus the formula also holds for $n = k+1$. By induction, it holds for all positive integers n.

42. The formula $\sum_{i=1}^{n} r^{i-1} = (r^n - 1)/(r-1)$ (for $r \neq 1$) holds for $n = 1$, since it says $1 = 1$ in this case. Now assume that it holds for $n =$ some number $k \geq 1$; that is, $\sum_{i=1}^{k} r^{i-1} = (r^k - 1)/(r-1)$. Then for $n = k+1$, we have

$$\sum_{i=1}^{k+1} r^{i-1} = \sum_{i=1}^{k} r^{i-1} + r^k = \frac{r^k - 1}{r - 1} + r^k = \frac{r^{k+1} - 1}{r - 1}.$$

Thus the formula also holds for $n = k+1$. By induction, it holds for all positive integers n.

44. To show that

$$\sum_{j=1}^{n} j^3 = 1^3 + 2^3 + 3^3 + \cdots + n^3 = \frac{n^2(n+1)^2}{4},$$

we write n copies of the identity

$$(k+1)^4 - k^4 = 4k^3 + 6k^2 + 4k + 1,$$

one for each k from 1 to n:

$$2^4 - 1^4 = 4(1)^3 + 6(1)^2 + 4(1) + 1$$
$$3^4 - 2^4 = 4(2)^3 + 6(2)^2 + 4(2) + 1$$
$$4^4 - 3^4 = 4(3)^3 + 6(3)^2 + 4(3) + 1$$
$$\vdots$$
$$(n+1)^4 - n^4 = 4(n)^3 + 6(n)^2 + 4(n) + 1.$$

Adding the left and right sides of these formulas we get

$$(n+1)^4 - 1^4 = 4\sum_{j=1}^{n} j^3 + 6\sum_{j=1}^{n} j^2 + 4\sum_{j=1}^{n} j + n$$

$$= 4\sum_{j=1}^{n} j^3 + \frac{6n(n+1)(2n+1)}{6} + \frac{4n(n+1)}{2} + n.$$

Hence,

$$4\sum_{j=1}^{n} j^3 = (n+1)^4 - 1 - n(n+1)(2n+1) - 2n(n+1) - n$$

$$= n^2(n+1)^2$$

so $\sum_{j=1}^{n} j^3 = \dfrac{n^2(n+1)^2}{4}$.

46. To find $\sum_{j=1}^{n} j^4 = 1^4 + 2^4 + 3^4 + \cdots + n^4$, we write n copies of the identity

$$(k+1)^5 - k^5 = 5k^4 + 10k^3 + 10k^2 + 5k + 1,$$

one for each k from 1 to n:

$$2^5 - 1^5 = 5(1)^4 + 10(1)^3 + 10(1)^2 + 5(1) + 1$$
$$3^5 - 2^5 = 5(2)^4 + 10(2)^3 + 10(2)^2 + 5(2) + 1$$
$$4^5 - 3^5 = 5(3)^4 + 10(3)^3 + 10(3)^2 + 5(3) + 1$$
$$\vdots$$
$$(n+1)^5 - n^5 = 5(n)^4 + 10(n)^3 + 10(n)^2 + 5(n) + 1.$$

Adding the left and right sides of these formulas we get

$$(n+1)^5 - 1^5 = 5\sum_{j=1}^{n} j^4 + 10\sum_{j=1}^{n} j^3 + 10\sum_{j=1}^{n} j^2 + 5\sum_{j=1}^{n} j + n.$$

Substituting the known formulas for all the sums except $\sum_{j=1}^{n} j^4$, and solving for this quantity, gives

$$\sum_{j=1}^{n} j^4 = \frac{n(n+1)(2n+1)(3n^2 + 3n - 1)}{30}.$$

Of course we got Maple to do the donkey work!

Section 5.2   Areas as Limits of Sums (page 314)

2. This is similar to #1; the rectangles now have width $3/n$ and the ith has height $2(3i/n) + 1$, the value of $2x + 1$ at $x = 3i/n$. The area is

$$A = \lim_{n \to \infty} \sum_{i=1}^{n} \frac{3}{n} \left(2\frac{3i}{n} + 1 \right)$$

$$= \lim_{n \to \infty} \frac{18}{n^2} \sum_{i=1}^{n} i + \frac{3}{n} n$$

$$= \lim_{n \to \infty} \frac{18}{n^2} \frac{n(n+1)}{2} + 3 = 9 + 3 = 12 \text{ sq. units.}$$

4. This is similar to #1; the rectangles have width $(2 - (-1))/n = 3/n$ and the ith has height the value of $3x + 4$ at $x = -1 + (3i/n)$. The area is

$$A = \lim_{n \to \infty} \sum_{i=1}^{n} \frac{3}{n} \left(-3 + 3\frac{3i}{n} + 4 \right)$$

$$= \lim_{n \to \infty} \frac{27}{n^2} \sum_{i=1}^{n} i + \frac{3}{n} n$$

$$= \lim_{n \to \infty} \frac{27}{n^2} \frac{n(n+1)}{2} + 3 = \frac{27}{2} + 3 = \frac{33}{2} \text{ sq. units.}$$

The region in question lies between $x = -1$ and $x = 1$ and is symmetric about the y-axis. We can therefore double the area between $x = 0$ and $x = 1$. If we divide this interval into n equal subintervals of width $1/n$ and use the distance $0 - (x^2 - 1) = 1 - x^2$ between $y = 0$ and $y = x^2 - 1$ for the heights of rectangles, we find that the required area is

$$A = 2 \lim_{n \to \infty} \sum_{i=1}^{n} \frac{1}{n}\left(1 - \frac{i^2}{n^2}\right)$$

$$= 2 \lim_{n \to \infty} \sum_{i=1}^{n} \left(\frac{1}{n} - \frac{i^2}{n^3}\right)$$

$$= 2 \lim_{n \to \infty} \left(\frac{n}{n} - \frac{n(n+1)(2n+1)}{6n^3}\right) = 2 - \frac{4}{6} = \frac{4}{3} \text{ sq. units.}$$

6. Divide $[0, a]$ into n equal subintervals of length $\Delta x = \dfrac{a}{n}$ by points $x_i = \dfrac{ia}{n}$, $(0 \le i \le n)$. Then

$$S_n = \sum_{i=1}^{n} \left(\frac{a}{n}\right)\left[\left(\frac{ia}{n}\right)^2 + 1\right]$$

$$= \left(\frac{a}{n}\right)^3 \sum_{i=1}^{n} i^2 + \frac{a}{n} \sum_{i=1}^{n} (1)$$

(Use Theorem 1(a) and 1(c).)

$$= \left(\frac{a}{n}\right)^3 \frac{n(n+1)(2n+1)}{6} + \frac{a}{n}(n)$$

$$= \frac{a^3}{6} \frac{(n+1)(2n+1)}{n^2} + a.$$

Area $= \lim_{n \to \infty} S_n = \dfrac{a^3}{3} + a$ sq. units.

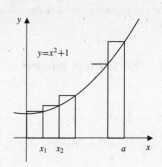

Fig. 5.2.6

8.

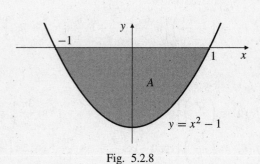

Fig. 5.2.8

10.

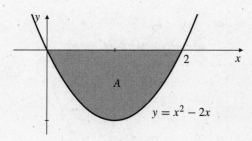

Fig. 5.2.10

The height of the region at position x is $0 - (x^2 - 2x) = 2x - x^2$. The "base" is an interval of length 2, so we approximate using n rectangles of width $2/n$. The shaded area is

$$A = \lim_{n \to \infty} \sum_{i=1}^{n} \frac{2}{n}\left(2\frac{2i}{n} - \frac{4i^2}{n^2}\right)$$

$$= \lim_{n \to \infty} \sum_{i=1}^{n} \left(\frac{8i}{n^2} - \frac{8i^2}{n^3}\right)$$

$$= \lim_{n \to \infty} \left(\frac{8}{n^2}\frac{n(n+1)}{2} - \frac{8}{n^3}\frac{n(n+1)(2n+1)}{6}\right)$$

$$= 4 - \frac{8}{3} = \frac{4}{3} \text{ sq. units.}$$

12. Divide $[0, b]$ into n equal subintervals of length $\Delta x = \dfrac{b}{n}$ by points $x_i = \dfrac{ib}{n}$, $(0 \le i \le n)$. Then

$$S_n = \sum_{i=1}^{n} \frac{b}{n}\left(e^{(ib/n)}\right) = \frac{b}{n} \sum_{i=1}^{n} \left(e^{(b/n)}\right)^i$$

$$= \frac{b}{n}e^{(b/n)}\sum_{i=1}^{n}\left(e^{(b/n)}\right)^{i-1} \quad \text{(Use Thm. 6.1.2(d).)}$$

$$= \frac{b}{n}e^{(b/n)}\frac{e^{(b/n)n}-1}{e^{(b/n)}-1}$$

$$= \frac{b}{n}e^{(b/n)}\frac{e^{b}-1}{e^{(b/n)}-1}.$$

Let $r = \dfrac{b}{n}$.

$$\text{Area} = \lim_{n\to\infty} S_n = (e^b - 1)\lim_{r\to 0+} e^r \lim_{r\to 0+} \frac{r}{e^r - 1} \quad \begin{bmatrix}0\\0\end{bmatrix}$$

$$= (e^b - 1)(1)\lim_{r\to 0+}\frac{1}{e^r} = e^b - 1\text{sq. units.}$$

14. $\quad \text{Area} = \lim_{n\to\infty}\frac{b}{n}\left[\left(\frac{b}{n}\right)^3 + \left(\frac{2b}{n}\right)^3 + \cdots + \left(\frac{nb}{n}\right)^3\right]$

$$= \lim_{n\to\infty}\frac{b^4}{n^4}(1^3 + 2^3 + 3^3 + \cdots + n^3)$$

$$= \lim_{n\to\infty}\frac{b^4}{n^4}\cdot\frac{n^2(n+1)^2}{4} = \frac{b^4}{4}\text{sq. units.}$$

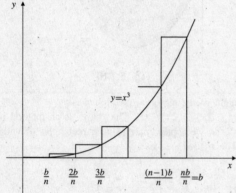

Fig. 5.2.14

16.

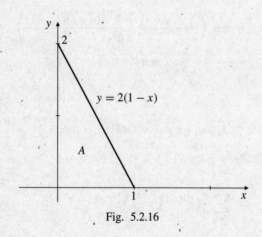

Fig. 5.2.16

$s_n = \displaystyle\sum_{i=1}^{n}\frac{2}{n}\left(1 - \frac{i}{n}\right)$ represents a sum of areas of n rectangles each of width $1/n$ and having heights equal to the height to the graph $y = 2(1 - x)$ at the points $x = i/n$. Thus $\lim_{n\to\infty} S_n$ is the area A of the triangle in the figure above, and therefore has the value 1.

18.

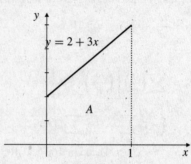

Fig. 5.2.18

$s_n = \displaystyle\sum_{i=1}^{n}\frac{2n + 3i}{n^2} = \sum_{i=1}^{n}\frac{1}{n}\left(2 + \frac{3i}{n}\right)$ represents a sum of areas of n rectangles each of width $1/n$ and having heights equal to the height to the graph $y = 2 + 3x$ at the points $x = i/n$. Thus $\lim_{n\to\infty} S_n$ is the area of the trapezoid in the figure above, and has the value $1(2 + 5)/2 = 7/2$.

Section 5.3 The Definite Integral (page 320)

2. $\quad f(x) = x^2$ on $[0, 4]$, $n = 4$.

$$L(f, P_4) = \left(\frac{4 - 0}{4}\right)[0 + (1)^2 + (2)^2 + (3)^2] = 14.$$

$$U(f, P_4) = \left(\frac{4 - 0}{4}\right)[(1)^2 + (2)^2 + (3)^2 + (4)^2] = 30.$$

4. $\quad f(x) = \ln x$ on $[1, 2]$, $n = 5$.

$$L(f, P_5) = \left(\frac{2 - 1}{5}\right)\left[\ln 1 + \ln\frac{6}{5} + \ln\frac{7}{5} + \ln\frac{8}{5} + \ln\frac{9}{5}\right]$$

$$\approx 0.3153168.$$

$$U(f, P_5) = \left(\frac{2 - 1}{5}\right)\left[\ln\frac{6}{5} + \ln\frac{7}{5} + \ln\frac{8}{5} + \ln\frac{9}{5} + \ln 2\right]$$

$$\approx 0.4539462.$$

6. $\quad f(x) = \cos x$ on $[0, 2\pi]$, $n = 4$.

$$L(f, P_4) = \left(\frac{2\pi}{4}\right)\left[\cos\frac{\pi}{2} + \cos\pi + \cos\pi + \cos\frac{3\pi}{2}\right] = -\pi.$$

$$U(f, P_4) = \left(\frac{2\pi}{4}\right)\left[\cos 0 + \cos\frac{\pi}{2} + \cos\frac{3\pi}{2} + \cos 2\pi\right] = \pi.$$

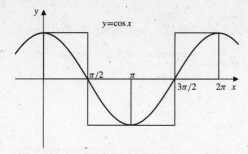

Fig. 5.3.6

8. $f(x) = 1 - x$ on $[0, 2]$. $P_n = \left\{0, \frac{2}{n}, \frac{4}{n}, \dots, \frac{2n-2}{n}, \frac{2n}{n}\right\}$. We have

$$L(f, P_n) = \frac{2}{n}\left(\left(1 - \frac{2}{n}\right) + \left(1 - \frac{4}{n}\right) + \cdots + \left(1 - \frac{2n}{n}\right)\right)$$

$$= \frac{2}{n}n - \frac{4}{n^2}\sum_{i=1}^{n} i$$

$$= 2 - \frac{4}{n^2}\frac{n(n+1)}{2} = -\frac{2}{n} \to 0 \text{ as } n \to \infty,$$

$$U(f, P_n) = \frac{2}{n}\left(\left(1 - \frac{0}{n}\right) + \left(1 - \frac{2}{n}\right) + \cdots + \left(1 - \frac{2n-2}{n}\right)\right)$$

$$= \frac{2}{n}n - \frac{4}{n^2}\sum_{i=0}^{n-1} i$$

$$= 2 - \frac{4}{n^2}\frac{(n-1)n}{2} = \frac{2}{n} \to 0 \text{ as } n \to \infty.$$

Thus $\int_0^2 (1 - x)\,dx = 0$.

10. $f(x) = e^x$ on $[0, 3]$. $P_n = \left\{0, \frac{3}{n}, \frac{6}{n}, \dots, \frac{3n-3}{n}, \frac{3n}{n}\right\}$. We have (using the result of Exercise 51 (or 52) of Section 6.1)

$$L(f, P_n) = \frac{3}{n}\left(e^{0/n} + e^{3/n} + e^{6/n} + \cdots + e^{3(n-1)/n}\right)$$

$$= \frac{3}{n}\frac{e^{3n/n} - 1}{e^{3/n} - 1} = \frac{3(e^3 - 1)}{n(e^{3/n} - 1)},$$

$$U(f, P_n) = \frac{3}{n}\left(e^{3/n} + e^{6/n} + e^{9/n} + \cdots + e^{3n/n}\right) = e^{3/n}L(f, P_n).$$

By l'Hôpital's Rule,

$$\lim_{n\to\infty} n(e^{3/n} - 1) = \lim_{n\to\infty}\frac{e^{3/n} - 1}{1/n}$$

$$= \lim_{n\to\infty}\frac{e^{3/n}(-3/n^2)}{-1/n^2} = \lim_{n\to\infty}\frac{3e^{3/n}}{1} = 3.$$

Thus

$$\lim_{n\to\infty} L(f, P_n) = \lim_{n\to\infty} U(f, P_n) = e^3 - 1 = \int_0^3 e^x\,dx.$$

12. $\displaystyle\lim_{n\to\infty}\sum_{i=1}^{n}\frac{1}{n}\sqrt{\frac{i-1}{n}} = \int_0^1 \sqrt{x}\,dx$

14. $\displaystyle\lim_{n\to\infty}\sum_{i=1}^{n}\frac{2}{n}\ln\left(1 + \frac{2i}{n}\right) = \int_0^2 \ln(1+x)\,dx$

16. $\displaystyle\lim_{n\to\infty}\sum_{i=1}^{n}\frac{n}{n^2 + i^2} = \lim_{n\to\infty}\sum_{i=1}^{n}\frac{1}{n}\frac{1}{1 + (i/n)^2} = \int_0^1 \frac{dx}{1 + x^2}$

18. $P = \{x_0 < x_1 < \cdots < x_n\}$,
$P' = \{x_0 < x_1 < \cdots < x_{j-1} < x' < x_j < \cdots < x_n\}$.
Let m_i and M_i be, respectively, the minimum and maximum values of $f(x)$ on the interval $[x_{i-1}, x_i]$, for $1 \le i \le n$. Then

$$L(f, P) = \sum_{i=1}^{n} m_i(x_i - x_{i-1}),$$

$$U(f, P) = \sum_{i=1}^{n} M_i(x_i - x_{i-1}).$$

If m_j' and M_j' are the minimum and maximum values of $f(x)$ on $[x_{j-1}, x']$, and if m_j'' and M_j'' are the corresponding values for $[x', x_j]$, then

$$m_j' \ge m_j, \quad m_j'' \ge m_j, \quad M_j' \le M_j, \quad M_j'' \le M_j.$$

Therefore we have

$$m_j(x_j - x_{j-1}) \le m_j'(x' - x_{j-1}) + m_j''(x_j - x'),$$
$$M_j(x_j - x_{j-1}) \ge M_j'(x' - x_{j-1}) + M_j''(x_j - x').$$

Hence $L(f, P) \le L(f, P')$ and $U(f, P) \ge U(f, P')$.

If P'' is any refinement of P we can add the new points in P'' to those in P one at a time, and thus obtain

$$L(f, P) \le L(f, P''), \qquad U(f, P'') \le U(f, P).$$

Section 5.4 Properties of the Definite Integral (page 327)

2. $\displaystyle\int_0^2 3f(x)\,dx + \int_1^3 3f(x)\,dx - \int_0^3 2f(x)\,dx$

$$-\int_1^2 3f(x)\,dx$$

$$= \int_0^1 (3-2)f(x)\,dx + \int_1^2 (3+3-2-3)f(x)\,dx$$

$$+ \int_2^3 (3-2)f(x)\,dx$$

$$= \int_0^3 f(x)\,dx$$

4. $\int_0^2 (3x + 1)\,dx = $ shaded area $= \frac{1}{2}(1 + 7)(2) = 8$

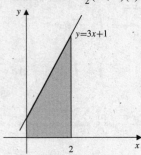

Fig. 5.4.4

6. $\int_{-1}^2 (1 - 2x)\,dx = A_1 - A_2 = 0$

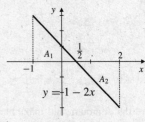

Fig. 5.4.6

8. $\int_{-\sqrt{2}}^0 \sqrt{2 - x^2}\,dx = $ quarter disk $= \frac{1}{4}\pi(\sqrt{2})^2 = \frac{\pi}{2}$

10. $\int_{-a}^a (a - |s|)\,ds = $ shaded area $= 2(\frac{1}{2}a^2) = a^2$

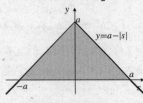

Fig. 5.4.10

12. Let $y = \sqrt{2x - x^2} \Rightarrow y^2 + (x - 1)^2 = 1$.

$\int_0^2 \sqrt{2x - x^2}\,dx = $ shaded area $= \frac{1}{2}\pi(1)^2 = \frac{\pi}{2}$.

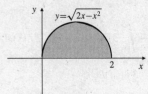

Fig. 5.4.12

14. $\int_{-3}^3 (2 + t)\sqrt{9 - t^2}\,dt = 2\int_{-3}^3 \sqrt{9 - t^2}\,dt + \int_{-3}^3 t\sqrt{9 - t^2}\,dt$

$$= 2\left(\frac{1}{2}\pi 3^2\right) + 0 = 9\pi$$

16. $\int_1^2 \sqrt{4 - x^2}\,dx = $ area A_2 in figure above

$= $ area sector POQ $-$ area triangle POR

$= \frac{1}{6}(\pi 2^2) - \frac{1}{2}(1)\sqrt{3}$

$= \frac{2\pi}{3} - \frac{\sqrt{3}}{2}$

18. $\int_2^3 (x^2 - 4)\,dx = \int_0^3 x^2\,dx - \int_0^2 x^2\,dx - 4(3 - 2)$

$= \frac{3^3}{3} - \frac{2^3}{3} - 4 = \frac{7}{3}$

20. $\int_0^2 (v^2 - v)\,dv = \frac{2^3}{3} - \frac{2^2}{2} = \frac{2}{3}$

22. $\int_{-6}^6 x^2(2 + \sin x)\,dx = \int_{-6}^6 2x^2\,dx + \int_{-6}^6 x^2 \sin x\,dx$

$= 4\int_0^6 x^2\,dx + 0 = \frac{4}{3}(6^3) = 288$

24. $\int_2^4 \frac{1}{t}\,dt = \int_1^4 \frac{1}{t}\,dt - \int_1^2 \frac{1}{t}\,dt$

$= \ln 4 - \ln 2 = \ln(4/2) = \ln 2$

26. $\int_{1/4}^3 \frac{1}{s}\,ds = \int_1^3 \frac{1}{s}\,ds - \int_1^{1/4} \frac{1}{s}\,ds$

$= \ln 3 - \ln \frac{1}{4} = \ln 3 + \ln 4 = \ln 12$

28. Average $= \frac{1}{b - a}\int_a^b (x + 2)\,dx$

$= \frac{1}{b - a}\left[\frac{1}{2}(b^2 - a^2) + 2(b - a)\right]$

$= \frac{1}{2}(b + a) + 2 = \frac{4 + a + b}{2}$

30. Average $= \frac{1}{3 - 0}\int_0^3 x^2\,dx = \frac{1}{3}\frac{3^3}{3} = 3$

32. Average value $= \frac{1}{2 - (1/2)}\int_{1/2}^2 \frac{1}{s}\,ds$

$= \frac{2}{3}\left(\ln 2 - \ln\frac{1}{2}\right) = \frac{4}{3}\ln 2$

34. Let

$$f(x) = \begin{cases} 1 + x & \text{if } x < 0 \\ 2 & \text{if } x \geq 0. \end{cases}$$

Then

$\int_{-3}^2 f(x)\,dx = \text{area}(1) + \text{area}(2) - \text{area}(3)$

$= (2 \times 2) + \frac{1}{2}(1)(1) - \frac{1}{2}(2)(2) = 2\frac{1}{2}.$

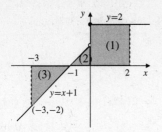

Fig. 5.4.34

36. $\int_0^3 |2-x|\, dx = \int_0^2 (2-x)\, dx + \int_2^3 (x-2)\, dx$

$$= \left(2x - \frac{x^2}{2}\right)\Big|_0^2 + \left(\frac{x^2}{2} - 2x\right)\Big|_2^3$$

$$= 4 - 2 - 0 + \frac{9}{2} - 6 - 2 + 4 = \frac{5}{2}$$

38. $\int_0^{3.5} \lfloor x \rfloor\, dx = $ shaded area $= 1 + 2 + 1.5 = 4.5.$

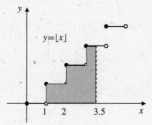

Fig. 5.4.38

40.

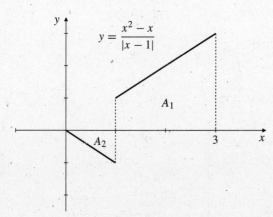

Fig. 5.4.40

$$\int_0^3 \frac{x^2 - x}{|x-1|}\, dx$$

$$= \text{area } A_1 - \text{area } A_2$$

$$= \frac{1+3}{2}(2) - \frac{1}{2}(1)(1) = \frac{7}{2}$$

42. $\int_a^b \left(f(x) - \bar{f}\right) dx = \int_a^b f(x)\, dx - \int_a^b \bar{f}\, dx$

$$= (b-a)\bar{f} - \bar{f} \int_a^b dx$$

$$= (b-a)\bar{f} - (b-a)\bar{f} = 0$$

Section 5.5 The Fundamental Theorem of Calculus (page 333)

2. $\int_0^4 \sqrt{x}\, dx = \frac{2}{3} x^{3/2}\Big|_0^4 = \frac{16}{3}$

4. $\int_{-2}^{-1} \left(\frac{1}{x^2} - \frac{1}{x^3}\right) dx = \left(-\frac{1}{x} + \frac{1}{2x^2}\right)\Big|_{-2}^{-1}$

$$= 1 + \frac{1}{2} - \left(\frac{1}{2} + \frac{1}{8}\right) = \frac{7}{8}$$

6. $\int_1^2 \left(\frac{2}{x^3} - \frac{x^3}{2}\right) dx = \left(-\frac{1}{x^2} - \frac{x^4}{8}\right)\Big|_1^2 = -9/8$

8. $\int_4^9 \left(\sqrt{x} - \frac{1}{\sqrt{x}}\right) dx = \frac{2}{3} x^{3/2} - 2\sqrt{x}\Big|_4^9$

$$= \left[\frac{2}{3}(9)^{3/2} - 2\sqrt{9}\right] - \left[\frac{2}{3}(4)^{3/2} - 2\sqrt{4}\right] = \frac{32}{3}$$

10. $\int_0^{\pi/3} \sec^2\theta\, d\theta = \tan\theta\Big|_0^{\pi/3} = \tan\frac{\pi}{3} = \sqrt{3}$

12. $\int_0^{2\pi} (1 + \sin u)\, du = (u - \cos u)\Big|_0^{2\pi} = 2\pi$

14. $\int_{-2}^2 (e^x - e^{-x})\, dx = 0$ (odd function, symmetric interval)

16. $\int_{-1}^1 2^x\, dx = \frac{2^x}{\ln 2}\Big|_{-1}^1 = \frac{2}{\ln 2} - \frac{1}{2\ln 2} = \frac{3}{2\ln 2}$

18. $\int_0^{1/2} \frac{dx}{\sqrt{1-x^2}} = \sin^{-1}x\Big|_0^{1/2} = \frac{\pi}{6}$

20. $\int_{-2}^0 \frac{dx}{4+x^2} = \frac{1}{2}\tan^{-1}\frac{x}{2}\Big|_{-2}^0 = 0 - \frac{1}{2}\tan^{-1}(-1) = \frac{\pi}{8}$

22. Area $= \int_e^{e^2} \frac{1}{x}\, dx = \ln x\Big|_e^{e^2}$

$$= \ln e^2 - \ln e = 2 - 1 = 1 \text{ sq. units.}$$

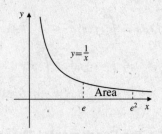

Fig. 5.5.22

24. Since $y = 5 - 2x - 3x^2 = (5 + 3x)(1 - x)$, therefore $y = 0$ at $x = -\frac{5}{3}$ and 1, and $y > 0$ if $-\frac{5}{3} < x < 1$. Thus, the area is

$$\int_{-1}^{1} (5 - 2x - 3x^2)\, dx = 2 \int_{0}^{1} (5 - 3x^2)\, dx$$

$$= 2(5x - x^3)\Big|_{0}^{1}$$

$$= 2(5 - 1) = 8 \text{ sq. units.}$$

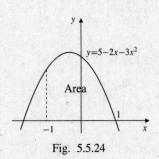

Fig. 5.5.24

26. Since $y = \sqrt{x}$ and $y = \dfrac{x}{2}$ intersect where $\sqrt{x} = \dfrac{x}{2}$, that is, at $x = 0$ and $x = 4$, thus,

$$\text{Area} = \int_{0}^{4} \sqrt{x}\, dx - \int_{0}^{4} \frac{x}{2}\, dx$$

$$= \frac{2}{3}x^{3/2}\Big|_{0}^{4} - \frac{x^2}{4}\Big|_{0}^{4}$$

$$= \frac{16}{3} - \frac{16}{4} = \frac{4}{3} \text{ sq. units.}$$

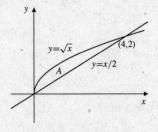

Fig. 5.5.26

28. The two graphs intersect at $(\pm 3, 3)$, thus

$$\text{Area} = 2 \int_{0}^{3} (12 - x^2)\, dx - 2 \int_{0}^{3} x\, dx$$

$$= 2\left(12x - \frac{1}{3}x^3\right)\Big|_{0}^{3} - 2\left(\frac{1}{2}x^2\right)\Big|_{0}^{3}$$

$$= 2(36 - 9) - 9 = 45 \text{ sq. units.}$$

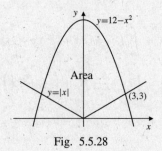

Fig. 5.5.28

30. $\text{Area} = \int_{-a}^{0} e^{-x}\, dx = -e^{-x}\Big|_{-a}^{0} = e^a - 1$ sq. units.

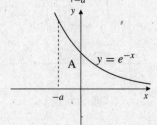

Fig. 5.5.30

32. $\text{Area} = \displaystyle\int_{1}^{27} x^{-1/3}\, dx = \frac{3}{2}x^{2/3}\Big|_{1}^{27}$

$$= \frac{3}{2}(27)^{2/3} - \frac{3}{2} = 12 \text{ sq. units.}$$

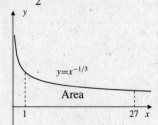

Fig. 5.5.32

34. $\displaystyle\int_{1}^{3} \frac{\text{sgn}\,(x - 2)}{x^2}\, dx = -\int_{1}^{2} \frac{dx}{x^2} + \int_{2}^{3} \frac{dx}{x^2}$

$$= \frac{1}{x}\Big|_{1}^{2} - \frac{1}{x}\Big|_{2}^{3} = -\frac{1}{3}$$

36. Average value $= \dfrac{1}{2-(-2)} \displaystyle\int_{-2}^{2} e^{3x}\, dx$

$\qquad = \dfrac{1}{4}\left(\dfrac{1}{3}e^{3x}\right)\Big|_{-2}^{2}$

$\qquad = \dfrac{1}{12}(e^{6} - e^{-6}).$

38. Since

$$g(t) = \begin{cases} 0, & \text{if } 0 \le t \le 1, \\ 1, & \text{if } 1 < t \le 3, \end{cases}$$

the average value of $g(t)$ over $[0,3]$ is

$$\dfrac{1}{3}\left[\int_0^1 (0)\, dt + \int_1^3 1\, dt\right] = \dfrac{1}{3}\left[0 + t\Big|_1^3\right]$$
$$= \dfrac{1}{3}(3-1) = \dfrac{2}{3}.$$

40. $\dfrac{d}{dt}\displaystyle\int_t^3 \dfrac{\sin x}{x}\, dx = \dfrac{d}{dt}\left[-\int_3^t \dfrac{\sin x}{x}\, dx\right] = -\dfrac{\sin t}{t}$

42. $\dfrac{d}{dx}x^2 \displaystyle\int_0^{x^2} \dfrac{\sin u}{u}\, du$

$= 2x\displaystyle\int_0^{x^2} \dfrac{\sin u}{u}\, du + x^2 \dfrac{d}{dx}\int_0^{x^2} \dfrac{\sin u}{u}\, du$

$= 2x\displaystyle\int_0^{x^2} \dfrac{\sin u}{u}\, du + x^2\left[\dfrac{2x \sin x^2}{x^2}\right]$

$= 2x\displaystyle\int_0^{x^2} \dfrac{\sin u}{u}\, du + 2x \sin(x^2)$

44. $\dfrac{d}{d\theta}\displaystyle\int_{\sin\theta}^{\cos\theta} \dfrac{1}{1-x^2}\, dx$

$= \dfrac{d}{d\theta}\left[\displaystyle\int_a^{\cos\theta} \dfrac{1}{1-x^2}\, dx - \int_a^{\sin\theta} \dfrac{1}{1-x^2}\, dx\right]$

$= \dfrac{-\sin\theta}{1-\cos^2\theta} - \dfrac{\cos\theta}{1-\sin^2\theta}$

$= \dfrac{-1}{\sin\theta} - \dfrac{1}{\cos\theta} = -\csc\theta - \sec\theta$

46. $H(x) = 3x \displaystyle\int_4^{x^2} e^{-\sqrt{t}}\, dt$

$H'(x) = 3\displaystyle\int_4^{x^2} e^{-\sqrt{t}}\, dt + 3x(2xe^{-|x|})$

$H'(2) = 3\displaystyle\int_4^4 e^{-\sqrt{t}}\, dt + 3(2)(4e^{-2})$

$\qquad = 3(0) + 24e^{-2} = \dfrac{24}{e^2}$

48. $f(x) = 1 - \displaystyle\int_0^x f(t)\, dt$

$f'(x) = -f(x) \implies f(x) = Ce^{-x}$

$1 = f(0) = C$

$f(x) = e^{-x}.$

50. If $F(x) = \displaystyle\int_{17}^x \dfrac{\sin t}{1+t^2}\, dt$, then $F'(x) = \dfrac{\sin x}{1+x^2}$ and $F(17) = 0.$

52. $\displaystyle\lim_{n\to\infty} \dfrac{1}{n}\left[\left(1+\dfrac{1}{n}\right)^5 + \left(1+\dfrac{2}{n}\right)^5 + \cdots + \left(1+\dfrac{n}{n}\right)^5\right]$

$\quad = $ area below $y = x^5$, above $y = 0$,

$\qquad$ between $x = 1$ and $x = 2$

$\quad = \displaystyle\int_1^2 x^5\, dx = \dfrac{1}{6}x^6\Big|_1^2 = \dfrac{1}{6}(2^6 - 1) = \dfrac{21}{2}$

54. $\displaystyle\lim_{n\to\infty}\left(\dfrac{n}{n^2+1} + \dfrac{n}{n^2+4} + \dfrac{n}{n^2+9} + \cdots + \dfrac{n}{2n^2}\right)$

$= \displaystyle\lim_{n\to\infty} \dfrac{1}{n}\left(\dfrac{n^2}{n^2+1} + \dfrac{n^2}{n^2+4} + \dfrac{n^2}{n^2+9} + \cdots + \dfrac{n^2}{2n^2}\right)$

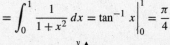

$= \displaystyle\lim_{n\to\infty} \dfrac{1}{n}\left(\dfrac{1}{1+\left(\dfrac{1}{n}\right)^2} + \dfrac{1}{1+\left(\dfrac{2}{n}\right)^2} + \cdots + \dfrac{1}{1+\left(\dfrac{n}{n}\right)^2}\right)$

$= $ area below $y = \dfrac{1}{1+x^2}$, above $y = 0$,

$\quad$ between $x = 0$ and $x = 1$

$= \displaystyle\int_0^1 \dfrac{1}{1+x^2}\, dx = \tan^{-1} x\Big|_0^1 = \dfrac{\pi}{4}$

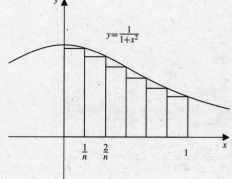

Fig. 5.5.54

Section 5.6 The Method of Substitution
(page 341)

2. $\displaystyle\int \cos(ax+b)\,dx$ Let $u = ax+b$
$$du = a\,dx$$
$$= \frac{1}{a}\int \cos u\,du = \frac{1}{a}\sin u + C$$
$$= \frac{1}{a}\sin(ax+b) + C.$$

4. $\displaystyle\int e^{2x}\sin(e^{2x})\,dx$ Let $u = e^{2x}$
$$du = 2e^{2x}\,dx$$
$$= \frac{1}{2}\int \sin u\,du = -\frac{1}{2}\cos u + C$$
$$= -\frac{1}{2}\cos(e^{2x}) + C.$$

6. $\displaystyle\int \frac{\sin\sqrt{x}}{\sqrt{x}}\,dx$ Let $u = \sqrt{x}$
$$du = \frac{dx}{2\sqrt{x}}$$
$$= 2\int \sin u\,du = -2\cos u + C$$
$$= -2\cos\sqrt{x} + C.$$

8. $\displaystyle\int x^2 2^{x^3+1}\,dx$ Let $u = x^3 + 1$
$$du = 3x^2\,dx$$
$$= \frac{1}{3}\int 2^u\,du = \frac{1}{3}\frac{2^u}{\ln 2} + C$$
$$= \frac{2^{x^3+1}}{3\ln 2} + C.$$

10. $\displaystyle\int \frac{\sec^2 x}{\sqrt{1-\tan^2 x}}\,dx$ Let $u = \tan x$
$$du = \sec^2 x\,dx$$
$$= \int \frac{du}{\sqrt{1-u^2}}$$
$$= \sin^{-1} u + C$$
$$= \sin^{-1}(\tan x) + C.$$

12. $\displaystyle\int \frac{\ln t}{t}\,dt$ Let $u = \ln t$
$$du = \frac{dt}{t}$$
$$= \int u\,du = \frac{1}{2}u^2 + C = \frac{1}{2}(\ln t)^2 + C.$$

14. $\displaystyle\int \frac{x+1}{\sqrt{x^2+2x+3}}\,dx$ Let $u = x^2 + 2x + 3$
$$du = 2(x+1)\,dx$$
$$= \frac{1}{2}\int \frac{1}{\sqrt{u}}\,du = \sqrt{u} + C = \sqrt{x^2+2x+3} + C$$

16. $\displaystyle\int \frac{x^2}{2+x^6}\,dx$ Let $u = x^3$
$$du = 3x^2\,dx$$
$$= \frac{1}{3}\int \frac{du}{2+u^2} = \frac{1}{3\sqrt{2}}\tan^{-1}\left(\frac{u}{\sqrt{2}}\right) + C$$
$$= \frac{1}{3\sqrt{2}}\tan^{-1}\left(\frac{x^3}{\sqrt{2}}\right) + C.$$

18. $\displaystyle\int \frac{dx}{e^x + e^{-x}} = \int \frac{e^x\,dx}{e^{2x}+1}$ Let $u = e^x$
$$du = e^x\,dx$$
$$= \int \frac{du}{u^2+1} = \tan^{-1}u + C$$
$$= \tan^{-1}e^x + C.$$

20. $\displaystyle\int \frac{x+1}{\sqrt{1-x^2}}\,dx$
$$= \int \frac{x\,dx}{\sqrt{1-x^2}} + \int \frac{dx}{\sqrt{1-x^2}}$$ Let $u = 1 - x^2$
$$du = -2x\,dx$$
in the first integral only
$$= -\frac{1}{2}\int \frac{du}{\sqrt{u}} + \sin^{-1}x = -\sqrt{u} + \sin^{-1}x + C$$
$$= -\sqrt{1-x^2} + \sin^{-1}x + C.$$

22. $\displaystyle\int \frac{dx}{\sqrt{4+2x-x^2}} = \frac{dx}{\sqrt{5-(1-x)^2}}$ Let $u = 1 - x$
$$du = -dx$$
$$= -\int \frac{du}{\sqrt{5-u^2}} = -\sin^{-1}\left(\frac{u}{\sqrt{5}}\right) + C$$
$$= -\sin^{-1}\left(\frac{1-x}{\sqrt{5}}\right) + C = \sin^{-1}\left(\frac{x-1}{\sqrt{5}}\right) + C.$$

24. $\displaystyle\int \sin^4 t \cos^5 t\,dt$
$$= \int \sin^4 t(1-\sin^2 t)^2 \cos t\,dt$$ Let $u = \sin t$
$$du = \cos t\,dt$$
$$= \int (u^4 - 2u^6 + u^8)\,du = \frac{u^5}{5} - \frac{2u^7}{7} + \frac{u^9}{9} + C$$
$$= \frac{1}{5}\sin^5 t - \frac{2}{7}\sin^7 t + \frac{1}{9}\sin^9 t + C.$$

26. $\displaystyle\int \sin^2 x \cos^2 x\,dx = \int \left(\frac{\sin 2x}{2}\right)^2 dx$
$$= \frac{1}{4}\int \frac{1-\cos 4x}{2}\,dx = \frac{x}{8} - \frac{\sin 4x}{32} + C.$$

28. $\displaystyle\int \cos^4 x\,dx = \int \frac{[1+\cos(2x)]^2}{4}\,dx$

$\displaystyle = \frac{1}{4}\int [1 + 2\cos(2x) + \cos^2(2x)]\,dx$

$\displaystyle = \frac{x}{4} + \frac{\sin(2x)}{4} + \frac{1}{8}\int 1 + \cos(4x)\,dx$

$\displaystyle = \frac{x}{4} + \frac{\sin(2x)}{4} + \frac{x}{8} + \frac{\sin(4x)}{32} + C$

$\displaystyle = \frac{3x}{8} + \frac{\sin(2x)}{4} + \frac{\sin(4x)}{32} + C.$

30. $\displaystyle\int \sec^6 x \tan^2 x\,dx$

$\displaystyle = \int \sec^2 x \tan^2 x (1+\tan^2 x)^2\,dx$ Let $u = \tan x$
$\displaystyle \qquad\qquad\qquad\qquad\qquad du = \sec^2 x\,dx$

$\displaystyle = \int (u^2 + 2u^4 + u^6)\,du = \frac{1}{3}u^3 + \frac{2}{5}u^5 + \frac{1}{7}u^7 + C$

$\displaystyle = \frac{1}{3}\tan^3 x + \frac{2}{5}\tan^5 x + \frac{1}{7}\tan^7 x + C.$

32. $\displaystyle\int \sin^{-2/3} x \cos^3 x\,dx$ Let $u = \sin x$
$\displaystyle \qquad\qquad\qquad\qquad\qquad du = \cos x\,dx$

$\displaystyle = \int \frac{1-u^2}{u^{2/3}}\,du = 3u^{1/3} - \frac{3}{7}u^{7/3} + C$

$\displaystyle = 3\sin^{1/3} x - \frac{3}{7}\sin^{7/3} x + C.$

34. $\displaystyle\int \frac{\sin^3(\ln x)\cos^3(\ln x)}{x}\,dx$ Let $u = \sin(\ln x)$
$\displaystyle \qquad\qquad\qquad\qquad\qquad\qquad du = \frac{\cos(\ln x)}{x}\,dx$

$\displaystyle = \int u^3(1-u^2)\,du = \frac{1}{4}u^4 - \frac{1}{6}u^6 + C$

$\displaystyle = \frac{1}{4}\sin^4(\ln x) - \frac{1}{6}\sin^6(\ln x) + C.$

36. $\displaystyle\int \frac{\sin^3 x}{\cos^4 x}\,dx = \int \tan^3 x \sec x\,dx$

$\displaystyle = \int (\sec^2 x - 1)\sec x \tan x\,dx$ Let $u = \sec x$
$\displaystyle \qquad\qquad\qquad\qquad\qquad\qquad du = \sec x \tan x\,dx$

$\displaystyle = \int (u^2 - 1)\,du = \tfrac{1}{3}u^3 - u + C$

$\displaystyle = \tfrac{1}{3}\sec^3 x - \sec x + C.$

38. $\displaystyle\int \frac{\cos^4 x}{\sin^8 x}\,dx = \int \cot^4 x \csc^4 x\,dx$

$\displaystyle = \int \cot^4 x (1 + \cot^2 x)\csc^2 x\,dx$ Let $u = \cot x$
$\displaystyle \qquad\qquad\qquad\qquad\qquad\qquad du = -\csc^2 x\,dx$

$\displaystyle = -\int u^4(1+u^2)\,du = -\frac{u^5}{5} - \frac{u^7}{7} + C$

$\displaystyle = -\frac{1}{5}\cot^5 x - \frac{1}{7}\cot^7 x + C.$

40. $\displaystyle\int_1^{\sqrt{e}} \frac{\sin(\pi \ln x)}{x}\,dx$ Let $u = \pi \ln x$
$\displaystyle \qquad\qquad\qquad\qquad\qquad du = \frac{\pi}{x}\,dx$

$\displaystyle = \frac{1}{\pi}\int_0^{\pi/2} \sin u\,du = \left.-\frac{1}{\pi}\cos u\right|_0^{\pi/2}$

$\displaystyle = -\frac{1}{\pi}(0 - 1) = \frac{1}{\pi}.$

42. $\displaystyle\int_{\pi/4}^{\pi} \sin^5 x\,dx$

$\displaystyle = \int_{\pi/4}^{\pi} (1 - \cos^2 x)^2 \sin x\,dx$ Let $u = \cos x$
$\displaystyle \qquad\qquad\qquad\qquad\qquad\qquad du = -\sin x\,dx$

$\displaystyle = -\int_{1/\sqrt{2}}^{-1} (1 - 2u^2 + u^4)\,du = \left. u - \frac{2}{3}u^3 + \frac{1}{5}u^5\right|_{-1}^{1/\sqrt{2}}$

$\displaystyle = \frac{1}{\sqrt{2}} - \frac{1}{3\sqrt{2}} + \frac{1}{20\sqrt{2}} - \left(-1 + \frac{2}{3} - \frac{1}{5}\right) = \frac{43}{60\sqrt{2}} + \frac{8}{15}.$

44. $\displaystyle\int_{\pi^2/16}^{\pi^2/9} \frac{2^{\sin\sqrt{x}}\cos\sqrt{x}}{\sqrt{x}}\,dx$ Let $u = \sin\sqrt{x}$
$\displaystyle \qquad\qquad\qquad\qquad\qquad\qquad du = \frac{\cos\sqrt{x}}{2\sqrt{x}}\,dx$

$\displaystyle = 2\int_{1/\sqrt{2}}^{\sqrt{3}/2} 2^u\,du = \left.\frac{2(2^u)}{\ln 2}\right|_{1/\sqrt{2}}^{\sqrt{3}/2}$

$\displaystyle = \frac{2}{\ln 2}(2^{\sqrt{3}/2} - 2^{1/\sqrt{2}}).$

46. $\displaystyle\text{Area} = \int_0^2 \frac{x}{x^2 + 16}\,dx$ Let $u = x^2 + 16$
$\displaystyle \qquad\qquad\qquad\qquad\qquad du = 2x\,dx$

$\displaystyle = \frac{1}{2}\int_{16}^{20} \frac{du}{u} = \left.\frac{1}{2}\ln u\right|_{16}^{20}$

$\displaystyle = \frac{1}{2}(\ln 20 - \ln 16) = \frac{1}{2}\ln\left(\frac{5}{4}\right)$ sq. units.

48. The area bounded by the ellipse $(x^2/a^2) + (y^2/b^2) = 1$ is

$\displaystyle 4\int_0^a b\sqrt{1 - \frac{x^2}{a^2}}\,dx$ Let $x = au$
$\displaystyle \qquad\qquad\qquad\qquad dx = a\,du$

$\displaystyle = 4ab\int_0^1 \sqrt{1 - u^2}\,du.$

The integral is the area of a quarter circle of radius 1. Hence

$$\text{Area} = 4ab\left(\frac{\pi(1)^2}{4}\right) = \pi ab \text{ sq. units.}$$

99

50. We have

$$\int \cos ax \cos bx \, dx$$

$$= \frac{1}{2} \int [\cos(ax - bx) + \cos(ax + bx)] \, dx$$

$$= \frac{1}{2} \int \cos[(a - b)x] \, dx + \frac{1}{2} \int \cos[(a + b)x] \, dx$$

Let $u = (a - b)x$, $du = (a - b) \, dx$ in the first integral;
let $v = (a + b)x$, $dv = (a + b) \, dx$ in the second integral.

$$= \frac{1}{2(a - b)} \int \cos u \, du + \frac{1}{2(a + b)} \int \cos v \, dv$$

$$= \frac{1}{2} \left[\frac{\sin[(a - b)x]}{(a - b)} + \frac{\sin[(a + b)x]}{(a + b)} \right] + C.$$

$$\int \sin ax \sin bx \, dx$$

$$= \frac{1}{2} \int [\cos(ax - bx) - \cos(ax + bx)] \, dx$$

$$= \frac{1}{2} \left[\frac{\sin[(a - b)x]}{(a - b)} - \frac{\sin[(a + b)x]}{(a + b)} \right] + C.$$

$$\int \sin ax \cos bx \, dx$$

$$= \frac{1}{2} \int [\sin(ax + bx) + \sin(ax - bx)] \, dx$$

$$= \frac{1}{2} \left[\int \sin[(a + b)x] \, dx + \int \sin[(a - b)x] \, dx \right]$$

$$= -\frac{1}{2} \left[\frac{\cos[(a + b)x]}{(a + b)} + \frac{\cos[(a - b)x]}{(a - b)} \right] + C.$$

52. If $1 \le m \le k$, we have

$$\int_{-\pi}^{\pi} f(x) \cos mx \, dx = \frac{a_0}{2} \int_{-\pi}^{\pi} \cos mx \, dx$$

$$+ \sum_{n=1}^{k} a_n \int_{-\pi}^{\pi} \cos nx \cos mx \, dx$$

$$+ \sum_{n=1}^{k} b_n \int_{-\pi}^{\pi} \sin nx \cos mx \, dx.$$

By the previous exercise, all the integrals on the right side are zero except the one in the first sum having $n = m$. Thus the whole right side reduces to

$$a_m \int_{-\pi}^{\pi} \cos^2(mx) \, dx = a_m \int_{-\pi}^{\pi} \frac{1 + \cos(2mx)}{2} \, dx$$

$$= \frac{a_m}{2}(2\pi + 0) = \pi a_m.$$

Thus

$$a_m = \frac{1}{\pi} \int_{-\pi}^{\pi} f(x) \cos mx \, dx.$$

A similar argument shows that

$$b_m = \frac{1}{\pi} \int_{-\pi}^{\pi} f(x) \sin mx \, dx.$$

For $m = 0$ we have

$$\int_{-\pi}^{\pi} f(x) \cos mx \, dx = \int_{-\pi}^{\pi} f(x) \, dx$$

$$= \frac{a_0}{2} \int_{-\pi}^{\pi} dx$$

$$+ \sum_{n=1}^{k} (a_n \cos(nx) + b_n \sin(nx)) \, dx$$

$$= \frac{a_0}{2}(2\pi) + 0 + 0 = a_0 \pi,$$

so the formula for a_m holds for $m = 0$ also.

Section 5.7   Areas of Plane Regions (page 346)

2. Area of $R = \displaystyle\int_0^1 (\sqrt{x} - x^2) \, dx$

$$= \left(\frac{2}{3} x^{3/2} - \frac{1}{3} x^3 \right) \Big|_0^1 = \frac{2}{3} - \frac{1}{3} = \frac{1}{3} \text{ sq. units.}$$

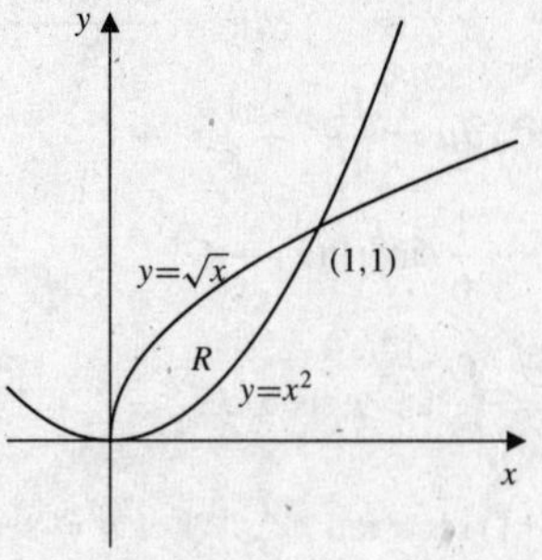

Fig. 5.7.2

4. For intersections:
$$x^2 - 2x = 6x - x^2 \Rightarrow 2x^2 - 8x = 0$$
i.e., $x = 0$ or 4.

Area of $R = \displaystyle\int_0^4 \left[6x - x^2 - (x^2 - 2x) \right] dx$

$$= \int_0^4 (8x - 2x^2) \, dx$$

$$= \left(4x^2 - \frac{2}{3} x^3 \right) \Big|_0^4 = \frac{64}{3} \text{ sq. units.}$$

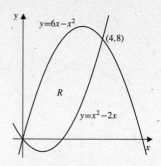

Fig. 5.7.4

6. For intersections:

$7 + y = 2y^2 - y + 3 \Rightarrow 2y^2 - 2y - 4 = 0$

$2(y - 2)(y + 1) = 0 \Rightarrow$ i.e., $y = -1$ or 2.

Area of $R = \int_{-1}^{2} [(7 + y) - (2y^2 - y + 3)]\, dy$

$= 2 \int_{-1}^{2} (2 + y - y^2)\, dy$

$= 2\left(2y + \frac{1}{2}y^2 - \frac{1}{3}y^3\right)\bigg|_{-1}^{2} = 9$ sq. units.

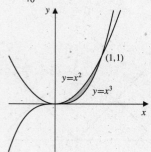

Fig. 5.7.6

8. Shaded area $= \int_{0}^{1} (x^2 - x^3)\, dx$

$= \left(\frac{1}{3}x^3 - \frac{1}{4}x^4\right)\bigg|_{0}^{1} = \frac{1}{12}$ sq. units.

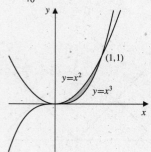

Fig. 5.7.8

10. For intersections:

$y^2 = 2y^2 - y - 2 \Rightarrow y^2 - y - 2 = 0$

$(y - 2)(y + 1) = 0 \Rightarrow$ i.e., $y = -1$ or 2.

Area of $R = \int_{-1}^{2} [y^2 - (2y^2 - y - 2)]\, dy$

$= \int_{-1}^{2} [2 + y - y^2]\, dy = \left(2y + \frac{1}{2}y^2 - \frac{1}{3}y^3\right)\bigg|_{-1}^{2}$

$= \frac{9}{2}$ sq. units.

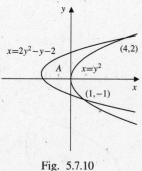

Fig. 5.7.10

12. Area of shaded region $= 2 \int_{0}^{1} [(1 - x^2) - (x^2 - 1)^2]\, dx$

$= 2 \int_{0}^{1} (x^2 - x^4)\, dx = 2\left(\frac{1}{3}x^3 - \frac{1}{5}x^5\right)\bigg|_{0}^{1} = \frac{4}{15}$ sq. units.

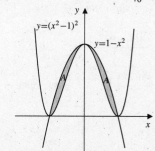

Fig. 5.7.12

14. For intersections:

$\frac{4x}{3 + x^2} = 1 \Rightarrow x^2 - 4x + 3 = 0$

i.e., $x = 1$ or 3.

Shaded area $= \int_{1}^{3} \left[\frac{4x}{3 + x^2} - 1\right] dx$

$= [2\ln(3 + x^2) - x]\bigg|_{1}^{3} = 2\ln 3 - 2$ sq. units.

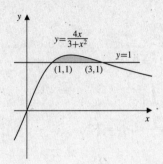

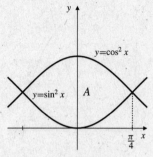

Fig. 5.7.14

Fig. 5.7.20

16. Area $A = \displaystyle\int_{-\pi}^{\pi} (\sin y - (y^2 - \pi^2))\, dy$

$= \left(-\cos y + \pi^2 y - \dfrac{y^3}{3} \right)\Big|_{-\pi}^{\pi} = \dfrac{4\pi^3}{3}$ sq. units.

22. For intersections: $x^{1/3} = \tan(\pi x/4)$. Thus $x = \pm 1$.

Area $A = 2\displaystyle\int_0^1 \left(x^{1/3} - \tan\dfrac{\pi x}{4} \right) dx$

$= 2\left(\dfrac{3}{4}x^{4/3} - \dfrac{4}{\pi}\ln\left|\sec\dfrac{\pi x}{4}\right| \right)\Big|_0^1$

$= \dfrac{3}{2} - \dfrac{8}{\pi}\ln\sqrt{2} = \dfrac{3}{2} - \dfrac{4}{\pi}\ln 2$ sq. units.

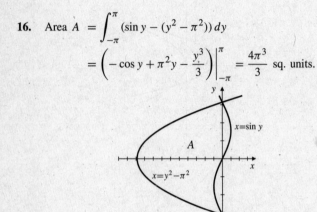

Fig. 5.7.16

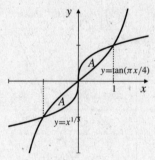

Fig. 5.7.22

18. Area $= \displaystyle\int_{-\pi/2}^{\pi/2} (1 - \sin^2 x)\, dx$

$= 2\displaystyle\int_0^{\pi/2} \dfrac{1 + \cos(2x)}{2}\, dx$

$= \left(x + \dfrac{\sin(2x)}{2} \right)\Big|_0^{\pi/2} = \dfrac{\pi}{2}$ sq. units.

24. For intersections: $|x| = \sqrt{2}\cos(\pi x/4)$. Thus $x = \pm 1$.

Area $A = 2\displaystyle\int_0^1 \left(\sqrt{2}\cos\dfrac{\pi x}{4} - x \right) dx$

$= \left(\dfrac{8\sqrt{2}}{\pi}\sin\dfrac{\pi x}{4} - x^2 \right)\Big|_0^1$

$= \dfrac{8}{\pi} - 1$ sq. units.

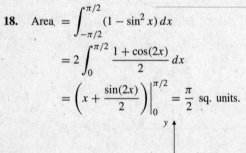

Fig. 5.7.18

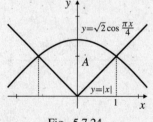

Fig. 5.7.24

20. Area $A = 2\displaystyle\int_0^{\pi/4} (\cos^2 x - \sin^2 x)\, dx$

$= 2\displaystyle\int_0^{\pi/4} \cos(2x)\, dx = \sin(2x)\Big|_0^{\pi/4} = 1$ sq. units.

26. For intersections: $e^x = x + 2$. There are two roots, both of which must be found numerically. We used a TI-85 solve routine to get $x_1 \approx -1.841406$ and $x_2 \approx 1.146193$. Thus

$$\text{Area } A = \int_{x_1}^{x_2} \left(x + 2 - e^x \right) dx$$

$$= \left(\frac{x^2}{2} + 2x - e^x \right) \Bigg|_{x_1}^{x_2}$$

$$\approx 1.949091 \text{ sq. units.}$$

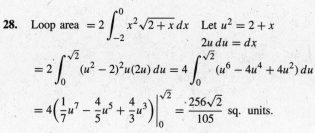

Fig. 5.7.26

28. Loop area $= 2 \displaystyle\int_{-2}^{0} x^2 \sqrt{2 + x}\, dx$ Let $u^2 = 2 + x$

$$2u\, du = dx$$

$$= 2 \int_{0}^{\sqrt{2}} (u^2 - 2)^2 u(2u)\, du = 4 \int_{0}^{\sqrt{2}} (u^6 - 4u^4 + 4u^2)\, du$$

$$= 4 \left(\frac{1}{7}u^7 - \frac{4}{5}u^5 + \frac{4}{3}u^3 \right) \Bigg|_0^{\sqrt{2}} = \frac{256\sqrt{2}}{105} \text{ sq. units.}$$

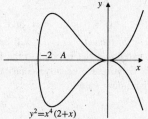

Fig. 5.7.28

30. The tangent line to $y = x^3$ at $(1, 1)$ is $y - 1 = 3(x - 1)$, or $y = 3x - 2$. The intersections of $y = x^3$ and this tangent line occur where $x^3 - 3x + 2 = 0$. Of course $x = 1$ is a (double) root of this cubic equation, which therefore factors to $(x - 1)^2(x + 2) = 0$. The other intersection is at $x = -2$. Thus

$$\text{Area of } R = \int_{-2}^{1} (x^3 - 3x + 2)\, dx$$

$$= \left(\frac{x^4}{4} - \frac{3x^2}{2} + 2x \right) \Bigg|_{-2}^{1}$$

$$= -\frac{15}{4} - \frac{3}{2} + 6 + 2 + 4 = \frac{27}{4} \text{ sq. units.}$$

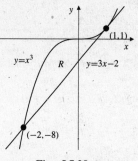

Fig. 5.7.30

Review Exercises 5   (page 347)

2. The number of balls is

$$40 \times 30 + 39 \times 29 + \cdots + 12 \times 2 + 11 \times 1$$

$$= \sum_{i=1}^{30} i(i + 10) = \frac{(30)(31)(61)}{6} + 10 \frac{(30)(31)}{2} = 14{,}105.$$

4. $R_n = \sum_{i=1}^{n} (1/n)\sqrt{1 + (i/n)}$ is a Riemann sum for $f(x) = \sqrt{1 + x}$ on the interval $[0, 1]$. Thus

$$\lim_{n \to \infty} R_n = \int_0^1 \sqrt{1 + x}\, dx$$

$$= \frac{2}{3}(1 + x)^{3/2} \Bigg|_0^1 = \frac{4\sqrt{2} - 2}{3}.$$

6. $\displaystyle\int_0^{\sqrt{5}} \sqrt{5 - x^2}\, dx = 1/4$ of the area of a circle of radius $\sqrt{5}$

$$= \frac{1}{4}\pi(\sqrt{5})^2 = \frac{5\pi}{4}$$

8. $\int_0^\pi \cos x\, dx = \text{area } A_1 - \text{area } A_2 = 0$

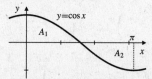

Fig. R-5.8

10. $\bar{h} = \dfrac{1}{3}\displaystyle\int_0^3 |x - 2|\, dx = \dfrac{1}{3}\dfrac{5}{2} = \dfrac{5}{6}$ (via #9)

12. $f(x) = \displaystyle\int_{-13}^{\sin x} \sqrt{1 + t^2}\, dt,$ $f'(x) = \sqrt{1 + \sin^2 x}\,(\cos x)$

14. $g(\theta) = \displaystyle\int_{e^{\sin \theta}}^{e^{\cos \theta}} \ln x\, dx$

$$g'(\theta) = (\ln(e^{\cos \theta}))e^{\cos \theta}(-\sin \theta) - (\ln(e^{\sin \theta}))e^{\sin \theta}\cos \theta$$

$$= -\sin \theta \cos \theta (e^{\cos \theta} + e^{\sin \theta})$$

16. $I = \displaystyle\int_0^\pi x f(\sin x)\, dx$ Let $x = \pi - u$
$$dx = -du$$

$$= -\int_\pi^0 (\pi - u) f(\sin(\pi - u))\, du \quad (\text{but } \sin(\pi - u) = \sin u)$$

$$= \pi \int_0^\pi f(\sin u)\, du - \int_0^\pi u f(\sin u)\, du$$

$$= \pi \int_0^\pi f(\sin x)\, dx - I.$$

Now, solving for I, we get

$$\int_0^\pi x f(\sin x)\, dx = I = \frac{\pi}{2}\int_0^\pi f(\sin x)\, dx.$$

18. The area bounded by $y = (x-1)^2$, $y = 0$, and $x = 0$ is

$$\int_0^1 (x-1)^2\, dx = \left.\frac{(x-1)^3}{3}\right|_0^1 = \frac{1}{3} \text{ sq. units..}$$

20. $y = 4x - x^2$ and $y = 3$ meet where $x^2 - 4x + 3 = 0$, that is, at $x = 1$ and $x = 3$. Since $4x - x^2 \geq 3$ on $[1, 3]$, the required area is

$$\int_1^3 (4x - x^2 - 3)\, dx = \left(2x^2 - \frac{x^3}{3} - 3x\right)\Big|_1^3 = \frac{4}{3} \text{ sq. units.}$$

22. $y = 5 - x^2$ and $y = 4/x^2$ meet where $5 - x^2 = 4/x^2$, that is, where
$$x^4 - 5x^2 + 4 = 0,$$
$$(x^2 - 1)(x^2 - 4) = 0.$$

There are four intersections: $x = \pm 1$ and $x = \pm 2$. By symmetry (see the figure) the total area bounded by the curves is

$$2\int_1^2 \left(5 - x^2 - \frac{4}{x^2}\right) dx = 2\left(5x - \frac{x^3}{3} + \frac{4}{x}\right)\Big|_1^2 = \frac{4}{3} \text{ sq. units.}$$

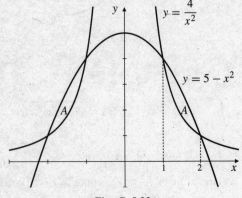

Fig. R-5.22

24. $\displaystyle\int_1^e \frac{\ln x}{x}\, dx$ Let $u = \ln x$
$$du = dx/x$$

$$= \int_0^1 u\, du = \left.\frac{u^2}{2}\right|_0^1 = \frac{1}{2}$$

26. $\displaystyle\int \sin^3(\pi x)\, dx$

$$= \int \sin(\pi x)\left(1 - \cos^2(\pi x)\right) dx \quad \text{Let } u = \cos(\pi x)$$
$$du = -\pi \sin(\pi x)\, dx$$

$$= -\frac{1}{\pi}\int (1 - u^2)\, du$$

$$= \frac{1}{\pi}\left(\frac{u^3}{3} - u\right) + C = \frac{1}{3\pi}\cos^3(\pi x) - \frac{1}{\pi}\cos(\pi x) + C$$

28. $\displaystyle\int_1^{\sqrt[4]{e}} \frac{\tan^2(\pi \ln x)}{x}\, dx$ Let $u = \pi \ln x$
$$du = (\pi/x)\, dx$$

$$= \frac{1}{\pi}\int_0^{\pi/4} \tan^2 u\, du = \frac{1}{\pi}\int_0^{\pi/4} (\sec^2 u - 1)\, du$$

$$= \frac{1}{\pi}(\tan u - u)\Big|_0^{\pi/4} = \frac{1}{\pi} - \frac{1}{4}$$

30. $\displaystyle\int \cos^2\frac{t}{5}\sin^2\frac{t}{5}\, dt = \frac{1}{4}\int \sin^2\frac{2t}{5}\, dt$

$$= \frac{1}{8}\int \left(1 - \cos\frac{4t}{5}\right) dt$$

$$= \frac{1}{8}\left(t - \frac{5}{4}\sin\frac{4t}{5}\right) + C$$

32. $f(x) = 4x - x^2 \geq 0$ if $0 \leq x \leq 4$, and $f(x) < 0$ otherwise. If $a < b$, then $\int_a^b f(x)\, dx$ will be maximum if $[a, b] = [0, 4]$; extending the interval to the left of 0 or to the right of 4 will introduce negative contributions to the integral. The maximum value is

$$\int_0^4 (4x - x^2)\, dx = \left(2x^2 - \frac{x^3}{3}\right)\Big|_0^4 = \frac{32}{3}.$$

34. If $y(t)$ is the distance the object falls in t seconds from its release time, then

$$y''(t) = g, \quad y(0) = 0, \quad \text{and } y'(0) = 0.$$

Antidifferentiating twice and using the initial conditions leads to

$$y(t) = \frac{1}{2}gt^2.$$

The average height during the time interval $[0, T]$ is

$$\frac{1}{T}\int_0^T \frac{1}{2}gt^2\, dt = \frac{g}{2T}\frac{T^3}{3} = \frac{gT^2}{6} = y\left(\frac{T}{\sqrt{3}}\right).$$

Challenging Problems 5   (page 347)

2.   a) $\cos\left((j+\tfrac{1}{2})t\right) - \cos\left((j-\tfrac{1}{2})t\right)$

$= \cos(jt)\cos(\tfrac{1}{2}t) - \sin(jt)\sin(\tfrac{1}{2}t)$

$\qquad - \cos(jt)\cos(\tfrac{1}{2}t) - \sin(jt)\sin(\tfrac{1}{2}t)$

$= -2\sin(jt)\sin(\tfrac{1}{2}t).$

Therefore, we obtain a telescoping sum:

$$\sum_{j=1}^{n} \sin(jt)$$

$$= -\frac{1}{2\sin(\tfrac{1}{2}t)} \sum_{j=1}^{n} \left[\cos\left((j+\tfrac{1}{2})t\right) - \cos\left((j-\tfrac{1}{2})t\right)\right]$$

$$= -\frac{1}{2\sin(\tfrac{1}{2}t)} \left[\cos\left((n+\tfrac{1}{2})t\right) - \cos(\tfrac{1}{2}t)\right]$$

$$= \frac{1}{2\sin(\tfrac{1}{2}t)} \left[\cos(\tfrac{1}{2}t) - \cos\left((n+\tfrac{1}{2})t\right)\right].$$

b)   Let $P_n = \{0, \frac{\pi}{2n}, \frac{2\pi}{2n}, \frac{3\pi}{2n}, \ldots \frac{n\pi}{2n}\}$ be the partition of $[0, \pi/2]$ into n subintervals of equal length $\Delta x = \pi/2n$. Using $t = \pi/2n$ in the formula obtained in part (a), we get

$$\int_0^{\pi/2} \sin x \, dx$$

$$= \lim_{n\to\infty} \sum_{j=1}^{n} \sin\left(\frac{j\pi}{2n}\right)\frac{\pi}{2n}$$

$$= \lim_{n\to\infty} \frac{\pi}{2n} \frac{1}{2\sin(\pi/(4n))}\left(\cos\frac{\pi}{4n} - \cos\frac{(2n+1)\pi}{4n}\right)$$

$$= \lim_{n\to\infty} \frac{\pi/(4n)}{\sin(\pi/(4n))} \lim_{n\to\infty}\left(\cos\frac{\pi}{4n} - \cos\frac{(2n+1)\pi}{4n}\right)$$

$$= 1 \times \left(\cos 0 - \cos\frac{\pi}{2}\right) = 1.$$

4. $f(x) = 1/x^2$, $1 = x_0 < x_1 < x_2 < \cdots < x_n = 2$. If $c_i = \sqrt{x_{i-1}x_i}$, then

$$x_{i-1}^2 < x_{i-1}x_i = c_i^2 < x_i^2,$$

so $x_{i-1} < c_i < x_i$. We have

$$\sum_{i=1}^{n} f(c_i)\,\Delta x_i = \sum_{i=1}^{n} \frac{1}{x_{i-1}x_i}(x_i - x_{i-1})$$

$$= \sum_{i=1}^{n}\left(\frac{1}{x_{i-1}} - \frac{1}{x_i}\right) \quad \text{(telescoping)}$$

$$= \frac{1}{x_0} - \frac{1}{x_n} = 1 - \frac{1}{2} = \frac{1}{2}.$$

Thus $\displaystyle\int_1^2 \frac{dx}{x^2} = \lim_{n\to\infty}\sum_{i=1}^{n} f(c_i)\,\Delta x_i = \frac{1}{2}.$

6.   Let $f(x) = ax^3 + bx^2 + cx + d$. We used Maple to calculate the following:

The tangent to $y = f(x)$ at $P = (p, f(p))$ has equation

$$y = g(x) = ap^3 + bp^2 + cp + d + (3ap^2 + 2bp + c)(x - p).$$

This line intersects $y = f(x)$ at $x = p$ (double root) and at $x = q$, where

$$q = -\frac{2ap + b}{a}.$$

Similarly, the tangent to $y = f(x)$ at $x = q$ has equation

$$y = h(x) = aq^3 + bq^2 + cq + d + (3aq^2 + 2bq + c)(x - q),$$

and intersects $y = f(x)$ at $x = q$ (double root) and $x = r$, where

$$r = -\frac{2aq + b}{a} = \frac{4ap + b}{a}.$$

The area between $y = f(x)$ and the tangent line at P is the absolute value of

$$\int_p^q (f(x) - g(x))\,dx$$

$$= -\frac{1}{12}\left(\frac{81a^4p^4 + 108a^3bp^3 + 54a^2b^2p^2 + 12ab^3p + b^4}{a^3}\right).$$

The area between $y = f(x)$ and the tangent line at $Q = (q, f(q))$ is the absolute value of

$$\int_q^r (f(x) - h(x))\,dx$$

$$= -\frac{4}{3}\left(\frac{81a^4p^4 + 108a^3bp^3 + 54a^2b^2p^2 + 12ab^3p + b^4}{a^3}\right),$$

which is 16 times the area between $y = f(x)$ and the tangent at P.

8.   Let $f(x) = ax^4 + bx^3 + cx^2 + dx + e$. The tangent to $y = f(x)$ at $P = (p, f(p))$ has equation

$$y = g(x) = ap^4 + bp^3 + cp^2 + dp + e + (4ap^3 + 3bp^2 + 2cp + d)(x - p),$$

and intersects $y = f(x)$ at $x = p$ (double root) and at the two points

$$x = \frac{-2ap - b \pm \sqrt{b^2 - 4ac - 4abp - 8a^2p^2}}{2a}.$$

If these latter two points coincide, then the tangent is a "double tangent." This happens if

$$8a^2p^2 + 4abp + 4ac - b^2 = 0,$$

which has two solutions, which we take to be p and q:

$$p = \frac{-b + \sqrt{3b^2 - 8ac}}{4a}$$

$$q = \frac{-b - \sqrt{3b^2 - 8ac}}{4a} = -p - \frac{b}{2a}.$$

(Both roots exist and are distinct provided $3b^2 > 8ac$.)
The point T corresponds to $x = t = (p+q)/2 = -b/4a$.
The tangent to $y = f(x)$ at $x = t$ has equation

$$y = h(x) = -\frac{3b^4}{256a^3} + \frac{b^2 c}{16a^2} - \frac{bd}{4a} + e + \left(\frac{b^3}{8a^2} - \frac{bc}{2a} + d \right) \left(x + \frac{b}{4a} \right)$$

and it intersects $y = f(x)$ at the points U and V with x-coordinates

$$u = \frac{-b - \sqrt{2}\sqrt{3b^2 - 8ac}}{4a},$$

$$v = \frac{-b + \sqrt{2}\sqrt{3b^2 - 8ac}}{4a}.$$

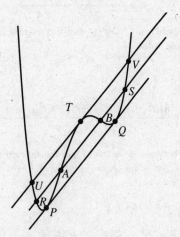

Fig. C-5.8

a) The areas between the curve $y = f(x)$ and the lines PQ and UV are, respectively, the absolute values of

$$A_1 = \int_p^q (f(x) - g(x))\,dx \quad \text{and} \quad A_2 = \int_u^v (h(x) - f(x))\,dx.$$

Maple calculates these two integrals and simplifies the ratio A_1/A_2 to be $1/\sqrt{2}$.

b) The two inflection points A and B of f have x-coordinates shown by Maple to be

$$\alpha = \frac{-3b - \sqrt{3(3b^2 - 8ac)}}{12a} \quad \text{and}$$

$$\beta = \frac{-3b + \sqrt{3(3b^2 - 8ac)}}{12a}.$$

It then determines the four points of intersection of the line $y = k(x)$ through these inflection points and the curve. The other two points have x-coordinates

$$r = \frac{-3b - \sqrt{15(3b^2 - 8ac)}}{12a} \quad \text{and}$$

$$s = \frac{-3b + \sqrt{15(3b^2 - 8ac)}}{12a}.$$

The region bounded by RS and the curve $y = f(x)$ is divided into three parts by A and B. The areas of these three regions are the absolute values of

$$A_1 = \int_r^\alpha (k(x) - f(x))\,dx$$

$$A_2 = \int_\alpha^\beta (f(x) - k(x))\,dx$$

$$A_3 = \int_\beta^s (k(x) - f(x))\,dx.$$

The expressions calculated by Maple for $k(x)$ and for these three areas are very complicated, but Maple simplifies the rations A_3/A_1 and A_2/A_1 to 1 and 2 respectively, as was to be shown.

CHAPTER 6. TECHNIQUES OF INTE-GRATION

Section 6.1 Integration by Parts (page 355)

2. $\displaystyle\int (x+3)e^{2x}\,dx$

$$U = x+3 \quad dV = e^{2x}\,dx$$
$$dU = dx \qquad V = \tfrac{1}{2}e^{2x}$$

$$= \frac{1}{2}(x+3)e^{2x} - \frac{1}{2}\int e^{2x}\,dx$$

$$= \frac{1}{2}(x+3)e^{2x} - \frac{1}{4}e^{2x} + C.$$

4. $\displaystyle\int (x^2 - 2x)e^{kx}\,dx$

$$U = x^2 - 2x \quad dV = e^{kx}$$
$$dU = (2x-2)\,dx \quad V = \frac{1}{k}e^{kx}$$

$$= \frac{1}{k}(x^2-2x)e^{kx} - \frac{1}{k}\int (2x-2)e^{kx}\,dx$$

$$U = x-1 \quad dV = e^{kx}\,dx$$
$$dU = dx \qquad V = \frac{1}{k}e^{kx}$$

$$= \frac{1}{k}(x^2-2x)e^{kx} - \frac{2}{k}\left[\frac{1}{k}(x-1)e^{kx} - \frac{1}{k}\int e^{kx}\,dx\right]$$

$$= \frac{1}{k}(x^2-2x)e^{kx} - \frac{2}{k^2}(x-1)e^{kx} + \frac{2}{k^3}e^{kx} + C.$$

6. $\displaystyle\int x(\ln x)^3\,dx = I_3$ where

$$I_n = \int x(\ln x)^n\,dx$$

$$U = (\ln x)^n \qquad dV = x\,dx$$
$$dU = \frac{n}{x}(\ln x)^{n-1}\,dx \quad V = \frac{1}{2}x^2$$

$$= \frac{1}{2}x^2(\ln x)^n - \frac{n}{2}\int x(\ln x)^{n-1}\,dx$$

$$= \frac{1}{2}x^2(\ln x)^n - \frac{n}{2}I_{n-1}$$

$$I_3 = \frac{1}{2}x^2(\ln x)^3 - \frac{3}{2}I_2$$

$$= \frac{1}{2}x^2(\ln x)^3 - \frac{3}{2}\left[\frac{1}{2}x^2(\ln x)^2 - \frac{2}{2}I_1\right]$$

$$= \frac{1}{2}x^2(\ln x)^3 - \frac{3}{4}x^2(\ln x)^2 + \frac{3}{2}\left[\frac{1}{2}x^2(\ln x) - \frac{1}{2}I_0\right]$$

$$= \frac{1}{2}x^2(\ln x)^3 - \frac{3}{4}x^2(\ln x)^2 + \frac{3}{4}x^2(\ln x) - \frac{3}{4}\int x\,dx$$

$$= \frac{x^2}{2}\left[(\ln x)^3 - \frac{3}{2}(\ln x)^2 + \frac{3}{2}(\ln x) - \frac{3}{4}\right] + C.$$

8. $\displaystyle\int x^2 \tan^{-1} x\,dx$

$$U = \tan^{-1} x \quad dV = x^2\,dx$$
$$dU = \frac{dx}{1+x^2} \quad V = \frac{x^3}{3}$$

$$= \frac{x^3}{3}\tan^{-1}x - \frac{1}{3}\int \frac{x^3}{1+x^2}\,dx$$

$$= \frac{x^3}{3}\tan^{-1}x - \frac{1}{3}\int \left(x - \frac{x}{1+x^2}\right)dx$$

$$= \frac{x^3}{3}\tan^{-1}x - \frac{x^2}{6} + \frac{1}{6}\ln(1+x^2) + C.$$

10. $\displaystyle\int x^5 e^{-x^2}\,dx = I_2$ where

$$I_n = \int x^{(2n+1)}e^{-x^2}\,dx$$

$$U = x^{2n} \qquad dV = xe^{-x^2}\,dx$$
$$dU = 2nx^{(2n-1)}\,dx \quad V = -\tfrac{1}{2}e^{-x^2}$$

$$= -\frac{1}{2}x^{2n}e^{-x^2} + n\int x^{(2n-1)}e^{-x^2}\,dx$$

$$= -\frac{1}{2}x^{2n}e^{-x^2} + nI_{n-1}$$

$$I_2 = -\frac{1}{2}x^4 e^{-x^2} + 2\left[-\frac{1}{2}x^2 e^{-x^2} + \int xe^{-x^2}\,dx\right]$$

$$= -\frac{1}{2}e^{-x^2}(x^4 + 2x^2 + 2) + C.$$

12. $\displaystyle I = \int \tan^2 x \sec x\,dx$

$$U = \tan x \qquad dV = \sec x \tan x\,dx$$
$$dU = \sec^2 x\,dx \quad V = \sec x$$

$$= \sec x \tan x - \int \sec^3 x\,dx$$

$$= \sec x \tan x - \int (1 + \tan^2 x)\sec x\,dx$$

$$= \sec x \tan x - \ln|\sec x + \tan x| - I$$

Thus, $I = \frac{1}{2}\sec x \tan x - \frac{1}{2}\ln|\sec x + \tan x| + C.$

14. $\displaystyle I = \int xe^{\sqrt{x}}\,dx$ Let $x = w^2$
$$dx = 2w\,dw$$

$$= 2\int w^3 e^w\,dw = 2I_3 \quad \text{where}$$

$$I_n = \int w^n e^w\,dw$$

$$U = w^n \qquad dV = e^w\,dw$$
$$dU = nw^{n-1}\,dw \quad V = e^w$$

$$= w^n e^w - nI_{n-1}.$$

$$I = 2I_3 = 2w^3 e^w - 6[w^2 e^w - 2(we^w - I_0)]$$

$$= e^{\sqrt{x}}(2x\sqrt{x} - 6x + 12\sqrt{x} - 12) + C.$$

16. $\displaystyle\int_0^1 \sqrt{x}\,\sin(\pi\sqrt{x})\,dx$ Let $x = w^2$

$$dx = 2w\,dw$$

$$= 2\int_0^1 w^2\sin(\pi w)\,dw$$

$$U = w^2 \qquad dV = \sin(\pi w)\,dw$$

$$dU = 2w\,dw \qquad V = -\frac{\cos(\pi w)}{\pi}$$

$$= -\frac{2}{\pi}w^2\cos(\pi w)\Big|_0^1 + \frac{4}{\pi}\int_0^1 w\cos(\pi w)\,dw$$

$$U = w \qquad dV = \cos(\pi w)\,dw$$

$$dU = dw \qquad V = \frac{\sin(\pi w)}{\pi}$$

$$= \frac{2}{\pi} + \frac{4}{\pi}\left[\frac{w}{\pi}\sin(\pi w)\right]\Big|_0^1 - \frac{4}{\pi^2}\int_0^1 \sin(\pi w)\,dw$$

$$= \frac{2}{\pi} + \frac{4}{\pi^3}\cos(\pi w)\Big|_0^1 = \frac{2}{\pi} + \frac{4}{\pi^3}(-2) = \frac{2}{\pi} - \frac{8}{\pi^3}.$$

18. $\displaystyle\int x\sin^2 x\,dx = \frac{1}{2}\int(x - x\cos 2x)\,dx$

$$= \frac{x^2}{4} - \frac{1}{2}\int x\cos 2x\,dx$$

$$U = x \qquad dV = \cos 2x\,dx$$

$$dU = dx \qquad V = \tfrac{1}{2}\sin 2x$$

$$= \frac{x^2}{4} - \frac{1}{2}\left[\frac{1}{2}x\sin 2x - \frac{1}{2}\int\sin 2x\,dx\right]$$

$$= \frac{x^2}{4} - \frac{x}{4}\sin 2x - \frac{1}{8}\cos 2x + C.$$

20. $\displaystyle I = \int_1^e \sin(\ln x)\,dx$

$$U = \sin(\ln x) \qquad dV = dx$$

$$dU = \frac{\cos(\ln x)}{x}\,dx \qquad V = x$$

$$= x\sin(\ln x)\Big|_1^e - \int_1^e \cos(\ln x)\,dx$$

$$U = \cos(\ln x) \qquad dV = dx$$

$$dU = -\frac{\sin(\ln x)}{x}\,dx \qquad V = x$$

$$= e\sin(1) - \left[x\cos(\ln x)\Big|_1^e + I\right]$$

Thus, $I = \dfrac{1}{2}[e\sin(1) - e\cos(1) + 1].$

22. $\displaystyle\int_0^4 \sqrt{x}\,e^{\sqrt{x}}\,dx$ Let $x = w^2$

$$dx = 2w\,dw$$

$$= 2\int_0^2 w^2 e^w\,dw = 2I_2$$

See solution #16 for the formula

$$I_n = \int w^n e^w\,dw = w^n e^w - nI_{n-1}.$$

$$= 2\left(w^2 e^w\Big|_0^2 - 2I_1\right) = 8e^2 - 4\left(we^w\Big|_0^2 - I_0\right)$$

$$= 8e^2 - 8e^2 + 4\int_0^2 e^w\,dw = 4(e^2 - 1).$$

24. $\displaystyle\int x\sec^{-1}x\,dx$

$$U = \sec^{-1}x \qquad dV = x\,dx$$

$$dU = \frac{dx}{|x|\sqrt{x^2 - 1}} \qquad V = \frac{1}{2}x^2$$

$$= \frac{1}{2}x^2\sec^{-1}x - \frac{1}{2}\int\frac{|x|}{\sqrt{x^2 - 1}}\,dx$$

$$= \frac{1}{2}x^2\sec^{-1}x - \frac{1}{2}\operatorname{sgn}(x)\sqrt{x^2 - 1} + C.$$

26. $\displaystyle\int(\sin^{-1}x)^2\,dx$ Let $x = \sin\theta$

$$dx = \cos\theta\,d\theta$$

$$= \int\theta^2\cos\theta\,d\theta$$

$$U = \theta^2 \qquad dV = \cos\theta\,d\theta$$

$$dU = 2\theta\,d\theta \qquad V = \sin\theta$$

$$= \theta^2\sin\theta - 2\int\theta\sin\theta\,d\theta$$

$$U = \theta \qquad dV = \sin\theta\,d\theta$$

$$dU = d\theta \qquad V = -\cos\theta$$

$$= \theta^2\sin\theta - 2\left(-\theta\cos\theta + \int\cos\theta\,d\theta\right)$$

$$= \theta^2\sin\theta + 2\theta\cos\theta - 2\sin\theta + C$$

$$= x(\sin^{-1}x)^2 + 2\sqrt{1 - x^2}(\sin^{-1}x) - 2x + C.$$

28. By the procedure used in Example 4 of Section 7.1,

$$\int e^x\cos x\,dx = \tfrac{1}{2}e^x(\sin x + \cos x) + C;$$

$$\int e^x\sin x\,dx = \tfrac{1}{2}e^x(\sin x - \cos x) + C.$$

Now

$$\int xe^x\cos x\,dx$$

$$U = x \qquad dV = e^x\cos x\,dx$$

$$dU = dx \qquad V = \tfrac{1}{2}e^x(\sin x + \cos x)$$

$$= \tfrac{1}{2}xe^x(\sin + \cos x) - \tfrac{1}{2}\int e^x(\sin x + \cos x)\,dx$$

$$= \tfrac{1}{2}xe^x(\sin + \cos x)$$
$$- \tfrac{1}{4}e^x(\sin x - \cos x + \sin x + \cos x) + C$$
$$= \tfrac{1}{2}xe^x(\sin x + \cos x) - \tfrac{1}{2}e^x \sin x + C.$$

30. The tangent line to $y = \ln x$ at $x = 1$ is $y = x - 1$, Hence,

$$\text{Shaded area} = \frac{1}{2}(1)(1) + (1)(e-2) - \int_1^e \ln x\, dx$$

$$= e - \frac{3}{2} - (x \ln x - x)\Big|_1^e$$

$$= e - \frac{3}{2} - e + e + 0 - 1 = e - \frac{5}{2} \text{ sq. units.}$$

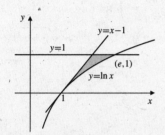

Fig. 6.1.30

32. $I_n = \displaystyle\int_0^{\pi/2} x^n \sin x\, dx$

$$U = x^n \qquad\qquad dV = \sin x\, dx$$
$$dU = nx^{n-1}\, dx \qquad V = -\cos x$$

$$= -x^n \cos x\Big|_0^{\pi/2} + n \int_0^{\pi/2} x^{n-1} \cos x\, dx$$

$$U = x^{n-1} \qquad\qquad dV = \cos x\, dx$$
$$dU = (n-1)x^{n-2}\, dx \qquad V = \sin x$$

$$= n\left[x^{n-1} \sin x\Big|_0^{\pi/2} - (n-1) \int_0^{\pi/2} x^{n-2} \sin x\, dx \right]$$

$$= n\left(\frac{\pi}{2}\right)^{n-1} - n(n-1)I_{n-2}, \qquad (n \ge 2).$$

$$I_0 = \int_0^{\pi/2} \sin x\, dx = -\cos x\Big|_0^{\pi/2} = 1.$$

$$I_6 = 6\left(\frac{\pi}{2}\right)^5 - 6(5)\left\{ 4\left(\frac{\pi}{2}\right)^3 - 4(3)\left[2\left(\frac{\pi}{2}\right) - 2(1)I_0\right] \right\}$$

$$= \frac{3}{16}\pi^5 - 15\pi^3 + 360\pi - 720.$$

34. We have

$$I_n = \int \sec^n x\, dx \qquad (n \ge 3)$$

$$U = \sec^{n-2} x \qquad\qquad dV = \sec^2 x\, dx$$
$$dU = (n-2)\sec^{n-2} x \tan x\, dx \qquad V = \tan x$$

$$= \sec^{n-2} x \tan x - (n-2) \int \sec^{n-2} x \tan^2 x\, dx$$

$$= \sec^{n-2} x \tan x - (n-2) \int \sec^{n-2} x(\sec^2 x - 1)\, dx$$

$$= \sec^{n-2} x \tan x - (n-2)I_n + (n-2)I_{n-2} + C$$

$$I_n = \frac{1}{n-1}(\sec^{n-2} x \tan x) + \frac{n-2}{n-1}I_{n-2} + C.$$

$$I_1 = \int \sec x\, dx = \ln|\sec x + \tan x| + C;$$

$$I_2 = \int \sec^2 x\, dx = \tan x + C.$$

$$I_6 = \frac{1}{5}(\sec^4 x \tan x) + \frac{4}{5}\left(\frac{1}{3}\sec^2 x \tan x + \frac{2}{3}I_2\right) + C$$

$$= \frac{1}{5}\sec^4 x \tan x + \frac{4}{15}\sec^2 x \tan x + \frac{8}{15}\tan x + C.$$

$$I_7 = \frac{1}{6}(\sec^5 x \tan x) + \frac{5}{6}\left[\frac{1}{4}\sec^3 x \tan x + \right.$$

$$\left. \frac{3}{4}\left(\frac{1}{2}\sec x \tan x + \frac{1}{2}I_1\right)\right] + C$$

$$= \frac{1}{6}\sec^5 x \tan x + \frac{5}{24}\sec^3 x \tan x + \frac{15}{48}\sec x \tan x +$$

$$\frac{15}{48}\ln|\sec x + \tan x| + C.$$

36. Given that $f(a) = f(b) = 0$.

$$\int_a^b (x-a)(b-x)f''(x)\, dx$$

$$U = (x-a)(b-x) \qquad dV = f''(x)\, dx$$
$$dU = (b+a-2x)\, dx \qquad V = f'(x)$$

$$= (x-a)(b-x)f'(x)\Big|_a^b - \int_a^b (b+a-2x)f'(x)\, dx$$

$$U = b+a-2x \qquad dV = f'(x)\, dx$$
$$dU = -2dx \qquad\qquad V = f(x)$$

$$= 0 - \left[(b+a-2x)f(x)\Big|_a^b + 2\int_a^b f(x)\, dx\right]$$

$$= -2\int_a^b f(x)\, dx.$$

38. $I_n = \displaystyle\int_0^{\pi/2} \cos^n x\, dx.$

a) For $0 \le x \le \pi/2$ we have $0 \le \cos x \le 1$, and so $0 \le \cos^{2n+2} x \le \cos^{2n+1} x \le \cos^{2n} x$. Therefore $0 \le I_{2n+2} \le I_{2n+1} \le I_{2n}$.

b) Since $I_n = \dfrac{n-1}{n}I_{n-2}$, we have $I_{2n+2} = \dfrac{2n+1}{2n+2}I_{2n}$. Combining this with part (a), we get

$$\frac{2n+1}{2n+2} = \frac{I_{2n+2}}{I_{2n}} \le \frac{I_{2n+1}}{I_{2n}} \le 1.$$

The left side approaches 1 as $n \to \infty$, so, by the Squeeze Theorem,

$$\lim_{n \to \infty} \frac{I_{2n+1}}{I_{2n}} = 1.$$

c) By Example 6 we have, since $2n+1$ is odd and $2n$ is even,

$$I_{2n+1} = \frac{2n}{2n+1} \cdot \frac{2n-2}{2n-1} \cdots \frac{4}{5} \cdot \frac{2}{3}$$

$$I_{2n} = \frac{2n-1}{2n} \cdot \frac{2n-3}{2n-2} \cdots \frac{3}{4} \cdot \frac{1}{2} \cdot \frac{\pi}{2}.$$

Multiplying the expression for I_{2n+1} by $\pi/2$ and dividing by the expression for I_{2n}, we obtain, by part (b),

$$\lim_{n \to \infty} \frac{\frac{2n}{2n+1} \cdot \frac{2n-2}{2n-1} \cdots \frac{4}{5} \cdot \frac{2}{3} \cdot \frac{\pi}{2}}{\frac{2n-1}{2n} \cdot \frac{2n-3}{2n-2} \cdots \frac{3}{4} \cdot \frac{1}{2} \cdot \frac{\pi}{2}} = \frac{\pi}{2} \times 1 = \frac{\pi}{2},$$

or, rearranging the factors on the left,

$$\lim_{n \to \infty} \frac{2}{1} \cdot \frac{2}{3} \cdot \frac{4}{3} \cdot \frac{4}{5} \cdots \frac{2n}{2n-1} \cdot \frac{2n}{2n+1} = \frac{\pi}{2}.$$

Section 6.2 Inverse Substitutions (page 363)

2. $\displaystyle\int \frac{x^2\, dx}{\sqrt{1-4x^2}}$ Let $2x = \sin u$

$\qquad\qquad\qquad\qquad 2\, dx = \cos u\, du$

$\displaystyle = \frac{1}{8} \int \frac{\sin^2 u \cos u\, du}{\cos u}$

$\displaystyle = \frac{1}{16} \int (1 - \cos 2u)\, du = \frac{u}{16} - \frac{\sin 2u}{32} + C$

$\displaystyle = \frac{1}{16} \sin^{-1} 2x - \frac{1}{16} \sin u \cos u + C$

$\displaystyle = \frac{1}{16} \sin^{-1} 2x - \frac{1}{8} x \sqrt{1 - 4x^2} + C.$

4. $\displaystyle\int \frac{dx}{x\sqrt{1-4x^2}}$ Let $x = \frac{1}{2} \sin \theta$

$\qquad\qquad\qquad\qquad dx = \frac{1}{2} \cos \theta\, d\theta$

$\displaystyle = \int \frac{\cos \theta\, d\theta}{\sin \theta \sqrt{1 - \sin^2 \theta}} = \int \csc \theta\, d\theta$

$\displaystyle = \ln|\csc \theta - \cot \theta| + C = \ln\left| \frac{1}{2x} - \frac{\sqrt{1-4x^2}}{2x} \right| + C$

$\displaystyle = \ln\left| \frac{1 - \sqrt{1-4x^2}}{x} \right| + C_1.$

6. $\displaystyle\int \frac{dx}{x\sqrt{9-x^2}}$ Let $x = 3 \sin \theta$

$\qquad\qquad\qquad\qquad dx = 3 \cos \theta\, d\theta$

$\displaystyle = \int \frac{3 \cos \theta\, d\theta}{3 \sin \theta\, 3 \cos \theta} = \frac{1}{3} \int \csc \theta\, d\theta$

$\displaystyle = \frac{1}{3} \ln|\csc \theta - \cot \theta| + C = \frac{1}{3} \ln\left| \frac{3}{x} - \frac{\sqrt{9-x^2}}{x} \right| + C$

$\displaystyle = \frac{1}{3} \ln\left| \frac{3 - \sqrt{9-x^2}}{x} \right| + C.$

8. $\displaystyle\int \frac{dx}{\sqrt{9+x^2}}$ Let $x = 3 \tan \theta$

$\qquad\qquad\qquad\qquad dx = 3 \sec^2 \theta\, d\theta$

$\displaystyle = \int \frac{3 \sec^2 \theta\, d\theta}{3 \sec \theta} = \int \sec \theta\, d\theta$

$\displaystyle = \ln|\sec \theta + \tan \theta| + C = \ln(x + \sqrt{9+x^2}) + C_1.$

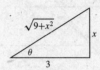

Fig. 6.2.8

10. $\displaystyle\int \frac{\sqrt{9+x^2}}{x^4}\, dx$ Let $x = 3 \tan \theta$

$\qquad\qquad\qquad\qquad\quad dx = 3 \sec^2 \theta\, d\theta$

$\displaystyle = \int \frac{(3 \sec \theta)(3 \sec^2 \theta)\, d\theta}{81 \tan^4 \theta}$

$\displaystyle = \frac{1}{9} \int \frac{\sec^3 \theta}{\tan^4 \theta}\, d\theta = \frac{1}{9} \int \frac{\cos \theta}{\sin^4 \theta}\, d\theta$ Let $u = \sin \theta$

$\qquad\qquad\qquad\qquad\qquad\qquad\qquad\qquad du = \cos \theta\, d\theta$

$\displaystyle = \frac{1}{9} \int \frac{du}{u^4} = -\frac{1}{27u^3} + C = -\frac{1}{27 \sin^3 \theta} + C$

$\displaystyle = -\frac{(9+x^2)^{3/2}}{27x^3} + C.$

12. $\displaystyle\int \frac{dx}{(a^2+x^2)^{3/2}}$ Let $x = a \tan \theta$

$\qquad\qquad\qquad\qquad\qquad dx = a \sec^2 \theta\, d\theta$

$\displaystyle = \int \frac{a \sec^2 \theta\, d\theta}{(a^2 + a^2 \tan^2 \theta)^{3/2}} = \int \frac{a \sec^2 \theta\, d\theta}{a^3 \sec^3 \theta}$

$\displaystyle = \frac{1}{a^2} \int \cos \theta\, d\theta = \frac{1}{a^2} \sin \theta + C = \frac{x}{a^2\sqrt{a^2+x^2}} + C.$

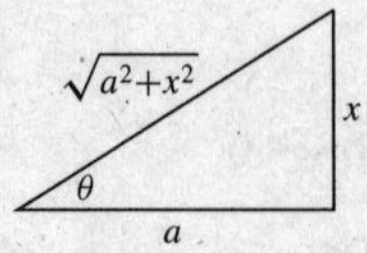

Fig. 6.2.12

14. $\displaystyle\int \frac{dx}{(1+2x^2)^{5/2}}$ Let $x = \dfrac{1}{\sqrt{2}}\tan\theta$

$$dx = \frac{1}{\sqrt{2}}\sec^2\theta\,d\theta$$

$$= \frac{1}{\sqrt{2}}\int \frac{\sec^2\theta\,d\theta}{(1+\tan^2\theta)^{5/2}} = \frac{1}{\sqrt{2}}\int \cos^3\theta\,d\theta$$

$$= \frac{1}{\sqrt{2}}\int (1-\sin^2\theta)\cos\theta\,d\theta \quad \text{Let } u = \sin\theta$$
$$du = \cos\theta\,d\theta$$

$$= \frac{1}{\sqrt{2}}\int (1-u^2)\,du = \frac{1}{\sqrt{2}}\left(u - \frac{1}{3}u^3\right) + C$$

$$= \frac{1}{\sqrt{2}}\sin\theta - \frac{1}{3\sqrt{2}}\sin^3\theta + C$$

$$= \frac{\sqrt{2}x}{\sqrt{2}\sqrt{1+2x^2}} - \frac{1}{3\sqrt{2}}\left(\frac{\sqrt{2}x}{\sqrt{1+2x^2}}\right)^3 + C$$

$$= \frac{4x^3 + 3x}{3(1+2x^2)^{3/2}} + C.$$

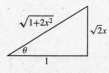

Fig. 6.2.14

16. $\displaystyle\int \frac{dx}{x^2\sqrt{x^2-a^2}}$ Let $x = a\sec\theta \ (a > 0)$

$$dx = a\sec\theta\,\tan\theta\,d\theta$$

$$= \int \frac{a\sec\theta\,\tan\theta\,d\theta}{a^2\sec^2\theta\,a\tan\theta}$$

$$= \frac{1}{a^2}\int \cos\theta\,d\theta = \frac{1}{a^2}\sin\theta + C$$

$$= \frac{1}{a^2}\frac{\sqrt{x^2-a^2}}{x} + C.$$

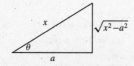

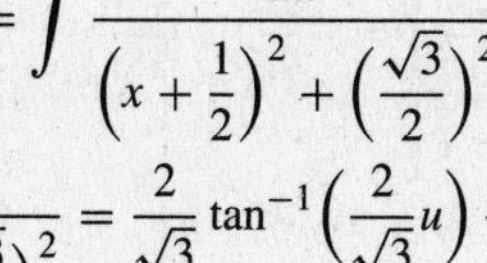

Fig. 6.2.16

18. $\displaystyle\int \frac{dx}{x^2+x+1} = \int \frac{dx}{\left(x+\frac{1}{2}\right)^2 + \left(\frac{\sqrt{3}}{2}\right)^2}$ Let $u = x + \frac{1}{2}$
$$du = dx$$

$$= \int \frac{du}{u^2 + \left(\frac{\sqrt{3}}{2}\right)^2} = \frac{2}{\sqrt{3}}\tan^{-1}\left(\frac{2}{\sqrt{3}}u\right) + C$$

$$= \frac{2}{\sqrt{3}}\tan^{-1}\left(\frac{2x+1}{\sqrt{3}}\right) + C.$$

20. $\displaystyle\int \frac{x\,dx}{x^2-2x+3} = \int \frac{(x-1)+1}{(x-1)^2+2}\,dx$ Let $u = x - 1$
$$du = dx$$

$$= \int \frac{u\,du}{u^2+2} + \int \frac{du}{u^2+2}$$

$$= \frac{1}{2}\ln(u^2+2) + \frac{1}{\sqrt{2}}\tan^{-1}\left(\frac{u}{\sqrt{2}}\right) + C$$

$$= \frac{1}{2}\ln(x^2-2x+3) + \frac{1}{\sqrt{2}}\tan^{-1}\left(\frac{x-1}{\sqrt{2}}\right) + C.$$

22. $\displaystyle\int \frac{dx}{(4x-x^2)^{3/2}}$

$$= \int \frac{dx}{[4-(2-x)^2]^{3/2}} \quad \text{Let } 2-x = 2\sin u$$
$$-dx = 2\cos u\,du$$

$$= -\int \frac{2\cos u\,du}{8\cos^3 u} = -\frac{1}{4}\int \sec^2 u\,du$$

$$= -\frac{1}{4}\tan u + C = \frac{1}{4}\frac{x-2}{\sqrt{4x-x^2}} + C.$$

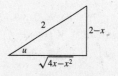

Fig. 6.2.22

24. $\displaystyle\int \frac{dx}{(x^2+2x+2)^2} = \int \frac{dx}{[(x+1)^2+1]^2}$ Let $x+1 = \tan u$
$$dx = \sec^2 u\,du$$

$$= \int \frac{\sec^2 u\,du}{\sec^4 u} = \int \cos^2 u\,du$$

$$= \frac{1}{2}\int (1 + \cos 2u)\,du = \frac{u}{2} + \frac{\sin 2u}{4} + C$$

$$= \frac{1}{2}\tan^{-1}(x+1) + \frac{1}{2}\sin u\,\cos u + C$$

$$= \frac{1}{2}\tan^{-1}(x+1) + \frac{1}{2}\frac{x+1}{x^2+2x+2} + C.$$

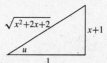

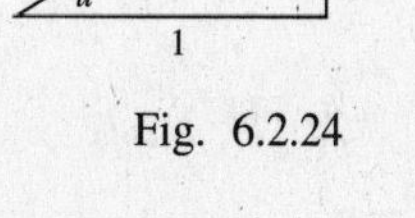

Fig. 6.2.24

26. $\displaystyle\int \frac{x^2\,dx}{(1+x^2)^2}$ Let $x = \tan u$
$$dx = \sec^2 u\,du$$

$$= \int \frac{\tan^2 u\,\sec^2 u\,du}{\sec^4 u} = \int \frac{\tan^2 u\,du}{\sec^2 u}$$

$$= \int \sin^2 u\,du = \frac{1}{2}\int (1 - \cos 2u)\,du$$

$$= \frac{u}{2} - \frac{\sin u\,\cos u}{2} + C$$

$$= \frac{1}{2}\tan^{-1}x - \frac{1}{2}\frac{x}{1+x^2} + C.$$

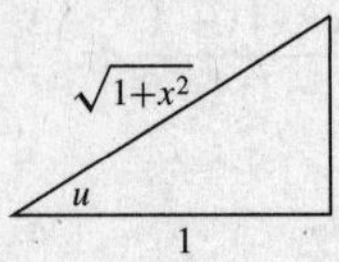

Fig. 6.2.26

28. $I = \displaystyle\int \sqrt{9 + x^2}\, dx$ Let $x = 3\tan\theta$
$$dx = 3\sec^2\theta\, d\theta$$

$= \displaystyle\int 3\sec\theta\, 3\sec^2\theta\, d\theta$

$= 9\displaystyle\int \sec^3\theta\, d\theta$

$\quad\quad U = \sec\theta \quad\quad\quad dV = \sec^2\theta\, d\theta$

$\quad\quad dU = \sec\theta\tan\theta\, d\theta \quad\quad V = \tan\theta$

$= 9\sec\theta\tan\theta - 9\displaystyle\int \sec\theta\tan^2\theta\, d\theta$

$= 9\sec\theta\tan\theta - 9\displaystyle\int \sec\theta(\sec^2\theta - 1)\, d\theta$

$= 9\sec\theta\tan\theta + 9\displaystyle\int \sec\theta\, d\theta - 9\displaystyle\int \sec^3\theta\, d\theta$

$= 9\sec\theta\tan\theta + 9\ln|\sec\theta + \tan\theta| - I$

$I = \dfrac{9}{2}\left[\left(\dfrac{\sqrt{9+x^2}}{3}\right)\left(\dfrac{x}{3}\right)\right] + \dfrac{9}{2}\ln\left|\dfrac{\sqrt{9+x^2}}{3} + \dfrac{x}{3}\right| + C$

$= \dfrac{1}{2}x\sqrt{9+x^2} + \dfrac{9}{2}\ln\left(\sqrt{9+x^2} + x\right) + C_1.$

$\quad\quad$ (where $C_1 = C - \dfrac{9}{2}\ln 3$)

30. $\displaystyle\int \dfrac{dx}{1 + x^{1/3}}$ Let $x = u^3$
$$dx = 3u^2\, du$$

$= 3\displaystyle\int \dfrac{u^2\, du}{1 + u}$ Let $v = 1 + u$
$$dv = du$$

$= 3\displaystyle\int \dfrac{v^2 - 2v + 1}{v}\, dv = 3\displaystyle\int \left(v - 2 + \dfrac{1}{v}\right) dv$

$= 3\left(\dfrac{v^2}{2} - 2v + \ln|v|\right) + C$

$= \dfrac{3}{2}(1 + x^{1/3})^2 - 6(1 + x^{1/3}) + 3\ln|1 + x^{1/3}| + C.$

32. $\displaystyle\int \dfrac{x\sqrt{2 - x^2}}{\sqrt{x^2 + 1}}\, dx$ Let $u^2 = x^2 + 1$
$$2u\, du = 2x\, dx$$

$= \displaystyle\int \dfrac{u\sqrt{3 - u^2}\, du}{u}$

$= \displaystyle\int \sqrt{3 - u^2}\, du$ Let $u = \sqrt{3}\sin v$
$$du = \sqrt{3}\cos v\, dv$$

$= \displaystyle\int (\sqrt{3}\cos v)\sqrt{3}\cos v\, dv = 3\displaystyle\int \cos^2 v\, dv$

$= \dfrac{3}{2}(v + \sin v\cos v) + C$

$= \dfrac{3}{2}\sin^{-1}\left(\dfrac{u}{\sqrt{3}}\right) + \dfrac{3}{2}\dfrac{u\sqrt{3 - u^2}}{3} + C$

$= \dfrac{3}{2}\sin^{-1}\left(\sqrt{\dfrac{x^2 + 1}{3}}\right) + \dfrac{1}{2}\sqrt{(x^2 + 1)(2 - x^2)} + C.$

34. $\displaystyle\int_0^{\pi/2} \dfrac{\cos x}{\sqrt{1 + \sin^2 x}}\, dx$ Let $u = \sin x$
$$du = \cos x\, dx$$

$= \displaystyle\int_0^1 \dfrac{du}{\sqrt{1 + u^2}}$ Let $u = \tan w$
$$du = \sec^2 w\, dw$$

$= \displaystyle\int_0^{\pi/4} \dfrac{\sec^2 w\, dw}{\sec w} = \displaystyle\int_0^{\pi/4} \sec w\, dw$

$= \ln|\sec w + \tan w|\Big|_0^{\pi/4}$

$= \ln|\sqrt{2} + 1| - \ln|1 + 0| = \ln(\sqrt{2} + 1).$

36. $\displaystyle\int_1^2 \dfrac{dx}{x^2\sqrt{9 - x^2}}$ Let $x = 3\sin u$
$$dx = 3\cos u\, du$$

$= \displaystyle\int_{x=1}^{x=2} \dfrac{3\cos u\, du}{9\sin^2 u(3\cos u)} = \dfrac{1}{9}\displaystyle\int_{x=1}^{x=2} \csc^2 u\, du$

$= \dfrac{1}{9}(-\cot u)\Big|_{x=1}^{x=2} = -\dfrac{1}{9}\left(\dfrac{\sqrt{9 - x^2}}{x}\right)\Big|_{x=1}^{x=2}$

$= -\dfrac{1}{9}\left(\dfrac{\sqrt{5}}{2} - \dfrac{\sqrt{8}}{1}\right) = \dfrac{2\sqrt{2}}{9} - \dfrac{\sqrt{5}}{18}.$

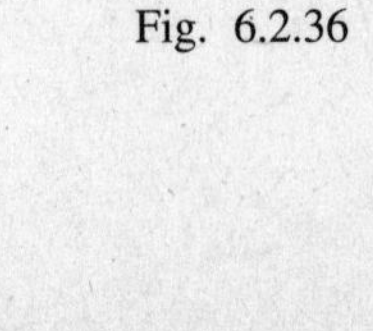

Fig. 6.2.36

38. $\displaystyle\int_0^{\pi/2} \frac{d\theta}{1+\cos\theta+\sin\theta}$ Let $x = \tan\dfrac{\theta}{2}, \quad d\theta = \dfrac{2}{1+x^2}\,dx,$

$$\cos\theta = \frac{1-x^2}{1+x^2}, \quad \sin\theta = \frac{2x}{1+x^2}.$$

$$= \int_0^1 \frac{\left(\dfrac{2}{1+x^2}\right)dx}{1+\left(\dfrac{1-x^2}{1+x^2}\right)+\left(\dfrac{2x}{1+x^2}\right)}$$

$$= 2\int_0^1 \frac{dx}{2+2x} = \int_0^1 \frac{dx}{1+x}$$

$$= \ln|1+x|\,\Big|_0^1 = \ln 2.$$

40. Area $\displaystyle= \int_{1/2}^1 \frac{dx}{\sqrt{2x-x^2}} = \int_{1/2}^1 \frac{dx}{\sqrt{1-(x-1)^2}}$

$$\text{Let } u = x-1$$
$$du = dx$$

$$= \int_{-1/2}^0 \frac{du}{\sqrt{1-u^2}} = \sin^{-1} u\,\Big|_{-1/2}^0$$

$$= 0 - \left(-\frac{\pi}{6}\right) = \frac{\pi}{6} \text{ sq. units.}$$

42. Average value $\displaystyle= \frac{1}{4}\int_0^4 \frac{dx}{(x^2-4x+8)^{3/2}}$

$$= \frac{1}{4}\int_0^4 \frac{dx}{[(x-2)^2+4]^{3/2}}$$

$$\text{Let } x-2 = 2\tan u$$
$$dx = 2\sec^2 u\,du$$

$$= \frac{1}{4}\int_{-\pi/4}^{\pi/4} \frac{2\sec^2 u\,du}{8\sec^3 u}$$

$$= \frac{1}{16}\int_{-\pi/4}^{\pi/4} \cos u\,du = \frac{1}{16}\sin u\,\Big|_{-\pi/4}^{\pi/4}$$

$$= \frac{1}{16}\left(\frac{1}{\sqrt2}+\frac{1}{\sqrt2}\right) = \frac{\sqrt2}{16}.$$

44. The circles intersect at $x = \frac14$, so the common area is

$A_1 + A_2$ where

$$A_1 = 2\int_{1/4}^1 \sqrt{1-x^2}\,dx \quad \text{Let } x = \sin u$$
$$dx = \cos u\,du$$

$$= 2\int_{x=1/4}^{x=1} \cos^2 u\,du$$

$$= (u + \sin u\cos u)\,\Big|_{x=1/4}^{x=1}$$

$$= (\sin^{-1} x + x\sqrt{1-x^2})\,\Big|_{x=1/4}^{x=1}$$

$$= \frac{\pi}{2} - \sin^{-1}\frac14 - \frac{\sqrt{15}}{16} \text{ sq. units.}$$

$$A_2 = 2\int_0^{1/4} \sqrt{4-(x-2)^2}\,dx \quad \text{Let } x-2 = 2\sin v$$
$$dx = 2\cos v\,dv$$

$$= 8\int_{x=0}^{x=1/4} \cos^2 v\,dv$$

$$= 4(v + \sin v\cos v)\,\Big|_{x=0}^{x=1/4}$$

$$= 4\left[\sin^{-1}\left(\frac{x-2}{2}\right) + \left(\frac{x-2}{2}\right)\frac{\sqrt{4x-x^2}}{2}\right]\Bigg|_{x=0}^{x=1/4}$$

$$= 4\left[\sin^{-1}\left(-\frac78\right) - \frac{7\sqrt{15}}{64} + \frac{\pi}{2}\right]$$

$$= -4\sin^{-1}\left(\frac78\right) - \frac{7\sqrt{15}}{16} + 2\pi \text{ sq. units.}$$

Hence, the common area is

$$A_1 + A_2 = \frac{5\pi}{2} - \frac{\sqrt{15}}{2}$$
$$- \sin^{-1}\left(\frac14\right) - 4\sin^{-1}\left(\frac78\right) \text{ sq. units.}$$

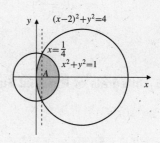

Fig. 6.2.44

46. Shaded area $= 2 \int_c^a b \sqrt{1 - \left(\frac{x}{a}\right)^2} \, dx$ Let $x = a \sin u$

$\qquad\qquad\qquad\qquad\qquad\qquad\qquad dx = a \cos u \, du$

$= 2ab \int_{x=c}^{x=a} \cos^2 u \, du$

$= ab(u + \sin u \cos u) \Big|_{x=c}^{x-a}$

$= \left(ab \sin^{-1} \frac{x}{a} + \frac{b}{a} x \sqrt{a^2 - x^2}\right) \Big|_c^a$

$= ab\left(\frac{\pi}{2} - \sin^{-1} \frac{c}{a}\right) - \frac{cb}{a} \sqrt{a^2 - c^2}$ sq. units.

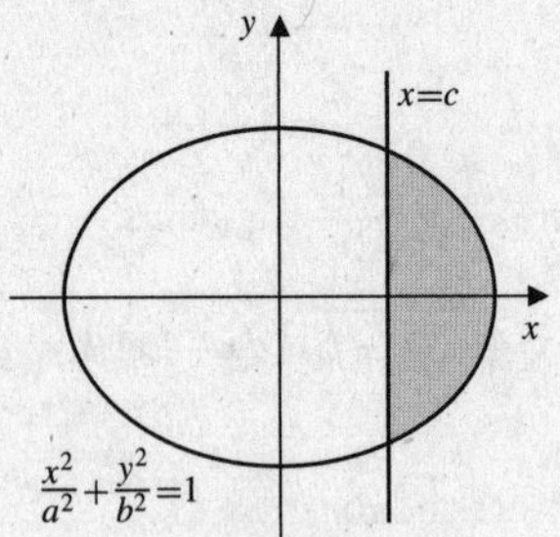

Fig. 6.2.46

48. $\int \frac{dx}{\sqrt{x^2 - a^2}}$ Let $x = a \cosh u$

$\qquad\qquad\qquad dx = a \sinh u \, du$

$= \int \frac{a \sinh u \, du}{a \sinh u} = u + C$

$= \cosh^{-1} \frac{x}{a} + C = \ln(x + \sqrt{x^2 - a^2}) + C, \quad (x \geq a).$

$\int \frac{dx}{x^2 \sqrt{x^2 - a^2}} = \int \frac{a \sinh u \, du}{a^2 \cosh^2 u \, a \sinh u}$

$= \frac{1}{a^2} \int \text{sech}^2 u \, du = \frac{1}{a^2} \tanh u + C$

$= \frac{1}{a^2} \tanh \left(\cosh^{-1} \frac{x}{a}\right) + C$

$= \frac{1}{a^2} \cdot \frac{\dfrac{x}{a} + \sqrt{\dfrac{x^2}{a^2} - 1} - \dfrac{1}{\dfrac{x}{a} - \sqrt{\dfrac{x^2}{a^2} - 1}}}{\dfrac{x}{a} + \sqrt{\dfrac{x^2}{a^2} - 1} + \dfrac{1}{\dfrac{x}{a} - \sqrt{\dfrac{x^2}{a^2} - 1}}} + C$

$= \frac{\sqrt{x^2 - a^2}}{a^2 x} + C_1.$

Section 6.3 Integrals of Rational Functions (page 372)

2. $\int \frac{dx}{5 - 4x} = -\frac{1}{4} \ln|5 - 4x| + C.$

4. $\int \frac{x^2}{x - 4} \, dx = \int \left(x + 4 + \frac{16}{x - 4}\right) dx$

$\qquad\qquad = \frac{x^2}{2} + 4x + 16 \ln|x - 4| + C.$

6. $\frac{1}{5 - x^2} = \frac{A}{\sqrt{5} - x} + \frac{B}{\sqrt{5} + x}$

$\qquad = \frac{(A + B)\sqrt{5} + (A - B)x}{5 - x^2}$

$\Rightarrow \begin{cases} A + B = \dfrac{1}{\sqrt{5}} \\ A - B = 0 \end{cases} \Rightarrow A = B = \dfrac{1}{2\sqrt{5}}.$

$\int \frac{1}{5 - x^2} \, dx = \frac{1}{2\sqrt{5}} \int \left(\frac{1}{\sqrt{5} - x} + \frac{1}{\sqrt{5} + x}\right) dx$

$\qquad = \frac{1}{2\sqrt{5}} \left(-\ln|\sqrt{5} - x| + \ln|\sqrt{5} + x|\right) + C$

$\qquad = \frac{1}{2\sqrt{5}} \ln \left|\frac{\sqrt{5} + x}{\sqrt{5} - x}\right| + C.$

8. $\frac{1}{b^2 - a^2 x^2} = \frac{A}{b - ax} + \frac{B}{b + ax}$

$\qquad = \frac{(A + B)b + (A - B)ax}{b^2 - a^2 x^2}$

$\Rightarrow A = B = \frac{1}{2b}$

$\int \frac{dx}{b^2 - a^2 x^2} = \frac{1}{2b} \int \left(\frac{1}{b - ax} + \frac{1}{b + ax}\right) dx$

$\qquad = \frac{1}{2b} \left(\frac{-\ln|b - ax|}{a} + \frac{\ln|b + ax|}{a}\right) + C$

$\qquad = \frac{1}{2ab} \ln \left|\frac{b + ax}{b - ax}\right| + C.$

10. $\frac{x}{3x^2 + 8x - 3} = \frac{A}{3x - 1} + \frac{B}{x + 3}$

$\qquad = \frac{(A + 3B)x + (3A - B)}{3x^2 + 8x - 3}$

$\Rightarrow \begin{cases} A + 3B = 1 \\ 3A - B = 0 \end{cases} \Rightarrow A = \frac{1}{10}, \ B = \frac{3}{10}.$

$\int \frac{x \, dx}{3x^2 + 8x - 3} = \frac{1}{10} \int \left(\frac{1}{3x - 1} + \frac{3}{x + 3}\right) dx$

$\qquad = \frac{1}{30} \ln|3x - 1| + \frac{3}{10} \ln|x + 3| + C.$

12. $\frac{1}{x^3 + 9x} = \frac{A}{x} + \frac{Bx + C}{x^2 + 9}$

$\qquad = \frac{Ax^2 + 9A + Bx^2 + Cx}{x^3 + 9x}$

$\Rightarrow \begin{cases} A + B = 0 \\ C = 0 \\ 9A = 1 \end{cases} \Rightarrow A = \frac{1}{9}, \ B = -\frac{1}{9}, \ C = 0.$

$\int \frac{dx}{x^3 + 9x} = \frac{1}{9} \int \left(\frac{1}{x} - \frac{x}{x^2 + 9}\right) dx$

$\qquad = \frac{1}{9} \ln|x| - \frac{1}{18} \ln(x^2 + 9) + K.$

14. $\displaystyle\int \frac{x}{2+6x+9x^2}\,dx = \int \frac{x}{(3x+1)^2+1}\,dx$ Let $u = 3x+1$

$$du = 3\,dx$$

$\displaystyle\frac{1}{9}\int \frac{u-1}{u^2+1}\,du = \frac{1}{9}\int \frac{u}{u^2+1}\,du - \frac{1}{9}\int \frac{1}{u^2+1}\,du$

$\displaystyle = \frac{1}{18}\ln(u^2+1) - \frac{1}{9}\tan^{-1}u + C$

$\displaystyle = \frac{1}{18}\ln(2+6x+9x^2) - \frac{1}{9}\tan^{-1}(3x+1) + C.$

16. First divide to obtain

$$\frac{x^3+1}{x^2+7x+12} = x-7 + \frac{37x+85}{(x+4)(x+3)}$$

$$\frac{37x+85}{(x+4)(x+3)} = \frac{A}{x+4} + \frac{B}{x+3}$$

$$= \frac{(A+B)x + 3A + 4B}{x^2+7x+12}$$

$$\Rightarrow \begin{cases} A+B = 37 \\ 3A+4B = 85 \end{cases} \Rightarrow A = 63, \ B = -26.$$

Now we have

$\displaystyle\int \frac{x^3+1}{12+7x+x^2}\,dx = \int\left(x-7 + \frac{63}{x+4} - \frac{26}{x+3}\right)dx$

$\displaystyle = \frac{x^2}{2} - 7x + 63\ln|x+4| - 26\ln|x+3| + C.$

18. The partial fraction decomposition is

$$\frac{1}{x^4-a^4} = \frac{A}{x-a} + \frac{B}{x+a} + \frac{Cx+D}{x^2+a^2}$$

$$= \frac{A(x^3+ax^2+a^2x+a^3) + B(x^3-ax^2+a^2x-a^3)}{x^4-a^4}$$

$$+ \frac{C(x^3-a^2x) + D(x^2-a^2)}{x^4-a^4}$$

$$\Rightarrow \begin{cases} A+B+C = 0 \\ aA - aB + D = 0 \\ a^2A + a^2B - a^2C = 0 \\ a^3A - a^3B - a^2D = 1 \end{cases}$$

$$\Rightarrow A = \frac{1}{4a^3}, \ B = -\frac{1}{4a^3}, \ C = 0, \ D = -\frac{1}{2a^2}.$$

$\displaystyle\int \frac{dx}{x^4-a^4} = \frac{1}{4a^3}\int\left(\frac{1}{x-a} - \frac{1}{x+a} - \frac{2a}{x^2+a^2}\right)dx$

$\displaystyle = \frac{1}{4a^3}\ln\left|\frac{x-a}{x+a}\right| - \frac{1}{2a^3}\tan^{-1}\left(\frac{x}{a}\right) + K.$

20. Here the expansion is

$$\frac{1}{x^3+2x^2+2x} = \frac{A}{x} + \frac{Bx+C}{x^2+2x+2}$$

$$= \frac{A(x^2+2x+2) + Bx^2 + Cx}{x^3+2x^2+2}$$

$$\Rightarrow \begin{cases} A+B = 0 \\ 2A+C = 0 \\ 2A = 1 \end{cases} \Rightarrow A = -B = \frac{1}{2}, \ C = -1,$$

so we have

$\displaystyle\int \frac{dx}{x^3+2x^2+2x} = \frac{1}{2}\int \frac{dx}{x} - \frac{1}{2}\int \frac{x+2}{x^2+2x+2}\,dx$

Let $u = x+1$

$$du = dx$$

$\displaystyle = \frac{1}{2}\ln|x| - \frac{1}{2}\int \frac{u+1}{u^2+1}\,du$

$\displaystyle = \frac{1}{2}\ln|x| - \frac{1}{4}\ln(u^2+1) - \frac{1}{2}\tan^{-1}u + K$

$\displaystyle = \frac{1}{2}\ln|x| - \frac{1}{4}\ln(x^2+2x+2) - \frac{1}{2}\tan^{-1}(x+1) + K.$

22. Here the expansion is

$$\frac{x^2+1}{x^3+8} = \frac{A}{x+2} + \frac{Bx+C}{x^2-2x+4}$$

$$= \frac{A(x^2-2x+4) + B(x^2+2x) + C(x+2)}{x^3+8}$$

$$\Rightarrow \begin{cases} A+B = 1 \\ -2A+2B+C = 0 \\ 4A+2C = 1 \end{cases} \Rightarrow A = \frac{5}{12}, \ B = \frac{7}{12}, \ C = -\frac{1}{3},$$

so we have

$\displaystyle\int \frac{x^2+1}{x^3+8}\,dx = \frac{5}{12}\int \frac{dx}{x+2} + \frac{1}{12}\int \frac{7x-4}{(x-1)^2+3}\,dx$

Let $u = x-1$

$$du = dx$$

$\displaystyle = \frac{5}{12}\ln|x+2| + \frac{1}{12}\int \frac{7u+3}{u^2+3}\,du$

$\displaystyle = \frac{5}{12}\ln|x+2| + \frac{7}{24}\ln(x^2-2x+4)$

$\displaystyle \quad + \frac{1}{4\sqrt{3}}\tan^{-1}\frac{x-1}{\sqrt{3}} + K.$

24. The expansion is

$$\frac{x^2}{(x^2-1)(x^2-4)} = \frac{A}{x-1} + \frac{B}{x+1} + \frac{C}{x-2} + \frac{D}{x+2}$$

$\displaystyle A = \lim_{x\to 1}\frac{x^2}{(x+1)(x^2-4)} = \frac{1}{2(-3)} = -\frac{1}{6}$

$\displaystyle B = \lim_{x\to -1}\frac{x^2}{(x-1)(x^2-4)} = \frac{1}{-2(-3)} = \frac{1}{6}$

$\displaystyle C = \lim_{x\to 2}\frac{x^2}{(x^2-1)(x+2)} = \frac{4}{3(4)} = \frac{1}{3}$

$\displaystyle D = \lim_{x\to -2}\frac{x^2}{(x^2-1)(x-2)} = \frac{4}{3(-4)} = -\frac{1}{3}.$

Therefore

$$\int \frac{x^2}{(x^2-1)(x^2-4)}\,dx = -\frac{1}{6}\ln|x-1| + \frac{1}{6}\ln|x+1| +$$
$$\frac{1}{3}\ln|x-2| - \frac{1}{3}\ln|x+2| + K.$$

26. We have

$$\int \frac{x\,dx}{(x^2-x+1)^2} = \int \frac{x\,dx}{\left[(x-\frac{1}{2})^2 + \frac{3}{4}\right]^2} \qquad \begin{array}{l} \text{Let } u = x - \frac{1}{2} \\ du = dx \end{array}$$

$$= \int \frac{u\,du}{(u^2+\frac{3}{4})^2} + \frac{1}{2}\int \frac{du}{(u^2+\frac{3}{4})^2}$$

Let $u = \dfrac{\sqrt{3}}{2}\tan v$,

$du = \dfrac{\sqrt{3}}{2}\sec^2 v\,dv$ in the second integral.

$$= -\frac{1}{2}\left(\frac{1}{u^2+\frac{3}{4}}\right) + \frac{1}{2}\int \frac{\frac{\sqrt{3}}{2}\sec^2 v\,dv}{\frac{9}{16}\sec^4 v}$$

$$= \frac{-1}{2(x^2-x+1)} + \frac{4}{3\sqrt{3}}\int \cos^2 v\,dv$$

$$= \frac{-1}{2(x^2-x+1)} + \frac{2}{3\sqrt{3}}(v + \sin v\cos v) + C$$

$$= \frac{-1}{2(x^2-x+1)} + \frac{2}{3\sqrt{3}}\tan^{-1}\frac{2x-1}{\sqrt{3}} + \frac{2}{3\sqrt{3}}\frac{2(x-\frac{1}{2})\sqrt{3}}{(2\sqrt{x^2-x+1})^2} + C$$

$$= \frac{2}{3\sqrt{3}}\tan^{-1}\frac{2x-1}{\sqrt{3}} + \frac{x-2}{3(x^2-x+1)} + C.$$

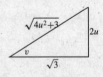

Fig. 6.3.26

28. We have

$$\int \frac{dt}{(t-1)(t^2-1)^2}$$

$$= \int \frac{dt}{(t-1)^3(t+1)^2} \qquad \begin{array}{l} \text{Let } u = t-1 \\ du = dt \end{array}$$

$$= \int \frac{du}{u^3(u+2)^2}$$

$$\frac{1}{u^3(u+2)^2} = \frac{A}{u} + \frac{B}{u^2} + \frac{C}{u^3} + \frac{D}{u+2} + \frac{E}{(u+2)^2}$$

$$= \frac{A(u^4+4u^3+4u^2) + B(u^3+4u^2+4u)}{u^3(u+2)^2}$$

$$\frac{C(u^2+4u+4) + D(u^4+2u^3) + Eu^3}{u^3(u+2)^2}$$

$$\Rightarrow \begin{cases} A + D = 0 \\ 4A + B + 2D + E = 0 \\ 4A + 4B + C = 0 \\ 4B + 4C = 0 \\ 4C = 1 \end{cases}$$

$$\Rightarrow A = \frac{3}{16},\ B = -\frac{1}{4},\ C = \frac{1}{4},\ D = -\frac{3}{16},\ E = -\frac{1}{8}.$$

$$\int \frac{du}{u^3(u+2)^2}$$

$$= \frac{3}{16}\int \frac{du}{u} - \frac{1}{4}\int \frac{du}{u^2} + \frac{1}{4}\int \frac{du}{u^3}$$

$$- \frac{3}{16}\int \frac{du}{u+2} - \frac{1}{8}\int \frac{du}{(u+2)^2}$$

$$= \frac{3}{16}\ln|t-1| + \frac{1}{4(t-1)} - \frac{1}{8(t-1)^2} -$$
$$\frac{3}{16}\ln|t+1| + \frac{1}{8(t+1)} + K.$$

30. $\displaystyle \int \frac{dx}{e^{2x}-4e^x+4} = \int \frac{dx}{(e^x-2)^2} \qquad \begin{array}{l} \text{Let } u = e^x \\ du = e^x\,dx \end{array}$

$$= \int \frac{du}{u(u-2)^2}$$

$$\frac{1}{u(u-2)^2} = \frac{A}{u} + \frac{B}{u-2} + \frac{C}{(u-2)^2}$$

$$= \frac{A(u^2-4u+4) + B(u^2-2u) + +Cu}{u(u-2)^2}$$

$$\Rightarrow \begin{cases} A + B = 0 \\ -4A - 2B + C = 0 \\ 4A = 1 \end{cases} \Rightarrow A = \frac{1}{4},\ B = -\frac{1}{4},\ C = \frac{1}{2}.$$

$$\int \frac{du}{u(u-2)^2} = \frac{1}{4}\int \frac{du}{u} - \frac{1}{4}\int \frac{du}{u-2} + \frac{1}{2}\int \frac{du}{(u-2)^2}$$

$$= \frac{1}{4}\ln|u| - \frac{1}{4}\ln|u-2| - \frac{1}{2}\frac{1}{(u-2)} + K$$

$$= \frac{x}{4} - \frac{1}{4}\ln|e^x-2| - \frac{1}{2(e^x-2)} + K.$$

32. We have

$$I = \int \frac{dx}{x(1-x^2)^{3/2}} \qquad \begin{array}{l} \text{Let } u^2 = 1-x^2 \\ 2u\,du = -2x\,dx \end{array}$$

$$= -\int \frac{u\,du}{(1-u^2)u^3} = -\int \frac{du}{(1-u^2)u^2}$$

$$\frac{1}{u^2(1-u^2)} = \frac{A}{u} + \frac{B}{u^2} + \frac{C}{1-u} + \frac{D}{1+u}$$

$$= \frac{A(u-u^3) + B(1-u^2) + C(u^2+u^3) + D(u^2-u^3)}{u^2(1-u^2)}$$

$$\Rightarrow \begin{cases} -A + C - D = 0 \\ -B + C + D = 0 \\ A = 0 \\ B = 1 \end{cases}$$

$$\Rightarrow A = 0, \ B = 1, \ C = \frac{1}{2}, \ D = \frac{1}{2}.$$

$$I = -\int \frac{du}{(1-u^2)u^2} = -\int \frac{du}{u^2} - \frac{1}{2}\int \frac{du}{1-u} - \frac{1}{2}\int \frac{du}{1+u}$$

$$= \frac{1}{u} + \frac{1}{2}\ln|1-u| - \frac{1}{2}\ln|1+u| + K$$

$$= \frac{1}{\sqrt{1-x^2}} + \frac{1}{2}\ln\left|\frac{1-\sqrt{1-x^2}}{1+\sqrt{1-x^2}}\right| + K$$

$$= \frac{1}{\sqrt{1-x^2}} + \ln\left(1-\sqrt{1-x^2}\right) - \ln|x| + K.$$

34. $\displaystyle\int \frac{d\theta}{\cos\theta(1+\sin\theta)}$ Let $u = \sin\theta$

$$du = \cos\theta\, d\theta$$

$$= \int \frac{du}{(1-u^2)(1+u)} = \int \frac{du}{(1-u)(1+u)^2}$$

$$\frac{1}{(1-u)(1+u)^2} = \frac{A}{1-u} + \frac{B}{1+u} + \frac{C}{(1+u)^2}$$

$$= \frac{A(1+2u+u^2) + B(1-u^2) + C(1-u)}{(1-u)(1+u)^2}$$

$$\Rightarrow \begin{cases} A - B = 0 \\ 2A - C = 0 \\ A + B + C = 1 \end{cases} \Rightarrow A = \frac{1}{4}, \ B = \frac{1}{4}, \ C = \frac{1}{2}.$$

$$\int \frac{du}{(1-u)(1+u)^2}$$

$$= \frac{1}{4}\int \frac{du}{1-u} + \frac{1}{4}\int \frac{du}{1+u} + \frac{1}{2}\int \frac{du}{(1+u)^2}$$

$$= \frac{1}{4}\ln\left|\frac{1+\sin\theta}{1-\sin\theta}\right| - \frac{1}{2(1+\sin\theta)} + C.$$

Section 6.4 Integration Using Computer Algebra or Tables (page 376)

2. According to Maple

$$\int \frac{1+x+x^2}{(x^4-1)(x^4-16)^2}\, dx$$

$$= \frac{\ln(x-1)}{300} - \frac{\ln(x+1)}{900} - \frac{7}{15{,}360(x-2)}$$

$$- \frac{613}{460{,}800}\ln(x-2) - \frac{1}{5{,}120(x+2)} + \frac{79}{153{,}600}\ln(x+2)$$

$$- \frac{\ln(x^2+1)}{900} + \frac{47}{115{,}200}\ln(x^2+4)$$

$$- \frac{23}{25{,}600}\tan^{-1}(x/2) - \frac{6x+8}{15{,}360(x^2+4)}$$

One suspects it has forgotten to use absolute values in some of the logarithms.

4. Maple, Mathematica, and Derive readily gave

$$\int_0^1 \frac{1}{(x^2+1)^3}\, dx = \frac{3\pi}{32} + \frac{1}{4}.$$

6. Use the last integral in the list involving $\sqrt{x^2 \pm a^2}$.

$$\int \sqrt{(x^2+4)^3}\, dx = \frac{x}{4}(x^2+10)\sqrt{x^2+4} + 6\ln|x+\sqrt{x^2+4}| + C$$

8. Use the 8th integral in the miscellaneous algebraic set.

$$\int \frac{dt}{t\sqrt{3t-5}} = \frac{2}{\sqrt{5}}\tan^{-1}\sqrt{\frac{3t-5}{5}} + C$$

10. We make a change of variable and then use the first two integrals in the exponential/logarithmic set.

$$\int x^7 e^{x^2}\, dx \quad \text{Let } u = x^2$$

$$du = 2x\, dx$$

$$= \frac{1}{2}\int u^3 e^u\, du$$

$$= \frac{1}{2}\left(u^3 e^u - 3\int u^2 e^u\, du\right)$$

$$= \frac{u^3 e^u}{2} - \frac{3}{2}\left(u^2 e^u - 2\int u e^u\, du\right)$$

$$= \left(\frac{u^3}{2} - \frac{3u^2}{2} + 3(u-1)\right)e^u + C$$

$$= \left(\frac{x^6}{2} - \frac{3x^4}{2} + 3x^2 - 3\right)e^{x^2} + C$$

12. Use integrals 17 and 16 in the miscellaneous algebraic set.

$$\int \frac{\sqrt{2x-x^2}}{x^2}\, dx$$

$$= -\frac{(2x-x^2)^{3/2}}{x^2} - \frac{1}{1}\int \frac{\sqrt{2x-x^2}}{x}\, dx$$

$$= -\frac{(2x-x^2)^{3/2}}{x^2} - \sqrt{2x-x^2} - \sin^{-1}(x-1) + C$$

14. Use the last integral in the miscellaneous algebraic set. Then complete the square, change variables, and use the second last integral in the elementary list.

$$\int \frac{dx}{(\sqrt{4x - x^2})^4}$$

$$= \frac{x - 2}{8}(\sqrt{4x - x^2})^{-2} + \frac{1}{8}\int \frac{dx}{4x - x^2}$$

$$= \frac{x - 2}{8(4x - x^2)} + \frac{1}{8}\int \frac{dx}{4 - (x - 2)^2} \quad \text{Let } u = x - 2$$
$$\qquad\qquad\qquad\qquad\qquad\qquad\qquad\qquad du = dx$$

$$= \frac{x - 2}{8(4x - x^2)} + \frac{1}{8}\int \frac{du}{4 - u^2}$$

$$= \frac{x - 2}{8(4x - x^2)} + \frac{1}{32}\ln\left|\frac{u + 2}{u - 2}\right| + C$$

$$= \frac{x - 2}{8(4x - x^2)} + \frac{1}{32}\ln\left|\frac{x}{x - 4}\right| + C$$

Section 6.5 Improper Integrals (page 384)

2. $\displaystyle\int_3^\infty \frac{1}{(2x - 1)^{2/3}}\,dx \quad \text{Let } u = 2x - 1$
$$\qquad\qquad\qquad\qquad\qquad\qquad du = 2\,dx$$

$$= \frac{1}{2}\int_5^\infty \frac{du}{u^{2/3}} = \frac{1}{2}\lim_{R\to\infty}\int_5^R u^{-2/3}\,du$$

$$= \frac{1}{2}\lim_{R\to\infty}\left.3u^{1/3}\right|_5^R = \infty \quad \text{(diverges)}$$

4. $\displaystyle\int_{-\infty}^{-1} \frac{dx}{x^2 + 1} = \lim_{R\to -\infty}\int_R^{-1} \frac{dx}{x^2 + 1}$

$$= \lim_{R\to -\infty}\left[\tan^{-1}(-1) - \tan^{-1}(R)\right]$$

$$= -\frac{\pi}{4} - \left(-\frac{\pi}{2}\right) = \frac{\pi}{4}.$$

This integral converges.

6. $\displaystyle\int_0^a \frac{dx}{a^2 - x^2} = \lim_{C\to a-}\int_0^C \frac{dx}{a^2 - x^2}$

$$= \lim_{C\to a-} \left.\frac{1}{2a}\ln\left|\frac{a + x}{a - x}\right|\right|_0^C$$

$$= \lim_{C\to a-} \frac{1}{2a}\ln\frac{a + C}{a - C} = \infty.$$

The integral diverges to infinity.

8. $\displaystyle\int_0^1 \frac{dx}{x\sqrt{1 - x}} \quad \text{Let } u^2 = 1 - x$
$$\qquad\qquad\qquad\qquad\quad 2u\,du = -dx$$

$$= \int_0^1 \frac{2u\,du}{(1 - u^2)u} = 2\lim_{c\to 1-}\int_0^c \frac{du}{1 - u^2}$$

$$= 2\lim_{c\to 1-} \left.\frac{1}{2}\ln\left|\frac{u + 1}{u - 1}\right|\right|_0^c = \infty \quad \text{(diverges)}$$

10. $\displaystyle\int_0^\infty xe^{-x}\,dx$

$$= \lim_{R\to\infty}\int_0^R xe^{-x}\,dx$$

$$\qquad U = x \qquad dV = e^{-x}\,dx$$
$$\qquad dU = dx \qquad V = -e^{-x}$$

$$= \lim_{R\to\infty}\left(\left.-xe^{-x}\right|_0^R + \int_0^R e^{-x}\,dx\right)$$

$$= \lim_{R\to\infty}\left(-\frac{R}{e^R} - \frac{1}{e^R} + 1\right) = 1.$$

The integral converges.

12. $\displaystyle\int_0^\infty \frac{x}{1 + 2x^2}\,dx = \lim_{R\to\infty}\int_0^R \frac{x}{1 + 2x^2}\,dx$

$$= \lim_{R\to\infty}\left.\frac{1}{4}\ln(1 + 2x^2)\right|_0^R$$

$$= \lim_{R\to\infty}\left[\frac{1}{4}\ln(1 + 2R^2) - \frac{1}{4}\ln 1\right] = \infty.$$

This integral diverges to infinity.

14. $\displaystyle\int_0^{\pi/2} \sec x\,dx = \lim_{C\to(\pi/2)-}\left.\ln|\sec x + \tan x|\right|_0^C$

$$= \lim_{C\to(\pi/2)-}\ln|\sec C + \tan C| = \infty.$$

This integral diverges to infinity.

16. $\displaystyle\int_e^\infty \frac{dx}{x(\ln x)} \quad \text{Let } u = \ln x$
$$\qquad\qquad\qquad\qquad\qquad du = \frac{dx}{x}$$

$$= \lim_{R\to\infty}\int_1^{\ln R} \frac{du}{u} = \lim_{R\to\infty}\left.\ln|u|\right|_1^{\ln R}$$

$$= \lim_{R\to\infty}\ln(\ln R) - \ln 1 = \infty.$$

This integral diverges to infinity.

18. $\displaystyle\int_e^\infty \frac{dx}{x(\ln x)^2} \quad \text{Let } u = \ln x$
$$\qquad\qquad\qquad\qquad\qquad du = \frac{dx}{x}$$

$$= \lim_{R\to\infty}\int_1^{\ln R} \frac{du}{u^2} = \lim_{R\to\infty}\left(-\frac{1}{\ln R} + 1\right) = 1.$$

The integral converges.

20. $\displaystyle I = \int_{-\infty}^\infty \frac{x\,dx}{1 + x^4} = \int_{-\infty}^0 + \int_0^\infty = I_1 + I_2$

$$I_2 = \int_0^\infty \frac{x\,dx}{1 + x^4} \quad \text{Let } u = x^2$$
$$\qquad\qquad\qquad\qquad\qquad du = 2x\,dx$$

$$= \frac{1}{2}\int_0^\infty \frac{du}{1 + u^2} = \frac{1}{2}\lim_{R\to\infty}\left.\tan^{-1}u\right|_0^R = \frac{\pi}{4}$$

Similarly, $I_1 = -\dfrac{\pi}{4}$. Therefore, $I = 0$.

22. $I = \displaystyle\int_{-\infty}^{\infty} e^{-|x|}\,dx = \int_{-\infty}^{0} e^x\,dx + \int_{0}^{\infty} e^{-x}\,dx = I_1 + I_2$

$I_2 = \displaystyle\int_{0}^{\infty} e^{-x}\,dx = 1$

Similarly, $I_1 = 1$. Therefore, $I = 2$.

24. Area of shaded region $= \displaystyle\int_{0}^{\infty} (e^{-x} - e^{-2x})\,dx$

$= \displaystyle\lim_{R \to \infty} \left(-e^{-x} + \frac{1}{2}e^{-2x}\right)\Bigg|_{0}^{R}$

$= \displaystyle\lim_{R \to \infty} \left(-e^{-R} + \frac{1}{2}e^{-2R} + 1 - \frac{1}{2}\right) = \frac{1}{2}$ sq. units.

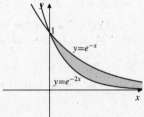

Fig. 6.5.24

26. The required area is

Area $= \displaystyle\int_{0}^{\infty} x^{-2} e^{-1/x}\,dx$

$= \displaystyle\int_{0}^{1} x^{-2} e^{-1/x}\,dx + \int_{1}^{\infty} x^{-2} e^{-1/x}\,dx$

$= I_1 + I_2$.

Then let $u = -\dfrac{1}{x}$ and $du = x^{-2}\,dx$ in both I_1 and I_2:

$I_1 = \displaystyle\lim_{C \to 0+} \int_{C}^{1} x^{-2} e^{-1/x}\,dx = \lim_{C \to 0+} \int_{-1/C}^{-1} e^u\,du$

$= \displaystyle\lim_{C \to 0+} (e^{-1} - e^{-1/C}) = \frac{1}{e}$.

$I_2 = \displaystyle\lim_{R \to \infty} \int_{1}^{R} x^{-2} e^{-1/x}\,dx = \lim_{R \to \infty} \int_{-1}^{-1/R} e^u\,du$

$= \displaystyle\lim_{R \to \infty} (e^{-1/R} - e^{-1}) = 1 - \frac{1}{e}$.

Hence, the total area is $I_1 + I_2 = 1$ square unit.

28. $\displaystyle\int_{-1}^{1} \frac{x\,\mathrm{sgn}\,x}{x+2}\,dx = \int_{-1}^{0} \frac{-x}{x+2}\,dx + \int_{0}^{1} \frac{x}{x+2}\,dx$

$= \displaystyle\int_{-1}^{0} \left(-1 + \frac{2}{x+2}\right)dx + \int_{0}^{1} \left(1 - \frac{2}{x+2}\right)dx$

$= (-x + 2\ln|x+2|)\Big|_{-1}^{0} + (x - 2\ln|x+2|)\Big|_{0}^{1} = \ln\frac{16}{9}$.

30. Since $\dfrac{x^2}{x^5+1} \le \dfrac{1}{x^3}$ for all $x \ge 0$, therefore

$I = \displaystyle\int_{0}^{\infty} \frac{x^2}{x^5+1}\,dx$

$= \displaystyle\int_{0}^{1} \frac{x^2}{x^5+1}\,dx + \int_{1}^{\infty} \frac{x^2}{x^5+1}\,dx$

$\le \displaystyle\int_{0}^{1} \frac{x^2}{x^5+1}\,dx + \int_{1}^{\infty} \frac{dx}{x^3}$

$= I_1 + I_2$.

Since I_1 is a proper integral (finite) and I_2 is a convergent improper integral, (see Theorem 2), therefore I converges.

32. Since $\dfrac{x\sqrt{x}}{x^2-1} \ge \dfrac{1}{\sqrt{x}}$ for all $x > 1$, therefore

$I = \displaystyle\int_{2}^{\infty} \frac{x\sqrt{x}}{x^2-1}\,dx \ge \int_{2}^{\infty} \frac{dx}{\sqrt{x}} = I_1 = \infty$.

Since I_1 is a divergent improper integral, I diverges.

34. On $[0,1]$, $\dfrac{1}{\sqrt{x} + x^2} \le \dfrac{1}{\sqrt{x}}$. On $[1,\infty]$, $\dfrac{1}{\sqrt{x} + x^2} \le \dfrac{1}{x^2}$. Thus,

$\displaystyle\int_{0}^{1} \frac{dx}{\sqrt{x} + x^2} \le \int_{0}^{1} \frac{dx}{\sqrt{x}}$

$\displaystyle\int_{1}^{\infty} \frac{dx}{\sqrt{x} + x^2} \le \int_{1}^{\infty} \frac{dx}{x^2}$.

Since both of these integrals are convergent, therefore so is their sum $\displaystyle\int_{0}^{\infty} \frac{dx}{\sqrt{x} + x^2}$.

36. Since $\sin x \le x$ for all $x \ge 0$, thus $\dfrac{\sin x}{x} \le 1$. Then

$I = \displaystyle\int_{0}^{\pi} \frac{\sin x}{x}\,dx = \lim_{\epsilon \to 0+} \int_{\epsilon}^{\pi} \frac{\sin x}{x}\,dx \le \int_{0}^{\pi} (1)\,dx = \pi$.

Hence, I converges.

38. Since $0 \le 1 - \cos\sqrt{x} = 2\sin^2\left(\dfrac{\sqrt{x}}{2}\right) \le 2\left(\dfrac{\sqrt{x}}{2}\right)^2 = \dfrac{x}{2}$,

for $x \ge 0$, therefore $\displaystyle\int_{0}^{\pi^2} \frac{dx}{1 - \cos\sqrt{x}} \ge 2\int_{0}^{\pi^2} \frac{dx}{x}$, which diverges to infinity.

40. Since $\ln x$ grows more slowly than any positive power of x, therefore we have $\ln x \le kx^{1/4}$ for some constant k and every $x \ge 2$. Thus, $\dfrac{1}{\sqrt{x}\ln x} \ge \dfrac{1}{kx^{3/4}}$ for $x \ge 2$

and $\displaystyle\int_{2}^{\infty} \frac{dx}{\sqrt{x}\ln x}$ diverges to infinity by comparison with

$\dfrac{1}{k}\displaystyle\int_{2}^{\infty} \frac{dx}{x^{3/4}}$.

42. We are given that $\int_0^\infty e^{-x^2}\,dx = \frac{1}{2}\sqrt{\pi}$.

a) First we calculate

$$\int_0^\infty x^2 e^{-x^2}\,dx = \lim_{R\to\infty} \int_0^R x^2 e^{-x^2}\,dx$$

$$U = x \qquad dV = xe^{-x^2}\,dx$$
$$dU = dx \qquad V = -\tfrac{1}{2}e^{-x^2}$$

$$= \lim_{R\to\infty}\left[-\frac{1}{2}xe^{-x^2}\Big|_0^R + \frac{1}{2}\int_0^R e^{-x^2}\,dx \right]$$

$$= -\frac{1}{2}\lim_{R\to\infty} Re^{-R^2} + \frac{1}{2}\int_0^\infty e^{-x^2}\,dx$$

$$= 0 + \frac{1}{4}\sqrt{\pi} = \frac{1}{4}\sqrt{\pi}.$$

b) Similarly,

$$\int_0^\infty x^4 e^{-x^2}\,dx = \lim_{R\to\infty}\int_0^R x^4 e^{-x^2}\,dx$$

$$U = x^3 \qquad dV = xe^{-x^2}\,dx$$
$$dU = 3x^2\,dx \qquad V = -\tfrac{1}{2}e^{-x^2}$$

$$= \lim_{R\to\infty}\left[-\frac{1}{2}x^3 e^{-x^2}\Big|_0^R + \frac{3}{2}\int_0^R x^2 e^{-x^2}\,dx \right]$$

$$= -\frac{1}{2}\lim_{R\to\infty} R^3 e^{-R^2} + \frac{3}{2}\int_0^\infty x^2 e^{-x^2}\,dx$$

$$= 0 + \frac{3}{2}\left(\frac{1}{4}\sqrt{\pi}\right) = \frac{3}{8}\sqrt{\pi}.$$

44. $\Gamma(x) = \int_0^\infty t^{x-1}e^{-t}\,dt$.

a) Since $\lim_{t\to\infty} t^{x-1}e^{-t/2} = 0$, there exists $T > 0$ such that $t^{x-1}e^{-t/2} \le 1$ if $t \ge T$. Thus

$$0 \le \int_T^\infty t^{x-1}e^{-t}\,dt \le \int_T^\infty e^{-t}\,dt = 2e^{-T/2}$$

and $\int_T^\infty t^{x-1}e^{-t}\,dt$ converges by the comparison theorem.

If $x > 0$, then

$$0 \le \int_0^T t^{x-1}e^{-t}\,dt < \int_0^T t^{x-1}\,dt$$

converges by Theorem 2(b). Thus the integral defining $\Gamma(x)$ converges.

b) $\Gamma(x+1) = \int_0^\infty t^x e^{-t}\,dt$

$$= \lim_{\substack{c\to 0+ \\ R\to\infty}} \int_c^R t^x e^{-t}\,dt$$

$$U = t^x \qquad dV = e^{-t}\,dt$$
$$dU = xt^{x-1}\,dx \qquad V = -e^{-t}$$

$$= \lim_{\substack{c\to 0+ \\ R\to\infty}} \left(-t^x e^{-t}\Big|_c^R + x\int_c^R t^{x-1}e^{-t}\,dt \right)$$

$$= 0 + x\int_0^\infty t^{x-1}e^{-t}\,dt = x\Gamma(x).$$

c) $\Gamma(1) = \int_0^\infty e^{-t}\,dt = 1 = 0!.$

By (b), $\Gamma(2) = 1\Gamma(1) = 1 \times 1 = 1 = 1!.$
In general, if $\Gamma(k+1) = k!$ for some positive integer k, then
$\Gamma(k+2) = (k+1)\Gamma(k+1) = (k+1)k! = (k+1)!.$
Hence $\Gamma(n+1) = n!$ for all integers $n \ge 0$, by induction.

d) $\Gamma\left(\dfrac{1}{2}\right) = \int_0^\infty t^{-1/2}e^{-t}\,dt$ Let $t = x^2$
$$dt = 2x\,dx$$

$$= \int_0^\infty \frac{1}{x}e^{-x^2}\,2x\,dx = 2\int_0^\infty e^{-x^2}\,dx = \sqrt{\pi}$$

$$\Gamma\left(\frac{3}{2}\right) = \frac{1}{2}\Gamma\left(\frac{1}{2}\right) = \frac{1}{2}\sqrt{\pi}.$$

Section 6.6 The Trapezoid and Midpoint Rules (page 392)

2. The exact value of I is

$$I = \int_0^1 e^{-x}\,dx = -e^{-x}\Big|_0^1$$

$$= 1 - \frac{1}{e} \approx 0.6321206.$$

The approximations are

$$T_4 = \frac{1}{4}\left(\frac{1}{2}e^0 + e^{-1/4} + e^{-1/2} + e^{-3/4} + \frac{1}{2}e^{-1}\right)$$
$$\approx 0.6354094$$

$$M_4 = \frac{1}{4}\left(e^{-1/8} + e^{-3/8} + e^{-5/8} + e^{-7/8}\right)$$
$$\approx 0.6304774$$

$$T_8 = \frac{1}{2}(T_4 + M_4) \approx 0.6329434$$

$$M_8 = \frac{1}{8}\left(e^{-1/16} + e^{-3/16} + e^{-5/16} + e^{-7/16} + \right.$$
$$\left. e^{-9/16} + e^{-11/16} + e^{-13/16} + e^{-15/16}\right)$$
$$\approx 0.6317092$$

$$T_{16} = \frac{1}{2}(T_8 + M_8) \approx 0.6323263.$$

The exact errors are

$$I - T_4 = -0.0032888; \quad I - M_4 = 0.0016432;$$
$$I - T_8 = -0.0008228; \quad I - M_8 = 0.0004114;$$
$$I - T_{16} = -0.0002057.$$

If $f(x) = e^{-x}$, then $f^{(2)}(x) = e^{-x}$. On $[0,1]$, $|f^{(2)}(x)| \leq 1$. Therefore, the error bounds are:

$$\text{Trapezoid} : |I - T_n| \leq \frac{1}{12}\left(\frac{1}{n}\right)^2$$

$$|I - T_4| \leq \frac{1}{12}\left(\frac{1}{16}\right) \approx 0.0052083;$$

$$|I - T_8| \leq \frac{1}{12}\left(\frac{1}{64}\right) \approx 0.001302;$$

$$|I - T_{16}| \leq \frac{1}{12}\left(\frac{1}{256}\right) \approx 0.0003255.$$

$$\text{Midpoint} : |I - M_n| \leq \frac{1}{24}\left(\frac{1}{n}\right)^2$$

$$|I - M_4| \leq \frac{1}{24}\left(\frac{1}{16}\right) \approx 0.0026041;$$

$$|I - M_8| \leq \frac{1}{24}\left(\frac{1}{64}\right) \approx 0.000651.$$

Note that the actual errors satisfy these bounds.

4. The exact value of I is

$$I = \int_0^1 \frac{dx}{1+x^2} = \tan^{-1} x \Big|_0^1 = \frac{\pi}{4} \approx 0.7853982.$$

The approximations are

$$T_4 = \frac{1}{4}\left[\frac{1}{2}(1) + \frac{16}{17} + \frac{4}{5} + \frac{16}{25} + \frac{1}{2}\left(\frac{1}{2}\right)\right]$$
$$\approx 0.7827941$$

$$M_4 = \frac{1}{4}\left[\frac{64}{65} + \frac{64}{73} + \frac{64}{89} + \frac{64}{113}\right]$$
$$\approx 0.7867001$$

$$T_8 = \frac{1}{2}(T_4 + M_4) \approx 0.7847471$$

$$M_8 = \frac{1}{8}\left[\frac{256}{257} + \frac{256}{265} + \frac{256}{281} + \frac{256}{305} + \right.$$
$$\left. \frac{256}{337} + \frac{256}{377} + \frac{256}{425} + \frac{256}{481}\right]$$
$$\approx 0.7857237$$

$$T_{16} = \frac{1}{2}(T_8 + M_8) \approx 0.7852354.$$

The exact errors are

$$I - T_4 = 0.0026041; \quad I - M_4 = -0.0013019;$$
$$I - T_8 = 0.0006511; \quad I - M_8 = -0.0003255;$$
$$I - T_{16} = 0.0001628.$$

Since $f(x) = \frac{1}{1+x^2}$, then $f'(x) = \frac{-2x}{(1+x^2)^2}$ and $f''(x) = \frac{6x^2 - 2}{(1+x^2)^3}$. On $[0,1]$, $|f''(x)| \leq 4$. Therefore, the error bounds are

$$\text{Trapezoid} : |I - T_n| \leq \frac{4}{12}\left(\frac{1}{n}\right)^2$$

$$|I - T_4| \leq \frac{4}{12}\left(\frac{1}{16}\right) \approx 0.0208333;$$

$$|I - T_8| \leq \frac{4}{12}\left(\frac{1}{64}\right) \approx 0.0052083;$$

$$|I - T_{16}| \leq \frac{4}{12}\left(\frac{1}{256}\right) \approx 0.001302.$$

$$\text{Midpoint} : |I - M_n| \leq \frac{4}{24}\left(\frac{1}{n}\right)^2$$

$$|I - M_4| \leq \frac{4}{24}\left(\frac{1}{16}\right) \approx 0.0104167;$$

$$|I - M_8| \leq \frac{4}{24}\left(\frac{1}{64}\right) \approx 0.0026042.$$

The exact errors are much smaller than these bounds. In part, this is due to very crude estimates made for $|f''(x)|$.

6. $M_4 = 2(3.8 + 6.7 + 8 + 5.2) = 47.4$

8. $M_4 = 100 \times 2(4 + 5.5 + 5.5 + 4) = 3,800 \text{ km}^2$

10. The approximations for $I = \int_0^1 e^{-x^2} \, dx$ are

$$M_8 = \frac{1}{8}\left(e^{-1/256} + e^{-9/256} + e^{-25/256} + e^{-49/256} + \right.$$
$$\left. e^{-81/256} + e^{-121/256} + e^{-169/256} + e^{-225/256}\right)$$
$$\approx 0.7473$$

$$T_{16} = \frac{1}{16}\left[\frac{1}{2}(1) + e^{-1/256} + e^{-1/64} + e^{-9/256} + e^{-1/16} + \right.$$
$$e^{-25/256} + e^{-9/64} + e^{-49/256} + e^{-1/4} + e^{-81/256} +$$
$$e^{-25/64} + e^{-121/256} + e^{-9/16} + e^{-169/256} + e^{-49/64} +$$
$$\left. e^{-225/256} + \frac{1}{2}e^{-1}\right]$$
$$\approx 0.74658.$$

Since $f(x) = e^{-x^2}$, we have $f'(x) = -2xe^{-x^2}$, $f''(x) = 2(2x^2 - 1)e^{-x^2}$, and $f'''(x) = 4x(3 - 2x^2)e^{-x^2}$. Since $f'''(x) \neq 0$ on $(0,1)$, therefore the maximum value of $|f''(x)|$ on $[0, 1]$ must occur at an endpoint of that interval. We have $f''(0) = -2$ and $f''(1) = 2/e$, so $|f''(x)| \leq 2$ on $[0, 1]$. The error bounds are

$$|I - M_n| \leq \frac{2}{24}\left(\frac{1}{n}\right)^2 \Rightarrow |I - M_8| \leq \frac{2}{24}\left(\frac{1}{64}\right)$$
$$\approx 0.00130.$$

$$|I - T_n| \leq \frac{2}{12}\left(\frac{1}{n}\right)^2 \Rightarrow |I - T_{16}| \leq \frac{2}{12}\left(\frac{1}{256}\right)$$
$$\approx 0.000651.$$

According to the error bounds,

$$\int_0^1 e^{-x^2}\, dx = 0.747,$$

accurate to two decimal places, with error no greater than 1 in the third decimal place.

12. The exact value of I is

$$I = \int_0^1 x^2\, dx = \frac{x^3}{3}\Big|_0^1 = \frac{1}{3}.$$

The approximation is

$$T_1 = (1)\left[\frac{1}{2}(0)^2 + \frac{1}{2}(1)^2\right] = \frac{1}{2}.$$

The actual error is $I - T_1 = -\frac{1}{6}$. However, since $f(x) = x^2$, then $f''(x) = 2$ on $[0,1]$, so the error estimate here gives

$$|I - T_1| \leq \frac{2}{12}(1)^2 = \frac{1}{6}.$$

Since this is the actual size of the error in this case, the constant "12" in the error estimate cannot be improved (i.e., cannot be made larger).

14. Let $y = f(x)$. We are given that m_1 is the midpoint of $[x_0, x_1]$ where $x_1 - x_0 = h$. By tangent line approximate in the subinterval $[x_0, x_1]$,

$$f(x) \approx f(m_1) + f'(m_1)(x - m_1).$$

The error in this approximation is

$$E(x) = f(x) - f(m_1) - f'(m_1)(x - m_1).$$

If $f''(t)$ exists for all t in $[x_0, x_1]$ and $|f''(t)| \leq K$ for some constant K, then by Theorem 4 of Section 3.5,

$$|E(x)| \leq \frac{K}{2}(x - m_1)^2.$$

Hence,

$$|f(x) - f(m_1) - f'(m_1)(x - m_1)| \leq \frac{K}{2}(x - m_1)^2.$$

We integrate both sides of this inequality. Noting that $x_1 - m_1 = m_1 - x_0 = \frac{1}{2}h$, we obtain for the left side

$$\left|\int_{x_0}^{x_1} f(x)\, dx - \int_{x_0}^{x_1} f(m_1)\, dx\right.$$
$$\left. - \int_{x_0}^{x_1} f'(m_1)(x - m_1)\, dx\right|$$
$$= \left|\int_{x_0}^{x_1} f(x)\, dx - f(m_1)h - f'(m_1)\frac{(x - m_1)^2}{2}\Big|_{x_0}^{x_1}\right|$$
$$= \left|\int_{x_0}^{x_1} f(x)\, dx - f(m_1)h\right|.$$

Integrating the right-hand side, we get

$$\int_{x_0}^{x_1} \frac{K}{2}(x - m_1)^2\, dx = \frac{K}{2}\frac{(x - m_1)^3}{3}\Big|_{x_0}^{x_1}$$
$$= \frac{K}{6}\left(\frac{h^3}{8} + \frac{h^3}{8}\right) = \frac{K}{24}h^3.$$

Hence,

$$\left|\int_{x_0}^{x_1} f(x)\, dx - f(m_1)h\right|$$
$$= \left|\int_{x_0}^{x_1}[f(x) - f(m_1) - f'(m_1)(x - m_1)]\, dx\right|$$
$$\leq \frac{K}{24}h^3.$$

A similar estimate holds on each subinterval $[x_{j-1}, x_j]$ for $1 \leq j \leq n$. Therefore,

$$\left|\int_a^b f(x)\, dx - M_n\right| = \left|\sum_{j=1}^{n}\left(\int_{x_{j-1}}^{x_j} f(x)\, dx - f(m_j)h\right)\right|$$
$$\leq \sum_{j=1}^{n}\left|\int_{x_{j-1}}^{x_j} f(x)\, dx - f(m_j)h\right|$$
$$\leq \sum_{j=1}^{n}\frac{K}{24}h^3 = \frac{K}{24}nh^3 = \frac{K(b - a)}{24}h^2.$$

because $nh = b - a$.

Section 6.7   Simpson's Rule   (page 397)

2. The exact value of I is

$$I = \int_0^1 e^{-x}\, dx = -e^{-x}\Big|_0^1$$

$$= 1 - \frac{1}{e} \approx 0.6321206.$$

The approximations are

$$S_4 = \frac{1}{12}(e^0 + 4e^{-1/4} + 2e^{-1/2} + 4e^{-3/4} + e^{-1})$$

$$\approx 0.6321342$$

$$S_8 = \frac{1}{24}(e^0 + 4e^{-1/8} + 2e^{-1/4} + 4e^{-3/8} +$$

$$2e^{-1/2} + 4e^{-5/8} + 2e^{-3/4} + 4e^{-7/8} + e^{-1})$$

$$\approx 0.6321214.$$

The actual errors are

$$I - S_4 = -0.0000136; \quad I - S_8 = -0.0000008.$$

These errors are evidently much smaller than the corresponding errors for the corresponding Trapezoid Rule approximations.

4. The exact value of I is

$$I = \int_0^1 \frac{dx}{1+x^2} = \tan^{-1} x\Big|_0^1 = \frac{\pi}{4} \approx 0.7853982.$$

The approximations are

$$S_4 = \frac{1}{12}\left[1 + 4\left(\frac{16}{17}\right) + 2\left(\frac{4}{5}\right) + 4\left(\frac{16}{25}\right) + \frac{1}{2}\right]$$

$$\approx 0.7853922$$

$$S_8 = \frac{1}{24}\left[1 + 4\left(\frac{64}{65}\right) + 2\left(\frac{16}{17}\right) + 4\left(\frac{64}{73}\right) + \right.$$

$$\left. 2\left(\frac{4}{5}\right) + 4\left(\frac{64}{89}\right) + 2\left(\frac{16}{25}\right) + 4\left(\frac{64}{113}\right) + \frac{1}{2}\right]$$

$$\approx 0.7853981.$$

The actual errors are

$$I - S_4 = 0.0000060; \quad I - S_8 = 0.0000001,$$

accurate to 7 decimal places. These errors are evidently much smaller than the corresponding errors for the corresponding Trapezoid Rule approximation.

6. $S_8 = 100 \times \frac{1}{3}[0 + 4(4 + 5.5 + 5.5 + 4) + 2(5.5 + 5 + 4.5) + 0]$

$\approx 3,533 \text{ km}^2$

8. Let $I = \int_a^b f(x)\, dx$, and the interval $[a, b]$ be subdivided into $2n$ subintervals of equal length $h = (b-a)/2n$. Let $y_j = f(x_j)$ and $x_j = a + jh$ for $0 \le j \le 2n$, then

$$S_{2n} = \frac{1}{3}\left(\frac{b-a}{2n}\right)\left[y_0 + 4y_1 + 2y_2 + \cdots\right.$$

$$\left. + 2y_{2n-2} + 4y_{2n-1} + y_{2n}\right]$$

$$= \frac{1}{3}\left(\frac{b-a}{2n}\right)\left[y_0 + 4\sum_{j=1}^{2n-1} y_j - 2\sum_{j=1}^{n-1} y_{2j} + y_{2n}\right]$$

and

$$T_{2n} = \frac{1}{2}\left(\frac{b-a}{2n}\right)\left(y_0 + 2\sum_{j=1}^{2n-1} y_j + y_{2n}\right)$$

$$T_n = \frac{1}{2}\left(\frac{b-a}{n}\right)\left(y_0 + 2\sum_{j=1}^{n-1} y_{2j} + y_{2n}\right).$$

Since $T_{2n} = \frac{1}{2}(T_n + M_n) \Rightarrow M_n = 2T_{2n} - T_n$, then

$$\frac{T_n + 2M_n}{3} = \frac{T_n + 2(2T_{2n} - T_n)}{3} = \frac{4T_{2n} - T_n}{3}$$

$$\frac{2T_{2n} + M_n}{3} = \frac{2T_{2n} + 2T_{2n} - T_n}{3} = \frac{4T_{2n} - T_n}{3}.$$

Hence,

$$\frac{T_n + 2M_n}{3} = \frac{2T_{2n} + M_n}{3} = \frac{4T_{2n} - T_n}{3}.$$

Using the formulas of T_{2n} and T_n obtained above,

$$\frac{4T_{2n} - T_n}{3}$$

$$= \frac{1}{3}\left[\frac{4}{2}\left(\frac{b-a}{2n}\right)\left(y_0 + 2\sum_{j=1}^{2n-1} y_j + y_{2n}\right)\right.$$

$$\left. - \frac{1}{2}\left(\frac{b-a}{n}\right)\left(y_0 + 2\sum_{j=1}^{n-1} y_{2j} + y_{2n}\right)\right]$$

$$= \frac{1}{3}\left(\frac{b-a}{2n}\right)\left[y_0 + 4\sum_{j=1}^{2n-1} y_j - 2\sum_{j=1}^{n-1} y_{2j} + y_{2n}\right]$$

$$= S_{2n}.$$

Hence,

$$S_{2n} = \frac{4T_{2n} - T_n}{3} = \frac{T_n + 2M_n}{3} = \frac{2T_{2n} + M_n}{3}.$$

10. The approximations for $I = \int_0^1 e^{-x^2}\,dx$ are

$$S_8 = \frac{1}{3}\left(\frac{1}{8}\right)\left[1 + 4\left(e^{-1/64} + e^{-9/64} + e^{-25/64} + \right.\right.$$

$$\left.e^{-49/64}\right) + 2\left(e^{-1/16} + e^{-1/4} + e^{-9/16}\right) + e^{-1}\Big]$$

$$\approx 0.7468261$$

$$S_{16} = \frac{1}{3}\left(\frac{1}{16}\right)\left[1 + 4\left(e^{-1/256} + e^{-9/256} + e^{-25/256} + \right.\right.$$

$$e^{-49/256} + e^{-81/256} + e^{-121/256} + e^{-169/256} +$$

$$\left.e^{-225/256}\right) + 2\left(e^{-1/64} + e^{-1/16} + e^{-9/64} + e^{-1/4} + \right.$$

$$\left.e^{-25/64} + e^{-9/16} + e^{-49/64}\right) + e^{-1}\Big]$$

$$\approx 0.7468243.$$

If $f(x) = e^{-x^2}$, then $f^{(4)}(x) = 4e^{-x^2}(4x^4 - 12x^2 + 3)$. On $[0,1]$, $|f^{(4)}(x)| \le 12$, and the error bounds are

$$|I - S_n| \le \frac{12(1)}{180}\left(\frac{1}{n}\right)^4$$

$$|I - S_8| \le \frac{12}{180}\left(\frac{1}{8}\right)^4 \approx 0.0000163$$

$$|I - S_{16}| \le \frac{12}{180}\left(\frac{1}{16}\right)^4 \approx 0.0000010.$$

Comparing the two approximations,

$$I = \int_0^1 e^{-x^2}\,dx = 0.7468,$$

accurate to 4 decimal places.

12. The exact value of I is

$$I = \int_0^1 x^3\,dx = \frac{x^4}{4}\bigg|_0^1 = \frac{1}{4}.$$

The approximation is

$$S_2 = \frac{1}{3}\left(\frac{1}{2}\right)\left[0^3 + 4\left(\frac{1}{2}\right)^3 + 1^3\right] = \frac{1}{4}.$$

The actual error is zero. Hence, Simpson's Rule is exact for the cubic function $f(x) = x^3$. Since it is evidently exact for quadratic functions $f(x) = Bx^2 + Cx + D$, it must also be exact for arbitrary cubics $f(x) = Ax^3 + Bx^2 + Cx + D$.

Section 6.8 Other Aspects of Approximate Integration   (page 403)

2. $\int_0^1 \dfrac{e^x}{\sqrt{1-x}}\,dx$ Let $t^2 = 1 - x$

$$2t\,dt = -dx$$

$$= -\int_1^0 \frac{e^{1-t^2}}{t}2t\,dt = 2\int_0^1 e^{1-t^2}\,dt.$$

4. $\int_1^\infty \dfrac{dx}{x^2 + \sqrt{x} + 1}$ Let $x = \dfrac{1}{t^2}$

$$dx = -\frac{2\,dt}{t^3}$$

$$= \int_1^0 \frac{1}{\left(\frac{1}{t^2}\right)^2 + \sqrt{\frac{1}{t^2}} + 1}\left(-\frac{2\,dt}{t^3}\right)$$

$$= 2\int_0^1 \frac{t\,dt}{t^4 + t^3 + 1}.$$

6. Let

$$\int_0^\infty \frac{dx}{x^4 + 1} = \int_0^1 \frac{dx}{x^4 + 1} + \int_1^\infty \frac{dx}{x^4 + 1} = I_1 + I_2.$$

Let $x = \dfrac{1}{t}$ and $dx = -\dfrac{dt}{t^2}$ in I_2, then

$$I_2 = \int_1^0 \frac{1}{\left(\frac{1}{t}\right)^4 + 1}\left(-\frac{dt}{t^2}\right) = \int_0^1 \frac{t^2}{1 + t^4}\,dt.$$

Hence,

$$\int_0^\infty \frac{dx}{x^4 + 1} = \int_0^1\left(\frac{1}{x^4 + 1} + \frac{x^2}{1 + x^4}\right)dx$$

$$= \int_0^1 \frac{x^2 + 1}{x^4 + 1}\,dx.$$

8. Let

$$I = \int_1^\infty e^{-x^2}\,dx \quad \text{Let } x = \frac{1}{t}$$

$$dx = -\frac{dt}{t^2}$$

$$= \int_1^0 e^{-(1/t)^2}\left(-\frac{1}{t^2}\right)dt = \int_0^1 \frac{e^{-1/t^2}}{t^2}\,dt.$$

Observe that

$$\lim_{t\to 0+} \frac{e^{-1/t^2}}{t^2} = \lim_{t\to 0+} \frac{t^{-2}}{e^{1/t^2}} \quad \left[\frac{\infty}{\infty}\right]$$

$$= \lim_{t\to 0+} \frac{-2t^{-3}}{e^{1/t^2}(-2t^{-3})}$$

$$= \lim_{t\to 0+} \frac{1}{e^{1/t^2}} = 0.$$

Hence,

$$S_2 = \frac{1}{3}\left(\frac{1}{2}\right)\left[0 + 4(4e^{-4}) + e^{-1}\right]$$

$$\approx 0.1101549$$

$$S_4 = \frac{1}{3}\left(\frac{1}{4}\right)\left[0 + 4(16e^{-16}) + 2(4e^{-4})\right.$$

$$\left. + 4\left(\frac{16}{9}e^{-16/9}\right) + e^{-1}\right]$$

$$\approx 0.1430237$$

$$S_8 = \frac{1}{3}\left(\frac{1}{8}\right)\left[0 + 4\left(64e^{-64} + \frac{64}{9}e^{-64/9} + \frac{64}{25}e^{-64/25} + \right.\right.$$

$$\left.\left. \frac{64}{49}e^{-64/49}\right) + 2\left(16e^{-16} + 4e^{-4} + \frac{16}{9}e^{-16/9}\right) + e^{-1}\right]$$

$$\approx 0.1393877.$$

Hence, $I \approx 0.14$, accurate to 2 decimal places. These approximations do not converge very quickly, because the fourth derivative of e^{-1/t^2} has very large values for some values of t near 0. In fact, higher and higher derivatives behave more and more badly near 0, so higher order methods cannot be expected to work well either.

10. We are given that $\int_0^\infty e^{-x^2}\,dx = \frac{1}{2}\sqrt{\pi}$ and from the previous exercise $\int_0^1 e^{-x^2}\,dx = 0.74684$. Therefore,

$$\int_1^\infty e^{-x^2}\,dx = \int_0^\infty e^{-x^2}\,dx - \int_0^1 e^{-x^2}\,dx$$

$$= \frac{1}{2}\sqrt{\pi} - 0.74684$$

$$= 0.139 \qquad \text{(to 3 decimal places).}$$

12. For any function f we use the approximation

$$\int_{-1}^1 f(x)\,dx \approx f(-1/\sqrt{3}) + f(1/\sqrt{3}).$$

We have

$$\int_{-1}^1 x^4\,dx \approx \left(-\frac{1}{\sqrt{3}}\right)^4 + \left(\frac{1}{\sqrt{3}}\right)^4 = \frac{2}{9}$$

$$\text{Error} = \int_{-1}^1 x^4\,dx - \frac{2}{9} = \frac{2}{5} - \frac{2}{9} \approx 0.17778$$

$$\int_{-1}^1 \cos x\,dx \approx \cos\left(-\frac{1}{\sqrt{3}}\right) + \cos\left(\frac{1}{\sqrt{3}}\right) \approx 1.67582$$

$$\text{Error} = \int_{-1}^1 \cos x\,dx - 1.67582 \approx 0.00712$$

$$\int_{-1}^1 e^x\,dx \approx e^{-1/\sqrt{3}} + e^{1/\sqrt{3}} \approx 2.34270$$

$$\text{Error} = \int_{-1}^1 e^x\,dx - 2.34270 \approx 0.00771.$$

14. For any function f we use the approximation

$$\int_{-1}^1 f(x)\,dx \approx \frac{5}{9}\left[f(-\sqrt{3/5}) + f(\sqrt{3/5})\right] + \frac{8}{9}f(0).$$

We have

$$\int_{-1}^1 x^6\,dx \approx \frac{5}{9}\left[\left(-\sqrt{\frac{3}{5}}\right)^6 + \left(\sqrt{\frac{3}{5}}\right)^6\right] + 0 = 0.24000$$

$$\text{Error} = \int_{-1}^1 x^6\,dx - 0.24000 \approx 0.04571$$

$$\int_{-1}^1 \cos x\,dx \approx \frac{5}{9}\left[\cos\left(-\sqrt{\frac{3}{5}}\right) + \cos\left(\sqrt{\frac{3}{5}}\right)\right] + \frac{8}{9}$$

$$\approx 1.68300$$

$$\text{Error} = \int_{-1}^1 \cos x\,dx - 1.68300 \approx 0.00006$$

$$\int_{-1}^1 e^x\,dx \approx e^{-\sqrt{3/5}} + e^{\sqrt{3/5}} \approx 2.35034$$

$$\text{Error} = \int_{-1}^1 e^x\,dx - 2.35034 \approx 0.00006.$$

16. From Exercise 9 in Section 7.6, for $I = \int_0^{1.6} f(x)\,dx$,

$$T_0^0 = T_1 = 1.9196$$

$$T_1^0 = T_2 = 2.00188$$

$$T_2^0 = T_4 = 2.02622$$

$$T_3^0 = T_8 = 2.02929.$$

Hence,

$$R_1 = T_1^1 = \frac{4T_1^0 - T_0^0}{3} = 2.0346684$$

$$T_2^1 = \frac{4T_2^0 - T_1^0}{3} = 2.0343333 = S_4$$

$$R_2 = T_2^2 = \frac{16T_2^1 - T_1^1}{15} = 2.0346684$$

$$T_3^1 = \frac{4T_3^0 - T_2^0}{3} = 2.0303133 = S_8$$

$$T_3^2 = \frac{16T_3^1 - T_2^1}{15} = 2.0300453$$

$$R_3 = T_3^3 = \frac{64T_3^2 - T_2^2}{63} = 2.0299719.$$

18. Let

$$I = \int_\pi^\infty \frac{\sin x}{1 + x^2}\, dx \quad \text{Let } x = \frac{1}{t}$$

$$dx = -\frac{dt}{t^2}$$

$$= \int_{1/\pi}^0 \frac{\sin\left(\frac{1}{t}\right)}{1 + \left(\frac{1}{t^2}\right)}\left(-\frac{1}{t^2}\right) dt$$

$$= \int_0^{1/\pi} \frac{\sin\left(\frac{1}{t}\right)}{1 + t^2}\, dt.$$

The transformation is not suitable because the derivative of $\sin\left(\frac{1}{t}\right)$ is $-\frac{1}{t^2}\cos\left(\frac{1}{t}\right)$, which has very large values at some points close to 0.

In order to approximate the integral I to an desired degree of accuracy, say with error less than ϵ in absolute value, we have to divide the integral into two parts:

$$I = \int_\pi^\infty \frac{\sin x}{1 + x^2}\, dx$$

$$= \int_\pi^t \frac{\sin x}{1 + x^2}\, dx + \int_t^\infty \frac{\sin x}{1 + x^2}\, dx$$

$$= I_1 + I_2.$$

If $t \geq \tan\dfrac{\pi - \epsilon}{2}$, then

$$\int_t^\infty \frac{\sin x}{1 + x^2}\, dx < \int_t^\infty \frac{dx}{1 + x^2}$$

$$= \tan^{-1}(x)\Big|_t^\infty = \frac{\pi}{2} - \tan^{-1}(t) \leq \frac{\epsilon}{2}.$$

Now let A be a numerical approximation to the proper integral $\int_\pi^t \frac{\sin x}{1 + x^2}\, dx$, having error less than $\epsilon/2$ in absolute value. Then

$$|I - A| = |I_1 + I_2 - A|$$
$$\leq |I_1 - A| + |I_2|$$
$$\leq \frac{\epsilon}{2} + \frac{\epsilon}{2} = \epsilon.$$

Hence, A is an approximation to the integral I with the desired accuracy.

Review Exercises on Techniques of Integration (page 404)

2. $\displaystyle\int \frac{x}{(x-1)^3}\, dx \quad \text{Let } u = x - 1$
$$du = dx$$

$$= \int \frac{u + 1}{u^3}\, du = \int \left(\frac{1}{u^2} + \frac{1}{u^3}\right) du$$

$$= -\frac{1}{u} - \frac{1}{2u^2} + C = -\frac{1}{x-1} - \frac{1}{2(x-1)^2} + C.$$

4. $\displaystyle\int \frac{(1 + \sqrt{x})^{1/3}}{\sqrt{x}}\, dx \quad \text{Let } u = 1 + \sqrt{x}$
$$du = \frac{dx}{2\sqrt{x}}$$

$$= 2\int u^{1/3}\, du = 2(\tfrac{3}{4})u^{4/3} + C$$

$$= \tfrac{3}{2}(1 + \sqrt{x})^{4/3} + C.$$

6. $\displaystyle\int (x^2 + x - 2)\sin 3x\, dx$

$$U = x^2 + x - 2 \qquad dV = \sin 3x$$
$$dU = (2x + 1)\, dx \qquad V = -\tfrac{1}{3}\cos 3x$$

$$= -\tfrac{1}{3}(x^2 + x - 2)\cos 3x + \tfrac{1}{3}\int (2x + 1)\cos 3x\, dx$$

$$U = 2x + 1 \qquad dV = \cos 3x\, dx$$
$$dU = 2\, dx \qquad V = \tfrac{1}{3}\sin 3x$$

$$= -\tfrac{1}{3}(x^2 + x - 2)\cos 3x + \tfrac{1}{9}(2x + 1)\sin 3x$$

$$\qquad - \tfrac{2}{9}\int \sin 3x\, dx$$

$$= -\tfrac{1}{3}(x^2 + x - 2)\cos 3x + \tfrac{1}{9}(2x + 1)\sin 3x$$

$$\qquad + \frac{2}{27}\cos 3x + C.$$

8. $\displaystyle\int x^3 \cos(x^2)\, dx$ Let $w = x^2$
$$dw = 2x\, dx$$
$$= \tfrac{1}{2}\int w \cos w\, dw$$

$$U = w \qquad dV = \cos w\, dw$$
$$dU = dw \qquad V = \sin w$$

$$= \tfrac{1}{2} w \sin w - \tfrac{1}{2}\int \sin w\, dw$$
$$= \tfrac{1}{2} x^2 \sin(x^2) + \tfrac{1}{2}\cos(x^2) + C.$$

10. $\displaystyle\frac{1}{x^2 + 2x - 15} = \frac{A}{x-3} + \frac{B}{x+5} = \frac{(A+B)x + (5A - 3B)}{x^2 + 2x - 15}$
$$\Rightarrow \begin{cases} A + B = 0 \\ 5A - 3B = 1 \end{cases} \Rightarrow A = \frac{1}{8},\ B = -\frac{1}{8}.$$
$$\int \frac{dx}{x^2 + 2x - 15} = \frac{1}{8}\int \frac{dx}{x-3} - \frac{1}{8}\int \frac{dx}{x+5}$$
$$= \frac{1}{8}\ln\left|\frac{x-3}{x+5}\right| + C.$$

12. $\displaystyle\int (\sin x + \cos x)^2\, dx = \int (1 + \sin 2x)\, dx$
$$= x - \tfrac{1}{2}\cos 2x + C.$$

14. $\displaystyle\int \frac{\cos x}{1 + \sin^2 x}\, dx$ Let $u = \sin x$
$$du = \cos x\, dx$$
$$= \int \frac{du}{1 + u^2} = \tan^{-1} u + C$$
$$= \tan^{-1}(\sin x) + C.$$

16. We have
$$\int \frac{x^2\, dx}{(3 + 5x^2)^{3/2}} \quad \text{Let } x = \sqrt{\tfrac{3}{5}}\tan u$$
$$dx = \sqrt{\tfrac{3}{5}}\sec^2 u\, du$$
$$= \int \frac{(\tfrac{3}{5}\tan^2 u)(\sqrt{\tfrac{3}{5}}\sec^2 u)\, du}{(3)^{3/2}\sec^3 u}$$
$$= \frac{1}{5\sqrt{5}}\int (\sec u - \cos u)\, du$$
$$= \frac{1}{5\sqrt{5}}(\ln|\sec u + \tan u| - \sin u) + C$$
$$= \frac{1}{5\sqrt{5}}\left(\ln\left|\frac{\sqrt{5x^2 + 3}}{\sqrt{3}} + \frac{\sqrt{5}x}{\sqrt{3}}\right| - \frac{\sqrt{5}x}{\sqrt{5x^2 + 3}}\right) + C$$
$$= \frac{1}{5\sqrt{5}}\ln\left(\sqrt{5x^2 + 3} + \sqrt{5}x\right) - \frac{x}{5\sqrt{5x^2 + 3}} + C_0,$$
$$\text{where } C_0 = C - \frac{1}{5\sqrt{5}}\ln\sqrt{3}.$$

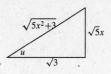

Fig. RT.16

18. $\displaystyle I = \int \frac{2x^2 + 4x - 3}{x^2 + 5x}\, dx = \int \frac{2x^2 + 10x - 6x - 3}{x^2 + 5x}\, dx$
$$= \int \left[2 - \frac{6x + 3}{x(x + 5)}\right] dx$$
$$\frac{6x + 3}{x(x + 5)} = \frac{A}{x} + \frac{B}{x+5} = \frac{(A+B)x + 5A}{x(x+5)}$$
$$\Rightarrow \begin{cases} A + B = 6 \\ 5A = 3 \end{cases} \Rightarrow A = \frac{3}{5},\ B = \frac{27}{5}.$$
$$I = \int 2\, dx - \frac{3}{5}\int \frac{dx}{x} - \frac{27}{5}\int \frac{dx}{x+5}$$
$$= 2x - \frac{3}{5}\ln|x| - \frac{27}{5}\ln|x + 5| + C.$$

20. $\displaystyle\frac{1}{4x^3 + x} = \frac{A}{x} + \frac{Bx + C}{4x^2 + 1}$
$$= \frac{A(4x^2 + 1) + Bx^2 + Cx}{4x^3 + x}$$
$$\Rightarrow \begin{cases} 4A + B = 0 \\ C = 0,\ A = 1 \end{cases} \Rightarrow B = -4.$$
$$\int \frac{1}{4x^3 + x}\, dx = \int \frac{dx}{x} - 4\int \frac{x\, dx}{4x^2 + 1}$$
$$= \ln|x| - \frac{1}{2}\ln(4x^2 + 1) + C.$$

22. $\displaystyle\int \sin^2 x \cos^4 x\, dx$
$$= \int \tfrac{1}{2}(1 - \cos 2x)[\tfrac{1}{2}(1 + \cos 2x)]^2\, dx$$
$$= \frac{1}{8}\int (1 + \cos 2x - \cos^2 2x - \cos^3 2x)\, dx$$
$$= \frac{1}{8}x + \frac{1}{16}\sin 2x - \frac{1}{16}\int (1 + \cos 4x)\, dx$$
$$\quad - \frac{1}{8}\int (1 - \sin^2 2x)\cos 2x\, dx$$
$$= \frac{x}{8} + \frac{1}{16}\sin 2x - \frac{x}{16} - \frac{1}{64}\sin 4x - \frac{1}{16}\sin 2x$$
$$\quad + \frac{1}{48}\sin^3 2x + C$$
$$= \frac{x}{16} - \frac{\sin 4x}{64} + \frac{\sin^3 2x}{48} + C.$$

24. We have

$$I = \int \tan^4 x \, \sec x \, dx$$

$$U = \tan^3 x \qquad\qquad dV = \tan x \, \sec x \, dx$$
$$dU = 3\tan^2 x \, \sec^2 x \, dx \qquad V = \sec x$$

$$= \tan^3 x \, \sec x - 3 \int \tan^2 x \, \sec^3 x \, dx$$

$$= \tan^3 x \, \sec x - 3 \int \tan^2 x (\tan^2 x + 1) \sec x \, dx$$

$$= \tan^3 x \, \sec x - 3I - 3J \quad \text{where}$$

$$J = \int \tan^2 x \, \sec x \, dx$$

$$U = \tan x \qquad dV = \tan x \, \sec x \, dx$$
$$dU = \sec^2 x \, dx \qquad V = \sec x$$

$$= \tan x \, \sec x - \int \sec^3 x \, dx$$

$$= \tan x \, \sec x - \int (\tan^2 x + 1) \sec x \, dx$$

$$= \tan x \, \sec x - J - \ln|\sec x + \tan x| + C$$

$$J = \tfrac{1}{2} \tan x \, \sec x - \tfrac{1}{2} \ln|\sec x + \tan x| + C.$$

$$I = \tfrac{1}{4} \tan^3 x \, \sec x - \tfrac{3}{8} \tan x \, \sec x$$
$$\quad + \tfrac{3}{8} \ln|\sec x + \tan x| + C.$$

26. We have

$$\int x \sin^{-1}\left(\frac{x}{2}\right) dx$$

$$U = \sin^{-1}\left(\frac{x}{2}\right) \qquad dV = x \, dx$$
$$dU = \frac{dx}{\sqrt{4 - x^2}} \qquad V = \frac{x^2}{2}$$

$$= \frac{x^2}{2} \sin^{-1}\left(\frac{x}{2}\right) - \frac{1}{2} \int \frac{x^2 \, dx}{\sqrt{4 - x^2}} \quad \begin{array}{l} \text{Let } x = 2\sin u \\ dx = 2\cos u \, du \end{array}$$

$$= \frac{x^2}{2} \sin^{-1}\left(\frac{x}{2}\right) - 2 \int \sin^2 u \, du$$

$$= \frac{x^2}{2} \sin^{-1}\left(\frac{x}{2}\right) - \int (1 - \cos 2u) \, du$$

$$= \frac{x^2}{2} \sin^{-1}\left(\frac{x}{2}\right) - u + \sin u \, \cos u + C$$

$$= \left(\frac{x^2}{2} - 1\right) \sin^{-1}\left(\frac{x}{2}\right) + \frac{1}{4} x\sqrt{4 - x^2} + C.$$

28. We have

$$I = \int \frac{dx}{x^5 - 2x^3 + x} = \int \frac{x \, dx}{x^6 - 2x^4 + x^2} \quad \begin{array}{l} \text{Let } u = x^2 \\ du = 2x \, dx \end{array}$$

$$= \frac{1}{2} \int \frac{du}{u^3 - 2u^2 + u} = \frac{1}{2} \int \frac{du}{u(u - 1)^2}$$

$$\frac{1}{u(u - 1)^2} = \frac{A}{u} + \frac{B}{u - 1} + \frac{C}{(u - 1)^2}$$

$$= \frac{A(u^2 - 2u + 1) + B(u^2 - u) + Cu}{u^3 - 2u^2 + u}$$

$$\Rightarrow \begin{cases} A + B = 0 \\ -2A - B + C = 0 \Rightarrow A = 1, \ B = -1, \ C = 1. \\ A = 1 \end{cases}$$

$$\frac{1}{2} \int \frac{du}{u^3 - 2u^2 + u} = \frac{1}{2} \int \frac{du}{u} - \frac{1}{2} \int \frac{du}{u - 1}$$
$$\qquad\qquad + \frac{1}{2} \int \frac{du}{(u - 1)^2}$$

$$= \frac{1}{2} \ln|u| - \frac{1}{2} \ln|u - 1| - \frac{1}{2}\frac{1}{u - 1} + K$$

$$= \frac{1}{2} \ln \frac{x^2}{|x^2 - 1|} - \frac{1}{2(x^2 - 1)} + K.$$

30. Let

$$I_n = \int x^n 3^x \, dx$$

$$U = x^n \qquad dV = 3^x \, dx$$
$$dU = nx^{n-1} \, dx \qquad V = \frac{3^x}{\ln 3}$$

$$= \frac{x^n 3^x}{\ln 3} - \frac{n}{\ln 3} I_{n-1}.$$

$$I_0 = \int 3^x \, dx = \frac{3^x}{\ln 3} + C.$$

Hence,

$$I_3 = \int x^3 3^x \, dx$$

$$= \frac{x^3 3^x}{\ln 3} - \frac{3}{\ln 3} \left[\frac{x^2 3^x}{\ln 3} - \frac{2}{\ln 3} \left(\frac{x 3^x}{\ln 3} - \frac{1}{\ln 3} I_0 \right) \right] + C_1$$

$$= 3^x \left[\frac{x^3}{\ln 3} - \frac{3x^2}{(\ln 3)^2} + \frac{6x}{(\ln 3)^3} - \frac{6}{(\ln 3)^4} \right] + C_1.$$

32. We have

$$\int \frac{x^2 + 1}{x^2 + 2x + 2} \, dx = \int \left(1 - \frac{2x + 1}{x^2 + 2x + 2} \right) dx$$

$$= x - \int \frac{2x + 1}{(x + 1)^2 + 1} \, dx \quad \begin{array}{l} \text{Let } u = x + 1 \\ du = dx \end{array}$$

$$= x - \int \frac{2u - 1}{u^2 + 1} \, du$$

$$= x - \ln|u^2 + 1| + \tan^{-1} u + C$$

$$= x - \ln(x^2 + 2x + 2) + \tan^{-1}(x + 1) + C.$$

34. We have

$$\int x^3 (\ln x)^2 \, dx$$

$$U = (\ln x)^2 \qquad dV = x^3 \, dx$$

$$dU = \frac{2}{x} \ln x \, dx \qquad V = \frac{1}{4} x^4$$

$$= \frac{1}{4} x^4 (\ln x)^2 - \frac{1}{2} \int x^3 \ln x \, dx$$

$$U = \ln x \qquad dV = x^3 \, dx$$

$$dU = \frac{1}{x} \, dx \qquad V = \frac{1}{4} x^4$$

$$= \frac{1}{4} x^4 (\ln x)^2 - \frac{1}{8} x^4 \ln x + \frac{1}{8} \int x^3 \, dx$$

$$= \frac{x^4}{4} \left[(\ln x)^2 - \frac{1}{2} \ln x + \frac{1}{8} \right] + C.$$

36. $\displaystyle \int \frac{e^{1/x}}{x^2} \, dx$ Let $u = \dfrac{1}{x}$

$$du = -\frac{1}{x^2} \, dx$$

$$= -\int e^u \, du = -e^u + C = -e^{1/x} + C.$$

38. $\displaystyle \int e^{(x^{1/3})}$ Let $x = u^3$

$$dx = 3u^2 \, du$$

$$= 3 \int u^2 e^u \, du = 3 I_2$$

See solution to #16 of Section 6.6 for

$$I_n = \int u^n e^u \, dx = u^n e^u - n I_{n-1}.$$

$$= 3 [u^2 e^u - 2(u e^u - e^u)] + C$$

$$= e^{(x^{1/3})} (3 x^{2/3} - 6 x^{1/3} + 6) + C.$$

40. $\displaystyle \int \frac{10^{\sqrt{x+2}} \, dx}{\sqrt{x+2}}$ Let $u = \sqrt{x+2}$

$$du = \frac{dx}{2\sqrt{x+2}}$$

$$= 2 \int 10^u \, du = \frac{2}{\ln 10} 10^u + C = \frac{2}{\ln 10} 10^{\sqrt{x+2}} + C.$$

42. Assume that $x \geq 1$ and let $x = \sec u$ and $dx = \sec u \tan u \, du.$ Then

$$\int \frac{x^2 \, dx}{\sqrt{x^2 - 1}}$$

$$= \int \frac{\sec^3 u \tan u \, du}{\tan u} = \int \sec^3 u \, du$$

$$= \frac{1}{2} \sec u \tan u + \frac{1}{2} \ln |\sec u + \tan u| + C$$

$$= \frac{1}{2} x \sqrt{x^2 - 1} + \frac{1}{2} \ln |x + \sqrt{x^2 - 1}| + C.$$

Differentiation shows that this solution is valid for $x \leq -1$ also.

44. $\displaystyle \int \frac{2x - 3}{\sqrt{4 - 3x + x^2}} \, dx$ Let $u = 4 - 3x + x^2$

$$du = (-3 + 2x) \, dx$$

$$= \int \frac{du}{\sqrt{u}} = 2\sqrt{u} + C = 2\sqrt{4 - 3x + x^2} + C.$$

46. Let $\sqrt{3} x = \sec u$ and $\sqrt{3} \, dx = \sec u \tan u \, du.$ Then

$$\int \frac{\sqrt{3x^2 - 1}}{x} \, dx$$

$$= \int \frac{\tan u \, \dfrac{1}{\sqrt{3}} \sec u \tan u \, du}{\dfrac{1}{\sqrt{3}} \sec u}$$

$$= \int \tan^2 u \, du = \int (\sec^2 u - 1) \, du$$

$$= \tan u - u + C = \sqrt{3x^2 - 1} - \sec^{-1}(\sqrt{3} x) + C$$

$$= \sqrt{3x^2 - 1} + \sin^{-1} \left(\frac{1}{\sqrt{3} x} \right) + C_1.$$

48. $\displaystyle \int \sqrt{x - x^2} \, dx$

$$= \int \sqrt{\frac{1}{4} - (x - \tfrac{1}{2})^2} \, dx$$ Let $x - \tfrac{1}{2} = \tfrac{1}{2} \sin u$

$$dx = \tfrac{1}{2} \cos u \, du$$

$$= \frac{1}{4} \int \cos^2 u \, du = \frac{1}{8} u + \frac{1}{8} \sin u \cos u + C$$

$$= \frac{1}{8} \sin^{-1}(2x - 1) + \frac{1}{4}(2x - 1)\sqrt{x - x^2} + C.$$

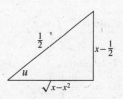

Fig. RT.48

50. $\displaystyle \int x \tan^{-1} \left(\frac{x}{3} \right) dx$

$$U = \tan^{-1} \left(\frac{x}{3} \right) \qquad dV = x \, dx$$

$$dU = \frac{3 \, dx}{9 + x^2} \qquad V = \frac{x^2}{2}$$

$$= \frac{x^2}{2} \tan^{-1} \left(\frac{x}{3} \right) - \frac{3}{2} \int \frac{x^2}{9 + x^2} \, dx$$

$$= \frac{x^2}{2} \tan^{-1} \left(\frac{x}{3} \right) - \frac{3}{2} \int \left(1 - \frac{9}{9 + x^2} \right) dx$$

$$= \frac{x^2}{2} \tan^{-1} \left(\frac{x}{3} \right) - \frac{3x}{2} + \frac{9}{2} \tan^{-1} \left(\frac{x}{3} \right) + C.$$

52. Let $u = x^2$ and $du = 2x\,dx$; then we have

$$I = \int \frac{dx}{x(x^2+4)^2} = \int \frac{x\,dx}{x^2(x^2+4)^2} = \frac{1}{2}\int \frac{du}{u(u+4)^2};$$

Since

$$\frac{1}{u(u+4)^2} = \frac{A}{u} + \frac{B}{u+4} + \frac{C}{(u+4)^2}$$

$$= \frac{A(u^2+8u+16) + B(u^2+4u) + Cu}{u(u+4)^2}$$

$$\Rightarrow \begin{cases} A+B=0 \\ 8A+4B+C=0 \Rightarrow A=\frac{1}{16}, \ B=-\frac{1}{16}, \ C=-\frac{1}{4}, \\ 16A=1 \end{cases}$$

therefore

$$I = \frac{1}{32}\int \frac{du}{u} - \frac{1}{32}\int \frac{du}{u+4} - \frac{1}{8}\int \frac{du}{(u+4)^2}$$

$$= \frac{1}{32}\ln\left|\frac{u}{u+4}\right| + \frac{1}{8}\frac{1}{u+4} + C$$

$$= \frac{1}{32}\ln\left|\frac{x^2}{x^2+4}\right| + \frac{1}{8(x^2+4)} + C.$$

54. Since

$$I = \int \frac{\sin(\ln x)}{x^2}\,dx$$

$$\begin{array}{ll} U = \sin(\ln x) & dV = \dfrac{dx}{x^2} \\[2mm] dU = \dfrac{\cos(\ln x)}{x}\,dx & V = -\dfrac{1}{x} \end{array}$$

$$= -\frac{\sin(\ln x)}{x} + \int \frac{\cos(\ln x)}{x^2}\,dx$$

$$\begin{array}{ll} U = \cos(\ln x) & dV = \dfrac{dx}{x^2} \\[2mm] dU = -\dfrac{\sin(\ln x)}{x}\,dx & V = \dfrac{-1}{x} \end{array}$$

$$= -\frac{\sin(\ln x)}{x} - \frac{\cos(\ln x)}{x} - I,$$

therefore

$$I = -\frac{1}{2x}\Big[\sin(\ln x) + \cos(\ln x)\Big] + C.$$

56. We have

$$I = \int \frac{x^3+x-2}{x^2-7}\,dx = \int \frac{x^3-7x+8x-2}{x^2-7}\,dx$$

$$= \int \left(x + \frac{8x-2}{x^2-7}\right)dx.$$

Since

$$\frac{8x-2}{x^2-7} = \frac{A}{x+\sqrt{7}} + \frac{B}{x-\sqrt{7}} = \frac{(A+B)x + (B-A)\sqrt{7}}{x^2-7}$$

$$\Rightarrow \begin{cases} A+B=8 \\ B-A=-\dfrac{2}{\sqrt{7}} \end{cases} \Rightarrow A=4+\frac{1}{\sqrt{7}}, \ B=4-\frac{1}{\sqrt{7}},$$

therefore

$$I = \int \left(x + \frac{8x-2}{x^2-7}\right)dx$$

$$= \frac{x^2}{2} + \left(4+\frac{1}{\sqrt{7}}\right)\int \frac{dx}{x+\sqrt{7}} + \left(4-\frac{1}{\sqrt{7}}\right)\int \frac{dx}{x-\sqrt{7}}$$

$$= \frac{x^2}{2} + \left(4+\frac{1}{\sqrt{7}}\right)\ln|x+\sqrt{7}| + \left(4-\frac{1}{\sqrt{7}}\right)\ln|x-\sqrt{7}| + C.$$

58.

$$\int \cos^7 x\,dx = \int (1-\sin^2 x)^3 \cos x\,dx \quad \begin{array}{l} \text{Let } u=\sin x \\ du=\cos x\,dx \end{array}$$

$$= \int (1-u^2)^3\,du = \int (1-3u^2+3u^4-u^6)\,du$$

$$= u - u^3 + \tfrac{3}{5}u^5 - \tfrac{1}{7}u^7 + C$$

$$= \sin x - \sin^3 x + \tfrac{3}{5}\sin^5 x - \tfrac{1}{7}\sin^7 x + C.$$

60. We have

$$\int \tan^4(\pi x)\,dx = \int \tan^2(\pi x)[\sec^2(\pi x) - 1]\,dx$$

$$= \int \tan^2(\pi x)\sec^2(\pi x)\,dx - \int [\sec^2(\pi x) - 1]\,dx$$

$$= \frac{1}{3\pi}\tan^3(\pi x) - \frac{1}{\pi}\tan(\pi x) + x + C.$$

62. $\displaystyle\int e^x(1-e^{2x})^{5/2}\,dx$ Let $e^x = \sin u$

$$e^x\,dx = \cos u\,du$$

$$= \int \cos^6 u\,du = \left(\frac{1}{2}\right)^3 \int (1+\cos 2u)^3\,du$$

$$= \frac{1}{8}\int (1 + 3\cos 2u + 3\cos^2 2u + \cos^3 2u)\,du$$

$$= \frac{u}{8} + \frac{3}{16}\sin 2u + \frac{3}{16}\int(1+\cos 4u)\,du +$$

$$\quad \frac{1}{8}\int(1-\sin^2 2u)\cos 2u\,du$$

$$= \frac{5u}{16} + \frac{3}{16}\sin 2u + \frac{3}{64}\sin 4u + \frac{\sin 2u}{16}$$

$$\quad - \frac{1}{48}\sin^3 2u + C$$

$$= \frac{5}{16}\sin^{-1}(e^x) + \frac{1}{4}\sin[2\sin^{-1}(e^x)] +$$

$$\quad \frac{3}{64}\sin[4\sin^{-1}(e^x)] - \frac{1}{48}\sin^3[2\sin^{-1}(e^x)] + C$$

$$= \frac{5}{16}\sin^{-1}(e^x) + \frac{1}{2}e^x\sqrt{1-e^{2x}}$$

$$\quad + \frac{3}{16}e^x\sqrt{1-e^{2x}}\left(1-2e^{2x}\right)$$

$$\quad - \frac{1}{6}e^{3x}\left(1-e^{2x}\right)^{3/2} + C.$$

64. $\displaystyle\int \frac{x^2}{2x^2-3}\,dx = \frac{1}{2}\int\left(1+\frac{3}{2x^2-3}\right)dx$

$$= \frac{x}{2} + \frac{\sqrt{3}}{4}\int\left(\frac{1}{\sqrt{2}x-\sqrt{3}} - \frac{1}{\sqrt{2}x+\sqrt{3}}\right)dx$$

$$= \frac{x}{2} + \frac{\sqrt{3}}{4\sqrt{2}}\ln\left|\frac{\sqrt{2}x-\sqrt{3}}{\sqrt{2}x+\sqrt{3}}\right| + C.$$

66. We have

$$\int \frac{dx}{x(x^2+x+1)^{1/2}}$$

$$= \int \frac{dx}{x[(x+\frac{1}{2})^2+\frac{3}{4}]^{1/2}}$$ Let $x + \dfrac{1}{2} = \dfrac{\sqrt{3}}{2}\tan\theta$

$$dx = \frac{\sqrt{3}}{2}\sec^2\theta\,d\theta$$

$$= \int \frac{\frac{\sqrt{3}}{2}\sec^2\theta\,d\theta}{\left(\frac{\sqrt{3}}{2}\tan\theta - \frac{1}{2}\right)\left(\frac{\sqrt{3}}{2}\sec\theta\right)}$$

$$= \int \frac{2\sec\theta\,d\theta}{\sqrt{3}\tan\theta - 1} = 2\int \frac{d\theta}{\sqrt{3}\sin\theta - \cos\theta}$$

$$= 2\int \frac{\sqrt{3}\sin\theta + \cos\theta}{3\sin^2\theta - \cos^2\theta}\,d\theta$$

$$= 2\sqrt{3}\int \frac{\sin\theta\,d\theta}{3\sin^2\theta - \cos^2\theta} + 2\int \frac{\cos\theta\,d\theta}{3\sin^2\theta - \cos^2\theta}$$

$$= 2\sqrt{3}\int \frac{\sin\theta\,d\theta}{3 - 4\cos^2\theta} + 2\int \frac{\cos\theta\,d\theta}{4\sin^2\theta - 1}$$

Let $u = \cos\theta$, $du = -\sin\theta\,d\theta$ in the first integral;
let $v = \sin\theta$, $dv = \cos\theta\,d\theta$ in the second integral.

$$= -2\sqrt{3}\int \frac{du}{3-4u^2} + 2\int \frac{dv}{4v^2-1}$$

$$= -\frac{\sqrt{3}}{2}\int \frac{du}{\frac{3}{4}-u^2} - \frac{1}{2}\int \frac{du}{\frac{1}{4}-v^2}$$

$$= -\frac{\sqrt{3}}{2}\left(\frac{1}{2}\right)\left(\frac{2}{\sqrt{3}}\right)\ln\left|\frac{\cos\theta + \frac{\sqrt{3}}{2}}{\cos\theta - \frac{\sqrt{3}}{2}}\right|$$

$$\quad -\frac{1}{2}\left(\frac{1}{2}\right)(2)\ln\left|\frac{\sin\theta + \frac{1}{2}}{\sin\theta - \frac{1}{2}}\right| + C$$

$$= \frac{1}{2}\ln\left|\frac{\left(\cos\theta - \frac{\sqrt{3}}{2}\right)\left(\sin\theta - \frac{1}{2}\right)}{\left(\cos\theta + \frac{\sqrt{3}}{2}\right)\left(\sin\theta + \frac{1}{2}\right)}\right| + C.$$

Since $\sin\theta = \dfrac{2x+1}{2\sqrt{x^2+x+1}}$ and $\cos\theta = \dfrac{\sqrt{3}}{2\sqrt{x^2+x+1}}$, therefore

$$\int \frac{dx}{x(x^2+x+1)^{1/2}} = \frac{1}{2}\ln\left|\frac{(x+2) - 2\sqrt{x^2+x+1}}{(x+2) + 2\sqrt{x^2+x+1}}\right| + C.$$

68. $\displaystyle\int \frac{x\,dx}{4x^4+4x^2+5}$ Let $u = x^2$

$$du = 2x\,dx$$

$$= \frac{1}{2}\int \frac{du}{4u^2+4u+5}$$

$$= \frac{1}{2}\int \frac{du}{(2u+1)^2+4}$$ Let $w = 2u+1$

$$dw = 2du$$

$$= \frac{1}{4}\int \frac{dw}{w^2+4} = \frac{1}{8}\tan^{-1}\left(\frac{w}{2}\right) + C$$

$$= \frac{1}{8}\tan^{-1}\left(x^2 + \frac{1}{2}\right) + C.$$

70. Use the partial fraction decomposition

$$\frac{1}{x^3+x^2+x} = \frac{A}{x} + \frac{Bx+C}{x^2+x+1}$$

$$= \frac{A(x^2+x+1) + Bx^2 + Cx}{x^3+x^2+x}$$

$$\Rightarrow \begin{cases} A+B=0 \\ A+C=0 \\ A=1 \end{cases} \Rightarrow A=1,\ B=-1,\ C=-1.$$

131

Therefore,

$$\int \frac{dx}{x^3 + x^2 + x}$$

$$= \int \frac{dx}{x} - \int \frac{x+1}{x^2 + x + 1}\, dx \quad \text{Let } u = x + \frac{1}{2}$$
$$\qquad\qquad\qquad\qquad\qquad\qquad du = dx$$

$$= \ln|x| - \int \frac{u + \frac{1}{2}}{u^2 + \frac{3}{4}}\, du$$

$$= \ln|x| - \frac{1}{2}\ln(x^2 + x + 1) - \frac{1}{\sqrt{3}}\tan^{-1}\left(\frac{2x+1}{\sqrt{3}}\right) + C.$$

72. $\displaystyle \int e^x \sec(e^x)\, dx \quad \text{Let } u = e^x$
$$\qquad\qquad\qquad\qquad\qquad du = e^x\, dx$$

$$= \int \sec u\, du = \ln|\sec u + \tan u| + C$$

$$= \ln|\sec(e^x) + \tan(e^x)| + C.$$

74. $\displaystyle \int \frac{dx}{x^{1/3} - 1} \quad \text{Let } x = (u+1)^3$
$$\qquad\qquad\qquad\qquad\qquad dx = 3(u+1)^2\, du$$

$$= 3\int \frac{(u+1)^2}{u}\, du = 3\int \left(u + 2 + \frac{1}{u}\right) du$$

$$= 3\left(\frac{u^2}{2} + 2u + \ln|u|\right) + C$$

$$= \frac{3}{2}(x^{1/3} - 1)^2 + 6(x^{1/3} - 1) + 3\ln|x^{1/3} - 1| + C.$$

76. $\displaystyle \int \frac{x\, dx}{\sqrt{3 - 4x - 4x^2}} = \int \frac{x\, dx}{\sqrt{4 - (2x+1)^2}} \quad \begin{array}{l}\text{Let } u = 2x + 1 \\ du = 2\, dx\end{array}$

$$= \frac{1}{4}\int \frac{u - 1}{\sqrt{4 - u^2}}\, du$$

$$= -\frac{1}{4}\sqrt{4 - u^2} - \frac{1}{4}\sin^{-1}\left(\frac{u}{2}\right) + C$$

$$= -\frac{1}{4}\sqrt{3 - 4x - 4x^2} - \frac{1}{4}\sin^{-1}\left(x + \frac{1}{2}\right) + C.$$

78. $\displaystyle \int \sqrt{1 + e^x}\, dx \quad \text{Let } u^2 = 1 + e^x$
$$\qquad\qquad\qquad\qquad\qquad 2u\, du = e^x\, dx$$

$$= \int \frac{2u^2\, du}{u^2 - 1} = \int \left(2 + \frac{2}{u^2 - 1}\right) du$$

$$= \int \left(2 + \frac{1}{u - 1} - \frac{1}{u + 1}\right) du$$

$$= 2u + \ln\left|\frac{u - 1}{u + 1}\right| + C$$

$$= 2\sqrt{1 + e^x} + \ln\left|\frac{\sqrt{1 + e^x} - 1}{\sqrt{1 + e^x} + 1}\right| + C.$$

80. By the procedure used in Example 4 of Section 7.1,

$$\int e^x \cos x\, dx = \frac{1}{2}e^x(\sin x + \cos x) + C,$$

$$\int e^x \sin x\, dx = \frac{1}{2}e^x(\sin x - \cos x) + C.$$

Now

$$\int x e^x \cos x\, dx$$

$$\qquad U = x \qquad dV = e^x \cos x\, dx$$
$$\qquad dU = dx \qquad V = \frac{1}{2}e^x(\sin x + \cos x)$$

$$= \frac{1}{2}x e^x(\sin + \cos x) - \frac{1}{2}\int e^x(\sin x + \cos x)\, dx$$

$$= \frac{1}{2}x e^x(\sin + \cos x)$$
$$\qquad - \frac{1}{4}e^x(\sin x - \cos x + \sin x + \cos x) + C$$

$$= \frac{1}{2}x e^x(\sin x + \cos x) - \frac{1}{2}e^x \sin x + C.$$

Other Review Exercises 6   (page 405)

2. $\displaystyle \int_0^\infty x^r e^{-x}\, dx$

$$= \lim_{\substack{c \to 0+ \\ R \to \infty}} \int_c^R x^r e^{-x}\, dx$$

$$\qquad U = x^r \qquad\qquad dV = e^{-x}\, dx$$
$$\qquad dU = rx^{r-1}\, dr \qquad V = -e^{-x}$$

$$= \lim_{\substack{c \to 0+ \\ R \to \infty}} -x^r e^{-x}\Big|_c^R + r\int_0^\infty x^{r-1} e^{-x}\, dx$$

$$= \lim_{c \to 0+} c^r e^{-c} + r\int_0^\infty x^{r-1} e^{-x}\, dx$$

because $\lim_{R\to\infty} R^r e^{-R} = 0$ for any r. In order to ensure that $\lim_{c\to 0+} c^r e^{-c} = 0$ we must have $\lim_{c\to 0+} c^r = 0$, so we need $r > 0$.

4. $\displaystyle \int_1^\infty \frac{dx}{x + x^3} = \lim_{R\to\infty} \int_1^R \left(\frac{1}{x} - \frac{x}{1 + x^2}\right) dx$

$$= \lim_{R\to\infty} \left(\ln|x| - \frac{1}{2}\ln(1 + x^2)\right)\Big|_1^R$$

$$= \lim_{R\to\infty} \frac{1}{2}\left(\ln \frac{R^2}{1 + R^2} + \ln 2\right) = \frac{\ln 2}{2}$$

6. $\displaystyle \int_0^1 \frac{dx}{x\sqrt{1 - x^2}} > \int_0^1 \frac{dx}{x} = \infty \text{ (diverges)}$

Therefore $\displaystyle \int_{-1}^1 \frac{dx}{x\sqrt{1 - x^2}}$ diverges.

8. Volume $= \int_0^{60} A(x)\,dx$. The approximation is

$$T_6 = \frac{10}{2}\big[10,200 + 2(9,200 + 8,000 + 7,100$$
$$+\, 4,500 + 2,400) + 100\big]$$
$$\approx 364,000 \text{ m}^3.$$

10. $I = \int_0^1 \sqrt{2 + \sin(\pi x)}\,dx$

$$T_4 = \frac{1}{8}\big[\sqrt{2} + 2(\sqrt{2 + \sin(\pi/4)} + \sqrt{2 + \sin(\pi/2)}$$
$$+\, \sqrt{2 + \sin(3\pi/4)} + \sqrt{2}\big]$$
$$\approx 1.609230$$

$$M_4 = \frac{1}{4}\big[\sqrt{2 + \sin(\pi/8)} + \sqrt{2 + \sin(3\pi/8)}$$
$$\sqrt{2 + \sin(5\pi/8)} + \sqrt{2 + \sin(7\pi/8)}\big]$$
$$\approx 1.626765$$
$$I \approx 1.6$$

12. $I = \int_{1/2}^{\infty} \frac{x^2}{x^5 + x^3 + 1}\,dx$ Let $x = 1/t$
$$\qquad\qquad\qquad\qquad dx = -(1/t^2)\,dt$$

$$= \int_0^2 \frac{(1/t^4)\,dt}{(1/t^5) + (1/t^3) + 1} = \int_0^2 \frac{t\,dt}{t^5 + t^2 + 1}$$

$T_4 \approx 0.4444 \qquad M_4 \approx 0.4799$

$T_8 \approx 0.4622 \qquad M_8 \approx 0.4708$

$S_8 \approx 0.4681 \qquad S_{16} \approx 0.4680$

$I \approx 0.468$ to 3 decimal places

Challenging Problems 6 (page 406)

2. a) $I_n = \int (1 - x^2)^n\,dx$

$$U = (1 - x^2)^n \qquad\qquad dV = dx$$
$$dU = -2nx(1 - x^2)^{n-1}\,dx \qquad V = x$$

$$= x(1 - x^2)^n + 2n \int x^2(1 - x^2)^{n-1}\,dx$$

$$= x(1 - x^2)^n - 2n \int (1 - x^2 - 1)(1 - x^2)^{n-1}\,dx$$

$$= x(1 - x^2)^n - 2nI_n + 2nI_{n-1}, \quad \text{so}$$

$$I_n = \frac{1}{2n+1}x(1 - x^2)^n + \frac{2n}{2n+1}I_{n-1}.$$

b) Let $J_n = \int_0^1 (1 - x^2)^n\,dx$. Observe that $J_0 = 1$. By
(a), if $n > 0$, then we have

$$J_n = \frac{x(1 - x^2)^n}{2n+1}\Big|_0^1 + \frac{2n}{2n+1}J_{n-1} = \frac{2n}{2n+1}J_{n-1}.$$

Therefore,

$$J_n = \frac{2n}{2n+1} \cdot \frac{2n-2}{2n-1} \cdots \frac{4}{5} \cdot \frac{2}{3} J_0$$
$$= \frac{[(2n)(2n-2)\cdots(4)(2)]^2}{(2n+1)!} = \frac{2^{2n}(n!)^2}{(2n+1)!}.$$

c) From (a):

$$I_{n-1} = \frac{2n+1}{2n}I_n - \frac{1}{2n}x(1 - x^2)^n.$$

Thus

$$\int (1 - x^2)^{-3/2}\,dx = I_{-3/2}$$
$$= \frac{2(-1/2)+1}{-1}I_{-1/2} - \frac{1}{-1}x(1 - x^2)^{-1/2}$$
$$= \frac{x}{\sqrt{1 - x^2}}.$$

4. $I_{m,n} = \int_0^1 x^m(\ln x)^n\,dx$ Let $x = e^{-t}$
$$\qquad\qquad\qquad\qquad dx = -e^{-t}\,dt$$

$$= \int_0^{\infty} e^{-mt}(-t)^n e^{-t}\,dt$$

$$= (-1)^n \int_0^{\infty} t^n e^{-(m+1)t}\,dt \quad \text{Let } u = (m+1)t$$
$$\qquad\qquad\qquad\qquad\qquad du = (m+1)\,dt$$

$$= \frac{(-1)^n}{(m+1)^n} \int_0^{\infty} u^n e^{-u}\,du$$

$$= \frac{(-1)^n}{(m+1)^n}\Gamma(n+1) \quad \text{(see \#50 in Section 7.5)}$$

$$= \frac{(-1)^n n!}{(m+1)^n}.$$

6. $I = \int_0^1 e^{-Kx}\,dx = \frac{e^{-Kx}}{-K}\Big|_0^1 = \frac{1}{K}\left(1 - \frac{1}{e^K}\right).$

For very large K, the value of I is very small ($I < 1/K$).
However,

$$T_{100} = \frac{1}{100}(1 + \cdots) > \frac{1}{100}$$
$$S_{100} = \frac{1}{300}(1 + \cdots) > \frac{1}{300}$$
$$M_{100} = \frac{1}{100}(e^{-K/200} + \cdots) < \frac{1}{100}.$$

In each case the $\cdots$ represent terms much less than the
first term (shown) in the sum. Evidently M_{100} is smallest
if k is much greater than 100, and is therefore the best
approximation. T_{100} appears to be the worst.

8. a) $f'(x) < 0$ on $[1, \infty)$, and $\lim_{x\to\infty} f(x) = 0$. Therefore

$$\int_1^\infty |f'(x)| \, dx = -\int_1^\infty f'(x) \, dx$$

$$= -\lim_{R\to\infty} \int_1^R f'(x) \, dx$$

$$= \lim_{R\to\infty} (f(1) - f(R)) = f(1).$$

Thus

$$\left| \int_R^\infty f'(x) \cos x \, dx \right| \le \int_R^\infty |f'(x)| \, dx \to 0 \text{ as } R \to \infty.$$

Thus $\lim_{R\to\infty} \int_1^R f'(x) \cos x \, dx$ exists.

b) $\displaystyle\int_1^\infty f(x) \sin x \, dx$

$$U = f(x) \qquad dV = \sin x \, dx$$
$$dU = f'(x) \, dx \qquad V = -\cos x$$

$$= \lim_{R\to\infty} f(x) \cos x \Big|_1^R + \int_1^\infty f'(x) \cos x \, dx$$

$$= -f(1) \cos(1) + \int_1^\infty f'(x) \cos x \, dx;$$

the integral converges.

c) $f(x) = 1/x$ satisfies the conditions of part (a), so

$$\int_1^\infty \frac{\sin x}{x} \, dx \quad \text{converges}$$

by part (b). Similarly, it can be shown that

$$\int_1^\infty \frac{\cos(2x)}{x} \, dx \quad \text{converges.}$$

But since $|\sin x| \ge \sin^2 x = \frac{1}{2}(1 - \cos(2x))$, we have

$$\int_1^\infty \frac{|\sin x|}{x} \, dx \ge \int_1^\infty \frac{1 - \cos(2x)}{2x}.$$

The latter integral diverges because $\int_1^\infty (1/x) \, dx$ diverges to infinity while $\int_1^\infty (\cos(2x))/(2x) \, dx$ converges. Therefore

$$\int_1^\infty \frac{|\sin x|}{x} \, dx \quad \text{diverges to infinity.}$$

CHAPTER 7. APPLICATIONS OF INTE-GRATION

Section 7.1 Volumes of Solids of Revolution (page 416)

2. Slicing:

$$V = \pi \int_0^1 (1 - y)\, dy$$

$$= \pi \left(y - \frac{1}{2}y^2 \right) \Big|_0^1 = \frac{\pi}{2} \text{ cu. units.}$$

Shells:

$$V = 2\pi \int_0^1 x^3\, dx$$

$$= 2\pi \left(\frac{x^4}{4} \right) \Big|_0^1 = \frac{\pi}{2} \text{ cu. units.}$$

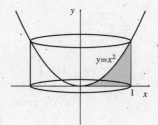

Fig. 7.1.2

4. Slicing:

$$V = \pi \int_0^1 (y - y^4)\, dy$$

$$= \pi \left(\frac{1}{2}y^2 - \frac{1}{5}y^5 \right) \Big|_0^1 = \frac{3\pi}{10} \text{ cu. units.}$$

Shells:

$$V = 2\pi \int_0^1 x(x^{1/2} - x^2)\, dx$$

$$= 2\pi \left(\frac{2}{5}x^{5/2} - \frac{1}{4}x^4 \right) \Big|_0^1 = \frac{3\pi}{10} \text{ cu. units.}$$

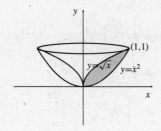

Fig. 7.1.4

6. Rotate about

a) the x-axis

$$V = \pi \int_0^1 (x^2 - x^4)\, dx$$

$$= \pi \left(\frac{1}{3}x^3 - \frac{1}{5}x^5 \right) \Big|_0^1 = \frac{2\pi}{15} \text{ cu. units.}$$

b) the y-axis

$$V = 2\pi \int_0^1 x(x - x^2)\, dx$$

$$= 2\pi \left(\frac{1}{3}x^3 - \frac{1}{4}x^4 \right) \Big|_0^1 = \frac{\pi}{6} \text{ cu. units.}$$

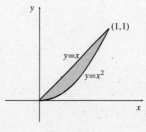

Fig. 7.1.6

8. Rotate about

a) the x-axis

$$V = \pi \int_0^\pi [(1 + \sin x)^2 - 1]\, dx$$

$$= \pi \int_0^\pi (2 \sin x + \sin^2 x)\, dx$$

$$= \left(-2\pi \cos x + \frac{\pi}{2}x - \frac{\pi}{4} \sin 2x \right) \Big|_0^\pi$$

$$= 4\pi + \frac{1}{2}\pi^2 \text{ cu. units.}$$

b) the y-axis

$$V = 2\pi \int_0^\pi x \sin x\, dx$$

$$U = x \qquad dV = \sin x\, dx$$

$$dU = dx \qquad V = -\cos x$$

$$= 2\pi \left[-x \cos x \Big|_0^\pi + \int_0^\pi \cos x\, dx \right]$$

$$= 2\pi^2 \text{ cu. units.}$$

10. By symmetry, rotation about the x-axis gives the same volume as rotation about the y-axis, namely

$$V = 2\pi \int_{1/3}^{3} x\left(\frac{10}{3} - x - \frac{1}{x}\right) dx$$

$$= 2\pi \left(\frac{5}{3}x^2 - \frac{1}{3}x^3 - x\right)\Bigg|_{1/3}^{3}$$

$$= \frac{512\pi}{81} \text{ cu. units.}$$

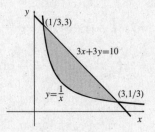

Fig. 7.1.10

12. $V = \pi \int_{-1}^{1} [(1)^2 - (x^2)^2] dx$

$$= \pi \left(x - \frac{1}{5}x^5\right)\Bigg|_{-1}^{1}$$

$$= \frac{8\pi}{5} \text{ cu. units.}$$

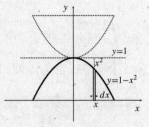

Fig. 7.1.12

14. The radius of the hole is $\sqrt{R^2 - \frac{1}{4}L^2}$. Thus, by slicing, the remaining volume is

$$V = \pi \int_{-L/2}^{L/2}\left[\left(R^2 - x^2\right) - \left(R^2 - \frac{L^2}{4}\right)\right] dx$$

$$= 2\pi \left(\frac{L^2}{4}x - \frac{1}{3}x^3\right)\Bigg|_{0}^{L/2}$$

$$= \frac{\pi}{6}L^3 \text{ cu. units (independent of } R).$$

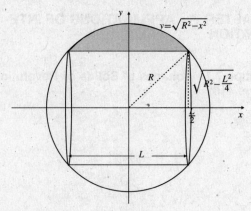

Fig. 7.1.14

16. Let a circular disk with radius a have centre at point $(a, 0)$. Then the disk is rotated about the y-axis which is one of its tangent lines. The volume is:

$$V = 2 \times 2\pi \int_{0}^{2a} x\sqrt{a^2 - (x - a)^2}\, dx \quad \text{Let } u = x - a$$
$$\qquad\qquad\qquad\qquad\qquad\qquad\qquad du = dx$$

$$= 4\pi \int_{-a}^{a} (u + a)\sqrt{a^2 - u^2}\, du$$

$$= 4\pi \int_{-a}^{a} u\sqrt{a^2 - u^2}\, du + 4\pi a \int_{-a}^{a} \sqrt{a^2 - u^2}\, du$$

$$= 0 + 4\pi a\left(\frac{1}{2}\pi a^2\right) = 2\pi^2 a^3 \text{ cu. units.}$$

(Note that the first integral is zero because the integrand is odd and the interval is symmetric about zero; the second integral is the area of a semicircle.)

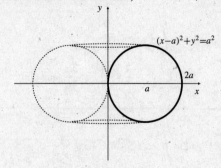

Fig. 7.1.16

18. Let the centre of the bowl be at $(0, 30)$. Then the volume of the water in the bowl is

$$V = \pi \int_{0}^{20} \left[30^2 - (y - 30)^2\right] dy$$

$$= \pi \int_{0}^{20} 60y - y^2\, dy$$

$$= \pi \left[30y^2 - \frac{1}{3}y^3\right]\Bigg|_{0}^{20}$$

$$\approx 29322 \text{ cm}^3.$$

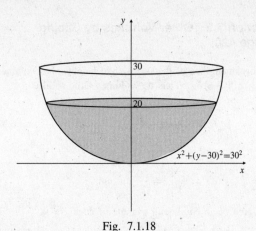

Fig. 7.1.18

20. The cross-section at height y is an annulus (ring) having inner radius $b - \sqrt{a^2 - y^2}$ and outer radius $b + \sqrt{a^2 - y^2}$. Thus the volume of the torus is

$$V = \pi \int_{-a}^{a} \left[(b + \sqrt{a^2 - y^2})^2 - (b - \sqrt{a^2 - y^2})^2 \right] dy$$

$$= 2\pi \int_{0}^{a} 4b\sqrt{a^2 - y^2} \, dy$$

$$= 8\pi b \frac{\pi a^2}{4} = 2\pi^2 a^2 b \text{ cu. units..}$$

We used the area of a quarter-circle of radius a to evaluate the last integral.

22. The volume is

$$V = \pi \int_{1}^{\infty} x^{-2k} \, dx = \pi \lim_{R \to \infty} \left. \frac{x^{1-2k}}{1-2k} \right|_{1}^{R}$$

$$= \pi \lim_{R \to \infty} \frac{R^{1-2k}}{1-2k} + \frac{\pi}{2k-1}.$$

In order for the solid to have finite volume we need

$$1 - 2k < 0, \quad \text{that is,} \quad k > \frac{1}{2}.$$

24. Using heights $f(x)$ estimated from the given graph, we obtain

$$V = \pi \int_{1}^{9} \left(f(x) \right)^2 dx$$

$$\approx \frac{\pi}{3} \left[3^2 + 4(3.8)^2 + 2(5)^2 + 4(6.7)^2 + 2(8)^2 \right.$$

$$\left. + 4(8)^2 + 2(7)^2 + 4(5.2)^2 + 3^2 \right] \approx 938 \text{ cu. units.}$$

26. Using heights $f(x)$ estimated from the given graph, we obtain

$$V = 2\pi \int_{1}^{9} (x+1) f(x) \, dx$$

$$\approx \frac{2\pi}{3} \left[2(3) + 4(3)(3.8) + 2(4)(5) + 4(5)(6.7) + 2(6)(8) \right.$$

$$\left. + 4(7)(8) + 2(8)(7) + 4(9)(5.2) + 10(3) \right] \approx 1832 \text{ cu. units.}$$

28. The volume of the ball is $\frac{4}{3}\pi R^3$. Expressing this volume as the "sum" (i.e., integral) of volume elements that are concentric spherical shells of radius r and thickness dr, and therefore surface area kr^2 and volume $kr^2 \, dr$, we obtain

$$\frac{4}{3}\pi R^3 = \int_{0}^{R} kr^2 \, dr = \frac{k}{3}R^3.$$

Thus $k = 4\pi$.

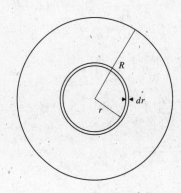

Fig. 7.1.28

30. Let P be the point $(t, \frac{5}{2} - t)$. The line through P perpendicular to AB has equation $y = x + \frac{5}{2} - 2t$, and meets the curve $xy = 1$ at point Q with x-coordinate s equal to the positive root of $s^2 + (\frac{5}{2} - 2t)s = 1$. Thus,

$$s = \frac{1}{2}\left[2t - \frac{5}{2} + \sqrt{\left(\frac{5}{2} - 2t\right)^2 + 4} \right].$$

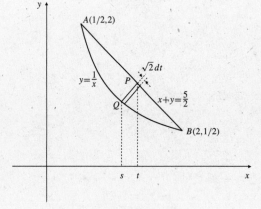

Fig. 7.1.30

The volume element at P has radius

$$PQ = \sqrt{2}(t-s)$$

$$= \sqrt{2}\left[\frac{5}{4} - \frac{1}{2}\sqrt{\left(\frac{5}{2} - 2t\right)^2 + 4}\,\right]$$

and thickness $\sqrt{2}\,dt$. Hence, the volume of the solid is

$$V = \pi \int_{1/2}^{2} \left[\sqrt{2}\left(\frac{5}{4} - \frac{1}{2}\sqrt{\left(\frac{5}{2} - 2t\right)^2 + 4}\,\right)\right]^2 \sqrt{2}\,dt$$

$$= 2\sqrt{2}\pi \int_{1/2}^{2} \left[\frac{25}{16} - \frac{5}{4}\left(\sqrt{\left(\frac{5}{2} - 2t\right)^2 + 4}\,\right) + \right.$$

$$\left. \frac{1}{4}\left[\left(\frac{5}{2} - 2t\right)^2 + 4\right]\right] dt \quad \text{Let } u = 2t - \frac{5}{2}$$
$$\qquad\qquad\qquad\qquad\qquad du = 2\,dt$$

$$= \sqrt{2}\pi \int_{-3/2}^{3/2} \left(\frac{41}{16} - \frac{5}{4}\sqrt{u^2 + 4} + \frac{u^2}{4}\right) du$$

$$= \sqrt{2}\pi \left(\frac{41}{16}u + \frac{1}{12}u^3\right)\Bigg|_{-3/2}^{3/2} -$$

$$\frac{5\sqrt{2}\pi}{4}\int_{-3/2}^{3/2} \sqrt{u^2 + 4}\,du \quad \text{Let } u = 2\tan v$$
$$\qquad\qquad\qquad\qquad\qquad du = 2\sec^2 v\,dv$$

$$= \frac{33\sqrt{2}\pi}{4} - 5\sqrt{2}\pi \int_{\tan^{-1}(-3/4)}^{\tan^{-1}(3/4)} \sec^3 v\,dv$$

$$= \frac{33\sqrt{2}\pi}{4} - 10\sqrt{2}\pi \int_{0}^{\tan^{-1}(3/4)} \sec^3 v\,dv$$

$$= \frac{33\sqrt{2}\pi}{4} - 5\sqrt{2}\pi\left(\sec v \tan v + \right.$$

$$\left. \ln|\sec v + \tan v|\right)\Bigg|_{0}^{\tan^{-1}(3/4)}$$

$$= \sqrt{2}\pi\left[\frac{33}{4} - 5\left(\frac{15}{16} + \ln 2 - 0 - \ln 1\right)\right]$$

$$= \sqrt{2}\pi\left(\frac{57}{16} - 5\ln 2\right) \text{ cu. units.}$$

Section 7.2 Other Volumes by Slicing (page 420)

2. A horizontal slice of thickness dz at height a has volume $dV = z(h - z)\,dz$. Thus the volume of the solid is

$$V = \int_{0}^{h} (z(h-z)\,dz = \left(\frac{hz^2}{2} - \frac{z^3}{3}\right)\Bigg|_{0}^{h} = \frac{h^3}{6} \text{ units}^3.$$

4. $V = \displaystyle\int_{1}^{3} x^2\,dx = \frac{x^3}{3}\Bigg|_{1}^{3} = \frac{26}{3}$ cu. units

6. The area of an equilateral triangle of edge $\sqrt{x}$ is $A(x) = \frac{1}{2}\sqrt{x}\left(\frac{\sqrt{3}}{2}\sqrt{x}\right) = \frac{\sqrt{3}}{4}x$ sq. units. The volume of the solid is

$$V = \int_{1}^{4} \frac{\sqrt{3}}{4}x\,dx = \frac{\sqrt{3}}{8}x^2\Bigg|_{1}^{4} = \frac{15\sqrt{3}}{8} \text{ cu. units.}$$

8. Since $V = 4$, we have

$$4 = \int_{0}^{2} kx^3\,dx = k\frac{x^4}{4}\Bigg|_{0}^{2} = 4k.$$

Thus $k = 1$.

10. This is similar to Exercise 7. We have $4z = \displaystyle\int_{0}^{z} A(t)\,dt$, so $A(z) = 4$. Thus the square cross-section at height z has side 2 units.

12. The area of an equilateral triangle of base $2y$ is $\frac{1}{2}(2y)(\sqrt{3}y) = \sqrt{3}y^2$. Hence, the solid has volume

$$V = 2\int_{0}^{r} \sqrt{3}(r^2 - x^2)\,dx$$

$$= 2\sqrt{3}\left(r^2 x - \frac{1}{3}x^3\right)\Bigg|_{0}^{r}$$

$$= \frac{4}{\sqrt{3}}r^3 \text{ cu. units.}$$

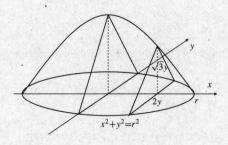

Fig. 7.2.12

14. The volume of a solid of given height h and given cross-sectional area $A(z)$ at height z above the base is given by

$$V = \int_0^h A(z)\,dz.$$

If two solids have the same height h and the same area function $A(z)$, then they must necessarily have the same volume.

16. The plane $z = k$ meets the ellipsoid in the ellipse

$$\left(\frac{x}{a}\right)^2 + \left(\frac{y}{b}\right)^2 = 1 - \left(\frac{k}{c}\right)^2$$

that is, $\dfrac{x^2}{a^2\left[1 - \left(\frac{k}{c}\right)^2\right]} + \dfrac{y^2}{b^2\left[1 - \left(\frac{k}{c}\right)^2\right]} = 1$

which has area

$$A(k) = \pi ab\left[1 - \left(\frac{k}{c}\right)^2\right].$$

The volume of the ellipsoid is found by summing volume elements of thickness dk:

$$V = \int_{-c}^c \pi ab\left[1 - \left(\frac{k}{c}\right)^2\right] dk$$

$$= \pi ab\left[k - \frac{1}{3c^2}k^3\right]\Bigg|_{-c}^c$$

$$= \frac{4}{3}\pi abc \text{ cu. units.}$$

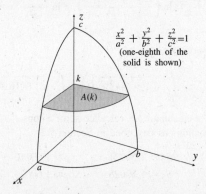

$\dfrac{x^2}{a^2} + \dfrac{y^2}{b^2} + \dfrac{z^2}{c^2} = 1$
(one-eighth of the solid is shown)

$A(k)$

Fig. 7.2.16

18. The solution is similar to that of Exercise 15 except that the legs of the right-triangular cross-sections are $y - 10$ instead of y, and x goes from $-10\sqrt{3}$ to $10\sqrt{3}$ instead of -20 to 20. The volume of the notch is

$$V = 2\int_0^{10\sqrt{3}} \frac{1}{2}(\sqrt{400 - x^2} - 10)^2\,dx$$

$$= \int_0^{10\sqrt{3}} \left(500 - x^2 - 20\sqrt{400 - x^2}\right)dx$$

$$= 3{,}000\sqrt{3} - \frac{4{,}000\pi}{3} \approx 1{,}007 \text{ cm}^3.$$

20. One eighth of the region lying inside both cylinders is shown in the figure. If the region is sliced by a horizontal plane at height z, then the intersection is a rectangle with area

$$A(z) = \sqrt{b^2 - z^2}\sqrt{a^2 - z^2}.$$

The volume of the whole region is

$$V = 8\int_0^b \sqrt{b^2 - z^2}\sqrt{a^2 - z^2}\,dz.$$

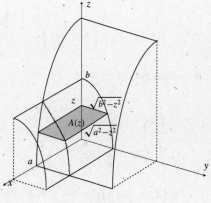

Fig. 7.2.20

Section 7.3 Arc Length and Surface Area (page 428)

2. $y = ax + b$, $A \leq x \leq B$, $y' = a$. The length is

$$L = \int_A^B \sqrt{1+a^2}\, dx = \sqrt{1+a^2}(B-A) \text{ units.}$$

4. $y^2 = (x-1)^3$, $\quad y = (x-1)^{3/2}$, $\quad y' = \dfrac{3}{2}\sqrt{x-1}$

$$L = \int_1^2 \sqrt{1 + \frac{9}{4}(x-1)}\, dx = \frac{1}{2}\int_1^2 \sqrt{9x-5}\, dx$$

$$= \frac{1}{27}(9x-5)^{3/2}\Big|_1^2 = \frac{13^{3/2}-8}{27} \text{ units.}$$

6. $2(x+1)^3 = 3(y-1)^2$, $\quad y = 1 + \sqrt{\frac{2}{3}}(x+1)^{3/2}$

$$y' = \sqrt{\frac{3}{2}}(x+1)^{1/2},$$

$$ds = \sqrt{1 + \frac{3x+3}{2}}\, dx = \sqrt{\frac{3x+5}{2}}\, dx$$

$$L = \frac{1}{\sqrt{2}}\int_{-1}^0 \sqrt{3x+5}\, dx = \frac{\sqrt{2}}{9}(3x+5)^{3/2}\Big|_{-1}^0$$

$$= \frac{\sqrt{2}}{9}\left(5^{3/2} - 2^{3/2}\right) \text{ units.}$$

8. $y = \dfrac{x^3}{3} + \dfrac{1}{4x}$, $\quad y' = x^2 - \dfrac{1}{4x^2}$

$$ds = \sqrt{1 + \left(x^2 - \frac{1}{4x^2}\right)^2}\, dx = \left(x^2 + \frac{1}{4x^2}\right) dx$$

$$L = \int_1^2 \left(x^2 + \frac{1}{4x^2}\right) dx = \left(\frac{x^3}{3} - \frac{1}{4x}\right)\Big|_1^2 = \frac{59}{24} \text{ units.}$$

10. If $y = x^2 - \dfrac{\ln x}{8}$ then $y' = 2x - \dfrac{1}{8x}$ and

$$1 + (y')^2 = \left(2x + \frac{1}{8x}\right)^2.$$

Thus the arc length is given by

$$s = \int_1^2 \sqrt{1 + \left(2x - \frac{1}{8x}\right)^2}\, dx$$

$$= \int_1^2 \left(2x + \frac{1}{8x}\right) dx$$

$$= \left(x^2 + \frac{1}{8}\ln x\right)\Big|_1^2 = 3 + \frac{1}{8}\ln 2 \text{ units.}$$

12. $s = \displaystyle\int_{\pi/6}^{\pi/4} \sqrt{1 + \tan^2 x}\, dx$

$$= \int_{\pi/6}^{\pi/4} \sec x\, dx = \ln|\sec x + \tan x|\Big|_{\pi/6}^{\pi/4}$$

$$= \ln(\sqrt{2}+1) - \ln\left(\frac{2}{\sqrt{3}} + \frac{1}{\sqrt{3}}\right)$$

$$= \ln\frac{\sqrt{2}+1}{\sqrt{3}} \text{ units.}$$

14. $y = \ln\dfrac{e^x - 1}{e^x + 1}$, $\qquad 2 \leq x \leq 4$

$$y' = \frac{e^x + 1}{e^x - 1}\frac{(e^x+1)e^x - (e^x-1)e^x}{(e^x+1)^2}$$

$$= \frac{2e^x}{e^{2x} - 1}.$$

The length of the curve is

$$L = \int_2^4 \sqrt{1 + \frac{4e^{2x}}{(e^{2x}-1)^2}}\, dx$$

$$= \int_2^4 \frac{e^{2x} + 1}{e^{2x} - 1}\, dx$$

$$= \int_2^4 \frac{e^x + e^{-x}}{e^x - e^{-x}}\, dx = \ln\left|e^x - e^{-x}\right|\Big|_2^4$$

$$= \ln\left(e^4 - \frac{1}{e^4}\right) - \ln\left(e^2 - \frac{1}{e^2}\right)$$

$$= \ln\left(\frac{e^8 - 1}{e^4}\frac{e^2}{e^4 - 1}\right) = \ln\frac{e^4 + 1}{e^2} \text{ units.}$$

16. The required length is

$$L = \int_0^1 \sqrt{1 + (4x^3)^2}\, dx = \int_0^1 \sqrt{1 + 16x^6}\, dx.$$

Using a calculator we calculate some Simpson's Rule approximations as described in Section 7.2:

$$S_2 \approx 1.59921 \qquad S_4 \approx 1.60110$$
$$S_8 \approx 1.60025 \qquad S_{16} \approx 1.60023.$$

To four decimal places the length is 1.6002 units.

18. For the ellipse $3x^2 + y^2 = 3$, we have $6x + 2yy' = 0$, so $y' = -3x/y$. Thus

$$ds = \sqrt{1 + \frac{9x^2}{3 - 3x^2}}\, dx = \sqrt{\frac{3 + 6x^2}{3 - 3x^2}}\, dx.$$

The circumference of the ellipse is

$$4\int_0^1 \sqrt{\frac{3 + 6x^2}{3 - 3x^2}}\, dx \approx 8.73775 \text{ units}$$

(with a little help from Maple's numerical integration routine.)

20. $S = 2\pi \int_0^2 |x| \sqrt{1 + 4x^2}\,dx$ Let $u = 1 + 4x^2$
$$du = 8x\,dx$$

$= \dfrac{\pi}{4} \int_1^{17} \sqrt{u}\,du = \dfrac{\pi}{4} \left(\dfrac{2}{3} u^{3/2}\right)\Big|_1^{17}$

$= \dfrac{\pi}{6} (17\sqrt{17} - 1)$ sq. units.

22. $y = x^{3/2}$, $0 \le x \le 1$. $ds = \sqrt{1 + \frac{9}{4}x}\,dx$.

The area of the surface of rotation about the x-axis is

$S = 2\pi \int_0^1 x^{3/2} \sqrt{1 + \dfrac{9x}{4}}\,dx$ Let $9x = 4u^2$
$$9\,dx = 8u\,du$$

$= \dfrac{128\pi}{243} \int_0^{3/2} u^4 \sqrt{1 + u^2}\,du$ Let $u = \tan v$
$$du = \sec^2 v\,dv$$

$= \dfrac{128\pi}{243} \int_0^{\tan^{-1}(3/2)} \tan^4 v \sec^3 v\,dv$

$= \dfrac{128\pi}{243} \int_0^{\tan^{-1}(3/2)} (\sec^7 v - 2\sec^5 v + \sec^3 v)\,dv.$

At this stage it is convenient to use the reduction formula

$$\int \sec^n v\,dv = \dfrac{1}{n-1}\sec^{n-2} v \tan v + \dfrac{n-2}{n-1} \int \sec^{n-2} v\,dv$$

(see Exercise 36 of Section 7.1) to reduce the powers of secant down to 3, and then use

$$\int_0^a \sec^3 v\,dv = \dfrac{1}{2}(\sec a \tan a + \ln|\sec a + \tan a|).$$

We have

$I = \int_0^a (\sec^7 v - 2\sec^5 v + \sec^3 v)\,dv$

$= \dfrac{\sec^5 v \tan v}{6}\Big|_0^a + \left(\dfrac{5}{6} - 2\right) \int_0^a \sec^5 v\,dv + \int_0^a \sec^3 v\,dv$

$= \dfrac{\sec^5 a \tan a}{6} - \dfrac{7}{6}\left[\dfrac{\sec^3 v \tan v}{4}\Big|_0^a + \dfrac{3}{4}\int_0^a \sec^3 v\,dv\right]$

$\quad + \int_0^a \sec^3 v\,dv$

$= \dfrac{\sec^5 a \tan a}{6} - \dfrac{7\sec^3 a \tan a}{24} + \dfrac{1}{8}\int_0^a \sec^3 v\,dv$

$= \dfrac{\sec^5 a \tan a}{6} - \dfrac{7\sec^3 a \tan a}{24} + \dfrac{\sec a \tan a + \ln|\sec a + \tan a|}{16}.$

Substituting $a = \arctan(3/2)$ now gives the following value for the surface area:

$$S = \dfrac{28\sqrt{13}\pi}{81} + \dfrac{8\pi}{243} \ln\left(\dfrac{3 + \sqrt{13}}{2}\right) \text{ sq. units.}$$

24. We have

$S = 2\pi \int_0^1 e^x \sqrt{1 + e^{2x}}\,dx$ Let $e^x = \tan\theta$
$$e^x\,dx = \sec^2\theta\,d\theta$$

$= 2\pi \int_{x=0}^{x=1} \sqrt{1 + \tan^2\theta}\,\sec^2\theta\,d\theta = 2\pi \int_{x=0}^{x=1} \sec^3\theta\,d\theta$

$= \pi\Big[\sec\theta \tan\theta + \ln|\sec\theta + \tan\theta|\Big]\Big|_{x=0}^{x=1}.$

Since

$$x = 1 \Rightarrow \tan\theta = e, \ \sec\theta = \sqrt{1 + e^2},$$
$$x = 0 \Rightarrow \tan\theta = 1, \ \sec\theta = \sqrt{2},$$

therefore

$S = \pi\Big[e\sqrt{1 + e^2} + \ln|\sqrt{1 + e^2} + e| - \sqrt{2} - \ln|\sqrt{2} + 1|\Big]$

$= \pi\left[e\sqrt{1 + e^2} - \sqrt{2} + \ln\dfrac{\sqrt{1 + e^2} + e}{\sqrt{2} + 1}\right]$ sq. units.

26. $1 + (y')^2 = 1 + \left(\dfrac{x^2}{4} - \dfrac{1}{x^2}\right)^2 = \left(\dfrac{x^2}{4} + \dfrac{1}{x^2}\right)^2$

$S = 2\pi \int_1^4 \left(\dfrac{x^3}{12} + \dfrac{1}{x}\right)\left(\dfrac{x^2}{4} + \dfrac{1}{x^2}\right)\,dx$

$= 2\pi \int_1^4 \left(\dfrac{x^5}{48} + \dfrac{x}{3} + \dfrac{1}{x^3}\right)\,dx$

$= 2\pi \left(\dfrac{x^6}{288} + \dfrac{x^2}{6} - \dfrac{1}{2x^2}\right)\Big|_1^4$

$= \dfrac{275}{8}\pi$ sq. units.

28. The area of the cone obtained by rotating the line $y = (h/r)x$, $0 \le x \le r$, about the y-axis is

$S = 2\pi \int_0^r x\sqrt{1 + (h/r)^2}\,dx = 2\pi \dfrac{\sqrt{r^2 + h^2}}{r} \dfrac{x^2}{2}\Big|_0^r$

$= \pi r\sqrt{r^2 + h^2}$ sq. units.

30. The top half of $x^2 + 4y^2 = 4$ is $y = \dfrac{1}{2}\sqrt{4 - x^2}$, so $\dfrac{dy}{dx} = \dfrac{-x}{2\sqrt{4 - x^2}}$, and

$$S = 2 \times 2\pi \int_0^2 \frac{\sqrt{4 - x^2}}{2}\sqrt{1 + \left(\frac{x}{2\sqrt{4 - x^2}}\right)^2}\, dx$$

$$= \pi \int_0^2 \sqrt{16 - 3x^2}\, dx \quad \text{Let } x = \sqrt{\frac{16}{3}}\sin\theta$$

$$dx = \sqrt{\frac{16}{3}}\cos\theta\, d\theta$$

$$= \pi \int_0^{\pi/3} (4\cos\theta)\frac{4}{\sqrt{3}}\cos\theta\, d\theta$$

$$= \frac{16\pi}{\sqrt{3}} \int_0^{\pi/3} \cos^2\theta\, d\theta$$

$$= \frac{8\pi}{\sqrt{3}} \left(\theta + \sin\theta\,\cos\theta\right)\Big|_0^{\pi/3}$$

$$= \frac{2\pi(4\pi + 3\sqrt{3})}{3\sqrt{3}} \text{ sq. units.}$$

32. As in Example 4, the arc length element for the ellipse is

$$ds = \sqrt{1 + \left(\frac{dy}{dx}\right)^2}\, dx = \sqrt{\frac{a^2 - \frac{a^2 - b^2}{a^2}x^2}{a^2 - x^2}}\, dx.$$

To get the area of the ellipsoid, we must rotate both the upper and lower semi-ellipses (see the figure for Exercise 20 of Section 8.1):

$$S = 2 \times 2\pi \int_0^a \left[\left(c - b\sqrt{1 - \left(\frac{x}{a}\right)^2}\right) + \left(c + b\sqrt{1 - \left(\frac{x}{a}\right)^2}\right)\right] ds$$

$$= 8\pi c \int_0^a \sqrt{\frac{a^2 - \frac{a^2 - b^2}{a^2}x^2}{a^2 - x^2}}\, dx$$

$$= 8\pi c \left[\frac{1}{4} \text{ of the circumference of the ellipse}\right]$$

$$= 8\pi c a E(\epsilon)$$

where $\epsilon = \dfrac{\sqrt{a^2 - b^2}}{a}$ and $E(\epsilon) = \int_0^{\pi/2}\sqrt{1 - \epsilon^2 \sin t}\, dt$ as defined in Example 4.

34. Let the equation of the sphere be $x^2 + y^2 = R^2$. Then the surface area between planes $x = a$ and $x = b$ $(-R \le a < b \le R)$ is

$$S = 2\pi \int_a^b \sqrt{R^2 - x^2}\sqrt{1 + \left(\frac{dy}{dx}\right)^2}\, dx$$

$$= 2\pi \int_a^b \sqrt{R^2 - x^2}\frac{R}{\sqrt{R^2 - x^2}}\, dx$$

$$= 2\pi R \int_a^b dx = 2\pi R(b - a) \text{ sq. units.}$$

Thus, the surface area depends only on the radius R of the sphere, and the distance $(b - a)$ between the parellel planes.

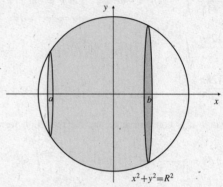

Fig. 7.3.34

36.
$$S = 2\pi \int_0^1 |x|\sqrt{1 + \frac{1}{x^2}}\, dx$$

$$= 2\pi \int_0^1 \sqrt{x^2 + 1}\, dx \quad \text{Let } x = \tan\theta$$

$$dx = \sec^2\theta\, d\theta$$

$$= 2\pi \int_0^{\pi/4} \sec^3\theta\, d\theta$$

$$= \pi \left(\sec\theta\,\tan\theta + \ln|\sec\theta + \tan\theta|\right)\Big|_0^{\pi/4}$$

$$= \pi[\sqrt{2} + \ln(\sqrt{2} + 1)] \text{ sq. units.}$$

Section 7.4 Mass, Moments, and Centre of Mass (page 436)

2. A slice of the wire of width dx at x has volume $dV = \pi(a + bx)^2\, dx$. Therefore the mass of the whole wire is

$$m = \int_0^L \delta_0 \pi (a + bx)^2\, dx$$

$$= \delta_0 \pi \int_0^L (a^2 + 2abx + b^2 x^2)\, dx$$

$$= \delta_0 \pi \left(a^2 L + abL^2 + \frac{1}{3}b^2 L^3\right).$$

Its moment about $x = 0$ is

$$M_{x=0} = \int_0^L x\delta_0\pi(a+bx)^2\,dx$$

$$= \delta_0\pi \int_0^L (a^2x + 2abx^2 + b^2x^3)\,dx$$

$$= \delta_0\pi\left(\frac{1}{2}a^2L^2 + \frac{2}{3}abL^3 + \frac{1}{4}b^2L^4\right).$$

Thus, the centre of mass is

$$\bar{x} = \frac{\delta_0\pi\left(\frac{1}{2}a^2L^2 + \frac{2}{3}abL^3 + \frac{1}{4}b^2L^4\right)}{\delta_0\pi\left(a^2L + abL^2 + \frac{1}{3}b^2L^3\right)}$$

$$= \frac{L\left(\frac{1}{2}a^2 + \frac{2}{3}abL + \frac{1}{4}b^2L^2\right)}{a^2 + abL + \frac{1}{3}b^2L^2}.$$

4. A vertical strip has area $dA = \sqrt{a^2 - x^2}\,dx$. Therefore, the mass of the quarter-circular plate is

$$m = \int_0^a (\delta_0 x)\sqrt{a^2 - x^2}\,dx \quad \text{Let } u = a^2 - x^2$$
$$du = -2x\,dx$$

$$= \frac{1}{2}\delta_0 \int_0^{a^2} \sqrt{u}\,du = \frac{1}{2}\delta_0\left(\frac{2}{3}u^{3/2}\right)\Big|_0^{a^2} = \frac{1}{3}\delta_0 a^3.$$

The moment about $x = 0$ is

$$M_{x=0} = \int_0^a \delta_0 x^2\sqrt{a^2 - x^2}\,dx \quad \text{Let } x = a\sin\theta$$
$$dx = a\cos\theta\,d\theta$$

$$= \delta_0 a^4 \int_0^{\pi/2} \sin^2\theta\cos^2\theta\,d\theta$$

$$= \frac{\delta_0 a^4}{4}\int_0^{\pi/2} \sin^2 2\theta\,d\theta$$

$$= \frac{\delta_0 a^4}{8}\int_0^{\pi/2}(1 - \cos 4\theta)\,d\theta = \frac{\pi\delta_0 a^4}{16}.$$

The moment about $y = 0$ is

$$M_{y=0} = \frac{1}{2}\delta_0 \int_0^a x(a^2 - x^2)\,dx$$

$$= \frac{1}{2}\delta_0\left(\frac{a^2x^2}{2} - \frac{x^4}{4}\right)\Big|_0^a = \frac{1}{8}a^4\delta_0.$$

Thus, $\bar{x} = \dfrac{3}{16}\pi a$ and $\bar{y} = \dfrac{3}{8}a$. Hence, the centre of mass is located at $(\dfrac{3}{16}\pi a, \dfrac{3}{8}a)$.

6. A vertical strip at h has area $dA = (2 - \frac{2}{3}h)\,dh$. Thus, the mass of the plate is

$$m = \int_0^3 (5h)\left(2 - \frac{2}{3}h\right)dh = 10\int_0^3\left(h - \frac{h^2}{3}\right)dh$$

$$= 10\left(\frac{h^2}{2} - \frac{h^3}{9}\right)\Big|_0^3 = 15 \text{ kg}.$$

The moment about $x = 0$ is

$$M_{x=0} = 10\int_0^3\left(h^2 - \frac{h^3}{3}\right)dh$$

$$= 10\left(\frac{h^3}{3} - \frac{h^4}{12}\right)\Big|_0^3 = \frac{45}{2} \text{ kg-m}.$$

The moment about $y = 0$ is

$$M_{y=0} = 10\int_0^3 \frac{1}{2}\left(2 - \frac{2}{3}h\right)\left(h - \frac{1}{3}h^2\right)dh$$

$$= 10\int_0^3\left(h - \frac{2}{3}h^2 + \frac{1}{9}h^3\right)dh$$

$$= 10\left(\frac{h^2}{2} - \frac{2h^3}{9} + \frac{h^4}{36}\right)\Big|_0^3 = \frac{15}{2} \text{ kg-m}.$$

Thus, $\bar{x} = \dfrac{\left(\frac{45}{2}\right)}{15} = \dfrac{3}{2}$ and $\bar{y} = \dfrac{\left(\frac{15}{2}\right)}{15} = \dfrac{1}{2}$. The centre of mass is located at $(\frac{3}{2}, \frac{1}{2})$.

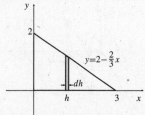

Fig. 7.4.6

8. A vertical strip has area $dA = 2\left(\dfrac{a}{\sqrt{2}} - r\right)dr$. Thus, the mass is

$$m = 2\int_0^{a/\sqrt{2}} kr\left[2\left(\frac{a}{\sqrt{2}} - r\right)\right]dr$$

$$= 4k\int_0^{a/\sqrt{2}}\left(\frac{a}{\sqrt{2}}r - r^2\right)dr = \frac{k}{3\sqrt{2}}a^3 \text{ g}.$$

Since the mass is symmetric about the y-axis, and the plate is symmetric about both the x- and y-axis, therefore the centre of mass must be located at the centre of the square.

143

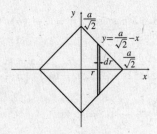

Fig. 7.4.8

10. The slice of the brick shown in the figure has volume $dV = 50\,dx$. Thus, the mass of the brick is

$$m = \int_0^{20} kx50\,dx = 25kx^2\Big|_0^{20} = 10000k \text{ g}.$$

The moment about $x = 0$, i.e., the yz-plane, is

$$M_{x=0} = 50k\int_0^{20} x^2\,dx = \frac{50}{3}kx^3\Big|_0^{20}$$
$$= \frac{50}{3}(8000)k \text{ g-cm}.$$

Thus, $\bar{x} = \dfrac{\frac{50}{3}(8000)k}{10000k} = \dfrac{40}{3}$. Since the density is independent of y and z, $\bar{y} = \dfrac{5}{2}$ and $\bar{z} = 5$. Hence, the centre of mass is located on the 20 cm long central axis of the brick, two-thirds of the way from the least dense 10×5 face to the most dense such face.

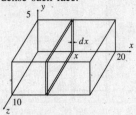

Fig. 7.4.10

12. A slice at height z has volume $dV = \pi y^2\,dz$ and density kz g/cm^3. Thus, the mass of the cone is

$$m = \int_0^b kz\pi y^2\,dz$$
$$= \pi ka^2\int_0^b z\left(1 - \frac{z}{b}\right)^2\,dz$$
$$= \pi ka^2\left(\frac{z^2}{2} - \frac{2z^3}{3b} + \frac{z^4}{4b^2}\right)\Big|_0^b$$
$$= \frac{1}{12}\pi ka^2 b^2 \text{ g}.$$

The moment about $z = 0$ is

$$M_{z=0} = \pi ka^2\int_0^b z^2\left(1 - \frac{z}{b}\right)^2\,dz = \frac{1}{30}\pi ka^2 b^3 \text{ g-cm}.$$

Thus, $\bar{z} = \dfrac{2b}{5}$. Hence, the centre of mass is on the axis of the cone at height $2b/5$ cm above the base.

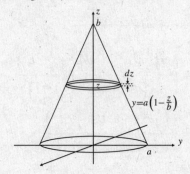

Fig. 7.4.12

14. Assume the cone has its base in the xy-plane and its vertex at height b on the z-axis. By symmetry, the centre of mass lies on the z-axis. A cylindrical shell of thickness dx and radius x about the z-axis has height $z = b(1 - (x/a))$. Since it's density is constant kx, its mass is

$$dm = 2\pi bkx^2\left(1 - \frac{x}{a}\right)\,dx.$$

Also its centre of mass is at half its height,

$$\bar{y}_{\text{shell}} = \frac{b}{2}\left(1 - \frac{x}{a}\right).$$

Thus its moment about $z = 0$ is

$$dM_{z=0} = \bar{y}_{\text{shell}}\,dm = \pi bkx^2\left(1 - \frac{x}{a}\right)^2\,dx.$$

Hence

$$m = \int_0^a 2\pi bkx^2\left(1 - \frac{x}{a}\right)\,dx = \frac{\pi kba^3}{6}$$
$$M_{z=0} = \int_0^a \pi bkx^2\left(1 - \frac{x}{a}\right)^2\,dx = \frac{\pi kb^2 a^3}{30}$$

and $\bar{z} = M_{z=0}/m = b/5$. The centre of mass is on the axis of the cone at height $b/5$ cm above the base.

16.

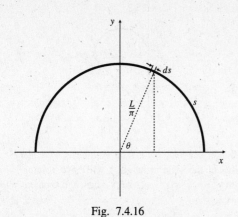

Fig. 7.4.16

The radius of the semicircle is $\dfrac{L}{\pi}$. Let s measure the distance along the wire from the point where it leaves the positive x-axis. Thus, the density at position s is $\delta\delta(s) = \sin\left(\dfrac{\pi s}{L}\right)$ g/cm. The mass of the wire is

$$m = \int_0^L \sin\frac{\pi s}{L}\,ds = -\frac{L}{\pi}\cos\frac{\pi s}{L}\bigg|_0^L = \frac{2L}{\pi}\ \text{g}.$$

Since an arc element ds at position s is at height $y = \dfrac{L}{\pi}\sin\theta = \dfrac{L}{\pi}\sin\dfrac{\pi s}{L}$, the moment of the wire about $y = 0$ is

$$M_{y=0} = \int_0^L \frac{L}{\pi}\sin^2\frac{\pi s}{L}\,ds \quad \text{Let } \theta = \pi s/L$$
$$\qquad\qquad\qquad\qquad\qquad\qquad d\theta = \pi ds/L$$
$$= \left(\frac{L}{\pi}\right)^2 \int_0^\pi \sin^2\theta\,d\theta$$
$$= \frac{L^2}{2\pi^2}\left(\theta - \sin\theta\cos\theta\right)\bigg|_0^\pi = \frac{L^2}{2\pi}\ \text{g-cm}.$$

Since the wire and the density function are both symmetric about the y-axis, we have $M_{x=0} = 0$.

Hence, the centre of mass is located at $\left(0, \dfrac{L}{4}\right)$.

18.
$$\bar{r} = \frac{1}{m}\int_0^\infty rCe^{-kr^2}(4\pi r^2)\,dr$$
$$= \frac{4\pi C}{C\pi^{3/2}k^{-3/2}}\int_0^\infty r^3 e^{-kr^2}\,dr \quad \text{Let } u = kr^2$$
$$\qquad\qquad\qquad\qquad\qquad\qquad\qquad\qquad du = 2kr\,dr$$
$$= \frac{4k^{3/2}}{\sqrt{\pi}}\frac{1}{2k^2}\int_0^\infty ue^{-u}\,du$$
$$\qquad U = u \qquad dV = e^{-u}\,du$$
$$\qquad dU = du \qquad V = -e^{-u}$$
$$= \frac{2}{\sqrt{\pi k}}\lim_{R\to\infty}\left(-ue^{-u}\bigg|_0^R + \int_0^R e^{-u}\,du\right)$$
$$= \frac{2}{\sqrt{\pi k}}\left(0 + \lim_{R\to\infty}(e^0 - e^{-R})\right) = \frac{2}{\sqrt{\pi k}}.$$

Section 7.5 Centroids (page 442)

2. By symmetry, $\bar{x} = 0$. A horizontal strip at y has mass $dm = 2\sqrt{9 - y}\,dy$ and moment $dM_{y=0} = 2y\sqrt{9 - y}\,dy$ about $y = 0$. Thus,

$$m = 2\int_0^9 \sqrt{9 - y}\,dy = -2\left(\frac{2}{3}\right)(9 - y)^{3/2}\bigg|_0^9 = 36$$

and

$$M_{y=0} = 2\int_0^9 y\sqrt{9 - y}\,dy \quad \text{Let } u^2 = 9 - y$$
$$\qquad\qquad\qquad\qquad\qquad\qquad 2u\,du = -dy$$
$$= 4\int_0^3 (9u^2 - u^4)\,du = 4(3u^3 - \tfrac{1}{5}u^5)\bigg|_0^3 = \frac{648}{5}.$$

Thus, $\bar{y} = \dfrac{648}{5 \times 36} = \dfrac{18}{5}$. Hence, the centroid is at $\left(0, \dfrac{18}{5}\right)$.

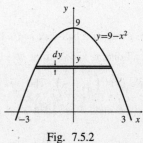

Fig. 7.5.2

4. The area of the sector is $A = \tfrac{1}{8}\pi r^2$. Its moment about $x = 0$ is

$$M_{x=0} = \int_0^{r/\sqrt{2}} x^2\,dx + \int_{r/\sqrt{2}}^r x\sqrt{r^2 - x^2}\,dx$$
$$= \frac{r^3}{6\sqrt{2}} - \frac{1}{3}(r^2 - x^2)^{3/2}\bigg|_{r/\sqrt{2}}^r = \frac{r^3}{3\sqrt{2}}.$$

Thus, $\bar{x} = \dfrac{r^3}{3\sqrt{2}} \times \dfrac{8}{\pi r^2} = \dfrac{8r}{3\sqrt{2}\pi}$. By symmetry, the centroid must lie on the line $y = x\left(\tan\dfrac{\pi}{8}\right) = x(\sqrt{2} - 1)$.

Thus, $\bar{y} = \dfrac{8r(\sqrt{2} - 1)}{3\sqrt{2}\pi}$.

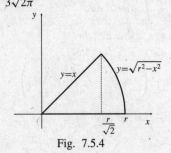

Fig. 7.5.4

6. By symmetry, $\bar{x} = 0$. The area is $A = \frac{1}{2}\pi ab$. The moment about $y = 0$ is

$$M_{y=0} = \frac{1}{2}\int_{-a}^{a} b^2\left[1 - \left(\frac{x}{a}\right)^2\right]dx = b^2\int_{0}^{a} 1 - \frac{x^2}{a^2}\,dx$$
$$= b^2\left(x - \frac{x^3}{3a^2}\right)\Big|_{0}^{a} = \frac{2}{3}ab^2.$$

Thus, $\bar{y} = \dfrac{2ab^2}{3} \times \dfrac{2}{\pi ab} = \dfrac{4b}{3\pi}$.

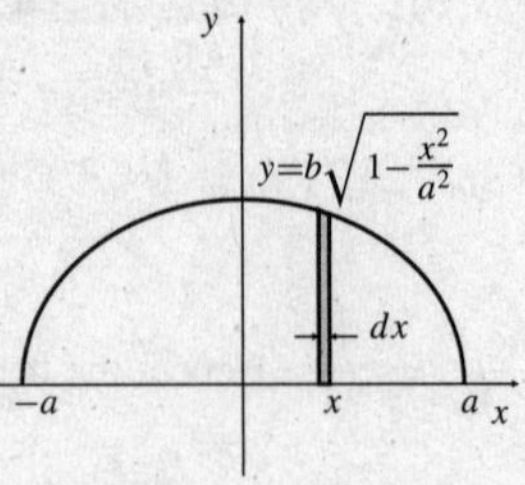

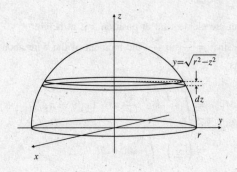

Fig. 7.5.6

8. The region is the union of a half-disk and a triangle. The centroid of the half-disk is known to be at $\left(1, \dfrac{4}{3\pi}\right)$ and that of the triangle is at $\left(\dfrac{2}{3}, -\dfrac{2}{3}\right)$. The area of the semicircle is $\dfrac{\pi}{2}$ and the triangle is 2. Hence,

$$M_{x=0} = \left(\frac{\pi}{2}\right)(1) + (2)\left(\frac{2}{3}\right) = \frac{3\pi + 8}{6};$$
$$M_{y=0} = \left(\frac{\pi}{2}\right)\left(\frac{4}{3\pi}\right) + (2)\left(-\frac{2}{3}\right) = -\frac{2}{3}.$$

Since the area of the whole region is $\dfrac{\pi}{2} + 2$, then

$$\bar{x} = \frac{3\pi + 8}{3(\pi + 4)} \quad \text{and} \quad \bar{y} = -\frac{4}{3(\pi + 4)}.$$

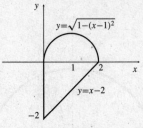

Fig. 7.5.8

10. By symmetry, $\bar{x} = \bar{y} = 0$. The volume is $V = \frac{2}{3}\pi r^3$. A thin slice of the solid at height z will have volume $dV = \pi y^2\,dz = \pi(r^2 - z^2)\,dz$. Thus, the moment about $z = 0$ is

$$M_{z=0} = \int_{0}^{r} z\pi(r^2 - z^2)\,dz$$
$$= \pi\left(\frac{r^2 z^2}{2} - \frac{z^4}{4}\right)\Big|_{0}^{r} = \frac{\pi r^4}{4}.$$

Thus, $\bar{z} = \dfrac{\pi r^4}{4} \times \dfrac{3}{2\pi r^3} = \dfrac{3r}{8}$. Hence, the centroid is on the axis of the hemisphere at distance $3r/8$ from the base.

Fig. 7.5.10

12. A band at height z with vertical width dz has radius $y = r\left(1 - \dfrac{z}{h}\right)$, and has actual (slant) width

$$ds = \sqrt{1 + \left(\frac{dy}{dz}\right)^2}\,dz = \sqrt{1 + \frac{r^2}{h^2}}\,dz.$$

Its area is

$$dA = 2\pi r\left(1 - \frac{z}{h}\right)\sqrt{1 + \frac{r^2}{h^2}}\,dz.$$

Thus the area of the conical surface is

$$A = 2\pi r\sqrt{1 + \frac{r^2}{h^2}}\int_{0}^{h}\left(1 - \frac{z}{h}\right)dz = \pi r\sqrt{r^2 + h^2}.$$

The moment about $z = 0$ is

$$M_{z=0} = 2\pi r \sqrt{1 + \frac{r^2}{h^2}} \int_0^h z\left(1 - \frac{z}{h}\right) dz$$

$$= 2\pi r \sqrt{1 + \frac{r^2}{h^2}} \left(\frac{z^2}{2} - \frac{z^3}{3h}\right)\Big|_0^h = \frac{1}{3}\pi rh\sqrt{r^2 + h^2}.$$

Thus, $\bar{z} = \dfrac{\pi rh\sqrt{r^2 + h^2}}{3} \times \dfrac{1}{\pi r\sqrt{r^2 + h^2}} = \dfrac{h}{3}$. By symmetry, $\bar{x} = \bar{y} = 0$. Hence, the centroid is on the axis of the conical surface, at distance $h/3$ from the base.

14. The region in figure (b) is the union of a square of area $(\sqrt{2})^2 = 2$ and centroid $(0, 0)$ and a triangle of area $1/2$ and centroid $(2/3, 2/3)$. Therefore its area is $5/2$ and its centroid is $(\bar{x}, \bar{y})$, where

$$\frac{5}{2}\bar{x} = 2(0) + \frac{1}{2}\left(\frac{2}{3}\right) = \frac{1}{3}.$$

Therefore, $\bar{x} = \bar{y} = 2/15$, and the centroid is $(2/15, 2/15)$.

16. The region in figure (d) is the union of three half-disks, one with area $\pi/2$ and centroid $(0, 4/(3\pi))$, and two with areas $\pi/8$ and centroids $(-1/2, -2/(3\pi))$ and $(1/2, -2/(3\pi))$. Therefore its area is $3\pi/4$ and its centroid is $(\bar{x}, \bar{y})$, where

$$\frac{3\pi}{4}(\bar{x}) = \frac{\pi}{2}(0) + \frac{\pi}{8}\left(\frac{-1}{2}\right) + \frac{\pi}{8}\left(\frac{1}{2}\right) = 0$$

$$\frac{3\pi}{4}(\bar{y}) = \frac{\pi}{2}\left(\frac{4}{3\pi}\right) + \frac{\pi}{8}\left(\frac{-2}{3\pi}\right) + \frac{\pi}{8}\left(\frac{-2}{3\pi}\right) = \frac{1}{2}.$$

Therefore, the centroid is $(0, 2/(3\pi))$.

18. The area of the region is

$$A = \int_0^{\pi/2} \cos x \, dx = \sin x \Big|_0^{\pi/2} = 1.$$

The moment about $x = 0$ is

$$M_{x=0} = \int_0^{\pi/2} x \cos x \, dx$$

$$U = x \qquad dV = \cos x \, dx$$
$$dU = dx \qquad V = \sin x$$

$$= x \sin x \Big|_0^{\pi/2} - \int_0^{\pi/2} \sin x \, dx = \frac{\pi}{2} - 1.$$

Thus, $\bar{x} = \dfrac{\pi}{2} - 1$. The moment about $y = 0$ is

$$M_{y=0} = \frac{1}{2}\int_0^{\pi/2} \cos^2 x \, dx$$

$$= \frac{1}{4}\left(x + \frac{1}{2}\sin 2x\right)\Big|_0^{\pi/2} = \frac{\pi}{8}.$$

Thus, $\bar{y} = \dfrac{\pi}{8}$. The centroid is $\left(\dfrac{\pi}{2} - 1, \dfrac{\pi}{8}\right)$.

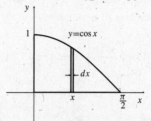

Fig. 7.5.18

20. The solid S in question consists of a solid cone C with vertex at the origin, height 1, and top a circular disk of radius 2, and a solid cylinder D of radius 2 and height 1 sitting on top of the cone. These solids have volumes $V_C = 4\pi/3$, $V_D = 4\pi$, and $V_S = V_C + V_D = 16\pi/3$.

By symmetry, the centroid of the solid lies on its vertical axis of symmetry; let us continue to call this the y-axis. We need only determine $\bar{y}_S$. Since D lies between $y = 1$ and $y = 2$, its centroid satisfies $\bar{y}_D = 3/2$. Also, by Exercise 11, the centroid of the solid cone satisfies $\bar{y}_C = 3/4$. Thus C and D have moments about $y = 0$:

$$M_{C,y=0} = \left(\frac{4\pi}{3}\right)\left(\frac{3}{4}\right) = \pi, \quad M_{D,y=0} = (4\pi)\left(\frac{3}{2}\right) = 6\pi.$$

Thus $M_{S,y=0} = \pi + 6\pi = 7\pi$, and $\bar{z}_S = 7\pi/(16\pi/3) = 21/16$. The centroid of the solid S is on its vertical axis of symmetry at height $21/16$ above the vertex of the conical part.

22. The line segment from $(1, 0)$ to $(0, 1)$ has centroid $(\frac{1}{2}, \frac{1}{2})$ and length $\sqrt{2}$. By Pappus's Theorem, the surface area of revolution about $x = 2$ is

$$A = 2\pi\left(2 - \frac{1}{2}\right)\sqrt{2} = 3\pi\sqrt{2} \text{ sq. units.}$$

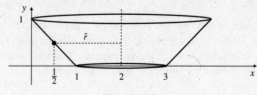

Fig. 7.5.22

24. The altitude h of the triangle is $\dfrac{s\sqrt{3}}{2}$. Its centroid is at

height $\dfrac{h}{3} = \dfrac{s}{2\sqrt{3}}$ above the base side. Thus, by Pappus's Theorem, the volume of revolution is

$$V = 2\pi \left(\frac{s}{2\sqrt{3}}\right)\left(\frac{s}{2} \times \frac{\sqrt{3}s}{2}\right) = \frac{\pi s^3}{4} \text{ cu. units.}$$

The centroid of one side is $\dfrac{h}{2} = \dfrac{s\sqrt{3}}{4}$ above the base. Thus, the surface area of revolution is

$$S = 2 \times 2\pi \left(\frac{\sqrt{3}s}{4}\right)(s) = s^2\pi\sqrt{3} \text{ sq. units.}$$

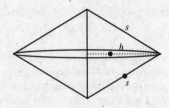

Fig. 7.5.24

26. The region bounded by $y = 0$ and $y = \ln(\sin x)$ between $x = 0$ and $x = \pi/2$ lies below the x-axis, so

$$A = -\int_0^{\pi/2} \ln(\sin x)\,dx \approx 1.088793$$

$$\bar{x} = \frac{-1}{A}\int_0^{\pi/2} x\ln(\sin x)\,dx \approx 0.30239$$

$$\bar{y} = \frac{-1}{2A}\int_0^{\pi/2} \left(\ln(\sin x)\right)^2 dx \approx -0.93986.$$

28. The surface area is given by
$$S = 2\pi \int_{-\infty}^{\infty} e^{-x^2}\sqrt{1 + 4x^2 e^{-2x^2}}\,dx. \text{ Since}$$

$\lim\limits_{x \to \pm\infty} 1 + 4x^2 e^{-2x^2} = 1$, this expression must be bounded

for all x, that is, $1 \le 1 + 4x^2 e^{-2x^2} \le K^2$ for some constant

K. Thus, $S \le 2\pi K \int_{-\infty}^{\infty} e^{-x^2}\,dx = 2K\pi\sqrt{\pi}$. The integral converges and the surface area is finite. Since the whole curve $y = e^{-x^2}$ lies above the x-axis, its centroid would have to satisfy $\bar{y} > 0$. However, Pappus's Theorem would then imply that the surface of revolution would have infinite area: $S = 2\pi\bar{y} \times$ (length of curve) $= \infty$. The curve cannot, therefore, have any centroid.

30. Let us take L to be the y-axis and suppose that a plane curve C lies between $x = a$ and $x = b$ where $0 < a < b$. Thus, $\bar{r} = \bar{x}$, the x-coordinate of the centroid of C. Let ds denote an arc length element of C at position x. This arc length element generates, on rotation about L, a circular band of surface area $dS = 2\pi x\,ds$, so the surface area of the surface of revolution is

$$S = 2\pi \int_{x=a}^{x=b} x\,ds = 2\pi M_{x=0} = 2\pi\bar{r}s.$$

Section 7.6 Other Physical Applications (page 449)

2. A vertical slice of water at position y with thickness dy is in contact with the botttom over an area $8\sec\theta\,dy = \frac{4}{5}\sqrt{101}\,dy$ m^2, which is at depth $x = \frac{1}{10}y + 1$ m. The force exerted on this area is then $dF = \rho g(\frac{1}{10}y + 1)\frac{4}{5}\sqrt{101}\,dy$. Hence, the total force exerted on the bottom is

$$F = \frac{4}{5}\sqrt{101}\,\rho g \int_0^{20} \left(\frac{1}{10}y + 1\right)dy$$

$$= \frac{4}{5}\sqrt{101}\,(1000)(9.8)\left(\frac{y^2}{20} + y\right)\Big|_0^{20}$$

$$\approx 3.1516 \times 10^6 \text{ N.}$$

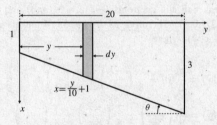

Fig. 7.6.2

4. The height of each triangular face is $2\sqrt{3}$ m and the height of the pyramid is $2\sqrt{2}$ m. Let the angle between the triangular face and the base be θ, then $\sin\theta = \sqrt{\dfrac{2}{3}}$ and $\cos\theta = \dfrac{1}{\sqrt{3}}$.

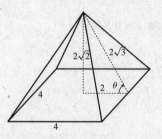

Fig. 7.6.4

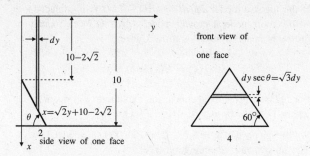

Fig. 7.6.4

A vertical slice of water with thickness dy at a distance y from the vertex of the pyramid exerts a force on the shaded strip shown in the front view, which has area $2\sqrt{3}y\,dy$ m^2 and which is at depth $\sqrt{2}y + 10 - 2\sqrt{2}$ m. Hence, the force exerted on the triangular face is

$$F = \rho g \int_0^2 (\sqrt{2}y + 10 - 2\sqrt{2})2\sqrt{3}y\,dy$$

$$= 2\sqrt{3}(9800)\left[\frac{\sqrt{2}}{3}y^3 + (5 - \sqrt{2})y^2\right]\Big|_0^2$$

$$\approx 6.1495 \times 10^5 \text{ N.}$$

6. The spring force is $F(x) = kx$, where x is the amount of compression. The work done to compress the spring 3 cm is

$$100 \text{ N·cm} = W = \int_0^3 kx\,dx = \frac{1}{2}kx^2\Big|_0^3 = \frac{9}{2}k.$$

Hence, $k = \dfrac{200}{9}$ N/cm. The work necessary to compress the spring a further 1 cm is

$$W = \int_3^4 kx\,dx = \left(\frac{200}{9}\right)\frac{1}{2}x^2\Big|_3^4 = \frac{700}{9} \text{ N·cm.}$$

8. The horizontal cross-sectional area of the pool at depth h is

$$A(h) = \begin{cases} 160, & \text{if } 0 \le h \le 1; \\ 240 - 80h, & \text{if } 1 < h \le 3. \end{cases}$$

The work done to empty the pool is

$$W = \rho g \int_0^3 hA(h)\,dh$$

$$= \rho g\left[\int_0^1 160h\,dh + \int_1^3 240h - 80h^2\,dh\right]$$

$$= 9800\left[80h^2\Big|_0^1 + \left(120h^2 - \frac{80}{3}h^3\right)\Big|_1^3\right]$$

$$= 3.3973 \times 10^6 \text{ N·m.}$$

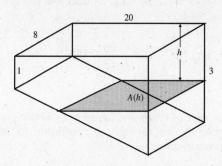

Fig. 7.6.8

10. Let the time required to raise the bucket to height h m be t minutes. Given that the velocity is 2 m/min, then $t = \dfrac{h}{2}$. The weight of the bucket at time t is

16 kg − (1 kg/min)(t min) = $16 - \dfrac{h}{2}$ kg. Therefore, the work done required to move the bucket to a height of 10 m is

$$W = g \int_0^{10}\left(16 - \frac{h}{2}\right)dh$$

$$= 9.8\left(16h - \frac{h^2}{4}\right)\Big|_0^{10} = 1323 \text{ N·m.}$$

Section 7.7 Applications in Business, Finance, and Ecology   (page 453)

2. The number of chips sold in the first year was

$$1,000 \int_0^{52} te^{-t/10}\,dt = 100,000 - 620,000e^{-26/5}$$

that is, about 96,580.

4. The price per kg at time t (years) is $10 + 5t$. Thus the revenue per year at time t is $400(10+5t)/(1+0.1t)$ $/year. The total revenue over the year is

$$\int_0^1 \frac{400(10 + 5t)}{1 + 0.1t}\,dt \approx \$4,750.37.$$

6. The present value of continuous payments of $1,000 per year for 10 years at a discount rate of 5% is

$$V = \int_0^{10} 1,000e^{-0.05t}\,dt = \frac{1,000}{-0.05}e^{-0.05t}\Big|_0^{10} = \$7,869.39.$$

8. The present value of continuous payments of $1,000 per year for 25 years beginning 10 years from now at a discount rate of 5% is

$$V = \int_{10}^{35} 1{,}000 e^{-0.05t}\, dt = \frac{1{,}000}{-0.05} e^{-0.05t}\Big|_{10}^{35} = \$8{,}655.13.$$

10. The present value of continuous payments of $1,000 per year beginning 10 years from now and continuing for all future time at a discount rate of 5% is

$$V = \int_{10}^{\infty} 1{,}000 e^{-0.05t}\, dt = \frac{1{,}000}{-0.05} e^{-0.5} = \$12{,}130.61.$$

12. After t years, money is flowing at $\$1{,}000(1.1)^t$ per year. The present value of 10 years of payments discounted at 5% is

$$V = 1{,}000 \int_{0}^{10} e^{t \ln(1.1)} e^{-0.05t}\, dt$$

$$= \frac{1{,}000}{\ln(1.1) - 0.05} e^{t(\ln(1.1)-0.05)}\Big|_{0}^{10} = \$12{,}650.23.$$

14. Let T be the time required for the account balance to reach $1,000,000. The $\$5{,}000(1.1)^t\, dt$ deposited in the time interval $[t, t+dt]$ grows for $T - t$ years, so the balance after T years is

$$\int_{0}^{T} 5{,}000(1.1)^t (1.06)^{T-t}\, dt = 1{,}000{,}000$$

$$(1.06)^T \int_{0}^{T} \left(\frac{1.1}{1.06}\right)^t dt = \frac{1{,}000{,}000}{5{,}000} = 200$$

$$\frac{(1.06)^T}{\ln(1.1/1.06)} \left[\left(\frac{1.1}{1.06}\right)^T - 1\right] = 200$$

$$(1.1)^T - (1.06)^T = 200 \ln \frac{1.1}{1.06}.$$

This equation can be solved by Newton's method or using a calculator "solve" routine. The solution is $T \approx 26.05$ years.

16. The analysis carried out in the text for the logistic growth model showed that the total present value of future harvests could be maximized by holding the population size x at a value that maximizes the quadratic expression

$$Q(x) = kx\left(1 - \frac{x}{L}\right) - \delta x.$$

If the logistic model $dx/dt = kx(1 - (x/L))$ is replaced with a more general growth model $dx/dt = F(x)$, exactly the same analysis leads us to maximize

$$Q(x) = F(x) - \delta x.$$

For realistic growth functions, the maximum will occur where $Q'(x) = 0$, that is, where $F'(x) = \delta$.

18. We are given that $k = 0.02$, $L = 150{,}000$, $p = \$10{,}000$. The growth rate at population level x is

$$\frac{dx}{dt} = 0.02x \left(1 - \frac{x}{150{,}000}\right).$$

a) The maximum sustainable annual harvest is

$$\frac{dx}{dt}\bigg|_{x=L/2} = 0.02(75{,}000)(0.5) = 750 \text{ whales.}$$

b) The resulting annual revenue is $\$750p = \$7{,}500{,}000$.

c) If the whole population of 75,000 is harvested and the proceeds invested at 2%, the annual interest will be

$$75{,}000(\$10{,}000)(0.02) = \$15{,}000{,}000.$$

d) At 5%, the interest would be $(5/2)(\$15{,}000) = \$37{,}500{,}000$.

e) The total present value of all future harvesting revenue if the population level is maintained at 75,000 and $\delta = 0.05$ is

$$\int_{0}^{\infty} e^{-0.05t} 7{,}500{,}000\, dt = \frac{7{,}500{,}000}{0.05} = \$150{,}000{,}000.$$

Section 7.8 Probability (page 464)

2. We have $f(x) = Cx$ on $[1, 2]$.

a) To find C, we have

$$1 = \int_{1}^{2} Cx\, dx = \frac{C}{2} x^2 \Big|_{1}^{2} = \frac{3}{2}C.$$

Hence, $C = \frac{2}{3}$.

b) The mean is

$$\mu = E(X) = \frac{2}{3} \int_{1}^{2} x^2\, dx = \frac{2}{9} x^3 \Big|_{1}^{2} = \frac{14}{9} \approx 1.556.$$

Since $E(X^2) = \frac{2}{3}\int_1^2 x^3\,dx = \frac{1}{6}x^4\Big|_1^2 = \frac{5}{2}$, the variance is

$$\sigma^2 = E(X^2) - \mu^2 = \frac{5}{2} - \frac{196}{81} = \frac{13}{162}$$

and the standard deviation is

$$\sigma = \sqrt{\frac{13}{162}} \approx 0.283.$$

c) We have

$$\Pr(\mu - \sigma \le X \le \mu + \sigma) = \frac{2}{3}\int_{\mu-\sigma}^{\mu+\sigma} x\,dx$$

$$= \frac{(\mu+\sigma)^2 - (\mu-\sigma)^2}{3} = \frac{4\mu\sigma}{3} \approx 0.5875.$$

4. We have $f(x) = C\sin x$ on $[0, \pi]$.

a) To find C, we calculate

$$1 = \int_0^\pi C\sin x\,dx = -C\cos x\Big|_0^\pi = 2C.$$

Hence, $C = \frac{1}{2}$.

b) The mean is

$$\mu = E(X) = \frac{1}{2}\int_0^\pi x\sin x\,dx$$

$$U = x \qquad dV = \sin x\,dx$$
$$dU = dx \qquad V = -\cos x$$

$$= \frac{1}{2}\left[-x\cos x\Big|_0^\pi + \int_0^\pi \cos x\,dx\right]$$

$$= \frac{\pi}{2} = 1.571.$$

Since

$$E(X^2) = \frac{1}{2}\int_0^\pi x^2\sin x\,dx$$

$$U = x^2 \qquad dV = \sin x\,dx$$
$$dU = 2x\,dx \qquad V = -\cos x$$

$$= \frac{1}{2}\left[-x^2\cos x\Big|_0^\pi + 2\int_0^\pi x\cos x\,dx\right]$$

$$U = x \qquad dV = \cos x\,dx$$
$$dU = dx \qquad V = \sin x$$

$$= \frac{1}{2}\left[\pi^2 + 2\left(x\sin x\Big|_0^\pi - \int_0^\pi \sin x\,dx\right)\right]$$

$$= \frac{1}{2}(\pi^2 - 4).$$

Hence, the variance is

$$\sigma^2 = E(X^2) - \mu^2 = \frac{\pi^2 - 4}{2} - \frac{\pi^2}{4} = \frac{\pi^2 - 8}{4} \approx 0.467$$

and the standard deviation is

$$\sigma = \sqrt{\frac{\pi^2 - 8}{4}} \approx 0.684.$$

c) Then

$$\Pr(\mu - \sigma \le X \le \mu + \sigma) = \frac{1}{2}\int_{\mu-\sigma}^{\mu+\sigma} \sin x\,dx$$

$$= -\frac{1}{2}\left[\cos(\mu+\sigma) - \cos(\mu-\sigma)\right]$$

$$= \sin\mu\,\sin\sigma = \sin\sigma \approx 0.632.$$

6. It was shown in Section 6.1 (p. 349) that

$$\int x^n e^{-x}\,dx = -x^n e^{-x} + n\int x^{n-1}e^{-x}\,dx.$$

If $I_n = \int_0^\infty x^n e^{-x}\,dx$, then

$$I_n = \lim_{R\to\infty} -R^n e^{-R} + nI_{n-1} = nI_{n-1} \qquad \text{if } n \ge 1.$$

Since $I_0 = \int_0^\infty e^{-x}\,dx = 1$, therefore $I_n = n!$ for $n \ge 1$.
Let $u = kx$; then

$$\int_0^\infty x^n e^{-kx}\,dx = \frac{1}{k^{n+1}}\int_0^\infty u^n e^{-u}\,du = \frac{1}{k^{n+1}}I_n = \frac{n!}{k^{n+1}}.$$

Now let $f(x) = Cxe^{-kx}$ on $[0, \infty)$.

a) To find C, observe that

$$1 = C\int_0^\infty xe^{-kx}\,dx = \frac{C}{k^2}.$$

Hence, $C = k^2$.

b) The mean is

$$\mu = E(X) = k^2\int_0^\infty x^2 e^{-kx}\,dx = k^2\left(\frac{2}{k^3}\right) = \frac{2}{k}.$$

Since $E(X^2) = k^2\int_0^\infty x^3 e^{-kx}\,dx = k^2\left(\frac{6}{k^4}\right) = \frac{6}{k^2}$,
then the variance is

$$\sigma^2 = E(X^2) - \mu^2 = \frac{6}{k^2} - \frac{4}{k^2} = \frac{2}{k^2}$$

and the standard deviation is $\sigma = \dfrac{\sqrt{2}}{k}$.

c) Finally,

$$\Pr(\mu - \sigma \le X \le \mu + \sigma)$$

$$= k^2 \int_{\mu-\sigma}^{\mu+\sigma} x e^{-kx}\, dx \quad \text{Let } u = kx$$
$$\qquad\qquad\qquad\qquad\qquad\quad du = k\, dx$$

$$= \int_{k(\mu-\sigma)}^{k(\mu+\sigma)} u e^{-u}\, du$$

$$= -u e^{-u}\Big|_{k(\mu-\sigma)}^{k(\mu+\sigma)} + \int_{k(\mu-\sigma)}^{k(\mu+\sigma)} e^{-u}\, du$$

$$= -(2+\sqrt{2})e^{-(2+\sqrt{2})} + (2-\sqrt{2})e^{-(2-\sqrt{2})}$$
$$\qquad - e^{-(2+\sqrt{2})} + e^{-(2-\sqrt{2})}$$

$$\approx 0.738.$$

8. No. The identity $\displaystyle\int_{-\infty}^{\infty} C\, dx = 1$ is not satisfied for any constant C.

10. Since $f(x) = \dfrac{2}{\pi(1+x^2)} > 0$ on $[0, \infty)$ and

$$\frac{2}{\pi}\int_0^\infty \frac{dx}{1+x^2} = \lim_{R\to\infty} \frac{2}{\pi}\tan^{-1}(R) = \frac{2}{\pi}\left(\frac{\pi}{2}\right) = 1,$$

therefore $f(x)$ is a probability density function on $[0, \infty)$. The expectation of X is

$$\mu = E(X) = \frac{2}{\pi}\int_0^\infty \frac{x\, dx}{1+x^2}$$

$$= \lim_{R\to\infty} \frac{1}{\pi}\ln(1+R^2) = \infty.$$

No matter what the cost per game, you should be willing to play (if you have an adequate bankroll). Your expected winnings per game in the long term is infinite.

12. The density function for T is $f(t) = k e^{-kt}$ on $[0, \infty)$, where $k = \dfrac{1}{\mu} = \dfrac{1}{20}$ (see Example 6). Then

$$\Pr(T \ge 12) = \frac{1}{20}\int_{12}^\infty e^{-t/20}\, dt = 1 - \frac{1}{20}\int_0^{12} e^{-t/20}\, dt$$

$$= 1 + e^{-t/20}\Big|_0^{12} = e^{-12/20} \approx 0.549.$$

The probability that the system will last at least 12 hours is about 0.549.

Section 7.9 First-Order Differential Equations (page 472)

2.
$$\frac{dy}{dx} = \frac{3y-1}{x}$$

$$\int \frac{dy}{3y-1} = \int \frac{dx}{x}$$

$$\frac{1}{3}\ln|3y-1| = \ln|x| + \frac{1}{3}\ln C$$

$$\frac{3y-1}{x^3} = C$$

$$\Rightarrow \quad y = \frac{1}{3}(1 + Cx^3).$$

4.
$$\frac{dy}{dx} = x^2 y^2$$

$$\int \frac{dy}{y^2} = \int x^2\, dx$$

$$-\frac{1}{y} = \frac{1}{3}x^3 + \frac{1}{3}C$$

$$\Rightarrow \quad y = -\frac{3}{x^3 + C}.$$

6.
$$\frac{dx}{dt} = e^x \sin t$$

$$\int e^{-x}\, dx = \int \sin t\, dt$$

$$-e^{-x} = -\cos t - C$$

$$\Rightarrow \quad x = -\ln(\cos t + C).$$

8.
$$\frac{dy}{dx} = 1 + y^2$$

$$\int \frac{dy}{1+y^2} = \int dx$$

$$\tan^{-1} y = x + C$$

$$\Rightarrow \quad y = \tan(x + C).$$

10. We have

$$\frac{dy}{dx} = y^2(1-y)$$

$$\int \frac{dy}{y^2(1-y)} = \int dx = x + K.$$

Expand the left side in partial fractions:

$$\frac{1}{y^2(1-y)} = \frac{A}{y} + \frac{B}{y^2} + \frac{C}{1-y}$$

$$= \frac{A(y-y^2) + B(1-y) + Cy^2}{y^2(1-y)}$$

$$\Rightarrow \begin{cases} -A + C = 0; \\ A - B = 0; \quad \Rightarrow A = B = C = 1. \\ B = 1. \end{cases}$$

Hence,

$$\int \frac{dy}{y^2(1-y)} = \int \left(\frac{1}{y} + \frac{1}{y^2} + \frac{1}{1-y}\right) dy$$

$$= \ln|y| - \frac{1}{y} - \ln|1-y|.$$

Therefore,

$$\ln\left|\frac{y}{1-y}\right| - \frac{1}{y} = x + K.$$

12. We have $\dfrac{dy}{dx} + \dfrac{2y}{x} = \dfrac{1}{x^2}$. Let

$\mu = \int \dfrac{2}{x}\, dx = 2\ln x = \ln x^2$, then $e^\mu = x^2$, and

$$\frac{d}{dx}(x^2 y) = x^2 \frac{dy}{dx} + 2xy$$

$$= x^2 \left(\frac{dy}{dx} + \frac{2y}{x}\right) = x^2 \left(\frac{1}{x^2}\right) = 1$$

$$\Rightarrow \quad x^2 y = \int dx = x + C$$

$$\Rightarrow \quad y = \frac{1}{x} + \frac{C}{x^2}.$$

14. We have $\dfrac{dy}{dx} + y = e^x$. Let $\mu = \int dx = x$, then $e^\mu = e^x$, and

$$\frac{d}{dx}(e^x y) = e^x \frac{dy}{dx} + e^x y = e^x \left(\frac{dy}{dx} + y\right) = e^{2x}$$

$$\Rightarrow \quad e^x y = \int e^{2x}\, dx = \frac{1}{2} e^{2x} + C.$$

Hence, $y = \dfrac{1}{2} e^x + C e^{-x}$.

16. We have $\dfrac{dy}{dx} + 2e^x y = e^x$. Let $\mu = \int 2e^x\, dx = 2e^x$, then

$$\frac{d}{dx}\left(e^{2e^x} y\right) = e^{2e^x} \frac{dy}{dx} + 2e^x e^{2e^x} y$$

$$= e^{2e^x}\left(\frac{dy}{dx} + 2e^x y\right) = e^{2e^x} e^x.$$

Therefore,

$$e^{2e^x} y = \int e^{2e^x} e^x\, dx \quad \text{Let } u = 2e^x$$
$$du = 2e^x\, dx$$
$$= \frac{1}{2}\int e^u\, du = \frac{1}{2} e^{2e^x} + C.$$

Hence, $y = \dfrac{1}{2} + C e^{-2e^x}$.

18. $y(x) = 1 + \displaystyle\int_0^x \frac{(y(t))^2}{1+t^2}\, dt \quad \Longrightarrow \quad y(0) = 1$

$$\frac{dy}{dx} = \frac{y^2}{1+x^2}, \quad \text{i.e. } dy/y^2 = dx/(1+x^2)$$

$$-\frac{1}{y} = \tan^{-1} x + C$$

$$-1 = 0 + C \quad \Longrightarrow \quad C = -1$$

$$y = 1/(1 - \tan^{-1} x).$$

20. $y(x) = 3 + \displaystyle\int_0^x e^{-y}\, dt \quad \Longrightarrow \quad y(0) = 3$

$$\frac{dy}{dx} = e^{-y}, \quad \text{i.e. } e^y\, dy = dx$$

$$e^y = x + C \quad \Longrightarrow \quad y = \ln(x + C)$$

$$3 = y(0) = \ln C \quad \Longrightarrow \quad C = e^3$$

$$y = \ln(x + e^3).$$

22. Given that $m\dfrac{dv}{dt} = mg - kv$, then

$$\int \frac{dv}{g - \dfrac{k}{m} v} = \int dt$$

$$-\frac{m}{k} \ln\left|g - \frac{k}{m} v\right| = t + C.$$

Since $v(0) = 0$, therefore $C = -\dfrac{m}{k} \ln g$. Also, $g - \dfrac{k}{m} v$ remains positive for all $t > 0$, so

$$\frac{m}{k} \ln \frac{g}{g - \dfrac{k}{m} v} = t$$

$$\frac{g - \dfrac{k}{m} v}{g} = e^{-kt/m}$$

$$\Rightarrow \quad v = v(t) = \frac{mg}{k}\left(1 - e^{-kt/m}\right).$$

Note that $\displaystyle\lim_{t \to \infty} v(t) = \frac{mg}{k}$. This limiting velocity can be obtained directly from the differential equation by setting $\dfrac{dv}{dt} = 0$.

24. The balance in the account after t years is $y(t)$ and $y(0) = 1000$. The balance must satisfy

$$\frac{dy}{dt} = 0.1y - \frac{y^2}{1,000,000}$$

$$\frac{dy}{dt} = \frac{10^5 y - y^2}{10^6}$$

$$\int \frac{dy}{10^5 y - y^2} = \int \frac{dt}{10^6}$$

$$\frac{1}{10^5} \int \left(\frac{1}{y} + \frac{1}{10^5 - y} \right) dy = \frac{t}{10^6} - \frac{C}{10^5}$$

$$\ln|y| - \ln|10^5 - y| = \frac{t}{10} - C$$

$$\frac{10^5 - y}{y} = e^{C - (t/10)}$$

$$y = \frac{10^5}{e^{C - (t/10)} + 1}.$$

Since $y(0) = 1000$, we have

$$1000 = y(0) = \frac{10^5}{e^C + 1} \quad \Rightarrow \quad C = \ln 99,$$

and

$$y = \frac{10^5}{99 e^{-t/10} + 1}.$$

The balance after 1 year is

$$y = \frac{10^5}{99 e^{-1/10} + 1} \approx \$1,104.01.$$

As $t \to \infty$, the balance can grow to

$$\lim_{t \to \infty} y(t) = \lim_{t \to \infty} \frac{10^5}{e^{(4.60 - 0.1t)} + 1} = \frac{10^5}{0 + 1} = \$100,000.$$

For the account to grow to $50,000, t must satisfy

$$50,000 = y(t) = \frac{100,000}{99 e^{-t/10} + 1}$$

$$\Rightarrow \quad 99 e^{-t/10} + 1 = 2$$

$$\Rightarrow \quad t = 10 \ln 99 \approx 46 \text{ years}.$$

26. Let $x(t)$ be the number of kg of salt in the solution in the tank after t minutes. Thus, $x(0) = 50$. Salt is coming into the tank at a rate of 10 g/L × 12 L/min = 0.12 kg/min. Since the contents flow out at a rate of 10 L/min, the volume of the solution is increasing at 2 L/min and thus, at any time t, the volume of the solution is $1000 + 2t$ L. Therefore the concentration of salt is $\dfrac{x(t)}{1000 + 2t}$ L. Hence, salt is being removed at a rate

$$\frac{x(t)}{1000 + 2t} \text{ kg/L} \times 10 \text{ L/min} = \frac{5x(t)}{500 + t} \text{ kg/min}.$$

Therefore,

$$\frac{dx}{dt} = 0.12 - \frac{5x}{500 + t}$$

$$\frac{dx}{dt} + \frac{5}{500 + t} x = 0.12.$$

Let $\mu = \displaystyle\int \frac{5}{500 + t} \, dt = 5 \ln|500 + t| = \ln(500 + t)^5$ for $t > 0$. Then $e^\mu = (500 + t)^5$, and

$$\frac{d}{dt}\left[(500 + t)^5 x\right] = (500 + t)^5 \frac{dx}{dy} + 5(500 + t)^4 x$$

$$= (500 + t)^5 \left(\frac{dx}{dy} + \frac{5x}{500 + t} \right)$$

$$= 0.12(500 + t)^5.$$

Hence,

$$(500 + t)^5 x = 0.12 \int (500 + t)^5 \, dt = 0.02(500 + t)^6 + C$$

$$\Rightarrow x = 0.02(500 + t) + C(500 + t)^{-5}.$$

Since $x(0) = 50$, we have $C = 1.25 \times 10^{15}$ and

$$x = 0.02(500 + t) + (1.25 \times 10^{15})(500 + t)^{-5}.$$

After 40 min, there will be

$$x = 0.02(540) + (1.25 \times 10^{15})(540)^{-5} = 38.023 \text{ kg}$$

of salt in the tank.

Review Exercises 7   (page 473)

2. Let $A(y)$ be the cross-sectional area of the bowl at height y above the bottom. When the depth of water in the bowl is Y, then the volume of water in the bowl is

$$V(Y) = \int_0^Y A(y) \, dy.$$

The water evaporates at a rate proportional to exposed surface area. Thus

$$\frac{dV}{dt} = kA(Y)$$

$$\frac{dV}{dY} \frac{dY}{dt} = kA(Y)$$

$$A(Y) \frac{dY}{dt} = kA(Y).$$

Hence $dY/dt = k$; the depth decreases at a constant rate.

4. A vertical slice parallel to the top ridge of the solid at distance x to the right of the centre is a rectangle of base $2\sqrt{100-x^2}$ cm^and height $\sqrt{3}(10-x)$ cm. Thus the solid has volume

$$V = 2\int_0^{10} \sqrt{3}(10-x)2\sqrt{100-x^2}\,dx$$

$$= 40\sqrt{3}\int_0^{10}\sqrt{100-x^2}\,dx - 4\sqrt{3}\int_0^{10} x\sqrt{100-x^2}\,dx$$

Let $u = 100 - x^2$

$du = -2x\,dx$

$$= 40\sqrt{3}\frac{100\pi}{4} - 2\sqrt{3}\int_0^{100}\sqrt{u}\,du$$

$$= 1{,}000\sqrt{3}\left(\pi - \frac{4}{3}\right)\ \text{cm}^3.$$

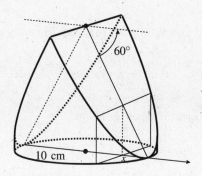

Fig. R-7.4

6. The area of revolution of $y = \sqrt{x}$, $(0 \le x \le 6)$, about the x-axis is

$$S = 2\pi\int_0^6 y\sqrt{1 + \left(\frac{dy}{dx}\right)^2}\,dx$$

$$= 2\pi\int_0^6 \sqrt{x}\sqrt{1 + \frac{1}{4x}}\,dx$$

$$= 2\pi\int_0^6 \sqrt{x + \frac{1}{4}}\,dx$$

$$= \frac{4\pi}{3}\left(x + \frac{1}{4}\right)^{3/2}\Big|_0^6 = \frac{4\pi}{3}\left[\frac{125}{8} - \frac{1}{8}\right] = \frac{62\pi}{3}\ \text{sq. units.}$$

8.

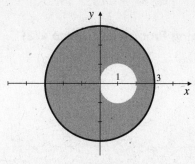

Fig. R-7.8

Let the disk have centre (and therefore centroid) at $(0,0)$. Its area is 9π. Let the hole have centre (and therefore centroid) at $(1,0)$. Its area is π. The remaining part has area 8π and centroid at $(\bar{x}, 0)$, where

$$(9\pi)(0) = (8\pi)\bar{x} + (\pi)(1).$$

Thus $\bar{x} = -1/8$. The centroid of the remaining part is $1/8$ ft from the centre of the disk on the side opposite the hole.

10. We are told that for any $a > 0$,

$$\pi\int_0^a \left[\big(f(x)\big)^2 - \big(g(x)\big)^2\right]dx = 2\pi\int_0^a x\big[f(x) - g(x)\big]\,dx.$$

Differentiating both sides of this equation with respect to a, we get

$$\big(f(a)\big)^2 - \big(g(a)\big)^2 = 2a\big[f(a) - g(a)\big],$$

or, equivalently, $f(a) + g(a) = 2a$. Thus f and g must satisfy

$$f(x) + g(x) = 2x \quad \text{for every } x > 0.$$

12. The ellipses $3x^2 + 4y^2 = C$ all satisfy the differential equation

$$6x + 8y\frac{dy}{dx} = 0, \quad \text{or} \quad \frac{dy}{dx} = -\frac{3x}{4y}.$$

A family of curves that intersect these ellipses at right angles must therefore have slopes given by $\dfrac{dy}{dx} = \dfrac{4y}{3x}$. Thus

$$3\int\frac{dy}{y} = 4\int\frac{dx}{x}$$

$$3\ln|y| = 4\ln|x| + \ln|C|.$$

The family is given by $y^3 = Cx^4$.

Challenging Problems 7 (page 473)

2.

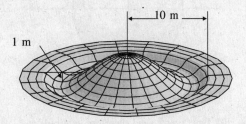

Fig. C-7.2

$h(r) = a(r^2 - 100)(r^2 - k^2)$, where $0 < k < 10$

$h'(r) = 2ar(r^2 - k^2) + 2ar(r^2 - 100) = 2ar(2r^2 - 100 - k^2)$.

The deepest point occurs where $2r^2 = 100 + k^2$, i.e., $r^2 = 50 + (k^2/2)$. Since this depth must be 1 m, we require

$$a\left(\frac{k^2}{2} - 50\right)\left(50 - \frac{k^2}{2}\right) = -1,$$

or, equivalently, $a(100 - k^2)^2 = 4$. The volume of the pool is

$$V_P = 2\pi a \int_k^{10} r(100 - r^2)(r^2 - k^2)\, dr$$

$$= 2\pi a\left(\frac{250,000}{3} - 2,500k^2 + 25k^4 - \frac{1}{12}k^6\right).$$

The volume of the hill is

$$V_H = 2\pi a \int_0^k r(r^2 - 100)(r^2 - k^2)\, dr = 2\pi a\left(25k^4 - \frac{1}{12}k^6\right).$$

These two volumes must be equal, so $k^2 = 100/3$ and $k \approx 5.77$ m. Thus $a = 4/(100 - k^2)^2 = 0.0009$. The volume of earth to be moved is V_H with these values of a and k, namely

$$2\pi(0.0009)\left[25\left(\frac{100}{3}\right)^2 - \frac{1}{12}\left(\frac{100}{3}\right)^4\right] \approx 140 \text{ m}^3.$$

4. a) If $f(x) = \begin{cases} a + bx + cx^2 & \text{for } 0 \le x \le 1 \\ p + qx + rx^2 & \text{for } 1 \le x \le 3 \end{cases}$, then

$f'(x) = \begin{cases} b + 2cx & \text{for } 0 < x < 1 \\ q + 2rx & \text{for } 1 < x < 3 \end{cases}$. We require that

$$\begin{array}{ll} a = 1 & p + 3q + 9r = 0 \\ a + b + c = 2 & p + q + r = 2 \\ b + 2c = m & q + 2r = m. \end{array}$$

The solutions of these systems are $a = 1$, $b = 2 - m$, $c = m - 1$, $p = \frac{3}{2}(1 - m)$, $q = 2m + 1$, and $r = -\frac{1}{2}(1 + m)$. $f(x, m)$ is $f(x)$ with these values of the six constants.

b) The length of the spline is

$$L(m) = \int_0^1 \sqrt{1 + (b + 2cx)^2}\, dx + \int_1^3 \sqrt{1 + (q + 2rx)^2}\, dx$$

with the values of b, c, q, and r determined above. A plot of the graph of $L(m)$ reveals a minimum value in the neighbourhood of $m = -0.3$. The derivative of $L(m)$ is a horrible expression, but Mathematica determined its zero to be about $m = -0.281326$, and the corresponding minimum value of L is about 4.41748. The polygonal line ABC has length $3\sqrt{2} \approx 4.24264$, which is only slightly shorter.

6. Starting with $V_1(r) = 2r$, and using repeatedly the formula

$$V_n(r) = \int_{-r}^r V_{n-1}(\sqrt{r^2 - x^2})\, dx,$$

Maple gave the following results:

$$\begin{array}{ll} V_1(r) = 2r & V_2(r) = \pi r^2 \\[4pt] V_3(r) = \frac{4}{3}\pi r^3 & V_4(r) = \frac{1}{2}\pi^2 r^4 \\[4pt] V_5(r) = \frac{8}{15}\pi^2 r^5 & V_6(r) = \frac{1}{6}\pi^3 r^6 \\[4pt] V_7(r) = \frac{16}{105}\pi^3 r^7 & V_8(r) = \frac{1}{24}\pi^4 r^8 \\[4pt] V_9(r) = \frac{32}{945}\pi^4 r^9 & V_{10}(r) = \frac{1}{120}\pi^5 r^{10} \end{array}$$

It appears that

$$V_{2n}(r) = \frac{1}{n!}\pi^n r^{2n}, \quad \text{and}$$

$$V_{2n-1}(r) = \frac{2^n}{1 \cdot 3 \cdot 5 \cdots (2n - 1)}\pi^{n-1} r^{2n-1}$$

$$= \frac{2^{2n-1}(n-1)!}{(2n-1)!}\pi^{n-1} r^{2n-1}.$$

These formulas predict that

$$V_{11}(r) = \frac{2^{11} 5!}{11!}\pi^5 r^{11} \quad \text{and} \quad V_{12}(r) = \frac{1}{6!}\pi^6 r^{12},$$

both of which Maple is happy to confirm.

8.

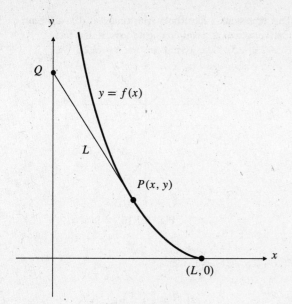

Fig. C-7.8

If $Q = (0, Y)$, then the slope of PQ is

$$\frac{y - Y}{x - 0} = f'(x) = \frac{dy}{dx}.$$

Since $|PQ| = L$, we have $(y - Y)^2 = L^2 - x^2$. Since the slope dy/dx is negative at P, $dy/dx = -\sqrt{L^2 - x^2}/x$. Thus

$$y = -\int \frac{\sqrt{L^2 - x^2}}{x} \, dx = L \ln\left(\frac{L + \sqrt{L^2 - x^2}}{x}\right) - \sqrt{L^2 - x^2} + C.$$

Since $y = 0$ when $x = L$, we have $C = 0$ and the equation of the tractrix is

$$y = L \ln\left(\frac{L + \sqrt{L^2 - x^2}}{x}\right) - \sqrt{L^2 - x^2}.$$

Note that the first term can be written in an alternate way:

$$y = L \ln\left(\frac{x}{L - \sqrt{L^2 - x^2}}\right) - \sqrt{L^2 - x^2}.$$

CHAPTER 8. CONICS, PARAMETRIC CURVES, AND POLAR CURVES

Section 8.1 Conics (page 487)

2. The ellipse with foci $(0, 1)$ and $(4, 1)$ has $c = 2$, centre $(2, 1)$, and major axis along $y = 1$. If $\epsilon = 1/2$, then $a = c/\epsilon = 4$ and $b^2 = 16 - 4 = 12$. The ellipse has equation

$$\frac{(x-2)^2}{16} + \frac{(y-1)^2}{12} = 1.$$

4. A parabola with focus at $(0, -1)$ and principal axis along $y = -1$ will have vertex at a point of the form $(v, -1)$. Its equation will then be of the form $(y + 1)^2 = \pm 4v(x - v)$. The origin lies on this curve if $1 = \pm 4(-v^2)$. Only the $-$ sign is possible, and in this case $v = \pm 1/2$. The possible equations for the parabola are $(y + 1)^2 = 1 \pm 2x$.

6. The hyperbola with foci at $(\pm 5, 1)$ and asymptotes $x = \pm(y - 1)$ is rectangular, has centre at $(0, 1)$ and has transverse axis along the line $y = 1$. Since $c = 5$ and $a = b$ (because the asymptotes are perpendicular to each other) we have $a^2 = b^2 = 25/2$. The equation of the hyperbola is

$$x^2 - (y - 1)^2 = \frac{25}{2}.$$

8. If $x^2 + 4y^2 - 4y = 0$, then

$$x^2 + 4\left(y^2 - y + \frac{1}{4}\right) = 1, \quad \text{or} \quad \frac{x^2}{1} + \frac{(y - \frac{1}{2})^2}{\frac{1}{4}} = 1.$$

This represents an ellipse with centre at $\left(0, \frac{1}{2}\right)$, semi-major axis 1, semi-minor axis $\frac{1}{2}$, and foci at $\left(\pm\frac{\sqrt{3}}{2}, \frac{1}{2}\right)$.

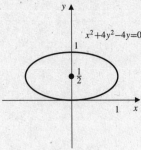

Fig. 8.1.8

10. If $4x^2 - y^2 - 4y = 0$, then

$$4x^2 - (y^2 + 4y + 4) = -4, \quad \text{or} \quad \frac{x^2}{1} - \frac{(y + 2)^2}{4} = -1.$$

This represents a hyperbola with centre at $(0, -2)$, semi-transverse axis 2, semi-conjugate axis 1, and foci at $(0, -2 \pm \sqrt{5})$. The asymptotes are $y = \pm 2x - 2$.

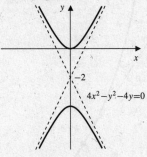

Fig. 8.1.10

12. If $x + 2y + 2y^2 = 1$, then

$$2\left(y^2 + y + \frac{1}{4}\right) = \frac{3}{2} - x$$

$$\Leftrightarrow \quad x = \frac{3}{2} - 2\left(y + \frac{1}{2}\right)^2.$$

This represents a parabola with vertex at $(\frac{3}{2}, -\frac{1}{2})$, focus at $(\frac{11}{8}, -\frac{1}{2})$ and directrix $x = \frac{13}{8}$.

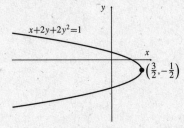

Fig. 8.1.12

14. If $9x^2 + 4y^2 - 18x + 8y = -13$, then

$$9(x^2 - 2x + 1) + 4(y^2 + 2y + 1) = 0$$

$$\Leftrightarrow 9(x - 1)^2 + 4(y + 1)^2 = 0.$$

This represents the single point $(1, -1)$.

16. The equation $(x - y)^2 - (x + y)^2 = 1$ simplifies to $4xy = -1$ and hence represents a rectangular hyperbola with centre at the origin, asymptotes along the coordinate axes, transverse axis along $y = -x$, conjugate axis along $y = x$, vertices at $\left(\frac{1}{2}, -\frac{1}{2}\right)$ and $\left(-\frac{1}{2}, \frac{1}{2}\right)$, semi-transverse and semi-conjugate axes equal to $1/\sqrt{2}$, semi-focal separation equal to $\sqrt{\frac{1}{2} + \frac{1}{2}} = 1$, and hence foci at the points $\left(\frac{1}{\sqrt{2}}, -\frac{1}{\sqrt{2}}\right)$ and $\left(-\frac{1}{\sqrt{2}}, \frac{1}{\sqrt{2}}\right)$. The eccentricity is $\sqrt{2}$.

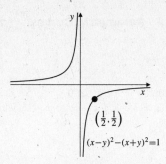

$\left(\frac{1}{2}, \frac{1}{2}\right)$

$(x-y)^2 - (x+y)^2 = 1$

Fig. 8.1.16

18. The foci of the ellipse are $(0,0)$ and $(3,0)$, so the centre is $(3/2, 0)$ and $c = 3/2$. The semi-axes a and b must satisfy $a^2 - b^2 = 9/4$. Thus the possible equations of the ellipse are

$$\frac{(x - (3/2))^2}{(9/4) + b^2} + \frac{y^2}{b^2} = 1.$$

20. We have $x^2 + 2xy + y^2 = 4x - 4y + 4$ and $A = 1$, $B = 2$, $C = 1$, $D = -4$, $E = 4$ and $F = -4$. We rotate the axes through angle θ satisfying $\tan 2\theta = B/(A - C) = \infty \Rightarrow \theta = \dfrac{\pi}{4}$. Then $A' = 2$, $B' = 0$, $C' = 0$, $D' = 0$, $E' = 4\sqrt{2}$ and the transformed equation is

$$2u^2 + 4\sqrt{2}v - 4 = 0 \quad \Rightarrow \quad u^2 = -2\sqrt{2}\left(v - \frac{1}{\sqrt{2}}\right)$$

which represents a parabola with vertex at $(u, v) = \left(0, \dfrac{1}{\sqrt{2}}\right)$ and principal axis along $u = 0$. The distance a from the focus to the vertex is given by $4a = 2\sqrt{2}$, so $a = 1/\sqrt{2}$ and the focus is at $(0, 0)$. The directrix is $v = \sqrt{2}$. Since $x = \dfrac{1}{\sqrt{2}}(u - v)$ and $y = \dfrac{1}{\sqrt{2}}(u + v)$, the vertex of the parabola in terms of xy-coordinates is $(-\frac{1}{2}, \frac{1}{2})$, and the focus is $(0, 0)$. The directrix is $x - y = 2$. The principal axis is $y = -x$.

$y = -x$

$x^2 + 2xy + y^2 = 4x - 4y + 4$

$(-1/2, 1/2)$

Fig. 8.1.20

22. We have $x^2 - 4xy + 4y^2 + 2x + y = 0$ and $A = 1$, $B = -4$, $C = 4$, $D = 2$, $E = 1$ and $F = 0$. We rotate the axes through angle θ satisfying $\tan 2\theta = B/(A - C) = \frac{4}{3}$. Then

$$\sec 2\theta = \sqrt{1 + \tan^2 2\theta} = \frac{5}{3} \quad \Rightarrow \quad \cos 2\theta = \frac{3}{5}$$

$$\Rightarrow \begin{cases} \cos\theta = \sqrt{\dfrac{1 + \cos 2\theta}{2}} = \sqrt{\dfrac{4}{5}} = \dfrac{2}{\sqrt{5}}; \\[2mm] \sin\theta = \sqrt{\dfrac{1 - \cos 2\theta}{2}} = \sqrt{\dfrac{1}{5}} = \dfrac{1}{\sqrt{5}}. \end{cases}$$

Then $A' = 0$, $B' = 0$, $C' = 5$, $D' = \sqrt{5}$, $E' = 0$ and the transformed equation is

$$5v^2 + \sqrt{5}u = 0 \quad \Rightarrow \quad v^2 = -\frac{1}{\sqrt{5}}u$$

which represents a parabola with vertex at $(u, v) = (0, 0)$, focus at $\left(-\dfrac{1}{4\sqrt{5}}, 0\right)$. The directrix is $u = \dfrac{1}{4\sqrt{5}}$ and the principal axis is $v = 0$. Since $x = \dfrac{2}{\sqrt{5}}u - \dfrac{1}{\sqrt{5}}v$ and $y = \dfrac{1}{\sqrt{5}}u + \dfrac{2}{\sqrt{5}}v$, in terms of the xy-coordinates, the vertex is at $(0, 0)$, the focus at $\left(-\dfrac{1}{10}, -\dfrac{1}{20}\right)$. The directrix is $2x + y = \frac{1}{4}$ and the principal axis is $2y - x = 0$.

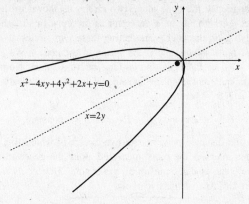

$x^2 - 4xy + 4y^2 + 2x + y = 0$

$x = 2y$

Fig. 8.1.22

24. Let the equation of the parabola be $y^2 = 4ax$. The focus F is at $(a, 0)$ and vertex at $(0, 0)$. Then the distance from the vertex to the focus is a. At $x = a$, $y = \sqrt{4a(a)} = \pm 2a$. Hence, $\ell = 2a$, which is twice the distance from the vertex to the focus.

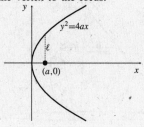

$y^2 = 4ax$

ℓ

$(a, 0)$

Fig. 8.1.24

26. Suppose the hyperbola has equation $\dfrac{x^2}{a^2} - \dfrac{y^2}{b^2} = 1$. The vertices are at $(\pm a, 0)$ and the foci are at $(\pm c, 0)$ where $c = \sqrt{a^2 + b^2}$. At $x = \sqrt{a^2 + b^2}$,

$$\frac{a^2 + b^2}{a^2} - \frac{y^2}{b^2} = 1$$
$$(a^2 + b^2)b^2 - a^2 y^2 = a^2 b^2$$
$$y = \pm \frac{b^2}{a}.$$

Hence, $\ell = \dfrac{b^2}{a}$.

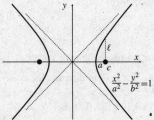

Fig. 8.1.26

28. Let F_1 and F_2 be the points where the plane is tangent to the spheres. Let P be an arbitrary point P on the hyperbola in which the plane intersects the cone. The spheres are tangent to the cone along two circles as shown in the figure. Let $PAVB$ be a generator of the cone (a straight line lying on the cone) intersecting these two circles at A and B as shown. (V is the vertex of the cone.) We have $PF_1 = PA$ because two tangents to a sphere from a point outside the sphere have equal lengths. Similarly, $PF_2 = PB$. Therefore

$$PF_2 - PF_1 = PB - PA = AB = \text{constant},$$

since the distance between the two circles in which the spheres intersect the cone, measured along the generators of the cone, is the same for all generators. Hence, F_1 and F_2 are the foci of the hyperbola.

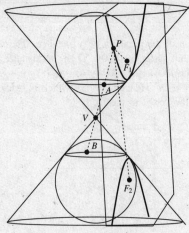

Fig. 8.1.28

Section 8.2 Parametric Curves (page 494)

2. If $x = 2 - t$ and $y = t + 1$ for $0 \le t < \infty$, then $y = 2 - x + 1 = 3 - x$ for $-\infty < x \le 2$, which is a half line.

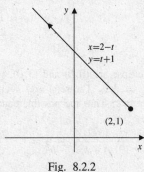

Fig. 8.2.2

4. If $x = \dfrac{1}{1 + t^2}$ and $y = \dfrac{t}{1 + t^2}$ for $-\infty < t < \infty$, then

$$x^2 + y^2 = \frac{1 + t^2}{(1 + t^2)^2} = \frac{1}{1 + t^2} = x$$
$$\Leftrightarrow \left(x - \frac{1}{2}\right)^2 + y^2 = \frac{1}{4}.$$

This curve consists of all points of the circle with centre at $(\frac{1}{2}, 0)$ and radius $\frac{1}{2}$ except the origin $(0, 0)$.

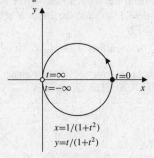

Fig. 8.2.4

6. If $x = a \sec t$ and $y = b \tan t$ for $-\dfrac{\pi}{2} < t < \dfrac{\pi}{2}$, then

$$\frac{x^2}{a^2} - \frac{y^2}{b^2} = \sec^2 t - \tan^2 t = 1.$$

The curve is one arch of this hyperbola.

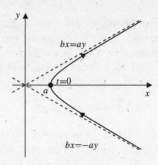

Fig. 8.2.6

8. If $x = \cos \sin s$ and $y = \sin \sin s$ for $-\infty < s < \infty$, then $x^2 + y^2 = 1$. The curve consists of the arc of this circle extending from $(a, -b)$ through $(1, 0)$ to (a, b) where $a = \cos(1)$ and $b = \sin(1)$, traversed infinitely often back and forth.

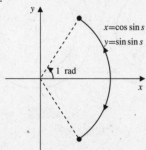

Fig. 8.2.8

10. If $x = 1 - \sqrt{4 - t^2}$ and $y = 2 + t$ for $-2 \le t \le 2$ then

$$(x - 1)^2 = 4 - t^2 = 4 - (y - 2)^2.$$

The parametric curve is the left half of the circle of radius 4 centred at $(1, 2)$, and is traced in the direction of increasing y.

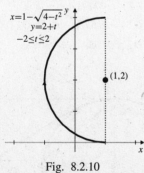

Fig. 8.2.10

12. $x = 2 - 3 \cosh t$, $y = -1 + 2 \sinh t$ represents the left half (branch) of the hyperbola

$$\frac{(x - 2)^2}{9} - \frac{(y + 1)^2}{4} = 1.$$

14. (i) If $x = \cos^4 t$ and $y = \sin^4 t$, then

$$
\begin{aligned}
(x - y)^2 &= (\cos^4 t - \sin^4 t)^2 \\
&= \left[(\cos^2 t + \sin^2 t)(\cos^2 t - \sin^2 t) \right]^2 \\
&= (\cos^2 t - \sin^2 t)^2 \\
&= \cos^4 t + \sin^4 t - 2 \cos^2 t \sin^2 t
\end{aligned}
$$

and

$$1 = (\cos^2 t + \sin^2 t)^2 = \cos^4 t + \sin^4 t + 2 \cos^2 t \sin^2 t.$$

Hence,

$$1 + (x - y)^2 = 2(\cos^4 t + \sin^4 t) = 2(x + y).$$

(ii) If $x = \sec^4 t$ and $y = \tan^4 t$, then

$$
\begin{aligned}
(x - y)^2 &= (\sec^4 t - \tan^4 t)^2 \\
&= (\sec^2 t + \tan^2 t)^2 \\
&= \sec^4 t + \tan^4 t + 2 \sec^2 t \tan^2 t
\end{aligned}
$$

and

$$1 = (\sec^2 t - \tan^2 t)^2 = \sec^4 t + \tan^4 t - 2 \sec^2 t \tan^2 t.$$

Hence,

$$1 + (x - y)^2 = 2(\sec^4 t + \tan^4 t) = 2(x + y).$$

(iii) Similarly, if $x = \tan^4 t$ and $y = \sec^4 t$, then

$$
\begin{aligned}
1 + (x - y)^2 &= 1 + (y - x)^2 \\
&= (\sec^2 t - \tan^2 t)^2 + (\sec^4 t - \tan^4 t)^2 \\
&= 2(\tan^4 t + \sec^4 t) \\
&= 2(x + y).
\end{aligned}
$$

These three parametric curves above correspond to different parts of the parabola $1 + (x - y)^2 = 2(x + y)$, as shown in the following diagram.

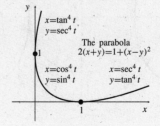

Fig. 8.2.14

16. If (x, y) is any point on the circle $x^2 + y^2 = R^2$ other than $(R, 0)$, then the line from (x, y) to $(R, 0)$ has slope $m = \dfrac{y}{x - R}$. Thus $y = m(x - R)$, and

$$x^2 + m^2(x - R)^2 = R^2$$
$$(m^2 + 1)x^2 - 2xRm^2 + (m^2 - 1)R^2 = 0$$
$$\left[(m^2 + 1)x - (m^2 - 1)R\right](x - R) = 0$$
$$\Rightarrow \quad x = \frac{(m^2 - 1)R}{m^2 + 1} \text{ or } x = R.$$

The parametrization of the circle in terms of m is given by

$$x = \frac{(m^2 - 1)R}{m^2 + 1}$$
$$y = m\left[\frac{(m^2 - 1)R}{m^2 + 1} - R\right] = -\frac{2Rm}{m^2 + 1}$$

where $-\infty < m < \infty$. This parametrization gives every point on the circle except $(R, 0)$.

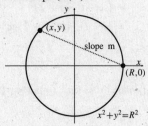

Fig. 8.2.16

18. The coordinates of P satisfy

$$x = a \sec t, \quad y = b \sin t.$$

The Cartesian equation is $\dfrac{y^2}{b^2} + \dfrac{a^2}{x^2} = 1$.

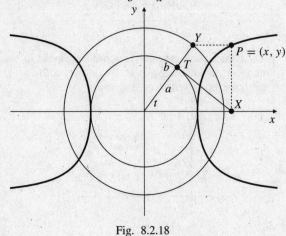

Fig. 8.2.18

20. Let C_0 and P_0 be the original positions of the centre of the wheel and a point at the bottom of the flange whose path is to be traced. The wheel is also shown in a subsequent position in which it makes contact with the rail at R. Since the wheel has been rotated by an angle θ,

$$OR = \text{arc } SR = a\theta.$$

Thus, the new position of the centre is $C = (a\theta, a)$. Let $P = (x, y)$ be the new position of the point; then

$$x = OR - PQ = a\theta - b\sin(\pi - \theta) = a\theta - b\sin\theta,$$
$$y = RC + CQ = a + b\cos(\pi - \theta) = a - b\cos\theta.$$

These are the parametric equations of the prolate cycloid.

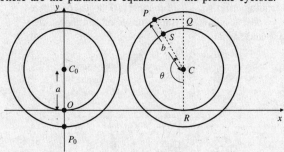

Fig. 8.2.20

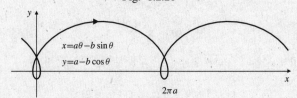

Fig. 8.2.20

22. a) From triangles in the figure,

$$x = |TX| = |OT|\tan t = \tan t$$
$$y = |OY| = \sin\left(\frac{\pi}{2} - t\right) = |OY|\cos t$$
$$= |OT|\cos t \cos t = \cos^2 t.$$

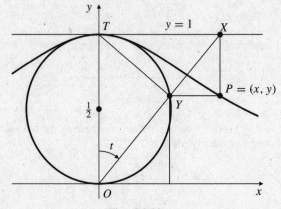

Fig. 8.2.22

b) $\dfrac{1}{y} = \sec^2 t = 1 + \tan^2 t = 1 + x^2$. Thus $y = \dfrac{1}{1+x^2}$.

24. $x = \sin t$, $y = \sin(3t)$

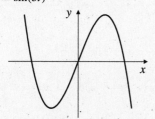

Fig. 8.2.24

26. $x = \sin(2t)$, $y = \sin(5t)$

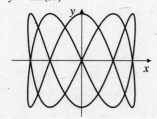

Fig. 8.2.26

28. $x = \left(1 + \dfrac{1}{n}\right)\cos t + \dfrac{1}{n}\cos((n-1)t)$

$y = \left(1 + \dfrac{1}{n}\right)\sin t - \dfrac{1}{n}\sin((n-1)t)$

represents a cycloid-like curve that is wound around the inside circle $x^2 + y^2 = \left(1 + (2/n)\right)^2$ and is externally tangent to $x^2 + y^2 = 1$. If $n \geq 2$ is an integer, the curve closes after one revolution and has n cusps. The figure shows the curve for $n = 7$. If n is a rational number but not an integer, the curve will wind around the circle more than once before it closes.

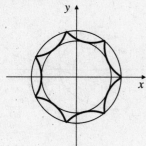

Fig. 8.2.28

Section 8.3 Smooth Parametric Curves and Their Slopes (page 499)

2. $x = t^2 - 2t \qquad y = t^2 + 2t$

$\dfrac{dx}{dt} = 2t - 2 \qquad \dfrac{dy}{dt} = 2t + 2$

Horizontal tangent at $t = -1$, i.e., at $(3, -1)$.
Vertical tangent at $t = 1$, i.e., at $(-1, 3)$.

4. $x = t^3 - 3t \qquad y = 2t^3 + 3t^2$

$\dfrac{dx}{dt} = 3(t^2 - 1) \qquad \dfrac{dy}{dt} = 6t(t+1)$

Horizontal tangent at $t = 0$, i.e., at $(0, 0)$.
Vertical tangent at $t = 1$, i.e., at $(-2, 5)$.
At $t = -1$ (i.e., at $(2, 1)$) both dx/dt and dy/dt change sign, so the curve is not smooth there. (It has a cusp.)

6. $x = \sin t \qquad y = \sin t - t\cos t$

$\dfrac{dx}{dt} = \cos t \qquad \dfrac{dy}{dt} = t\sin t$

Horizontal tangent at $t = n\pi$, i.e., at $(0, -(-1)^n n\pi)$ (for integers n).
Vertical tangent at $t = (n + \tfrac{1}{2})\pi$, i.e. at $(1, 1)$ and $(-1, -1)$.

8. $x = \dfrac{3t}{1+t^3} \qquad y = \dfrac{3t^2}{1+t^3}$

$\dfrac{dx}{dt} = \dfrac{3(1 - 2t^3)}{(1+t^3)^2} \qquad \dfrac{dy}{dt} = \dfrac{3t(2 - t^3)}{(1+t^3)^2}$

Horizontal tangent at $t = 0$ and $t = 2^{1/3}$, i.e., at $(0, 0)$ and $(2^{1/3}, 2^{2/3})$.
Vertical tangent at $t = 2^{-1/3}$, i.e., at $(2^{2/3}, 2^{1/3})$. The curve also approaches $(0, 0)$ vertically as $t \to \pm\infty$.

10. $x = t^4 - t^2 \qquad y = t^3 + 2t$

$\dfrac{dx}{dt} = 4t^3 - 2t \qquad \dfrac{dy}{dt} = 3t^2 + 2$

At $t = -1$; $\dfrac{dy}{dx} = \dfrac{3(-1)^2 + 2}{4(-1)^3 - 2(-1)} = -\dfrac{5}{2}$.

12. $x = e^{2t} \qquad y = te^{2t}$

$\dfrac{dx}{dt} = 2e^{2t} \qquad \dfrac{dy}{dt} = e^{2t}(1 + 2t)$

At $t = -2$; $\dfrac{dy}{dx} = \dfrac{e^{-4}(1-4)}{2e^{-4}} = -\dfrac{3}{2}$.

14.
$$x = t - \cos t = \frac{\pi}{4} - \frac{1}{\sqrt{2}}$$

$$\frac{dx}{dt} = 1 + \sin t = 1 + \frac{1}{\sqrt{2}}$$

$$y = 1 - \sin t = 1 - \frac{1}{\sqrt{2}} \quad \text{at } t = \frac{\pi}{4}$$

$$\frac{dy}{dt} = -\cos t = -\frac{1}{\sqrt{2}} \quad \text{at } t = \frac{\pi}{4}$$

Tangent line: $x = \frac{\pi}{4} - \frac{1}{\sqrt{2}} + \left(1 + \frac{1}{\sqrt{2}}\right) t,$

$$y = 1 - \frac{1}{\sqrt{2}} - \frac{t}{\sqrt{2}}.$$

16. $x = \sin t$, $y = \sin(2t)$ is at $(0,0)$ at $t = 0$ and $t = \pi$.
Since
$$\frac{dy}{dx} = \frac{2\cos(2t)}{\cos t} = \begin{cases} 2 & \text{if } t = 0 \\ -2 & \text{if } t = \pi, \end{cases}$$
the tangents at $(0,0)$ at $t = 0$ and $t = \pi$ have slopes 2 and -2, respectively.

18. $x = (t-1)^4 \qquad y = (t-1)^3$

$$\frac{dx}{dt} = 4(t-1)^3 \quad \frac{dy}{dt} = 3(t-1)^2 \quad \text{both vanish at } t = 1.$$

Since $\dfrac{dx}{dy} = \dfrac{4(t-1)}{3} \to 0$ as $t \to 1$, and dy/dt does not change sign at $t = 1$, the curve is smooth at $t = 1$ and therefore everywhere.

20. $x = t^3 \qquad y = t - \sin t$

$$\frac{dx}{dt} = 3t^2 \quad \frac{dy}{dt} = 1 - \cos t \quad \text{both vanish at } t = 0.$$

$$\lim_{t\to 0} \frac{dx}{dy} = \lim_{t\to 0} \frac{3t^2}{1 - \cos t} = \lim_{t\to 0} \frac{6t}{\sin t} = 6 \text{ and } dy/dt \text{ does}$$
not change sign at $t = 0$. Thus the curve is smooth at $t = 0$, and hence everywhere.

22. If $x = f(t) = t^3$ and $y = g(t) = 3t^2 - 1$, then

$$f'(t) = 3t^2, \ f''(t) = 6t;$$
$$g'(t) = 6t, \quad g''(t) = 6.$$

Both $f'(t)$ and $g'(t)$ vanish at $t = 0$. Observe that

$$\frac{dy}{dx} = \frac{6t}{3t^2} = \frac{2}{t}.$$

Thus,

$$\lim_{t\to 0+} \frac{dy}{dx} = \infty, \qquad \lim_{t\to 0-} \frac{dy}{dx} = -\infty.$$

and the curve has a cusp at $t = 0$, i.e., at $(0, -1)$. Since

$$\frac{d^2y}{dx^2} = \frac{(3t^2)(6) - (6t)(6t)}{(3t^2)^3} = -\frac{2}{3t^4} < 0$$

for all t, the curve is concave down everywhere.

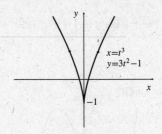

Fig. 8.3.22

24. If $x = f(t) = t^3 - 3t - 2$ and $y = g(t) = t^2 - t - 2$, then

$$f'(t) = 3t^2 - 3, \ f''(t) = 6t;$$
$$g'(t) = 2t - 1, \quad g''(t) = 2.$$

The tangent is horizontal at $t = \dfrac{1}{2}$, i.e., at $\left(-\dfrac{27}{8}, -\dfrac{9}{4}\right)$.
The tangent is vertical at $t = \pm 1$, i.e., $(-4, -2)$ and $(0, 0)$.
Directional information is as follows:

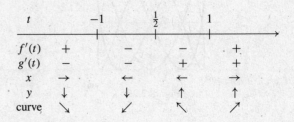

t		-1		$\frac{1}{2}$		1	
$f'(t)$	$+$		$-$		$-$		$+$
$g'(t)$	$-$		$-$		$+$		$+$
x	$\rightarrow$		$\leftarrow$		$\leftarrow$		$\rightarrow$
y	$\downarrow$		$\downarrow$		$\uparrow$		$\uparrow$
curve	$\searrow$		$\swarrow$		$\nwarrow$		$\nearrow$

For concavity,

$$\frac{d^2y}{dx^2} = \frac{3(t^2 - 1)(2) - (2t - 1)(6t)}{[3(t^2 - 1)]^3} = -\frac{2(t^2 - t + 1)}{9(t^2 - 1)^3}$$

which is undefined at $t = \pm 1$, therefore

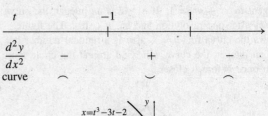

t		-1		1	
$\frac{d^2y}{dx^2}$	$-$		$+$		$-$
curve	$\frown$		$\smile$		$\frown$

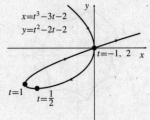

Fig. 8.3.24

Section 8.4 Arc Lengths and Areas for Parametric Curves (page 504)

2. If $x = 1 + t^3$ and $y = 1 - t^2$ for $-1 \le t \le 2$, then the arc length is

$$s = \int_{-1}^{2} \sqrt{(3t^2)^2 + (-2t)^2}\, dt$$

$$= \int_{-1}^{2} |t| \sqrt{9t^2 + 4}\, dt$$

$$= \left(\int_{0}^{1} + \int_{0}^{2} \right) t\sqrt{9t^2 + 4}\, dt \quad \text{Let } u = 9t^2 + 4$$
$$du = 18t\, dt$$

$$= \frac{1}{18} \left(\int_{4}^{13} + \int_{4}^{40} \right) \sqrt{u}\, du$$

$$= \frac{1}{27} \left(13\sqrt{13} + 40\sqrt{40} - 16 \right) \text{ units.}$$

4. If $x = \ln(1 + t^2)$ and $y = 2\tan^{-1} t$ for $0 \le t \le 1$, then

$$\frac{dx}{dt} = \frac{2t}{1 + t^2}; \qquad \frac{dy}{dt} = \frac{2}{1 + t^2}.$$

The arc length is

$$s = \int_{0}^{1} \sqrt{\frac{4t^2 + 4}{(1 + t^2)^2}}\, dt$$

$$= 2 \int_{0}^{1} \frac{dt}{\sqrt{1 + t^2}} \quad \begin{array}{l} \text{Let } t = \tan\theta \\ dt = \sec^2\theta\, d\theta \end{array}$$

$$= 2 \int_{0}^{\pi/4} \sec\theta\, d\theta$$

$$= 2\ln|\sec\theta + \tan\theta|\Big|_{0}^{\pi/4} = 2\ln(1 + \sqrt{2}) \text{ units.}$$

6. $\quad x = \cos t + t\sin t \qquad y = \sin t - t\cos t \quad (0 \le t \le 2\pi)$

$$\frac{dx}{dt} = t\cos t \qquad \frac{dy}{dt} = t\sin t$$

$$\text{Length } = \int_{0}^{2\pi} \sqrt{t^2 \cos^2 t + t^2 \sin^2 t}\, dt$$

$$= \int_{0}^{2\pi} t\, dt = \frac{t^2}{2}\Big|_{0}^{2\pi} = 2\pi^2 \text{ units.}$$

8. $\quad x = \sin^2 t \qquad y = 2\cos t \quad (0 \le t \le \pi/2)$

$$\frac{dx}{dt} = 2\sin t\cos t \qquad \frac{dy}{dt} = -2\sin t$$

Length
$$= \int_{0}^{\pi/2} \sqrt{4\sin^2 t\cos^2 t + 4\sin^2 t}\, dt$$

$$= 2 \int_{0}^{\pi/2} \sin t \sqrt{1 + \cos^2 t}\, dt \quad \begin{array}{l} \text{Let } \cos t = \tan u \\ -\sin t\, dt = \sec^2 u\, du \end{array}$$

$$= 2 \int_{0}^{\pi/4} \sec^3 u\, du$$

$$= \Big(\sec u\tan u + \ln(\sec u + \tan u) \Big)\Big|_{0}^{\pi/4}$$

$$= \sqrt{2} + \ln(1 + \sqrt{2}) \text{ units.}$$

10. If $x = at - a\sin t$ and $y = a - a\cos t$ for $0 \le t \le 2\pi$, then

$$\frac{dx}{dt} = a - a\cos t, \qquad \frac{dy}{dt} = a\sin t;$$

$$ds = \sqrt{(a - a\cos t)^2 + (a\sin t)^2}\, dt$$

$$= a\sqrt{2}\sqrt{1 - \cos t}\, dt = a\sqrt{2}\sqrt{2\sin^2\left(\frac{t}{2}\right)}\, dt$$

$$= 2a\sin\left(\frac{t}{2}\right) dt.$$

a) The surface area generated by rotating the arch about the x-axis is

$$S_x = 2\pi \int_{0}^{2\pi} |y|\, ds$$

$$= 4\pi \int_{0}^{\pi} (a - a\cos t) 2a\sin\left(\frac{t}{2}\right) dt$$

$$= 16\pi a^2 \int_{0}^{\pi} \sin^3\left(\frac{t}{2}\right) dt$$

$$= 16\pi a^2 \int_{0}^{\pi} \left[1 - \cos^2\left(\frac{t}{2}\right)\right] \sin\left(\frac{t}{2}\right) dt$$

$$\text{Let } u = \cos\left(\frac{t}{2}\right)$$
$$du = -\frac{1}{2}\sin\left(\frac{t}{2}\right) dt$$

$$= -32\pi a^2 \int_{1}^{0} (1 - u^2)\, du$$

$$= 32\pi a^2 \left[u - \frac{1}{3}u^3\right]\Big|_{0}^{1}$$

$$= \frac{64}{3}\pi a^3 \text{ sq. units.}$$

b) The surface area generated by rotating the arch about the y-axis is

$$S_y = 2\pi \int_0^{2\pi} |x|\, ds$$

$$= 2\pi \int_0^{2\pi} (at - a\sin t) 2a \sin\left(\frac{t}{2}\right) dt$$

$$= 4\pi a^2 \int_0^{2\pi} \left[t - 2\sin\left(\frac{t}{2}\right) \cos\left(\frac{t}{2}\right) \right] \sin\left(\frac{t}{2}\right) dt$$

$$= 4\pi a^2 \int_0^{2\pi} t \sin\left(\frac{t}{2}\right) dt$$

$$\quad - 8\pi a^2 \int_0^{2\pi} \sin^2\left(\frac{t}{2}\right) \cos\left(\frac{t}{2}\right) dt$$

$$= 4\pi a^2 \left[-2t \cos\left(\frac{t}{2}\right) \Big|_0^{2\pi} + 2\int_0^{2\pi} \cos\left(\frac{t}{2}\right) dt \right] - 0$$

$$= 4\pi a^2 [4\pi + 0] = 16\pi^2 a^2 \text{ sq. units.}$$

12. The area of revolution of the curve in Exercise 11 about the y-axis is

$$\int_{t=0}^{t=\pi/2} 2\pi x\, ds = 2\sqrt{2}\pi \int_0^{\pi/2} e^{2t} \cos t\, dt$$

$$= 2\sqrt{2}\pi \frac{e^{2t}}{5} (2\cos t + \sin t) \Big|_0^{\pi/2}$$

$$= \frac{2\sqrt{2}\pi}{5} (e^\pi - 2) \text{ sq. units.}$$

14. The area of revolution of the curve of Exercise 13 about the x-axis is

$$\int_{t=0}^{t=1} 2\pi y\, ds = 24\pi \int_0^1 t^4 \sqrt{1+t^2}\, dt \quad \text{Let } t = \tan u$$
$$\qquad\qquad\qquad\qquad\qquad\qquad\qquad dt = \sec^2 u\, du$$

$$= 24\pi \int_0^{\pi/4} \tan^4 u \sec^3 u\, du$$

$$= 24\pi \int_0^{\pi/4} (\sec^7 u - 2\sec^5 u + \sec^3 u)\, du$$

$$= \frac{\pi}{2}\left(7\sqrt{2} + 3\ln(1 + \sqrt{2})\right) \text{ sq. units.}$$

We have omitted the details of evaluation of the final integral. See Exercise 24 of Section 8.3 for a similar evaluation.

16. Area of $R = 4 \times \int_{\pi/2}^0 (a\sin^3 t)(-3a \sin t \cos^2 t)\, dt$

$$= -12a^2 \int_{\pi/2}^0 \sin^4 t \cos^2 t\, dt$$

$$= 12a^2 \left[\frac{t}{16} - \frac{\sin(4t)}{64} - \frac{\sin^3(2t)}{48} \right]\Big|_0^{\pi/2}$$

(See Exercise 34 of Section 6.4.)

$$= \frac{3}{8}\pi a^2 \text{ sq. units.}$$

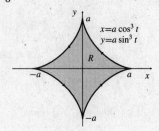

Fig. 8.4.16

18. If $x = \cos s \sin s = \frac{1}{2}\sin 2s$ and $y = \sin^2 s = \frac{1}{2} - \frac{1}{2}\cos 2s$ for $0 \le s \le \frac{1}{2}\pi$, then

$$x^2 + \left(y - \frac{1}{2}\right)^2 = \frac{1}{4}\sin^2 2s + \frac{1}{4}\cos^2 2s = \frac{1}{4}$$

which is the right half of the circle with radius $\frac{1}{2}$ and centre at $(0, \frac{1}{2})$. Hence, the area of R is

$$\frac{1}{2}\left[\pi \left(\frac{1}{2}\right)^2\right] = \frac{\pi}{8} \text{ sq. units.}$$

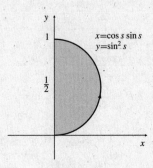

Fig. 8.4.18

20. To find the shaded area we subtract the area under the upper half of the hyperbola from that of a right triangle:

Shaded area = Area $\triangle ABC$ − Area sector ABC

$$= \frac{1}{2}\sec t_0 \tan t_0 - \int_0^{t_0} \tan t (\sec t \tan t)\,dt$$

$$= \frac{1}{2}\sec t_0 \tan t_0 - \int_0^{t_0} (\sec^3 t - \sec t)\,dt$$

$$= \frac{1}{2}\sec t_0 \tan t_0 - \left[\frac{1}{2}\sec t \tan t + \right.$$

$$\left. \frac{1}{2}\ln|\sec t + \tan t| - \ln|\sec t + \tan t|\right]\Big|_0^{t_0}$$

$$= \frac{1}{2}\ln|\sec t_0 + \tan t_0| \text{ sq. units.}$$

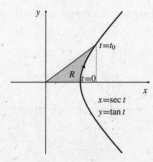

Fig. 8.4.20

22. If $x = f(t) = at - a\sin t$ and $y = g(t) = a - a\cos t$, then the volume of the solid obtained by rotating about the x-axis is

$$V = \int_{t=0}^{t=2\pi} \pi y^2\,dx = \pi \int_{t=0}^{t=2\pi} [g(t)]^2 f'(t)\,dt$$

$$= \pi \int_0^{2\pi} (a - a\cos t)^2 (a - a\cos t)\,dt$$

$$= \pi a^3 \int_0^{2\pi} (1 - \cos t)^3\,dt$$

$$= \pi a^3 \int_0^{2\pi} (1 - 3\cos t + 3\cos^2 t - \cos^3 t)\,dt$$

$$= \pi a^3 \left[2\pi - 0 + \frac{3}{2}\int_0^{2\pi}(1 + \cos 2t)\,dt - 0\right]$$

$$= \pi a^3 \left[2\pi + \frac{3}{2}(2\pi)\right] = 5\pi^2 a^3 \text{ cu. units.}$$

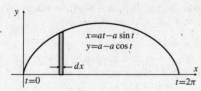

$$x = at - a\sin t$$
$$y = a - a\cos t$$

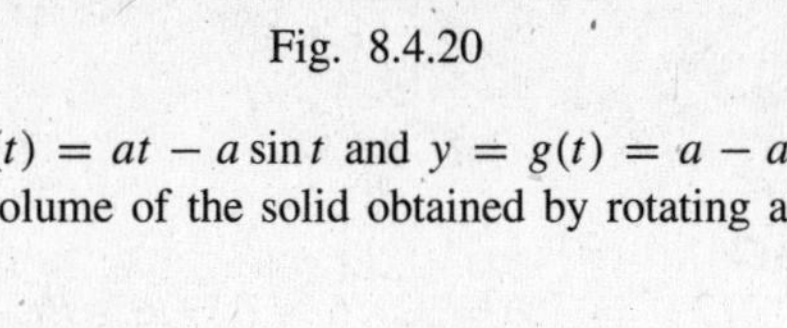

Fig. 8.4.22

Section 8.5 Polar Coordinates and Polar Curves (page 511)

2. $r = -2\csc\theta \Rightarrow r\sin\theta = -2$

$\Leftrightarrow y = -2$ a horizontal line.

4. $r = \sin\theta + \cos\theta$

$r^2 = r\sin\theta + r\cos\theta$

$x^2 + y^2 = y + x$

$\left(x - \frac{1}{2}\right)^2 + \left(y - \frac{1}{2}\right)^2 = \frac{1}{2}$

a circle with centre $\left(\frac{1}{2}, \frac{1}{2}\right)$ and radius $\frac{1}{\sqrt{2}}$.

6. $r = \sec\theta \tan\theta \Rightarrow r\cos\theta = \dfrac{r\sin\theta}{r\cos\theta}$

$x^2 = y$ a parabola.

8. $r = \dfrac{2}{\sqrt{\cos^2\theta + 4\sin^2\theta}}$

$r^2\cos^2\theta + 4r^2\sin^2\theta = 4$

$x^2 + 4y^2 = 4$ an ellipse.

10. $r = \dfrac{2}{2 - \cos\theta}$

$2r - r\cos\theta = 2$

$4r^2 = (2 + x)^2$

$4x^2 + 4y^2 = 4 + 4x + x^2$

$3x^2 + 4y^2 - 4x = 4$ an ellipse.

12. $r = \dfrac{2}{1 + \sin\theta}$

$r + r\sin\theta = 2$

$r^2 = (2 - y)^2$

$x^2 + y^2 = 4 - 4y + y^2$

$x^2 = 4 - 4y$ a parabola.

14. If $r = 1 - \cos\left(\theta + \dfrac{\pi}{4}\right)$, then $r = 0$ at $\theta = -\dfrac{\pi}{4}$ and $\dfrac{7\pi}{4}$. This is a cardioid.

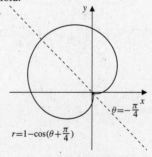

$r = 1 - \cos(\theta + \frac{\pi}{4})$

$\theta = -\frac{\pi}{4}$

Fig. 8.5.14

16. If $r = 1 - 2\sin\theta$, then $r = 0$ at $\theta = \dfrac{\pi}{6}$ and $\dfrac{5\pi}{6}$.

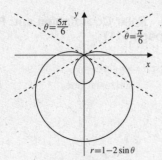

Fig. 8.5.16

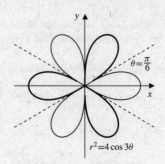

Fig. 8.5.22

24. If $r = \ln\theta$, then $r = 0$ at $\theta = 1$. Note that

$$y = r\sin\theta = \ln\theta\sin\theta = (\theta\ln\theta)\left(\frac{\sin\theta}{\theta}\right) \to 0$$

as $\theta \to 0+$. Therefore, the (negative) x-axis is an asymptote of the curve.

18. If $r = 2\sin 2\theta$, then $r = 0$ at $\theta = 0$, $\pm\dfrac{\pi}{2}$ and π.

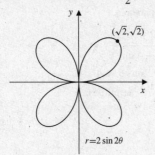

Fig. 8.5.18

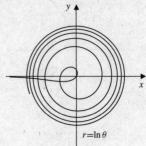

Fig. 8.5.24

26. $r^2 = 2\cos(2\theta)$, $r = 1$.
$\cos(2\theta) = 1/2 \quad \Rightarrow \quad \theta = \pm\pi/6$ or $\theta = \pm 5\pi/6$.
Intersections: $[1, \pm\pi/6]$ and $[1, \pm 5\pi/6]$.

28. Let $r_1(\theta) = \theta$ and $r_2(\theta) = \theta + \pi$. Although the equation $r_1(\theta) = r_2(\theta)$ has no solutions, the curves $r = r_1(\theta)$ and $r = r_2(\theta)$ can still intersect if $r_1(\theta_1) = -r_2(\theta_2)$ for two angles θ_1 and θ_2 having the opposite directions in the polar plane. Observe that $\theta_1 = -n\pi$ and $\theta_2 = (n-1)\pi$ are two such angles provided n is any integer. Since

$$r_1(\theta_1) = -n\pi = -r_2((n-1)\pi),$$

the curves intersect at any point of the form $[n\pi, 0]$ or $[n\pi, \pi]$.

20. If $r = 2\cos 4\theta$, then $r = 0$ at $\theta = \pm\dfrac{\pi}{8}$, $\pm\dfrac{3\pi}{8}$, $\pm\dfrac{5\pi}{8}$ and $\pm\dfrac{7\pi}{8}$. (an eight leaf rosette)

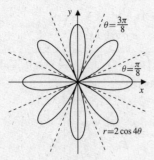

Fig. 8.5.20

30. The graph of $r = \cos n\theta$ has $2n$ leaves if n is an even integer and n leaves if n is an odd integer. The situation for $r^2 = \cos n\theta$ is reversed. The graph has $2n$ leaves if n is an odd integer (provided negative values of r are allowed), and it has n leaves if n is even.

32. $r = \cos\theta\cos(m\theta)$
For odd m this flower has $2m$ petals, 2 large ones and 4 each of $(m-1)/2$ smaller sizes.
For even m the flower has $m+1$ petals, one large and 2 each of $m/2$ smaller sizes.

22. If $r^2 = 4\cos 3\theta$, then $r = 0$ at $\theta = \pm\dfrac{\pi}{6}$, $\pm\dfrac{\pi}{2}$ and $\pm\dfrac{5\pi}{6}$. This equation defines two functions of r, namely $r = \pm 2\sqrt{\cos 3\theta}$. Each contributes 3 leaves to the graph.

34. $r = \sin(2\theta)\sin(m\theta)$
For odd m there are $m + 1$ petals, 2 each of $(m + 1)/2$ different sizes.
For even m there are always $2m$ petals. They are of n different sizes if $m = 4n - 2$ or $m = 4n$.

36. $r = C + \cos\theta\cos(2\theta)$
The curve always has 3 bulges, one larger than the other two. For $C = 0$ these are 3 distinct petals. For $0 < C < 1$ there is a fourth supplementary petal inside the large one. For $C = 1$ the curve has a cusp at the origin. For $C > 1$ the curve does not approach the origin, and the petals become less distinct as C increases.

38.

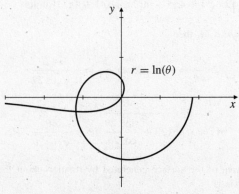

Fig. 8.5.38

We will have $[\ln\theta_1, \theta_1] = [\ln\theta_2, \theta_2]$ if

$$\theta_2 = \theta_1 + \pi \quad \text{and} \quad \ln\theta_1 = -\ln\theta_2,$$

that is, if $\ln\theta_1 + \ln(\theta_1 + \pi) = 0$. This equation has solution $\theta_1 \approx 0.29129956$. The corresponding intersection point has Cartesian coordinates $(\ln\theta_1\cos\theta_1, \ln\theta_1\sin\theta_1) \approx (-1.181442, -0.354230)$.

Section 8.6 Slopes, Areas, and Arc Lengths for Polar Curves (page 515)

2. Area $= \dfrac{1}{2}\displaystyle\int_0^{2\pi}\theta^2\,d\theta = \dfrac{\theta^3}{6}\bigg|_0^{2\pi} = \dfrac{4}{3}\pi^3$ sq. units.

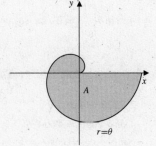

Fig. 8.6.2

4. Area $= \dfrac{1}{2}\displaystyle\int_0^{\pi/3}\sin^2 3\theta\,d\theta = \dfrac{1}{4}\displaystyle\int_0^{\pi/3}(1-\cos 6\theta)\,d\theta$

$= \dfrac{1}{4}\left(\theta - \dfrac{1}{6}\sin 6\theta\right)\bigg|_0^{\pi/3} = \dfrac{\pi}{12}$ sq. units.

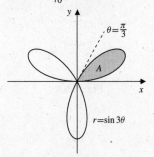

Fig. 8.6.4

6. The circles $r = a$ and $r = 2a\cos\theta$ intersect at $\theta = \pm\pi/3$. By symmetry, the common area is $4 \times$ (area of sector − area of right triangle) (see the figure), i.e.,

$$4\times\left[\left(\dfrac{1}{6}\pi a^2\right) - \left(\dfrac{1}{2}\dfrac{a}{2}\dfrac{\sqrt{3}a}{2}\right)\right] = \dfrac{4\pi - 3\sqrt{3}}{6}a^2 \text{ sq. units.}$$

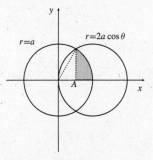

Fig. 8.6.6

8. Area $= \dfrac{1}{2}\pi a^2 + 2\times\dfrac{1}{2}\displaystyle\int_0^{\pi/2}a^2(1-\sin\theta)^2\,d\theta$

$= \dfrac{\pi a^2}{2} + a^2\displaystyle\int_0^{\pi/2}\left(1 - 2\sin\theta + \dfrac{1-\cos 2\theta}{2}\right)d\theta$

$= \dfrac{\pi a^2}{2} + a^2\left(\dfrac{3}{2}\theta + 2\cos\theta - \dfrac{1}{4}\sin 2\theta\right)\bigg|_0^{\pi/2}$

$= \left(\dfrac{5\pi}{4} - 2\right)a^2$ sq. units.

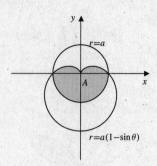

Fig. 8.6.8

14.

$$s = \int_0^{2\pi} \sqrt{a^2 + a^2\theta^2}\, d\theta$$

$$= a \int_0^{2\pi} \sqrt{1+\theta^2}\, d\theta \quad \text{Let } \theta = \tan u$$
$$\qquad\qquad d\theta = \sec^2 u\, d\theta$$

$$= a \int_{\theta=0}^{\theta=2\pi} \sec^3 u\, du$$

$$= \frac{a}{2}\left(\sec u\, \tan u + \ln|\sec u + \tan u|\right)\Big|_{\theta=0}^{\theta=2\pi}$$

$$= \frac{a}{2}\left[\theta\sqrt{1+\theta^2} + \ln|\sqrt{1+\theta^2} + \theta|\right]\Big|_{\theta=0}^{\theta=2\pi}$$

$$= \frac{a}{2}\left[2\pi\sqrt{1+4\pi^2} + \ln(2\pi + \sqrt{1+4\pi^2})\right] \text{ units.}$$

16. If $r^2 = \cos 2\theta$, then

$$2r\frac{dr}{d\theta} = -2\sin 2\theta \Rightarrow \frac{dr}{d\theta} = -\frac{\sin 2\theta}{\sqrt{\cos 2\theta}}$$

and

$$ds = \sqrt{\cos 2\theta + \frac{\sin^2 2\theta}{\cos 2\theta}}\, d\theta = \frac{d\theta}{\sqrt{\cos 2\theta}}.$$

a) Area of the surface generated by rotation about the x-axis is

$$S_x = 2\pi \int_0^{\pi/4} r\sin\theta\, ds$$

$$= 2\pi \int_0^{\pi/4} \sqrt{\cos 2\theta}\,\sin\theta\,\frac{d\theta}{\sqrt{\cos 2\theta}}$$

$$= -2\pi \cos\theta\Big|_0^{\pi/4} = (2 - \sqrt{2})\pi \text{ sq. units.}$$

b) Area of the surface generated by rotation about the y-axis is

$$S_y = 2\pi \int_{-\pi/4}^{\pi/4} r\cos\theta\, ds$$

$$= 4\pi \int_0^{\pi/4} \sqrt{\cos 2\theta}\,\cos\theta\,\frac{d\theta}{\sqrt{\cos 2\theta}}$$

$$= 4\pi \sin\theta\Big|_0^{\pi/4} = 2\sqrt{2}\pi \text{ sq. units.}$$

10. Since $r^2 = 2\cos 2\theta$ meets $r = 1$ at $\theta = \pm\frac{\pi}{6}$ and $\pm\frac{5\pi}{6}$, the area inside the lemniscate and outside the circle is

$$4 \times \frac{1}{2}\int_0^{\pi/6}\left[2\cos 2\theta - 1^2\right] d\theta$$

$$= 2\sin 2\theta\Big|_0^{\pi/6} \quad -\frac{\pi}{3} = \sqrt{3} - \frac{\pi}{3} \text{ sq. units.}$$

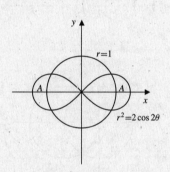

Fig. 8.6.10

12.

$$s = \int_0^\pi \sqrt{\left(\frac{dr}{d\theta}\right)^2 + r^2}\, d\theta = \int_0^\pi \sqrt{4\theta^2 + \theta^4}\, d\theta$$

$$= \int_0^\pi \theta\sqrt{4+\theta^2}\, d\theta \quad \text{Let } u = 4 + \theta^2$$
$$\qquad\qquad du = 2\theta\, d\theta$$

$$= \frac{1}{2}\int_4^{4+\pi^2} \sqrt{u}\, du = \frac{1}{3}u^{3/2}\Big|_4^{4+\pi^2}$$

$$= \frac{1}{3}\left[(4+\pi^2)^{3/2} - 8\right] \text{ units.}$$

18. The two curves $r^2 = 2\sin 2\theta$ and $r = 2\cos\theta$ intersect where

$$2\sin 2\theta = 4\cos^2\theta$$
$$4\sin\theta\cos\theta = 4\cos^2\theta$$
$$(\sin\theta - \cos\theta)\cos\theta = 0$$
$$\Leftrightarrow \quad \sin\theta = \cos\theta \text{ or } \cos\theta = 0,$$

i.e., at $P_1 = \left[\sqrt{2}, \dfrac{\pi}{4}\right]$ and $P_2 = (0, 0)$.

For $r^2 = 2\sin 2\theta$ we have $2r\dfrac{dr}{d\theta} = 4\cos 2\theta$. At P_1 we have $r = \sqrt{2}$ and $dr/d\theta = 0$. Thus the angle ψ between the curve and the radial line $\theta = \pi/4$ is $\psi = \pi/2$.
For $r = 2\cos\theta$ we have $dr/d\theta = -2\sin\theta$, so the angle between this curve and the radial line $\theta = \pi/4$ satisfies $\tan\psi = \dfrac{r}{dr/d\theta}\bigg|_{\theta=\pi/4} = -1$, and $\psi = 3\pi/4$. The two curves intersect at P_1 at angle $\dfrac{3\pi}{4} - \dfrac{\pi}{2} = \dfrac{\pi}{4}$.
The Figure shows that at the origin, P_2, the circle meets the lemniscate twice, at angles 0 and $\pi/2$.

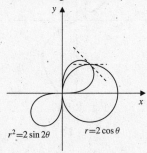

Fig. 8.6.18

20. We have $r = \cos\theta + \sin\theta$. For horizontal tangents:

$$0 = \frac{dy}{d\theta} = \frac{d}{d\theta}\left(\cos\theta\sin\theta + \sin^2\theta\right)$$
$$= \cos^2\theta - \sin^2\theta + 2\sin\theta\cos\theta$$
$$\Leftrightarrow \quad \cos 2\theta = -\sin 2\theta \quad \Leftrightarrow \quad \tan 2\theta = -1.$$

Thus $\theta = -\dfrac{\pi}{8}$ or $\dfrac{3\pi}{8}$. The tangents are horizontal at $\left[\cos\left(\dfrac{\pi}{8}\right) - \sin\left(\dfrac{\pi}{8}\right), -\dfrac{\pi}{8}\right]$ and $\left[\cos\left(\dfrac{3\pi}{8}\right) + \sin\left(\dfrac{3\pi}{8}\right), \dfrac{3\pi}{8}\right]$.
For vertical tangent:

$$0 = \frac{dx}{d\theta} = \frac{d}{d\theta}\left(\cos^2\theta + \cos\theta\sin\theta\right)$$
$$= -2\cos\theta\sin\theta + \cos^2\theta - \sin^2\theta$$
$$\Leftrightarrow \quad \sin 2\theta = \cos 2\theta \quad \Leftrightarrow \quad \tan 2\theta = 1.$$

Thus $\theta = \pi/8$ of $5\pi/8$. There are vertical tangents at $\left[\cos\left(\dfrac{\pi}{8}\right) + \sin\left(\dfrac{\pi}{8}\right), \dfrac{\pi}{8}\right]$ and $\left[\cos\left(\dfrac{5\pi}{8}\right) + \sin\left(\dfrac{5\pi}{8}\right), \dfrac{5\pi}{8}\right]$.

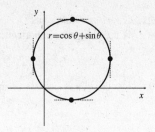

Fig. 8.6.20

22. We have $r^2 = \cos 2\theta$, and $2r\dfrac{dr}{d\theta} = -2\sin 2\theta$. For horizontal tangents:

$$0 = \frac{d}{d\theta}r\sin\theta = r\cos\theta + \sin\theta\left(-\frac{\sin 2\theta}{r}\right)$$
$$\Leftrightarrow \quad \cos 2\theta\cos\theta = \sin 2\theta\sin\theta$$
$$\Leftrightarrow \quad (\cos^2\theta - \sin^2\theta)\cos\theta = 2\sin^2\theta\cos\theta$$
$$\Leftrightarrow \quad \cos\theta = 0 \quad \text{or} \quad \cos^2\theta = 3\sin^2\theta.$$

There are no points on the curve where $\cos\theta = 0$. Therefore, horizontal tangents occur only where $\tan^2\theta = 1/3$. There are horizontal tangents at $\left[\dfrac{1}{\sqrt{2}}, \pm\dfrac{\pi}{6}\right]$ and $\left[\dfrac{1}{\sqrt{2}}, \pm\dfrac{5\pi}{6}\right]$.
For vertical tangents:

$$0 = \frac{d}{d\theta}r\cos\theta = -r\sin\theta + \cos\theta\left(-\frac{\sin 2\theta}{r}\right)$$
$$\Leftrightarrow \quad \cos 2\theta\sin\theta = -\sin 2\theta\cos\theta$$
$$\Leftrightarrow \quad (\cos^2\theta - \sin^2\theta)\sin\theta = -2\sin\theta\cos^2\theta$$
$$\Leftrightarrow \quad \sin\theta = 0 \quad \text{or} \quad 3\cos^2\theta = \sin^2\theta.$$

There are no points on the curve where $\tan^2\theta = 3$, so the only vertical tangents occur where $\sin\theta = 0$, that is, at the points with polar coordinates $[1, 0]$ and $[1, \pi]$.

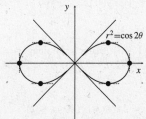

Fig. 8.6.22

24. We have $r = e^\theta$ and $\dfrac{dr}{d\theta} = e^\theta$. For horizontal tangents:

$$0 = \frac{d}{d\theta}r\sin\theta = e^\theta\cos\theta + e^\theta\sin\theta$$
$$\Leftrightarrow \quad \tan\theta = -1 \quad \Leftrightarrow \quad \theta = -\frac{\pi}{4} + k\pi,$$

where $k = 0, \pm 1, \pm 2, \ldots$. At the points $[e^{k\pi - \pi/4}, k\pi - \pi/4]$ the tangents are horizontal. For vertical tangents:

$$0 = \frac{d}{d\theta} r \cos\theta = e^{\theta} \cos\theta - e^{\theta} \sin\theta$$

$$\Leftrightarrow \quad \tan\theta = 1 \quad \Leftrightarrow \quad \theta = \frac{\pi}{4} + k\pi.$$

At the points $[e^{k\pi + \pi/4}, k\pi + \pi/4]$ the tangents are vertical.

26. $x = r\cos\theta = f(\theta)\cos\theta$, $y = r\sin\theta = f(\theta)\sin\theta$.

$$\frac{dx}{d\theta} = f'(\theta)\cos\theta - f(\theta)\sin\theta, \quad \frac{dy}{d\theta} = f'(\theta)\sin\theta + f(\theta)\cos\theta$$

$$ds = \sqrt{\left(f'(\theta)\cos\theta - f(\theta)\sin\theta\right)^2 + \left(f'(\theta)\sin\theta + f(\theta)\cos\theta\right)^2}\, d\theta$$

$$= \left[\left(f'(\theta)\right)^2 \cos^2\theta - 2f'(\theta)f(\theta)\cos\theta\sin\theta + \left(f(\theta)\right)^2 \sin^2\theta\right.$$

$$\left. + \left(f'(\theta)\right)^2 \sin^2\theta + 2f'(\theta)f(\theta)\sin\theta\cos\theta + \left(f(\theta)\right)^2 \cos^2\theta\right]^{1/2} d\theta$$

$$= \sqrt{\left(f'(\theta)\right)^2 + \left(f(\theta)\right)^2}\, d\theta.$$

Review Exercises 8   (page 516)

2. $9x^2 - 4y^2 = 36 \quad \Leftrightarrow \quad \dfrac{x^2}{4} - \dfrac{y^2}{9} = 1$
Hyperbola, transverse axis along the x-axis.
Semi-transverse axis $a = 2$, semi-conjugate axis $b = 3$.
$c^2 = a^2 + b^2 = 13$. Foci: $(\pm\sqrt{13}, 0)$.
Asymptotes: $3x \pm 2y = 0$.

4. $2x^2 + 8y^2 = 4x - 48y$
$2(x^2 - 2x + 1) + 8(y^2 + 6y + 9) = 74$

$$\frac{(x-1)^2}{37} + \frac{(y+3)^2}{37/4} = 1.$$

Ellipse, centre $(1, -3)$, major axis along $y = -3$.
$a = \sqrt{37}$, $b = \sqrt{37}/2$, $c^2 = a^2 - b^2 = 111/4$.
Foci: $(1 \pm \sqrt{111}/2, -3)$.

6. $x = 2\sin(3t)$, $y = 2\cos(3t)$, $(0 \le t \le 2)$
Part of a circle of radius 2 centred at the origin from the point $(0, 2)$ clockwise to $(2\sin 6, 2\cos 6)$.

8. $x = e^t$, $y = e^{-2t}$, $(-1 \le t \le 1)$.
Part of the curve $x^2 y = 1$ from $(1/e, e^2)$ to $(e, 1/e^2)$.

10. $x = \cos t + \sin t$, $y = \cos t - \sin t$, $(0 \le t \le 2\pi)$
The circle $x^2 + y^2 = 2$, traversed clockwise, starting and ending at $(1, 1)$.

12. $x = t^3 - 3t \qquad y = t^3 + 3t$

$$\frac{dx}{dt} = 3(t^2 - 1) \quad \frac{dy}{dt} = 3(t^2 + 1)$$

Horizontal tangent: none.
Vertical tangent at $t = \pm 1$, i.e., at $(2, -4)$ and $(-2, 4)$.

Slope $\dfrac{dy}{dx} = \dfrac{t^2 + 1}{t^2 - 1} \quad \begin{cases} > 0 & \text{if } |t| > 1 \\ < 0 & \text{if } |t| < 1 \end{cases}$
Slope $\to 1$ as $t \to \pm\infty$.

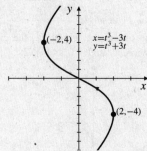

Fig. R-8.12

14. $x = t^3 - 3t \qquad y = t^3 - 12t$

$$\frac{dx}{dt} = 3(t^2 - 1) \quad \frac{dy}{dt} = 3(t^2 - 4)$$

Horizontal tangent at $t = \pm 2$, i.e., at $(2, -16)$ and $(-2, 16)$.
Vertical tangent at $t = \pm 1$, i.e., at $(2, 11)$ and $(-2, -11)$.

Slope $\dfrac{dy}{dx} = \dfrac{t^2 - 4}{t^2 - 1} \quad \begin{cases} > 0 & \text{if } |t| > 2 \text{ or } |t| < 1 \\ < 0 & \text{if } 1 < |t| < 2 \end{cases}$
Slope $\to 1$ as $t \to \pm\infty$.

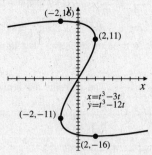

Fig. R-8.14

16.   The volume of revolution about the y-axis is

$$V = \pi \int_{t=0}^{t=1} x^2 \, dy$$

$$= \pi \int_0^1 (t^6 - 2t^4 + t^2) 3t^2 \, dt$$

$$= 3\pi \int_0^1 (t^8 - 2t^6 + t^4) \, dt$$

$$= 3\pi \left(\frac{1}{9} - \frac{2}{7} + \frac{1}{5}\right) = \frac{8\pi}{105} \text{ cu. units.}$$

18. Area of revolution about the x-axis is

$$S = 2\pi \int 4e^{t/2}(e^t + 1)\, dt$$

$$= 8\pi \left(\frac{2}{3}e^{3t/2} + 2e^{t/2} \right)\Bigg|_0^2$$

$$= \frac{16\pi}{3}(e^3 + 3e - 4) \text{ sq. units.}$$

20. $r = |\theta|, \quad (-2\pi \le \theta \le 2\pi)$

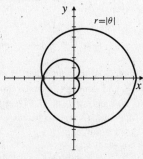

Fig. R-8.20

22. $r = 2 + \cos(2\theta)$

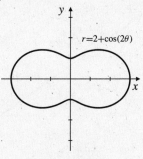

Fig. R-8.22

24. $r = 1 - \sin(3\theta)$

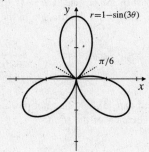

Fig. R-8.24

26. Area of a small loop:

$$A = 2 \times \frac{1}{2} \int_{\pi/3}^{\pi/2} (1 + 2\cos(2\theta))^2\, d\theta$$

$$= \int_{\pi/3}^{\pi/2} [1 + 4\cos(2\theta) + 2(1 + \cos(4\theta))]\, d\theta$$

$$= \left(3\theta + 2\sin(2\theta) + \frac{1}{2}\sin(4\theta) \right)\Bigg|_{\pi/3}^{\pi/2}$$

$$= \frac{\pi}{2} - \frac{3\sqrt{3}}{4} \text{ sq. units.}$$

28. $r\cos\theta = x = 1/4$ and $r = 1 + \cos\theta$ intersect where

$$1 + \cos\theta = \frac{1}{4\cos\theta}$$

$$4\cos^2\theta + 4\cos\theta - 1 = 0$$

$$\cos\theta = \frac{-4 \pm \sqrt{16 + 16}}{8} = \frac{\pm\sqrt{2} - 1}{2}.$$

Only $(\sqrt{2} - 1)/2$ is between -1 and 1, so is a possible value of $\cos\theta$. Let $\theta_0 = \cos^{-1}\dfrac{\sqrt{2} - 1}{2}$. Then

$$\sin\theta_0 = \sqrt{1 - \left(\frac{\sqrt{2} - 1}{2} \right)^2} = \frac{\sqrt{1 + 2\sqrt{2}}}{2}.$$

By symmetry, the area inside $r = 1 + \cos\theta$ to the left of the line $x = 1/4$ is

$$A = 2 \times \frac{1}{2} \int_{\theta_0}^{\pi} \left(1 + 2\cos\theta + \frac{1 + \cos(2\theta)}{2} \right) d\theta + \cos\theta_0 \sin\theta_0$$

$$= \frac{3}{2}(\pi - \theta_0) + \left(2\sin\theta + \frac{1}{4}\sin(2\theta) \right)\Bigg|_{\theta_0}^{\pi}$$

$$\quad + \frac{(\sqrt{2} - 1)\sqrt{1 + 2\sqrt{2}}}{4}$$

$$= \frac{3}{2}\left(\pi - \cos^{-1}\frac{\sqrt{2} - 1}{2} \right) + \sqrt{1 + 2\sqrt{2}}\left(\frac{\sqrt{2} - 9}{8} \right) \text{ sq. units.}$$

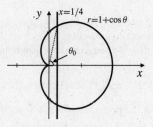

Fig. R-8.28

173

Challenging Problems 8   (page 516)

2. Let S_1 and S_2 be two spheres inscribed in the cylinder, one on each side of the plane that intersects the cylinder in the curve C that we are trying to show is an ellipse. Let the spheres be tangent to the cylinder around the circles C_1 and C_2, and suppose they are also tangent to the plane at the points F_1 and F_2, respectively, as shown in the figure.

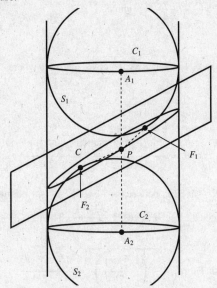

Fig. C-8.2

Let P be any point on C. Let A_1A_2 be the line through P that lies on the cylinder, with A_1 on C_1 and A_2 on C_2. Then $PF_1 = PA_1$ because both lengths are of tangents drawn to the sphere S_1 from the same exterior point P. Similarly, $PF_2 = PA_2$. Hence

$$PF_1 + PF_2 = PA_1 + PA_2 = A_1A_2,$$

which is constant, the distance between the centres of the two spheres. Thus C must be an ellipse, with foci at F_1 and F_2.

4. Without loss of generality, choose the axes and axis scales so that the parabola has equation $y = x^2$. If P is the point (x_0, x_0^2) on it, then the tangent to the parabola at P has equation

$$y = x_0^2 + 2x_0(x - x_0),$$

which intersects the principal axis $x = 0$ at $(0, -x_0^2)$. Thus $R = (0, -x_0^2)$ and $Q = (0, x_0^2)$. Evidently the vertex $V = (0, 0)$ bisects RQ.

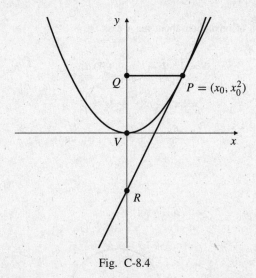

Fig. C-8.4

To construct the tangent at a given point P on a parabola with given vertex V and principal axis L, drop a perpendicular from P to L, meeting L at Q. Then find R on L on the side of V opposite Q and such that $QV = VR$. Then PR is the desired tangent.

6.

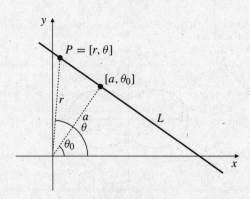

Fig. C-8.6

a) Let L be a line not passing through the origin, and let $[a, \theta_0]$ be the polar coordinates of the point on L that is closest to the origin. If $P = [r, \theta]$ is any point on the line, then, from the triangle in the figure,

$$\frac{a}{r} = \cos(\theta - \theta_0), \quad \text{or} \quad r = \frac{a}{\cos(\theta - \theta_0)}.$$

b) As shown in part (a), any line not passing through the origin has equation of the form

$$r = g(\theta) = \frac{a}{\cos(\theta - \theta_0)} = a\sec(\theta - \theta_0),$$

for some constants a and θ_0. We have

$$g'(\theta) = a\sec(\theta - \theta_0)\tan(\theta - \theta_0)$$
$$g''(\theta) = a\sec(\theta - \theta_0)\tan^2(\theta - \theta_0)$$
$$+ a\sec^3(\theta - \theta_0)$$
$$\big(g(\theta)\big)^2 + 2\big(g'(\theta)\big)^2 - g(\theta)g''(\theta)$$
$$= a^2\sec^2(\theta - \theta_0) + 2a^2\sec^2(\theta - \theta_0)\tan^2(\theta - \theta_0)$$
$$- a^2\sec^2(\theta - \theta_0)\tan^2(\theta - \theta_0) - a^2\sec^4(\theta - \theta_0)$$
$$= a^2\big[\sec^2(\theta - \theta_0)\big(1 + \tan^2(\theta - \theta_0)\big) - \sec^4(\theta - \theta_0)\big]$$
$$= 0.$$

c) If $r = g(\theta)$ is the polar equation of the tangent to $r = f(\theta)$ at $\theta = \alpha$, then $g(\alpha) = f(\alpha)$ and $g'(\alpha) = f'(\alpha)$. Suppose that

$$\big(f(\alpha)\big)^2 + 2\big(f'(\alpha)\big)^2 - f(\alpha)f''(\alpha) > 0.$$

By part (b) we have

$$\big(g(\alpha)\big)^2 + 2\big(g'(\alpha)\big)^2 - g(\alpha)g''(\alpha) = 0.$$

Subtracting, and using $g(\alpha) = f(\alpha)$ and $g'(\alpha) = f'(\alpha)$, we get $f''(\alpha) < g''(\alpha)$. It follows that $f(\theta) < g(\theta)$ for values of θ near α; that is, the graph of $r = f(\theta)$ is curving to the origin side of its tangent at α. Similarly, if

$$\big(f(\alpha)\big)^2 + 2\big(f'(\alpha)\big)^2 - f(\alpha)f''(\alpha) < 0,$$

then the graph is curving to the opposite side of the tangent, away from the origin.

8. Take the origin at station O as shown in the figure. Both of the lines L_1 and L_2 pass at distance $100\cos\epsilon$ from the origin. Therefore, by Problem 6(a), their equations are

$$L_1: \quad r = \frac{100\cos\epsilon}{\cos\big[\theta - \big(\frac{\pi}{2} - \epsilon\big)\big]} = \frac{100\cos\epsilon}{\sin(\theta + \epsilon)}$$

$$L_2: \quad r = \frac{100\cos\epsilon}{\cos\big[\theta - \big(\frac{\pi}{2} + \epsilon\big)\big]} = \frac{100\cos\epsilon}{\sin(\theta - \epsilon)}.$$

The search area $A(\epsilon)$ is, therefore,

$$A(\epsilon) = \frac{1}{2}\int_{\frac{\pi}{4} - \epsilon}^{\frac{\pi}{4} + \epsilon}\left(\frac{100^2\cos^2\epsilon}{\sin^2(\theta - \epsilon)} - \frac{100^2\cos^2\epsilon}{\sin^2(\theta + \epsilon)}\right)d\theta$$

$$= 5{,}000\cos^2\epsilon\int_{\frac{\pi}{4} - \epsilon}^{\frac{\pi}{4} + \epsilon}\big(\csc^2(\theta - \epsilon) - \csc^2(\theta + \epsilon)\big)d\theta$$

$$= 5{,}000\cos^2\epsilon\left[\cot\big(\tfrac{\pi}{4} + 2\epsilon\big) - 2\cot\tfrac{\pi}{4} + \cot\big(\tfrac{\pi}{4} - 2\epsilon\big)\right]$$

$$= 5{,}000\cos^2\epsilon\left[\frac{\cos\big(\frac{\pi}{4} + 2\epsilon\big)}{\sin\big(\frac{\pi}{4} + 2\epsilon\big)} + \frac{\sin\big(\frac{\pi}{4} + 2\epsilon\big)}{\cos\big(\frac{\pi}{4} + 2\epsilon\big)} - 2\right]$$

$$= 10{,}000\cos^2\epsilon\left[\csc\big(\tfrac{\pi}{2} + 4\epsilon\big) - 1\right]$$

$$= 10{,}000\cos^2\epsilon(\sec(4\epsilon) - 1)\ \text{mi}^2.$$

For $\epsilon = 3° = \pi/60$, we have $A(\epsilon) \approx 222.8$ square miles. Also

$$A'(\epsilon) = -20{,}000\cos\epsilon\sin\epsilon(\sec(4\epsilon) - 1)$$
$$+ 40{,}000\cos^2\epsilon\sec(4\epsilon)\tan(4\epsilon)$$
$$A'(\pi/60) \approx 8645.$$

When $\epsilon = 3°$, the search area increases at about $8645(\pi/180) \approx 151$ square miles per degree increase in ϵ.

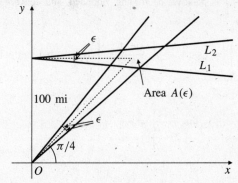

Fig. C-8.8

CHAPTER 9. SEQUENCES, SERIES, AND POWER SERIES

Section 9.1 Sequences and Convergence (page 526)

2. $\left\{ \dfrac{2n}{n^2+1} \right\} = \left\{ 1, \dfrac{4}{5}, \dfrac{3}{5}, \dfrac{8}{17}, \ldots \right\}$ is bounded, positive, decreasing, and converges to 0.

4. $\left\{ \sin \dfrac{1}{n} \right\} = \left\{ \sin 1, \sin\left(\dfrac{1}{2}\right), \sin\left(\dfrac{1}{3}\right), \ldots \right\}$ is bounded, positive, decreasing, and converges to 0.

6. $\left\{ \dfrac{e^n}{\pi^n} \right\} = \left\{ \dfrac{e}{\pi}, \left(\dfrac{e}{\pi}\right)^2, \left(\dfrac{e}{\pi}\right)^3, \ldots \right\}$ is bounded, positive, decreasing, and converges to 0, since $e < \pi$.

8. $\left\{ \dfrac{(-1)^n n}{e^n} \right\} = \left\{ \dfrac{-1}{e}, \dfrac{2}{e^2}, \dfrac{-3}{e^3}, \ldots \right\}$ is bounded, alternating, and converges to 0.

10. $\dfrac{(n!)^2}{(2n)!} = \dfrac{1}{n+1} \dfrac{2}{n+2} \dfrac{3}{n+3} \cdots \dfrac{n}{2n} \leq \left(\dfrac{1}{2}\right)^n$.

Also, $\dfrac{a_{n+1}}{a_n} = \dfrac{(n+1)^2}{(2n+2)(2n+1)} < \dfrac{1}{2}$. Thus the sequence $\left\{ \dfrac{(n!)^2}{(2n)!} \right\}$ is positive, decreasing, bounded, and convergent to 0.

12. $\left\{ \dfrac{\sin n}{n} \right\} = \left\{ \sin 1, \dfrac{\sin 2}{2}, \dfrac{\sin 3}{3}, \ldots \right\}$ is bounded and converges to 0.

14. $\lim \dfrac{5-2n}{3n-7} = \lim \dfrac{\dfrac{5}{n}-2}{3-\dfrac{7}{n}} = -\dfrac{2}{3}$.

16. $\lim \dfrac{n^2}{n^3+1} = \lim \dfrac{\dfrac{1}{n}}{1+\dfrac{1}{n^3}} = 0$.

18. $\lim \dfrac{n^2-2\sqrt{n}+1}{1-n-3n^2} = \lim \dfrac{1-\dfrac{2}{n\sqrt{n}}+\dfrac{1}{n^2}}{\dfrac{1}{n^2}-\dfrac{1}{n}-3} = -\dfrac{1}{3}$.

20. $\lim n \sin \dfrac{1}{n} = \lim_{x\to 0+} \dfrac{\sin x}{x} = \lim_{x\to 0+} \dfrac{\cos x}{1} = 1$.

22. $\lim \dfrac{n}{\ln(n+1)} = \lim_{x\to\infty} \dfrac{x}{\ln(x+1)}$

$= \lim_{x\to\infty} \dfrac{1}{\left(\dfrac{1}{x+1}\right)} = \lim_{x\to\infty} x+1 = \infty$.

24. $\lim\left(n - \sqrt{n^2-4n}\right) = \lim \dfrac{n^2 - (n^2-4n)}{n+\sqrt{n^2-4n}}$

$= \lim \dfrac{4n}{n+\sqrt{n^2-4n}} = \lim \dfrac{4}{1+\sqrt{1-\dfrac{4}{n}}} = 2$.

26. If $a_n = \left(\dfrac{n-1}{n+1}\right)^n$, then

$\lim a_n = \lim \left(\dfrac{n-1}{n}\right)^n \left(\dfrac{n}{n+1}\right)^n$

$= \lim \left(1-\dfrac{1}{n}\right)^n \bigg/ \lim \left(1+\dfrac{1}{n}\right)^n$

$= \dfrac{e^{-1}}{e} = e^{-2}$ (by Theorem 6 of Section 3.4).

28. We have $\lim \dfrac{n^2}{2^n} = 0$ since 2^n grows much faster than n^2 and $\lim \dfrac{4^n}{n!} = 0$ by Theorem 3(b). Hence,

$\lim \dfrac{n^2 2^n}{n!} = \lim \dfrac{n^2}{2^n} \cdot \dfrac{2^{2n}}{n!} = \left(\lim \dfrac{n^2}{2^n}\right)\left(\lim \dfrac{4^n}{n!}\right) = 0$.

30. Let $a_1 = 1$ and $a_{n+1} = \sqrt{1+2a_n}$ for $n = 1, 2, 3, \ldots$. Then we have $a_2 = \sqrt{3} > a_1$. If $a_{k+1} > a_k$ for some k, then

$$a_{k+2} = \sqrt{1+2a_{k+1}} > \sqrt{1+2a_k} = a_{k+1}.$$

Thus, $\{a_n\}$ is increasing by induction. Observe that $a_1 < 3$ and $a_2 < 3$. If $a_k < 3$ then

$$a_{k+1} = \sqrt{1+2a_k} < \sqrt{1+2(3)} = \sqrt{7} < \sqrt{9} = 3.$$

Therefore, $a_n < 3$ for all n, by induction. Since $\{a_n\}$ is increasing and bounded above, it converges. Let $\lim a_n = a$. Then

$$a = \sqrt{1+2a} \Rightarrow a^2 - 2a - 1 = 0 \Rightarrow a = 1 \pm \sqrt{2}.$$

Since $a = 1 - \sqrt{2} < 0$, it is not appropriate. Hence, we must have $\lim a_n = 1 + \sqrt{2}$.

32. Let $a_n = \left(1+\dfrac{1}{n}\right)^n$ so $\ln a_n = n \ln\left(1+\dfrac{1}{n}\right)$.

a) If $f(x) = x \ln\left(1 + \dfrac{1}{x}\right) = x \ln(x+1) - x \ln x$, then

$$f'(x) = \ln(x+1) + \frac{x}{x+1} - \ln x - 1$$

$$= \ln\left(\frac{x+1}{x}\right) - \frac{1}{x+1}$$

$$= \int_x^{x+1} \frac{dt}{t} - \frac{1}{x+1}$$

$$> \frac{1}{x+1} \int_x^{x+1} dt - \frac{1}{x+1}$$

$$= \frac{1}{x+1} - \frac{1}{x+1} = 0.$$

Since $f'(x) > 0$, $f(x)$ must be an increasing function. Thus, $\{a_n\} = \{e^{f(x_n)}\}$ is increasing.

b) Since $\ln x \leq x - 1$,

$$\ln a_k = k \ln\left(1 + \frac{1}{k}\right) \leq k\left(1 + \frac{1}{k} - 1\right) = 1$$

which implies that $a_k \leq e$ for all k. Since $\{a_n\}$ is increasing, e is an upper bound for $\{a_n\}$.

34. If $\{|a_n|\}$ is bounded then it is bounded above, and there exists a constant K such that $|a_n| \leq K$ for all n. Therefore, $-K \leq a_n \leq K$ for all n, and so $\{a_n\}$ is bounded above and below, and is therefore bounded.

36.
a) "If $\lim a_n = \infty$ and $\lim b_n = L > 0$, then $\lim a_n b_n = \infty$" is TRUE. Let R be an arbitrary, large positive number. Since $\lim a_n = \infty$, and $L > 0$, it must be true that $a_n \geq \dfrac{2R}{L}$ for n sufficiently large. Since $\lim b_n = L$, it must also be that $b_n \geq \dfrac{L}{2}$ for n sufficiently large. Therefore $a_n b_n \geq \dfrac{2R}{L}\dfrac{L}{2} = R$ for n sufficiently large. Since R is arbitrary, $\lim a_n b_n = \infty$.

b) "If $\lim a_n = \infty$ and $\lim b_n = -\infty$, then $\lim(a_n + b_n) = 0$" is FALSE. Let $a_n = 1 + n$ and $b_n = -n$; then $\lim a_n = \infty$ and $\lim b_n = -\infty$ but $\lim(a_n + b_n) = 1$.

c) "If $\lim a_n = \infty$ and $\lim b_n = -\infty$, then $\lim a_n b_n = -\infty$" is TRUE. Let R be an arbitrary, large positive number. Since $\lim a_n = \infty$ and $\lim b_n = -\infty$, we must have $a_n \geq \sqrt{R}$ and $b_n \leq -\sqrt{R}$, for all sufficiently large n. Thus $a_n b_n \leq -R$, and $\lim a_n b_n = -\infty$.

d) "If neither $\{a_n\}$ nor $\{b_n\}$ converges, then $\{a_n b_n\}$ does not converge" is FALSE. Let $a_n = b_n = (-1)^n$; then $\lim a_n$ and $\lim b_n$ both diverge. But $a_n b_n = (-1)^{2n} = 1$ and $\{a_n b_n\}$ does converge (to 1).

e) "If $\{|a_n|\}$ converges, then $\{a_n\}$ converges" is FALSE. Let $a_n = (-1)^n$. Then $\lim_{n\to\infty} |a_n| = \lim_{n\to\infty} 1 = 1$, but $\lim_{n\to\infty} a_n$ does not exist.

Section 9.2 Infinite Series (page 534)

2. $3 - \dfrac{3}{4} + \dfrac{3}{16} - \dfrac{3}{64} + \cdots = \displaystyle\sum_{n=1}^{\infty} 3\left(-\dfrac{1}{4}\right)^{n-1} = \dfrac{3}{1 + \frac{1}{4}} = \dfrac{12}{5}.$

4. $\displaystyle\sum_{n=0}^{\infty} \dfrac{5}{10^{3n}} = 5\left[1 + \dfrac{1}{1000} + \left(\dfrac{1}{1000}\right)^2 + \cdots\right]$

$$= \dfrac{5}{1 - \dfrac{1}{1000}} = \dfrac{5000}{999}.$$

6. $\displaystyle\sum_{n=0}^{\infty} \dfrac{1}{e^n} = 1 + \dfrac{1}{e} + \left(\dfrac{1}{e}\right)^2 + \cdots = \dfrac{1}{1 - \dfrac{1}{e}} = \dfrac{e}{e-1}.$

8. $\displaystyle\sum_{j=1}^{\infty} \pi^{j/2} \cos(j\pi) = \sum_{j=2}^{\infty} (-1)^j \pi^{j/2}$ diverges because $\lim_{j\to\infty} (-1)^j \pi^{j/2}$ does not exist.

10. $\displaystyle\sum_{n=0}^{\infty} \dfrac{3 + 2^n}{3^{n+2}} = \dfrac{1}{3}\sum_{n=0}^{\infty}\left(\dfrac{1}{3}\right)^n + \dfrac{1}{9}\sum_{n=0}^{\infty}\left(\dfrac{2}{3}\right)^n$

$$= \dfrac{1}{3} \cdot \dfrac{1}{1 - \dfrac{1}{3}} + \dfrac{1}{9} \cdot \dfrac{1}{1 - \dfrac{2}{3}} = \dfrac{1}{2} + \dfrac{1}{3} = \dfrac{5}{6}.$$

12. Let

$$\sum_{n=1}^{\infty} \frac{1}{(2n-1)(2n+1)} = \frac{1}{1 \times 3} + \frac{1}{3 \times 5} + \frac{1}{5 \times 7} + \cdots.$$

Since $\dfrac{1}{(2n-1)(2n+1)} = \dfrac{1}{2}\left(\dfrac{1}{2n-1} - \dfrac{1}{2n+1}\right)$, the partial sum is

$$s_n = \frac{1}{2}\left(1 - \frac{1}{3}\right) + \frac{1}{2}\left(\frac{1}{3} - \frac{1}{5}\right) + \cdots$$
$$+ \frac{1}{2}\left(\frac{1}{2n-3} - \frac{1}{2n-1}\right) + \frac{1}{2}\left(\frac{1}{2n-1} - \frac{1}{2n+1}\right)$$
$$= \frac{1}{2}\left(1 - \frac{1}{2n+1}\right).$$

Hence,

$$\sum_{n=1}^{\infty} \frac{1}{(2n-1)(2n+1)} = \lim s_n = \frac{1}{2}.$$

14. Since

$$\frac{1}{n(n+1)(n+2)} = \frac{1}{2}\left[\frac{1}{n} - \frac{2}{n+1} + \frac{1}{n+2}\right],$$

the partial sum is

$$s_n = \frac{1}{2}\left(1 - \frac{2}{2} + \frac{1}{3}\right) + \frac{1}{2}\left(\frac{1}{2} - \frac{2}{3} + \frac{1}{4}\right) + \cdots$$
$$+ \frac{1}{2}\left(\frac{1}{n-1} - \frac{2}{n} + \frac{1}{n+1}\right) + \frac{1}{2}\left(\frac{1}{n} - \frac{2}{n+1} + \frac{1}{n+2}\right)$$
$$= \frac{1}{2}\left(\frac{1}{2} - \frac{1}{n+1} + \frac{1}{n+2}\right).$$

Hence,

$$\sum_{n=1}^{\infty} \frac{1}{n(n+1)(n+2)} = \lim s_n = \frac{1}{4}.$$

16. $\displaystyle\sum_{n=1}^{\infty} \frac{n}{n+2}$ diverges to infinity since $\displaystyle\lim \frac{n}{n+2} = 1 > 0$.

18. $\displaystyle\sum_{n=1}^{\infty} \frac{2}{n+1} = 2\left(\frac{1}{2} + \frac{1}{3} + \frac{1}{4} + \cdots\right)$ diverges to infinity since it is just twice the harmonic series with the first term omitted.

20. Since $1 + 2 + 3 + \cdots + n = \dfrac{n(n+1)}{2}$, the given series is $\sum_{n=1}^{\infty} \dfrac{2}{n(n+1)}$ which converges to 2 by the result of Example 3 of this section.

22. The balance at the end of 8 years is

$$s_n = 1000\left[(1.1)^8 + (1.1)^7 + \cdots + (1.1)^2 + (1.1)\right]$$
$$= 1000(1.1)\left(\frac{(1.1)^8 - 1}{1.1 - 1}\right) \approx \$12,579.48.$$

24. If $\{a_n\}$ is ultimately positive, then the sequence $\{s_n\}$ of partial sums of the series must be ultimately increasing. By Theorem 2, if $\{s_n\}$ is ultimately increasing, then either it is bounded above, and therefore convergent, or else it is not bounded above and diverges to infinity. Since $\sum a_n = \lim s_n$, $\sum a_n$ must either converge when $\{s_n\}$ converges and $\lim s_n = s$ exists, or diverge to infinity when $\{s_n\}$ diverges to infinity.

26. "If $a_n = 0$ for every n, then $\sum a_n$ converge" is TRUE because $s_n = \sum_{k=0}^{n} 0 = 0$, for every n, and so $\sum a_n = \lim s_n = 0$.

28. "If $\sum a_n$ and $\sum b_n$ both diverge, then so does $\sum(a_n + b_n)$" is FALSE. Let $a_n = \dfrac{1}{n}$ and $b_n = -\dfrac{1}{n}$, then $\sum a_n = \infty$ and $\sum b_n = -\infty$ but $\sum(a_n + b_n) = \sum(0) = 0$.

30. "If $\sum a_n$ diverges and $\{b_n\}$ is bounded, then $\sum a_n b_n$ diverges" is FALSE. Let $a_n = \dfrac{1}{n}$ and $b_n = \dfrac{1}{n+1}$. Then $\sum a_n = \infty$ and $0 \le b_n \le 1/2$. But $\sum a_n b_n = \sum \dfrac{1}{n(n+1)}$ which converges by Example 3.

Section 9.3 Convergence Tests for Positive Series (page 545)

2. $\displaystyle\sum_{n=1}^{\infty} \frac{n}{n^4 - 2}$ converges by comparison with $\displaystyle\sum_{n=1}^{\infty} \frac{1}{n^3}$ since

$$\lim \frac{\left(\dfrac{n}{n^4 - 2}\right)}{\left(\dfrac{1}{n^3}\right)} = 1, \quad \text{and} \quad 0 < 1 < \infty.$$

4. $\displaystyle\sum_{n=1}^{\infty} \frac{\sqrt{n}}{n^2 + n + 1}$ converges by comparison with $\displaystyle\sum_{n=1}^{\infty} \frac{1}{n^{3/2}}$ since

$$\lim \frac{\left(\dfrac{\sqrt{n}}{n^2 + n + 1}\right)}{\left(\dfrac{1}{n^{3/2}}\right)} = 1, \quad \text{and} \quad 0 < 1 < \infty.$$

6. $\displaystyle\sum_{n=8}^{\infty} \frac{1}{\pi^n + 5}$ converges by comparison with the geometric series $\displaystyle\sum_{n=8}^{\infty} \left(\frac{1}{\pi}\right)^n$ since $0 < \dfrac{1}{\pi^n + 5} < \dfrac{1}{\pi^n}$.

8. $\displaystyle\sum_{n=1}^{\infty} \frac{1}{\ln(3n)}$ diverges to infinity by comparison with the harmonic series $\displaystyle\sum_{n=1}^{\infty} \frac{1}{3n}$ since $\dfrac{1}{\ln(3n)} > \dfrac{1}{3n}$ for $n \ge 1$.

10. $\displaystyle\sum_{n=0}^{\infty} \frac{1+n}{2+n}$ diverges to infinity since $\displaystyle\lim \frac{1+n}{2+n} = 1 > 0$.

12. $\displaystyle\sum_{n=1}^{\infty} \frac{n^2}{1 + n\sqrt{n}}$ diverges to infinity since $\displaystyle\lim \frac{n^2}{1 + n\sqrt{n}} = \infty$.

14. $\displaystyle\sum_{n=2}^{\infty} \frac{1}{n \ln n (\ln \ln n)^2}$ converges by the integral test:

$$\int_a^{\infty} \frac{dt}{t \ln t (\ln \ln t)^2} = \int_{\ln \ln a}^{\infty} \frac{du}{u^2} < \infty \quad \text{if } \ln \ln a > 0.$$

16. The series

$$\sum_{n=1}^{\infty} \frac{1+(-1)^n}{\sqrt{n}} = 0 + \frac{2}{\sqrt{2}} + 0 + \frac{2}{\sqrt{4}} + 0 + \frac{2}{\sqrt{6}} + \cdots$$

$$= 2\sum_{k=1}^{\infty} \frac{1}{\sqrt{2k}} = \sqrt{2}\sum_{k=1}^{\infty} \frac{1}{\sqrt{k}}$$

diverges to infinity.

18. $\sum_{n=1}^{\infty} \frac{n^4}{n!}$ converges by the ratio test since

$$\lim \frac{\frac{(n+1)^4}{(n+1)!}}{\frac{n^4}{n!}} = \lim \left(\frac{n+1}{n}\right)^4 \frac{1}{n+1} = 0.$$

20. $\sum_{n=1}^{\infty} \frac{(2n)!6^n}{(3n)!}$ converges by the ratio test since

$$\lim \frac{(2n+2)!6^{n+1}}{(3n+3)!} \bigg/ \frac{(2n)!6^n}{(3n)!}$$

$$= \lim \frac{(2n+2)(2n+1)6}{(3n+3)(3n+2)(3n+1)} = 0.$$

22. $\sum_{n=0}^{\infty} \frac{n^{100}2^n}{\sqrt{n!}}$ converges by the ratio test since

$$\lim \frac{(n+1)^{100}2^{n+1}}{\sqrt{(n+1)!}} \bigg/ \frac{n^{100}2^n}{\sqrt{n!}}$$

$$= \lim 2\left(\frac{n+1}{n}\right)^{100} \frac{1}{\sqrt{n+1}} = 0.$$

24. $\sum_{n=1}^{\infty} \frac{1+n!}{(1+n)!}$ diverges by comparison with the harmonic

series $\sum_{n=1}^{\infty} \frac{1}{n+1}$ since $\frac{1+n!}{(1+n)!} > \frac{n!}{(1+n)!} = \frac{1}{n+1}$.

26. $\sum_{n=1}^{\infty} \frac{n^n}{\pi^n n!}$ converges by the ratio test since

$$\lim \frac{(n+1)^{n+1}}{\pi^{(n+1)}(n+1)!} \bigg/ \frac{n^n}{\pi^n n!} = \frac{1}{\pi} \lim\left(1+\frac{1}{n}\right)^n = \frac{e}{\pi} < 1.$$

28. Since $f(x) = \frac{1}{x^3}$ is positive, continuous and decreasing on $[1, \infty)$, for any $n = 1, 2, 3, \ldots$, we have

$$s_n + A_{n+1} \le s \le s_n + A_n$$

where $s_n = \sum_{k=1}^{n} \frac{1}{k^3}$ and $A_n = \int_n^{\infty} \frac{dx}{x^3} = \frac{1}{2n^2}$. If $s_n^* = s_n + \frac{1}{2}(A_{n+1} + A_n)$, then

$$|s_n - s_n^*| \le \frac{A_n - A_{n+1}}{2} = \frac{1}{4}\left[\frac{1}{n^2} - \frac{1}{(n+1)^2}\right]$$

$$= \frac{1}{4}\frac{2n+1}{n^2(n+1)^2} < 0.001$$

if $n = 8$. Thus, the error in the approximation $s \approx s_8^*$ is less than 0.001.

30. Again, we have $s_n + A_{n+1} \le s \le s_n + A_n$ where $s_n = \sum_{k=1}^{n} \frac{1}{k^2+4}$ and

$$A_n = \int_n^{\infty} \frac{dx}{x^2+4} = \frac{1}{2}\tan^{-1}\left(\frac{x}{2}\right)\bigg|_n^{\infty} = \frac{\pi}{4} - \frac{1}{2}\tan^{-1}\left(\frac{n}{2}\right).$$

If $s_n^* = s_n + \frac{1}{2}(A_{n+1} + A_n)$, then

$$|s_n - s_n^*| \le \frac{A_n - A_{n+1}}{2}$$

$$= \frac{1}{2}\left[\frac{\pi}{4} - \frac{1}{2}\tan^{-1}\left(\frac{n}{2}\right) - \frac{\pi}{4} + \frac{1}{2}\tan^{-1}\left(\frac{n+1}{2}\right)\right]$$

$$= \frac{1}{4}\left[\tan^{-1}\left(\frac{n+1}{2}\right) - \tan^{-1}\left(\frac{n}{2}\right)\right] = \frac{1}{4}(a-b),$$

where $a = \tan^{-1}\left(\frac{n+1}{2}\right)$ and $b = \tan^{-1}\left(\frac{n}{2}\right)$. Now

$$\tan(a-b) = \frac{\tan a - \tan b}{1 + \tan a \tan b}$$

$$= \frac{\left(\frac{n+1}{2}\right) - \left(\frac{n}{2}\right)}{1 + \left(\frac{n+1}{2}\right)\left(\frac{n}{2}\right)}$$

$$= \frac{2}{n^2+n+4}$$

$$\Leftrightarrow \quad a - b = \tan^{-1}\left(\frac{2}{n^2+n+4}\right).$$

We want error less than 0.001:

$$\frac{1}{4}(a-b) = \frac{1}{4}\tan^{-1}\left(\frac{2}{n^2+n+4}\right) < 0.001$$

$$\Leftrightarrow \quad \frac{2}{n^2+n+4} < \tan 0.004$$

$$\Leftrightarrow \quad n^2 + n > 2\cot(0.004) - 4 \approx 496.$$

$n = 22$ will do. The approximation $s \approx s_{22}^*$ has error less than 0.001.

32. We have $s = \sum_{k=1}^{\infty} \dfrac{1}{(2k-1)!}$ and

$$s_n = \sum_{k=1}^{n} \frac{1}{(2k-1)!} = \frac{1}{1!} + \frac{1}{3!} + \frac{1}{5!} + \cdots + \frac{1}{(2n-1)!}.$$

Then

$$0 < s - s_n = \frac{1}{(2n+1)!} + \frac{1}{(2n+3)!} + \frac{1}{(2n+5)!} + \cdots$$

$$= \frac{1}{(2n+1)!}\left[1 + \frac{1}{(2n+2)(2n+3)} + \frac{1}{(2n+2)(2n+3)(2n+4)(2n+5)} + \cdots \right]$$

$$< \frac{1}{(2n+1)!}\left[1 + \frac{1}{(2n+2)(2n+3)} + \frac{1}{[(2n+2)(2n+3)]^2} + \cdots \right]$$

$$= \frac{1}{(2n+1)!}\left[\frac{1}{1 - \dfrac{1}{(2n+2)(2n+3)}} \right]$$

$$= \frac{1}{(2n+1)!} \frac{4n^2+10n+6}{4n^2+10n+5} < 0.001$$

if $n = 3$. Thus, $s \approx s_3 = 1 + \dfrac{1}{3!} + \dfrac{1}{5!} = 1.175$ with error less than 0.001.

34. We have $s = \sum_{k=1}^{\infty} \dfrac{1}{k^k}$ and

$$s_n = \sum_{k=1}^{n} \frac{1}{k^k} = \frac{1}{1} + \frac{1}{2^2} + \frac{1}{3^3} + \cdots + \frac{1}{n^n}.$$

Then

$$0 < s - s_n = \frac{1}{(n+1)^{n+1}} + \frac{1}{(n+2)^{n+2}} + \frac{1}{(n+3)^{n+3}} + \cdots$$

$$< \frac{1}{(n+1)^{n+1}}\left[1 + \frac{1}{n+1} + \frac{1}{(n+1)^2} + \cdots \right]$$

$$= \frac{1}{(n+1)^{n+1}}\left[\frac{1}{1 - \dfrac{1}{n+1}} \right]$$

$$= \frac{1}{n(n+1)^n} < 0.001$$

if $n = 4$. Thus, $s \approx s_4 = 1 + \dfrac{1}{2^2} + \dfrac{1}{3^3} + \dfrac{1}{4^4} = 1.291$ with error less than 0.001.

36. Let $u = \ln\ln t$, $du = \dfrac{dt}{t \ln t}$ and $\ln\ln a > 0$; then

$$\int_a^{\infty} \frac{dt}{t \ln t (\ln\ln t)^p} = \int_{\ln\ln a}^{\infty} \frac{du}{u^p}$$

will converge if and only if $p > 1$. Thus,

$\sum_{n=3}^{\infty} \dfrac{1}{n \ln n (\ln\ln n)^p}$ will converge if and only if $p > 1$.

Similarly,

$$\sum_{n=N}^{\infty} \frac{1}{n(\ln n)(\ln\ln n)\cdots(\ln_j n)(\ln_{j+1} n)^p}$$

converges if and only if $p > 1$, where N is large enough that $\ln_j N > 1$.

38. Let $a_n = 2^{n+1}/n^n$. Then

$$\lim_{n\to\infty} \sqrt[n]{a_n} = \lim_{n\to\infty} \frac{2 \times 2^{1/n}}{n} = 0.$$

Since this limit is less than 1, $\sum_{n=1}^{\infty} a_n$ converges by the root test.

40. Let $a_n = \dfrac{2^{n+1}}{n^n}$. Then

$$\frac{a_{n+1}}{a_n} = \frac{2^{n+2}}{(n+1)^{n+1}} \cdot \frac{n^n}{2^{n+1}}$$

$$= \frac{2}{(n+1)\left(\dfrac{n}{n+1}\right)^n} = \frac{2}{n+1} \cdot \frac{1}{\left(1 + \dfrac{1}{n}\right)^n}$$

$$\to 0 \times \frac{1}{e} = 0 \text{ as } n \to \infty.$$

Thus $\sum_{n=1}^{\infty} a_n$ converges by the ratio test.
(Remark: the question contained a typo. It was intended to ask that #33 be repeated, using the ratio test. That is a little harder.)

42. We have

$$a_n = \frac{(2n)!}{2^{2n}(n!)^2} = \frac{1 \times 2 \times 3 \times 4 \times \cdots \times 2n}{(2 \times 4 \times 6 \times 8 \times \cdots \times 2n)^2}$$

$$= \frac{1 \times 3 \times 5 \times \cdots \times (2n-1)}{2 \times 4 \times 6 \times \cdots \times (2n-2) \times 2n}$$

$$= 1 \times \frac{3}{2} \times \frac{5}{4} \times \frac{7}{6} \times \cdots \times \frac{2n-1}{2n-2} \times \frac{1}{2n} > \frac{1}{2n}.$$

Therefore $\sum_{n=1}^{\infty} \dfrac{(2n)!}{2^{2n}(n!)^2}$ diverges to infinity by comparison

with the harmonic series $\sum_{n=1}^{\infty} \dfrac{1}{2n}$.

44. If $s = \sum_{k=1}^{\infty} c_k = \sum_{k=1}^{\infty} \frac{1}{k^2(k+1)}$, then we have

$$s_n + A_{n+1} \leq s \leq s_n + A_n$$

where $s_n = \sum_{k=1}^{n} \frac{1}{k^2(k+1)}$ and

$$A_n = \int_n^{\infty} \frac{dx}{x^2(x+1)} = \int_n^{\infty} \left(\frac{-1}{x} + \frac{1}{x^2} + \frac{1}{x+1} \right) dx$$

$$= -\ln x - \frac{1}{x} + \ln(x+1) \Big|_n^{\infty}$$

$$= \ln\left(1 + \frac{1}{x}\right) - \frac{1}{x} \Big|_n^{\infty}$$

$$= \frac{1}{n} - \ln\left(1 + \frac{1}{n}\right).$$

If $s_n^* = s_n + \frac{1}{2}(A_{n+1} + A_n)$, then

$$|s_n - s_n^*| \leq \frac{A_n - A_{n+1}}{2}$$

$$= \frac{1}{2}\left[\frac{1}{n} - \ln\left(1 + \frac{1}{n}\right) - \frac{1}{n+1} + \ln\left(1 + \frac{1}{n+1}\right)\right]$$

$$= \frac{1}{2}\left[\frac{1}{n(n+1)} + \ln\left(\frac{n^2 + 2n}{n^2 + 2n + 1}\right)\right]$$

$$\leq \frac{1}{2}\left[\frac{1}{n(n+1)} + \left(\frac{n^2 + 2n}{n^2 + 2n + 1} - 1\right)\right]$$

$$= \frac{1}{2n(n+1)^2} < 0.001$$

if $n = 8$. Thus,

$$\sum_{n=1}^{\infty} \frac{1}{n^2} = 1 + s_8^* = 1 + s_8 + \frac{1}{2}(A_9 + A_8)$$

$$= 1 + \left[\frac{1}{2} + \frac{1}{2^2(3)} + \frac{1}{3^2(4)} + \cdots + \frac{1}{8^2(9)}\right] +$$

$$\frac{1}{2}\left[\left(\frac{1}{9} - \ln\frac{10}{9}\right) + \left(\frac{1}{8} - \ln\frac{9}{8}\right)\right]$$

$$= 1.6450$$

with error less than 0.001.

Section 9.4 Absolute and Conditional Convergence (page 553)

2. $\sum_{n=1}^{\infty} \frac{(-1)^n}{n^2 + \ln n}$ converges absolutely since $\left|\frac{(-1)^n}{n^2 + \ln n}\right| \leq \frac{1}{n^2}$

and $\sum_{n=1}^{\infty} \frac{1}{n^2}$ converges.

4. $\sum_{n=1}^{\infty} \frac{(-1)^{2n}}{2^n} = \sum_{n=1}^{\infty} \frac{1}{2^n}$ is a positive, convergent geometric series so must converge absolutely.

6. $\sum_{n=1}^{\infty} \frac{(-2)^n}{n!}$ converges absolutely by the ratio test since

$$\lim \left|\frac{(-2)^{n+1}}{(n+1)!} \cdot \frac{n!}{(-2)^n}\right| = 2\lim \frac{1}{n+1} = 0.$$

8. $\sum_{n=0}^{\infty} \frac{-n}{n^2 + 1}$ diverges to $-\infty$ since all terms are negative

and $\sum_{n=0}^{\infty} \frac{n}{n^2 + 1}$ diverges to infinity by comparison with

$\sum_{n=0}^{\infty} \frac{1}{n}$.

10. $\sum_{n=1}^{\infty} \frac{100\cos(n\pi)}{2n+3} = \sum_{n=1}^{\infty} \frac{100(-1)^n}{2n+3}$ converges by the alternating series test but only conditionally since

$$\left|\frac{100(-1)^n}{2n+3}\right| = \frac{100}{2n+3}$$

and $\sum_{n=1}^{\infty} \frac{100}{2n+3}$ diverges to infinity.

12. $\sum_{n=10}^{\infty} \frac{\sin(n + \frac{1}{2})\pi}{\ln\ln n} = \sum_{n=10}^{\infty} \frac{(-1)^n}{\ln\ln n}$ converges by the alternating series test but only conditionally since $\sum_{n=10}^{\infty} \frac{1}{\ln\ln n}$

diverges to infinity by comparison with $\sum_{n=10}^{\infty} \frac{1}{n}$.

($\ln\ln n < n$ for $n \geq 10$.)

14. Since the terms of the series $s = \sum_{n=0}^{\infty} \frac{(-1)^n}{(2n)!}$ are alternating in sign and decreasing in size, the size of the error in the approximation $s \approx s_n$ does not exceed that of the first omitted term:

$$|s - s_n| \leq \frac{1}{(2n+2)!} < 0.001$$

if $n = 3$. Hence $s \approx 1 - \frac{1}{2!} + \frac{1}{4!} - \frac{1}{6!}$; four terms will approximate s with error less than 0.001 in absolute value.

16. Since the terms of the series $s = \sum_{n=0}^{\infty}(-1)^n \frac{3^n}{n!}$ are alternating in sign and ultimately decreasing in size (they decrease after the third term), the size of the error in the approximation $s \approx s_n$ does not exceed that of the first omitted term (provided $n \geq 3$): $|s - s_n| \leq \frac{3^{n+1}}{(n+1)!} < 0.001$ if $n = 12$. Thus twelve terms will suffice to approximate s with error less than 0.001 in absolute value.

18. Let $a_n = \frac{(x-2)^n}{n^2 2^{2n}}$. Apply the ratio test

$$\rho = \lim \left| \frac{(x-2)^{n+1}}{(n+1)^2 2^{2n+2}} \times \frac{n^2 2^{2n}}{(x-2)^n} \right| = \frac{|x-2|}{4} < 1$$

if and only if $|x-2| < 4$, that is $-2 < x < 6$. If $x = -2$, then $\sum_{n=1}^{\infty} a_n = \sum_{n=1}^{\infty} \frac{(-1)^n}{n^2}$, which converges absolutely. If $x = 6$, then $\sum_{n=1}^{\infty} a_n = \sum_{n=1}^{\infty} \frac{1}{n^2}$, which also converges absolutely. Thus, the series converges absolutely if $-2 \leq x \leq 6$ and diverges elsewhere.

20. Let $a_n = \frac{1}{2n-1} \left(\frac{3x+2}{-5} \right)^n$. Apply the ratio test

$$\rho = \lim \left| \frac{1}{2n+1} \left(\frac{3x+2}{-5} \right)^{n+1} \times \frac{2n-1}{1} \left(\frac{3x+2}{-5} \right)^{-n} \right|$$

$$= \left| \frac{3x+2}{5} \right| < 1$$

if and only if $\left| x + \frac{2}{3} \right| < \frac{5}{3}$, that is $-\frac{7}{3} < x < 1$. If $x = -\frac{7}{3}$, then $\sum_{n=1}^{\infty} a_n = \sum_{n=1}^{\infty} \frac{1}{2n-1}$, which diverges. If $x = 1$, then $\sum_{n=1}^{\infty} a_n = \sum_{n=1}^{\infty} \frac{(-1)^n}{2n-1}$, which converges conditionally. Thus, the series converges absolutely if $-\frac{7}{3} < x < 1$, converges conditionally if $x = 1$ and diverges elsewhere.

22. Let $a_n = \frac{(4x+1)^n}{n^3}$. Apply the ratio test

$$\rho = \lim \left| \frac{(4x+1)^{n+1}}{(n+1)^3} \times \frac{n^3}{(4x+1)^n} \right| = |4x+1| < 1$$

if and only if $-\frac{1}{2} < x < 0$. If $x = -\frac{1}{2}$, then $\sum_{n=1}^{\infty} a_n = \sum_{n=1}^{\infty} \frac{(-1)^n}{n^3}$, which converges absolutely. If $x = 0$, then $\sum_{n=1}^{\infty} a_n = \sum_{n=1}^{\infty} \frac{1}{n^3}$, which also converges absolutely. Thus, the series converges absolutely if $-\frac{1}{2} \leq x \leq 0$ and diverges elsewhere.

24. Let $a_n = \frac{1}{n} \left(1 + \frac{1}{x} \right)^n$. Apply the ratio test

$$\rho = \lim \left| \frac{1}{n+1} \left(1 + \frac{1}{x} \right)^{n+1} \times \frac{n}{1} \left(1 + \frac{1}{x} \right)^{-n} \right| = \left| 1 + \frac{1}{x} \right| < 1$$

if and only if $|x+1| < |x|$, that is, $-2 < \frac{1}{x} < 0 \Rightarrow x < -\frac{1}{2}$. If $x = -\frac{1}{2}$, then $\sum_{n=1}^{\infty} a_n = \sum_{n=1}^{\infty} \frac{(-1)^n}{n}$, which converges conditionally.

Thus, the series converges absolutely if $x < -\frac{1}{2}$, converges conditionally if $x = -\frac{1}{2}$ and diverges elsewhere. It is undefined at $x = 0$.

26. If

$$a_n = \begin{cases} \dfrac{10}{n^2}, & \text{if } n \text{ is even;} \\ \dfrac{-1}{10n^3}, & \text{if } n \text{ is odd;} \end{cases}$$

then $|a_n| \leq \frac{10}{n^2}$ for every $n \geq 1$. Hence, $\sum_{n=1}^{\infty} a_n$ converges absolutely by comparison with $\sum_{n=1}^{\infty} \frac{10}{n^2}$.

28. a) We have

$$\ln(n!) = \ln 1 + \ln 2 + \ln 3 + \cdots + \ln n$$
$$= \text{sum of area of the shaded rectangles}$$
$$> \int_1^n \ln t \, dt = (t \ln t - t) \Big|_1^n$$
$$= n \ln n - n + 1.$$

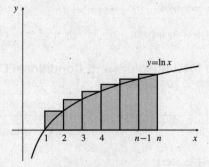

Fig. 9.4.28

b) Let $a_n = \dfrac{n!x^n}{n^n}$. Apply the ratio test

$$\rho = \lim \left| \frac{(n+1)!x^{n+1}}{(n+1)^{n+1}} \times \frac{n^n}{n!x^n} \right|$$

$$= \lim \frac{|x|}{\left(1 + \dfrac{1}{n}\right)^n} = \frac{|x|}{e} < 1$$

if and only if $-e < x < e$. If $x = \pm e$, then, by (a),

$$\ln \left| \frac{n!e^n}{n^n} \right| = \ln(n!) + \ln e^n - \ln n^n$$

$$> (n \ln n - n + 1) + n - n \ln n = 1.$$

$$\Rightarrow \left| \frac{n!e^n}{n^n} \right| > e.$$

Hence, $\displaystyle\sum_{n=1}^{\infty} a_n$ converges absolutely if $-e < x < e$ and diverges elsewhere.

30. Let $p_n = \dfrac{1}{2n-1}$ and $q_n = -\dfrac{1}{2n}$. Then $\sum p_n$ diverges to ∞ and $\sum q_n$ diverges to $-\infty$. Also, the alternating harmonic series is the sum of all the p_ns and q_ns in a specific order:

$$\sum_{n=1}^{\infty} \frac{(-1)^{n-1}}{n} = \sum_{n=1}^{\infty} (p_n + q_n).$$

a) Rearrange the terms as follows: first add terms of $\sum p_n$ until the sum exceeds 2. Then add q_1. Then add more terms of $\sum p_n$ until the sum exceeds 3. Then add q_2. Continue in this way; at the nth stage, add new terms from $\sum p_n$ until the sum exceeds $n + 1$, and then add q_n. All partial sums after the nth stage exceed n, so the rearranged series diverges to infinity.

b) Rearrange the terms of the original alternating harmonic series as follows: first add terms of $\sum q_n$ until the sum is less than -2. Then add p_1. The sum will now be greater than -2. (Why?) Then resume adding new terms from $\sum q_n$ until the sum is less than -2 again, and add p_2, which will raise the sum above -2 again. Continue in this way. After the nth stage, all succeeding partial sums will differ from -2 by less than $1/n$, so the rearranged series will converge to -2.

Section 9.5 Power Series (page 564)

2. We have $\displaystyle\sum_{n=0}^{\infty} 3n(x+1)^n$. The centre of convergence is $x = -1$. The radius of convergence is

$$R = \lim \frac{3n}{3(n+1)} = 1.$$

The series converges absolutely on $(-2, 0)$ and diverges on $(-\infty, -2)$ and $(0, \infty)$. At $x = -2$, the series is $\displaystyle\sum_{n=0}^{\infty} 3n(-1)^n$, which diverges. At $x = 0$, the series is $\displaystyle\sum_{n=0}^{\infty} 3n$, which diverges to infinity. Hence, the interval of convergence is $(-2, 0)$.

4. We have $\displaystyle\sum_{n=1}^{\infty} \frac{(-1)^n}{n^4 2^{2n}} x^n$. The centre of convergence is $x = 0$. The radius of convergence is

$$R = \lim \left| \frac{(-1)^n}{n^4 2^{2n}} \cdot \frac{(n+1)^4 2^{2n+2}}{(-1)^{n+1}} \right|$$

$$= \lim \left| \left(\frac{n+1}{n} \right)^4 \cdot 4 \right| = 4.$$

At $x = 4$, the series is $\displaystyle\sum_{n=1}^{\infty} \frac{(-1)^n}{n^4}$, which converges.

At $x = -4$, the series is $\displaystyle\sum_{n=1}^{\infty} \frac{1}{n^4}$, which also converges. Hence, the interval of convergence is $[-4, 4]$.

6. We have $\displaystyle\sum_{n=1}^{\infty} \frac{e^n}{n^3}(4-x)^n$. The centre of convergence is $x = 4$. The radius of convergence is

$$R = \lim \frac{e^n}{n^3} \cdot \frac{(n+1)^3}{e^{n+1}} = \frac{1}{e}.$$

At $x = 4 + \dfrac{1}{e}$, the series is $\displaystyle\sum_{n=1}^{\infty} \frac{(-1)^n}{n^3}$, which converges.

At $x = 4 - \dfrac{1}{e}$, the series is $\displaystyle\sum_{n=1}^{\infty} \frac{1}{n^3}$, which also converges.

Hence, the interval of convergence is $\left[4 - \dfrac{1}{e}, 4 + \dfrac{1}{e} \right]$.

8. We have $\displaystyle\sum_{n=1}^{\infty} \frac{(4x-1)^n}{n^n} = \sum_{n=1}^{\infty}\left(\frac{4}{n}\right)^n \left(x-\frac{1}{4}\right)^n$. The centre of convergence is $x=\frac{1}{4}$. The radius of convergence is

$$R = \lim \frac{4^n}{n^n} \cdot \frac{(n+1)^{n+1}}{4^{n+1}}$$

$$= \frac{1}{4}\lim\left(\frac{n+1}{n}\right)^n (n+1) = \infty.$$

Hence, the interval of convergence is $(-\infty, \infty)$.

10. We have

$$1 + x + x^2 + x^3 + \cdots = \frac{1}{1-x} = \sum_{n=0}^{\infty} x^n$$

and

$$1 - x + x^2 - x^3 + \cdots = \frac{1}{1+x} = \sum_{n=0}^{\infty} (-1)^n x^n$$

holds for $-1 < x < 1$. Since $a_n = 1$ and $b_n = (-1)^n$ for $n = 0, 1, 2, \ldots$, we have

$$C_n = \sum_{j=0}^{n} (-1)^{n-j} = \begin{cases} 0, & \text{if } n \text{ is odd;} \\ 1, & \text{if } n \text{ is even.} \end{cases}$$

Then the Cauchy product is

$$1 + x^2 + x^4 + \cdots = \sum_{n=0}^{\infty} x^{2n} = \frac{1}{1-x} \cdot \frac{1}{1+x} = \frac{1}{1-x^2}$$

for $-1 < x < 1$.

12. $\displaystyle \frac{1}{2-x} = \frac{1}{2}\frac{1}{\left(1-\dfrac{x}{2}\right)} = \frac{1}{2}\sum_{n=0}^{\infty}\left(\frac{x}{2}\right)^n$

$$= \frac{1}{2} + \frac{x}{2^2} + \frac{x^2}{2^3} + \frac{x^3}{2^4} + \cdots \quad (-2 < x < 2).$$

14. $\displaystyle \frac{1}{1+2x} = \sum_{n=0}^{\infty}(-2x)^n$

$$= 1 - 2x + 2^2 x^2 - 2^3 x^3 + \cdots \quad \left(-\frac{1}{2} < x < \frac{1}{2}\right).$$

16. Let $y = x - 1$. Then $x = 1 + y$ and

$$\frac{1}{x} = \frac{1}{1+y} = \sum_{n=0}^{\infty}(-y)^n \quad (-1 < y < 1)$$

$$= \sum_{n=0}^{\infty}[-(x-1)]^n$$

$$= 1 - (x-1) + (x-1)^2 - (x-1)^3 + (x-1)^4 - \cdots$$
$$(\text{for } 0 < x < 2).$$

18. $\displaystyle \frac{1-x}{1+x} = \frac{2}{1+x} - 1$

$$= 2(1 - x + x^2 - x^3 + \cdots) - 1$$

$$= 1 + 2\sum_{n=1}^{\infty}(-x)^n \quad (-1 < x < 1).$$

20. Let $y = x - 4$. Then $x = 4 + y$ and

$$\frac{1}{x} = \frac{1}{4+y} = \frac{1}{4}\frac{1}{\left(1+\dfrac{y}{4}\right)} = \frac{1}{4}\sum_{n=0}^{\infty}\left(-\frac{y}{4}\right)^n$$

$$= \frac{1}{4}\sum_{n=0}^{\infty}\left[-\frac{(x-4)}{4}\right]^n$$

$$= \frac{1}{4} - \frac{(x-4)}{4^2} + \frac{(x-4)^2}{4^3} - \frac{(x-4)^3}{4^4} + \cdots$$

for $0 < x < 8$. Therefore,

$$\ln x = \int_1^x \frac{dt}{t} = \int_1^4 \frac{dt}{t} + \int_4^x \frac{dt}{t}$$

$$= \ln 4 + \int_4^x \left[\frac{1}{4} - \frac{(t-4)}{4^2} + \frac{(t-4)^2}{4^3} - \frac{(t-4)^3}{4^4} + \cdots\right] dt$$

$$= \ln 4 + \frac{x-4}{4} - \frac{(x-4)^2}{2 \cdot 4^2} + \frac{(x-4)^3}{3 \cdot 4^3} - \frac{(x-4)^4}{4 \cdot 4^4} + \cdots$$
$$(\text{for } 0 < x \leq 8).$$

22. We differentiate the series

$$\sum_{n=0}^{\infty} x^n = 1 + x + x^2 + x^3 + \cdots = \frac{1}{1-x}$$

and multiply by x to get

$$\sum_{n=0}^{\infty} n x^n = x + 2x^2 + 3x^3 + \cdots = \frac{x}{(1-x)^2}$$

for $-1 < x < 1$. Therefore,

$$\sum_{n=0}^{\infty}(n+3)x^n = \sum_{n=0}^{\infty} nx^n + 3\sum_{n=0}^{\infty} x^n$$

$$= \frac{x}{(1-x)^2} + \frac{3}{1-x}$$

$$= \frac{3-2x}{(1-x)^2} \quad (-1 < x < 1).$$

24. We start with

$$1 - x + x^2 - x^3 + x^4 - \cdots = \frac{1}{1+x}$$

and differentiate to get

$$-1 + 2x - 3x^3 + 4x^3 - \cdots = -\frac{1}{(1+x)^2}.$$

Now we multiply by $-x^3$:

$$x^3 - 2x^4 + 3x^5 - 4x^6 + \cdots = \frac{x^3}{(1+x)^2}.$$

Differentiating again we get

$$3x^2 - 2 \times 4x^3 + 3 \times 5x^4 - 4 \times 6x^5 + \cdots = \frac{x^3 + 3x^2}{(1+x)^3}.$$

Finally, we remove the factor x^2:

$$3 - 2 \times 4x + 3 \times 5x^2 - 4 \times 6x^3 + \cdots = \frac{x+3}{(1+x)^3}.$$

All steps are valid for $-1 < x < 1$.

26. Since $x - \dfrac{x^2}{2} + \dfrac{x^3}{3} - \dfrac{x^4}{4} + \cdots = \ln(1+x)$ for $-1 < x \le 1$, therefore

$$x^2 - \frac{x^4}{2} + \frac{x^6}{3} - \frac{x^8}{4} + \cdots = \ln(1+x^2)$$

for $-1 \le x \le 1$, and, dividing by x^2,

$$1 - \frac{x^2}{2} + \frac{x^4}{3} - \frac{x^6}{4} + \cdots = \begin{cases} \dfrac{\ln(1+x^2)}{x^2} & \text{if } -1 \le x \le 1, x \ne 0 \\ 1 & \text{if } x = 0. \end{cases}$$

28. From Example 5(a) with $x = 1/2$,

$$\sum_{n=0}^{\infty} \frac{n+1}{2^n} = \sum_{k=1}^{\infty} k \left(\frac{1}{2}\right)^{k-1} = \frac{1}{\left(1 - \frac{1}{2}\right)^2} = 4.$$

30. From Example 5(a),

$$\sum_{n=1}^{\infty} n x^{n-1} = \frac{1}{(1-x)^2}, \quad (-1 < x < 1).$$

Differentiate with respect to x and then replace n by $n+1$:

$$\sum_{n=2}^{\infty} n(n-1)x^{n-2} = \frac{2}{(1-x)^3}, \quad (-1 < x < 1)$$

$$\sum_{n=1}^{\infty} (n+1)n x^{n-1} = \frac{2}{(1-x)^3}, \quad (-1 < x < 1).$$

Now let $x = -1/2$:

$$\sum_{n=1}^{\infty} (-1)^{n-1} \frac{n(n+1)}{2^{n-1}} = \frac{16}{27}.$$

Finally, multiply by $-1/2$:

$$\sum_{n=1}^{\infty} (-1)^n \frac{n(n+1)}{2^n} = -\frac{8}{27}.$$

32. In the series for $\ln(1+x)$ in Example 5(c), put $x = -1/2$ to get

$$\sum_{n=1}^{\infty} (-1) \frac{1}{n2^n} = \sum_{k=0}^{\infty} \frac{(-1)^k}{k+1} \left(-\frac{1}{2}\right)^{k+1} = \ln\left(1 - \frac{1}{2}\right) = -\ln 2.$$

Therefore

$$\sum_{n=1}^{\infty} \frac{1}{n2^n} = \ln 2$$

$$\sum_{n=3}^{\infty} \frac{1}{n2^n} = \ln 2 - \frac{1}{2} - \frac{1}{8} = \ln 2 - \frac{5}{8}.$$

Section 9.6 Taylor and Maclaurin Series (page 572)

2. $\cos(2x^3) = 1 - \dfrac{(2x^3)^2}{2!} + \dfrac{(2x^3)^4}{4!} - \dfrac{(2x^3)^6}{6!} + \cdots$

$\qquad = 1 - \dfrac{2^2 x^6}{2!} + \dfrac{2^4 x^{12}}{4!} - \dfrac{2^6 x^{18}}{6!} + \cdots$

$\qquad = \displaystyle\sum_{n=0}^{\infty} \frac{(-1)^n 4^n}{(2n)!} x^{6n} \quad$ (for all x).

4. $\cos(2x - \pi) = -\cos(2x)$

$\qquad = -1 + \dfrac{2^2 x^2}{2!} - \dfrac{2^4 x^4}{4!} + \dfrac{2^6 x^6}{6!} - \cdots$

$\qquad = -\displaystyle\sum_{n=0}^{\infty} \frac{(-1)^n}{(2n)!} (2x)^{2n}$

$\qquad = \displaystyle\sum_{n=0}^{\infty} \frac{(-1)^{n+1}}{(2n)!} 4^n (x)^{2n} \quad$ (for all x).

6. $\cos^2 \left(\dfrac{x}{2}\right) = \dfrac{1}{2}(1 + \cos x)$

$\qquad = \dfrac{1}{2}\left(1 + 1 - \dfrac{x^2}{2!} + \dfrac{x^4}{4!} - \dfrac{x^6}{6!} + \cdots\right)$

$\qquad = 1 + \dfrac{1}{2} \displaystyle\sum_{n=1}^{\infty} \frac{(-1)^n}{(2n)!} x^{2n} \quad$ (for all x).

8. $\tan^{-1}(5x^2) = (5x^2) - \dfrac{(5x^2)^3}{3} + \dfrac{(5x^2)^5}{5} - \dfrac{(5x^2)^7}{7} + \cdots$

$$= \sum_{n=0}^{\infty} \frac{(-1)^n}{(2n+1)}(5x^2)^{2n+1}$$

$$= \sum_{n=0}^{\infty} \frac{(-1)^n 5^{2n+1}}{(2n+1)} x^{4n+2}$$

$$\left(\text{for } -\frac{1}{\sqrt{5}} \le x \le \frac{1}{\sqrt{5}}\right).$$

10. $\ln(2 + x^2) = \ln 2\left(1 + \dfrac{x^2}{2}\right)$

$$= \ln 2 + \ln\left(1 + \frac{x^2}{2}\right)$$

$$= \ln 2 + \left[\frac{x^2}{2} - \frac{1}{2}\left(\frac{x^2}{2}\right)^2 + \frac{1}{3}\left(\frac{x^2}{2}\right)^3 - \cdots\right]$$

$$= \ln 2 + \sum_{n=1}^{\infty} \frac{(-1)^{n-1}}{n} \cdot \frac{x^{2n}}{2^n}$$

$$\left(\text{for } -\sqrt{2} \le x \le \sqrt{2}\right).$$

12. $\dfrac{e^{2x^2} - 1}{x^2} = \dfrac{1}{x^2}\left(e^{2x^2} - 1\right)$

$$= \frac{1}{x^2}\left(1 + 2x^2 + \frac{(2x^2)^2}{2!} + \frac{(2x^2)^3}{3!} + \cdots - 1\right)$$

$$= 2 + \frac{2^2 x^2}{2!} + \frac{2^3 x^4}{3!} + \frac{2^4 x^6}{4!} + \cdots$$

$$= \sum_{n=0}^{\infty} \frac{2^{n+1}}{(n+1)!} x^{2n} \quad (\text{for all } x \ne 0).$$

14. $\sinh x - \sin x = \displaystyle\sum_{n=0}^{\infty}\left[1 - (-1)^n\right]\frac{x^{2n+1}}{(2n+1)!}$

$$= 2\left(\frac{x^2}{2!} + \frac{x^6}{6!} + \frac{x^{10}}{10!} + \cdots\right)$$

$$= 2\sum_{n=0}^{\infty} \frac{x^{4n+3}}{(4n+3)!} \quad (\text{for all } x).$$

16. Let $y = x - \dfrac{\pi}{2}$; then $x = y + \dfrac{\pi}{2}$. Hence,

$$\sin x = \sin\left(y + \frac{\pi}{2}\right) = \cos y$$

$$= 1 - \frac{y^2}{2!} + \frac{y^4}{4!} - \cdots \quad (\text{for all } y)$$

$$= 1 - \frac{1}{2!}\left(x - \frac{\pi}{2}\right)^2 + \frac{1}{4!}\left(x - \frac{\pi}{2}\right)^4 - \cdots$$

$$= \sum_{n=0}^{\infty} \frac{(-1)^n}{(2n)!}\left(x - \frac{\pi}{2}\right)^{2n} \quad (\text{for all } x).$$

18. Let $y = x - 3$; then $x = y + 3$. Hence,

$$\ln x = \ln(y + 3) = \ln 3 + \ln\left(1 + \frac{y}{3}\right)$$

$$= \ln 3 + \frac{y}{3} - \frac{1}{2}\left(\frac{y}{3}\right)^2 + \frac{1}{3}\left(\frac{y}{3}\right)^3 - \frac{1}{4}\left(\frac{y}{3}\right)^4 + \cdots$$

$$= \ln 3 + \frac{(x-3)}{3} - \frac{(x-3)^2}{2 \cdot 3^2} + \frac{(x-3)^3}{3 \cdot 3^3} - \frac{(x-3)^4}{4 \cdot 3^4} + \cdots$$

$$= \ln 3 + \sum_{n=1}^{\infty} \frac{(-1)^{n-1}}{n \cdot 3^n}(x-3)^n \quad (0 < x \le 6).$$

20. Let $t = x + 1$. Then $x = t - 1$, and

$$e^{2x+3} = e^{2t+1} = e\, e^{2t}$$

$$= e\sum_{n=0}^{\infty} \frac{2^n t^n}{n!} \quad (\text{for all } t)$$

$$= \sum_{n=0}^{\infty} \frac{e 2^n (x+1)^n}{n!} \quad (\text{for all } x).$$

22. Let $y = x - \dfrac{\pi}{8}$; then $x = y + \dfrac{\pi}{8}$. Thus,

$$\cos^2 x = \cos^2\left(y + \frac{\pi}{8}\right)$$

$$= \frac{1}{2}\left[1 + \cos\left(2y + \frac{\pi}{4}\right)\right]$$

$$= \frac{1}{2}\left[1 + \frac{1}{\sqrt{2}}\cos(2y) - \frac{1}{\sqrt{2}}\sin(2y)\right]$$

$$= \frac{1}{2} + \frac{1}{2\sqrt{2}}\left[1 - \frac{(2y)^2}{2!} + \frac{(2y)^4}{4!} - \cdots\right]$$

$$\quad - \frac{1}{2\sqrt{2}}\left[2y - \frac{(2y)^3}{3!} + \frac{(2y)^5}{5!} - \cdots\right]$$

$$= \frac{1}{2} + \frac{1}{2\sqrt{2}}\left[1 - 2y - \frac{(2y)^2}{2!} + \frac{(2y)^3}{3!}\right.$$

$$\left. + \frac{(2y)^4}{4!} - \frac{(2y)^5}{5!} - \cdots\right]$$

$$= \frac{1}{2} + \frac{1}{2\sqrt{2}}\left[1 - 2\left(x - \frac{\pi}{8}\right) - \frac{2^2}{2!}\left(x - \frac{\pi}{8}\right)^2\right.$$

$$\left. + \frac{2^3}{3!}\left(x - \frac{\pi}{8}\right)^3 + \frac{2^4}{4!}\left(x - \frac{\pi}{8}\right)^4 - \frac{2^5}{5!}\left(x - \frac{\pi}{8}\right)^5 - \cdots\right]$$

$$= \frac{1}{2} + \frac{1}{2\sqrt{2}} + \frac{1}{2\sqrt{2}}\sum_{n=1}^{\infty}(-1)^n\left[\frac{2^{2n-1}}{(2n-1)!}\left(x - \frac{\pi}{8}\right)^{2n-1}\right.$$

$$\left. + \frac{2^{2n}}{(2n)!}\left(x - \frac{\pi}{8}\right)^{2n}\right] \quad (\text{for all } x).$$

24. Let $y = x - 1$; then $x = y + 1$. Thus,

$$\frac{x}{1+x} = \frac{1+y}{2+y} = 1 - \frac{1}{2\left(1+\frac{y}{2}\right)}$$

$$= 1 - \frac{1}{2}\left[1 - \frac{y}{2} + \left(\frac{y}{2}\right)^2 - \left(\frac{y}{2}\right)^3 + \cdots\right]$$

$$= \frac{1}{2}\left[1 + \frac{y}{2} - \frac{y^2}{2^2} + \frac{y^3}{2^3} - \frac{y^4}{2^4} + \cdots\right] \quad (-1 < y < 1)$$

$$= \frac{1}{2} + \frac{1}{2^2}(x-1) - \frac{1}{2^3}(x-1)^2 + \frac{1}{2^4}(x-1)^3 - \cdots$$

$$= \frac{1}{2} + \sum_{n=1}^{\infty} \frac{(-1)^{n-1}}{2^{n+1}}(x-1)^n \quad \text{(for } 0 < x < 2\text{)}.$$

26. Let $u = x + 2$. Then $x = u - 2$, and

$$xe^x = (u-2)e^{u-2}$$

$$= (u-2)e^{-2}\sum_{n=0}^{\infty}\frac{u^n}{n!} \quad \text{(for all } u\text{)}$$

$$= \sum_{n=0}^{\infty}\frac{e^{-2}u^{n+1}}{n!} - \sum_{n=0}^{\infty}\frac{2e^{-2}u^n}{n!}.$$

In the first sum replace n by $n-1$.

$$xe^x = \sum_{n=1}^{\infty}\frac{e^{-2}u^n}{(n-1)!} - \sum_{n=0}^{\infty}\frac{2e^{-2}u^n}{n!}$$

$$= -\frac{2}{e^2} + \sum_{n=1}^{\infty}\frac{1}{e^2}\left(\frac{1}{(n-1)!} - \frac{2}{n!}\right)u^n$$

$$= -\frac{2}{e^2} + \sum_{n=1}^{\infty}\frac{1}{e^2}\left(\frac{1}{(n-1)!} - \frac{2}{n!}\right)(x+2)^n \quad \text{(for all } x\text{)}.$$

28. If we divide the first four terms of the series

$$\cos x = 1 - \frac{x^2}{2} + \frac{x^4}{24} - \frac{x^6}{720} + \cdots$$

into 1 we obtain

$$\sec x = 1 + \frac{x^2}{2} + \frac{5x^4}{24} + \frac{61x^6}{720} + \cdots.$$

Now we can differentiate and obtain

$$\sec x \tan x = x + \frac{5x^3}{6} + \frac{61x^5}{120} + \cdots.$$

(Note: the same result can be obtained by multiplying the first three nonzero terms of the series for $\sec x$ (from Exercise 25) and $\tan x$ (from Example 6(b)).)

30. We have

$$e^{\tan^{-1} x} - 1 = \exp\left[x - \frac{x^3}{3} + \frac{x^5}{5} - \frac{x^7}{7} + \cdots\right] - 1$$

$$= 1 + \left(x - \frac{x^3}{3} + \frac{x^5}{5} - \cdots\right) + \frac{1}{2!}\left(x - \frac{x^3}{3} + \cdots\right)^2$$

$$+ \frac{1}{3!}(x - \cdots)^3 + \cdots - 1$$

$$= x - \frac{x^3}{3} + \frac{x^2}{2} + \frac{x^3}{6} + \text{higher degree terms}$$

$$= x + \frac{x^2}{2} - \frac{x^3}{6} + \cdots.$$

32. $\csc x$ does not have a Maclaurin series because $\lim_{x\to 0} \csc x$ does not exist.
Let $y = x - \frac{\pi}{2}$. Then $x = y + \frac{\pi}{2}$ and $\sin x = \cos y$.
Therefore, using the result of Exercise 25,

$$\csc x = \sec y = 1 + \frac{y^2}{2} + \frac{5y^4}{24} + \cdots$$

$$= 1 + \frac{1}{2}\left(x - \frac{\pi}{2}\right)^2 + \frac{5}{24}\left(x - \frac{\pi}{2}\right)^4 + \cdots.$$

34. $x^3 - \frac{x^9}{3! \times 4} + \frac{x^{15}}{5! \times 16} - \frac{x^{21}}{7! \times 64} + \frac{x^{27}}{9! \times 256} - \cdots$

$$= 2\left[\frac{x^3}{2} - \frac{1}{3!}\left(\frac{x^3}{2}\right)^3 + \frac{1}{5!}\left(\frac{x^3}{2}\right)^5 - \cdots\right]$$

$$= 2\sin\left(\frac{x^3}{2}\right) \quad \text{(for all } x\text{)}.$$

36. $1 + \frac{1}{2 \times 2!} + \frac{1}{4 \times 3!} + \frac{1}{8 \times 4!} + \cdots$

$$= 2\left[\frac{1}{2} + \frac{1}{2!}\left(\frac{1}{2}\right)^2 + \frac{1}{3!}\left(\frac{1}{2}\right)^3 + \cdots\right]$$

$$= 2\left(e^{1/2} - 1\right).$$

38. If $a \neq 0$ and $|x - a| < |a|$, then

$$\frac{1}{x} = \frac{1}{a + (x-a)} = \frac{1}{a}\frac{1}{1 + \frac{x-a}{a}}$$

$$= \frac{1}{a}\left[1 - \frac{x-a}{a} + \frac{(x-a)^2}{a^2} - \frac{(x-a)^3}{a^3} + \cdots\right].$$

The radius of convergence of this series is $|a|$, and the series converges to $1/x$ throughout its interval of convergence. Hence, $1/x$ is analytic at $x = a$.

40. If

$$f(x) = \begin{cases} e^{-1/x^2}, & \text{if } x \neq 0; \\ 0, & \text{if } x = 0; \end{cases}$$

then the Maclaurin series for $f(x)$ is the identically zero series $0 + 0x + 0x^2 + \cdots$ since $f^{(k)}(0) = 0$ for every k. The series converges for every x, but converges to $f(x)$ only at $x = 0$, since $f(x) \neq 0$ if $x \neq 0$. Hence, f cannot be analytic at $x = 0$.

Section 9.7 Applications of Taylor and Maclaurin Series (page 576)

2. We have

$$\frac{1}{e} = e^{-1} = 1 - \frac{1}{1!} + \frac{1}{2!} - \frac{1}{3!} + \frac{1}{4!} - \cdots$$

which satisfies the conditions for the alternating series test, and the error incurred in using a partial sum to approximate e^{-1} is less than the first omitted term in absolute value. Now $\frac{1}{(n+1)!} < 5 \times 10^{-5}$ if $n = 7$, so

$$\frac{1}{e} \approx \frac{1}{2} - \frac{1}{6} + \frac{1}{24} - \frac{1}{120} + \frac{1}{720} - \frac{1}{5040} \approx 0.36786$$

with error less than 5×10^{-5} in absolute value.

4. We have

$$\sin(0.1) = 0.1 - \frac{(0.1)^3}{3!} + \frac{(0.1)^5}{5!} - \frac{(0.1)^7}{7!} + \cdots.$$

Since $\frac{(0.1)^5}{5!} = 8.33 \times 10^{-8} < 5 \times 10^{-5}$, therefore

$$\sin(0.1) = 0.1 - \frac{(0.1)^3}{3!} \approx 0.09983$$

with error less than 5×10^{-5} in absolute value.

6. We have

$$\ln\left(\frac{6}{5}\right) = \ln\left(1 + \frac{1}{5}\right)$$
$$= \frac{1}{5} - \frac{1}{2}\left(\frac{1}{5}\right)^2 + \frac{1}{3}\left(\frac{1}{5}\right)^3 - \frac{1}{4}\left(\frac{1}{5}\right)^4 + \cdots.$$

Since $\frac{1}{n}\left(\frac{1}{5}\right)^n < 5 \times 10^{-5}$ if $n = 6$, therefore

$$\ln\left(\frac{6}{5}\right) \approx \frac{1}{5} - \frac{1}{2}\left(\frac{1}{5}\right)^2 + \frac{1}{3}\left(\frac{1}{5}\right)^3 - \frac{1}{4}\left(\frac{1}{5}\right)^4 + \frac{1}{5}\left(\frac{1}{5}\right)^5$$
$$\approx 0.18233$$

with error less than 5×10^{-5} in absolute value.

8. We have

$$\sin 80° = \cos 10° = \cos\left(\frac{\pi}{18}\right)$$
$$= 1 - \frac{1}{2!}\left(\frac{\pi}{18}\right)^2 + \frac{1}{4!}\left(\frac{\pi}{18}\right)^4 - \cdots.$$

Since $\frac{1}{4!}\left(\frac{\pi}{18}\right)^4 < 5 \times 10^{-5}$, therefore

$$\sin 80° \approx 1 - \frac{1}{2!}\left(\frac{\pi}{18}\right)^2 \approx 0.98477$$

with error less than 5×10^{-5} in absolute value.

10. We have

$$\tan^{-1}(0.2) = 0.2 - \frac{(0.2)^3}{3} + \frac{(0.2)^5}{5} - \frac{(0.2)^7}{7} + \cdots.$$

Since $\frac{(0.2)^7}{7} < 5 \times 10^{-5}$, therefore

$$\tan^{-1}(0.2) \approx 0.2 - \frac{(0.2)^3}{3} + \frac{(0.2)^5}{5} \approx 0.19740$$

with error less than 5×10^{-5} in absolute value.

12. We have

$$\ln\left(\frac{3}{2}\right) = \ln\left(1 + \frac{1}{2}\right)$$
$$= \frac{1}{2} - \frac{1}{2}\left(\frac{1}{2}\right)^2 + \frac{1}{3}\left(\frac{1}{2}\right)^3 - \frac{1}{4}\left(\frac{1}{2}\right)^4 + \cdots.$$

Since $\frac{1}{n}\left(\frac{1}{2}\right)^n < \frac{1}{20000}$ if $n = 11$, therefore

$$\ln\left(\frac{3}{2}\right) \approx \frac{1}{2} - \frac{1}{2}\left(\frac{1}{2}\right)^2 + \frac{1}{3}\left(\frac{1}{2}\right)^3 - \cdots - \frac{1}{10}\left(\frac{1}{2}\right)^{10}$$
$$\approx 0.40543$$

with error less than 5×10^{-5} in absolute value.

14.
$$J(x) = \int_0^x \frac{e^t - 1}{t}\, dt$$
$$= \int_0^x \left(1 + \frac{t}{2!} + \frac{t^2}{3!} + \frac{t^3}{4!} + \cdots\right) dt$$
$$= x + \frac{x^2}{2! \cdot 2} + \frac{x^3}{3! \cdot 3} + \frac{x^4}{4! \cdot 4} + \cdots$$
$$= \sum_{n=1}^{\infty} \frac{x^n}{n! \cdot n}.$$

16. $L(x) = \int_0^x \cos(t^2)\,dt$

$$= \int_0^x \left(1 - \frac{t^4}{2!} + \frac{t^8}{4!} - \frac{t^{12}}{6!} + \cdots\right) dt$$

$$= x - \frac{x^5}{2! \cdot 5} + \frac{x^9}{4! \cdot 9} - \frac{x^{13}}{6! \cdot 13} + \cdots$$

$$= \sum_{n=0}^{\infty} (-1)^n \frac{x^{4n+1}}{(2n)! \cdot (4n+1)}.$$

18. We have

$$L(0.5) = 0.5 - \frac{(0.5)^5}{2! \cdot 5} + \frac{(0.5)^9}{4! \cdot 9} - \frac{(0.5)^{13}}{6! \cdot 13} + \cdots.$$

Since $\dfrac{(0.5)^{4n+1}}{(2n)! \cdot (4n+1)} < 5 \times 10^{-4}$ if $n = 2$, therefore

$$L(0.5) \approx 0.5 - \frac{(0.5)^5}{2! \cdot 5} \approx 0.497$$

rounded to three decimal places.

20. $\displaystyle\lim_{x \to 0} \frac{\sin(x^2)}{\sinh x} = \lim_{x \to 0} \dfrac{x^2 - \dfrac{x^6}{3!} + \dfrac{x^{10}}{5!} - \cdots}{x + \dfrac{x^3}{3!} + \dfrac{x^5}{5!} + \cdots}$

$$= \lim_{x \to 0} \dfrac{x - \dfrac{x^5}{3!} + \dfrac{x^9}{5!} - \cdots}{1 + \dfrac{x^2}{3!} + \dfrac{x^4}{5!} + \cdots} = 0.$$

22. We have

$$\lim_{x \to 0} \frac{(e^x - 1 - x)^2}{x^2 - \ln(1 + x^2)} = \lim_{x \to 0} \dfrac{\left(\dfrac{x^2}{2!} + \dfrac{x^3}{3!} + \dfrac{x^4}{4!} + \cdots\right)^2}{\dfrac{x^4}{2} - \dfrac{x^6}{3} + \dfrac{x^8}{4} - \cdots}$$

$$= \lim_{x \to 0} \dfrac{\dfrac{x^4}{4}\left(1 + \dfrac{x}{3} + \dfrac{x^2}{12} + \cdots\right)^2}{\dfrac{x^4}{2} - \dfrac{x^6}{3} + \dfrac{x^8}{4} - \cdots} = \dfrac{\left(\dfrac{1}{4}\right)}{\left(\dfrac{1}{2}\right)} = \frac{1}{2}.$$

24. We have

$$\lim_{x \to 0} \frac{\sin(\sin x) - x}{x[\cos(\sin x) - 1]}$$

$$= \lim_{x \to 0} \dfrac{\left(\sin x - \dfrac{1}{3!}\sin^3 x + \dfrac{1}{5!}\sin^5 x - \cdots\right) - x}{x\left[1 - \dfrac{1}{2!}\sin^2 x + \dfrac{1}{4!}\sin^4 x - \cdots - 1\right]}$$

$$= \lim_{x \to 0} \dfrac{\left(x - \dfrac{x^3}{3!} + \cdots\right) - \dfrac{1}{3!}\left(x - \dfrac{x^3}{3!} + \cdots\right)^3 + \dfrac{1}{5!}\left(x - \cdots\right)^5 - \cdots - x}{x\left[-\dfrac{1}{2!}\left(x - \dfrac{x^3}{3!} + \cdots\right)^2 + \dfrac{1}{4!}\left(x - \cdots\right)^4 - \cdots\right]}$$

$$= \lim_{x \to 0} \dfrac{-\dfrac{2}{3!}x^3 + \text{higher degree terms}}{-\dfrac{1}{2!}x^3 + \text{higher degree terms}} = \dfrac{\dfrac{2}{3!}}{\dfrac{1}{2!}} = \frac{2}{3}.$$

Section 9.8 Taylor's Formula Revisited (page 580)

2. Let $f(x) = \cos x$, then

$$f'(x) = -\sin x \quad f''(x) = -\cos x \quad f'''(x) = \sin x$$
$$f^{(4)}(x) = \cos x \quad f^{(5)}(x) = -\sin x \quad f^{(6)}(x) = -\cos x$$
$$f^{(7)}(x) = \sin x.$$

Since $P_6(1)$ is used to approximate $f(1)$, the error is

$$R_6(1) = \frac{f^{(7)}(X)}{7!}(1)^7 = \frac{\sin(X)}{7!}$$

for some X between 0 and 1. Since $1 < \dfrac{\pi}{3}$, we have

$$\sin X < \sin \frac{\pi}{3} = \frac{\sqrt{3}}{2}. \text{ Thus,}$$

$$|R_6(1)| < \frac{\sqrt{3}}{2 \cdot 7!} \approx 0.000172.$$

4. Let $f(x) = \sec x$, then

$$f'(x) = \sec x \tan x,$$
$$f''(x) = \sec x \tan^2 x + \sec^3 x,$$
$$f'''(x) = \sec x \tan^3 x + 5\sec^3 x \tan x.$$

If $P_2(0.2)$ is used to approximate $f(0.2)$, then

$$R_2(0.2) = \frac{f'''(X)}{3!}(0.2)^3$$

for some X between 0 and 0.2. Since $0.2 < \pi/6$,

$$0 \le \sec X \le \sec(0.2) < \sec\frac{\pi}{6} = \frac{2}{\sqrt{3}};$$
$$0 \le \tan X \le \tan(0.2) < \tan\frac{\pi}{6} = \frac{1}{\sqrt{3}}.$$

Thus,

$$|f'''(X)| = \sec X \tan^3 X + 5\sec^3 X \tan X$$

$$< \left(\frac{2}{\sqrt{3}}\right)\left(\frac{1}{\sqrt{3}}\right)^3 + 5\left(\frac{2}{\sqrt{3}}\right)^3\left(\frac{1}{\sqrt{3}}\right) = \frac{42}{9}$$

and

$$|R_2(0.2)| < \frac{\left(\dfrac{42}{9}\right)}{3!}(0.2)^3 \approx 0.0062.$$

6. If $f(x) = \tan^{-1}(x)$, then $f^{(4)}(x) = \dfrac{24(x - x^3)}{(x^2 + 1)^4}$. If $P_3(0.99)$ is used to approximate $f(0.99)$, the error will be

$$R_3(0.99) = \frac{f^{(4)}(X)}{4!}(0.99 - 1)^4$$

for some X between 0.99 and 1. For such X,

$$|F^{(4)}(x)| \le \frac{24[1 - (0.99)^3]}{[(0.99)^2 + 1]^4} \approx 0.0464.$$

Hence,

$$|R_3(0.99)| \le \frac{0.0464}{4!}(0.99-1)^4 = 1.93 \times 10^{-11}.$$

8. If $f(x) = e^{-x}$, then $f^{(k)}(x) = (-1)^k e^{-x}$ and $f^{(k)}(0) = (-1)^k$ for $k = 0, 1, 2, \ldots$. The Lagrange remainder in Taylor's Formula is

$$R_n(x) = \frac{f^{(n+1)}(X)}{(n+1)!}x^{(n+1)} = (-1)^{n+1}e^{-X}\frac{x^{(n+1)}}{(n+1)!}$$

for some X between 0 and x. Clearly,

$$|R_n(x)| \le e^{|x|}\frac{|x|^{(n+1)}}{(n+1)!} \to 0$$

as $n \to \infty$ for all real x. Thus,

$$e^{-x} = 1 - x + \frac{x^2}{2!} - \frac{x^3}{3!} + \cdots = \sum_{k=0}^{\infty} \frac{(-1)^k x^k}{k!}.$$

10. If $f(x) = \cos x$, then for $k = 0, 1, 2, \ldots$,

$$f^{(n)}(0) = \begin{cases} 1, & \text{if } n = 4k; \\ 0, & \text{if } n = 4k+1 \text{ or } 4k+3; \\ -1, & \text{if } n = 4k+2. \end{cases}$$

The Taylor Formula with Lagrange remainder is

$$f(x) = 1 - \frac{x^2}{2!} + \frac{x^4}{4!} + \cdots + (-1)^n\frac{x^{2n}}{2n!} + R_{2n}(x)$$

where

$$R_{2n}(x) = \frac{f^{(2n+1)}(X)}{(2n+1)!}x^{2n+1} = (-1)^{n+1}\sin X \frac{x^{2n+1}}{(2n+1)!}$$

for some X between 0 and x. Since

$$|R_{2n}(x)| < \frac{|x|^{2n+1}}{(2n+1)!} \to 0 \text{ as } n \to \infty,$$

therefore

$$\cos x = 1 - \frac{x^2}{2!} + \frac{x^4}{4!} - \frac{x^6}{6!} + \cdots = \sum_{n=0}^{\infty}\frac{(-1)^n x^{2n}}{(2n)!}$$

for all real x.

12. If $f(x) = \sin^2 x = \dfrac{1 - \cos 2x}{2}$, then, for $k = 0, 1, 2, \ldots$,

$$f^{(n)}(x) = \begin{cases} 2^{(n-1)}\sin 2x, & \text{if } n = 4k+1, \\ 2^{(n-1)}\cos 2x, & \text{if } n = 4k+2, \\ -2^{(n-1)}\sin 2x, & \text{if } n = 4k+3, \\ -2^{(n-1)}\cos 2x, & \text{if } n = 4k+4; \end{cases}$$

and

$$f^{(n)}(0) = \begin{cases} 2^{(n-1)}, & \text{if } n = 4k+2, \\ 0, & \text{if } n = 4k+1 \text{ or } 4k+3, \\ -2^{(n-1)}, & \text{if } n = 4k+4. \end{cases}$$

The Taylor Formula is

$$f(x) = \frac{2}{2!}x^2 - \frac{2^3}{4!}x^4 + \frac{2^5}{6!}x^6 - \frac{2^7}{8!}x^8 + \cdots +$$
$$(-1)^{n-1}\frac{2^{2n-1}}{(2n)!}x^{2n} + R_{2n}$$

where

$$R_{2n}(x) = \frac{f^{(2n+1)}(X)}{(2n+1)!}x^{2n+1} = (-1)^n \sin 2X \frac{x^{2n+1}}{(2n+1)!}$$

for some X between 0 and x. Since

$$|R_{2n}(x)| < \frac{|x|^{2n+1}}{(2n+1)!} \to 0 \text{ as } n \to \infty,$$

therefore

$$f(x) = \frac{2}{2!}x^2 - \frac{2^3}{4!}x^4 + \frac{2^5}{6!}x^6 - \cdots$$
$$= \sum_{n=1}^{\infty}\frac{(-1)^{n-1}2^{2n-1}}{(2n)!}x^{2n}.$$

14. If $f(x) = \ln(1+x)$, then

$$f'(x) = \frac{1}{1+x}, \quad f''(x) = \frac{-1}{(1+x)^2}, \quad f'''(x) = \frac{2}{(1+x)^3},$$
$$f^{(4)}(x) = \frac{-3!}{(1+x)^4}, \quad \ldots, \quad f^{(n)} = \frac{(-1)^{n-1}(n-1)!}{(1+x)^n}$$

and

$$f(0) = 0, \quad f'(0) = 1, \quad f''(0) = -1, \quad f'''(0) = 2,$$
$$f^{(4)}(0) = -3!, \quad \ldots, \quad f^{(n)}(0) = (-1)^{n-1}(n-1)!.$$

Therefore, the Taylor Formula is

$$f(x) = x + \frac{-1}{2!}x^2 + \frac{2}{3!}x^3 + \frac{-3!}{4!}x^4 + \cdots +$$
$$\frac{(-1)^{n-1}(n-1)!}{n!}x^n + R_n(x)$$

where

$$R_n(x) = \frac{1}{n!}\int_0^x (x-t)^n f^{(n+1)}(t)\, dt$$
$$= \frac{1}{n!}\int_0^x (x-t)^n \frac{(-1)^n n!}{(1+t)^{n+1}}\, dt$$
$$= (-1)^n \int_0^x \frac{(x-t)^n}{(1+t)^{n+1}}\, dt.$$

If $0 \leq t \leq x \leq 1$, then $1 + t \geq 1$ and

$$|R_n(x)| \leq \int_0^x (x-t)^n \, dt = \frac{x^{n+1}}{n+1} \leq \frac{1}{n+1} \to 0$$

as $n \to \infty$.
If $-1 < x \leq t \leq 0$, then

$$\left| \frac{x-t}{1+t} \right| = \frac{t-x}{1+t} \leq |x|,$$

because $\dfrac{t-x}{1+t}$ increases from 0 to $-x = |x|$ as t increases from x to 0. Thus,

$$|R_n(x)| < \frac{1}{1+x} \int_0^{|x|} |x|^n \, dt = \frac{|x|^{n+1}}{1+x} \to 0$$

as $n \to \infty$ since $|x| < 1$. Therefore,

$$f(x) = x - \frac{x^2}{2} + \frac{x^3}{3} - \frac{x^4}{4} + \cdots = \sum_{n=1}^{\infty} (-1)^{n-1} \frac{x^n}{n},$$

for $-1 < x \leq 1$.

16. If $f(x) = e^x$, then $f^{(k)}(x) = e^x$ and $f^{(k)}(a) = e^a$ for all k. The Lagrange remainder is

$$R_n(x) = \frac{f^{(n+1)}(X)}{(n+1)!}(x-a)^{n+1} = e^X \frac{(x-a)^{n+1}}{(n+1)!}$$

for some X between a and x. Since $e^{|X|} < e^{|x|}$, clearly,

$$|R_n(x)| < e^{|x|} \frac{|x-a|^{n+1}}{(n+1)!} \to 0 \text{ as } n \to \infty.$$

Therefore,

$$f(x) = e^a + e^a (x-a) + \frac{e^a}{2!}(x-a)^2 + \cdots = e^a \sum_{n=0}^{\infty} \frac{(x-a)^n}{n!}.$$

18. If $f(x) = \cos x$, then for $k = 0, 1, 2, \ldots$,

$$f^{(n)}(x) = \begin{cases} \cos x, & \text{if } n = 4k; \\ -\sin x, & \text{if } n = 4k+1; \\ -\cos x, & \text{if } n = 4k+2; \\ \sin x, & \text{if } n = 4k+3. \end{cases}$$

Therefore,

$$f^{(n)}\left(\frac{\pi}{4}\right) = \begin{cases} \dfrac{1}{\sqrt{2}}, & \text{if } n = 4k \text{ or } 4k+3; \\ -\dfrac{1}{\sqrt{2}}, & \text{if } n = 4k+1 \text{ or } 4k+2. \end{cases}$$

The Lagrange remainder is

$$|R_n(x)| = \left| \frac{f^{(n+1)}(X)}{(n+1)!}\left(x - \frac{\pi}{4}\right)^{n+1} \right| \leq \frac{\left(x - \dfrac{\pi}{4}\right)^{n+1}}{(n+1)!},$$

since $|f^{(n+1)}(X)| \leq 1$ for any X. Clearly, $|R_n(x)| \to 0$ as $n \to \infty$ for all real x. Hence,

$$f(x) = \frac{1}{\sqrt{2}} - \frac{1}{\sqrt{2}}\left(x - \frac{\pi}{4}\right) - \frac{1}{\sqrt{2}}\left(\frac{1}{2!}\right)\left(x - \frac{\pi}{4}\right)^2$$
$$+ \frac{1}{\sqrt{2}}\left(\frac{1}{3!}\right)\left(x - \frac{\pi}{4}\right)^3 + \cdots$$
$$= \frac{1}{\sqrt{2}} \sum_{n=0}^{\infty} (-1)^n \left[\frac{1}{(2n)!}\left(x - \frac{\pi}{4}\right)^{2n}\right.$$
$$\left. - \frac{1}{(2n+1)!}\left(x - \frac{\pi}{4}\right)^{2n+1}\right].$$

20. Let $f(x) = \ln x$, then $f^{(k)}(x) = (-1)^{k-1}(k-1)! x^{-k}$. The Taylor Formula in powers of $x - 2$ is

$$f(x) = \ln 2 + \frac{x-2}{2} - \frac{1}{2}\left(\frac{x-2}{2}\right)^2 + \frac{1}{3}\left(\frac{x-2}{2}\right)^3 + \cdots +$$
$$\frac{(-1)^{n-1}}{n}\left(\frac{x-2}{2}\right)^n + R_n(x)$$

where

$$R_n(x) = \frac{1}{n!} \int_2^x (x-t)^n (-1)^n (n!) t^{-(n+1)} \, dt$$
$$= (-1)^n \int_2^x \frac{(x-t)^n}{t^{n+1}} \, dt.$$

If $2 \leq t \leq x \leq 4$ then $\left| \dfrac{x-t}{t}\right|^n \leq \left(\dfrac{4-t}{2}\right)^n$, so

$$|R_n| \leq \frac{1}{2^{n+1}} \int_2^4 (4-t)^n \, dt = \frac{1}{n+1} \to 0$$

as $n \to \infty$.
If $0 < x \leq t \leq 2$ then

$$\left| \frac{x-t}{t} \right| = \frac{t-x}{t} \leq \frac{2-x}{2},$$

since $\dfrac{t-x}{t}$ increases from 0 to $\dfrac{2-x}{2}$ as t increases from x to 2. Therefore

$$|R_n| \leq \frac{1}{x}\left(\frac{2-x}{2}\right)^n \int_x^2 dt = \frac{(2-x)^{n+1}}{2^n x} \to 0$$

as $n \to \infty$ (since $2 - x < 2$). Therefore

$$\ln x = \ln 2 + \frac{x-2}{2} - \frac{1}{2}\left(\frac{x-2}{2}\right)^2 + \frac{1}{3}\left(\frac{x-2}{2}\right)^3 - \cdots$$

$$= \ln 2 + \sum_{n=1}^{\infty} \frac{(-1)^{n-1}}{n}\left(\frac{x-2}{2}\right)^n,$$

for $0 < x \le 4$.

Section 9.9 The Binomial Theorem and Binomial Series (page 584)

2. $x\sqrt{1-x} = x(1-x)^{1/2}$

$$= x - \frac{x^2}{2} + \frac{1}{2}\left(-\frac{1}{2}\right)\frac{(-1)^2 x^3}{2!}$$

$$+ \frac{1}{2}\left(-\frac{1}{2}\right)\left(-\frac{3}{2}\right)\frac{(-1)^3 x^4}{3!} + \cdots$$

$$= x - \frac{x^2}{2} - \sum_{n=2}^{\infty} \frac{1 \cdot 3 \cdot 5 \cdots (2n-3)}{2^n n!} x^{n+1}$$

$$= x - \frac{x^2}{2} - \sum_{n=2}^{\infty} (-1)^{n-1} \frac{(2n-2)!}{2^{2n-1}(n-1)!n!} x^{n+1} \quad (-1 < x < 1).$$

4.

$$\frac{1}{\sqrt{4+x^2}} = \frac{1}{2\sqrt{1+\left(\frac{x}{2}\right)^2}} = \frac{1}{2}\left[1 + \left(\frac{x}{2}\right)^2\right]^{-1/2}$$

$$= \frac{1}{2}\left[1 + \left(-\frac{1}{2}\right)\left(\frac{x}{2}\right)^2 + \frac{1}{2!}\left(-\frac{1}{2}\right)\left(-\frac{3}{2}\right)\left(\frac{x}{2}\right)^4 + \right.$$

$$\left. \frac{1}{3!}\left(-\frac{1}{2}\right)\left(-\frac{3}{2}\right)\left(-\frac{5}{2}\right)\left(\frac{x}{2}\right)^6 + \cdots\right]$$

$$= \frac{1}{2} - \frac{1}{2^4}x^2 + \frac{3}{2^7 2!}x^4 - \frac{3 \times 5}{2^{10} 3!}x^6 + \cdots$$

$$= \frac{1}{2} + \sum_{n=1}^{\infty} (-1)^n \frac{1 \times 2 \times 3 \times \cdots \times (2n-1)}{2^{3n+1} n!} x^{2n}$$

$$(-2 \le x \le 2).$$

6. $(1+x)^{-3} = 1 - 3x + \frac{(-3)(-4)}{2!}x^2 + \frac{(-3)(-4)(-5)}{3!}x^3 + \cdots$

$$= 1 - 3x + \frac{(3)(4)}{2}x^2 - \frac{(4)(5)}{2}x^3 + \cdots$$

$$= \sum_{n=0}^{\infty} (-1)^n \frac{(n+2)(n+1)}{2}x^n \quad (-1 < x < 1).$$

8. The formula $(a+b)^n = \sum_{k=0}^{n}\binom{n}{k}a^{n-k}b^k$

holds for $n = 1$; it says $a + b = a + b$ in this case. Suppose the formula holds for $n = m$, where m is some positive integer. Then

$$(a+b)^{m+1} = (a+b)\sum_{k=0}^{m}\binom{m}{k}a^{m-k}b^k$$

$$= \sum_{k=0}^{m}\binom{m}{k}a^{m+1-k}b^k + \sum_{k=0}^{m}\binom{m}{k}a^{m-k}b^{k+1}$$

(replace k by $k-1$ in the latter sum)

$$= \sum_{k=0}^{m}\binom{m}{k}a^{m+1-k}b^k + \sum_{k=1}^{m+1}\binom{m}{k-1}a^{m+1-k}b^k$$

$$= a^{m+1} + \sum_{k=1}^{m}\left[\binom{m}{k} + \binom{m}{k-1}\right]a^{m+1-k}b^k + b^{m+1}$$

(by #13(i))

$$= a^{m+1} + \sum_{k=1}^{m}\binom{m+1}{k}a^{m+1-k}b^k + b^{m+1} \quad \text{(by #13(ii))}$$

$$= \sum_{k=0}^{m+1}\binom{m+1}{k}a^{m+1-k}b^k \quad \text{(by #13(i) again)}.$$

Thus the formula holds for $n = m + 1$. By induction it holds for all positive integers n.

Section 9.10 Series Solutions of Differential Equations (page 589)

2. $y'' = xy$. Try $\sum_{n=0}^{\infty} a_n x^n$. Then

$$y' = \sum_{n=0}^{\infty} na_n x^{n-1} = \sum_{n=1}^{\infty} na_n x^{n-1}$$

$$y'' = \sum_{n=2}^{\infty} n(n-1)a_n x^{n-2} = \sum_{n=0}^{\infty} (n+2)(n+1)a_{n+2}x^n.$$

Thus we have

$$0 = y'' - xy$$

$$= \sum_{n=0}^{\infty} (n+2)(n+1)a_{n+2}x^n - \sum_{n=0}^{\infty} a_n x^{n+1}$$

$$= \sum_{n=0}^{\infty} (n+2)(n+1)a_{n+2}x^n - \sum_{n=1}^{\infty} a_{n-1}x^n$$

$$= 2a_2 + \sum_{n=1}^{\infty}\left[(n+2)(n+1)a_{n+2} - a_{n-1}\right]x^n.$$

Thus $a_2 = 0$ and $a_{n+2} = \dfrac{a_{n-1}}{(n+2)(n+1)}$ for $n \geq 1$. Given a_0 and a_1, we have

$$a_3 = \frac{a_0}{2 \times 3}$$

$$a_6 = \frac{a_3}{5 \times 6} = \frac{a_0}{2 \times 3 \times 5 \times 6} = \frac{1 \times 4 \times a_0}{6!}$$

$$a_9 = \frac{a_6}{8 \times 9} = \frac{1 \times 4 \times 7 \times a_0}{9!}$$

$$\vdots$$

$$a_{3n} = \frac{1 \times 4 \times \cdots \times (3n-2)a_0}{(3n)!}$$

$$a_4 = \frac{a_1}{3 \times 4} = \frac{2 \times a_1}{4!}$$

$$a_7 = \frac{a_4}{6 \times 7} = \frac{2 \times 5 \times a_1}{7!}$$

$$\vdots$$

$$a_{3n+1} = \frac{2 \times 5 \times \cdots \times (3n-1)a_1}{(3n+1)!}$$

$$0 = a_2 = a_5 = a_8 = \cdots = a_{3n+2}.$$

Thus the general solution of the given equation is

$$y = a_0 \left(1 + \sum_{n=1}^{\infty} \frac{1 \times 4 \times \cdots \times (3n-2)}{(3n)!} x^{3n} \right)$$

$$+ a_1 \sum_{n=1}^{\infty} \frac{2 \times 5 \times \cdots \times (3n-1)}{(3n+1)!} x^{3n+1}.$$

4. If $y = \displaystyle\sum_{n=0}^{\infty} a_n x^n$, then $y' = \sum_{n=1}^{\infty} n a_n x^{n-1}$ and

$$y'' = \sum_{n=2}^{\infty} n(n-1)a_n x^{n-2} = \sum_{n=0}^{\infty} (n+2)(n+1)a_{n+2} x^n.$$

Thus,

$$0 = y'' + xy' + y$$

$$= \sum_{n=0}^{\infty} (n+2)(n+1)a_{n+2}x^n + x \sum_{n=1}^{\infty} n a_n x^{n-1} + \sum_{n=0}^{\infty} a_n x^n$$

$$= 2a_2 + a_0 + \sum_{n=1}^{\infty} \left[(n+2)(n+1)a_{n+2} + (n+1)a_n \right] x^n.$$

Since coefficients of all powers of x must vanish, therefore $2a_2 + a_0 = 0$ and, for $n \geq 1$,

$$(n+2)(n+1)a_{n+2} + (n+1)a_n = 0,$$

that is, $a_{n+2} = \dfrac{-a_n}{n+2}$.

If $y(0) = 1$, then $a_0 = 1$, $a_2 = \dfrac{-1}{2}$, $a_4 = \dfrac{1}{2^2 \cdot 2!}$, $a_6 = \dfrac{-1}{2^3 \cdot 3!}$, $a_8 = \dfrac{1}{2^4 \cdot 4!}, \ldots$. If $y'(0) = 0$, then $a_1 = a_3 = a_5 = \cdots = 0$. Hence,

$$y = 1 - \frac{1}{2}x^2 + \frac{1}{8}x^4 - \frac{1}{48}x^6 + \cdots = \sum_{n=0}^{\infty} \frac{(-1)^n}{2^n \cdot n!} x^{2n}.$$

6. $(1-x^2)y'' - xy' + 9y = 0$, $y(0) = 0$, $y'(0) = 1$. Try

$$y = \sum_{n=0}^{\infty} a_n x^n.$$

Then $a_0 = 0$ and $a_1 = 1$. We have

$$y' = \sum_{n=1}^{\infty} n a_n x^{n-1}$$

$$y'' = \sum_{n=2}^{\infty} n(n-1)a_n x^{n-2}$$

$$0 = (1-x^2)y'' - xy' + 9y$$

$$= \sum_{n=0}^{\infty} (n+2)(n+1)a_{n+2}x^n - \sum_{n=2}^{\infty} n(n-1)a_n x^n$$

$$- \sum_{n=1}^{\infty} n a_n x^n + 9 \sum_{n=0}^{\infty} a_n x^n$$

$$= 2a_2 + 9a_0 + (6a_3 + 8a_1)x$$

$$+ \sum_{n=2}^{\infty} \left[(n+2)(n+1)a_{n+2} - (n^2 - 9)a_n \right] x^n.$$

Thus $2a_2 + 9a_0 = 0$, $6a_3 + 8a_1 = 0$, and

$$a_{n+2} = \frac{(n^2 - 9)a_n}{(n+1)(n+2)}.$$

Therefore we have

$$a_2 = a_4 = a_6 = \cdots = 0$$

$$a_3 = -\frac{4}{3}, \quad a_5 = 0 = a_7 = a_9 = \cdots.$$

The initial-value problem has solution

$$y = x - \frac{4}{3}x^3.$$

8. $xy'' + y' + xy = 0$.
Since $x = 0$ is a regular singular point of this equation, try

$$y = \sum_{n=0}^{\infty} a_n x^{n+\mu} \qquad (a_0 = 1)$$

$$y' = \sum_{n=0}^{\infty} (n+\mu)a_n x^{n+\mu-1}$$

$$y'' = \sum_{n=0}^{\infty} (n+\mu)(n+\mu-1)a_n x^{n+\mu-2}.$$

Then we have

$$0 = xy'' + y' + xy$$

$$= \sum_{n=0}^{\infty}\big[(n+\mu)(n+\mu-1)+(n+\mu)\big]a_n x^{n+\mu-1}$$

$$+ \sum_{n=0}^{\infty} a_n x^{n+\mu+1}$$

$$= \sum_{n=0}^{\infty}(n+\mu)^2 a_n x^{n+\mu-1} + \sum_{n=2}^{\infty} a_{n-2}x^{n+\mu-1}$$

$$= \mu^2 x^{\mu-1} + (1+\mu)^2 a_1 x^{\mu}$$

$$+ \sum_{n=2}^{\infty}\big[(n+\mu)^2 a_n + a_{n-2}\big]x^{n+\mu-1}.$$

Thus $\mu = 0$, $a_1 = 0$, and $a_n = -\dfrac{a_{n-2}}{n^2}$ for $n \geq 2$.
It follows that $0 = a_1 = a_3 = a_5 = \cdots$, and, since $a_0 = 1$,

$$a_2 = -\frac{1}{2^2}; \qquad a_4 = \frac{1}{2^2 4^2}, \ \cdots$$

$$a_{2n} = \frac{(-1)^n}{2^2 4^2 \cdots (2n)^2} = \frac{(-1)^n}{2^{2n}(n!)^2}.$$

One series solution is

$$y = 1 + \sum_{n=1}^{\infty} \frac{(-1)^n x^{2n}}{2^{2n}(n!)^2}.$$

Review Exercises 9   (page 589)

2. $\lim\limits_{n\to\infty} \dfrac{n^{100} + 2^n \pi}{2^n} = \lim\limits_{n\to\infty}\left(\pi + \dfrac{n^{100}}{2^n}\right) = \pi.$
The sequence converges.

4. $\lim\limits_{n\to\infty} \dfrac{(-1)^n n^2}{\pi n(n-\pi)} = \lim\limits_{n\to\infty} \dfrac{(-1)^n}{1-(\pi/n)}$ does not exist.
The sequence diverges (oscillates).

6. By l'Hôpital's Rule,

$$\lim_{x\to\infty} \frac{\ln(x+1)}{\ln x} = \lim_{x\to\infty} \frac{1/(x+1)}{1/x} = \lim_{x\to\infty} \frac{x}{x+1} = 1.$$

Thus

$$\lim_{n\to\infty}\big(\ln\ln(n+1) - \ln\ln n\big) = \lim_{n\to\infty} \ln \frac{\ln(n+1)}{\ln n} = \ln 1 = 0.$$

8. $\displaystyle\sum_{n=0}^{\infty} \frac{4^{n-1}}{(\pi-1)^{2n}} = \frac{1}{4}\sum_{n=0}^{\infty}\left(\frac{4}{(\pi-1)^2}\right)^n$

$$= \frac{1}{4}\cdot\frac{1}{1-\dfrac{4}{(\pi-1)^2}} = \frac{(\pi-1)^2}{4(\pi-1)^2-16},$$

since $(\pi-1)^2 > 4$.

10. $\displaystyle\sum_{n=1}^{\infty} \frac{1}{n^2 - \frac{9}{4}} = \sum_{n=1}^{\infty} \frac{1}{3}\left(\frac{1}{n-\frac{3}{2}} - \frac{1}{n+\frac{3}{2}}\right)$ (telescoping)

$$= \frac{1}{3}\Bigg[\frac{1}{-1/2} - \frac{1}{5/2} + \frac{1}{1/2} - \frac{1}{7/2}$$

$$+ \frac{1}{3/2} - \frac{1}{9/2} + \frac{1}{5/2} - \frac{1}{11/2} + \cdots\Bigg]$$

$$= \frac{1}{3}\left[-2 + 2 + \frac{2}{3}\right] = \frac{2}{9}.$$

12. $\displaystyle\sum_{n=1}^{\infty} \frac{n+2^n}{1+3^n}$ converges by comparison with the convergent

geometric series $\displaystyle\sum_{n=1}^{\infty}\left(\frac{2}{3}\right)^n$ because

$$\lim_{n\to\infty} \frac{\dfrac{n+2^n}{1+3^n}}{(2/3)^n} = \lim_{n\to\infty} \frac{(n/2^n)+1}{(1/3^n)+1} = 1.$$

14. $\displaystyle\sum_{n=1}^{\infty} \frac{n^2}{(1+2^n)(1+n\sqrt{n})}$ converges by comparison with the

convergent series $\displaystyle\sum_{n=1}^{\infty} \frac{\sqrt{n}}{2^n}$ (which converges by the ratio

test) because

$$\lim_{n\to\infty} \frac{\dfrac{n^2}{(1+2^n)(1+n\sqrt{n})}}{\dfrac{\sqrt{n}}{2^n}} = \lim_{n\to\infty} \frac{1}{\left(\dfrac{1}{2^n}+1\right)\left(\dfrac{1}{n^{3/2}}+1\right)} = 1.$$

16. $\displaystyle\sum_{n=1}^{\infty} \frac{n!}{(n+2)!+1}$ converges by comparison with the con-

vergent p-series $\displaystyle\sum_{n=1}^{\infty} \frac{1}{n^2}$, because

$$0 \leq \frac{n!}{(n+2)!+1} < \frac{n!}{(n+2)!} = \frac{1}{(n+2)(n+1)} < \frac{1}{n^2}.$$

18. $\displaystyle\sum_{n=1}^{\infty}\frac{(-1)^n}{2^n-n}$ converges absolutely by comparison with the

convergent geometric series $\displaystyle\sum_{n=1}^{\infty}\frac{1}{2^n}$, because

$$\lim_{n\to\infty}\frac{\left|\dfrac{(-1)^n}{2^n-n}\right|}{\dfrac{1}{2^n}}=\lim_{n\to\infty}\frac{1}{1-\dfrac{n}{2^n}}=1.$$

20. $\displaystyle\sum_{n=1}^{\infty}\frac{n^2\cos(n\pi)}{1+n^3}$ converges by the alternating series test
(note that $\cos(n\pi)=(-1)^n$), but the convergence is only
conditional because

$$\left|\frac{n^2\cos(n\pi)}{1+n^3}\right|=\frac{n^2}{1+n^3}\ge\frac{1}{2n}$$

for $n\ge 1$, and $\displaystyle\sum_{n=1}^{\infty}\frac{1}{2n}$ is a divergent harmonic series.

22. $\displaystyle\lim_{n\to\infty}\left|\frac{\dfrac{(5-2x)^{n+1}}{n+1}}{\dfrac{(5-2x)^n}{n}}\right|=\lim_{n\to\infty}|5-2x|\frac{n}{n+1}=|5-2x|.$

$\displaystyle\sum_{n=1}^{\infty}\frac{(5-2x)^n}{n}$ converges absolutely if $|5-2x|<1$, that is,
if $2<x<3$, and diverges if $x<2$ or $x>3$.

If $x=2$ the series is $\displaystyle\sum\frac{1}{n}$, which diverges.

If $x=3$ the series is $\displaystyle\sum\frac{(-1)^n}{n}$, which converges conditionally.

24. Let $\displaystyle s=\sum_{k=1}^{\infty}\frac{1}{4+k^2}$ and $\displaystyle s_n=\sum_{k=1}^{n}\frac{1}{4+k^2}$. Then

$$\int_{n+1}^{\infty}\frac{dt}{4+t^2}<s-s_n<\int_{n}^{\infty}\frac{dt}{4+t^2}$$

$$s_n+\frac{\pi}{4}-\frac{1}{2}\tan^{-1}\frac{n+1}{2}<s<s_n+\frac{\pi}{4}-\frac{1}{2}\tan^{-1}\frac{n}{2}.$$

Let

$$s_n^*=s_n+\frac{\pi}{4}-\frac{1}{4}\left[\tan^{-1}\frac{n+1}{2}+\tan^{-1}\frac{n}{2}\right].$$

Then $s\approx s_n^*$ with error satisfying

$$|s-s_n^*|<\frac{1}{4}\left[\tan^{-1}\frac{n+1}{2}-\tan^{-1}\frac{n}{2}\right].$$

This error is less than 0.001 if $n\ge 22$. Hence

$$s\approx\sum_{k=1}^{22}\frac{1}{4+k^2}+\frac{\pi}{4}-\frac{1}{4}\left[\tan^{-1}\frac{23}{2}+\tan^{-1}(11)\right]\approx 0.6605$$

with error less than 0.001.

26. Replace x with x^2 in Exercise 25 and multiply by x to get

$$\frac{x}{3-x^2}=\sum_{n=0}^{\infty}\frac{x^{2n+1}}{3^{n+1}}\quad(-\sqrt{3}<x<\sqrt{3}).$$

28. $\displaystyle\frac{1-e^{-2x}}{x}=\frac{1}{x}\left(1-1-\sum_{n=1}^{\infty}\frac{(-2x)^n}{n!}\right)$

$$=\sum_{n=1}^{\infty}(-1)^{n-1}\frac{2^n x^{n-1}}{n!}\quad\text{(for all }x\ne 0\text{)}.$$

30. $\displaystyle\sin\left(x+\frac{\pi}{3}\right)=\sin x\cos\frac{\pi}{3}+\cos x\sin\frac{\pi}{3}$

$$=\frac{1}{2}\sum_{n=0}^{\infty}(-1)^n\frac{x^{2n+1}}{(2n+1)!}+\frac{\sqrt{3}}{2}\sum_{n=0}^{\infty}(-1)^n\frac{x^{2n}}{(2n)!}$$

$$=\sum_{n=0}^{\infty}\frac{(-1)^n}{2}\left(\frac{\sqrt{3}x^{2n}}{(2n)!}+\frac{x^{2n+1}}{(2n+1)!}\right)\quad\text{(for all }x\text{)}.$$

32. $\displaystyle(1+x)^{1/3}=1+\frac{1}{3}x+\frac{\left(\dfrac{1}{3}\right)\left(-\dfrac{2}{3}\right)}{2!}x^2$

$$+\frac{\left(\dfrac{1}{3}\right)\left(-\dfrac{2}{3}\right)\left(-\dfrac{5}{3}\right)}{3!}x^3+\cdots$$

$$=1+\frac{x}{3}+\sum_{n=2}^{\infty}(-1)^{n-1}\frac{2\cdot 5\cdot 8\cdots(3n-4)}{3^n n!}x^n\quad(-1<x<1).$$

(Remark: the series also converges at $x=1$.)

34. Let $u=x-(\pi/4)$, so $x=u+(\pi/4)$. Then

$$\sin x+\cos x=\sin\left(u+\frac{\pi}{4}\right)+\cos\left(u+\frac{\pi}{4}\right)$$

$$=\frac{1}{\sqrt{2}}\left((\sin u+\cos u)+(\cos u-\sin u)\right)$$

$$=\sqrt{2}\cos u=\sqrt{2}\sum_{n=0}^{\infty}(-1)^n\frac{u^{2n}}{(2n)!}$$

$$=\sqrt{2}\sum_{n=0}^{\infty}\frac{(-1)^n}{(2n)!}\left(x-\frac{\pi}{4}\right)^{2n}\quad\text{(for all }x\text{)}.$$

36. $\sin(1+x)=\sin(1)\cos x+\cos(1)\sin x$

$$=\sin(1)\left(1-\frac{x^2}{2!}+\cdots\right)+\cos(1)\left(x-\frac{x^3}{3!}+\cdots\right)$$

$$P_3(x)=\sin(1)+\cos(1)x-\frac{\sin(1)}{2}x^2-\frac{\cos(1)}{6}x^3.$$

38. $\sqrt{1 + \sin x} = 1 + \dfrac{1}{2}\sin x + \dfrac{\left(\dfrac{1}{2}\right)\left(-\dfrac{1}{2}\right)}{2!}(\sin x)^2$

$\quad + \dfrac{\left(\dfrac{1}{2}\right)\left(-\dfrac{1}{2}\right)\left(-\dfrac{3}{2}\right)}{3!}(\sin x)^3$

$\quad + \dfrac{\left(\dfrac{1}{2}\right)\left(-\dfrac{1}{2}\right)\left(-\dfrac{3}{2}\right)\left(-\dfrac{5}{2}\right)}{4!}(\sin x)^4 + \cdots$

$\quad = 1 + \dfrac{1}{2}\left(x - \dfrac{x^3}{6} + \cdots\right) - \dfrac{1}{8}\left(x - \dfrac{x^3}{6} + \cdots\right)^2$

$\quad + \dfrac{1}{16}(x - \cdots)^3 - \dfrac{5}{128}(x - \cdots)^4 + \cdots$

$\quad = 1 + \dfrac{x}{2} - \dfrac{x^3}{12} - \dfrac{x^2}{8} + \dfrac{x^4}{24} + \dfrac{x^3}{16} - \dfrac{5x^4}{128} + \cdots$

$\quad P_4(x) = 1 + \dfrac{x}{2} - \dfrac{x^2}{8} - \dfrac{x^3}{48} + \dfrac{x^4}{384}.$

40. Since

$$1 + \sum_{n=1}^{\infty} \frac{x^{2n}}{n^2} = \sum_{k=0}^{\infty} \frac{f^{(k)}(0)}{k!} x^k$$

for x near 0, we have, for $n = 1, 2, 3, \ldots$

$$f^{(2n)}(0) = \frac{(2n)!}{n^2}, \quad f^{(2n-1)}(0) = 0.$$

42. $\displaystyle\sum_{n=0}^{\infty} n x^n = \frac{x}{(1-x)^2}$ as in Exercise 23

$\displaystyle\sum_{n=0}^{\infty} n^2 x^{n-1} = \frac{d}{dx} \frac{x}{(1-x)^2} = \frac{1+x}{(1-x)^3}$

$\displaystyle\sum_{n=0}^{\infty} n^2 x^n = \frac{x(1+x)}{(1-x)^3}$

$\displaystyle\sum_{n=0}^{\infty} \frac{n^2}{\pi^n} = \frac{\dfrac{1}{\pi}\left(1 + \dfrac{1}{\pi}\right)}{\left(1 - \dfrac{1}{\pi}\right)^3} = \frac{\pi(\pi + 1)}{(\pi - 1)^3}.$

44. $\displaystyle\sum_{n=1}^{\infty} \frac{(-1)^{n-1} x^{2n-1}}{(2n-1)!} = \sin x$

$\displaystyle\sum_{n=1}^{\infty} \frac{(-1)^n \pi^{2n-1}}{(2n-1)!} = -\sin \pi = 0$

$\displaystyle\sum_{n=2}^{\infty} \frac{(-1)^n \pi^{2n-4}}{(2n-1)!} = \frac{1}{\pi^3}\left(0 - \frac{(-1)\pi}{1!}\right) = \frac{1}{\pi^2}.$

46. $\displaystyle\lim_{x \to 0} \frac{(x - \tan^{-1} x)(e^{2x} - 1)}{2x^2 - 1 + \cos(2x)}$

$\quad = \displaystyle\lim_{x \to 0} \frac{\left(x - x + \dfrac{x^3}{3} - \dfrac{x^5}{5} + \cdots\right)\left(2x + \dfrac{4x^2}{2!} + \cdots\right)}{2x^2 - 1 + 1 - \dfrac{4x^2}{2!} + \dfrac{16x^4}{4!} - \cdots}$

$\quad = \displaystyle\lim_{x \to 0} \frac{x^4\left(\dfrac{2}{3} + \cdots\right)}{x^4\left(\dfrac{2}{3} + \cdots\right)} = 1.$

48. If $f(x) = \ln(\sin x)$, then calculation of successive derivatives leads to

$$f^{(5)}(x) = 24 \csc^4 x \cot x - 8 \csc^2 \cot x.$$

Observe that $1.5 < \pi/2 \approx 1.5708$, that $\csc x \geq 1$ and $\cot x \geq 0$, and that both functions are decreasing on that interval. Thus

$$|f^{(5)}(x)| \leq 24 \csc^4(1.5) \cot(1.5) \leq 2$$

for $1.5 \leq x \leq \pi/2$. Therefore, the error in the approximation

$$\ln(\sin 1.5) \approx P_4(x),$$

where P_4 is the 4th degree Taylor polynomial for $f(x)$ about $x = \pi/2$, satisfies

$$|\text{error}| \leq \frac{2}{5!}\left|1.5 - \frac{\pi}{2}\right|^5 \leq 3 \times 10^{-8}.$$

Challenging Problems 9   (page 590)

2. a) If $s_n = \sum_{k=1}^n v_k$ for $n \geq 1$, and $s_0 = 0$, then $v_k = s_k - s_{k-1}$ for $k \geq 1$, and

$$\sum_{k=1}^n u_k v_k = \sum_{k=1}^n u_k s_k - \sum_{k=1}^n u_k s_{k-1}.$$

In the second sum on the right replace k with $k + 1$:

$$\sum_{k=1}^n u_k v_k = \sum_{k=1}^n u_k s_k - \sum_{k=0}^{n-1} u_{k+1} s_k$$

$$= \sum_{k=1}^n (u_k - u_{k+1}) s_k - u_1 s_0 + u_{n+1} s_n$$

$$= u_{n+1} s_n + \sum_{k=1}^n (u_k - u_{k+1}) s_k.$$

b) If $\{u_n\}$ is positive and decreasing, and $\lim_{n \to \infty} u_n = 0$, then

$$\sum_{k=1}^n (u_k - u_{k+1}) = u_1 - u_2 + u_2 - u_3 + \cdots + u_n - u_{n+1}$$

$$= u_1 - u_{n+1} \to u_1 \text{ as } n \to \infty.$$

Thus $\sum_{k=1}^{n}(u_k - u_{k+1})$ is a convergent, positive, telescoping series.

If the partial sums s_n of $\{v_n\}$ are bounded, say $|s_n| \le K$ for all n, then

$$|(u_n - u_{n+1})s_n| \le K(u_n - u_{n+1}),$$

so $\sum_{n=1}^{\infty}(u_n - u_{n+1})s_n$ is absolutely convergent (and therefore convergent) by the comparison test. Therefore, by part (a),

$$\sum_{k=1}^{\infty} u_k v_k = \lim_{n\to\infty}\left(u_{n+1}s_n + \sum_{k=1}^{n}(u_k - u_{k+1})s_k\right)$$

$$= \sum_{k=1}^{\infty}(u_k - u_{k+1})s_k$$

converges.

4. Let a_n be the nth integer that has no zeros in its decimal representation. The number of such integers that have m digits is 9^m. (There are nine possible choices for each of the m digits.) Also, each such m-digit number is greater than 10^{m-1} (the smallest m-digit number). Therefore the sum of all the terms $1/a_n$ for which a_n has m digits is less than $9^m/(10^{m-1})$. Therefore,

$$\sum_{n=1}^{\infty}\frac{1}{a_n} < 9\sum_{m=1}^{\infty}\left(\frac{9}{10}\right)^{m-1} = 90.$$

6. a) Since $e = \sum_{j=0}^{\infty}\frac{1}{j!}$, we have

$$0 < e - \sum_{j=0}^{n}\frac{1}{j!} = \sum_{j=n+1}^{\infty}\frac{1}{j!}$$

$$= \frac{1}{(n+1)!}\left(1 + \frac{1}{n+2} + \frac{1}{(n+2)(n+3)} + \cdots\right)$$

$$\le \frac{1}{(n+1)!}\left(1 + \frac{1}{n+2} + \frac{1}{(n+2)^2} + \cdots\right)$$

$$= \frac{1}{(n+1)!}\cdot\frac{1}{1 - \frac{1}{n+2}} = \frac{n+2}{(n+1)!(n+1)} < \frac{1}{n!n}.$$

The last inequality follows from $\dfrac{n+2}{(n+1)^2} < \dfrac{1}{n}$, that is, $n^2 + 2n < n^2 + 2n + 1$.

b) Suppose e is rational, say $e = M/N$ where M and N are positive integers. Then $N!e$ is an integer and $N!\sum_{j=0}^{N}(1/j!)$ is an integer (since each $j!$ is a factor of $N!$). Therefore the number

$$Q = N!\left(e - \sum_{j=0}^{N}\frac{1}{j!}\right)$$

is a difference of two integers and so is an integer.

c) By part (a), $0 < Q < \dfrac{1}{N} \le 1$. By part (b), Q is an integer. This is not possible; there are no integers between 0 and 1. Therefore e cannot be rational.

8. Let f be a polynomial and let

$$g(x) = \sum_{j=0}^{\infty}(-1)^j f^{(2j)}(x).$$

This "series" is really just a polynomial since sufficiently high derivatives of f are all identically zero.

a) By replacing j with $j - 1$, observe that

$$g''(x) = \sum_{j=0}^{\infty}(-1)^j f^{(2j+2)}(x)$$

$$= \sum_{j=1}^{\infty}(-1)^{j-1} f^{(2j)}(x) = -\big(g(x) - f(x)\big).$$

Also

$$\frac{d}{dx}\big(g'(x)\sin x - g(x)\cos x\big)$$
$$= g''(x)\sin x + g'(x)\cos x - g'(x)\cos x + g(x)\sin x$$
$$= \big(g''(x) + g(x)\big)\sin x = f(x)\sin x.$$

Thus

$$\int_0^{\pi} f(x)\sin x\,dx = \big(g'(x)\sin x - g(x)\cos x\big)\Big|_0^{\pi} = g(\pi) + g(0).$$

b) Suppose that $\pi = m/n$, where m and n are positive integers. Since $\lim_{k\to\infty} x^k/k! = 0$ for any x, there exists an integer k such that $(\pi m)^k/k! < 1/2$. Let

$$f(x) = \frac{x^k(m - nx)^k}{k!} = \frac{1}{k!}\sum_{j=0}^{k}\binom{k}{j}m^{k-j}(-n)^j x^{j+k}.$$

The sum is just the binomial expansion. For $0 < x < \pi = m/n$ we have

$$0 < f(x) < \frac{\pi^k m^k}{k!} < \frac{1}{2}.$$

Thus $0 < \int_0^{\pi} f(x)\sin x\,dx < \frac{1}{2}\int_0^{\pi}\sin x\,dx = 1$, and so $0 < g(\pi) + g(0) < 1$.

c) $f^{(i)}(x) = \dfrac{1}{k!} \displaystyle\sum_{j=0}^{k} \binom{k}{j} m^{k-j}(-n)^j$

$$\times (j+k)(j+k-1)\cdots(j+k-i+1)x^{j+k-i}$$

$$= \frac{1}{k!} \sum_{j=0}^{k} \binom{k}{j} m^{k-j}(-n)^j \frac{(j+k)!}{(j+k-i)!} x^{j+k-i}.$$

d) Evidently $f^{(i)}(0) = 0$ if $i < k$ or if $i > 2k$.
   If $k \le i \le 2k$, the only term in the sum for $f^{(i)}(0)$
   that is not zero is the term for which $j = i - k$. This
   term is the constant

$$\frac{1}{k!} \binom{k}{i-k} m^{k-j}(-n)^j \frac{i!}{0!}.$$

This constant is an integer because the binomial coefficient $\binom{k}{i-k}$ is an integer and $i!/k!$ is an integer.
(The other factors are also integers.) Hence $f^{(i)}(0)$ is
an integer, and so $g(0)$ is an integer.

e) Observe that $f(\pi - x) = f((m/n) - x) = f(x)$ for all
   x. Therefore $f^{(i)}(\pi)$ is an integer (for each i), and
   so $g(\pi)$ is an integer. Thus $g(\pi) + g(0)$ is an integer,
   which contradicts the conclusion of part (b). (There
   is no integer between 0 and 1.) Therefore, π cannot
   be rational.

CHAPTER 10. VECTORS AND COORDI-NATE GEOMETRY IN 3-SPACE

Section 10.1 Analytic Geometry in 3 Dimensions (page 599)

2. The distance between $(-1, -1, -1)$ and $(1, 1, 1)$ is

$$\sqrt{(1+1)^2 + (1+1)^2 + (1+1)^2} = 2\sqrt{3} \text{ units.}$$

4. The distance between $(3, 8, -1)$ and $(-2, 3, -6)$ is

$$\sqrt{(-2-3)^2 + (3-8)^2 + (-6+1)^2} = 5\sqrt{3} \text{ units.}$$

6. If $A = (1, 2, 3)$, $B = (4, 0, 5)$, and $C = (3, 6, 4)$, then

$$|AB| = \sqrt{3^2 + (-2)^2 + 2^2} = \sqrt{17}$$
$$|AC| = \sqrt{2^2 + 4^2 + 1^2} = \sqrt{21}$$
$$|BC| = \sqrt{(-1)^2 + 6^2 + (-1)^2} = \sqrt{38}.$$

Since $|AB|^2 + |AC|^2 = 17 + 21 = 38 = |BC|^2$, the triangle ABC has a right angle at A.

8. If $A = (1, 2, 3)$, $B = (1, 3, 4)$, and $C = (0, 3, 3)$, then

$$|AB| = \sqrt{(1-1)^2 + (3-2)^2 + (4-3)^2} = \sqrt{2}$$
$$|AC| = \sqrt{(0-1)^2 + (3-2)^2 + (3-3)^2} = \sqrt{2}$$
$$|BC| = \sqrt{(0-1)^2 + (3-3)^2 + (3-4)^2} = \sqrt{2}.$$

All three sides being equal, the triangle is equilateral.

10. The distance from the origin to $(1, 1, 1, \ldots, 1)$ in $\mathbb{R}^n$ is

$$\sqrt{1^2 + 1^2 + 1^2 + \cdots + 1} = \sqrt{n} \text{ units.}$$

12. $z = 2$ is a plane, perpendicular to the z-axis at $(0, 0, 2)$.

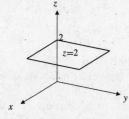

Fig. 10.1.12

14. $z = x$ is a plane containing the y-axis and making $45°$ angles with the positive directions of the x- and z-axes.

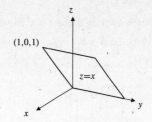

Fig. 10.1.14

16. $x^2 + y^2 + z^2 = 4$ is a sphere centred at the origin and having radius 2 (i.e., all points at distance 2 from the origin).

18. $x^2 + y^2 + z^2 = 2z$ can be rewritten

$$x^2 + y^2 + (z - 1)^2 = 1,$$

and so it represents a sphere with radius 1 and centre at $(0, 0, 1)$. It is tangent to the xy-plane at the origin.

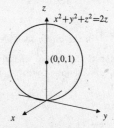

Fig. 10.1.18

20. $x^2 + z^2 = 4$ is a circular cylindrical surface of radius 2 with axis along the y-axis.

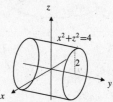

Fig. 10.1.20

22. $z \geq \sqrt{x^2 + y^2}$ represents every point whose distance above the xy-plane is not less than its horizontal distance from the z-axis. It therefore consists of all points inside and on a circular cone with axis along the positive z-axis, vertex at the origin, and semi-vertical angle $45°$.

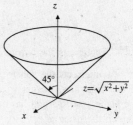

Fig. 10.1.22

24. $\begin{cases} x = 1 \\ y = 2 \end{cases}$ represents the vertical straight line in which the plane $x = 1$ intersects the plane $y = 2$.

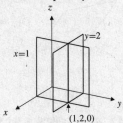

Fig. 10.1.24

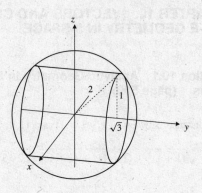

Fig. 10.1.28

26. $\begin{cases} x^2 + y^2 + z^2 = 4 \\ z = 1 \end{cases}$ is the circle in which the horizontal plane $z = 1$ intersects the sphere of radius 2 centred at the origin. The circle has centre $(0, 0, 1)$ and radius $\sqrt{4 - 1} = \sqrt{3}$.

30. $\begin{cases} y \geq x \\ z \leq y \end{cases}$ is the quarter-space consisting of all points lying on or on the same side of the planes $y = x$ and $z = y$ as does the point $(0, 1, 0)$.

32. $\begin{cases} x^2 + y^2 + z^2 \leq 1 \\ \sqrt{x^2 + y^2} \leq z \end{cases}$ represents all points which are inside or on the sphere of radius 1 centred at the origin and which are also inside or on the upper half of the circular cone with axis along the z-axis, vertex at the origin, and semi-vertical angle $45°$.

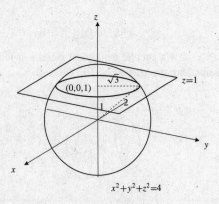

Fig. 10.1.26

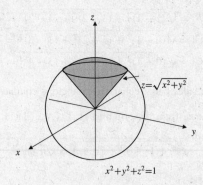

Fig. 10.1.32

34. $S = \{(x, y) : x \geq 0, \ y < 0\}$
The boundary of S consists of points $(x, 0)$ where $x \geq 0$, and points $(0, y)$ where $y \leq 0$.
The interior of S consists of all points of S that are not on the y-axis, that is, all points (x, y) satisfying $x > 0$ and $y < 0$.
S is neither open nor closed; it contains some, but not all, of its boundary points.
S is not bounded; $(x, -1)$ belongs to S for $0 < x < \infty$.

28. $\begin{cases} x^2 + y^2 + z^2 = 4 \\ x^2 + z^2 = 1 \end{cases}$ represents the two circles in which the cylinder $x^2 + z^2 - 1$ intersects the sphere $x^2 + y^2 + z^2 = 4$. Subtracting the two equations, we get $y^2 = 3$. Thus, one circle lies in the plane $y = \sqrt{3}$ and has centre $(0, \sqrt{3}, 0)$ and the other lies in the plane $y = -\sqrt{3}$ and has centre $(0, -\sqrt{3}, 0)$. Both circles have radius 1.

36. $S = \{(x, y) : |x| + |y| \le 1\}$
The boundary of S consists of all points on the edges of the square with vertices $(\pm 1, 0)$ and $(0, \pm 1)$.
The interior of S consists of all points inside that square. S is closed since it contains all its boundary points. It is bounded since all points in it are at distance not greater than 1 from the origin.

38. $S = \{(x, y, z) : x \ge 0, \ y > 1, \ z < 2\}$
Boundary: the quarter planes $x = 0$, $(y \ge 1, \ z \le 2)$, $y = 1$, $(x \ge 0, \ z \le 2)$, and $z = 2$, $(x \ge 0, \ y \ge 1)$.
Interior: the set of points (x, y, z) such that $x > 0$, $y > 1$, $z < 2$.
S is neither open nor closed.

40. $S = \{(x, y, z) : x^2 + y^2 < 1, \ y + z > 2\}$
Boundary: the part of the cylinder $x^2 + y^2 = 1$ that lies on or above the plane $y + z = 2$ together with the part of that plane that lies inside the cylinder.
Interior: all points that are inside the cylinder $x^2 + y^2 = 1$ and above the plane $y + z = 2$. S is open.

Section 10.2   Vectors   (page 609)

2. $\mathbf{u} = \mathbf{i} - \mathbf{j}$
$\mathbf{v} = \mathbf{j} + 2\mathbf{k}$

a) $\mathbf{u} + \mathbf{v} = \mathbf{i} + 2\mathbf{k}$
   $\mathbf{u} - \mathbf{v} = \mathbf{i} - 2\mathbf{j} - 2\mathbf{k}$
   $2\mathbf{u} - 3\mathbf{v} = 2\mathbf{i} - 5\mathbf{j} - 6\mathbf{k}$

b) $|\mathbf{u}| = \sqrt{1+1} = \sqrt{2}$
   $|\mathbf{v}| = \sqrt{1+4} = \sqrt{5}$

c) $\hat{\mathbf{u}} = \dfrac{1}{\sqrt{2}}(\mathbf{i} - \mathbf{j})$
   $\hat{\mathbf{v}} = \dfrac{1}{\sqrt{5}}(\mathbf{j} + 2\mathbf{k})$

d) $\mathbf{u} \bullet \mathbf{v} = 0 - 1 + 0 = -1$

e)   The angle between $\mathbf{u}$ and $\mathbf{v}$ is
   $\cos^{-1} \dfrac{-1}{\sqrt{10}} \approx 108.4°.$

f)   The scalar projection of $\mathbf{u}$ in the direction of $\mathbf{v}$ is
   $\dfrac{\mathbf{u} \bullet \mathbf{v}}{|\mathbf{v}|} = \dfrac{-1}{\sqrt{5}}.$

g)   The vector projection of $\mathbf{v}$ along $\mathbf{u}$ is
   $\dfrac{(\mathbf{v} \bullet \mathbf{u})\mathbf{u}}{|\mathbf{u}|^2} = -\dfrac{1}{2}(\mathbf{i} - \mathbf{j}).$

4. If $a = (-1, 1)$, $B = (2, 5)$ and $C = (10, -1)$, then $\overrightarrow{AB} = 3\mathbf{i} + 4\mathbf{j}$ and $\overrightarrow{BC} = 8\mathbf{i} - 6\mathbf{j}$. Since $\overrightarrow{AB} \bullet \overrightarrow{BC} = 0$, therefore, $\overrightarrow{AB} \perp \overrightarrow{BC}$. Hence, $\triangle ABC$ has a right angle at B.

6. We have
$$\overrightarrow{PQ} = \overrightarrow{PB} + \overrightarrow{BQ} = \tfrac{1}{2}\overrightarrow{AB} + \tfrac{1}{2}\overrightarrow{BC} = \tfrac{1}{2}\overrightarrow{AC};$$
$$\overrightarrow{SR} = \overrightarrow{SD} + \overrightarrow{DR} = \tfrac{1}{2}\overrightarrow{AD} + \tfrac{1}{2}\overrightarrow{DC} = \tfrac{1}{2}\overrightarrow{AC}.$$

Therefore, $\overrightarrow{PQ} = \overrightarrow{SR}$. Similarly,
$$\overrightarrow{QR} = \overrightarrow{QC} + \overrightarrow{CR} = \tfrac{1}{2}\overrightarrow{BD};$$
$$\overrightarrow{PS} = \overrightarrow{PA} + \overrightarrow{AS} = \tfrac{1}{2}\overrightarrow{BD}.$$

Therefore, $\overrightarrow{QR} = \overrightarrow{PS}$. Hence, $PQRS$ is a parallelogram.

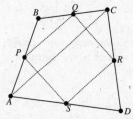

Fig. 10.2.6

8. Let X be the point of intersection of the medians AQ and BP as shown. We must show that CX meets AB in the midpoint of AB. Note that $\overrightarrow{PX} = \alpha \overrightarrow{PB}$ and $\overrightarrow{QX} = \beta \overrightarrow{QA}$ for certain real numbers α and β. Then
$$\overrightarrow{CX} = \frac{1}{2}\overrightarrow{CB} + \beta \overrightarrow{QA} = \frac{1}{2}\overrightarrow{CB} + \beta\left(\frac{1}{2}\overrightarrow{CB} + \overrightarrow{BA}\right)$$
$$= \frac{1+\beta}{2}\overrightarrow{CB} + \beta \overrightarrow{BA};$$
$$\overrightarrow{CX} = \frac{1}{2}\overrightarrow{CA} + \alpha \overrightarrow{PB} = \frac{1}{2}\overrightarrow{CA} + \alpha\left(\frac{1}{2}\overrightarrow{CA} + \overrightarrow{AB}\right)$$
$$= \frac{1+\alpha}{2}\overrightarrow{CA} + \alpha \overrightarrow{AB}.$$

Thus,
$$\frac{1+\beta}{2}\overrightarrow{CB} + \beta \overrightarrow{BA} = \frac{1+\alpha}{2}\overrightarrow{CA} + \alpha \overrightarrow{AB}$$
$$(\beta + \alpha)\overrightarrow{BA} = \frac{1+\alpha}{2}\overrightarrow{CA} - \frac{1+\beta}{2}\overrightarrow{CB}$$
$$(\beta + \alpha)(\overrightarrow{CA} - \overrightarrow{CB}) = \frac{1+\alpha}{2}\overrightarrow{CA} - \frac{1+\beta}{2}\overrightarrow{CB}$$
$$\left(\beta + \alpha - \frac{1+\alpha}{2}\right)\overrightarrow{CA} = \left(\beta + \alpha - \frac{1+\beta}{2}\right)\overrightarrow{CB}.$$

Since $\overrightarrow{CA}$ is not parallel to $\overrightarrow{CB}$,
$$\beta + \alpha - \frac{1+\alpha}{2} = \beta + \alpha - \frac{1+\beta}{2} = 0$$
$$\Rightarrow \alpha = \beta = \frac{1}{3}.$$

Since $\alpha = \beta$, x divides AQ and BP in the same ratio. By symmetry, the third median CM must also divide the other two in this ratio, and so must pass through X and $MX = \tfrac{1}{3}MC$.

201

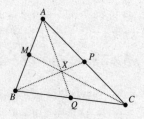

Fig. 10.2.8

10. Let the x-axis point east and the y-axis north. The velocity of the water is

$$\mathbf{v}_{\text{water}} = 3\mathbf{i}.$$

If you row through the water with speed 5 in the direction making angle θ west of north, then your velocity relative to the water will be

$$\mathbf{v}_{\text{boat rel water}} = -5\sin\theta\mathbf{i} + 5\cos\theta\mathbf{j}.$$

Therefore, your velocity relative to the land will be

$$\mathbf{v}_{\text{boat rel land}} = \mathbf{v}_{\text{boat rel water}} + \mathbf{v}_{\text{water}}$$
$$= (3 - 5\sin\theta)\mathbf{i} + 5\cos\theta\mathbf{j}.$$

To make progress in the direction $\mathbf{j}$, choose θ so that $3 = 5\sin\theta$. Thus $\theta = \sin^{-1}(3/5) \approx 36.87°$. In this case, your actual speed relative to the land will be

$5\cos\theta = \dfrac{4}{5} \times 5 = 4$ km/h.

To row from A to B, head in the direction $36.87°$ west of north. The 1/2 km crossing will take $(1/2)/4 = 1/8$ of an hour, or about $7\frac{1}{2}$ minutes.

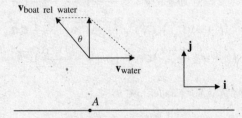

Fig. 10.2.10

12. Let $\mathbf{i}$ point east and $\mathbf{j}$ point north. If the aircraft heads in a direction θ north of east, then its velocity relative to the air is

$$750\cos\theta\mathbf{i} + 750\sin\theta\mathbf{j}.$$

The velocity of the air relative to the ground is

$$-\frac{100}{\sqrt{2}}\mathbf{i} + -\frac{100}{\sqrt{2}}\mathbf{j}.$$

Thus the velocity of the aircraft relative to the ground is

$$\left(750\cos\theta - \frac{100}{\sqrt{2}}\right)\mathbf{i} + \left(750\sin\theta - \frac{100}{\sqrt{2}}\right)\mathbf{j}.$$

If this velocity is true easterly, then

$$750\sin\theta = \frac{100}{\sqrt{2}},$$

so $\theta \approx 5.41°$. The speed relative to the ground is

$$750\cos\theta - \frac{100}{\sqrt{2}} \approx 675.9 \text{ km/h}.$$

The time for the 1500 km trip is $\dfrac{1500}{675.9} \approx 2.22$ hours.

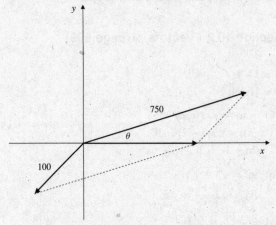

Fig. 10.2.12

14. The cube with edges $\mathbf{i}$, $\mathbf{j}$, and $\mathbf{k}$ has diagonal $\mathbf{i}+\mathbf{j}+\mathbf{k}$. The angle between $\mathbf{i}$ and the diagonal is

$$\cos^{-1}\frac{\mathbf{i}\bullet(\mathbf{i}+\mathbf{j}+\mathbf{k})}{\sqrt{3}} = \cos^{-1}\frac{1}{\sqrt{3}} \approx 54.7°.$$

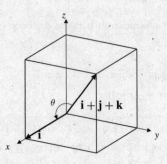

Fig. 10.2.14

16. If $\mathbf{u} = u_1\mathbf{i} + u_2\mathbf{j} + u_3\mathbf{k}$, then $\cos\alpha \dfrac{\mathbf{u}\bullet\mathbf{i}}{|\mathbf{u}|} = \dfrac{u_1}{|\mathbf{u}|}$.

Similarly, $\cos\beta = \dfrac{u_2}{|\mathbf{u}|}$ and $\cos\gamma = \dfrac{u_3}{|\mathbf{u}|}$.

Thus, the unit vector in the direction of $\mathbf{u}$ is

$$\hat{\mathbf{u}} = \frac{\mathbf{u}}{|\mathbf{u}|} = \cos\alpha\,\mathbf{i} + \cos\beta\,\mathbf{j} + \cos\gamma\,\mathbf{k},$$

and so $\cos^2\alpha + \cos^2\beta + \cos^2\gamma = |\hat{\mathbf{u}}|^2 = 1$.

18. If $A = (1,0,0)$, $B = (0,2,0)$, and $C = (0,0,3)$, then

$$\angle ABC = \cos^{-1}\frac{\overrightarrow{BA}\bullet\overrightarrow{BC}}{|BA||BC|} = \cos^{-1}\frac{4}{\sqrt{5}\sqrt{13}} \approx 60.26°$$

$$\angle BCA = \cos^{-1}\frac{\overrightarrow{CB}\bullet\overrightarrow{CA}}{|CB||CA|} = \cos^{-1}\frac{9}{\sqrt{10}\sqrt{13}} \approx 37.87°$$

$$\angle CAB = \cos^{-1}\frac{\overrightarrow{AC}\bullet\overrightarrow{AB}}{|AC||AB|} = \cos^{-1}\frac{1}{\sqrt{10}\sqrt{5}} \approx 81.87°.$$

20. If $\mathbf{a} \neq \mathbf{0}$, then $\mathbf{a}\bullet\mathbf{r} = 0$ implies that the position vector $\mathbf{r}$ is perpendicular to $\mathbf{a}$. Thus the equation is satisfied by all points on the plane through the origin that is normal (perpendicular) to $\mathbf{a}$.

In Exercises 22–24, $\mathbf{u} = 2\mathbf{i} + \mathbf{j} - 2\mathbf{k}$, $\mathbf{v} = \mathbf{i} + 2\mathbf{j} - 2\mathbf{k}$, and $\mathbf{w} = 2\mathbf{i} - 2\mathbf{j} + \mathbf{k}$.

22. Vector $\mathbf{x} = x\mathbf{i} + y\mathbf{j} + z\mathbf{k}$ is perpendicular to both $\mathbf{u}$ and $\mathbf{v}$ if

$$\mathbf{u}\bullet\mathbf{x} = 0 \quad\Leftrightarrow\quad 2x + y - 2z = 0$$
$$\mathbf{v}\bullet\mathbf{x} = 0 \quad\Leftrightarrow\quad x + 2y - 2z = 0.$$

Subtracting these equations, we get $x - y = 0$, so $x = y$. The first equation now gives $3x = 2z$. Now $\mathbf{x}$ is a unit vector if $x^2 + y^2 + z^2 = 1$, that is, if $x^2 + x^2 + \frac{9}{4}x^2 = 1$, or $x = \pm 2/\sqrt{17}$. The two unit vectors are

$$\mathbf{x} = \pm\left(\frac{2}{\sqrt{17}}\mathbf{i} + \frac{2}{\sqrt{17}}\mathbf{j} + \frac{3}{\sqrt{17}}\mathbf{k}\right).$$

24. Since $\mathbf{u}$, $\mathbf{v}$, and $\mathbf{w}$ all have the same length (3), a vector $\mathbf{x} = x\mathbf{i} + y\mathbf{j} + z\mathbf{k}$ will make equal angles with all three if it has equal dot products with all three, that is, if

$$2x + y - 2z = x + 2y - 2z \quad\Leftrightarrow\quad x = y = 0$$
$$2x + y - 2z = 2x - 2y + z \quad\Leftrightarrow\quad 3y - 3z = 0.$$

Thus $x = y = z$. Two unit vectors satisfying this condition are

$$\mathbf{x} = \pm\left(\frac{1}{\sqrt{3}}\mathbf{i} + \frac{1}{\sqrt{3}}\mathbf{j} + \frac{1}{\sqrt{3}}\mathbf{k}\right).$$

26. If $\mathbf{u}$ and $\mathbf{v}$ are not parallel, then neither is the zero vector, and the origin and the two points with position vectors $\mathbf{u}$ and $\mathbf{v}$ lie on a unique plane. The equation $\mathbf{r} = \lambda\mathbf{u} + \mu\mathbf{v}$ (λ, μ real) gives the position vector of an arbitrary point on that plane.

28. a) $\mathbf{u}$, $\mathbf{v}$, and $\mathbf{u} + \mathbf{v}$ are the sides of a triangle. The triangle inequality says that the length of one side cannot exceed the sum of the lengths of the other two sides.

   b) If $\mathbf{u}$ and $\mathbf{v}$ are parallel and point in the *same direction*, (or if at least one of them is the zero vector), then $|\mathbf{u} + \mathbf{v}| = |\mathbf{u}| + |\mathbf{v}|$.

30. Suppose $|\mathbf{u}| = |\mathbf{v}| = |\mathbf{w}| = 1$, and $\mathbf{u}\bullet\mathbf{v} = \mathbf{u}\bullet\mathbf{w} = \mathbf{v}\bullet\mathbf{w} = 0$, and let $\mathbf{r} = a\mathbf{u} + b\mathbf{v} + w\mathbf{w}$. Then

$$\mathbf{r}\bullet\mathbf{u} = a\mathbf{u}\bullet\mathbf{u} + b\mathbf{v}\bullet\mathbf{u} + c\mathbf{w}\bullet\mathbf{u} = a|\mathbf{u}|^2 + 0 + 0 = a.$$

Similarly, $\mathbf{r}\bullet\mathbf{v} = b$ and $\mathbf{r}\bullet\mathbf{w} = c$.

32. Let $\hat{\mathbf{n}}$ be a unit vector that is perpendicular to $\mathbf{u}$ and lies in the plane containing the origin and the points U, V, and P. Then $\hat{\mathbf{u}} = \mathbf{u}/|\mathbf{u}|$ and $\hat{\mathbf{n}}$ constitute a standard basis in that plane, so each of the vectors $\mathbf{v}$ and $\mathbf{r}$ can be expressed in terms of them:

$$\mathbf{v} = s\hat{\mathbf{u}} + t\hat{\mathbf{n}}$$
$$\mathbf{r} = x\hat{\mathbf{u}} + y\hat{\mathbf{n}}.$$

Since $\mathbf{v}$ is not parallel to $\mathbf{u}$, we have $t \neq 0$. Thus $\hat{\mathbf{n}} = (1/t)(\mathbf{v} - s\hat{\mathbf{u}})$ and

$$\mathbf{r} = x\hat{\mathbf{u}} + \frac{y}{t}(\mathbf{v} - s\hat{\mathbf{u}}) = \lambda\mathbf{u} + \mu\mathbf{v},$$

where $\lambda = (tx - ys)/(t|\mathbf{u}|)$ and $\mu = y/t$.

34. The derivation of the equation of the hanging cable given in the text needs to be modified by replacing $\mathbf{W} = -\delta gs\mathbf{j}$ with $\mathbf{W} = -\delta gx\mathbf{j}$. Thus $T_v = \delta gx$, and the slope of the cable satisfies

$$\frac{dy}{dx} = \frac{\delta gx}{H} = ax$$

where $a = \delta g/H$. Thus

$$y = \frac{1}{2}ax^2 + C;$$

the cable hangs in a parabola.

36. The cable hangs along the curve $y = \dfrac{1}{a}\cosh(ax)$, and its length from the lowest point at $x = 0$ to the support tower at $x = 45$ m is 50 m. Thus

$$50 = \int_0^{45}\sqrt{1 + \sinh^2(ax)}\,dx = \frac{1}{a}\sinh(45a).$$

The equation $\sinh(45a) = 50a$ has approximate solution $a \approx 0.0178541$. The vertical distance between the lowest point on the cable and the support point is

$$\frac{1}{a}\big(\cosh(45a) - 1\big) \approx 19.07 \text{ m}.$$

Section 10.3   The Cross Product in 3-Space (page 618)

2. $(\mathbf{j} + 2\mathbf{k}) \times (-\mathbf{i} - \mathbf{j} + \mathbf{k}) = 3\mathbf{i} - 2\mathbf{j} + \mathbf{k}$

4. A vector perpendicular to the plane containing the three given points is

$$(-a\mathbf{i} + b\mathbf{j}) \times (-a\mathbf{i} + c\mathbf{k}) = bc\mathbf{i} + ac\mathbf{j} + ab\mathbf{k}.$$

A unit vector in this direction is

$$\frac{bc\mathbf{i} + ac\mathbf{j} + ab\mathbf{k}}{\sqrt{b^2c^2 + a^2c^2 + a^2b^2}}.$$

The triangle has area $\frac{1}{2}\sqrt{b^2c^2 + a^2c^2 + a^2b^2}$.

6. A vector perpendicular to $\mathbf{u} = 2\mathbf{i} - \mathbf{j} - 2\mathbf{k}$ and to $\mathbf{v} = 2\mathbf{i} - 3\mathbf{j} + \mathbf{k}$ is the cross product

$$\mathbf{u} \times \mathbf{v} = \begin{vmatrix} \mathbf{i} & \mathbf{j} & \mathbf{k} \\ 2 & -1 & -2 \\ 2 & -3 & 1 \end{vmatrix} = -7\mathbf{i} - 6\mathbf{j} - 4\mathbf{k},$$

which has length $\sqrt{101}$. A unit vector with positive $\mathbf{k}$ component that is perpenducular to $\mathbf{u}$ and $\mathbf{v}$ is

$$\frac{-1}{\sqrt{101}}\mathbf{u} \times \mathbf{v} = \frac{1}{\sqrt{101}}(7\mathbf{i} + 6\mathbf{j} + 4\mathbf{k}).$$

8.
$$\mathbf{u} \times \mathbf{v} = \begin{vmatrix} \mathbf{i} & \mathbf{j} & \mathbf{k} \\ u_1 & u_2 & u_3 \\ v_1 & v_2 & v_3 \end{vmatrix}$$
$$= -\begin{vmatrix} \mathbf{i} & \mathbf{j} & \mathbf{k} \\ v_1 & v_2 & v_3 \\ u_1 & u_2 & u_3 \end{vmatrix} = -\mathbf{v} \times \mathbf{u}.$$

10.
$$(t\mathbf{u}) \times \mathbf{v} = \begin{vmatrix} \mathbf{i} & \mathbf{j} & \mathbf{k} \\ tu_1 & tu_2 & tu_3 \\ v_1 & v_2 & v_3 \end{vmatrix}$$
$$= t\begin{vmatrix} \mathbf{i} & \mathbf{j} & \mathbf{k} \\ u_1 & u_2 & u_3 \\ v_1 & v_2 & v_3 \end{vmatrix} = t(\mathbf{u} \times \mathbf{v}),$$
$$\mathbf{u} \times (t\mathbf{v}) = -(t\mathbf{v}) \times \mathbf{u}$$
$$= -t(\mathbf{v} \times \mathbf{u}) = t(\mathbf{u} \times \mathbf{v}).$$

12. Both $\mathbf{u} = \cos\beta\,\mathbf{i} + \sin\beta\,\mathbf{j}$ and $\mathbf{v} = \cos\alpha\,\mathbf{i} + \sin\alpha\,\mathbf{j}$ are unit vectors. They make angles β and α, respectively, with the positive x-axis, so the angle between them is $|\alpha - \beta| = \alpha - \beta$, since we are told that $0 \le \alpha - \beta \le \pi$. They span a parallelogram (actually a rhombus) having area

$$|\mathbf{u} \times \mathbf{v}| = |\mathbf{u}||\mathbf{v}|\sin(\alpha - \beta) = \sin(\alpha - \beta).$$

But

$$\mathbf{u} \times \mathbf{v} = \begin{vmatrix} \mathbf{i} & \mathbf{j} & \mathbf{k} \\ \cos\beta & \sin\beta & 0 \\ \cos\alpha & \sin\alpha & 0 \end{vmatrix} = (\sin\alpha\cos\beta - \cos\alpha\sin\beta)\mathbf{k}.$$

Because $\mathbf{v}$ is displaced counterclockwise from $\mathbf{u}$, the cross product above must be in the positive k direction. Therefore its length is the k component. Therefore

$$\sin(\alpha - \beta) = \sin\alpha\cos\beta - \cos\alpha\sin\beta.$$

14. The base of the tetrahedron is a triangle spanned by $\mathbf{v}$ and $\mathbf{w}$, which has area

$$A = \frac{1}{2}|\mathbf{v} \times \mathbf{w}|.$$

The altitude h of the tetrahedron (measured perpendicular to the plane of the base) is equal to the length of the projection of $\mathbf{u}$ onto the vector $\mathbf{v} \times \mathbf{w}$ (which is perpendicular to the base). Thus

$$h = \frac{|\mathbf{u} \bullet (\mathbf{v} \times \mathbf{w})|}{|\mathbf{v} \times \mathbf{w}|}.$$

The volume of the tetrahedron is

$$V = \frac{1}{3}Ah = \frac{1}{6}|\mathbf{u} \bullet (\mathbf{v} \times \mathbf{w})|$$
$$= \frac{1}{6}\left|\begin{vmatrix} u_1 & u_2 & u_3 \\ v_1 & v_2 & v_3 \\ w_1 & w_2 & w_3 \end{vmatrix}\right|.$$

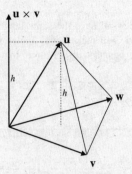

Fig. 10.3.14

16. Let the cube be as shown in the figure. The required parallelepiped is spanned by $a\mathbf{i}+a\mathbf{j}$, $a\mathbf{j}+a\mathbf{k}$, and $a\mathbf{i}+a\mathbf{k}$. Its volume is

$$V = \left| \begin{vmatrix} a & a & 0 \\ 0 & a & a \\ a & 0 & a \end{vmatrix} \right| = 2a^3 \text{ cu. units.}$$

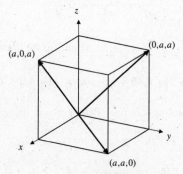

Fig. 10.3.16

18.
$$\mathbf{u} \bullet (\mathbf{v} \times \mathbf{w}) = \begin{vmatrix} u_1 & u_2 & u_3 \\ v_1 & v_2 & v_3 \\ w_1 & w_2 & w_3 \end{vmatrix}$$

$$= -\begin{vmatrix} v_1 & v_2 & v_3 \\ u_1 & u_2 & u_3 \\ w_1 & w_2 & w_3 \end{vmatrix}$$

$$= \begin{vmatrix} v_1 & v_2 & v_3 \\ w_1 & w_2 & w_3 \\ u_1 & u_2 & u_3 \end{vmatrix}$$

$$= \mathbf{v} \bullet (\mathbf{w} \times \mathbf{u})$$

$$= \mathbf{w} \bullet (\mathbf{u} \times \mathbf{v}) \qquad \text{(by symmetry).}$$

20. If $\mathbf{v} \times \mathbf{w} \neq \mathbf{0}$, then $(\mathbf{v} \times \mathbf{w}) \bullet (\mathbf{v} \times \mathbf{w}) \neq 0$. By the previous exercise, there exist constants λ, μ and ν such that

$$\mathbf{u} = \lambda \mathbf{v} + \mu \mathbf{w} + \nu(\mathbf{v} \times \mathbf{w}).$$

But $\mathbf{v} \times \mathbf{w}$ is perpendicular to both $\mathbf{v}$ and $\mathbf{w}$, so

$$\mathbf{u} \bullet (\mathbf{v} \times \mathbf{w}) = 0 + 0 + \nu(\mathbf{v} \times \mathbf{w}) \bullet (\mathbf{v} \times \mathbf{w}).$$

If $\mathbf{u} \bullet (\mathbf{v} \times \mathbf{w}) = 0$, then $\nu = 0$, and

$$\mathbf{u} = \lambda \mathbf{v} + \mu \mathbf{w}.$$

22. $\mathbf{u} \bullet \mathbf{v} \times \mathbf{w}$ makes sense in that it must mean $\mathbf{u} \bullet (\mathbf{v} \times \mathbf{w})$. $((\mathbf{u} \bullet \mathbf{v}) \times \mathbf{w}$ makes no sense since it is the cross product of a scalar and a vector.)

$\mathbf{u} \times \mathbf{v} \times \mathbf{w}$ makes no sense. It is ambiguous, since $(\mathbf{u} \times \mathbf{v}) \times \mathbf{w}$ and $\mathbf{u} \times (\mathbf{v} \times \mathbf{w})$ are not in general equal.

24. If $\mathbf{u}$, $\mathbf{v}$, and $\mathbf{w}$ are mutually perpendicular, then $\mathbf{v} \times \mathbf{w}$ is parallel to $\mathbf{u}$, so $\mathbf{u} \times (\mathbf{v} \times \mathbf{w}) = \mathbf{0}$. In this case, $\mathbf{u} \bullet (\mathbf{v} \times \mathbf{w}) = \pm |\mathbf{u}||\mathbf{v}||\mathbf{w}|$; the sign depends on whether $\mathbf{u}$ and $\mathbf{v} \times \mathbf{w}$ are in the same or opposite directions.

26. If $\mathbf{a} = -\mathbf{i} + 2\mathbf{j} + 3\mathbf{k}$ and $\mathbf{x} = x\mathbf{i} + y\mathbf{j} + z\mathbf{k}$, then

$$\mathbf{a} \times \mathbf{x} = \begin{vmatrix} \mathbf{i} & \mathbf{j} & \mathbf{k} \\ -1 & 2 & 3 \\ x & y & z \end{vmatrix}$$

$$= (2z - 3y)\mathbf{i} + (3x + z)\mathbf{y} - (y + 2x)\mathbf{k}$$

$$= \mathbf{i} + 5\mathbf{j} - 3\mathbf{k},$$

provided $2z - 3y = 1$, $3x + z = 5$, and $-y - 2x = -3$. This system is satisfied by $x = t$, $y = 3 - 2t$, $z = 5 - 3t$, for any real number t. Thus

$$x = t\mathbf{i} + (3 - 2t)\mathbf{j} + (5 - 3t)\mathbf{k}$$

gives a solution of $\mathbf{a} \times \mathbf{x} = \mathbf{i} + 5\mathbf{j} - 3\mathbf{k}$ for any t. These solutions constitute a line parallel to $\mathbf{a}$.

28. The equation $\mathbf{a} \times \mathbf{x} = \mathbf{b}$ can be solved for $\mathbf{x}$ if and only if $\mathbf{a} \bullet \mathbf{b} = 0$. The "only if" part is demonstrated in the previous solution. For the "if" part, observe that if $\mathbf{a} \bullet \mathbf{b} = 0$ and $\mathbf{x}_0 = (\mathbf{b} \times \mathbf{a})/|\mathbf{a}|^2$, then by Exercise 23,

$$\mathbf{a} \times \mathbf{x}_0 = \frac{1}{|\mathbf{a}|^2} \mathbf{a} \times (\mathbf{b} \times \mathbf{a}) = \frac{(\mathbf{a} \bullet \mathbf{a})\mathbf{b} - (\mathbf{a} \bullet \mathbf{b})\mathbf{a}}{|\mathbf{a}|^2} = \mathbf{b}.$$

The solution $\mathbf{x}_0$ is not unique; as suggested by the example in Exercise 26, any multiple of $\mathbf{a}$ can be added to it and the result will still be a solution. If $\mathbf{x} = \mathbf{x}_0 + t\mathbf{a}$, then

$$\mathbf{a} \times \mathbf{x} = \mathbf{a} \times \mathbf{x}_0 + t\mathbf{a} \times \mathbf{a} = \mathbf{b} + 0 = \mathbf{b}.$$

Section 10.4   Planes and Lines   (page 627)

2. The plane through $(0, 2, -3)$ normal to $4\mathbf{i} - \mathbf{j} - 2\mathbf{k}$ has equation

$$4(x - 0) - (y - 2) - 2(z + 3) = 0,$$

or $4x - y - 2z = 4$.

4. The plane passing through $(1, 2, 3)$, parallel to the plane $3x + y - 2z = 15$, has equation $3z + y - 2z = 3 + 2 - 6$, or $3x + y - 2z = -1$.

6. The plane passing through $(-2, 0, 0)$, $(0, 3, 0)$, and $(0, 0, 4)$ has equation

$$\frac{x}{-2} + \frac{y}{3} + \frac{z}{4} = 1,$$

or $6x - 4y - 3z = -12$.

8. Since $(-2, 0, -1)$ does not lie on $x - 4y + 2z = -5$, the required plane will have an equation of the form

$$2x + 3y - z + \lambda(x - 4y + 2z + 5) = 0$$

for some λ. Thus

$$-4 + 1 + \lambda(-2 - 2 + 5) = 0,$$

so $\lambda = 3$. The required plane is $5x - 9y + 5z = -15$.

10. Three distinct points will not determine a unique plane through them if they all lie on a straight line. If the points have position vectors $\mathbf{r}_1$, $\mathbf{r}_2$, and $\mathbf{r}_3$, then they will all lie on a straight line if

$$(\mathbf{r}_2 - \mathbf{r}_1) \times (\mathbf{r}_3 - \mathbf{r}_1) = \mathbf{0}.$$

12. $x + y + z = \lambda$ is the family of all (parallel) planes normal to the vector $\mathbf{i} + \mathbf{j} + \mathbf{k}$.

14. The distance from the planes

$$\lambda x + \sqrt{1 - \lambda^2}\, y = 1$$

to the origin is $1/\sqrt{\lambda^2 + 1 - \lambda^2} = 1$. Hence the equation represents the family of all vertical planes at distance 1 from the origin. All such planes are tangent to the cylinder $x^2 + y^2 = 1$.

16. The line through $(-1, 0, 1)$ perpendicular to the plane $2x - y + 7z = 12$ is parallel to the normal vector $2\mathbf{i} - \mathbf{j} + 7\mathbf{k}$ to that plane. The equations of the line are, in vector parametric form,

$$\mathbf{r} = (-1 + 2t)\mathbf{i} - t\mathbf{j} + (1 + 7t)\mathbf{k},$$

or in scalar parametric form,

$$x = -1 + 2t, \quad y = -t, \quad z = 1 + 7t,$$

or in standard form

$$\frac{x+1}{2} = \frac{y}{-1} = \frac{z-1}{7}.$$

18. A line parallel to $x + y = 0$ and to $x - y + 2z = 0$ is parallel to the cross product of the normal vectors to these two planes, that is, to the vector

$$(\mathbf{i} + \mathbf{j}) \times (\mathbf{i} - \mathbf{j} + 2\mathbf{k}) = 2(\mathbf{i} - \mathbf{j} - \mathbf{k}).$$

Since the line passes through $(2, -1, -1)$, its equations are, in vector parametric form

$$\mathbf{r} = (2 + t)\mathbf{i} - (1 + t)\mathbf{j} - (1 + t)\mathbf{k},$$

or in scalar parametric form

$$x = 2 + t, \quad y = -(1 + t), \quad z = -(1 + t),$$

or in standard form

$$x - 2 = -(y + 1) = -(z + 1).$$

20. The line $\mathbf{r} = (1 - 2t)\mathbf{i} + (4 + 3t)\mathbf{j} + (9 - 4t)\mathbf{k}$ has standard form

$$\frac{x-1}{-2} = \frac{y-4}{3} = \frac{z-9}{-4}.$$

22. The line $\begin{cases} x - 2y + 3z = 0 \\ 2x + 3y - 4z = 4 \end{cases}$ is parallel to the vector

$$(\mathbf{i} - 2\mathbf{j} + 3\mathbf{k}) \times (2\mathbf{i} + 3\mathbf{j} - 4\mathbf{k}) = -\mathbf{i} + 10\mathbf{j} + 7\mathbf{k}.$$

We need a point on this line. Putting $z = 0$, we get

$$x - 2y = 0, \quad 2x + 3y = 4.$$

The solution of this system is $y = 4/7$, $x = 8/7$. A possible standard form for the given line is

$$\frac{x - \dfrac{8}{7}}{-1} = \frac{y - \dfrac{4}{7}}{10} = \frac{z}{7},$$

though, of course, this answer is not unique as the coordinates of any point on the line could have been used.

24. The point on the line corresponding to $t = -1$ is the point P_3 such that P_1 is midway between P_3 and P_2.
The point on the line corresponding to $t = 1/2$ is the midpoint between P_1 and P_2.
The point on the line corresponding to $t = 2$ is the point P_4 such that P_2 is the midpoint between P_1 and P_4.

26. The distance from $(0, 0, 0)$ to $x + 2y + 3z = 4$ is

$$\frac{4}{\sqrt{1^2 + 2^2 + 3^2}} = \frac{4}{\sqrt{14}} \text{ units.}$$

28. A vector parallel to the line $x + y + z = 0$, $2x - y - 5z = 1$ is

$$\mathbf{a} = (\mathbf{i} + \mathbf{j} + \mathbf{k}) \times (2\mathbf{i} - \mathbf{j} - 5\mathbf{k}) = -4\mathbf{i} + 7\mathbf{j} - 3\mathbf{k}.$$

We need a point on this line: if $z = 0$ then $x + y = 0$ and $2x - y = 1$, so $x = 1/3$ and $y = -1/3$. The position vector of this point is

$$\mathbf{r}_1 = \frac{1}{3}\mathbf{i} - \frac{1}{3}\mathbf{j}.$$

The distance from the origin to the line is

$$s = \frac{|\mathbf{r}_1 \times \mathbf{a}|}{|\mathbf{a}|} = \frac{|\mathbf{i} + \mathbf{j} + \mathbf{k}|}{\sqrt{74}} = \sqrt{\frac{3}{74}} \text{ units.}$$

30. The line $x - 2 = \dfrac{y+3}{2} = \dfrac{z-1}{4}$ passes through the point $(2, -3, 1)$, and is parallel to $\mathbf{a} = \mathbf{i} + 2\mathbf{j} + 4\mathbf{k}$.

The plane $2y - z = 1$ has normal $\mathbf{n} = 2\mathbf{j} - \mathbf{k}$.

Since $\mathbf{a} \bullet \mathbf{n} = 0$, the line is parallel to the plane.

The distance from the line to the plane is equal to the distance from $(2, -3, 1)$ to the plane $2y - z = 1$, so is

$$D = \frac{|-6 - 1 - 1|}{\sqrt{4+1}} = \frac{8}{\sqrt{5}} \text{ units.}$$

32. $\dfrac{x - x_0}{\sqrt{1 - \lambda^2}} = \dfrac{y - y_0}{\lambda} = z - z_0$ represents all lines through (x_0, y_0, z_0) parallel to the vectors

$$\mathbf{a} = \sqrt{1 - \lambda^2}\,\mathbf{i} + \lambda\mathbf{j} + \mathbf{k}.$$

All such lines are generators of the circular cone

$$(z - z_0)^2 = (x - x_0)^2 + (y - y_0)^2,$$

so the given equations specify all straight lines lying on that cone.

Section 10.5 Quadric Surfaces (page 631)

2. $x^2 + y^2 + 4z^2 = 4$ represents an oblate spheroid, that is, an ellipsoid with its two longer semi-axes equal. In this case the longer semi-axes have length 2, and the shorter one (in the z direction) has length 1. Cross-sections in planes perpendicular to the z-axis between $z = -1$ and $z = 1$ are circles.

4. $x^2 + 4y^2 + 9z^2 + 4x - 8y = 8$
$(x + 2)^2 + 4(y - 1)^2 + 9z^2 = 8 + 8 = 16$
$\dfrac{(x+2)^2}{4^2} + \dfrac{(y-1)^2}{2^2} + \dfrac{z^2}{(4/3)^2} = 1$
This is an ellipsoid with centre $(-2, 1, 0)$ and semi-axes 4, 2, and 4/3.

6. $z = x^2 - 2y^2$ represents a hyperbolic paraboloid.

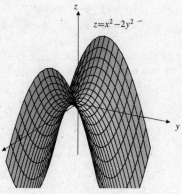

$z = x^2 - 2y^2$

Fig. 10.5.6

8. $-x^2 + y^2 + z^2 = 4$ represents a hyperboloid of one sheet, with circular cross-sections in all planes perpendicular to the x-axis.

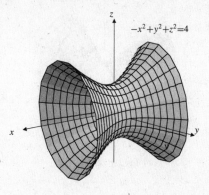

$-x^2 + y^2 + z^2 = 4$

Fig. 10.5.8

10. $x^2 + 4z^2 = 4$ represents an elliptic cylinder with axis along the y-axis.

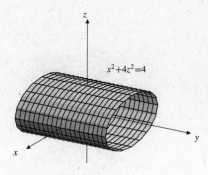

$x^2 + 4z^2 = 4$

Fig. 10.5.10

12. $y = z^2$ represents a parabolic cylinder with vertex line along the x-axis.

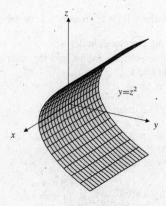

$y = z^2$

Fig. 10.5.12

14. $x^2 = y^2 + 2z^2$ represents an elliptic cone with vertex at the origin and axis along the x-axis.

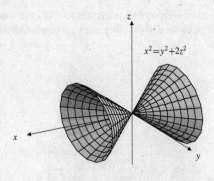

$x^2 = y^2 + 2z^2$

Fig. 10.5.14

16. $(z-1)^2 = (x-2)^2 + (y-3)^2 + 4$ represents a hyperboloid of two sheets with centre at $(2,3,1)$, axis along the line $x = 2$, $y = 3$, and vertices at $(2,3,-1)$ and $(2,3,3)$.

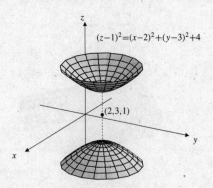

$(z-1)^2 = (x-2)^2 + (y-3)^2 + 4$

$(2,3,1)$

Fig. 10.5.16

18. $\begin{cases} x^2 + y^2 = 1 \\ z = x + y \end{cases}$ is the ellipse of intersection of the plane $z = x + y$ and the circular cylinder $x^2 + y^2 = 1$. The centre of the ellipse is at the origin, and the ends of the major axis are $\pm(1/\sqrt{2}, 1/\sqrt{2}, \sqrt{2})$.

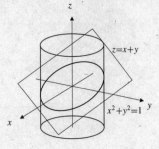

$z = x + y$

$x^2 + y^2 = 1$

Fig. 10.5.18

20. $\begin{cases} x^2 + 2y^2 + 3z^2 = 6 \\ y = 1 \end{cases}$ is an ellipse in the plane $y = 1$. Its projection onto the xz-plane is the ellipse $x^2 + 3z^2 = 4$. One quarter of the ellipse is shown in the figure.

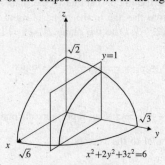

$\sqrt{2}$ $y=1$

$\sqrt{3}$

$\sqrt{6}$ $x^2 + 2y^2 + 3z^2 = 6$

Fig. 10.5.20

22. $z = xy$

Family 1: $\begin{cases} z = \lambda x \\ \lambda = y. \end{cases}$

Family 2: $\begin{cases} z = \mu y \\ \mu = x. \end{cases}$

24. The plane $z = cx + k$ intersects the elliptic cone $z^2 = 2x^2 + y^2$ on the cylinder

$$c^2 x^2 + 2ckx + k^2 = 2x^2 + y^2$$
$$(2 - c^2)x^2 - 2ckx + y^2 = k^2$$
$$(2 - c^2)\left(x - \frac{ck}{2 - c^2}\right)^2 + y^2 = k^2 + \frac{c^2 k^2}{2 - c^2} = \frac{2k^2}{2 - c^2}$$
$$\frac{(x - x_0)^2}{a^2} + \frac{y^2}{b^2} = 1,$$

where $x_0 = \dfrac{ck}{2 - c^2}$, $a^2 = \dfrac{2k^2}{(2 - c^2)^2}$, and $b^2 = \dfrac{2k^2}{2 - c^2}$.

As in the previous exercise, $z = cx + k$ intersects the cylinder (and hence the cone) in an ellipse with principal axes joining the points

$$(x_0 - a, 0, c(x_0 - a) + k) \quad \text{to} \quad (x_0 + a, 0, c(x_0 + a) + k),$$
$$\text{and} \quad (x_0, -b, cx_0 + k) \quad \text{to} \quad (x_0, b, cx_0 + k).$$

The centre of this ellipse is $(x_0, 0, cx_0 + k)$. The ellipse is a circle if its two semi-axes have equal lengths, that is, if

$$a^2 + c^2 a^2 = b^2,$$

that is,

$$(1 + c^2)\frac{2k^2}{(2 - c^2)^2} = \frac{2k^2}{2 - c^2},$$

or $1 + c^2 = 2 - c^2$. Thus $c = \pm 1/\sqrt{2}$. A vector normal to the plane $z = \pm(x/\sqrt{2}) + k$ is $\mathbf{a} = \mathbf{i} \pm \sqrt{2}\mathbf{k}$.

Section 10.6 A Little Linear Algebra (page 641)

2. $\begin{pmatrix} 1 & 1 & 1 \\ 0 & 1 & 1 \\ 0 & 0 & 1 \end{pmatrix}\begin{pmatrix} 1 & 1 & 1 \\ 0 & 1 & 1 \\ 0 & 0 & 1 \end{pmatrix} = \begin{pmatrix} 1 & 2 & 3 \\ 0 & 1 & 2 \\ 0 & 0 & 1 \end{pmatrix}$

4. $\begin{pmatrix} w & x \\ y & z \end{pmatrix}\begin{pmatrix} a & b \\ c & d \end{pmatrix} = \begin{pmatrix} aw+cx & bw+dx \\ ay+cz & by+dz \end{pmatrix}$

6. $\mathbf{x} = \begin{pmatrix} x \\ y \\ z \end{pmatrix}, \qquad \mathcal{A} = \begin{pmatrix} a & p & q \\ p & b & r \\ q & r & c \end{pmatrix}$

$\mathbf{x}\mathbf{x}^T = \begin{pmatrix} x \\ y \\ z \end{pmatrix}(x,y,z) = \begin{pmatrix} x^2 & xy & xz \\ xy & y^2 & yz \\ xz & yz & z^2 \end{pmatrix}$

$\mathbf{x}^T\mathbf{x} = (x,y,z)\begin{pmatrix} x \\ y \\ z \end{pmatrix} = (x^2+y^2+z^2)$

$\mathbf{x}^T\mathcal{A}\mathbf{x} = (x,y,z)\begin{pmatrix} a & p & q \\ p & b & r \\ q & r & c \end{pmatrix}\begin{pmatrix} x \\ y \\ z \end{pmatrix}$

$= (x,y,z)\begin{pmatrix} ax+py+qz \\ px+by+rz \\ qx+ry+cz \end{pmatrix}$

$= ax^2+by^2+cz^2+2pxy+2qxz+2ryz$

8. $\begin{vmatrix} 1 & 1 & 1 & 1 \\ 1 & 2 & 3 & 4 \\ -2 & 0 & 2 & 4 \\ 3 & -3 & 2 & -2 \end{vmatrix}$

$= -2\begin{vmatrix} 1 & 1 & 1 \\ 2 & 3 & 4 \\ -3 & 2 & -2 \end{vmatrix} + 2\begin{vmatrix} 1 & 1 & 1 \\ 1 & 2 & 4 \\ 3 & -3 & -2 \end{vmatrix}$

$\qquad - 4\begin{vmatrix} 1 & 1 & 1 \\ 1 & 2 & 3 \\ 3 & -3 & 2 \end{vmatrix}$

$= -2\begin{vmatrix} 1 & 1 & 1 \\ 0 & 1 & 2 \\ 0 & 5 & 1 \end{vmatrix} + 2\begin{vmatrix} 1 & 1 & 1 \\ 0 & 1 & 3 \\ 0 & -6 & -5 \end{vmatrix}$

$\qquad - 4\begin{vmatrix} 1 & 1 & 1 \\ 0 & 1 & 2 \\ 0 & -6 & -1 \end{vmatrix}$

$= -2\begin{vmatrix} 1 & 2 \\ 5 & 1 \end{vmatrix} + 2\begin{vmatrix} 1 & 3 \\ -6 & -5 \end{vmatrix} - 4\begin{vmatrix} 1 & 2 \\ -6 & -1 \end{vmatrix}$

$= -2(-9) + 2(13) - 4(11) = 0$

10. $\begin{vmatrix} 1 & 1 \\ x & y \end{vmatrix} = y - x$. If

$$f(x,y,z) = \begin{vmatrix} 1 & 1 & 1 \\ x & y & z \\ x^2 & y^2 & z^2 \end{vmatrix},$$

then f is a polynomial of degree 2 in z.
Since $f(x,y,x) = 0$ and $f(x,y,y) = 0$, we must have
$f(x,y,z) = A(z-x)(z-y)$ for some A independent of z.
But

$$Axy = f(x,y,0) = \begin{vmatrix} 1 & 1 & 1 \\ x & y & 0 \\ x^2 & y^2 & 0 \end{vmatrix} = xy(y-x),$$

so $A = y - x$ and

$$f(x,y,z) = (y-x)(z-x)(z-y).$$

Generalization:

$$\begin{vmatrix} 1 & 1 & 1 & \cdots & 1 \\ x_1 & x_2 & x_3 & \cdots & x_n \\ x_1^2 & x_2^2 & x_3^2 & \cdots & x_n^2 \\ \vdots & \vdots & \vdots & \ddots & \vdots \\ x_1^{n-1} & x_2^{n-1} & x_3^{n-1} & \cdots & x_n^{n-1} \end{vmatrix} = \prod_{1\le i<j\le n}(x_j - x_i).$$

12. If $\mathcal{A} = \begin{pmatrix} a & b \\ c & d \end{pmatrix}$, then $\mathcal{A}^T = \begin{pmatrix} a & c \\ b & d \end{pmatrix}$, and

$$\det(\mathcal{A}) = ad - bc = \det(\mathcal{A}^T).$$

We generalize this by induction.
Suppose $\det(\mathcal{B}^T) = \det(\mathcal{B})$ for any $(n-1) \times (n-1)$ matrix, where $n \ge 3$. Let

$$\mathcal{A} = \begin{pmatrix} a_{11} & a_{12} & \cdots & a_{1n} \\ a_{21} & a_{22} & \cdots & a_{2n} \\ \vdots & \vdots & \ddots & \vdots \\ a_{n1} & a_{n2} & \cdots & a_{nn} \end{pmatrix}$$

be an $n \times n$ matrix. If $\det(\mathcal{A})$ is expanded in minors about the first row, and $\det(\mathcal{A}^T)$ is expanded in minors about the first column, the corresponding terms in these expansions are equal by the induction hypothesis. (The $(n-1) \times (n-1)$ matrices whose determinants appear in one expansion are the transposes of those in the other expansion.) Therefore $\det(\mathcal{A}^T) = \det(\mathcal{A})$ for any square matrix $\mathcal{A}$.

14. If $\mathcal{A}_\theta = \begin{pmatrix} \cos\theta & \sin\theta \\ -\sin\theta & \cos\theta \end{pmatrix}$, then

$$\mathcal{A}_{-\theta} = \begin{pmatrix} \cos(-\theta) & \sin(-\theta) \\ -\sin(-\theta) & \cos(-\theta) \end{pmatrix} = \begin{pmatrix} \cos\theta & -\sin\theta \\ \sin\theta & \cos\theta \end{pmatrix},$$

and

$$\mathcal{A}_\theta \mathcal{A}_{-\theta} = \begin{pmatrix} 1 & 0 \\ 0 & 1 \end{pmatrix} = I.$$

Thus $\mathcal{A}_{-\theta} = (\mathcal{A}_\theta)^{-1}$.

16. Let $\mathcal{A} = \begin{pmatrix} 1 & 0 & -1 \\ -1 & 1 & 0 \\ 2 & 1 & 3 \end{pmatrix}$, $\mathcal{A}^{-1} = \begin{pmatrix} a & b & c \\ d & e & f \\ g & h & i \end{pmatrix}$. Since $\mathcal{A}\mathcal{A}^{-1} = I$ we must have

$a - g = 1$	$b - h = 0$	$c - i = 0$
$-a + d = 0$	$-b + e = 1$	$-c + f = 0$
$2a + d + 3g = 0$	$2b + e + 3h = 0$	$2c + f + 3i = 1.$

Solving these three systems of equations, we get

$$A^{-1} = \begin{pmatrix} \frac{1}{2} & -\frac{1}{6} & \frac{1}{6} \\ \frac{1}{2} & \frac{5}{6} & \frac{1}{6} \\ -\frac{1}{2} & -\frac{1}{6} & \frac{1}{6} \end{pmatrix}.$$

18. If A is the matrix of Exercises 16 and 17 then $\det(A) = 6$. By Cramer's Rule,

$$x = \frac{1}{6} \begin{vmatrix} -2 & 0 & -1 \\ 1 & 1 & 0 \\ 13 & 1 & 3 \end{vmatrix} = \frac{6}{6} = 1$$

$$y = \frac{1}{6} \begin{vmatrix} 1 & -2 & -1 \\ -1 & 1 & 0 \\ 2 & 13 & 3 \end{vmatrix} = \frac{12}{6} = 2$$

$$z = \frac{1}{6} \begin{vmatrix} 1 & 0 & -2 \\ -1 & 1 & 1 \\ 2 & 1 & 13 \end{vmatrix} = \frac{18}{6} = 3.$$

20. Let $F(x_1, x_2) = \mathcal{F}\begin{pmatrix} x_1 \\ x_2 \end{pmatrix}$, where $\mathcal{F} = \begin{pmatrix} a & b \\ c & d \end{pmatrix}$.
Let $G(y_1, y_2) = \mathcal{G}\begin{pmatrix} y_1 \\ y_2 \end{pmatrix}$, where $\mathcal{G} = \begin{pmatrix} p & q \\ r & s \end{pmatrix}$.
If $y_1 = ax_1 + bx_2$ and $y_2 = cx_1 + dx_2$, then

$$
\begin{aligned}
G \circ F(x_1, x_2) &= G(y_1, y_2) \\
&= \begin{pmatrix} p & q \\ r & s \end{pmatrix}\begin{pmatrix} ax_1 + bx_2 \\ cx_1 + dx_2 \end{pmatrix} \\
&= \begin{pmatrix} pax_1 + pbx_2 + qcx_1 + qdx_2 \\ rax_1 + rbx_2 + scx_1 + sdx_2 \end{pmatrix} \\
&= \begin{pmatrix} pa + qc & pb + qd \\ ra + sc & rb + sd \end{pmatrix}\begin{pmatrix} x_1 \\ x_2 \end{pmatrix} \\
&= \begin{pmatrix} p & q \\ r & s \end{pmatrix}\begin{pmatrix} a & b \\ c & d \end{pmatrix}\begin{pmatrix} x_1 \\ x_2 \end{pmatrix} \\
&= \mathcal{G}\mathcal{F}\begin{pmatrix} x_1 \\ x_2 \end{pmatrix}.
\end{aligned}
$$

Thus, $G \circ F$ is represented by the matrix $\mathcal{G}\mathcal{F}$.

22. $A = \begin{pmatrix} 1 & 2 & 0 \\ 2 & 1 & 0 \\ 0 & 0 & 1 \end{pmatrix}$. Use Theorem 8.

$$D_1 = 1 > 0, \quad D_2 = \begin{vmatrix} 1 & 2 \\ 2 & 1 \end{vmatrix} = -3 < 0,$$

$$D_3 = \begin{vmatrix} 1 & 2 & 0 \\ 2 & 1 & 0 \\ 0 & 0 & 1 \end{vmatrix} = -3 < 0.$$

Thus A is indefinite.

24. $A = \begin{pmatrix} 1 & 1 & 0 \\ 1 & 1 & 0 \\ 0 & 0 & 1 \end{pmatrix}$. Since $D_2 = \begin{vmatrix} 1 & 1 \\ 1 & 1 \end{vmatrix} = 0$, we cannot use Theorem 8. The corresponding quadratic form is

$$Q(x, y, z) = x^2 + y^2 + 2xy + z^2 = (x + y)^2 + z^2,$$

which is positive semidefinite. ($Q(1, -1, 0) = 0$.). Thus A is positive semidefinite.

26. $A = \begin{pmatrix} 2 & 0 & 1 \\ 0 & 4 & 11 \\ 1 & -1 & 1 \end{pmatrix}$. Use Theorem 8.

$$D_1 = 2 > 0, \quad D_2 = \begin{vmatrix} 2 & 0 \\ 0 & 4 \end{vmatrix} = 8 > 0,$$

$$D_3 = \begin{vmatrix} 2 & 0 & 1 \\ 0 & 4 & 11 \\ 1 & -1 & 1 \end{vmatrix} = 2 > 0.$$

Thus A is positive definite.

Section 10.7 Using Maple for Vector and Matrix Calculations (page 648)

It is assumed that the Maple package "vecops.def" has been loaded for all the calculations in this section.

2. The plane P through the origin containing the vectors $\mathbf{v}_1 = \mathbf{i} - 2\mathbf{j} - 3\mathbf{k}$ and $\mathbf{v}2 = 2\mathbf{i} + 3\mathbf{j} + 4\mathbf{k}$ has normal $\mathbf{n} = \mathbf{v}_1 \times \mathbf{v}_2$.

```
>  v1 := [1,-2,-3]: v2 := [2,3,4]:
>  n := v1 &x v2;
              n := [1, -10, 7]
```

The angle between $\mathbf{v} = \mathbf{i} - \mathbf{j} + 2\mathbf{k}$ and $\mathbf{n}$ (in degrees) is calculated as follows:

```
>  v := [1, -1, 2];
>  angvn := evalf((180/Pi)*
>    arccos((v &. n)/(len(v)*len(n))));
              angvn := 33.55730975
```

Since this angle is acute, the angle between $\mathbf{v}$ and the plane P is its complement.

```
>  angle := 90 - angvn;
              angle := 56.44269025
```

4. These calculations verify the identity:

```
>  u := vector(3): v := vector(3):
>  w := vector(3):
>  evl(u &x v) &x (u &x w)
>      - (u &. (v &x w)) * u);
              [0, 0, 0]
```

6. `vp := (u,v) -> ((u &. v)/(len(v))^2)*v`

8. `unitn := (u,v) -> (u &x v)/len(u &x v)`

10. `dist := (a,b) -> len(a-b)`

```
>  dist([1,1,1,1], [3,-1,2,5]);
              5
```

12. We use `linsolve`.

```
>  B := [[1,1,1,1,1],
>        [1,-2,3,-4,5],[1,-1,2,-2,5],
>        [2,-3,5,-6,8],[2,-1,4,-3,6]]:
>  linsolve(B,[6,-5,-1,-6,1]);
       [4 - _t[1], 3 - _t[1], -1 + _t[1], _t[1], 0]
```

There is a one-parameter family of solutions: $u = 4 - t$, $v = 3 - t$, $x = -1 + t$, $y = t$, $z = 0$, for arbitrary t.

14.
```
>  B := [[1,1,1,1,1],
>        [1,-2,3,-4,5],[1,-1,2,-2,5],
>        [2,-3,5,-6,8],[2,-1,4,-3,6]]:
>  eigenvals(B);
     0, RootOf(-20 * _Z + 4 - _Z^3 - 36 * _Z^2 + _Z^4)
```

One of the eigenvalues is 0. The other four are roots of a polynomial equation of degree 4. To obtain these to five decimal places we apply the Maple procedure all-values to the second element in the above list, and then evalf to the result:

```
>  evalf(allvalues(%[2]));
```

The output is a list of four complex numbers with real and imaginary parts each involving 10 significant digits. For instance, one of them is

$$6.754851669 - .1644910056 \times 10^{-10} I.$$

All the imaginary parts are of the order of 10^{-10} in size, and are due to round-off errors in the obtaining the solutions of the quartic equation. B is a real symmetric matrix, and so has real eigenvalues. The remaining four eigenvalues, rounded to 5 decimal places are 6.5485, 0.15602, -0.73300, and -5.17787.

16.
```
>  A := [[1,1/2,1/3],
>        [1/2,1/3,1/4],[1/3,1/4;1/5]]:
>  Ainv := evalm(A^(-1)):
>  evalf(eigenvals(A));
   1.408318927 + .1 10^{-10}I, .00268734042, .1223270660
>  evalf(eigenvals(Ainv));
372.1151278, .710066449 - .1 10^{-8}I, 8.174805751 - .1 10^{-8}I
```

The small imaginary parts are due to round-off errors in the solution process. The eigenvalues are real since the matrix and its inverse are real and symmetric.

Review Exercises 10   (page 648)

2. $y - z \geq 1$ represents all points on or below the plane parallel to the x-axis that passes through the points $(0, 1, 0)$ and $(0, 0, -1)$.

4. $x - 2y - 4z = 8$ represents all points on the plane passing through the three points $(8, 0, 0)$, $(0, -4, 0)$, and $(0, 0, -2)$.

6. $y = z^2$ represents the parabolic cylinder parallel to the x-axis containing the curve $y = z^2$ in the yz-plane.

8. $z = xy$ is the hyperbolic paraboloid containing the x- and y-axes that results from rotating the hyperbolic paraboloid $z = (x^2 - y^2)/2$ through $45°$ about the z-axis.

10. $x^2 + y^2 - 4z^2 = 4$ represents a hyperboloid of one sheet with circular cross-sections in planes perpendicular to the z-axis, and asymptotic to the cone obtained by rotating the line $x = 2z$ about the z-axis.

12. $x^2 - y^2 - 4z^2 = 4$ represents a hyperboloid of two sheets asymptotic to the cone of the previous exercise.

14. $(x - z)^2 + y^2 = z^2$ represents an elliptic cone with oblique axis along the line $z = x$ in the xz-plane, having circular cross-sections of radius $|k|$ in horizontal planes $z = k$. The z-axis lies on the cone.

16. $x + y + 2z = 1$, $x + y + z = 0$ together represent the straight line through the points $(-1, 0, 1)$ and $(0, -1, 1)$.

18. $x^2 + z^2 \leq 1$, $x - y \geq 0$ together represent all points that lie inside or on the circular cylinder of radius 1 and axis along the y-axis and also either on the vertical plane $x - y = 0$ or on the side of that plane containing the positive x-axis.

20. A plane through $(2, -1, 1)$ and $(1, 0, -1)$ is parallel to $\mathbf{b} = (2 - 1)\mathbf{i} + (-1 - 0)\mathbf{j} + (1 - (-1))\mathbf{k} = \mathbf{i} - \mathbf{j} + 2\mathbf{k}$. If it is also parallel to the vector $\mathbf{a}$ in the previous solution, then it is normal to

$$\mathbf{a} \times \mathbf{b} = \begin{vmatrix} \mathbf{i} & \mathbf{j} & \mathbf{k} \\ 2 & -1 & 3 \\ 1 & -1 & 2 \end{vmatrix} = \mathbf{i} - \mathbf{j} - \mathbf{k}.$$

The plane has equation $(x - 1) - (y - 0) - (z + 1) = 0$, or $x - y - z = 2$.

22. The plane through $A = (-1, 1, 0)$, $B = (0, 4, -1)$ and $C = (2, 0, 0)$ has normal

$$\overrightarrow{AC} \times \overrightarrow{AB} = \begin{vmatrix} \mathbf{i} & \mathbf{j} & \mathbf{k} \\ 3 & -1 & 0 \\ 1 & 3 & -1 \end{vmatrix} = \mathbf{i} + 3\mathbf{j} + 10\mathbf{k}.$$

Its equation is $(x - 2) + 3y + 10z = 0$, or $x + 3y + 10z = 2$.

24. A plane containing the line of intersection of the planes $x + y + z = 0$ and $2x + y - 3z = 2$ has equation

$$2x + y - 3z - 2 + \lambda(x + y + z - 0) = 0.$$

This plane is perpendicular to $x - 2y - 5z = 17$ if their normals are perpendicular, that is, if

$$1(2 + \lambda) - 2(1 + \lambda) - 5(-3 + \lambda) = 0,$$

or $9x + 7y - z = 4$.

26. A vector parallel to the planes $x - y = 3$ and $x + 2y + z = 1$ is $(\mathbf{i} - \mathbf{j}) \times (\mathbf{i} + 2\mathbf{j} + \mathbf{k}) = -\mathbf{i} - \mathbf{j} + 3\mathbf{k}$. A line through $(1, 0, -1)$ parallel to this vector is

$$\frac{x - 1}{-1} = \frac{y}{-1} = \frac{z + 1}{3}.$$

28. The vector

$$\mathbf{a} = (1 + t)\mathbf{i} - t\mathbf{j} - (2 + 2t)\mathbf{k} - \big(2s\mathbf{i} + (s - 2)\mathbf{j} - (1 + 3s)\mathbf{k}\big)$$
$$= (1 + t - 2s)\mathbf{i} - (t + s - 2)\mathbf{j} - (1 + 2t - 3s)\mathbf{k}$$

joins points on the two lines and is perpendicular to both lines if $\mathbf{a} \bullet (\mathbf{i} - \mathbf{j} - 2\mathbf{k}) = 0$ and $\mathbf{a} \bullet (2\mathbf{i} + \mathbf{j} - 3\mathbf{k}) = 0$, that is, if

$$1 + t - 2s + t + s - 2 + 2 + 4t - 6s = 0$$
$$2 + 2t - 4s - t - s + 2 + 3 + 6t - 9s = 0,$$

or, on simplification,

$$6t - 7s = -1$$
$$7t - 14s = -7.$$

This system has solution $t = 1$, $s = 1$. We would expect to use $\mathbf{a}$ as a vector perpendicular to both lines, but, as it happens, $\mathbf{a} = \mathbf{0}$ if $t = s = 1$, because the two given lines intersect at $(2, -1, -4)$. A nonzero vector perpendicular to both lines is

$$\begin{vmatrix} \mathbf{i} & \mathbf{j} & \mathbf{k} \\ 1 & -1 & -2 \\ 2 & 1 & -3 \end{vmatrix} = 5\mathbf{i} - \mathbf{j} + 3\mathbf{k}.$$

Thus the required line is parallel to this vector and passes through $(2, -1, -4)$, so its equation is

$$\mathbf{r} = (2 + 5t)\mathbf{i} - (1 + t)\mathbf{j} + (-4 + 3t)\mathbf{k}.$$

30. The points with position vectors $\mathbf{r}_1$, $\mathbf{r}_2$, $\mathbf{r}_3$, and $\mathbf{r}_4$ are coplanar if the tetrahedron having these points as vertices has zero volume, that is, if

$$\big[(\mathbf{r}_2 - \mathbf{r}_1) \times (\mathbf{r}_3 - \mathbf{r}_1)\big] \bullet (\mathbf{r}_4 - \mathbf{r}_1) = 0.$$

(Any permutation of the subscripts 1, 2, 3, and 4 in the above equation will do as well.)

32. The tetrahedron with vertices $A = (1, 2, 1)$, $B = (4, -1, 1)$, $C = (3, 4, -2)$, and $D = (2, 2, 2)$ has volume

$$\frac{1}{6}|(\overrightarrow{AB} \times \overrightarrow{AC}) \bullet \overrightarrow{AD}| = \frac{1}{6}|(9\mathbf{i} + 9\mathbf{j} + 12\mathbf{k}) \bullet (\mathbf{i} + \mathbf{k})|$$
$$= \frac{9 + 12}{6} = \frac{7}{2} \text{ cu. units.}$$

34. Let $\mathcal{A} = \begin{pmatrix} 1 & 1 & 1 \\ 2 & 1 & 0 \\ 1 & 0 & -1 \end{pmatrix}$, $\mathbf{x} = \begin{pmatrix} x_1 \\ x_2 \\ x_3 \end{pmatrix}$, and $\mathbf{b} = \begin{pmatrix} b_1 \\ b_2 \\ b_3 \end{pmatrix}$.

Then

$$\mathcal{A}\mathbf{x} = \mathbf{b} \quad \Leftrightarrow \quad \begin{aligned} x_1 + x_2 + x_3 &= b_1 \\ 2x_1 + x_2 \quad\;\; &= b_2 \\ x_1 \quad\quad - x_3 &= b_3. \end{aligned}$$

The sum of the first and third equations is $2x_1 + x_2 = b_1 + b_3$, which is incompatible with the second equation unless $b_2 = b_1 + b_3$, that is, unless

$$\mathbf{b} \bullet (\mathbf{i} - \mathbf{j} + \mathbf{k}) = 0.$$

If $\mathbf{b}$ satisfies this condition then there will be a line of solutions; if $x_1 = t$, then $x_2 = b_2 - 2t$, and $x_3 = t - b_3$, so

$$\mathbf{x} = \begin{pmatrix} t \\ b_2 - 2t \\ t - b_3 \end{pmatrix}$$

is a solution for any t.

Challenging Problems 10   (page 649)

2. By the formula for the vector triple product given in Exercise 23 of Section 1.3,

$$(\mathbf{u} \times \mathbf{v}) \times (\mathbf{w} \times \mathbf{x}) = [(\mathbf{u} \times \mathbf{v}) \bullet \mathbf{x}]\mathbf{w} - [(\mathbf{u} \times \mathbf{v}) \bullet \mathbf{w}]\mathbf{x}$$
$$(\mathbf{u} \times \mathbf{v}) \times (\mathbf{w} \times \mathbf{x}) = -(\mathbf{w} \times \mathbf{x}) \times (\mathbf{u} \times \mathbf{v})$$
$$= -[(\mathbf{w} \times \mathbf{x}) \bullet \mathbf{v}]\mathbf{u} + [(\mathbf{w} \times \mathbf{x}) \bullet \mathbf{u}]\mathbf{v}.$$

In particular, if $\mathbf{w} = \mathbf{u}$, then, since $(\mathbf{u} \times \mathbf{v}) \bullet \mathbf{u} = 0$, we have

$$(\mathbf{u} \times \mathbf{v}) \times (\mathbf{u} \times \mathbf{x}) = [(\mathbf{u} \times \mathbf{v}) \bullet \mathbf{x}]\mathbf{u},$$

or, replacing x with w,

$$(\mathbf{u} \times \mathbf{v}) \times (\mathbf{u} \times \mathbf{w}) = [(\mathbf{u} \times \mathbf{v}) \bullet \mathbf{w}]\mathbf{u}.$$

4. a) Let Q_1 and Q_2 be the points on lines L_1 and L_2, respectively, that are closest together. As observed in Example 9 of Section 1.4, $\overrightarrow{Q_1Q_2}$ is perpendicular to both lines.
Therefore, the plane P_1 through Q_1 having normal $\overrightarrow{Q_1Q_2}$ contains the line L_1. Similarly, the plane P_2 through Q_2 having normal $\overrightarrow{Q_1Q_2}$ contains the line L_2. These planes are parallel since they have the same normal. They are different planes because $Q_1 \neq Q_2$ (because the lines are skew).

b) Line L_1 through $(1, 1, 0)$ and $(2, 0, 1)$ is parallel to $\mathbf{i} - \mathbf{j} + \mathbf{k}$, and has parametric equation

$$\mathbf{r}_1 = (1 + t)\mathbf{i} + (1 - t)\mathbf{j} + t\mathbf{k}.$$

Line L_2 through $(0, 1, 1)$ and $(1, 2, 2)$ is parallel to $\mathbf{i} + \mathbf{j} + \mathbf{k}$, and has parametric equation

$$\mathbf{r}_2 = s\mathbf{i} + (1 + s)\mathbf{j} + (1 + s)\mathbf{k}.$$

Now $\mathbf{r}_2 - \mathbf{r}_1 = (s - t - 1)\mathbf{i} + (s + t)\mathbf{j} + (1 + s - t)\mathbf{k}$.

To find the points Q_1 on L_1 and Q_2 on L_2 for which $\overrightarrow{Q_1Q_2}$ is perpendicular to both lines, we solve

$$(s - t - 1) - (s + t) + (1 + s - t) = 0$$
$$(s - t - 1) + (s + t) + (1 + s - t) = 0.$$

Subtracting these equations gives $s + t = 0$, so $t = -s$. Then substituting into either equation gives $2s - 1 + 1 + 2s = 0$, so $s = -t = 0$. Thus $Q_1 = (1, 1, 0)$ and $Q_2 = (0, 1, 1)$, and $\overrightarrow{Q_1Q_2} = -\mathbf{i} + \mathbf{k}$. The required planes are $x - z = 1$ (containing L_1) and $x - z = -1$ (containing L_2).

CHAPTER 11. VECTOR FUNCTIONS AND CURVES

Section 11.1 Vector Functions of One Variable (page 657)

2. Position: $\mathbf{r} = t^2\mathbf{i} + \mathbf{k}$
Velocity: $\mathbf{v} = 2t\mathbf{i}$
Speed: $v = 2|t|$
Acceleration : $\mathbf{a} = 2\mathbf{i}$
Path: the line $z = 1$, $y = 0$.

4. Position: $\mathbf{r} = \mathbf{i} + t\mathbf{j} + t\mathbf{k}$
Velocity: $\mathbf{v} = \mathbf{j} + \mathbf{k}$
Speed: $v = \sqrt{2}$
Acceleration : $\mathbf{a} = 0$
Path: the straight line $x = 1$, $y = z$.

6. Position: $\mathbf{r} = t\mathbf{i} + t^2\mathbf{j} + t^2\mathbf{k}$
Velocity: $\mathbf{v} = \mathbf{i} + 2t\mathbf{j} + 2t\mathbf{k}$
Speed: $v = \sqrt{1 + 8t^2}$
Acceleration: $\mathbf{a} = 2\mathbf{j} + 2\mathbf{k}$
Path: the parabola $y = z = x^2$.

8. Position: $\mathbf{r} = a\cos\omega t\mathbf{i} + b\mathbf{j} + a\sin\omega t\mathbf{k}$
Velocity: $\mathbf{v} = -a\omega\sin\omega t\mathbf{i} + a\omega\cos\omega t\mathbf{k}$
Speed: $v = |a\omega|$
Acceleration: $\mathbf{a} = -a\omega^2\cos\omega t\mathbf{i} - a\omega^2\sin\omega t\mathbf{k}$
Path: the circle $x^2 + z^2 = a^2$, $y = b$.

10. Position: $\mathbf{r} = 3\cos t\mathbf{i} + 4\sin t\mathbf{j} + t\mathbf{k}$
Velocity: $\mathbf{v} = -3\sin t\mathbf{i} + 4\cos t\mathbf{j} + \mathbf{k}$
Speed: $v = \sqrt{9\sin^2 t + 16\cos^2 t + 1} = \sqrt{10 + 7\cos^2 t}$
Acceleration : $\mathbf{a} = -3\cos t\mathbf{i} - 4\sin t\mathbf{j} = t\mathbf{k} - \mathbf{r}$
Path: a helix (spiral) wound around the elliptic cylinder $(x^2/9) + (y^2/16) = 1$.

12. Position: $\mathbf{r} = at\cos\omega t\mathbf{i} + at\sin\omega t\mathbf{j} + b\ln t\mathbf{k}$
Velocity: $\mathbf{v} = a(\cos\omega t - \omega t\sin\omega t)\mathbf{i}$
$\qquad + a(\sin\omega t + \omega t\cos\omega t)\mathbf{j} + (b/t)\mathbf{k}$
Speed: $v = \sqrt{a^2(1 + \omega^2 t^2) + (b^2/t^2)}$
Acceleration: $\mathbf{a} = -a\omega(2\sin\omega t + \omega\cos\omega t)\mathbf{i}$
$\qquad + a\omega(2\cos\omega t - \omega\sin\omega t)\mathbf{j} - (b/t^2)\mathbf{k}$
Path: a spiral on the surface $x^2 + y^2 = a^2 e^{z/b}$.

14. Position: $\mathbf{r} = a\cos t\sin t\mathbf{i} + a\sin^2 t\mathbf{j} + a\cos t\mathbf{k}$
$\qquad = \dfrac{a}{2}\sin 2t\mathbf{i} + \dfrac{a}{2}(1 - \cos 2t)\mathbf{j} + a\cos t\mathbf{k}$
Velocity: $\mathbf{v} = a\cos 2t\mathbf{i} + a\sin 2t\mathbf{j} - a\sin t\mathbf{k}$
Speed: $v = a\sqrt{1 + \sin^2 t}$
Acceleration: $\mathbf{a} = -2a\sin 2t\mathbf{i} + 2a\cos 2t\mathbf{j} - a\cos t\mathbf{k}$
Path: the path lies on the sphere $x^2 + y^2 + z^2 = a^2$, on the surface defined in terms of spherical polar coordinates by $\phi = \theta$, on the circular cylinder $x^2 + y^2 = ay$, and on the parabolic cylinder $ay + z^2 = a^2$. Any two of these surfaces serve to pin down the shape of the path.

16. When its x-coordinate is x, the particle is at position $\mathbf{r} = x\mathbf{i} + (3/x)\mathbf{j}$, and its velocity and speed are

$$\mathbf{v} = \frac{d\mathbf{r}}{dt} = \frac{dx}{dt}\mathbf{i} - \frac{3}{x^2}\frac{dx}{dt}\mathbf{j}$$

$$v = \left|\frac{dx}{dt}\right|\sqrt{1 + \frac{9}{x^4}}.$$

We know that $dx/dt > 0$ since the particle is moving to the right. When $x = 2$, we have
$10 = v = (dx/dt)\sqrt{1 + (9/16)} = (5/4)(dx/dt)$. Thus $dx/dt = 8$. The velocity at that time is $\mathbf{v} = 8\mathbf{i} - 6\mathbf{j}$.

18. The position of the object when its x-coordinate is x is

$$\mathbf{r} = x\mathbf{i} + x^2\mathbf{j} + x^3\mathbf{k},$$

so its velocity is $\mathbf{v} = \dfrac{dx}{dt}\left[\mathbf{i} + 2x\mathbf{j} + 3x^2\mathbf{k}\right]$. Since $dz/dt = 3x^2\,dx/dt = 3$, when $x = 2$ we have $12\,dx/dt = 3$, so $dx/dt = 1/4$. Thus

$$\mathbf{v} = \frac{1}{4}\mathbf{i} + \mathbf{j} + 3\mathbf{k}.$$

20. $\mathbf{r} = x\mathbf{i} - x^2\mathbf{j} + +x^2\mathbf{k}$

$\mathbf{v} = \dfrac{dx}{dt}(\mathbf{i} - 2x\mathbf{j} + 2x\mathbf{k})$

$\mathbf{a} = \dfrac{d^2 x}{dt^2}(\mathbf{i} - 2x\mathbf{j} + 2x\mathbf{k}) + \left(\dfrac{dx}{dt}\right)^2(-2\mathbf{j} + 2\mathbf{k}).$

Thus $|\mathbf{v}| = \left|\dfrac{dx}{dt}\right|\sqrt{1 + 4x^4 + 4x^4} = \sqrt{1 + 8x^4}\dfrac{dx}{dt}$,
since x is increasing. At $(1, -1, 1)$, $x = 1$ and $|\mathbf{v}| = 9$, so $dx/dt = 3$, and the velocity at that point is $\mathbf{v} = 3\mathbf{i} - 6\mathbf{j} + 6\mathbf{k}$. Now

$$\frac{d}{dt}|\mathbf{v}| = \sqrt{1 + 8x^4}\frac{d^2 x}{dt^2} + \frac{16x^3}{\sqrt{1 + 8x^4}}\left(\frac{dx}{dt}\right)^2.$$

The left side is 3 when $x = 1$, so $3(d^2 x/dt^2) + 48 = 3$, and $d^2 x/dt^2 = -15$ at that point, and the acceleration there is

$$\mathbf{a} = -15(\mathbf{i} - 2\mathbf{j} + 2\mathbf{k}) + 9(-2\mathbf{j} + 2\mathbf{k}) = -15\mathbf{i} + 12\mathbf{j} - 12\mathbf{k}.$$

22. If $\mathbf{u}(t) = u_1(t)\mathbf{i} + u_2(t)\mathbf{j} + u_3(t)\mathbf{k}$
$\quad \mathbf{v}(t) = v_1(t)\mathbf{i} + v_2(t)\mathbf{j} + v_3(t)\mathbf{k}$
then $\mathbf{u} \bullet \mathbf{v} = u_1 v_1 + u_2 v_2 + u_3 v_3$, so

$$\frac{d}{dt}\mathbf{u} \bullet \mathbf{v} = \frac{du_1}{dt}v_1 + u_1\frac{dv_1}{dt} + \frac{du_2}{dt}v_2 + u_2\frac{dv_2}{dt}$$

$$+ \frac{du_3}{dt}v_3 + u_3\frac{dv_3}{dt}$$

$$= \frac{d\mathbf{u}}{dt} \bullet \mathbf{v} + \mathbf{u} \bullet \frac{d\mathbf{v}}{dt}.$$

24. $\frac{d}{dt}|\mathbf{r}|^2 = \frac{d}{dt}\mathbf{r}\bullet\mathbf{r} = 2\mathbf{r}\bullet\mathbf{v} = 0$ implies that $|\mathbf{r}|$ is constant.

Thus $\mathbf{r}(t)$ lies on a sphere centred at the origin.

26. If $\mathbf{r}\bullet\mathbf{v} > 0$ then $|\mathbf{r}|$ is increasing. (See Exercise 16 above.) Thus $\mathbf{r}$ is moving farther away from the origin. If $\mathbf{r}\bullet\mathbf{v} < 0$ then $\mathbf{r}$ is moving closer to the origin.

28. $\frac{d}{dt}\big(\mathbf{u}\bullet(\mathbf{v}\times\mathbf{w})\big)$

$= \mathbf{u}'\bullet(\mathbf{v}\times\mathbf{w}) + \mathbf{u}\bullet(\mathbf{v}'\times\mathbf{w}) + \mathbf{u}\bullet(\mathbf{v}\times\mathbf{w}')$.

30. $\frac{d}{dt}\left(\mathbf{u}\times\left(\frac{d\mathbf{u}}{dt}\times\frac{d^2\mathbf{u}}{dt^2}\right)\right)$

$= \frac{d\mathbf{u}}{dt}\times\left(\frac{d\mathbf{u}}{dt}\times\frac{d^2\mathbf{u}}{dt^2}\right) + \mathbf{u}\times\left(\frac{d^2\mathbf{u}}{dt^2}\times\frac{d^2\mathbf{u}}{dt^2}\right)$

$\quad + \mathbf{u}\times\left(\frac{d\mathbf{u}}{dt}\times\frac{d^3\mathbf{u}}{dt^3}\right)$

$= \frac{d\mathbf{u}}{dt}\times\left(\frac{d\mathbf{u}}{dt}\times\frac{d^2\mathbf{u}}{dt^2}\right) + \mathbf{u}\times\left(\frac{d\mathbf{u}}{dt}\times\frac{d^3\mathbf{u}}{dt^3}\right)$.

32. $\frac{d}{dt}\big[(\mathbf{u}\times\mathbf{u}')\bullet(\mathbf{u}'\times\mathbf{u}'')\big]$

$= (\mathbf{u}'\times\mathbf{u}')\bullet(\mathbf{u}'\times\mathbf{u}'') + (\mathbf{u}\times\mathbf{u}'')\bullet(\mathbf{u}'\times\mathbf{u}'')$

$\quad + (\mathbf{u}\times\mathbf{u}')\bullet(\mathbf{u}''\times\mathbf{u}'') + (\mathbf{u}\times\mathbf{u}')\bullet(\mathbf{u}'\times\mathbf{u}''')$

$= (\mathbf{u}\times\mathbf{u}'')\bullet(\mathbf{u}'\times\mathbf{u}'') + (\mathbf{u}\times\mathbf{u}')\bullet(\mathbf{u}'\times\mathbf{u}''')$.

34. $\mathbf{r} = \mathbf{r}_0\cos\omega t + \left(\frac{\mathbf{v}_0}{\omega}\right)\sin\omega t$

$\frac{d\mathbf{r}}{dt} = -\omega\mathbf{r}_0\sin\omega t + \mathbf{v}_0\cos\omega t$

$\frac{d^2\mathbf{r}}{dt^2} = -\omega^2\mathbf{r}_0\cos\omega t - \omega\mathbf{v}_0\sin\omega t = -\omega^2\mathbf{r}$

$\mathbf{r}(0) = \mathbf{r}_0, \qquad \left.\frac{d\mathbf{r}}{dt}\right|_{t=0} = \mathbf{v}_0.$

Observe that $\mathbf{r}\bullet(\mathbf{r}_0\times\mathbf{v}_0) = 0$ for all t. Therefore the path lies in a plane through the origin having normal $\mathbf{N} = \mathbf{r}_0\times\mathbf{v}_0$.

Let us choose our coordinate system so that $\mathbf{r}_0 = a\mathbf{i}$ ($a > 0$) and $\mathbf{v}_0 = \omega b\mathbf{i} + \omega c\mathbf{j}$ ($c > 0$). Therefore, $\mathbf{N}$ is in the direction of $\mathbf{k}$. The path has parametric equations

$$x = a\cos\omega t + b\sin\omega t$$
$$y = c\sin\omega t.$$

The curve is a conic section since it has a quadratic equation:

$$\frac{1}{a^2}\left(x - \frac{by}{c}\right)^2 + \frac{y^2}{c^2} = 1.$$

Since the path is bounded ($|\mathbf{r}(t)| \le |\mathbf{r}_0| + (|\mathbf{v}_0|/\omega)$), it must be an ellipse.

If $\mathbf{r}_0$ is perpendicular to $\mathbf{v}_0$, then $b = 0$ and the path is the ellipse $(x/a)^2 + (y/c)^2 = 1$ having semi-axes $a = |\mathbf{r}_0|$ and $c = |\mathbf{v}_0|/\omega$.

Section 11.2 Some Applications of Vector Differentiation (page 666)

2. Let $v(t)$ be the speed of the tank car at time t seconds. The mass of the car at time t is $m(t) = M - kt$ kg. At full power, the force applied to the car is $F = Ma$ (since the motor can accelerate the full car at a m/s^2). By Newton's Law, this force is the rate of change of the momentum of the car. Thus

$$\frac{d}{dt}\big[(M-kt)v\big] = Ma$$
$$(M-kt)\frac{dv}{dt} - kv = Ma$$
$$\frac{dv}{Ma+kv} = \frac{dt}{M-kt}$$
$$\frac{1}{k}\ln(Ma+kv) = -\frac{1}{k}\ln(M-kt) + \frac{1}{k}\ln C$$
$$Ma + kv = \frac{C}{M-kt}.$$

At $t = 0$ we have $v = 0$, so $Ma = C/M$. Thus $C = M^2a$ and

$$kv = \frac{M^2a}{M-kt} - Ma = \frac{Makt}{M-kt}.$$

The speed of the tank car at time t (before it is empty) is

$$v(t) = \frac{Mat}{M-kt} \text{ m/s}.$$

4. First observe that

$$\frac{d}{dt}|\mathbf{r}-\mathbf{b}|^2 = 2(\mathbf{r}-\mathbf{b})\bullet\frac{d\mathbf{r}}{dt} = 2(\mathbf{r}-\mathbf{b})\bullet\big(\mathbf{a}\times(\mathbf{r}-\mathbf{b})\big) = 0,$$

so $|\mathbf{r}-\mathbf{b}|$ is constant; for all t the object lies on the sphere centred at the point with position vector $\mathbf{b}$ having radius $\mathbf{r}_0 - \mathbf{b}$.

Next, observe that

$$\frac{d}{dt}(\mathbf{r}-\mathbf{r}_0)\bullet\mathbf{a} = \big(\mathbf{a}\times(\mathbf{r}-\mathbf{b})\big)\bullet\mathbf{a} = 0,$$

so $\mathbf{r}-\mathbf{r}_0 \perp \mathbf{a}$; for all t the object lies on the plane through $\mathbf{r}_0$ having normal $\mathbf{a}$. Hence the path of the object lies on the circle in which this plane intersects the sphere described above. The angle between $\mathbf{r}-\mathbf{b}$ and $\mathbf{a}$ must therefore also be constant, and so the object's speed $|d\mathbf{r}/dt|$ is constant. Hence the path must be the whole circle.

6. We use the fixed and rotating frames as described in the text. Assume the satellite is in an orbit in the plane spanned by the fixed basis vectors $\mathbf{I}$ and $\mathbf{K}$. When the satellite passes overhead an observer at latitude 45°, its position is

$$\mathbf{R} = R\frac{\mathbf{I}+\mathbf{K}}{\sqrt{2}},$$

where R is the radius of the earth, and since it circles the earth in 2 hours, its velocity at that point is

$$\mathbf{V} = \pi R \frac{\mathbf{I} - \mathbf{K}}{\sqrt{2}}.$$

The angular velocity of the earth is $\boldsymbol{\Omega} = (\pi/12)\mathbf{K}$.

The rotating frame with origin at the observer's position has, at the instant in question, its basis vectors satisfying

$$\mathbf{I} = -\frac{1}{\sqrt{2}}\mathbf{j} + \frac{1}{\sqrt{2}}\mathbf{k}$$

$$\mathbf{J} = \mathbf{i}$$

$$\mathbf{K} = \frac{1}{\sqrt{2}}\mathbf{j} + \frac{1}{\sqrt{2}}\mathbf{k}.$$

As shown in the text, the velocity $\mathbf{v}$ of the satellite as it appears to the observer is given by $\mathbf{V} = \mathbf{v} + \boldsymbol{\Omega} \times \mathbf{R}$. Thus

$$\mathbf{v} = \mathbf{V} - \boldsymbol{\Omega} \times \mathbf{R}$$
$$= \frac{\pi R}{\sqrt{2}}(\mathbf{I} - \mathbf{K}) - \frac{pi}{12}\mathbf{K} \times \frac{R}{\sqrt{2}}(\mathbf{I} + \mathbf{K})$$
$$= \frac{\pi R}{\sqrt{2}}(\mathbf{I} - \mathbf{K}) - \frac{\pi R}{12\sqrt{2}}\mathbf{J}$$
$$= -\pi R\mathbf{j} - \frac{\pi R}{12\sqrt{2}}\mathbf{i}.$$

$\mathbf{v}$ makes angle

$$\tan^{-1}\left(\frac{\pi R/12\sqrt{2}}{\pi R}\right) = \tan^{-1}(1/(12\sqrt{2}) \approx 3.37°\text{ with}$$

the southward direction. Thus the satellite appears to the observer to be moving in a direction 3.37° west of south.

The apparent Coriolis force is

$$-2\boldsymbol{\Omega} \times \mathbf{v} = -2\frac{\pi}{12}\mathbf{K} \times \left(\frac{\pi R}{\sqrt{2}}(\mathbf{I} - \mathbf{K} - \frac{\pi R}{12\sqrt{2}}\mathbf{J}\right)$$
$$= -\frac{\pi^2 R}{6\sqrt{2}}\left(\mathbf{J} + \frac{1}{12}\mathbf{I}\right)$$
$$= -\frac{\pi^2 R}{6\sqrt{2}}\left(\mathbf{i} + \frac{1}{12\sqrt{2}}(-\mathbf{j} + \mathbf{k})\right).$$

8. We continue with the same notation as in Example 4. Since $\mathbf{j}$ points northward at the observer's position, the angle μ between the direction vector of the sun, $\mathbf{S} = \cos\sigma\mathbf{I} + \sin\sigma\mathbf{J}$ and north satisfies

$$\cos\mu = \mathbf{S} \bullet \mathbf{j} = -\cos\sigma\cos\phi\cos\theta + \sin\sigma\sin\phi.$$

For the sun, $\theta = 0$ and at sunrise and sunset we have, by Example 4, $\cos\theta = -\tan\sigma/\tan\phi$, so that

$$\cos\mu = \cos\sigma\cos\phi\frac{\tan\sigma}{\tan\phi} + \sin\sigma\sin\phi$$
$$= \sin\sigma\frac{\cos^2\phi}{\sin\phi} + \sin\sigma\sin\phi$$
$$= \frac{\sin\sigma}{\sin\phi}.$$

10. At Umeå, $\phi = 90° - 63.5° = 26.5°$. On June 21st, $\sigma = 23.3°$. By Example 4 there will be

$$\frac{24}{\pi}\cos^{-1}\left(-\frac{\tan 23.3°}{\tan 26.5°}\right) \approx 20$$

hours between sunrise and sunset. By Exercise 8, the sun will rise and set at an angle

$$\cos^{-1}\left(\frac{\sin 23.3°}{\sin 26.5°}\right) \approx 27.6°$$

to the east and west of north.

Section 11.3   Curves and Parametrizations (page 673)

2. On the first quadrant part of the circle $x^2 + y^2 = a^2$ we have $y = \sqrt{a^2 - x^2}$, $0 \le x \le a$. The required parametrization is

$$\mathbf{r} = \mathbf{r}(x) = x\mathbf{i} + \sqrt{a^2 - x^2}\mathbf{j}, \quad (0 \le x \le a).$$

4. $x = a\sin\dfrac{s}{a}$, $y = a\cos\dfrac{s}{a}$, $0 \le \dfrac{s}{a} \le \dfrac{\pi}{2}$

$\mathbf{r} = a\sin\dfrac{s}{a}\mathbf{i} + a\cos\dfrac{s}{a}\mathbf{j}$, $\left(0 \le s \le \dfrac{a\pi}{2}\right)$.

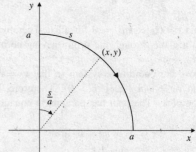

Fig. 11.3.4

6. $z = x^2$, $x + y + z = 1$. If $t = x$, then $z = t^2$ and $y = 1 - t - t^2$. The parametrization is $\mathbf{r} = t\mathbf{i} + (1 - t - t^2)\mathbf{j} + t^2\mathbf{k}$.

8. $x + y = 1$, $z = \sqrt{1 - x^2 - y^2}$. If $x = t$, then $y = 1 - t$ and

$$z = \sqrt{1 - t^2 - (1 - t)^2} = \sqrt{2(t - t^2)}.$$ One possible parametrization is

$$\mathbf{r} = t\mathbf{i} + (1 - t)\mathbf{j} + \sqrt{2(t - t^2)}\mathbf{k}.$$

10. $yz + x = 1$, $xz - x = 1$. One possible parametrization is $x = t$, $z = (1+t)/t$, and $y = (1-t)/z = (1-t)t/(1+t)$, that is,

$$\mathbf{r} = t\mathbf{i} + \frac{t - t^2}{1+t}\mathbf{j} + \frac{1+t}{t}\mathbf{k}.$$

12. By symmetry, the centre of the circle $\mathcal{C}$ of intersection of the plane $x + y + z = 1$ and the sphere $x^2 + y^2 + z^2 = 1$ must lie on the plane and must have its three coordinates equal. Thus the centre has position vector

$$\mathbf{r}_0 = \frac{1}{3}(\mathbf{i} + \mathbf{j} + \mathbf{k}).$$

Since $\mathcal{C}$ passes through the point $(0, 0, 1)$, its radius is

$$\sqrt{\left(0 - \frac{1}{3}\right)^2 + \left(0 - \frac{1}{3}\right)^2 + \left(1 - \frac{1}{3}\right)^2} = \sqrt{\frac{2}{3}}.$$

Any vector $\mathbf{v}$ that satisfies $\mathbf{v} \bullet (\mathbf{i} + \mathbf{j} + \mathbf{k}) = 0$ is parallel to the plane $x + y + z = 1$ containing $\mathcal{C}$. One such vector is $\mathbf{v}_1 = \mathbf{i} - \mathbf{j}$. A second one, perpendicular to $\mathbf{v}_1$, is

$$\mathbf{v}_2 = (\mathbf{i} + \mathbf{j} + \mathbf{k}) \times (\mathbf{i} - \mathbf{j}) = \mathbf{i} + \mathbf{j} - 2\mathbf{k}.$$

Two perpendicular unit vectors that are parallel to the plane of $\mathcal{C}$ are

$$\hat{\mathbf{v}}_1 = \frac{\mathbf{i} - \mathbf{j}}{\sqrt{2}}, \quad \hat{\mathbf{v}}_2 = \frac{\mathbf{i} + \mathbf{j} - 2\mathbf{k}}{\sqrt{6}}.$$

Thus one possible parametrization of $\mathcal{C}$ is

$$\mathbf{r} = \mathbf{r}_0 + \sqrt{\frac{2}{3}}(\cos t\,\hat{\mathbf{v}}_1 + \sin t\,\hat{\mathbf{v}}_2)$$

$$= \frac{\mathbf{i} + \mathbf{j} + \mathbf{k}}{3} + \frac{\cos t}{\sqrt{3}}(\mathbf{i} - \mathbf{j}) + \frac{\sin t}{3}(\mathbf{i} + \mathbf{j} - 2\mathbf{k}).$$

14. $\mathbf{r} = t\mathbf{i} + \lambda t^2\mathbf{j} + t^3\mathbf{k}$, $(0 \le t \le T)$

$v = \sqrt{1 + (2\lambda t)^2 + 9t^4} = \sqrt{(1 + 3t^2)^2}$

if $4\lambda^2 = 6$, that is, if $\lambda = \pm\sqrt{3/2}$. In this case, the length of the curve is

$$s(T) = \int_0^T (1 + 3t^2)\, dt = T + T^3.$$

16. $x = a\cos t\sin t = \dfrac{a}{2}\sin 2t$,

$y = a\sin^2 t = \dfrac{a}{2}(1 - \cos 2t)$,

$z = bt$.

The curve is a circular helix lying on the cylinder

$$x^2 + \left(y - \frac{a}{2}\right)^2 = \frac{a^2}{4}.$$

Its length, from $t = 0$ to $t = T$, is

$$L = \int_0^T \sqrt{a^2\cos^2 2t + a^2\sin^2 2t + b^2}\, dt$$

$$= T\sqrt{a^2 + b^2} \text{ units.}$$

18. One-eighth of the curve $\mathcal{C}$ lies in the first octant. That part can be parametrized

$$x = \cos t, \quad z = \frac{1}{\sqrt{2}}\sin t, \quad (0 \le t \le \pi/2)$$

$$y = \sqrt{1 - \cos^2 t - \frac{1}{2}\sin^2 t} = \frac{1}{\sqrt{2}}\sin t.$$

Since the first octant part of $\mathcal{C}$ lies in the plane $y = z$, it must be a quarter of a circle of radius 1. Thus the length of all of $\mathcal{C}$ is $8 \times (\pi/2) = 4\pi$ units.

If you wish to use an integral, the length is

$$8\int_0^{\pi/2} \sqrt{\sin^2 t + \frac{1}{2}\cos^2 t + \frac{1}{2}\cos^2 t}\, dt$$

$$= 8\int_0^{\pi/2} dt = 4\pi \text{ units.}$$

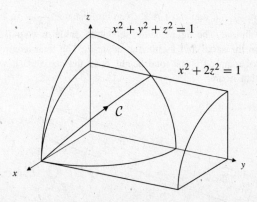

Fig. 11.3.18

20. $\mathbf{r} = t^3\mathbf{i} + t^2\mathbf{j}$

$\mathbf{v} = 3t^2\mathbf{i} + 2t\mathbf{j}$

$v = |\mathbf{v}| = \sqrt{9t^4 + 4t^2} = |t|\sqrt{9t^2 + 4}$

The length L between $t = -1$ and $t = 2$ is

$$L = \int_{-1}^0 (-t)\sqrt{9t^2 + 4}\, dt + \int_0^2 t\sqrt{9t^2 + 4}\, dt.$$

Making the substitution $u = 9t^2 + 4$ in each integral, we obtain

$$L = \frac{1}{18}\left[\int_4^{13} u^{1/2}\,du + \int_4^{40} u^{1/2}\,du\right]$$

$$= \frac{1}{27}\left(13^{3/2} + 40^{3/2} - 16\right) \quad \text{units.}$$

22. (Solution due to Roland Urbanek, a student at Okanagan College.) Suppose the spool is vertical and the cable windings make angle θ with the horizontal at each point.

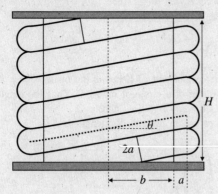

Fig. 11.3.22

The centreline of the cable is wound around a cylinder of radius $a + b$ and must rise a vertical distance $\dfrac{2a}{\cos\theta}$ in one revolution. The figure below shows the cable unwound from the spool and inclined at angle θ. The total length of spool required is the total height H of the cable as shown in that figure.

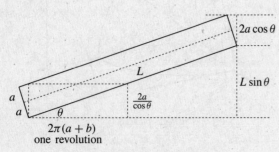

$2\pi(a + b)$
one revolution

Fig. 11.3.22

Observe that $\tan\theta = \dfrac{2a}{\cos\theta} \times \dfrac{1}{2\pi(a + b)}$. Therefore

$$\sin\theta = \frac{a}{\pi(a + b)}$$

$$\cos\theta = \sqrt{1 - \frac{a^2}{\pi^2(a + b)^2}} = \frac{\sqrt{\pi^2(a + b)^2 - a^2}}{\pi(a + b)}.$$

The total length of spool required is

$$H = L\sin\theta + 2a\cos\theta$$

$$= \frac{a}{\pi(a + b)}\left(L + 2\sqrt{\pi^2(a + b)^2 - a^2}\right) \text{ units.}$$

24. $\mathbf{r} = e^t\mathbf{i} + \sqrt{2}t\mathbf{j} - e^{-t}\mathbf{k}$
    $\mathbf{v} = e^t\mathbf{i} + \sqrt{2}\mathbf{j} + e^{-t}\mathbf{k}$
    $v = |\mathbf{v}| = \sqrt{e^{2t} + 2 + e^{-2t}} = e^t + e^{-t}.$
    The arc length from the point where $t = 0$ to the point corresponding to arbitrary t is

$$s = s(t) = \int_0^t (e^u + e^{-u})\,du = e^t - e^{-t} = 2\sinh t.$$

Thus $t = \sinh^{-1}(s/2) = \ln\left(\dfrac{s + \sqrt{s^2 + 4}}{2}\right)$,

and $e^t = \dfrac{s + \sqrt{s^2 + 4}}{2}$. The required parametrization is

$$\mathbf{r} = \frac{s + \sqrt{s^2 + 4}}{2}\mathbf{i} + \sqrt{2}\ln\left(\frac{s + \sqrt{s^2 + 4}}{2}\right)\mathbf{j} - \frac{2\mathbf{k}}{s + \sqrt{s^2 + 4}}.$$

26. $\mathbf{r} = 3t\cos t\,\mathbf{i} + 3t\sin t\,\mathbf{j} + 2\sqrt{2}t^{3/2}\mathbf{k}, \quad (t \ge 0)$
    $\mathbf{v} = 3(\cos t - t\sin t)\mathbf{i} + 3(\sin t + t\cos t)\mathbf{j} + 3\sqrt{2}\sqrt{t}\,\mathbf{k}$
    $v = |\mathbf{v}| = 3\sqrt{1 + t^2 + 2t} = 3(1 + t)$

$$s = \int_0^t 3(1 + u)\,du = 3\left(t + \frac{t^2}{2}\right)$$

Thus $t^2 + 2t = \dfrac{2s}{3}$, so $t = -1 + \sqrt{1 + \dfrac{2s}{3}}$ since $t \ge 0$. The required parametrization is the given one with t replaced by $-1 + \sqrt{1 + (2s)/3}$.

28. If $\mathbf{r} = \mathbf{r}(t)$ has nonvanishing velocity $\mathbf{v} = d\mathbf{r}/dt$ on $[a, b]$, then for any t_0 in $[a, b]$, the function

$$s = g(t) = \int_{t_0}^t |\mathbf{v}(u)|\,du,$$

which gives the (signed) arc length s measured from $\mathbf{r}(t_0)$ along the curve, is an increasing function:

$$\frac{ds}{dt} = g'(t) = |\mathbf{v}(t)| > 0$$

on $[a, b]$, by the Fundamental Theorem of Calculus. Hence g is invertible, and defines t as a function of arc length s:

$$t = g^{-1}(s) \Leftrightarrow s = g(t).$$

Then

$$\mathbf{r} = \mathbf{r}_2(s) = \mathbf{r}\big(g^{-1}(s)\big)$$

is a parametrization of the curve $\mathbf{r} = \mathbf{r}(t)$ in terms of arc length.

Section 11.4   Curvature, Torsion, and the Frenet Frame   (page 682)

2. $\mathbf{r} = a \sin \omega t \mathbf{i} + a \cos \omega t \mathbf{k}$

$\mathbf{v} = a\omega \cos \omega t \mathbf{i} - a\omega \sin \omega t \mathbf{k}, \quad v = |a\omega|$

$\hat{\mathbf{T}} = \operatorname{sgn}(a\omega)\left[\cos \omega t \mathbf{i} - \sin \omega t \mathbf{k}\right].$

4. $\mathbf{r} = a \cos t \mathbf{i} + b \sin t \mathbf{j} + t \mathbf{k}$

$\mathbf{v} = -a \sin t \mathbf{i} + b \cos t \mathbf{j} + \mathbf{k}$

$v = \sqrt{a^2 \sin^2 t + b^2 \cos^2 t + 1}$

$\hat{\mathbf{T}} = \dfrac{\mathbf{v}}{v} = \dfrac{-a \sin t \mathbf{i} + b \cos t \mathbf{j} + \mathbf{k}}{\sqrt{a^2 \sin^2 t + b^2 \cos^2 t + 1}}.$

6.   If $\tau(s) = 0$ for all s, then

$\dfrac{d\hat{\mathbf{B}}}{ds} = -\tau \hat{\mathbf{N}} = 0$, so $\hat{\mathbf{B}}(s) = \hat{\mathbf{B}}(0)$ is constant. Therefore,

$$\frac{d}{ds}\left(\mathbf{r}(s) - \mathbf{r}(0)\right) \bullet \hat{\mathbf{B}}(s) = \frac{d\mathbf{r}}{ds} \bullet \hat{\mathbf{B}}(s) = \hat{\mathbf{T}}(s) \bullet \hat{\mathbf{B}}(s) = 0.$$

It follows that

$$\left(\mathbf{r}(s) - \mathbf{r}(0)\right) \bullet \hat{\mathbf{B}}(0) = \left(\mathbf{r}(s) - \mathbf{r}(0)\right) \bullet \hat{\mathbf{B}}(s) = 0$$

for all s. This says that $\mathbf{r}(s)$ lies in the plane through $\mathbf{r}(0)$ having normal $\hat{\mathbf{B}}(0)$.

8.   The circular helix $\mathcal{C}_1$ given by

$$\mathbf{r} = a \cos t \mathbf{i} + a \sin t \mathbf{j} + bt \mathbf{k}$$

has curvature and torsion given by

$$\kappa(s) = \frac{a}{a^2 + b^2}, \qquad \tau(s) = \frac{b}{a^2 + b^2},$$

by Example 3.

if a curve $\mathcal{C}$ has constant curvature $\kappa(s) = C > 0$, and constant torsion $\tau(s) = T \neq 0$, then we can choose a and b so that

$$\frac{a}{a^2 + b^2} = C, \qquad \frac{b}{a^2 + b^2} = T.$$

(Specifically, $a = \dfrac{C}{C^2 + T^2}$, and $b = \dfrac{T}{C^2 + T^2}$.) By Theorem 3, $\mathcal{C}$ is itself a circular helix, congruent to $\mathcal{C}_1$.

Section 11.5   Curvature and Torsion for General Parametrizations   (page 689)

2.   For $y = \cos$ we have

$$\kappa(x) = \frac{|d^2y/dx^2|}{(1 + (dy/dx)^2)^{3/2}} = \frac{|\cos x|}{(1 + \sin^2 x)^{3/2}}.$$

Hence $\kappa(0) = 1$ and $\kappa(\pi/2) = 0$. The radius of curvature at $x = 0$ is 1. The radius of curvature at $x = \pi/2$ is infinite.

4. $\mathbf{r} = t^3 \mathbf{i} + t^2 \mathbf{j} + t \mathbf{k}$

$\mathbf{v} = 3t^2 \mathbf{i} + 2t \mathbf{j} + \mathbf{k}$

$\mathbf{a} = 6t \mathbf{i} + 2 \mathbf{j}$

$\mathbf{v}(1) = 3\mathbf{i} + 2\mathbf{j} + \mathbf{k}, \quad \mathbf{a}(1) = 6\mathbf{i} + 2\mathbf{j}$

$\mathbf{v}(1) \times \mathbf{a}(1) = -2\mathbf{i} + 6\mathbf{j} - 6\mathbf{k}$

$\kappa(1) = \dfrac{\sqrt{4 + 36 + 36}}{(9 + 4 + 1)^{3/2}} = \dfrac{2\sqrt{19}}{14^{3/2}}$

At $t = 1$ the radius of curvature is $14^{3/2}/(2\sqrt{19})$.

6. $\mathbf{r} = t\mathbf{i} + t^2 \mathbf{j} + t \mathbf{k}$

$\mathbf{v} = \mathbf{i} + 2t \mathbf{j} + \mathbf{k}$

$\mathbf{a} = 2\mathbf{j}$

$\mathbf{v} \times \mathbf{a} = -2\mathbf{i} + 2\mathbf{k}$

At $(1, 1, 1)$, where $t = 1$, we have

$\hat{\mathbf{T}} = \mathbf{v}/|\mathbf{v}| = (\mathbf{i} + 2\mathbf{j} + \mathbf{k})/\sqrt{6}$

$\hat{\mathbf{B}} = (\mathbf{v} \times \mathbf{a})/|\mathbf{v} \times \mathbf{a}| = -(\mathbf{i} - \mathbf{k})/\sqrt{2}$

$\hat{\mathbf{N}} = \hat{\mathbf{B}} \times \hat{\mathbf{T}} = -(\mathbf{i} - \mathbf{j} + \mathbf{k})/\sqrt{3}.$

8. $\mathbf{r} = e^t \cos t \mathbf{i} + e^t \sin t \mathbf{j} + e^t \mathbf{k}$

$\mathbf{v} = e^t(\cos t - \sin t)\mathbf{i} + e^t(\sin t + \cos t)\mathbf{j} + e^t \mathbf{k}$

$\mathbf{a} = -2e^t \sin t \mathbf{i} + 2e^t \cos t \mathbf{j} + e^t \mathbf{k}$

$\dfrac{d\mathbf{a}}{dt} = -2e^t(\cos t + \sin t)\mathbf{i} + 2e^t(\cos t - \sin t)\mathbf{j} + e^t \mathbf{k}$

$\mathbf{v} \times \mathbf{a} = e^{2t}(\sin t - \cos t)\mathbf{i} - e^{2t}(\cos t + \sin t)\mathbf{j} + 2e^{2t} \mathbf{k}$

$v = |\mathbf{v}| = \sqrt{3}e^t, \qquad |\mathbf{v} \times \mathbf{a}| = \sqrt{6}e^{2t}$

$(\mathbf{v} \times \mathbf{a}) \bullet \dfrac{d\mathbf{a}}{dt} = 2e^{3t}$

$\hat{\mathbf{T}} = \dfrac{\mathbf{v}}{v} = \dfrac{(\cos t - \sin t)\mathbf{i} + (\cos t + \sin t)\mathbf{j} + \mathbf{k}}{\sqrt{3}}$

$\hat{\mathbf{B}} = \dfrac{\mathbf{v} \times \mathbf{a}}{|\mathbf{v} \times \mathbf{a}|} = \dfrac{(\sin t - \cos t)\mathbf{i} - (\cos t + \sin t)\mathbf{j} + 2\mathbf{k}}{\sqrt{6}}$

$\hat{\mathbf{N}} = \hat{\mathbf{B}} \times \hat{\mathbf{T}} = -\dfrac{(\cos t + \sin t)\mathbf{i} - (\cos t - \sin t)\mathbf{j}}{\sqrt{2}}$

$\kappa = \dfrac{|\mathbf{v} \times \mathbf{a}|}{v^3} = \dfrac{\sqrt{2}}{3e^t}$

$\tau = \dfrac{(\mathbf{v} \times \mathbf{a}) \bullet \dfrac{d\mathbf{a}}{dt}}{|\mathbf{v} \times \mathbf{a}|^2} = \dfrac{1}{3e^t}.$

10. $\mathbf{r} = x\mathbf{i} + \sin x\mathbf{j}$

$\mathbf{v} = \dfrac{dx}{dt}\mathbf{i} + \cos x\,\dfrac{dx}{dt}\mathbf{j} = k(\mathbf{i} + \cos x\mathbf{j})$

$v = k\sqrt{1 + \cos^2 x}$

$\mathbf{a} = -k\sin x\,\dfrac{dx}{dt}\mathbf{j} = -k^2\sin x\mathbf{j}$

$\mathbf{v} \times \mathbf{a} = -k^3\sin x\mathbf{k}$

$\kappa = \dfrac{|\mathbf{v} \times \mathbf{a}|}{v^3} = \dfrac{|\sin x|}{(1 + \cos^2 x)^{3/2}}.$

The tangential and normal components of acceleration are

$\dfrac{dv}{dt} = \dfrac{k}{2\sqrt{1 + \cos^2 x}}2\cos x)(-\sin x)\dfrac{dx}{dt} = -\dfrac{k^2\cos x\sin x}{\sqrt{1 + \cos^2 x}}$

$v^2\kappa = \dfrac{k^2|\sin x|}{\sqrt{1 + \cos^2 x}}.$

12. $\mathbf{r} = a\cos t\mathbf{i} + b\sin t\mathbf{j}$

$\mathbf{v} = -a\sin t\mathbf{i} + b\cos t\mathbf{j}$

$\mathbf{a} = -a\cos t\mathbf{i} - b\sin t\mathbf{j}$

$\mathbf{v} \times \mathbf{a} = ab\mathbf{k}$

$v = \sqrt{a^2\sin^2 t + b^2\cos^2 t}.$

The tangential component of acceleration is

$$\dfrac{dv}{dt} = \dfrac{(a^2 - b^2)\sin t\cos t}{\sqrt{a^2\sin^2 t + b^2\cos^2 t}},$$

which is zero if t is an integer multiple of $\pi/2$, that is, at the ends of the major and minor axes of the ellipse. The normal component of acceleration is

$$v^2\kappa = v^2\dfrac{|\mathbf{v} \times \mathbf{a}|}{v^3} = \dfrac{ab}{\sqrt{a^2\sin^2 t + b^2\cos^2 t}}.$$

14. By Example 2, the curvature of $y = x^2$ at $(1, 1)$ is

$$\kappa = \left.\dfrac{2}{(1 + 4x^2)^{3/2}}\right|_{x=1} = \dfrac{2}{5\sqrt{5}}.$$

Thus the magnitude of the normal acceleration of the bead at that point is $v^2\kappa = 2v^2/(5\sqrt{5})$.

The rate of change of the speed, dv/dt, is the tangential component of the acceleration, and is due entirely to the tangential component of the gravitational force since there is no friction:

$$\dfrac{dv}{dt} = g\cos\theta = g(-\mathbf{j}) \bullet \hat{\mathbf{T}},$$

where θ is the angle between $\hat{\mathbf{T}}$ and $-\mathbf{j}$. (See the figure.) Since the slope of $y = x^2$ at $(1, 1)$ is 2, we have $\hat{\mathbf{T}} = -(\mathbf{i} + 2\mathbf{j})/\sqrt{5}$, and therefore $dv/dt = 2g/\sqrt{5}$.

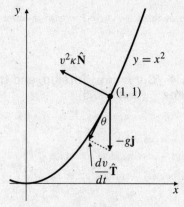

Fig. 11.5.14

16. The curve with polar equation $r = f(\theta)$ is given parametrically by

$$\mathbf{r} = f(\theta)\cos\theta\mathbf{i} + f(\theta)\sin\theta\mathbf{j}.$$

Thus we have

$\mathbf{v} = \big(f'(\theta)\cos\theta - f(\theta)\sin\theta\big)\mathbf{i}$
$\qquad + \big(f'(\theta)\sin\theta + f(\theta)\cos\theta\big)\mathbf{j}$

$\mathbf{a} = \big(f''(\theta)\cos\theta - 2f'(\theta)\sin\theta - f(\theta)\cos\theta\big)\mathbf{i}$
$\qquad + \big(f''(\theta)\sin\theta + 2f'(\theta)\cos\theta - f(\theta)\sin\theta\big)\mathbf{j}$

$v = |\mathbf{v}| = \sqrt{\big(f'(\theta)\big)^2 + \big(f(\theta)\big)^2}$

$\mathbf{v} \times \mathbf{a} = \Big[2\big(f'(\theta)\big)^2 + \big(f(\theta)\big)^2 - f(\theta)f''(\theta)\Big]\mathbf{k}.$

The curvature is, therefore,

$$\dfrac{\big|2\big(f'(\theta)\big)^2 + \big(f(\theta)\big)^2 - f(\theta)f''(\theta)\big|}{\Big[\big(f'(\theta)\big)^2 + \big(f(\theta)\big)^2\Big]^{3/2}}.$$

18. By Exercise 8 of Section 2.4, the required curve must be a circular helix with parameters $a = 1/2$ (radius), and $b = 1/2$. Its equation will be

$$\mathbf{r} = \dfrac{1}{2}\cos t\mathbf{i}_1 + \dfrac{1}{2}\sin t\mathbf{j}_1 + \dfrac{1}{2}t\mathbf{k}_1 + \mathbf{r}_0$$

for some right-handed basis $\{\mathbf{i}_1, \mathbf{j}_1, \mathbf{k}_1\}$, and some constant vector $\mathbf{r}_0$. Example 3 of Section 2.4 provides values for $\hat{\mathbf{T}}(0)$, $\hat{\mathbf{N}}(0)$, and $\hat{\mathbf{B}}(0)$, which we can equate to the given values of these vectors:

$$\mathbf{i} = \hat{\mathbf{T}}(0) = \dfrac{1}{\sqrt{2}}\mathbf{j}_1 + \dfrac{1}{\sqrt{2}}\mathbf{k}_1$$

$$\mathbf{j} = \hat{\mathbf{N}}(0) = -\mathbf{i}_1$$

$$\mathbf{k} = \hat{\mathbf{B}}(0) = -\dfrac{1}{\sqrt{2}}\mathbf{j}_1 + \dfrac{1}{\sqrt{2}}\mathbf{k}_1.$$

Solving these equations for $\mathbf{i}_1$, $\mathbf{j}_1$, and $\mathbf{k}_1$ in terms of the given basis vectors, we obtain

$$\mathbf{i}_1 = -\mathbf{j}$$
$$\mathbf{j}_1 = \frac{1}{\sqrt{2}}\mathbf{i} - \frac{1}{\sqrt{2}}\mathbf{k}.$$
$$\mathbf{k}_1 = \frac{1}{\sqrt{2}}\mathbf{i} + \frac{1}{\sqrt{2}}\mathbf{k}.$$

Therefore

$$\mathbf{r}(t) = \frac{t + \sin t}{2\sqrt{2}}\mathbf{i} - \frac{\cos t}{2}\mathbf{j} + \frac{t - \sin t}{2\sqrt{2}}\mathbf{k} + \mathbf{r}_0.$$

We also require that $\mathbf{r}(0) = \mathbf{i}$, so $\mathbf{r}_0 = \mathbf{i} + \frac{1}{2}\mathbf{j}$. The required equation is, therefore,

$$\mathbf{r}(t) = \left(\frac{t + \sin t}{2\sqrt{2}} + 1\right)\mathbf{i} + \frac{1 - \cos t}{2}\mathbf{j} + \frac{t - \sin t}{2\sqrt{2}}\mathbf{k}.$$

20. For $\mathbf{r} = a\cos t\,\mathbf{i} + a\sin t\,\mathbf{j} + bt\,\mathbf{k}$, we have, by Example 3 of Section 2.4,

$$\hat{\mathbf{N}} = -\cos t\,\mathbf{i} - \sin t\,\mathbf{j}, \qquad \kappa = \frac{a}{a^2 + b^2}.$$

The centre of curvature $\mathbf{r}_c$ is given by

$$\mathbf{r}_c = \mathbf{r} + \rho\hat{\mathbf{N}} = \mathbf{r} + \frac{1}{\kappa}\hat{\mathbf{N}}.$$

Thus the evolute has equation

$$\mathbf{r} = a\cos t\,\mathbf{i} + a\sin t\,\mathbf{j} + bt\,\mathbf{k}$$
$$\qquad - \frac{a^2 + b^2}{a}(\cos t\,\mathbf{i} + \sin t\,\mathbf{j})$$
$$= -\frac{b^2}{a}\cos t\,\mathbf{i} - \frac{b^2}{a}\sin t\,\mathbf{j} + bt\,\mathbf{k}.$$

The evolute is also a circular helix.

22. For the ellipse $\mathbf{r} = 2\cos t\,\mathbf{i} + \sin t\,\mathbf{j}$, we have

$$\mathbf{v} = -2\sin t\,\mathbf{i} + \cos t\,\mathbf{j}$$
$$\mathbf{a} = -2\cos t\,\mathbf{i} - \sin t\,\mathbf{j}$$
$$\mathbf{v} \times \mathbf{a} = 2\mathbf{k}$$
$$v = \sqrt{4\sin^2 t + \cos^2 t} = \sqrt{3\sin^2 t + 1}.$$

The curvature is $\kappa = \dfrac{2}{(3\sin^2 t + 1)^{3/2}}$, so the radius of curvature is $\rho = \dfrac{(3\sin^2 t + 1)^{3/2}}{2}$. We have

$$\hat{\mathbf{T}} = \frac{-2\sin t\,\mathbf{i} + \cos t\,\mathbf{j}}{\sqrt{3\sin^2 t + 1}}, \qquad \hat{\mathbf{B}} = \mathbf{k}$$
$$\hat{\mathbf{N}} = -\frac{\cos t\,\mathbf{i} + 2\sin t\,\mathbf{j}}{\sqrt{3\sin^2 t + 1}}.$$

Therefore the evolute has equation

$$\mathbf{r} = 2\cos t\,\mathbf{i} + \sin t\,\mathbf{j} - \frac{3\sin^2 t + 1}{2}(\cos t\,\mathbf{i} + 2\sin t\,\mathbf{j})$$
$$= \frac{3}{2}\cos^3\mathbf{i} - 3\sin^3 t\,\mathbf{j}.$$

24. We require

$$f(0) = 1, \qquad f'(0) = 0, \qquad f''(0) = -1,$$
$$f(-1) = 1, \qquad f'(-1) = 0, \qquad f''(-1) = 0.$$

The condition $f''(0) = -1$ follows from the fact that

$$\frac{d^2}{dx^2}\sqrt{1 - x^2}\,\bigg|_{x=0} = -1.$$

As in Example 5, we try

$$f(x) = A + Bx + Cx^2 + Dx^3 + Ex^4 + Fx^5$$
$$f'(x) = B + 2Cx + 3Dx^2 + 4Ex^3 + 5Fx^4$$
$$f'' = 2C + 6Dx + 12Ex^2 + 20Fx^3.$$

The required conditions force the coefficients to satisfy the system of equations

$$A - B + C - D + E - F = 1$$
$$B - 2C + 3D - 4E + 5F = 0$$
$$2C - 6D + 12E - 20F = 0$$
$$A = 1$$
$$B = 0$$
$$2C = -1$$

which has solution $A = 1$, $B = 0$, $C = -1/2$, $D = -3/2$, $E = -3/2$, $F = -1/2$. Thus we can use a track section in the shape of the graph of

$$f(x) = 1 - \frac{1}{2}x^2 - \frac{3}{2}x^3 - \frac{3}{2}x^4 - \frac{1}{2}x^5 = 1 - \frac{1}{2}x^2(1 + x)^3.$$

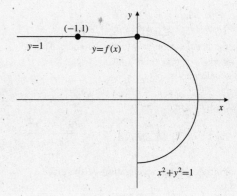

Fig. 11.5.24

221

26. Redefining `x := cos(t); y := 2*sin(t); z := cos(t);` in the `curve-3d.mws` worksheet, and reexecuting the sheet, leads to the following values for the curvature and torsion:

$$\text{curv}(t) = \frac{1}{\left(\cos^2 t + 1\right)^{3/2}}$$

$$\text{tors}(t) = 0.$$

The torsion is zero because the curve is lies in the plane $z = x$. It is the ellipse in which this plane intersects the ellipsoid $2x^2 + y^2 + 2z^2 = 4$.

28. Redefining `x := cos(t)*cos(2*t); y := cos(t)*sin(2*t); z := sin(t);` in the `curve-3d.mws` worksheet, and reexecuting the sheet, leads to the following values for the curvature and torsion:

$$\text{curv}(t) = \frac{\sqrt{48\cos^4 t + 60\cos^2 t + 17}}{\left(4\cos^2 t + 1\right)^{3/2}}$$

$$\text{tors}(t) = \frac{12\cos t (2\cos^2 t + 3)}{48\cos^4 t + 60\cos^2 t + 17}.$$

Plotting the curvature as a function of t, (`plot(curv(t),t=-2*Pi..2*Pi)`), shows that the minimum curvature occurs at $t = 0$ (and any integer multiple of π). The minimum curvature is $\sqrt{125}/5^{3/2} = 1$.

30. ```
evolute :=
    t -> evl(R(t)+(1/curv(t))*N(t));
```

## Section 11.6   Kepler's Laws of Planetary Motion   (page 699)

**2.** Position: $\mathbf{r} = r\hat{\mathbf{r}} = k\hat{\mathbf{r}}$.
Velocity: $\mathbf{v} = k\dot{\hat{\mathbf{r}}} = k\dot{\theta}\hat{\boldsymbol{\theta}}$; speed: $v = k\dot{\theta}$.
Acceleration: $k\ddot{\theta}\hat{\boldsymbol{\theta}} + k\dot{\theta}\dot{\hat{\boldsymbol{\theta}}} = -k\dot{\theta}^2\hat{\mathbf{r}} + k\ddot{\theta}\hat{\boldsymbol{\theta}}$.
Radial component of acceleration: $-k\dot{\theta}^2$.
Transverse component of acceleration: $k\ddot{\theta} = \dot{v}$ (the rate of change of the speed).

**4.** Path: $r = \theta$. Thus $\dot{r} = \dot{\theta}$, $\ddot{r} = \ddot{\theta}$.
Speed: $v = \sqrt{(\dot{r})^2 + (r\dot{\theta})^2} = \dot{\theta}\sqrt{1 + r^2}$.
Transverse acceleration = 0 (central force). Thus $r\ddot{\theta} + 2\dot{r}\dot{\theta} = 0$, or $\ddot{\theta} = -2\dot{\theta}^2/r$.
Radial acceleration:

$$\ddot{r} - r\dot{\theta}^2 = \ddot{\theta} - r\dot{\theta}^2$$
$$= -\left(\frac{2}{r} + r\right)\dot{\theta}^2 = -\frac{(2 + r^2)v^2}{r(1 + r^2)}.$$

The magnitude of the acceleration is, therefore,
$$\frac{(2 + r^2)v^2}{r(1 + r^2)}.$$

**6.** Let the period and the semi-major axis of the orbit of Halley's comet be $T_H = 76$ years and $a_H$ km respectively. Similar parameters for the earth's orbit are $T_E = 1$ year and $a_E = 150 \times 10^6$ km. By Kepler's third law

$$\frac{T_H^2}{a_H^3} = \frac{T_E^2}{a_E^3}.$$

Thus

$$a_H = 150 \times 10^6 \times 76^{2/3} \approx 2.69 \times 10^9.$$

The major axis of Halley's comet's orbit is $2a_H \approx 5.38 \times 10^9$ km.

**8.** The period $T$ (in years) and radius $R$ (in km) of the asteroid's orbit satisfies

$$\frac{T^2}{R^3} = \frac{T_{\text{earth}}^2}{R_{\text{earth}}^3} = \frac{1^2}{(150 \times 10^6)^3}.$$

Thus the radius of the asteroid's orbit is $R \approx 150 \times 10^6 T^{2/3}$ km.

**10.** At perihelion, $r = a - c = (1 - \epsilon)a$.
At aphelion $r = a + c = (1 + \epsilon)a$.
Since $\dot{r} = 0$ at perihelion and aphelion, the speed is $v = r\dot{\theta}$ at each point. Since $r^2\dot{\theta} = h$ is constant over the orbit, $v = h/r$. Therefore

$$v_{\text{perihelion}} = \frac{h}{a(1 - \epsilon)}, \qquad v_{\text{aphelion}} = \frac{h}{a(1 + \epsilon)}.$$

If $v_{\text{perihelion}} = 2v_{\text{aphelion}}$ then

$$\frac{h}{a(1 - \epsilon)} = \frac{2h}{a(1 + \epsilon)}.$$

Hence $1 + \epsilon = 2(1 - \epsilon)$, and $\epsilon = 1/3$. The eccentricity of the orbit is 1/3.

**12.** Since $r^2\dot{\theta} = h$ = constant for the planet's orbit, and since the speed is $v = r\dot{\theta}$ at perihelion and at aphelion (the radial velocity is zero at these points), we have

$$r_p v_p = r_a v_a,$$

where the subscripts $p$ and $a$ refer to perihelion and aphelion, respectively. Since $r_p/r_a = 8/10$, we must have $v_p/v_a = 10/8 = 1.25$. Also,

$$r_p = \frac{\ell}{1 + \epsilon\cos 0} = \frac{\ell}{1 + \epsilon}, \qquad r_a = \frac{\ell}{1 + \epsilon\cos\pi} = \frac{\ell}{1 - \epsilon}.$$

Thus $\ell/(1+\epsilon) = (8/10)\ell/(1-\epsilon)$, and so $10 - 10\epsilon = 8 + 8\epsilon$. Hence $2 = 18\epsilon$. The eccentricity of the orbit is $\epsilon = 1/9$.

**14.** As in Exercise 12, $r_P v_P = r_A v_A$, where $r_A = \ell/(1 - \epsilon)$ and $r_P = \ell/(1 + \epsilon)$, $\epsilon$ being the eccentricity of the orbit. Thus

$$\frac{v_P}{v_A} = \frac{r_A}{r_P} = \frac{1 + \epsilon}{1 - \epsilon}.$$
```

Solving this equation for ϵ in terms of v_P and v_A, we get

$$\epsilon = \frac{v_P - v_A}{v_P + v_A}.$$

By conservation of energy the speed v at the ends of the minor axis of the orbit (where $r = a$) satisfies

$$\frac{v^2}{2} - \frac{k}{a} = \frac{v_P^2}{2} - \frac{k}{r_P} = \frac{v_A^2}{2} - \frac{k}{r_A}.$$

The latter equality shows that

$$v_P^2 - v_A^2 = 2k\left(\frac{1}{r_P} - \frac{1}{r_A}\right) = \frac{4k\epsilon}{\ell}.$$

Using this result and the parameters of the orbit given in the text, we obtain

$$\begin{aligned} v^2 &= v_P^2 + 2k\left(\frac{1}{a} - \frac{1}{r_P}\right) \\ &= v_P^2 + \frac{2k}{\ell}\left(1 - \epsilon^2 - (1 + \epsilon)\right) \\ &= v_P^2 - \frac{2k\epsilon}{\ell}(1 + \epsilon) \\ &= v_P^2 - \frac{v_P^2 - v_A^2}{2}\left(1 + \frac{v_P - v_A}{v_P + v_A}\right) \\ &= v_P^2 - \frac{v_P - v_A}{2}(2v_P) = v_P v_A. \end{aligned}$$

Thus $v = \sqrt{v_P v_A}$.

16. By conservation of energy, we have

$$\frac{k}{r} - \frac{1}{2}\left(\dot{r}^2 + \frac{h^2}{r^2}\right) = -K$$

where K is a constant for the orbit (the total energy). The term in the parentheses is v^2, the square of the speed. Thus

$$\frac{k}{r} - \frac{1}{2}v^2 = -K = \frac{k}{r_0} - \frac{1}{2}v_0^2,$$

where r_0 and v_0 are the given distance and speed. We evaluate $-K$ at perihelion.
The parameters of the orbit are

$$\ell = \frac{h^2}{k}, \quad a = \frac{h^2}{k(1 - \epsilon^2)}, \quad b = \frac{h^2}{k\sqrt{1 - \epsilon^2}}, \quad c = \epsilon a.$$

At perihelion P we have

$$r = a - c = (1 - \epsilon)a = \frac{h^2}{k(1 + \epsilon)}.$$

Since $\dot{r} = 0$ at perihelion, the speed there is $v = r\dot{\theta}$. By Kepler's second law, $r^2\dot{\theta} = h$, so $v = h/r = k(1 + \epsilon)/h$. Thus

$$\begin{aligned} -K &= \frac{k}{r} - \frac{v^2}{2} \\ &= \frac{k^2}{h^2}(1 + \epsilon) - \frac{1}{2}\frac{k^2}{h^2}(1 + \epsilon)^2 \\ &= \frac{k^2}{2h^2}(1 + \epsilon)\left[2 - (1 + \epsilon)\right] \\ &= \frac{k^2}{2h^2}(1 - \epsilon^2) = \frac{k}{2a}. \end{aligned}$$

Thus $a = \dfrac{k}{-2K}$. By Kepler's third law,

$$T^2 = \frac{4\pi^2}{k}a^3 = \frac{4\pi^2}{k}\left(\frac{k}{-2K}\right)^3.$$

Thus $T = \dfrac{2\pi}{\sqrt{k}}\left(\dfrac{2}{r_0} - \dfrac{v_0^2}{k}\right)^{-3/2}$

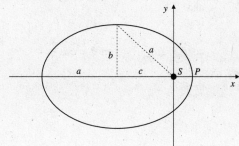

Fig. 11.6.16

18. Start with

$$\ddot{r} - \frac{h^2}{r^3} = -\frac{k}{r^2}.$$

Let $r(t) = \dfrac{1}{u(\theta)}$, where $\theta = \theta(t)$. Since $r^2\dot{\theta} = h$ (constant), we have

$$\dot{r} = -\frac{1}{u^2}\frac{du}{d\theta}\dot{\theta} = -r^2\frac{du}{d\theta}\frac{h}{r^2} = -h\frac{du}{d\theta}$$

$$\ddot{r} = -h\frac{d^2u}{d\theta^2}\dot{\theta} = -\frac{h^2}{r^2}\frac{d^2u}{d\theta^2} = -h^2u^2\frac{d^2u}{d\theta^2}.$$

Thus $-h^2u^2\dfrac{d^2u}{d\theta^2} - h^2u^3 = -ku^2$, or

$$\frac{d^2u}{d\theta^2} + u = \frac{k}{h^2}.$$

This is the DE for simple harmonic motion with a constant forcing term (nonhomogeneous term) on the right-hand side. It is easily verified that

$$u = \frac{k}{h^2}\left(1 + \epsilon\cos(\theta - \theta_0)\right)$$

is a solution for any choice of the constants ϵ and θ_0. Expressing the solution in terms of r, we have

$$r = \frac{h^2/k}{1 + \epsilon \cos(\theta - \theta_0)},$$

which is an ellipse if $|\epsilon| < 1$.

20. Since $\dfrac{k}{r} = \dfrac{1}{2}v^2 - K$ by conservation of energy, if $K < 0$, then

$$\frac{k}{r} \geq -K > 0,$$

so $r \leq -\dfrac{k}{K}$. The orbit is, therefore, bounded.

22. By Exercise 17, the asymptotes make angle $\theta = \cos^{-1}(1/\epsilon)$ with the transverse axis, as shown in the figure. The angle of deviation δ satisfies $2\theta + \delta = \pi$, so $\theta = \dfrac{\pi}{2} - \dfrac{\delta}{2}$, and

$$\cos\theta = \sin\frac{\delta}{2}, \qquad \sin\theta = \cos\frac{\delta}{2}.$$

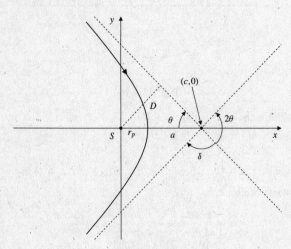

Fig. 11.6.22

By conservation of energy,

$$\frac{v^2}{2} - \frac{k}{r} = \text{constant} = \frac{v_\infty^2}{2}$$

for all points on the orbit. At perihelion,

$$r = r_p = c - a = (\epsilon - 1)a = \frac{\ell}{\epsilon + 1},$$

$$v = v_p = r_p\dot\theta = \frac{h}{r_p} = \frac{h(\epsilon + 1)}{\ell}.$$

Since $h^2 = k\ell$, we have

$$\begin{aligned}
v_\infty^2 &= v_p^2 - \frac{2k}{r_p} \\
&= \frac{h^2}{\ell^2}(\epsilon + 1)^2 - \frac{2k}{\ell}(\epsilon + 1) \\
&= \frac{k}{\ell}\left[(\epsilon + 1)^2 - 2(\epsilon + 1)\right] \\
&= \frac{k}{\ell}(\epsilon^2 - 1) = \frac{k}{a}.
\end{aligned}$$

Thus $av_\infty^2 = k$.

If D is the perpendicular distance from the sun S to an asymptote of the orbit (see the figure) then

$$\begin{aligned}
D &= c\sin\theta = \epsilon a\sin\theta = a\frac{\sin\theta}{\cos\theta} \\
&= a\frac{\cos(\delta/2)}{\sin(\delta/2)} = a\cot\frac{\delta}{2}.
\end{aligned}$$

Therefore

$$\frac{Dv_\infty^2}{k} = \frac{v_\infty^2 a}{k}\cot\frac{\delta}{2} = \cot\frac{\delta}{2}.$$

Review Exercises 11   (page 701)

2. $\mathbf{r} = t\cos t\,\mathbf{i} + t\sin t\,\mathbf{j} + (2\pi - t)\mathbf{k}$, $(0 \leq t \leq 2\pi)$ is a conical helix wound around the cone $z = 2\pi - \sqrt{x^2 + y^2}$ starting at the vertex $(0, 0, 2\pi)$, and completing one revolution to end up at $(2\pi, 0, 0)$. Since

$$\mathbf{v} = (\cos t - t\sin t)\mathbf{i} + (\sin t + t\cos t)\mathbf{j} - \mathbf{k},$$

the length of the curve is

$$L = \int_0^{2\pi} \sqrt{2 + t^2}\,dt = \pi\sqrt{2 + 4\pi^2} + \ln\left(\frac{2\pi + \sqrt{2 + 4\pi^2}}{\sqrt{2}}\right)$$

units.

4. The position, velocity, speed, and acceleration of the particle are given by

$$\mathbf{r} = x\mathbf{i} + x^2\mathbf{j}$$

$$\mathbf{v} = \frac{dx}{dt}(\mathbf{i} + 2x\mathbf{j}), \qquad v = \left|\frac{dx}{dt}\right|\sqrt{1 + 4x^2}$$

$$\mathbf{a} = \frac{d^2x}{dt^2}(\mathbf{i} + 2x\mathbf{j}) + 2\left(\frac{dx}{dt}\right)^2\mathbf{j}.$$

Let us assume that the particle is moving to the right, so that $dx/dt > 0$. Since the speed is t, we have

$$\frac{dx}{dt} = \frac{t}{\sqrt{1 + 4x^2}}$$

$$\frac{d^2x}{dt^2} = \frac{\sqrt{1 + 4x^2} - \dfrac{4tx}{\sqrt{1 + 4x^2}}\dfrac{dx}{dt}}{1 + 4x^2}.$$

If the particle is at $(\sqrt{2}, 2)$ at $t = 3$, then $dx/dt = 1$ at that time, and

$$\frac{d^2x}{dt^2} = \frac{3 - 4\sqrt{2}}{9}.$$

Hence the acceleration is

$$\mathbf{a} = \frac{3 - 4\sqrt{2}}{9}(\mathbf{i} + 2\sqrt{2}\mathbf{j}) + 2\mathbf{j}.$$

If the particle is moving to the left, so that $dx/dt < 0$, a similar calculation shows that at $t = 3$ its acceleration is

$$\mathbf{a} = -\frac{3 + 4\sqrt{2}}{9}(\mathbf{i} + 2\sqrt{2}\mathbf{j}) + 2\mathbf{j}.$$

6. Tangential acceleration: $dv/dt = e^t - e^{-t}$.
Normal acceleration: $v^2\kappa = \sqrt{2}$.
Since $v = 2\cosh t$, the minimum speed is 2 at time $t = 0$.

8. If $r = e^{-\theta}$, and $\dot{\theta} = k$, then $\dot{r} = -e^{-\theta}\dot{\theta} = -kr$, and $\ddot{r} = k^2 r$. Since $\mathbf{r} = r\hat{\mathbf{r}}$, we have

$$\mathbf{v} = \dot{r}\hat{\mathbf{r}} + r\dot{\theta}\hat{\boldsymbol{\theta}} = -kr\hat{\mathbf{r}} + kr\hat{\boldsymbol{\theta}}$$
$$\mathbf{a} = (\ddot{r} - r\dot{\theta}^2)\hat{\mathbf{r}} + (r\ddot{\theta} + 2\dot{r}\dot{\theta})\hat{\boldsymbol{\theta}}$$
$$= (k^2r - k^2r)\hat{\mathbf{r}} + (0 - 2k^2r)\hat{\boldsymbol{\theta}} = -2k^2r\hat{\boldsymbol{\theta}}.$$

10. $s = 4a\left(1 - \cos\dfrac{t}{2}\right) \Rightarrow t = 2\cos^{-1}\left(1 - \dfrac{s}{4a}\right) = t(s).$

The required arc length parametrization of the cycloid is

$$\mathbf{r} = a\big(t(s) - \sin t(s)\big)\mathbf{i} + a\big(1 - \cos t(s)\big)\mathbf{j}.$$

12. Let P be the point with position vector $\mathbf{r}(t)$ on the cycloid. By Exercise 9, the arc OP has length $4a - 4a\cos(t/2)$, and so PQ has length $4a$ - arc $OP = 4a\cos(t/2)$ units. Thus

$$\overrightarrow{PQ} = 4a\cos\frac{t}{2}\hat{\mathbf{T}}(t)$$
$$= 4a\cos\frac{t}{2}\left(\sin\frac{t}{2}\mathbf{i} + \cos\frac{t}{2}\mathbf{j}\right)$$
$$= 2a\sin t\mathbf{i} + 2a(1 + \cos t)\mathbf{j}.$$

It follows that Q has position vector

$$\mathbf{r}_Q = \mathbf{r} + \overrightarrow{PQ}$$
$$= a(t - \sin t)\mathbf{i} + a(1 - \cos t)\mathbf{j} + 2a\sin t\mathbf{i} + 2a(1 + \cos t)\mathbf{j}$$
$$= a(t + \sin t)\mathbf{i} + a(1 + \cos t + 2)\mathbf{j} \quad (\text{let } t = u + \pi)$$
$$= a(u - \sin u + \pi)\mathbf{i} + a(1 - \cos u + 2)\mathbf{j}.$$

Thus $\mathbf{r}_Q(t)$ represents the same cycloid as $\mathbf{r}(t)$, but translated πa units to the left and $2a$ units upward. From Exercise 11, the given cycloid is the evolute of its involute.

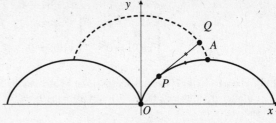

Fig. R-11.12

14. By Kepler's Second Law the position vector $\mathbf{r}$ from the origin (the sun) to the planet sweeps out area at a constant rate, say $h/2$:

$$\frac{dA}{dt} = \frac{h}{2}.$$

As observed in the text, $dA/dt = r^2\dot{\theta}/2$, so $r^2\dot{\theta} = h$, and

$$\mathbf{r} \times \mathbf{v} = (r\hat{\mathbf{r}}) \times (\dot{r}\hat{\mathbf{r}} + r\dot{\theta}\hat{\boldsymbol{\theta}}) = r^2\dot{\theta}\hat{\mathbf{r}} \times \hat{\boldsymbol{\theta}} = h\mathbf{k} = \mathbf{h}$$

is a constant vector.

16. By Exercise 15, $\mathbf{F}(\mathbf{r}) = m(\ddot{r} - r\dot{\theta}^2)\hat{\mathbf{r}} = -f(\mathbf{r})\hat{\mathbf{r}}$. We are given that $r = \ell/(1 + \epsilon\cos\theta)$. Thus

$$\dot{r} = -\frac{\ell}{(1 + \epsilon\cos\theta)^2}(-\epsilon\sin\theta)\dot{\theta}$$
$$= \frac{\epsilon\ell\sin\theta}{(1 + \epsilon\cos\theta)^2}\dot{\theta}$$
$$= \frac{\epsilon\sin\theta}{\ell}r^2\dot{\theta} = \frac{h\epsilon}{\ell}\sin\theta$$
$$\ddot{r} = \frac{h\epsilon}{\ell}(\cos\theta)\dot{\theta} = \frac{h^2\epsilon\cos\theta}{\ell r^2}.$$

It follows that

$$\ddot{r} - r\dot{\theta}^2 = \frac{h^2\epsilon\cos\theta}{\ell r^2} - \frac{h^2}{r^3}$$
$$= \frac{h^2}{\ell r^2}\left(\epsilon\cos\theta - \frac{\ell}{r}\right) = -\frac{h^2}{\ell r^2},$$

(because $(\ell/r) = 1 + \epsilon\cos\theta$). Hence

$$f(\mathbf{r}) = \frac{mh^2}{\ell r^2}.$$

This says that the magnitude of the force on the planet is inversely proportional to the square of its distance from the sun. Thus Newton's law of gravitation follows from Kepler's laws and the second law of motion.

Challenging Problems 11   (page 702)

2.
$$\begin{cases} \dfrac{d\mathbf{v}}{dt} = \mathbf{k} \times \mathbf{v} - 32\mathbf{k} \\ \mathbf{v}(0) = 70\mathbf{i} \end{cases}$$

a) If $\mathbf{v} = v_1\mathbf{i} + v_2\mathbf{j} + v_3\mathbf{k}$, then $\mathbf{k} \times \mathbf{v} = v_1\mathbf{j} - v_2\mathbf{i}$. Thus the initial-value problem breaks down into component equations as

$$\begin{cases} \dfrac{dv_1}{dt} = -v_2 \\ v_1(0) = 70 \end{cases} \quad \begin{cases} \dfrac{dv_2}{dt} = v_1 \\ v_2(0) = 0 \end{cases} \quad \begin{cases} \dfrac{dv_3}{dt} = -32 \\ v_3(0) = 0. \end{cases}$$

b) If $\mathbf{r} = x\mathbf{i} + y\mathbf{j} + z\mathbf{k}$ denotes the position of the baseball t s after it is thrown, then $x(0) = y(0) = z(0) = 0$ and we have

$$\frac{dz}{dt} = v_3 = -32t \implies z = -16t^2.$$

Also, $\dfrac{d^2v_1}{dt^2} = -\dfrac{dv_2}{dt} = -v_1$ (the equation of simple harmonic motion), so

$$v_1(t) = A\cos t + B\sin t, \quad v_2(t) = A\sin t - B\cos t.$$

Since $v_1(0) = 70$, $v_2(0) = 0$, $x(0) = 0$, and $y(0) = 0$, we have

$$\frac{dx}{dt} = v_1 = 70\cos t \qquad \frac{dy}{dt} = v_2 = 70\sin t$$
$$x(t) = 70\sin t \qquad\qquad y(t) = 70(1 - \cos t).$$

At time t seconds after it is thrown, the ball is at position

$$\mathbf{r} = 70\sin t\,\mathbf{i} + 70(1 - \cos t)\mathbf{j} - 16t^2\mathbf{k}.$$

c) At $t = 1/5$ s, the ball is at about $(13.9, 1.40, -0.64)$. If it had been thrown without the vertical spin, its position at time t would have been

$$\mathbf{r} = 70t\mathbf{i} - 16t^2\mathbf{k},$$

so its position at $t = 1/5$ s would have been $(14, 0, -0.64)$. Thus the spin has deflected the ball approximately 1.4 ft to the left (as seen from above) of what would have been its parabolic path had it not been given the spin.

4. The arc length element on $x = a(\theta - \sin\theta)$, $y = a(\cos\theta - 1)$ is (for $\theta \le \pi$)

$$ds = a\sqrt{(1 - \cos\theta)^2 + \sin^2\theta}\,d\theta$$
$$= a\sqrt{2(1 - \cos\theta)}\,d\theta = 2a\sin(\theta/2)\,d\theta.$$

If the bead slides downward from rest at height $y(\theta_0)$ to height $y(\theta)$, its gravitational potential energy has decreased by

$$mg\big[y(\theta_0) - y(\theta)\big] = mga(\cos\theta_0 - \cos\theta).$$

Since there is no friction, all this potential energy is converted to kinetic energy, so its speed v at height $y(\theta)$ is given by

$$\frac{1}{2}mv^2 = mga(\cos\theta_0 - \cos\theta),$$

and so $v = \sqrt{2ga(\cos\theta_0 - \cos\theta)}$. The time required for the bead to travel distance ds at speed v is $dt = ds/v$, so the time T required for the bead to slide from its starting position at $\theta = \theta_0$ to the lowest point on the wire, $\theta = \pi$, is

$$\begin{aligned} T &= \int_{\theta=\theta_0}^{\theta=\pi} \frac{ds}{v} = \int_{\theta_0}^{\pi} \frac{1}{v}\frac{ds}{d\theta}\,d\theta \\ &= \sqrt{\frac{2a}{g}} \int_{\theta_0}^{\pi} \frac{\sin(\theta/2)}{\sqrt{\cos\theta_0 - \cos\theta}}\,d\theta \\ &= \sqrt{\frac{2a}{g}} \int_{\theta_0}^{\pi} \frac{\sin(\theta/2)}{\sqrt{2\cos^2(\theta_0/2) - 2\cos^2(\theta/2)}}\,d\theta \end{aligned}$$

Let $u = \cos(\theta/2)$
$$du = -\tfrac{1}{2}\sin(\theta/2)\,d\theta$$

$$\begin{aligned} &= 2\sqrt{\frac{a}{g}} \int_0^{\cos(\theta_0/2)} \frac{du}{\sqrt{\cos^2(\theta_0/2) - u^2}} \\ &= 2\sqrt{\frac{a}{g}} \sin^{-1}\left(\frac{u}{\cos(\theta_0/2)}\right)\Bigg|_0^{\cos(\theta_0/2)} \\ &= \pi\sqrt{ag} \end{aligned}$$

which is independent of θ_0.

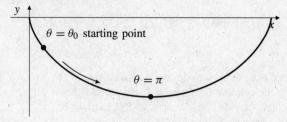

Fig. C-11.4

6. a) At time t, the hare is at $P = (0, vt)$ and the fox is at $Q = \big(x(t), y(t)\big)$, where x and y are such that the slope dy/dx of the fox's path is the slope of the line PQ:

$$\frac{dy}{dx} = \frac{y - vt}{x}.$$

b) Since $\dfrac{d}{dt}\dfrac{dy}{dx} = \dfrac{d^2y}{dx^2}\dfrac{dx}{dt}$, we have

$$\dfrac{dx}{dt}\dfrac{d^2y}{dx^2} = \dfrac{d}{dt}\left(\dfrac{y - vt}{x}\right)$$

$$= \dfrac{x\left(\dfrac{dy}{dt} - v\right) - (y - vt)\dfrac{dx}{dt}}{x^2}$$

$$= \dfrac{1}{x}\left(\dfrac{dy}{dx}\dfrac{dx}{dt} - v\right) - \dfrac{1}{x^2}(y - vt)\dfrac{dx}{dt}$$

$$= \dfrac{1}{x^2}(y - vt)\dfrac{dx}{dt} - \dfrac{v}{x} - \dfrac{1}{x^2}(y - vt)\dfrac{dx}{dt}$$

$$= -\dfrac{v}{x}.$$

Thus $x\dfrac{d^2y}{dx^2} = -\dfrac{v}{dx/dt}$.

Since the fox's speed is also v, we have

$$\left(\dfrac{dx}{dt}\right)^2 + \left(\dfrac{dy}{dt}\right)^2 = v^2.$$

Also, the fox is always running to the left (towards the y-axis from points where $x > 0$), so $dx/dt < 0$. Hence

$$\dfrac{v}{-\left(\dfrac{dx}{dt}\right)} = \sqrt{1 + \dfrac{(dy/dt)^2}{(dx/dt)^2}} = \sqrt{1 + \left(\dfrac{dy}{dx}\right)^2},$$

and so the fox's path $y = y(x)$ satisfies the DE

$$x\dfrac{d^2y}{dx^2} = \sqrt{1 + \left(\dfrac{dy}{dx}\right)^2}.$$

c) If $u = dy/dx$, then $u = 0$ and $y = 0$ when $x = a$, and

$$x\dfrac{du}{dx} = \sqrt{1 + u^2}$$

$$\int \dfrac{du}{\sqrt{1 + u^2}} = \int \dfrac{dx}{x} \quad \begin{array}{l} \text{Let } u = \tan\theta \\ du = \sec^2\theta\, d\theta \end{array}$$

$$\int \sec\theta\, d\theta = \ln x + \ln C$$

$$\ln(\tan\theta + \sec\theta) = \ln(Cx)$$

$$u + \sqrt{1 + u^2} = Cx.$$

Since $u = 0$ when $x = a$, we have $C = 1/a$.

$$\sqrt{1 + u^2} = \dfrac{x}{a} - u$$

$$1 + u^2 = \dfrac{x^2}{a^2} - \dfrac{2xu}{a} + u^2$$

$$\dfrac{2xu}{a} = \dfrac{x^2}{a^2} - 1$$

$$\dfrac{dy}{dx} = u = \dfrac{x}{2a} - \dfrac{a}{2x}$$

$$y = \dfrac{x^2}{4a} - \dfrac{a}{2}\ln x + C_1.$$

Since $y = 0$ when $x = a$, we have $C_1 = -\dfrac{a}{4} + \dfrac{a}{2}\ln a$, so

$$y = \dfrac{x^2 - a^2}{4} - \dfrac{a}{2}\ln\dfrac{x}{a}$$

is the path of the fox.

CHAPTER 12. PARTIAL DIFFERENTIATION

Section 12.1 Functions of Several Variables (page 712)

2. $f(x, y) = \sqrt{xy}$.
Domain is the set of points (x, y) for which $xy \geq 0$, that is, points on the coordinate axes and in the first and third quadrants.

4. $f(x, y) = \dfrac{xy}{x^2 - y^2}$.
The domain consists of all points not on the lines $x = \pm y$.

6. $f(x, y) = 1/\sqrt{x^2 - y^2}$.
The domain consists of all points in the part of the plane where $|x| > |y|$.

8. $f(x, y) = \sin^{-1}(x + y)$.
The domain consists of all points in the strip $-1 \leq x + y \leq 1$.

10. $f(x, y, z) = \dfrac{e^{xyz}}{\sqrt{xyz}}$.
The domain consists of all points (x, y, z) where $xyz > 0$, that is, all points in the four octants $x > 0$, $y > 0$, $z > 0$; $x > 0$, $y < 0$, $z < 0$; $x < 0$, $y > 0$, $z < 0$; and $x < 0$, $y < 0$, $z > 0$.

12. $f(x, y) = \sin x$, $0 \leq x \leq 2\pi$, $0 \leq y \leq 1$

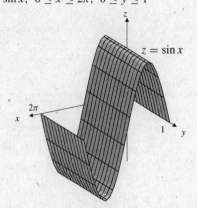

Fig. 12.1.12

14. $f(x, y) = 4 - x^2 - y^2$, $(x^2 + y^2 \leq 4$, $x \geq 0$, $y \geq 0)$

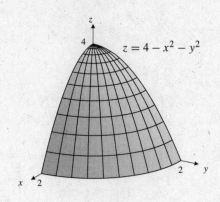

Fig. 12.1.14

16. $f(x, y) = 4 - x^2$

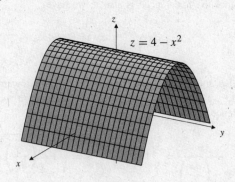

Fig. 12.1.16

18. $f(x, y) = 6 - x - 2y$

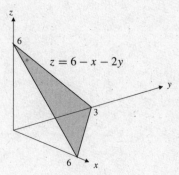

Fig. 12.1.18

20. $f(x, y) = x^2 + 2y^2 = C$, a family of similar ellipses centred at the origin.

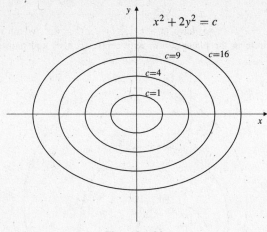

Fig. 12.1.20

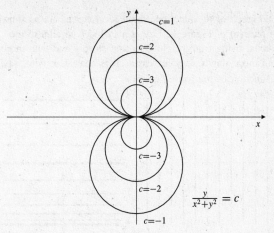

Fig. 12.1.24

26. $f(x, y) = \sqrt{\dfrac{1}{y} - x^2} = C \ \Rightarrow \ y = \dfrac{1}{x^2 + C^2}$.

22. $f(x, y) = \dfrac{x^2}{y} = C$, a family of parabolas, $y = x^2/C$, with vertices at the origin and vertical axes.

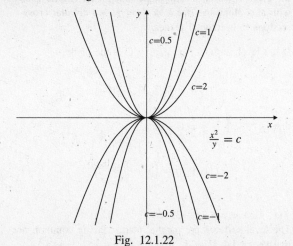

Fig. 12.1.22

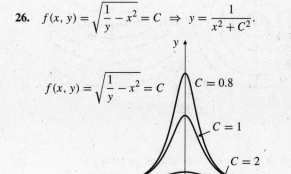

Fig. 12.1.26

28. C is a "pass" between two peaks to the east and west. The land is level at C and rises as you move to the east or west, but falls as you move to the north or south.

24. $f(x, y) = \dfrac{y}{x^2 + y^2} = C$.

This is the family $x^2 + \left(y - \dfrac{1}{2C}\right)^2 = \dfrac{1}{4C^2}$ of circles passing through the origin and having centres on the y-axis. The origin itself is, however, not on any of the level curves.

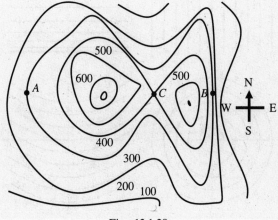

Fig. 12.1.28

30. The graph of the function whose level curves are as shown in part (b) of Figure 12.1.29 is a cylinder parallel to the x-axis, rising from height zero first steeply and then more and more slowly as y increases. It is consistent with, say, a function of the form $f(x, y) = \sqrt{y + 5}$.

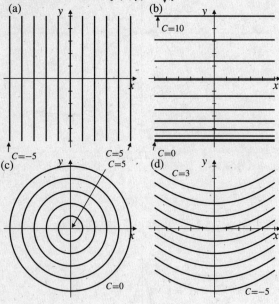

Fig. 12.1.30

32. The graph of the function whose level curves are as shown in part (d) of Figure 12.1.29 is a cylinder (possibly parabolic) with axis in the yz-plane, sloping upwards in the direction of increasing y. It is consistent with, say, a function of the form $f(x, y) = y - x^2$.

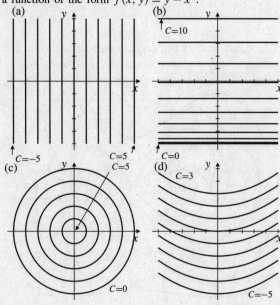

Fig. 12.1.32

34. $4z^2 = (x - z)^2 + (y - z)^2$.
If $z = c > 0$, we have $(x - c)^2 + (y - c)^2 = 4c^2$, which is a circle in the plane $z = c$, with centre (c, c, c) and radius $2c$.

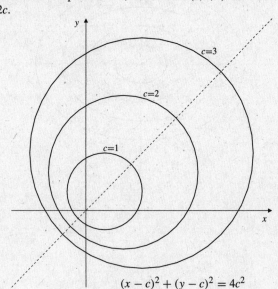

$$(x - c)^2 + (y - c)^2 = 4c^2$$

Fig. 12.1.34

The graph of the function $z = z(x, y) \geq 0$ defined by the given equation is (the upper half of) an elliptic cone with axis along the line $x = y = z$, and circular cross-sections in horizontal planes.

36. If the level surface $f(x, y, z) = C$ is the plane

$$\frac{x}{C^3} + \frac{y}{2C^3} + \frac{z}{3C^3} = 1,$$

that is, $x + \dfrac{y}{2} + \dfrac{z}{3} = C^3$, then

$$f(x, y, z) = \left(x + \frac{y}{2} + \frac{z}{3} \right)^{1/3}.$$

38. $f(x, y, z) = x + 2y + 3z$.
The level surfaces are parallel planes having common normal vector $\mathbf{i} + 2\mathbf{j} + 3\mathbf{k}$.

40. $f(x, y, z) = \dfrac{x^2 + y^2}{z^2}$.
The equation $f(x, y, z) = c$ can be rewritten $x^2 + y^2 = C^2 z^2$. The level surfaces are circular cones with vertices at the origin and axes along the z-axis.

42. $f(x, y, z, t) = x^2 + y^2 + z^2 + t^2$.
The "level hypersurface" $f(x, y, z, t) = c > 0$ is the "4-sphere" of radius $\sqrt{c}$ centred at the origin in $\mathbb{R}^4$. That is, it consists of all points in $\mathbb{R}^4$ at distance $\sqrt{c}$ from the origin.

44.

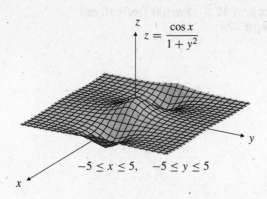

$$z = \frac{\cos x}{1 + y^2}$$

$$-5 \le x \le 5, \quad -5 \le y \le 5$$

Fig. 12.1.44

46.

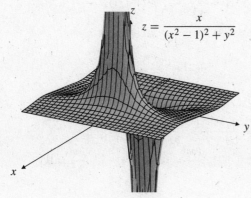

$$z = \frac{x}{(x^2 - 1)^2 + y^2}$$

Fig. 12.1.46

48. The graph is asymptotic to the coordinate planes.

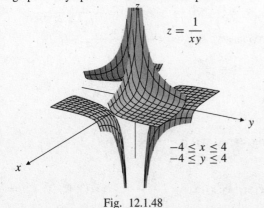

$$z = \frac{1}{xy}$$

$$-4 \le x \le 4$$
$$-4 \le y \le 4$$

Fig. 12.1.48

Section 12.2   Limits and Continuity (page 717)

2. $\lim\limits_{(x,y)\to(0,0)} \sqrt{x^2 + y^2} = 0$

4. Let $f(x, y) = \dfrac{x}{x^2 + y^2}$.

Then $|f(x, 0)| = |1/x| \to \infty$ as $x \to 0$.
But $|f(0, y)| = 0 \to 0$ as $y \to 0$.
Thus $\lim_{(x,y)\to(0,0)} f(x, y)$ does not exist.

6. $\lim\limits_{(x,y)\to(0,1)} \dfrac{x^2(y-1)^2}{x^2 + (y-1)^2} = 0$, because

$$0 \le \left| \frac{x^2(y-1)^2}{x^2 + (y-1)^2} \right| \le x^2$$

and $x^2 \to 0$ as $(x, y) \to (0, 1)$.

8. $\lim\limits_{(x,y)\to(0,0)} \dfrac{\sin(x-y)}{\cos(x+y)} = \dfrac{\sin 0}{\cos 0} = 0$.

10. The fraction is not defined at points of the line $y = 2x$ and so cannot have a limit at $(1, 2)$ by Definition 4. However, if we use the extended Definition 6, then, cancelling the common factor $2x - y$, we get

$$\lim_{(x,y)\to(1,2)} \frac{2x^2 - xy}{4x^2 - y^2} = \lim_{(x,y)\to(1,2)} \frac{x}{2x + y} = \frac{1}{4}.$$

12. If $x = 0$ and $y \ne 0$, then $\dfrac{x^2 y^2}{2x^4 + y^4} = 0$.

If $x = y \ne 0$, then $\dfrac{x^2 y^2}{2x^4 + y^4} = \dfrac{x^4}{2x^4 + x^4} = \dfrac{1}{3}$.

Therefore $\lim\limits_{(x,y)\to(0,0)} \dfrac{x^2 y^2}{2x^4 + y^4}$ does not exist.

14. For $x \ne y$, we have

$$f(x, y) = \frac{x^3 - y^3}{x - y} = x^2 + xy + y^2.$$

The latter expression has the value $3x^2$ at points of the line $x = y$. Therefore, if we extend the definition of $f(x, y)$ so that $f(x, x) = 3x^2$, then the resulting function will be equal to $x^2 + xy + y^2$ everywhere, and so continuous everywhere.

16. Let f be the function of Example 3 of Section 3.2:

$$f(x, y) = \begin{cases} \dfrac{2xy}{x^2 + y^2} & \text{if } (x, y) \ne (0, 0) \\ 0 & \text{if } (x, y) = (0, 0). \end{cases}$$

Let $a = b = 0$. If $g(x) = f(x, 0)$ and $h(y) = f(0, y)$, then $g(x) = 0$ for all x, and $h(y) = 0$ for all y, so g and h are continuous at 0. But, as shown in Example 3 of Section 3.2, f is not continuous at $(0, 0)$.

If $f(x, y)$ is continuous at (a, b), then $g(x) = f(x, b)$ is continuous at $x = a$ because

$$\lim_{x \to a} g(x) = \lim_{\substack{x \to a \\ y = b}} f(x, y) = f(a, b).$$

Similarly, $h(y) = f(a, y)$ is continuous at $y = b$.

231

18. Since $|x| \le \sqrt{x^2 + y^2}$ and $|y| \le \sqrt{x^2 + y^2}$, we have

$$\left| \frac{x^m y^n}{(x^2 + y^2)^p} \right| \le \frac{(x^2 + y^2)^{(m+n)/2}}{(x^2 + y^2)^p} = (x^2 + y^2)^{-p+(m+n)/2}.$$

The expression on the right $\to 0$ as $(x, y) \to (0, 0)$, provided $m + n > 2p$. In this case

$$\lim_{(x,y)\to(0,0)} \frac{x^m y^n}{(x^2 + y^2)^p} = 0.$$

20. $f(x, y) = \dfrac{\sin x \sin^3 y}{1 - \cos(x^2 + y^2)}$ cannot be defined at $(0, 0)$ so as to become continuous there, because $f(x, y)$ has no limit as $(x, y) \to (0, 0)$. To see this, observe that $f(x, 0) = 0$, so the limit must be 0 if it exists at all. However,

$$f(x, x) = \frac{\sin^4 x}{1 - \cos(2x^2)} = \frac{\sin^4 x}{2\sin^2(x^2)}$$

which approaches $1/2$ as $x \to 0$ by l'Hôpital's Rule or by using Maclaurin series.

22. The graphing software is unable to deal effectively with the discontinuity at $(x, y) = (0, 0)$ so it leaves some gaps and rough edges near the z-axis. The surface lies between a ridge along $y = x^2$, $z = 1$, and a ridge along $y = -x^2$, $z = -1$. It appears to be creased along the z-axis. The level curves are parabolas $y = kx^2$ through the origin. One of the families of rulings on the surface is the family of contours corresponding to level curves.

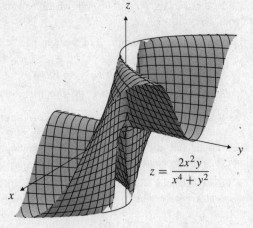

$$z = \frac{2x^2 y}{x^4 + y^2}$$

Fig. 12.2.22

Section 12.3 Partial Derivatives
(page 724)

2. $f(x, y) = xy + x^2$,
$f_1(x, y) = y + 2x$, $f_2(x, y) = x$,
$f_1(2, 0) = 4$, $f_2(2, 0) = 2$.

4. $g(x, y, z) = \dfrac{xz}{y + z}$,
$g_1(x, y, z) = \dfrac{z}{y + z}$, $g_1(1, 1, 1) = \dfrac{1}{2}$,
$g_2(x, y, z) = \dfrac{-xz}{(y + z)^2}$, $g_2(1, 1, 1) = -\dfrac{1}{4}$,
$g_3(x, y, z) = \dfrac{xy}{(y + z)^2}$, $g_3(1, 1, 1) = \dfrac{1}{4}$.

6. $w = \ln(1 + e^{xyz})$, $\dfrac{\partial w}{\partial x} = \dfrac{yze^{xyz}}{1 + e^{xyz}}$,
$\dfrac{\partial w}{\partial y} = \dfrac{xze^{xyz}}{1 + e^{xyz}}$, $\dfrac{\partial w}{\partial z} = \dfrac{xye^{xyz}}{1 + e^{xyz}}$,
At $(2, 0, -1)$: $\dfrac{\partial w}{\partial x} = 0$, $\dfrac{\partial w}{\partial y} = -1$, $\dfrac{\partial w}{\partial z} = 0$.

8. $f(x, y) = \dfrac{1}{\sqrt{x^2 + y^2}}$,
$f_1(x, y) = -\dfrac{1}{2}(x^2 + y^2)^{-3/2}(2x) = -\dfrac{x}{(x^2 + y^2)^{3/2}}$,
By symmetry, $f_2(x, y) = -\dfrac{y}{(x^2 + y^2)^{3/2}}$,
$f_1(-3, 4) = \dfrac{3}{125}$, $f_2(-3, 4) = -\dfrac{4}{125}$.

10. If $g(x_1, x_2, x_3, x_4) = \dfrac{x_1 - x_2^2}{x_3 + x_4^2}$, then

$$g_1(x_1, x_2, x_3, x_4) = \frac{1}{x_3 + x_4^2} \qquad g_1(3, 1, -1, -2) = \frac{1}{3}$$

$$g_2(x_1, x_2, x_3, x_4) = \frac{-2x_2}{x_3 + x_4^2} \qquad g_2(3, 1, -1, -2) = -\frac{2}{3}$$

$$g_3(x_1, x_2, x_3, x_4) = \frac{x_2^2 - x_1}{(x_3 + x_4^2)^2} \qquad g_3(3, 1, -1, -2) = -\frac{2}{9}$$

$$g_4(x_1, x_2, x_3, x_4) = \frac{(x_2^2 - x_1)2x_4}{(x_3 + x_4^2)^2} \qquad g_4(3, 1, -1, -2) = \frac{8}{9}.$$

12. $f(x, y) = \begin{cases} \dfrac{x^2 - 2y^2}{x - y} & \text{if } x \neq y \\ 0 & \text{if } x = y \end{cases}$

$f_1(0, 0) = \lim_{h \to 0} \dfrac{f(h, 0) - f(0, 0)}{h} = \lim_{h \to 0} \dfrac{h - 0}{h} = 1,$

$f_2(0, 0) = \lim_{k \to 0} \dfrac{f(0, k) - f(0, 0)}{k} = \lim_{k \to 0} \dfrac{2k}{k} = 2.$

14. $f(x, y) = \dfrac{x - y}{x + y}, \quad f(1, 1) = 0,$

$f_1(x, y) = \dfrac{(x + y) - (x - y)}{(x + y)^2}, \quad f_1(1, 1) = \dfrac{1}{2}$

$f_2(x, y) = \dfrac{(x + y)(-1) - (x - y)}{(x + y)^2}, \quad f_2(1, 1) = -\dfrac{1}{2}.$

Tangent plane to $z = f(x, y)$ at $(1,1)$ has equation

$z = \dfrac{x - 1}{2} - \dfrac{y - 1}{2}$, or $2z = x - y$.

Normal line: $2(x - 1) = -2(y - 1) = -z$.

16. $f(x, y) = e^{xy}, \quad f_1(x, y) = ye^{xy}, \quad f_2(x, y) = xe^{xy},$

$f(2, 0) = 1, \quad f_1(2, 0) = 0, \quad f_2(2, 0) = 2.$

Tangent plane to $z = e^{xy}$ at $(2,0)$ has equation $z = 1 + 2y$.

Normal line: $x = 2, \ y = 2 - 2z$.

18. $f(x, y) = ye^{-x^2}, \quad f_1 = -2xye^{-x^2}, \quad f_2 = e^{-x^2},$

$f(0, 1) = 1, \quad f_1(0, 1) = 0, \quad f_2(0, 1) = 1.$

Tangent plane to $z = f(x, y)$ at $(0, 1)$ has equation

$z = 1 + 1(y - 1)$, or $z = y$.

Normal line: $x = 0, \ y + z = 2$.

20. $f(x, y) = \dfrac{2xy}{x^2 + y^2}, \qquad f(0, 2) = 0$

$f_1(x, y) = \dfrac{(x^2 + y^2)2y - 2xy(2x)}{(x^2 + y^2)^2} = \dfrac{2y(y^2 - x^2)}{(x^2 + y^2)^2}$

$f_2(x, y) = \dfrac{2x(x^2 - y^2)}{(x^2 + y^2)^2}$ (by symmetry)

$f_1(0, 2) = 1, \qquad f_2(0, 2) = 0.$

Tangent plane at $(0, 2)$: $z = x$.

Normal line: $z + x = 0, \ y = 2$.

22. $f(x, y) = \sqrt{1 + x^3 y^2} \qquad f(2, 1) = 3$

$f_1(x, y) = \dfrac{3x^2 y^2}{2\sqrt{1 + x^3 y^2}} \qquad f_1(2, 1) = 2$

$f_2(x, y) = \dfrac{2x^3 y}{2\sqrt{1 + x^3 y^2}} \qquad f_2(2, 1) = \dfrac{8}{3}$

Tangent plane: $z = 3 + 2(x - 2) + \frac{8}{3}(y - 1)$, or

$6x + 8y - 3z = 11.$

Normal line: $\dfrac{x - 2}{2} = \dfrac{y - 1}{8/3} = \dfrac{z - 3}{-1}.$

24. $z = xye^{-(x^2 + y^2)/2}$

$\dfrac{\partial z}{\partial x} = ye^{-(x^2 + y^2)/2} - x^2 ye^{-(x^2 + y^2)/2} = y(1 - x^2)e^{-(x^2 + y^2)/2}$

$\dfrac{\partial z}{\partial y} = x(1 - y^2)e^{-(x^2 + y^2)/2}$ (by symmetry)

The tangent planes are horizontal at points where both of these first partials are zero, that is, points satisfying

$$y(1 - x^2) = 0 \quad \text{and} \quad x(1 - y^2) = 0.$$

These points are $(0, 0)$, $(1, 1)$, $(-1, -1)$, $(1, -1)$ and $(-1, 1)$.

At $(0,0)$ the tangent plane is $z = 0$.

At $(1, 1)$ and $(-1, -1)$ the tangent plane is $z = 1/e$.

At $(1, -1)$ and $(-1, 1)$ the tangent plane is $z = -1/e$.

26. $z = \dfrac{x + y}{x - y},$

$\dfrac{\partial z}{\partial x} = \dfrac{(x - y)(1) - (x + y)(1)}{(x - y)^2} = \dfrac{-2y}{(x - y)^2},$

$\dfrac{\partial z}{\partial y} = \dfrac{(x - y)(1) - (x + y)(-1)}{(x - y)^2} = \dfrac{2x}{(x - y)^2}.$

Therefore

$x\dfrac{\partial z}{\partial x} + y\dfrac{\partial z}{\partial y} = -\dfrac{2xy}{(x - y)^2} + \dfrac{2xy}{(x - y)^2} = 0.$

28. $w = x^2 + yz, \quad \dfrac{\partial w}{\partial x} = 2x, \quad \dfrac{\partial w}{\partial y} = z, \quad \dfrac{\partial w}{\partial z} = y.$

Therefore

$x\dfrac{\partial w}{\partial x} + y\dfrac{\partial w}{\partial y} + z\dfrac{\partial w}{\partial z}$

$\qquad\qquad = 2x^2 + yz + yz$

$\qquad\qquad = 2(x^2 + yz) = 2w.$

30. $z = f(x^2 + y^2),$

$\dfrac{\partial z}{\partial x} = f'(x^2 + y^2)(2x), \qquad \dfrac{\partial z}{\partial y} = f'(x^2 + y^2)(2y).$

Thus $y\dfrac{\partial z}{\partial x} - x\dfrac{\partial z}{\partial y} = 2xyf'(x^2 + y^2) - 2xyf'(x^2 + y^2) = 0.$

32. $f_1(x, y, z) = \lim_{h \to 0} \dfrac{f(x + h, y, z) - f(x, y, z)}{h}$

$f_2(x, y, z) = \lim_{k \to 0} \dfrac{f(x, y + k, z) - f(x, y, z)}{k}$

$f_3(x, y, z) = \lim_{\ell \to 0} \dfrac{f(x, y, z + \ell) - f(x, y, z)}{\ell}$

34. If $Q = (X, Y, Z)$ is the point on the surface $z = x^2 + y^2$ that is closest to $P = (1, 1, 0)$, then

$$\vec{PQ} = (X - 1)\mathbf{i} + (Y - 1)\mathbf{j} + Z\mathbf{k}$$

must be normal to the surface at Q, and hence must be parallel to $\mathbf{n} = 2X\mathbf{i} + 2Y\mathbf{j} - \mathbf{k}$. Hence $\overrightarrow{PQ} = t\mathbf{n}$ for some real number t, so

$$X - 1 = 2tX, \qquad Y - 1 = 2tY, \qquad Z = -t.$$

Thus $X = Y = \dfrac{1}{1 - 2t}$, and, since $Z = X^2 + Y^2$, we must have

$$-t = \frac{2}{(1 - 2t)^2}.$$

Evidently this equation is satisfied by $t = -\dfrac{1}{2}$. Since the left and right sides of the equation have graphs similar to those in Figure 12.18(b) (in the text), the equation has only this one real solution. Hence $X = Y = \dfrac{1}{2}$, and so $Z = \dfrac{1}{2}$.

The distance from $(1, 1, 0)$ to $z = x^2$ is the distance from $(1, 1, 0)$ to $\left(\frac{1}{2}, \frac{1}{2}, \frac{1}{2}\right)$, which is $\sqrt{3}/2$ units.

36. $f(x, y) = \dfrac{2xy}{x^2 + y^2}$ if $(x, y) \neq (0, 0)$, $\qquad f(0, 0) = 0$

$f_1(0, 0) = \lim_{h \to 0} \dfrac{f(h, 0) - f(0, 0)}{h} = \lim_{h \to 0} \dfrac{0 - 0}{h} = 0$

$f_2(0, 0) = \lim_{k \to 0} \dfrac{f(0, k) - f(0, 0)}{h} = \lim_{k \to 0} \dfrac{0 - 0}{k} = 0$

Thus $f_1(0, 0)$ and $f_2(0, 0)$ both exist even though f is not continuous at $(0, 0)$ (as shown in Example 2 of Section 3.2).

38. If $(x, y) \neq (0, 0)$, then

$$f_1(x, y) = 3x^2 \sin \frac{1}{x^2 + y^2} - \frac{(x^3 + y)2x}{(x^2 + y^2)^2} \cos \frac{1}{x^2 + y^2}.$$

The first term on the right $\to 0$ as $(x, y) \to (0, 0)$, but the second term has no limit at $(0, 0)$. (It is 0 along $x = 0$, but along $x = y$ it is

$$-\frac{2x^4 + 2x^2}{4x^4} \cos \frac{1}{2x^2} = -\frac{1}{2}\left(1 + \frac{1}{x^2}\right) \cos \frac{1}{2x^2},$$

which has no limit as $x \to 0$.) Thus $f_1(x, y)$ has no limit at $(0, 0)$ and is not continuous there.

40. $f(x, y, z) = \begin{cases} \dfrac{xy^2z}{x^4 + y^4 + z^4} & \text{if } (x, y, z) \neq (0, 0, 0) \\ 0 & \text{if } (x, y, z) = (0, 0, 0). \end{cases}$

By symmetry we have

$$f_3(0, 0, 0) = f_1(0, 0, 0) = \lim_{h \to 0} \frac{0}{h^5} = 0.$$

Also,

$$f_2(0, 0, 0) = \lim_{k \to 0} \frac{0}{k^5} = 0.$$

f is not continuous at $(0, 0, 0)$; it has different limits as $(x, y, z) \to (0, 0, 0)$ along $x = 0$ and along $x = y = z$. None of f_1, f_2, and f_3 is continuous at $(0, 0, 0)$ either. For example,

$$f_1(x, y, z) = \frac{(y^4 + z^4 - 3x^4)y^2z}{(x^4 + y^4 + z^4)^2},$$

which has no limit as $(x, y, z) \to (0, 0, 0)$ along the line $x = y = z$.

Section 12.4 Higher-Order Derivatives (page 731)

2. $f(x, y) = x^2 + y^2$, $\quad f_1(x, y) = 2x$, $\quad f_2(x, y) = 2y$, $f_{11}(x, y) = f_{22}(x, y) = 2$, $\quad f_{12}(x, y) = f_{21}(x, y) = 0$.

4. $z = \sqrt{3x^2 + y^2}$,

$\dfrac{\partial z}{\partial x} = \dfrac{3x}{\sqrt{3x^2 + y^2}}$, $\quad \dfrac{\partial z}{\partial y} = \dfrac{y}{\sqrt{3x^2 + y^2}}$,

$\dfrac{\partial^2 z}{\partial x^2} = \dfrac{\sqrt{3x^2 + y^2}\,(3) - 3x\dfrac{3x}{\sqrt{3x^2 + y^2}}}{3x^2 + y^2} = \dfrac{3y^2}{(3x^2 + y^2)^{3/2}}$,

$\dfrac{\partial^2 z}{\partial y^2} = \dfrac{\sqrt{3x^2 + y^2} - y\dfrac{y}{\sqrt{3x^2 + y^2}}}{3x^2 + y^2} = \dfrac{3x^2}{(3x^2 + y^2)^{3/2}}$,

$\dfrac{\partial^2 z}{\partial x \partial y} = \dfrac{\partial^2 z}{\partial y \partial x} = -\dfrac{3xy}{(3x^2 + y^2)^{3/2}}$.

6. $f(x, y) = \ln(1 + \sin(xy))$

$f_1(x, y) = \dfrac{y\cos(xy)}{1 + \sin(xy)}$, $\quad f_2(x, y) = \dfrac{x\cos(xy)}{1 + \sin(xy)}$

$f_{11}(x, y)$

$= \dfrac{(1 + \sin(xy))(-y^2 \sin(xy)) - (y\cos(xy))(y\cos(xy))}{(1 + \sin(xy))^2}$

$= -\dfrac{y^2}{1 + \sin(xy)}$

$f_{22}(x, y) = -\dfrac{x^2}{1 + \sin(xy)} \quad \text{(by symmetry)}$

$f_{12}(x, y) =$

$\dfrac{(1 + \sin(xy))(\cos(xy) - xy\sin(xy)) - (y\cos(xy))(x\cos(xy))}{(1 + \sin(xy))^2}$

$= \dfrac{\cos(xy) - xy}{1 + \sin(xy)} = f_{21}(x, y).$

8. $f(x, y) = A(x^2 - y^2) + Bxy$, $\quad f_1 = 2Ax + By$,
$f_2 = -2Ay + Bx$,
$f_{11} = 2A$, $\qquad f_{22} = -2A$,
Thus $f_{11} + f_{22} = 0$, and f is harmonic.

10. $f(x, y) = \dfrac{x}{x^2 + y^2}$

$$f_1(x, y) = \frac{x^2 + y^2 - 2x^2}{(x^2 + y^2)^2} = \frac{y^2 - x^2}{(x^2 + y^2)^2}$$

$$f_2(x, y) = -\frac{2xy}{(x^2 + y^2)^2}$$

$$f_{11}(x, y) = \frac{(x^2 + y^2)^2(-2x) - (y^2 - x^2)2(x^2 + y^2)(2x)}{(x^2 + y^2)^4}$$

$$= \frac{2x^3 - 6xy^2}{(x^2 + y^2)^3}$$

$$f_{22}(x, y) = -\frac{(x^2 + y^2)^2(2x) - 2xy2(x^2 + y^2)(2y)}{(x^2 + y^2)^4}$$

$$= \frac{-2x^3 + 6xy^2}{(x^2 + y^2)^3}.$$

Evidently $f_{11}(x, y) + f_{22}(x, y) = 0$ for $(x, y) \neq (0, 0)$. Hence f is harmonic except at the origin.

12. $f(x, y) = \tan^{-1}\left(\dfrac{y}{x}\right), \quad (x \neq 0).$

$$f_1(x, y) = \frac{1}{1 + \dfrac{y^2}{x^2}}\left(-\frac{y}{x^2}\right) = -\frac{y}{x^2 + y^2},$$

$$f_2(x, y) = \frac{1}{1 + \dfrac{y^2}{x^2}}\left(\frac{1}{x}\right) = \frac{x}{x^2 + y^2},$$

$$f_{11} = \frac{2xy}{(x^2 + y^2)^2}, \qquad f_{22} = -\frac{2xy}{(x^2 + y^2)^2}.$$

Thus $f_{11} + f_{22} = 0$ and f is harmonic.

14. Let $g(x, y, z) = zf(x, y)$. Then

$$g_1(x, y, z) = zf_1(x, y), \qquad g_{11}(x, y, z) = zf_{11}(x, y)$$
$$g_2(x, y, z) = zf_2(x, y), \qquad g_{22}(x, y, z) = zf_{22}(x, y)$$
$$g_3(x, y, z) = f(x, y), \qquad g_{33}(x, y, z) = 0.$$

Thus $g_{11} + g_{22} + g_{33} = z(f_{11} + f_{22}) = 0$ and g is harmonic because f is harmonic. This proves (a). The proofs of (b) and (c) are similar.

If $h(x, y, z) = f(ax + by, cz)$, then $h_{11} = a^2 f_{11}$, $h_{22} = b^2 f_{11}$ and $h_{33} = c^2 f_{22}$. If $a^2 + b^2 = c^2$ and f is harmonic then

$$h_{11} + h_{22} + h_{33} = c^2(f_{11} + f_{22}) = 0,$$

so h is harmonic.

16. Let

$$f(x, y) = \begin{cases} \dfrac{2xy}{x^2 + y^2} & \text{if } (x, y) \neq (0, 0) \\ 0 & \text{if } (x, y) = (0, 0). \end{cases}$$

For $(x, y) \neq (0, 0)$, we have

$$f_1(x, y) = \frac{(x^2 + y^2)2y - 2xy(2x)}{(x^2 + y^2)^2} = \frac{2y(y^2 - x^2)}{(x^2 + y^2)^2}$$

$$f_2(x, y) = \frac{2x(x^2 - y^2)}{(x^2 + y^2)^2} \qquad \text{(by symmetry)}.$$

Let $F(x, y) = (x^2 - y^2)f(x, y)$. Then we calculate

$$F_1(x, y) = 2xf(x, y) + (x^2 - y^2)f_1(x, y)$$

$$= 2xf(x, y) - \frac{2y(y^2 - x^2)^2}{(x^2 + y^2)^2}$$

$$F_2(x, y) = -2yf(x, y) + (x^2 - y^2)f_2(x, y)$$

$$= -2yf(x, y) + \frac{2x(x^2 - y^2)^2}{(x^2 + y^2)^2}$$

$$F_{12}(x, y) = \frac{2(x^6 + 9x^4y^2 - 9x^2y^4 - y^6)}{(x^2 + y^2)^3} = F_{21}(x, y).$$

For the values at $(0, 0)$ we revert to the definition of derivative to calculate the partials:

$$F_1(0, 0) = \lim_{h \to 0} \frac{F(h, 0) - F(0, 0)}{h} = 0 = F_2(0, 0)$$

$$F_{12}(0, 0) = \lim_{k \to 0} \frac{F_1(0, k) - F_1(0, 0)}{k} = \lim_{k \to 0} \frac{-2k(k^4)}{k(k^4)} = -2$$

$$F_{21}(0, 0) = \lim_{h \to 0} \frac{F_2(h, 0) - F_2(0, 0)}{h} = \lim_{h \to 0} \frac{2h(h^4)}{h(h^4)} = 2$$

This does not contradict Theorem 1 since the partials F_{12} and F_{21} are not continuous at $(0, 0)$. (Observe, for instance, that $F_{12}(x, x) = 0$, while $F_{12}(x, 0) = 2$ for $x \neq 0$.)

18. $u(x, y, t) = t^{-1}e^{-(x^2+y^2)/4t}$

$$\frac{\partial u}{\partial t} = -\frac{1}{t^2}e^{-(x^2+y^2)/4t} + \frac{x^2 + y^2}{4t^3}e^{-(x^2+y^2)/4t}$$

$$\frac{\partial u}{\partial x} = -\frac{x}{2t^2}e^{-(x^2+y^2)/4t}$$

$$\frac{\partial^2 u}{\partial x^2} = -\frac{1}{2t^2}e^{-(x^2+y^2)/4t} + \frac{x^2}{4t^3}e^{-(x^2+y^2)/4t}$$

$$\frac{\partial^2 u}{\partial y^2} = -\frac{1}{2t^2}e^{-(x^2+y^2)/4t} + \frac{y^2}{4t^3}e^{-(x^2+y^2)/4t}$$

Thus $\dfrac{\partial u}{\partial t} = \dfrac{\partial^2 u}{\partial x^2} + \dfrac{\partial^2 u}{\partial y^2}$.

20. $u(x, y)$ is biharmonic $\Leftrightarrow \dfrac{\partial^2 u}{\partial x^2} + \dfrac{\partial^2 u}{\partial y^2}$ is harmonic

$$\Leftrightarrow \left(\frac{\partial^2}{\partial x^2} + \frac{\partial^2}{\partial y^2}\right)\left(\frac{\partial^2 u}{\partial x^2} + \frac{\partial^2 u}{\partial y^2}\right) = 0$$

$$\Leftrightarrow \frac{\partial^4 u}{\partial x^4} + 2\frac{\partial^4 u}{\partial x^2 \partial y^2} + \frac{\partial^4 u}{\partial y^4} = 0$$

by the equality of mixed partials.

22. If u is harmonic, then $\dfrac{\partial^2 u}{\partial x^2} + \dfrac{\partial^2 u}{\partial y^2} = 0$. If
$v(x, y) = xu(x, y)$, then

$$\frac{\partial^2 v}{\partial x^2} = \frac{\partial}{\partial x}\left(u + x\frac{\partial u}{\partial x}\right) = 2\frac{\partial u}{\partial x} + x\frac{\partial^2 u}{\partial x^2}$$

$$\frac{\partial^2 v}{\partial y^2} = \frac{\partial}{\partial y}\left(x\frac{\partial u}{\partial y}\right) = x\frac{\partial^2 u}{\partial y^2}$$

$$\frac{\partial^2 v}{\partial x^2} + \frac{\partial^2 v}{\partial y^2} = 2\frac{\partial u}{\partial x} + x\left(\frac{\partial^2 u}{\partial x^2} + \frac{\partial^2 u}{\partial y^2}\right) = 2\frac{\partial u}{\partial x}.$$

Since u is harmonic, so is $\partial u/\partial x$:

$$\left(\frac{\partial^2}{\partial x^2} + \frac{\partial^2}{\partial y^2}\right)\frac{\partial u}{\partial x} = \frac{\partial}{\partial x}\left(\frac{\partial^2 u}{\partial x^2} + \frac{\partial^2 u}{\partial y^2}\right) = \frac{\partial}{\partial x}(0) = 0.$$

Thus $\dfrac{\partial^2 v}{\partial x^2} + \dfrac{\partial^2 v}{\partial y^2}$ is harmonic, and so v is biharmonic. The proof that $w(x, y) = yu(x, y)$ is biharmonic is similar.

24. By Exercise 11, $\ln(x^2 + y^2)$ is harmonic (except at the origin). Therefore $y\ln(x^2 + y^2)$ is biharmonic by Exercise 22.

26. $u(x, y, z)$ is biharmonic $\Leftrightarrow \dfrac{\partial^2 u}{\partial x^2} + \dfrac{\partial^2 u}{\partial y^2} + \dfrac{\partial^2 u}{\partial z^2}$ is harmonic

$$\Leftrightarrow \left(\frac{\partial^2}{\partial x^2} + \frac{\partial^2}{\partial y^2} + \frac{\partial^2}{\partial z^2}\right)\left(\frac{\partial^2 u}{\partial x^2} + \frac{\partial^2 u}{\partial y^2} + \frac{\partial^2 u}{\partial z^2}\right) = 0$$

$$\Leftrightarrow \frac{\partial^4 u}{\partial x^4} + \frac{\partial^4 u}{\partial y^4} + \frac{\partial^4 u}{\partial z^4} + 2\left(\frac{\partial^4 u}{\partial x^2 \partial y^2} + \frac{\partial^4 u}{\partial x^2 \partial z^2} + \frac{\partial^4 u}{\partial y^2 \partial z^2}\right) = 0$$

by the equality of mixed partials.

If $u(x, y, z)$ is harmonic then the functions $xu(x, y, z)$, $yu(x, y, z)$, and $zu(x, y, z)$ are all biharmonic. The proof is almost identical to that given in Exercise 22.

Section 12.5 The Chain Rule (page 741)

2. If $w = f(x, y, z)$ where $x = g(s)$, $y = h(s, t)$ and $z = k(t)$, then

$$\frac{\partial w}{\partial t} = f_2(x, y, z)h_2(s, t) + f_3(x, y, z)k'(t).$$

4. If $w = f(x, y)$ where $x = g(r, s)$, $y = h(r, t)$, $r = k(s, t)$ and $s = m(t)$, then

$$\begin{aligned}\frac{dw}{dt} = &\; f_1(x, y)\left[g_1(r, s)\left(k_1(s, t)m'(t)\right.\right.\\ &\left.+ k_2(s, t)\right) + g_2(r, s)m'(t)\Big]\\ &+ f_2(x, y)\Big[h_1(r, t)\left(k_1(s, t)m'(t)\right.\\ &\left.+ k_2(s, t)\right) + h_2(r, t)\Big].\end{aligned}$$

6. If $u = \sqrt{x^2 + y^2}$, where $x = e^{st}$ and $y = 1 + s^2\cos t$, then
Method I.

$$\begin{aligned}\frac{\partial u}{\partial t} &= \frac{x}{\sqrt{x^2 + y^2}}se^{st} + \frac{y}{\sqrt{x^2 + y^2}}(-s^2\sin t)\\ &= \frac{xse^{st} - ys^2\sin t}{\sqrt{x^2 + y^2}}.\end{aligned}$$

Method II.

$$\begin{aligned}u &= \sqrt{e^{2st} + (1 + s^2\cos t)^2}\\ \frac{\partial u}{\partial t} &= \frac{2se^{2st} - 2s^2\sin t(1 + s^2\cos t)}{2\sqrt{e^{2st} + (1 + s^2\cos t)^2}}\\ &= \frac{x^2 s - ys^2\sin t}{\sqrt{x^2 + y^2}}.\end{aligned}$$

8. If $z = txy^2$, where $x = t + \ln(y + t^2)$ and $y = e^t$, then
Method I.

$$\begin{aligned}\frac{dz}{dt} &= \frac{\partial z}{\partial t} + \frac{\partial z}{\partial x}\left(\frac{\partial x}{\partial t} + \frac{\partial x}{\partial y}\frac{\partial y}{\partial t}\right)\\ &\quad + \frac{\partial z}{\partial y}\frac{\partial y}{\partial t}\\ &= xy^2 + ty^2\left(1 + \frac{y + 2t}{y + t^2}\right) + 2txy^2.\end{aligned}$$

Method II.

$$\begin{aligned}z &= t\left(t + \ln(e^t + t^2)\right)e^{2t}\\ \frac{\partial z}{\partial t} &= \left(t + \ln(e^t + t^2)\right)e^{2t} + te^{2t}\left(1 + \frac{e^t + 2t}{e^t + t^2}\right)\\ &\quad + 2te^{2t}\left(t + \ln(e^t + t^2)\right)\\ &= xy^2 + ty^2\left(1 + \frac{y + 2t}{y + t^2}\right) + 2txy^2.\end{aligned}$$

10. $\dfrac{\partial}{\partial x}f(2y, 3x) = 3f_2(2y, 3x).$

12. $\dfrac{\partial}{\partial y}f\left(yf(x, t), f(y, t)\right)$

$$\begin{aligned}&= f(x, t)f_1\left(yf(x, t), f(y, t)\right)\\ &\quad + f_1(y, t)f_2\left(yf(x, t), f(y, t)\right).\end{aligned}$$

14. If $E = f(x, y, z, t)$, where $x = \sin t$, $y = \cos t$ and $z = t$, then the rate of change of E is

$$\frac{dE}{dt} = \frac{\partial E}{\partial x}\cos t - \frac{\partial E}{\partial y}\sin t + \frac{\partial E}{\partial z} + \frac{\partial E}{\partial t}.$$

16. Let $u = \dfrac{x}{x^2 + y^2}$, $v = -\dfrac{y}{x^2 + y^2}$. Then

$$\frac{\partial u}{\partial x} = \frac{y^2 - x^2}{(x^2 + y^2)^2} \qquad \frac{\partial v}{\partial x} = \frac{2xy}{(x^2 + y^2)^2}$$

$$\frac{\partial u}{\partial y} = -\frac{2xy}{(x^2 + y^2)^2} \qquad \frac{\partial v}{\partial y} = \frac{y^2 - x^2}{(x^2 + y^2)^2}.$$

We have

$$\frac{\partial}{\partial x} f(u, v) = f_1(u, v) \frac{\partial u}{\partial x} + f_2(u, v) \frac{\partial v}{\partial x}$$

$$\frac{\partial}{\partial y} f(u, v) = f_1(u, v) \frac{\partial u}{\partial y} + f_2(u, v) \frac{\partial v}{\partial y}$$

$$\frac{\partial^2}{\partial x^2} f(u, v) = f_{11} \left(\frac{\partial u}{\partial x} \right)^2 + f_{12} \frac{\partial u}{\partial x} \frac{\partial v}{\partial x} + f_1 \frac{\partial^2 u}{\partial x^2}$$
$$+ f_{21} \frac{\partial u}{\partial x} \frac{\partial v}{\partial x} + f_{22} \left(\frac{\partial v}{\partial x} \right)^2 + f_2 \frac{\partial^2 v}{\partial x^2}$$

$$\frac{\partial^2}{\partial y^2} f(u, v) = f_{11} \left(\frac{\partial u}{\partial y} \right)^2 + f_{12} \frac{\partial u}{\partial y} \frac{\partial v}{\partial y} + f_1 \frac{\partial^2 u}{\partial y^2}$$
$$+ f_{21} \frac{\partial u}{\partial y} \frac{\partial v}{\partial y} + f_{22} \left(\frac{\partial v}{\partial y} \right)^2 + f_2 \frac{\partial^2 v}{\partial y^2}.$$

Noting that

$$\left(\frac{\partial u}{\partial x} \right)^2 + \left(\frac{\partial u}{\partial y} \right)^2 = \frac{1}{(x^2 + y^2)^2} = \left(\frac{\partial v}{\partial x} \right)^2 + \left(\frac{\partial v}{\partial y} \right)^2$$

$$\frac{\partial u}{\partial x} \frac{\partial v}{\partial x} + \frac{\partial u}{\partial y} \frac{\partial v}{\partial y} = 0,$$

we have

$$\frac{\partial^2}{\partial x^2} f(u, v) + \frac{\partial^2}{\partial y^2} f(u, v)$$

$$= f_{11} \left[\left(\frac{\partial u}{\partial x} \right)^2 + \left(\frac{\partial u}{\partial y} \right)^2 \right]$$

$$+ f_{22} \left[\left(\frac{\partial v}{\partial x} \right)^2 + \left(\frac{\partial v}{\partial y} \right)^2 \right]$$

$$+ 2f_{12} \left[\frac{\partial u}{\partial x} \frac{\partial v}{\partial x} + \frac{\partial u}{\partial y} \frac{\partial v}{\partial y} \right]$$

$$+ f_1 \left[\frac{\partial^2 u}{\partial x^2} + \frac{\partial^2 u}{\partial y^2} \right] + f_2 \left[\frac{\partial^2 v}{\partial x^2} + \frac{\partial^2 v}{\partial y^2} \right]$$

$$= f_1 \left[\frac{\partial^2 u}{\partial x^2} + \frac{\partial^2 u}{\partial y^2} \right] + f_2 \left[\frac{\partial^2 v}{\partial x^2} + \frac{\partial^2 v}{\partial y^2} \right],$$

because we are given that f is harmonic, that is,
$f_{11}(u, v) + f_{22}(u, v) = 0$.

Finally, u is harmonic by Exercise 10 of Section 3.4, and, by symmetry, so is v. Thus

$$\frac{\partial^2}{\partial x^2} f(u, v) + \frac{\partial^2}{\partial y^2} f(u, v) = 0$$

and $f\left(\dfrac{x}{x^2 + y^2}, -\dfrac{y}{x^2 + y^2} \right)$ is harmonic for $(x, y) \neq (0, 0)$.

18.
$$\frac{\partial^3}{\partial x \partial y^2} f(2x + 3y, xy) = \frac{\partial^2}{\partial x \partial y} (3f_1 + x f_2)$$

$$= \frac{\partial}{\partial x} (9f_{11} + 3x f_{12} + 3x f_{21} + x^2 f_{22})$$

$$= \frac{\partial}{\partial x} (9f_{11} + 6x f_{12} + x^2 f_{22})$$

$$= 18 f_{111} + 9y f_{112} + 6 f_{12} + 12x f_{121} + 6xy f_{122}$$
$$+ 2x f_{22} + 2x^2 f_{221} + x^2 y f_{222}$$

$$= 18 f_{111} + (12x + 9y) f_{112} + (6xy + 2x^2) f_{122} + x^2 y f_{222}$$
$$+ 6 f_{12} + 2x f_{22},$$

where all partials are evaluated at $(2x + 3y, xy)$.

20.
$$\frac{\partial^3}{\partial t^2 \partial s} f(s^2 - t, s + t^2) = \frac{\partial^2}{\partial t^2} (2s f_1 + f_2)$$

$$= \frac{\partial}{\partial t} (-2s f_{11} + 4st f_{12} - f_{21} + 2t f_{22})$$

$$= \frac{\partial}{\partial t} (-2s f_{11} + (4st - 1) f_{12} + 2t f_{22})$$

$$= 2s f_{111} - 4st f_{112} + 4s f_{12} - (4st - 1) f_{121}$$
$$+ 2t(4st - 1) f_{122} + 2 f_{22} - 2t f_{221} + 4t^2 f_{222}$$

$$= 2s f_{111} + (1 - 8st) f_{112} + 4t(2st - 1) f_{122} + 4t^2 f_{222}$$
$$+ 4s f_{12} + 2 f_{22},$$

where all partials are evaluated at $(s^2 - t, s + t^2)$.

22. If $r^2 = x^2 + y^2 + z^2$, then $2r \dfrac{\partial r}{\partial x} = 2x$, so $\dfrac{\partial r}{\partial x} = \dfrac{x}{r}$. Similarly, $\dfrac{\partial r}{\partial y} = \dfrac{y}{r}$ and $\dfrac{\partial r}{\partial z} = \dfrac{z}{r}$. If $u = \dfrac{1}{r}$, then

$$\frac{\partial u}{\partial x} = -\frac{1}{r^2} \frac{\partial r}{\partial x} = -\frac{x}{r^3}$$

$$\frac{\partial^2 u}{\partial x^2} = -\frac{1}{r^3} + \frac{3x}{r^4} \frac{x}{r} = \frac{3x^2 - r^2}{r^5}.$$

Similarly,

$$\frac{\partial^2 u}{\partial y^2} = \frac{3y^2 - r^2}{r^5}, \qquad \frac{\partial^2 u}{\partial z^2} = \frac{3z^2 - r^2}{r^5}.$$

Adding these three expressions, we get

$$\frac{\partial^2 u}{\partial x^2} + \frac{\partial^2 u}{\partial y^2} + \frac{\partial^2 u}{\partial z^2} = 0,$$

so u is harmonic except at $r = 0$.

237

24. If $x = r\cos\theta$ and $y = r\sin\theta$, then $r^2 = x^2 + y^2$ and $\tan\theta = y/x$. Thus $2r\dfrac{\partial r}{\partial x} = 2x$, so $\dfrac{\partial r}{\partial x} = \dfrac{x}{r} = \cos\theta$, and

similarly, $\dfrac{\partial r}{\partial y} = \dfrac{y}{r} = \sin\theta$. Also

$$\sec^2\theta\,\frac{\partial\theta}{\partial x} = -\frac{y}{x^2} \qquad \sec^2\theta\,\frac{\partial\theta}{\partial y} = \frac{1}{x}$$

$$\frac{\partial\theta}{\partial x} = -\frac{y}{x^2+y^2} \qquad \frac{\partial\theta}{\partial x} = \frac{x}{x^2+y^2}$$

$$= -\frac{\sin\theta}{r} \qquad\qquad = \frac{\cos\theta}{r}.$$

Now

$$\frac{\partial u}{\partial x} = \frac{\partial u}{\partial r}\frac{\partial r}{\partial x} + \frac{\partial u}{\partial\theta}\frac{\partial\theta}{\partial x} = \cos\theta\,\frac{\partial u}{\partial r} - \frac{\sin\theta}{r}\frac{\partial u}{\partial\theta}$$

$$\frac{\partial u}{\partial y} = \frac{\partial u}{\partial r}\frac{\partial r}{\partial y} + \frac{\partial u}{\partial\theta}\frac{\partial\theta}{\partial y} = \sin\theta\,\frac{\partial u}{\partial r} + \frac{\cos\theta}{r}\frac{\partial u}{\partial\theta}$$

$$\frac{\partial^2 u}{\partial x^2} = \left(\frac{\partial}{\partial x}\cos\theta\right)\frac{\partial u}{\partial r} + \cos\theta\left(\cos\theta\,\frac{\partial^2 u}{\partial r^2} - \frac{\sin\theta}{r}\frac{\partial^2 u}{\partial\theta\partial r}\right)$$

$$\quad -\left(\frac{\partial}{\partial x}\frac{\sin\theta}{r}\right)\frac{\partial u}{\partial\theta} - \frac{\sin\theta}{r}\left(\cos\theta\,\frac{\partial^2 u}{\partial r\partial\theta} - \frac{\sin\theta}{r}\frac{\partial^2 u}{\partial\theta^2}\right)$$

$$= \frac{\sin^2\theta}{r}\frac{\partial u}{\partial r} + \frac{2\sin\theta\cos\theta}{r^2}\frac{\partial u}{\partial\theta} + \cos^2\theta\,\frac{\partial^2 u}{\partial r^2}$$

$$\quad - \frac{2\sin\theta\cos\theta}{r}\frac{\partial^2 u}{\partial r\partial\theta} + \frac{\sin^2\theta}{r^2}\frac{\partial^2 u}{\partial\theta^2}$$

$$\frac{\partial^2 u}{\partial y^2} = \left(\frac{\partial}{\partial y}\sin\theta\right)\frac{\partial u}{\partial r} + \sin\theta\left(\sin\theta\,\frac{\partial^2 u}{\partial r^2} + \frac{\cos\theta}{r}\frac{\partial^2 u}{\partial\theta\partial r}\right)$$

$$\quad + \left(\frac{\partial}{\partial y}\frac{\cos\theta}{r}\right)\frac{\partial u}{\partial\theta} + \frac{\cos\theta}{r}\left(\sin\theta\,\frac{\partial^2 u}{\partial r\partial\theta} + \frac{\cos\theta}{r}\frac{\partial^2 u}{\partial\theta^2}\right)$$

$$= \frac{\cos^2\theta}{r}\frac{\partial u}{\partial r} - \frac{2\sin\theta\cos\theta}{r^2}\frac{\partial u}{\partial\theta} + \sin^2\theta\,\frac{\partial^2 u}{\partial r^2}$$

$$\quad + \frac{2\sin\theta\cos\theta}{r}\frac{\partial^2 u}{\partial r\partial\theta} + \frac{\cos^2\theta}{r^2}\frac{\partial^2 u}{\partial\theta^2}.$$

Therefore

$$\frac{\partial^2 u}{\partial x^2} + \frac{\partial^2 u}{\partial y^2} = \frac{\partial^2 u}{\partial r^2} + \frac{1}{r}\frac{\partial u}{\partial r} + \frac{1}{r^2}\frac{\partial^2 u}{\partial\theta^2},$$

as was to be shown.

26. $f(tx, ty) = t^k f(x, y)$

$$xf_1(tx, ty) + yf_2(tx, ty) = kt^{k-1}f(x, y)$$

$$x\big(xf_{11}(tx, ty) + yf_{12}(tx, ty)\big)$$

$$\quad + y\big(xf_{21}(tx, ty) + yf_{22}(tx, ty)\big)$$

$$\quad = k(k-1)t^{k-2}f(x, y)$$

Put $t = 1$ and get

$$x^2 f_{11}(x, y) + 2xyf_{12}(x, y) + y^2 f_{22}(x, y) = k(k-1)f(x, y).$$

28. If $f(x_1, \cdots, x_n)$ is positively homogeneous of degree k and has continuous partial derivatives of mth order, then

$$\sum_{i_1,\ldots,i_m=1}^{n} x_{i_1}\cdots x_{i_m} f_{i_1\ldots i_m}(x_1, \cdots, x_n)$$

$$= k(k-1)\cdots(k-m+1)f(x_1, \cdots, x_n).$$

The proof is identical to those of Exercises 26 or 27, except that you differentiate m times before putting $t = 1$.

30. a) Since $F_{12}(x, y) = -F_{21}(y, x)$ for $(x, y) \neq (0, 0)$, we have $F_{12}(x, x) = -F_{21}(x, x)$ for $x \neq 0$. However, all partial derivatives of the rational function F are continuous except possibly at the origin. Thus $F_{12}(x, x) = F_{21}(x, x)$ for $x \neq 0$. Therefore, $F_{12}(x, x) = 0$ for $x \neq 0$.

   b) F_{12} cannot be continuous at $(0, 0)$ because its value there (which is -2) differs from the value of $F_{21}(0, 0)$ (which is 2). Alternatively, $F_{12}(0, 0)$ is not the limit of $F_{12}(x, x)$ as $x \to 0$.

32. If $w(r) = f(r) + g(s)$, where f and g are arbitrary twice differentiable functions, then

$$\frac{\partial^2 w}{\partial r\partial s} = \frac{\partial}{\partial r}g'(s) = 0.$$

34. By Exercise 39, the DE $u_t = c^2 u_{xx}$ has solution

$$u(x, t) = f(x + ct) + g(x - ct),$$

for arbitrary sufficiently smooth functions f and g. The initial conditions imply that

$$p(x) = u(x, 0) = f(x) + g(x)$$

$$q(x) = u_t(x, 0) = cf'(x) - cg'(x).$$

Integrating the second of these equations, we get

$$f(x) - g(x) = \frac{1}{c}\int_a^x q(s)\,ds,$$

where a is a constant. Solving the two equations for f and g we obtain

$$f(x) = \frac{1}{2}p(x) + \frac{1}{2c}\int_a^x q(s)\,ds$$

$$g(x) = \frac{1}{2}p(x) - \frac{1}{2c}\int_a^x q(s)\,ds.$$

Thus the solution to the initial-value problem is

$$u(x, t) = \frac{p(x + ct) + p(x - ct)}{2} + \frac{1}{2c}\int_{x-ct}^{x+ct} q(s)\,ds.$$

36.
```
> g := f(x/(x^2+y^2),y/(x^2+y^2)):
> simplify(diff(g,x$2)+diff(g,y$2));
```

$$\frac{D_{1,1}(f)\left(\dfrac{x}{x^2+y^2},\dfrac{y}{x^2+y^2}\right) + D_{2,2}(f)\left(\dfrac{x}{x^2+y^2},\dfrac{y}{x^2+y^2}\right)}{(x^2+y^2)^2}.$$

If f is harmonic, then the numerator is zero so g is harmonic.

38.
```
> simplify(diff(diff
>     (f(s^2-t,s+t^2),s),t$2));
```
$$2s\,D_{1,1,1}(f) - 8st\,D_{1,1,2}(f) + 8st^2 D_{1,2,2}(f) + 4s\,D_{1,2}(f)$$
$$+ D_{1,1,2}(f) - 4t\,D_{1,2,2}(f) + 4t^2 D_{2,2,2}(f) + 2D_{2,2}(f)$$

where all terms are evaluated at $(s^2 - t, s + t^2)$.

40.
```
> u := (x,t) -> (p(x-c*t)+p(x+c*t))/2
>       +(1/((2*c))*int(q(s),x=x-
c*t..x+c*t):
> simplify(diff(u(x,t),t$2)·
>   -c^2*diff(u(x,t),x$2));
                0
> simplify(u(x,0));
                p(x)
> simplify(subs(t=0,diff(u(x,t),t)));
                q(x)
```

so u satisfies the PDE and initial conditions given in Exercise 34.

Section 12.6 Linear Approximations, Differentiability, and Differentials (page 750)

2.
$$f(x, y) = \tan^{-1}\frac{y}{x} \qquad f(3, 3) = \frac{\pi}{4}$$
$$f_1(x, y) = -\frac{y}{x^2 + y^2} \qquad f_1(3, 3) = -\frac{1}{6}$$
$$f_2(x, y) = \frac{x}{x^2 + y^2} \qquad f_2(3, 3) = \frac{1}{6}$$
$$f(3.01, 2.99) = f(3 + 0.01, 3 - 0.01)$$
$$\approx f(3, 3) + 0.01 f_1(3, 3) - 0.01 f_2(3, 3)$$
$$= \frac{\pi}{4} - \frac{0.01}{6} - \frac{0.01}{6} = \frac{\pi}{4} - \frac{0.01}{3}$$
$$\approx 0.7820648$$

4.
$$f(x, y) = \frac{24}{x^2 + xy + y^2}$$
$$f_1(x, y) = \frac{-24(2x + y)}{(x^2 + xy + y^2)^2}, \qquad f_2(x, y) = \frac{-24(x + 2y)}{(x^2 + xy + y^2)^2}$$
$$f(2, 2) = 2, \qquad f_1(2, 2) = -1, \qquad f_2(2, 2) = -1$$
$$f(2.1, 1.8) \approx f(2, 2) + 0.1 f_1(2, 2) - 0.2 f_2(2, 2)$$
$$= 2 - 0.1 + 0.2 = 2.1$$

6.
$$f(x, y) = xe^{y+x^2} \qquad f(2, -4) = 2$$
$$f_1(x, y) = e^{y+x^2}(1 + 2x^2) \qquad f_1(2, -4) = 9$$
$$f_2(x, y) = xe^{y+x^2} \qquad f_2(2, -4) = 2$$
$$f(2.05, -3.92) \approx f(2, -4) + 0.05 f_1(2, -4) + 0.08 f_2(2, -4)$$
$$= 2 + 0.45 + 0.16 = 2.61$$

8. $V = \frac{1}{3}\pi r^2 h \Rightarrow dV = \frac{2}{3}\pi rh\,dr + \frac{1}{3}\pi r^2\,dh$. If $r = 25$ ft, $h = 21$ ft, and $dr = dh = 0.5/12$ ft, then

$$dV = \frac{\pi}{3}(2 \times 25 \times 21 + 25^2)\frac{0.5}{12} \approx 73.08.$$

The calculated volume can be in error by about 73 cubic feet.

10. If the sides and contained angle of the triangle are x and y m and θ radians, then its area A satisfies

$$A = \frac{1}{2}xy\sin\theta$$
$$dA = \frac{1}{2}y\sin\theta\,dx + \frac{1}{2}x\sin\theta\,dy + \frac{1}{2}xy\cos\theta\,d\theta$$
$$\frac{dA}{A} = \frac{dx}{x} + \frac{dy}{y} + \cot\theta\,d\theta.$$

For $x = 224$, $y = 158$, $\theta = 64° = 64\pi/180$, $dx = dy = 0.4$, and $d\theta = 2° = 2\pi/180$, we have

$$\frac{dA}{A} = \frac{0.4}{224} + \frac{0.4}{158} + (\cot 64°)\frac{2\pi}{180} \approx 0.0213.$$

The calculated area of the plot can be in error by a little over 2%.

12.
$$w = \frac{x^2 y^3}{z^4} \qquad\qquad \frac{\partial w}{\partial x} = \frac{2xy^3}{z^4} = \frac{2w}{x}$$
$$\frac{\partial w}{\partial y} = \frac{3x^2 y^2}{z^4} = \frac{3w}{y} \qquad \frac{\partial w}{\partial z} = -\frac{4x^2 y^3}{z^5} = -\frac{4w}{x}.$$
$$dw = \frac{\partial w}{\partial x}dx + \frac{\partial w}{\partial y}dy + \frac{\partial w}{\partial z}dz$$
$$\frac{dw}{w} = 2\frac{dx}{x} + 3\frac{dy}{y} - 4\frac{dz}{z}.$$

Since x increases by 1%, then $\frac{dx}{x} = \frac{1}{100}$. Similarly, $\frac{dy}{y} = \frac{2}{100}$ and $\frac{dz}{z} = \frac{3}{100}$. Therefore

$$\frac{\Delta w}{w} \approx \frac{dw}{w} = \frac{2 + 6 - 12}{100} = -\frac{4}{100},$$

and w decreases by about 4%.

14.
$$\mathbf{f}(\rho, \phi, \theta) = (\rho\sin\phi\cos\theta, \rho\sin\phi\sin\theta, \rho\cos\phi)$$
$$D\mathbf{f}(\rho, \phi, \theta) = \begin{pmatrix} \sin\phi\cos\theta & \rho\cos\phi\cos\theta & -\rho\sin\phi\sin\theta \\ \sin\phi\sin\theta & \rho\cos\phi\sin\theta & \rho\sin\phi\cos\theta \\ \cos\phi & -\rho\sin\phi & 0 \end{pmatrix}$$

16.
$$g(r, s, t) = \begin{pmatrix} r^2 s \\ r^2 t \\ s^2 - t^2 \end{pmatrix}$$

$$Dg(r, s, t) = \begin{pmatrix} 2rs & r^2 & 0 \\ 2rt & 0 & r^2 \\ 0 & 2s & -2t \end{pmatrix}$$

$$Dg(1, 3, 3) = \begin{pmatrix} 6 & 1 & 0 \\ 6 & 0 & 1 \\ 0 & 6 & -6 \end{pmatrix}$$

$$g(0.99, 3.02, 2.97) \approx g(1, 3, 3) + Dg(1, 3, 3) \begin{pmatrix} -0.01 \\ 0.02 \\ -0.03 \end{pmatrix}$$

$$= \begin{pmatrix} 3 \\ 3 \\ 0 \end{pmatrix} + \begin{pmatrix} -0.04 \\ -0.09 \\ 0.30 \end{pmatrix} = \begin{pmatrix} 2.96 \\ 2.91 \\ 0.30 \end{pmatrix}$$

18. Let $g(t) = f(a + th, b + tk)$. Then

$$g'(t) = hf_1(a + th, b + tk) + kf_2(a + th, b + tk).$$

If h and k are small enough that $(a + h, b + k)$ belongs to the disk referred to in the statement of the problem, then we can apply the (one-variable) Mean-Value Theorem to $g(t)$ on $[0, 1]$ and obtain

$$g(1) = g(0) + g'(\theta),$$

for some θ satisfying $0 < \theta < 1$, i.e.,

$$f(a + h, b + k) = f(a, b) + hf_1(a + \theta h, b + \theta k)$$
$$+ kf_2(a + \theta h, b + \theta k).$$

Section 12.7   Gradients and Directional Derivatives   (page 761)

2. $f(x, y) = \dfrac{x - y}{x + y}$, $\qquad f(1, 1) = 0$.

$\nabla f = \dfrac{2y\mathbf{i} - 2x\mathbf{j}}{(x + y)^2}$,

$\nabla f(1, 1) = \dfrac{1}{2}(\mathbf{i} - \mathbf{j})$. Tangent plane to $z = f(x, y)$ at $(1, 1, 0)$ has equation $\frac{1}{2}(x - 1) - \frac{1}{2}(y - 1) = z$, or $x - y - 2z = 0$.
Tangent line to $f(x, y) = 0$ at $(1, 1)$ has equation $\frac{1}{2}(x - 1) - \frac{1}{2}(y - 1)$, or $x = y$.

4. $f(x, y) = e^{xy}$, $\quad \nabla f = ye^{xy}\mathbf{i} + xe^{xy}\mathbf{j}$,
$\nabla f(2, 0) = 2\mathbf{j}$. Tangent plane to $z = f(x, y)$ at $(2, 0, 1)$ has equation $2y = z - 1$, or $2y - z = -1$.
Tangent line to $f(x, y) = 1$ at $(2, 0)$ has equation $y = 0$.

6. $f(x, y) = \sqrt{1 + xy^2}$, $\qquad f(2, -2) = 3$.
$\nabla f(x, y) = \dfrac{y^2\mathbf{i} + 2xy\mathbf{j}}{2\sqrt{1 + xy^2}}$,

$\nabla f(2, -2) = \dfrac{2}{3}\mathbf{i} - \dfrac{4}{3}\mathbf{j}$.
Tangent plane to $z = f(x, y)$ at $(2, -2, 3)$ has equation
$\frac{2}{3}(x - 2) - \frac{4}{3}(y + 2) = z - 3$, or $2x - 4y - 3z = 3$.
Tangent line to $f(x, y) = 3$ at $(2, -2)$ has equation
$\frac{2}{3}(x - 2) - \frac{4}{3}(y + 2) = 0$, or $x - 2y = 6$.

8. $f(x, y, z) = \cos(x + 2y + 3z)$,
$f\left(\dfrac{\pi}{2}, \pi, \pi\right) = \cos\dfrac{11\pi}{2} = 0$.
$\nabla f(x, y, z) = -\sin(x + 2y + 3z)(\mathbf{i} + 2\mathbf{j} + 3\mathbf{k})$,
$\nabla f\left(\dfrac{\pi}{2}, \pi, \pi\right) = -\sin\dfrac{11\pi}{2}(\mathbf{i} + 2\mathbf{j} + 3\mathbf{k}) = \mathbf{i} + 2\mathbf{j} + 3\mathbf{k}$.
Tangent plane to $f(x, y, z) = 0$ at $\left(\dfrac{\pi}{2}, \pi, \pi\right)$ has equation

$$x - \dfrac{\pi}{2} + 2(y - \pi) + 3(z - \pi) = 0,$$

or $x + 2y + 3z = \dfrac{11\pi}{2}$.

10. $f(x, y) = 3x - 4y$, $\qquad \nabla f(0, 2) = \nabla f(x, y) = 3\mathbf{i} - 4\mathbf{j}$,
$D_{-\mathbf{i}}f(0, 2) = -\mathbf{i} \bullet (3\mathbf{i} - 4\mathbf{j}) = -3$.

12. $f(x, y) = \dfrac{x}{1 + y}$, $\quad \nabla f(x, y) = \dfrac{1}{1 + y}\mathbf{i} - \dfrac{x}{(1 + y)^2}\mathbf{j}$,

$\nabla f(0, 0) = \mathbf{i}$, $\qquad \mathbf{u} = \dfrac{\mathbf{i} - \mathbf{j}}{\sqrt{2}}$,

$D_{\mathbf{u}}f(0, 0) = \mathbf{i} \bullet \left(\dfrac{\mathbf{i} - \mathbf{j}}{\sqrt{2}}\right) = \dfrac{1}{\sqrt{2}}$.

14. $f(x, y) = \ln|\mathbf{r}|$, where $\mathbf{r} = x\mathbf{i} + y\mathbf{j}$. Since $|\mathbf{r}| = \sqrt{x^2 + y^2}$, we have

$$\nabla f(x, y) = \dfrac{1}{|\mathbf{r}|}\left(\dfrac{x}{|\mathbf{r}|}\mathbf{i} + \dfrac{y}{|\mathbf{r}|}\mathbf{j}\right) = \dfrac{\mathbf{r}}{|\mathbf{r}|^2}.$$

16. Since $x = r\cos\theta$ and $y = r\sin\theta$, we have

$$\dfrac{\partial f}{\partial r} = \cos\theta\dfrac{\partial f}{\partial x} + \sin\theta\dfrac{\partial f}{\partial y}$$

$$\dfrac{\partial f}{\partial \theta} = -r\sin\theta\dfrac{\partial f}{\partial x} + r\cos\theta\dfrac{\partial f}{\partial y}.$$

Also,

$$\hat{\mathbf{r}} = \dfrac{x\mathbf{i} + y\mathbf{j}}{r} = (\cos\theta)\mathbf{i} + (\sin\theta)\mathbf{j}$$

$$\hat{\theta} = \dfrac{-y\mathbf{i} + x\mathbf{j}}{r} = -(\sin\theta)\mathbf{i} + (\cos\theta)\mathbf{j}.$$

Therefore,

$$\frac{\partial f}{\partial r}\hat{\mathbf{r}} + \frac{1}{r}\frac{\partial f}{\partial \theta}\hat{\boldsymbol{\theta}}$$

$$= \left(\cos^2\theta\,\frac{\partial f}{\partial x} + \sin\theta\cos\theta\,\frac{\partial f}{\partial y}\right)\mathbf{i}$$

$$+ \left(\cos\theta\sin\theta\,\frac{\partial f}{\partial x} + \sin^2\theta\,\frac{\partial f}{\partial y}\right)\mathbf{j}$$

$$+ \left(\sin^2\theta\,\frac{\partial f}{\partial x} - \sin\theta\cos\theta\,\frac{\partial f}{\partial y}\right)\mathbf{i}$$

$$+ \left(-\cos\theta\sin\theta\,\frac{\partial f}{\partial x} + \cos^2\theta\,\frac{\partial f}{\partial y}\right)\mathbf{j}$$

$$= \frac{\partial f}{\partial x}\mathbf{i} + \frac{\partial f}{\partial y}\mathbf{j} = \nabla f.$$

18. $f(x, y, z) = x^2 + y^2 - z^2$.
$\nabla f(a, b, c) = 2a\mathbf{i} + 2b\mathbf{j} - 2c\mathbf{k}$. The maximum rate of change of f at (a, b, c) is in the direction of $\nabla f(a, b, c)$, and is equal to $|\nabla f(a, b, c)|$.
Let $\mathbf{u}$ be a unit vector making an angle θ with $\nabla f(a, b, c)$. The rate of change of f at (a, b, c) in the direction of $\mathbf{u}$ will be half of the maximum rate of change of f at that point provided

$$\frac{1}{2}|\nabla f(a, b, c)| = \mathbf{u} \bullet \nabla f(a, b, c) = |\nabla f(a, b, c)|\cos\theta,$$

that is, if $\cos\theta = \frac{1}{2}$, which means $\theta = 60°$. At (a, b, c), f increases at half its maximal rate in all directions making $60°$ angles with the direction $a\mathbf{i} + b\mathbf{j} - c\mathbf{k}$.

20. Given the values $D_{\phi_1}f(a, b)$ and $D_{\phi_2}f(a, b)$, we can solve the equations

$$f_1(a, b)\cos\phi_1 + f_2(a, b)\sin\phi_1 = D_{\phi_1}f(a, b)$$
$$f_1(a, b)\cos\phi_2 + f_2(a, b)\sin\phi_2 = D_{\phi_2}f(a, b)$$

for unique values of $f_1(a, b)$ and $f_2(a, b)$ (and hence determine $\nabla f(a, b)$ uniquely), provided the coefficients satisfy

$$0 \neq \begin{vmatrix} \cos\phi_1 & \sin\phi_1 \\ \cos\phi_2 & \sin\phi_2 \end{vmatrix} = \sin(\phi_2 - \phi_1).$$

Thus ϕ_1 and ϕ_2 must not differ by an integer multiple of π.

22. Let the curve be $y = g(x)$. At (x, y) this curve has normal $\nabla(g(x) - y) = g'(x)\mathbf{i} - \mathbf{j}$.
A curve of the family $x^4 + y^2 = C$ has normal $\nabla(x^4 + y^2) = 4x^3\mathbf{i} + 2y\mathbf{j}$.
These curves will intersect at right angles if their normals are perpendicular. Thus we require that

$$0 = 4x^3 g'(x) - 2y = 4x^3 g'(x) - 2g(x),$$

or, equivalently,

$$\frac{g'(x)}{g(x)} = \frac{1}{2x^3}.$$

Integration gives $\ln|g(x)| = -\dfrac{1}{4x^2} + \ln|C|$,

or $g(x) = Ce^{-(1/4x^2)}$.
Since the curve passes through $(1, 1)$, we must have $1 = g(1) = Ce^{-1/4}$, so $C = e^{1/4}$.
The required curve is $y = e^{(1/4)-(1/4x^2)}$.

24. Let $f(x, y) = e^{-(x^2+y^2)}$. Then

$$\nabla f(x, y) = -2e^{-(x^2+y^2)}(x\mathbf{i} + y\mathbf{j}).$$

The vector $\mathbf{u} = \dfrac{a\mathbf{i} + b\mathbf{j}}{\sqrt{a^2 + b^2}}$ is a unit vector in the direction directly away from the origin at (a, b).
The first directional derivative of f at (x, y) in the direction of $\mathbf{u}$ is

$$\mathbf{u} \bullet \nabla f(x, y) = -\frac{2}{\sqrt{a^2 + b^2}}(ax + by)e^{-(x^2+y^2)}.$$

The second directional derivative is

$$\mathbf{u} \bullet \nabla\left(-\frac{2}{\sqrt{a^2 + b^2}}(ax + by)e^{-(x^2+y^2)}\right)$$

$$= -\frac{2}{a^2 + b^2}(a\mathbf{i} + b\mathbf{j}) \bullet e^{-(x^2+y^2)}$$

$$\left[\left(a - 2x(ax + by)\right)\mathbf{i} + \left(b - 2y(ax + by)\right)\mathbf{j}\right].$$

At (a, b) this second directional derivative is

$$-\frac{2e^{-(a^2+b^2)}}{a^2 + b^2}\left(a^2 - 2a^4 - 2a^2b^2 + b^2 - 2a^2b^2 - 2b^4\right)$$

$$= \frac{2}{a^2 + b^2}\left(2(a^2 + b^2)^2 - a^2 - b^2\right)e^{-(a^2+b^2)}$$

$$= 2\left(2(a^2 + b^2) - 1\right)e^{-(a^2+b^2)}.$$

Remark: Since $f(x, y) = e^{-r^2}$ (expressed in terms of polar coordinates), the second directional derivative of f at (a, b) in the direction directly away from the origin (i.e., the direction of increasing r) can be more easily calculated as

$$\left.\frac{d^2}{dr^2}e^{-r^2}\right|_{r^2=a^2+b^2}.$$

26. At $(1, -1, 1)$ the surface $x^2 + y^2 = 2$ has normal

$$\mathbf{n}_1 = \nabla(x^2 + y^2)\Big|_{(1,-1,1)} = 2\mathbf{i} - 2\mathbf{j},$$

and $y^2 + z^2 = 2$ has normal

$$\mathbf{n}_2 = \boldsymbol{\nabla}(y^2 + z^2)\Big|_{(1,-1,1)} = -2\mathbf{j} + 2\mathbf{k}.$$

A vector tangent to the curve of intersection of the two surfaces at $(1, -1, 1)$ must be perpendicular to both these normals. Since

$$(\mathbf{i} - \mathbf{j}) \times (-\mathbf{j} + \mathbf{k}) = -(\mathbf{i} + \mathbf{j} + \mathbf{k}),$$

the vector $\mathbf{i} + \mathbf{j} + \mathbf{k}$, or any scalar multiple of this vector, is tangent to the curve at the given point.

28. A vector tangent to the path of the fly at $(1, 1, 2)$ is given by

$$\mathbf{v} = \boldsymbol{\nabla}(3x^2 - y^2 - z) \times \boldsymbol{\nabla}(2x^2 + 2y^2 - z^2)\Big|_{(1,1,2)}$$

$$= (6x\mathbf{i} - 2y\mathbf{j} - \mathbf{k}) \times (4x\mathbf{i} + 4y\mathbf{j} - 2z\mathbf{k})\Big|_{(1,1,2)}$$

$$= (6\mathbf{i} - 2\mathbf{j} - \mathbf{k}) \times (4\mathbf{i} + 4\mathbf{j} - 4\mathbf{k})$$

$$= 4\begin{vmatrix} \mathbf{i} & \mathbf{j} & \mathbf{k} \\ 6 & -2 & -1 \\ 1 & 1 & -1 \end{vmatrix} = 4(3\mathbf{i} + 5\mathbf{j} + 8\mathbf{k}).$$

The temperature $T = x^2 - y^2 + z^2 + xz^2$ has gradient at $(1, 1, 2)$ given by

$$\boldsymbol{\nabla}T(1, 1, 2) = (2x + z^2)\mathbf{i} - 2y\mathbf{j} + 2z(1 + x)\mathbf{k}\Big|_{(1,1,2)}$$

$$= 6\mathbf{i} - 2\mathbf{j} + 8\mathbf{k}.$$

Thus the fly, passing through $(1, 1, 2)$ with speed 7, experiences temperature changing at rate

$$7 \times \frac{\mathbf{v}}{|\mathbf{v}|} \bullet \boldsymbol{\nabla}T(1, 1, 2) = 7\frac{3\mathbf{i} + 5\mathbf{j} + 8\mathbf{k}}{\sqrt{98}} \bullet (6\mathbf{i} - 2\mathbf{j} + 8\mathbf{k})$$

$$= \frac{1}{\sqrt{2}}(18 - 10 + 64) = \frac{72}{\sqrt{2}}.$$

We don't know which direction the fly is moving along the curve, so all we can say is that it experiences temperature changing at rate $36\sqrt{2}$ degrees per unit time.

30. The level surface of $f(x, y, z) = \cos(x + 2y + 3z)$ through (π, π, π) has equation $\cos(x + 2y + 3z) = \cos(6\pi) = 1$, which simplifies to $x + 2y + 3z = 6\pi$. This level surface is a plane, and is therefore its own tangent plane. We cannot determine this plane by the method used to find the tangent plane to the level surface of f through $(\pi/2, \pi, \pi)$ in Exercise 10, because $\boldsymbol{\nabla}f(\pi, \pi, \pi) = \mathbf{0}$, so the gradient does not provide a usable normal vector to define the tangent plane.

32. Let $f(x, y) = x^3 - y^2$. Then $\boldsymbol{\nabla}f(x, y) = 3x^2\mathbf{i} - 2y\mathbf{j}$ exists everywhere, but equals $\mathbf{0}$ at $(0, 0)$. The level curve of f passing through $(0, 0)$ is $y^2 = x^3$, which has a cusp at $(0, 0)$, so is not smooth there.

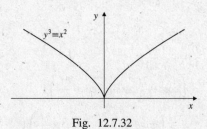

Fig. 12.7.32

34. $T = T(x, y, z)$. As measured by the observer,

$$\frac{dT}{dt} = D_{\mathbf{v}(t)}T = \mathbf{v}(t) \bullet \boldsymbol{\nabla}T$$

$$\frac{d^2T}{dt^2} = \mathbf{a}(t) \bullet \boldsymbol{\nabla}T + \mathbf{v}(t) \bullet \frac{d}{dt}\boldsymbol{\nabla}T$$

$$= D_{\mathbf{a}(t)}T + \left(v_1(t)\frac{d}{dt}\frac{\partial T}{\partial x} + \cdots\right)$$

$$= D_{\mathbf{a}(t)}T + \left(v_1(t)\mathbf{v}(t) \bullet \boldsymbol{\nabla}\frac{\partial T}{\partial x} + \cdots\right)$$

$$= D_{\mathbf{a}(t)}T + \left(\left(v_1(t)\right)^2\frac{\partial^2 T}{\partial x^2} + v_1(t)v_2(t)\frac{\partial^2 T}{\partial y \partial x} + \cdots\right)$$

$$= D_{\mathbf{a}(t)}T + D_{\mathbf{v}(t)}(D_{\mathbf{v}(t)}T)$$

(as in Exercise 37 above).

36. $f(x, y) = \begin{cases} \dfrac{\sin(xy)}{\sqrt{x^2 + y^2}} & \text{if } (x, y) \neq (0, 0) \\ 0 & \text{if } (x, y) = (0, 0) \end{cases}$.

a) $f_1(0, 0) = \lim\limits_{h \to 0}\dfrac{0 - 0}{h} = 0 = f_2(0, 0)$. Thus $\boldsymbol{\nabla}f(0, 0) = \mathbf{0}$.

b) If $\mathbf{u} = (\mathbf{i} + \mathbf{j})/\sqrt{2}$, then

$$D_{\mathbf{u}}f(0, 0) = \lim_{h \to 0+}\frac{1}{h}\frac{\sin(h^2/2)}{\sqrt{h^2}} = \frac{1}{2}.$$

c) f cannot be differentiable at $(0, 0)$; if it were, then the directional derivative obtained in part (b) would have been $\mathbf{u} \bullet \boldsymbol{\nabla}f(0, 0) = 0$.

38. After typing in the definition of Grad given in the problem statement, we can calculate

```
>  f := (x,y,z,t) -> x*y^2*z^3*t^4:
>  Grad(f,4)(1,1,1,1);
                [1, 2, 3, 4]
```

which is the expected result.

Section 12.8 Implicit Functions   (page 772)

2. $xy^3 = y - z$: $x = x(y, z)$

$y^3 \dfrac{\partial x}{\partial y} + 3xy^2 = 1$

$\dfrac{\partial x}{\partial y} = \dfrac{1 - 3xy^2}{y^3}$.

The given equation has a solution $x = x(y, z)$ with this partial derivative near any point where $y \neq 0$.

4. $e^{yz} - x^2 z \ln y = \pi$: $y = y(x, z)$

$e^{yz}\left(z\dfrac{\partial y}{\partial z} + y\right) - x^2 \ln y - \dfrac{x^2 z}{y}\dfrac{\partial y}{\partial z} = 0$

$\dfrac{\partial y}{\partial z} = \dfrac{x^2 \ln y - ye^{yz}}{ze^{yz} - \dfrac{x^2 z}{y}} = \dfrac{x^2 y \ln y - y^2 e^{yz}}{yze^{yz} - x^2 z}$.

The given equation has a solution $y = y(x, z)$ with this derivative near any point where $y > 0$, $z \neq 0$, and $ye^{yz} \neq x^2$.

6. $F(x, y, x^2 - y^2) = 0$: $y = y(x)$

$F_1 + F_2\dfrac{dy}{dx} + F_3\left(2x - 2y\dfrac{dy}{dx}\right) = 0$

$\dfrac{dy}{dx} = \dfrac{F_1(x, y, x^2 - y^2) + 2xF_3(x, y, x^2 - y^2)}{2yF_3(x, y, x^2 - y^2) - F_2(x, y, x^2 - y^2)}$.

The given equation has a solution with this derivative near any point where $F_2(x, y, x^2 - y^2) \neq 2yF_3(x, y, x^2 - y^2)$.

8. $F(x^2 - z^2, y^2 + xz) = 0$: $z = z(x, y)$

$F_1\left(2x - 2z\dfrac{\partial z}{\partial x}\right) + F_2\left(x\dfrac{\partial z}{\partial x} + z\right) = 0$

$\dfrac{\partial z}{\partial x} = \dfrac{2xF_1(x^2 - z^2, y^2 + xz) + zF_2(x^2 - z^2, y^2 + xz)}{2zF_1(x^2 - z^2, y^2 + xz) - xF_2(x^2 - z^2, y^2 + xz)}$.

The given equation has a solution with this derivative near any point where $xF_2(x^2 - z^2, y^2 + xz) \neq 2zF_1(x^2 - z^2, y^2 + xz)$.

10. $\begin{cases} xyuv = 1 \\ x + y + u + v = 0 \end{cases} \Rightarrow \begin{cases} y = y(x, u) \\ v = v(x, u) \end{cases}$

Differentiate the given equations with respect to x:

$yuv + xuv\dfrac{\partial y}{\partial x} + xyu\dfrac{\partial v}{\partial x} = 0$

$1 + \dfrac{\partial y}{\partial x} + \dfrac{\partial v}{\partial x} = 0$

Multiply the last equation by xyu and subtract the two equations:

$yuv - xyu + (xuv - xyu)\dfrac{\partial y}{\partial x} = 0$

$\left(\dfrac{\partial y}{\partial x}\right)_u = \dfrac{y(x - v)}{x(v - y)}$.

The given equations have a solution of the indicated form with this derivative near any point where $u \neq 0$, $x \neq 0$ and $y \neq v$.

12. $\begin{cases} x^2 y + y^2 u - u^3 = 0 \\ x^2 + yu = 1 \end{cases} \Rightarrow \begin{cases} u = u(x) \\ y = y(x) \end{cases}$

$2xy + (x^2 + 2yu)\dfrac{dy}{dx} + (y^2 - 3u^2)\dfrac{du}{dx} = 0$

$2x + u\dfrac{dy}{dx} + y\dfrac{du}{dx} = 0$

Multiply the first equation by u and the second by $x^2 + 2yu$ and subtract:

$2x(x^2 + yu) + (x^2 y + y^2 u + 3u^3)\dfrac{du}{dx} = 0$

$\dfrac{du}{dx} = -\dfrac{2x(x^2 + yu)}{3u^3 + x^2 y + y^2 u} = -\dfrac{x}{2u^3}$.

The given equations have a solution with the indicated derivative near any point where $u \neq 0$.

14. $\begin{cases} x = r^2 + 2s \\ y = s^2 - 2r \end{cases}$

$\dfrac{\partial(x, y)}{\partial(r, s)} = \begin{vmatrix} 2r & 2 \\ -2 & 2s \end{vmatrix} = 4(rs + 1)$.

The given system can be solved for r and s as functions of x and y near any point (r, s) where $rs \neq -1$.
We have

$1 = 2r\dfrac{\partial r}{\partial x} + 2\dfrac{\partial s}{\partial x}$

$0 = -2\dfrac{\partial r}{\partial x} + 2s\dfrac{\partial s}{\partial x}$

$0 = 2r\dfrac{\partial r}{\partial y} + 2\dfrac{\partial s}{\partial y}$

$1 = -2\dfrac{\partial r}{\partial y} + 2s\dfrac{\partial s}{\partial y}$.

Thus

$\dfrac{\partial r}{\partial x} = \dfrac{s}{2(rs + 1)}$ $\dfrac{\partial r}{\partial y} = -\dfrac{1}{2(rs + 1)}$

$\dfrac{\partial s}{\partial x} = \dfrac{1}{2(rs + 1)}$ $\dfrac{\partial s}{\partial y} = \dfrac{r}{2(rs + 1)}$.

16. $x = \rho\sin\phi\cos\theta$, $y = \rho\sin\phi\sin\theta$, $z = \rho\cos\phi$.

$\dfrac{\partial(x, y, z)}{\partial(\rho, \phi, \theta)} = \begin{vmatrix} \sin\phi\cos\theta & \rho\cos\phi\cos\theta & -\rho\sin\phi\sin\theta \\ \sin\phi\sin\theta & \rho\cos\phi\sin\theta & \rho\sin\phi\cos\theta \\ \cos\phi & -\rho\sin\phi & 0 \end{vmatrix}$

$= \cos\phi\begin{vmatrix} \rho\cos\phi\cos\theta & -\rho\sin\phi\sin\theta \\ \rho\cos\phi\sin\theta & \rho\sin\phi\cos\theta \end{vmatrix}$

$+ \rho\sin\phi\begin{vmatrix} \sin\phi\cos\theta & -\rho\sin\phi\sin\theta \\ \sin\phi\sin\theta & \rho\sin\phi\cos\theta \end{vmatrix}$

$= \rho^2\cos\phi\left[\cos\phi\sin\phi\cos^2\theta + \sin\phi\cos\phi\sin^2\theta\right]$

$+ \rho^2\sin\phi\left[\sin^2\phi\cos^2\theta + \sin^2\phi\sin^2\theta\right]$

$= \rho^2\cos^2\phi\sin\phi + \rho^2\sin^3\phi = \rho^2\sin\phi$.

243

The transformation is one-to-one (and invertible) near any point where $\rho^2 \sin \phi \neq 0$, that is, near any point not on the z-axis.

18. Let $F(x, y, z, u, v) = xe^y + uz - \cos v - 2$

$G(x, y, z, u, v) = u \cos y + x^2 v - yz^2 - 1$.

If P_0 is the point where $(x, y, z) = (2, 0, 1)$ and $(u, v) = (1, 0)$, then

$$\frac{\partial(F, G)}{\partial(u, v)}\bigg|_{P_0} = \begin{vmatrix} z & \sin v \\ \cos y & x^2 \end{vmatrix}\bigg|_{P_0}$$

$$= \begin{vmatrix} 1 & 0 \\ 1 & 4 \end{vmatrix} = 4.$$

Since this Jacobian is not zero, the equations $F = G = 0$ can be solved for u, and v in terms of x, y and z near P_0. Also,

$$\left(\frac{\partial u}{\partial z}\right)_{x,y}\bigg|_{(2,0,1)} = -\frac{1}{4}\frac{\partial(F, G)}{\partial(z, v)}\bigg|_{P_0}$$

$$= -\frac{1}{4}\begin{vmatrix} u & \sin v \\ -2yz & x^2 \end{vmatrix}\bigg|_{P_0}$$

$$= -\frac{1}{4}\begin{vmatrix} 1 & 0 \\ 0 & 4 \end{vmatrix} = -1.$$

20. $F(x, y, z, u, v) = 0$

$G(x, y, z, u, v) = 0$

$H(x, y, z, u, v) = 0$

To calculate $\dfrac{\partial x}{\partial y}$ we require that x be one of three dependent variables, and y be one of two independent variables. The other independent variable can be z or u or v. The possible interpretations for this partial, and their values, are

$$\left(\frac{\partial x}{\partial y}\right)_z = -\frac{\dfrac{\partial(F, G, H)}{\partial(y, u, v)}}{\dfrac{\partial(F, G, H)}{\partial(x, u, v)}}$$

$$\left(\frac{\partial x}{\partial y}\right)_u = -\frac{\dfrac{\partial(F, G, H)}{\partial(y, z, v)}}{\dfrac{\partial(F, G, H)}{\partial(x, z, v)}}$$

$$\left(\frac{\partial x}{\partial y}\right)_v = -\frac{\dfrac{\partial(F, G, H)}{\partial(y, z, u)}}{\dfrac{\partial(F, G, H)}{\partial(x, z, u)}}.$$

22. If $F(x, y, z) = 0 \implies z = z(x, y)$, then

$$F_1 + F_3 \frac{\partial z}{\partial x} = 0, \qquad F_2 + F_3 \frac{\partial z}{\partial y} = 0,$$

$$F_{11} + F_{13}\frac{\partial z}{\partial x} + F_{31}\frac{\partial z}{\partial x} + F_{33}\left(\frac{\partial z}{\partial x}\right)^2 + F_3 \frac{\partial^2 z}{\partial x^2} = 0.$$

Thus

$$\frac{\partial^2 z}{\partial x^2} = -\frac{1}{F_3}\left[F_{11} + 2F_{13}\left(-\frac{F_1}{F_3}\right) + F_{33}\left(-\frac{F_1}{F_3}\right)^2\right]$$

$$= -\frac{1}{F_3^3}\left[F_{11}F_3^2 - 2F_1 F_3 F_{13} + F_1^2 F_{33}\right].$$

Similarly,

$$\frac{\partial^2 z}{\partial y^2} = -\frac{1}{F_3^3}\left[F_{22}F_3^2 - 2F_2 F_3 F_{23} + F_2^2 F_{33}\right].$$

Also,

$$F_{12} + F_{13}\frac{\partial z}{\partial y} + \left(F_{32} + F_{33}\frac{\partial z}{\partial y}\right)\frac{\partial z}{\partial x} + F_3 \frac{\partial^2 z}{\partial y \partial x}.$$

Therefore

$$\frac{\partial^2 z}{\partial x \partial y} = -\frac{1}{F_3}\left[F_{12} + F_{13}\left(-\frac{F_2}{F_3}\right) + F_{23}\left(-\frac{F_1}{F_3}\right)\right.$$

$$\left. + F_{33}\left(\frac{F_1 F_2}{F_3^2}\right)\right]$$

$$= -\frac{1}{F_3^2}\left[F_3^2 F_{12} - F_2 F_3 F_{13} - F_1 F_3 F_{23} + F_1 F_2 F_{33}\right].$$

24. $pV = T - \dfrac{4p}{T^2}, \qquad T = T(p, V)$

a) $V = \dfrac{\partial T}{\partial p} - \dfrac{4}{T^2} + \dfrac{8p}{T^3}\dfrac{\partial T}{\partial p}$

$p = \dfrac{\partial T}{\partial V} + \dfrac{8p}{T^3}\dfrac{\partial T}{\partial V}$.

Putting $p = V = 1$ and $T = 2$, we obtain

$$2\frac{\partial T}{\partial p} = 2, \qquad 2\frac{\partial T}{\partial V} = 1,$$

so $\dfrac{\partial T}{\partial p} = 1$ and $\dfrac{\partial T}{\partial V} = \dfrac{1}{2}$.

b) $dT = \dfrac{\partial T}{\partial p}\,dp + \dfrac{\partial T}{\partial V}\,dV$.

If $p = 1$, $|dp| \leq 0.001$, $V = 1$, and $|dV| \leq 0.002$, then $T = 2$ and

$$|dT| \leq (1)(0.001) + \frac{1}{2}(0.002) = 0.002.$$

The approximate maximum error in T is 0.002.

26. Given $F(x, y, u, v) = 0, \qquad G(x, y, u, v) = 0$, let

$$\Delta = \frac{\partial(F, G)}{\partial(x, y)} = \frac{\partial F}{\partial x}\frac{\partial G}{\partial y} - \frac{\partial F}{\partial y}\frac{\partial G}{\partial x}.$$

Then, regarding the given equations as defining x and y as functions of u and v, we have

$$\frac{\partial x}{\partial u} = -\frac{1}{\Delta}\frac{\partial(F, G)}{\partial(u, y)} \qquad \frac{\partial y}{\partial u} = -\frac{1}{\Delta}\frac{\partial(F, G)}{\partial(x, u)}$$

$$\frac{\partial x}{\partial v} = -\frac{1}{\Delta}\frac{\partial(F, G)}{\partial(v, y)} \qquad \frac{\partial y}{\partial v} = -\frac{1}{\Delta}\frac{\partial(F, G)}{\partial(x, v)}.$$

Therefore,

$$\frac{\partial(x, y)}{\partial(u, v)} = \frac{1}{\Delta^2}\left(\frac{\partial(F, G)}{\partial(u, y)}\frac{\partial(F, G)}{\partial(x, v)} - \frac{\partial(F, G)}{\partial(v, y)}\frac{\partial(F, G)}{\partial(x, u)}\right)$$

$$= \frac{1}{\Delta^2}\left[\left(\frac{\partial F}{\partial u}\frac{\partial G}{\partial y} - \frac{\partial F}{\partial y}\frac{\partial G}{\partial u}\right)\left(\frac{\partial F}{\partial x}\frac{\partial G}{\partial v} - \frac{\partial F}{\partial v}\frac{\partial G}{\partial x}\right)\right.$$

$$\left. - \left(\frac{\partial F}{\partial v}\frac{\partial G}{\partial y} - \frac{\partial F}{\partial y}\frac{\partial G}{\partial v}\right)\left(\frac{\partial F}{\partial x}\frac{\partial G}{\partial u} - \frac{\partial F}{\partial u}\frac{\partial G}{\partial x}\right)\right]$$

$$= \frac{1}{\Delta^2}\left[\frac{\partial F}{\partial u}\frac{\partial G}{\partial y}\frac{\partial F}{\partial x}\frac{\partial G}{\partial v} - \frac{\partial F}{\partial y}\frac{\partial G}{\partial u}\frac{\partial F}{\partial x}\frac{\partial G}{\partial v}\right.$$

$$- \frac{\partial F}{\partial u}\frac{\partial G}{\partial y}\frac{\partial F}{\partial v}\frac{\partial G}{\partial x} + \frac{\partial F}{\partial y}\frac{\partial G}{\partial u}\frac{\partial F}{\partial v}\frac{\partial G}{\partial x}$$

$$- \frac{\partial F}{\partial v}\frac{\partial G}{\partial y}\frac{\partial F}{\partial x}\frac{\partial G}{\partial u} + \frac{\partial F}{\partial v}\frac{\partial G}{\partial y}\frac{\partial F}{\partial u}\frac{\partial G}{\partial x}$$

$$\left. + \frac{\partial F}{\partial y}\frac{\partial G}{\partial v}\frac{\partial F}{\partial x}\frac{\partial G}{\partial u} - \frac{\partial F}{\partial y}\frac{\partial G}{\partial v}\frac{\partial F}{\partial u}\frac{\partial G}{\partial x}\right]$$

$$= \frac{1}{\Delta^2}\left[\frac{\partial F}{\partial u}\frac{\partial G}{\partial y}\frac{\partial F}{\partial x}\frac{\partial G}{\partial v} + \frac{\partial F}{\partial y}\frac{\partial G}{\partial u}\frac{\partial F}{\partial v}\frac{\partial G}{\partial x}\right.$$

$$\left. - \frac{\partial F}{\partial v}\frac{\partial G}{\partial y}\frac{\partial F}{\partial x}\frac{\partial G}{\partial u} - \frac{\partial F}{\partial y}\frac{\partial G}{\partial v}\frac{\partial F}{\partial u}\frac{\partial G}{\partial x}\right]$$

$$= \frac{1}{\Delta^2}\left(\frac{\partial F}{\partial x}\frac{\partial G}{\partial y} - \frac{\partial F}{\partial y}\frac{\partial G}{\partial x}\right)\left(\frac{\partial F}{\partial u}\frac{\partial G}{\partial v} - \frac{\partial F}{\partial v}\frac{\partial G}{\partial u}\right)$$

$$= \frac{1}{\Delta^2}\frac{\partial(F, G)}{\partial(x, y)}\frac{\partial(F, G)}{\partial(u, v)}$$

$$= \frac{1}{\Delta}\frac{\partial(F, G)}{\partial(u, v)} = \frac{\partial(F, G)}{\partial(u, v)}\left/\frac{\partial(F, G)}{\partial(x, y)}\right..$$

28. By the Chain Rule,

$$\begin{pmatrix} \dfrac{\partial x}{\partial r} & \dfrac{\partial x}{\partial s} \\[2mm] \dfrac{\partial y}{\partial r} & \dfrac{\partial y}{\partial s} \end{pmatrix}$$

$$= \begin{pmatrix} \dfrac{\partial x}{\partial u}\dfrac{\partial u}{\partial r} + \dfrac{\partial x}{\partial v}\dfrac{\partial v}{\partial r} & \dfrac{\partial x}{\partial u}\dfrac{\partial u}{\partial s} + \dfrac{\partial x}{\partial v}\dfrac{\partial v}{\partial s} \\[2mm] \dfrac{\partial y}{\partial u}\dfrac{\partial u}{\partial r} + \dfrac{\partial y}{\partial v}\dfrac{\partial v}{\partial r} & \dfrac{\partial y}{\partial u}\dfrac{\partial u}{\partial s} + \dfrac{\partial y}{\partial v}\dfrac{\partial v}{\partial s} \end{pmatrix}$$

$$= \begin{pmatrix} \dfrac{\partial x}{\partial u} & \dfrac{\partial x}{\partial v} \\[2mm] \dfrac{\partial y}{\partial u} & \dfrac{\partial y}{\partial v} \end{pmatrix}\begin{pmatrix} \dfrac{\partial u}{\partial r} & \dfrac{\partial u}{\partial s} \\[2mm] \dfrac{\partial v}{\partial r} & \dfrac{\partial v}{\partial s} \end{pmatrix}.$$

Since the determinant of a product of matrices is the product of their determinants, we have

$$\frac{\partial(x, y)}{\partial(r, s)} = \frac{\partial(x, y)}{\partial(u, v)}\frac{\partial(u, v)}{\partial(r, s)}.$$

30. Let $u = f(x, y)$ and $v = g(x, y)$, and suppose that

$$\frac{\partial(u, v)}{\partial(x, y)} = \frac{\partial(f, g)}{\partial(x, y)} = 0$$

for all (x, y). Thus

$$\frac{\partial f}{\partial x}\frac{\partial g}{\partial y} - \frac{\partial f}{\partial y}\frac{\partial g}{\partial x} = 0.$$

Now consider the equations $u = f(x, y)$ and $v = g(x, y)$ as defining u and y as functions of x and v. Holding v constant and differentiating with respect to x, we get

$$\frac{\partial g}{\partial x} + \frac{\partial g}{\partial y}\frac{\partial y}{\partial x} = 0,$$

and

$$\left(\frac{\partial u}{\partial x}\right)_v = \frac{\partial f}{\partial x} + \frac{\partial f}{\partial y}\frac{\partial y}{\partial x}$$

$$= \frac{1}{\dfrac{\partial g}{\partial y}}\left(\frac{\partial f}{\partial x}\frac{\partial g}{\partial y} - \frac{\partial f}{\partial y}\frac{\partial g}{\partial x}\right) = 0.$$

This says that $u = u(x, v)$ is independent of x, and so depends only on v: $u = k(v)$ for some function k of one variable. Thus $f(x, y) = k\big(g(x, y)\big)$, so f and g are functionally dependent.

Section 12.9   Taylor Series and Approximations   (page 780)

2. Since $f(x, y) = \ln(1 + x + y + xy)$

$$= \ln\big((1 + x)(1 + y)\big)$$

$$= \ln(1 + x) + \ln(1 + y),$$

the Taylor series for f about $(0, 0)$ is

$$\sum_{n=1}^{\infty}(-1)^{n-1}\frac{x^n + y^n}{n}.$$

4. Let $u = x - 1$, $v = y + 1$. Thus

$$f(x, y) = x^2 + xy + y^3$$

$$= (u + 1)^2 + (u + 1)(v - 1) + (v - 1)^3$$

$$= 1 + 2u + u^2 - 1 + v - u + uv + v^3 - 3v^2 + 3v - 1$$

$$= -1 + u + 4v + u^2 + uv - 3v^2 + v^3$$

$$= -1 + (x - 1) + 4(y + 1) + (x - 1)^2$$

$$+ (x - 1)(y + 1) - 3(y + 1)^2 + (y + 1)^3.$$

This is the Taylor series for f about $(1, -1)$.

6. $f(x, y) = \sin(2x + 3y) = \sum_{n=0}^{\infty} (-1)^n \dfrac{(2x + 3y)^{2n+1}}{(2n + 1)!}$

$= \sum_{n=0}^{\infty} \dfrac{(-1)^n}{(2n + 1)!} \sum_{j=0}^{2n+1} \dfrac{(2n + 1)!}{j!(2n + 1 - j)!} (2x)^j (3y)^{2n+1-j}$

$= \sum_{n=0}^{\infty} \sum_{j=0}^{2n+1} \dfrac{(-1)^n 2^j 3^{2n+1-j}}{j!(2n + 1 - j)!} x^j y^{2n+1-j}.$

This is the Taylor series for f about $(0, 0)$.

8. Let $u = x - 1$. Then

$f(x, y) = \ln(x^2 + y^2) = \ln(1 + 2u + u^2 + y^2)$

$= (2u + u^2 + y^2) - \dfrac{(2u + u^2 + y^2)^2}{2}$

$\qquad + \dfrac{(2u + u^2 + y^2)^3}{3} - \cdots$

$= 2u + u^2 + y^2 - 2u^2 - 2u^3 - 2uy^2 + \dfrac{8u^3}{3} + \cdots.$

The Taylor polynomial of degree 3 for f near $(1, 0)$ is

$$2(x - 1) - (x - 1)^2 + y^2 - 2(x - 1)^3$$
$$- 2(x - 1)y^2 + \dfrac{8}{3}(x - 1)^3.$$

10. $f(x, y) = \cos(x + \sin y)$

$= 1 - \dfrac{(x + \sin y)^2}{2!} + \dfrac{(x + \sin y)^4}{4!} - \cdots$

$= 1 - \dfrac{\left(x + y - \dfrac{y^3}{6} + \cdots\right)^2}{2} + \dfrac{(x + y - \cdots)^4}{4} - \cdots$

$= 1 - \dfrac{1}{2}\left(x^2 + y^2 + 2xy - \dfrac{xy^3}{3} - \dfrac{y^4}{3} + \cdots\right)$

$\qquad + \dfrac{1}{4}(x^4 + 4x^3 y + 6x^2 y^2 + 4xy^3 + y^4 + \cdots).$

The Taylor polynomial of degree 4 for f near $(0, 0)$ is

$$1 - \dfrac{x^2}{2} - xy - \dfrac{y^2}{2} + \dfrac{x^4}{4} + x^3 y$$
$$+ \dfrac{3x^2 y^2}{2} + \dfrac{7xy^3}{6} + \dfrac{5y^4}{12}.$$

12. $f(x, y) = \dfrac{1 + x}{1 + x^2 + y^4}$

$= (1 + x)\left(1 - (x^2 + y^4) + \cdots\right)$

$= 1 + x - x^2 - \cdots.$

The Taylor polynomial of degree 2 for f near $(0, 0)$ is

$$1 + x - x^2.$$

14. The equation $\sqrt{1 + xy} = 1 + x + \ln(1 + y)$ can be rewritten $F(x, y) = 0$, where $F(x, y) = \sqrt{1 + xy} - 1 - x - \ln(1 + y)$. Since $F(0, 0) = 0$ and $F_2(0, 0) = -1 \neq 0$, the given equation has a solution of the form $y = f(x)$ where $f(0) = 0$.

Try $y = a_1 x + a_2 x^2 + a_3 x^3 + a_4 x^4 + \cdots$. We have

$\sqrt{1 + xy}$

$\quad = \sqrt{1 + a_1 x^2 + a_2 x^3 + a_3 x_4 + \cdots}$

$\quad = 1 + \dfrac{1}{2}(a_1 x^2 + a_2 x^3 + a_3 x^4 + \cdots)$

$\qquad - \dfrac{1}{8}(a_1 x^2 + \cdots)^2 + \cdots$

$1 + x + \ln(1 + y)$

$\quad = 1 + x + (a_1 x + a_2 x^2 + a_3 x^3 + a_4 x^4 + \cdots)$

$\qquad - \dfrac{1}{2}(a_1 x + a_2 x^2 + a_3 x^3 + \cdots)^2 + \dfrac{1}{3}(a_1 x + a_2 x^2 \cdots)^3 - \cdots$

Thus we must have

$$0 = 1 + a_1$$
$$\dfrac{1}{2}a_1 = a_2 - \dfrac{1}{2}a_1^2$$
$$\dfrac{1}{2}a_2 = a_3 - a_1 a_2 + \dfrac{1}{3}a_1^3$$
$$\dfrac{1}{2}a_3 - \dfrac{1}{8}a_1^2 = a_4 - \dfrac{1}{2}a_2^2 - a_1 a_3 + a_1^2 a_2,$$

and $a_1 = -1$, $a_2 = 0$, $a_3 = \dfrac{1}{3}$, $a_4 = -\dfrac{7}{24}$. The required solution is

$$y = -x + \dfrac{1}{3}x^3 - \dfrac{7}{24}x^4 + \cdots.$$

16. The coefficient of $x^2 y$ in the Taylor series for $f(x, y) = \tan^{-1}(x + y)$ about $(0, 0)$ is

$$\dfrac{1}{2!1!} f_{112}(0, 0) = \dfrac{1}{2} f_{112}(0, 0).$$

But

$\tan^{-1}(x + y) = x + y - \dfrac{1}{3}(x + y)^3 + \cdots$

$\qquad = x + y - \dfrac{1}{3}(x^3 + 3x^2 y + 3xy^2 + y^3) + \cdots$

so the coefficient of $x^2 y$ is -1. Hence $f_{112}(0, 0) = -2$.

Review Exercises 12   (page 780)

2. $T = \dfrac{140 + 30x^2 - 60x + 120y^2}{8 + x^2 - 2x + 4y^2}$

$\quad = 30 - \dfrac{100}{(x - 1)^2 + 4y^2 + 7}$

Ellipses: centre $(1, 0)$, values of T between $30 - (100/7)$ (minimum) at $(1, 0)$ and 30 (at infinite distance from $(1, 0)$).

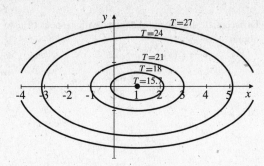

Fig. R-12.2

4. $f(x, y) = \begin{cases} x^3/(x^2 + y^2) & \text{if } (x, y) \neq (0, 0) \\ 0 & \text{if } (x, y) = (0, 0) \end{cases}$.

$$f_1(0, 0) = \lim_{h \to 0} \frac{(h^3 - 0)/h^2}{h} = 1$$

$$f_2(0, 0) = \lim_{k \to 0} \frac{0 - 0}{k} = 0.$$

For $(x, y) \neq (0, 0)$, we have

$$f_1(x, y) = \frac{x^4 + 3x^2 y^2}{(x^2 + y^2)^2}$$

$$f_2(x, y) = -\frac{2x^3 y}{(x^2 + y^2)^2}$$

$$f_{12}(0, 0) = \lim_{k \to 0} \frac{f_1(0, k) - f_1(0, 0)}{k} = \lim_{k \to 0} \frac{0 - 1}{k} \text{ does not exist}$$

$$f_{21}(0, 0) = \lim_{h \to 0} \frac{f_2(h, 0) - f_2(0, 0)}{h} = \lim_{h \to 0} \frac{0 - 0}{h} = 0.$$

6. $f(x, y) = e^{x^2 - 2x - 4y^2 + 5}$ $f(1, -1) = 1$
$f_1(x, y) = 2(x - 1)e^{x^2 - 2x - 4y^2 + 5}$ $f_1(1, -1) = 0$
$f_2(x, y) = -8y e^{x^2 - 2x - 4y^2 + 5}$ $f_2(1, -1) = 8$.

a) The tangent plane to $z = f(x, y)$ at $(1, -1, 1)$ has equation $z = 1 + 8(y + 1)$, or $z = 8y + 9$.

b) $f(x, y) = C \Rightarrow (x - 1)^2 - 4y^2 + 4 = \ln C$
$$\Rightarrow (x - 1)^2 - 4y^2 = \ln C - 4.$$
These are hyperbolas with centre $(1, 0)$ and asymptotes $x = 1 \pm 2y$.

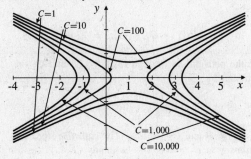

Fig. R-12.6

8. $\dfrac{1}{R} = \dfrac{1}{R_1} + \dfrac{1}{R_2}$

$-\dfrac{1}{R^2} dR = -\dfrac{1}{R_1^2} dR_1 - \dfrac{1}{R_2^2} dR_2$

If $R_1 = 100$ and $R_2 = 25$, so that $R = 20$, and if $|dR_1/R_1| = 5/100$ and $|dR_2/R_2| = 2/100$, then

$$\frac{1}{20}\left|\frac{dR}{R}\right| \leq \frac{1}{100} \cdot \frac{5}{100} + \frac{1}{25} \cdot \frac{2}{100} = \frac{13}{100^2}.$$

Thus $|dR/R| \leq 13/500$; R can be in error by about 2.6%.

10. $T = x^3 y + y^3 z + z^3 x$.

a) $\nabla T = (3x^2 y + z^3)\mathbf{i} + (3y^2 z + x^3)\mathbf{j} + (3z^2 x + y^3)\mathbf{k}$
$\nabla T(2, -1, 0) = -12\mathbf{i} + 8\mathbf{j} - \mathbf{k}$.
A unit vector in the direction from $(2, -1, 0)$ towards $(1, 1, 2)$ is $\mathbf{u} = (-\mathbf{i} + 2\mathbf{j} + 2\mathbf{k})/3$. The directional derivative of T at $(2, -1, 0)$ in the direction of $\mathbf{u}$ is

$$\mathbf{u} \bullet \nabla T(2, -1, 0) = \frac{12 + 16 - 2}{3} = \frac{26}{3}.$$

b) Since $\nabla(2x^2 + 3y^2 + z^2) = 4x\mathbf{i} + 6y\mathbf{j} + 2z\mathbf{k}$, at $t = 0$ the fly is at $(2, -1, 0)$ and is moving in the direction $\pm(8\mathbf{i} - 6\mathbf{j})$, so its velocity is

$$\pm 5\frac{8\mathbf{i} - 6\mathbf{j}}{10} = \pm(4\mathbf{i} - 3\mathbf{j}).$$

Since the fly is moving in the direction of increasing T, the rate at which it experiences T increasing is

$$\frac{dT}{dt} = |(4\mathbf{i} - 3\mathbf{j}) \bullet (-12\mathbf{i} + 8\mathbf{j} - \mathbf{k})| = 48 + 24 = 72.$$

12. $f(x, y, z) = (x^2 + z^2) \sin \dfrac{\pi x y}{2} + yz^2$, $P_0 = (1, 1, -1)$.

a) $\nabla f = \left(2x \sin \dfrac{\pi x y}{2} + \dfrac{\pi y}{2}(x^2 + z^2) \cos \dfrac{\pi x y}{2}\right)\mathbf{i}$
$\quad + \left(\dfrac{\pi x}{2}(x^2 + z^2) \cos \dfrac{\pi x y}{2} + z^2\right)\mathbf{j}$
$\quad + 2z\left(\sin \dfrac{\pi x y}{2} + y\right)\mathbf{k}$

$\nabla f(P_0) = 2\mathbf{i} + \mathbf{j} - 4\mathbf{k}$.

b) Since $f(P_0) = 2 + 1 = 3$, the linearization of f at P_0 is

$$L(x, y, z) = 3 + 2(x - 1) + (y - 1) - 4(z + 1).$$

c) The tangent plane at P_0 to the level surface of f through P_0 has equation

$$\nabla f(P_0) \bullet \big((x - 1)\mathbf{i} + (y - 1)\mathbf{j} + (z + 1)\mathbf{k}\big) = 0$$
$$2(x - 1) + (y - 1) - 4(z + 1) = 0$$
$$2x + y - 4z = 7.$$

d) The bird is flying in direction

$$(2-1)\mathbf{i} + (-1-1)\mathbf{j} + (1+1)\mathbf{k} = \mathbf{i} - 2\mathbf{j} + 2\mathbf{k},$$

a vector of length 3. Since the bird's speed is 5, its velocity is

$$\mathbf{v} = \frac{5}{3}(\mathbf{i} - 2\mathbf{j} + 2\mathbf{k}).$$

The rate of change of f as experienced by the bird is

$$\frac{df}{dt} = \mathbf{v} \bullet \nabla f(P_0) = \frac{5}{3}(2 - 2 - 8) = -\frac{40}{3}.$$

e) To experience the greatest rate of increase of f while flying through P_0 at speed 5, the bird should fly in the direction of $\nabla f(P_0)$, that is, $2\mathbf{i} + \mathbf{j} - 4\mathbf{k}$.

14. If $F(x, y, z) = 0$, $G(x, y, z) = 0$ are solved for $x = x(y)$, $z = z(y)$, then

$$F_1 \frac{dx}{dy} + F_2 + F_3 \frac{dz}{dy} = 0$$
$$G_1 \frac{dx}{dy} + G_2 + G_3 \frac{dz}{dy} = 0.$$

Eliminating dz/dy from these equations, we obtain

$$\frac{dx}{dy} = -\frac{F_2 G_3 - F_3 G_2}{F_1 G_3 - F_3 G_1}.$$

Similarly, if the equations are solved for $x = x(z)$, $y = y(z)$, then

$$\frac{dy}{dz} = -\frac{F_3 G_1 - F_1 G_3}{F_2 G_1 - F_1 G_2},$$

and if the equations are solved for $y = y(x)$, $z = z(x)$, then

$$\frac{dz}{dx} = -\frac{F_1 G_2 - F_2 G_1}{F_3 G_2 - F_2 G_3}.$$

Hence

$$\frac{dx}{dy} \cdot \frac{dy}{dz} \cdot \frac{dz}{dx}$$
$$= -\frac{F_2 G_3 - F_3 G_2}{F_1 G_3 - F_3 G_1} \cdot \frac{F_3 G_1 - F_1 G_3}{F_2 G_1 - F_1 G_2} \cdot \frac{F_1 G_2 - F_2 G_1}{F_3 G_2 - F_2 G_3} = 1.$$

16. $u = x^2 + y^2$
$v = x^2 - 2xy^2$
Assume these equations define $x = x(u, v)$ and $y = y(u, v)$ near the point $(u, v) = (5, -7)$, with $x = 1$ and $y = 2$ at that point.

a) Differentiate the given equations with respect to u to obtain

$$1 = 2x \frac{\partial x}{\partial u} + 2y \frac{\partial y}{\partial u}$$
$$0 = 2(x - y^2) \frac{\partial x}{\partial u} - 4xy \frac{\partial y}{\partial u}.$$

At $x = 1$, $y = 2$,

$$2\frac{\partial x}{\partial u} + 4\frac{\partial y}{\partial u} = 1$$
$$-6\frac{\partial x}{\partial u} - 8\frac{\partial y}{\partial u} = 0,$$

from which we obtain $\partial x/\partial u = -1$ and $\partial y/\partial u = 3/4$ at $(5, -7)$.

b) If $z = \ln(y^2 - x^2)$, then

$$\frac{\partial z}{\partial u} = \frac{1}{y^2 - x^2}\left[-2x \frac{\partial x}{\partial u} + 2y \frac{\partial y}{\partial u}\right].$$

At $(u, v) = (5, -7)$, we have $(x, y) = (1, 2)$, and so

$$\frac{\partial z}{\partial u} = \frac{1}{3}\left[-2(-1) + 4\left(\frac{3}{4}\right)\right] = \frac{5}{3}.$$

Challenging Problems 12 (page 781)

2. Let the position vector of the particle at time t be $\mathbf{r} = x(t)\mathbf{i} + y(t)\mathbf{j} + z(t)\mathbf{k}$. Then the velocity of the particle is

$$\mathbf{v} = \frac{dx}{dt}\mathbf{i} + \frac{dy}{dt}\mathbf{j} + \frac{dz}{dt}\mathbf{k}.$$

This velocity must be parallel to

$$\nabla f(x, y, z) = -2x\mathbf{i} - 4y\mathbf{j} + 6z\mathbf{k}$$

at every point of the path, that is,

$$\frac{dx}{dt} = -2tx, \quad \frac{dy}{dt} = -4ty, \quad \frac{dz}{dt} = 6tz,$$

so that $\dfrac{dx}{-2x} = \dfrac{dy}{-4y} = \dfrac{dz}{6z}$. Integrating these equations, we get

$$\ln|y| = 2\ln|x| + C_1, \quad \ln|z| = -3\ln|x| + C_2.$$

Since the path passes through $(1, 1, 8)$, C_1 and C_2 are determined by

$$\ln 1 = 2\ln 1 + C_1, \quad \ln 8 = -3\ln 1 + C_2.$$

Thus $C_1 = 0$ and $C_2 = \ln 8$. The path therefore has equations $y = x^2$, $z = 8/x^3$. Evidently $(2, 4, 1)$ lies on the path, and $(3, 7, 0)$ does not.

4. If $u(x, y, z, t) = v(\rho, t) = \dfrac{f(\rho - ct)}{\rho}$ is independent of θ and ϕ, then

$$\frac{\partial^2 u}{\partial x^2} + \frac{\partial^2 u}{\partial y^2} + \frac{\partial^2 u}{\partial z^2} = \frac{\partial^2 v}{\partial \rho^2} + \frac{2}{\rho} \frac{\partial v}{\partial \rho}$$

by Problem 3. We have

$$\frac{\partial v}{\partial \rho} = \frac{f'(\rho - ct)}{\rho} - \frac{f(\rho - ct)}{\rho^2}$$

$$\frac{\partial^2 v}{\partial \rho^2} = \frac{f''(\rho - ct)}{\rho} - \frac{2f'(\rho - ct)}{\rho^2} + \frac{2f(\rho - ct)}{\rho^3}$$

$$\frac{\partial v}{\partial t} = -\frac{cf'(\rho - ct)}{\rho}$$

$$\frac{\partial^2 v}{\partial t^2} = \frac{c^2 f''(\rho - ct)}{\rho}$$

$$\frac{\partial^2 v}{\partial \rho^2} + \frac{2}{\rho} \frac{\partial v}{\partial \rho}$$

$$= \frac{f''(\rho - ct)}{\rho} - \frac{2f'(\rho - ct)}{\rho^2} + \frac{2f(\rho - ct)}{\rho^3}$$

$$\quad + \frac{2f'(\rho - ct)}{\rho^2} - \frac{2f(\rho - ct)}{\rho^3}$$

$$= \frac{f''(\rho - ct)}{\rho}$$

$$= \frac{1}{c^2} \frac{\partial^2 v}{\partial t^2} = \frac{1}{c^2} \frac{\partial^2 u}{\partial t^2}.$$

The function $f(\rho - ct)/\rho$ represents the shape of a symmetrical wave travelling uniformly away from the origin at speed c. Its amplitude at distance ρ from the origin decreases as ρ increases; it is proportional to the reciprocal of ρ.

CHAPTER 13. APPLICATIONS OF PARTIAL DERIVATIVES

Section 13.1 Extreme Values (page 791)

2. $f(x, y) = xy - x + y$, $\quad f_1 = y - 1$, $\quad f_2 = x + 1$
$A = f_{11} = 0$, $\quad B = f_{12} = 1$, $\quad C = f_{22} = 0$.
Critical point $(-1, 1)$ is a saddle point since $B^2 - AC > 0$.

4. $f(x, y) = x^4 + y^4 - 4xy$, $\quad f_1 = 4(x^3 - y)$, $\quad f_2 = 4(y^3 - x)$
$A = f_{11} = 12x^2$, $\quad B = f_{12} = -4$, $\quad C = f_{22} = 12y^2$.
For critical points: $x^3 = y$ and $y^3 = x$. Thus $x^9 = x$, or
$x(x^8 - 1) = 0$, and $x = 0$, 1, or -1. The critical points are
$(0, 0)$, $(1, 1)$ and $(-1, -1)$.
At $(0, 0)$, $B^2 - AC = 16 - 0 > 0$, so $(0, 0)$ is a saddle
point.
At $(1, 1)$ and $(-1, -1)$, $B^2 - AC = 16 - 144 < 0$, $A > 0$,
so f has local minima at these points.

6. $f(x, y) = \cos(x + y)$, $\quad f_1 = -\sin(x + y) = f_2$.
All points on the lines $x + y = n\pi$ (n is an integer) are
critical points. If n is even, $f = 1$ at such points; if n is
odd, $f = -1$ there. Since $-1 \le f(x, y) \le 1$ at all points
in $\mathbb{R}^2$, f must have local and absolute maximum values
at points $x + y = n\pi$ with n even, and local and absolute
minimum values at such points with n odd.

8. $f(x, y) = \cos x + \cos y$, $\quad f_1 = -\sin x$, $\quad f_2 = -\sin y$
$A = f_{11} = -\cos x$, $\quad B = f_{12} = 0$, $\quad C = f_{22} = -\cos y$.
The critical points are points $(m\pi, n\pi)$, where m and n
are integers.
Here $B^2 - AC = -\cos(m\pi)\cos(n\pi) = (-1)^{m+n+1}$ which
is negative if $m + n$ is even, and positive if $m + n$ is odd.
If $m + n$ is odd then f has a saddle point at $(m\pi, n\pi)$.
If $m + n$ is even and m is odd then f has a local (and
absolute) minimum value, -2, at $(m\pi, n\pi)$. If $m + n$
is even and m is even then f has a local (and absolute)
maximum value, 2, at $(m\pi, n\pi)$.

10. $f(x, y) = \dfrac{xy}{2 + x^4 + y^4}$
$f_1 = \dfrac{(2 + x^4 + y^4)y - xy4x^3}{(2 + x^4 + y^4)^2} = \dfrac{y(2 + y^4 - 3x^4)}{(2 + x^4 + y^4)^2}$
$f_2 = \dfrac{x(2 + x^4 - 3y^4)}{(2 + x^4 + y^4)^2}$.
For critical points, $y(2 + y^4 - 3x^4) = 0$ and
$x(2 + x^4 - 3y^4) = 0$.
One critical point is $(0, 0)$. Since $f(0, 0) = 0$ but
$f(x, y) > 0$ in the first quadrant and $f(x, y) < 0$ in
the second quadrant, $(0, 0)$ must be a saddle point of f.
Any other critical points must satisfy $2 + y^4 - 3x^4 = 0$
and $2 + x^4 - 3y^4 = 0$, that is, $y^4 = x^4$, or $y = \pm x$. Thus
$2 - 2x^4 = 0$ and $x = \pm 1$. Therefore there are four other
critical points: $(1, 1)$, $(-1, -1)$, $(1, -1)$ and $(-1, 1)$. f is
positive at the first two of these, and negative at the other
two. Since $f(x, y) \to 0$ as $x^2 + y^2 \to \infty$, f must have
maximum values at $(1, 1)$ and $(-1, -1)$, and minimum
values at $(1, -1)$ and $(-1, 1)$.

12. $f(x, y) = \dfrac{1}{1 - x + y + x^2 + y^2}$
$= \dfrac{1}{\left(x - \dfrac{1}{2}\right)^2 + \left(y + \dfrac{1}{2}\right)^2 + \dfrac{1}{2}}$.
Evidently f has absolute maximum value 2 at $\left(\dfrac{1}{2}, -\dfrac{1}{2}\right)$.
Since
$$f_1(x, y) = \frac{1 - 2x}{(1 - x + y + x^2 + y^2)^2}$$
$$f_2(x, y) = -\frac{1 + 2y}{(1 - x + y + x^2 + y^2)^2},$$
$\left(\dfrac{1}{2}, -\dfrac{1}{2}\right)$ is the only critical point of f.

14. $f(x, y, z) = xyz - x^2 - y^2 - z^2$. For critical points we
have
$$0 = f_1 = yz - 2x, \quad 0 = f_2 = xz - 2y, \quad 0 = f_3 = xy - 2z.$$
Thus $xyz = 2x^2 = 2y^2 = 2z^2$, so $x^2 = y^2 = z^2$. Hence
$x^3 = \pm 2x^2$, and $x = \pm 2$ or 0. Similarly for y and z.
The only critical points are $(0, 0, 0)$, $(2, 2, 2)$, $(-2, -2, 2)$,
$(-2, 2, -2)$, and $(2, -2, -2)$.
Let $\mathbf{u} = u\mathbf{i} + v\mathbf{j} + w\mathbf{k}$, where $u^2 + v^2 + w^2 = 1$. Then
$$D_{\mathbf{u}} f(x, y, z) = (yz - 2x)u + (xz - 2y)v + (xy - 2z)w$$
$$D_{\mathbf{u}}\big(D_{\mathbf{u}} f(x, y, z)\big) = (-2u + zv + yw)u$$
$$+ (zu - 2v + xw)v + (yu + xv - 2w)w.$$
At $(0, 0, 0)$, $D_{\mathbf{u}}\big(D_{\mathbf{u}} f(0, 0, 0)\big) = -2u^2 - 2v^2 - 2w^2 < 0$
for $\mathbf{u} \ne \mathbf{0}$, so f has a local maximum value at $(0, 0, 0)$.

At $(2, 2, 2)$, we have

$$D_{\mathbf{u}}\big(D_{\mathbf{u}}f(2,2,2)\big) = (-2u + 2v + 2w)u + (2u - 2v + 2w)v$$
$$+ (2u + 2v - 2w)w$$
$$= -2(u^2 + v^2 + w^2) + 4(uv + vw + wu)$$
$$= -2[(u - v - w)^2 - 4vw]$$
$$\begin{cases} < 0 & \text{if } v = w = 0, \; u \neq 0 \\ > 0 & \text{if } v = w \neq 0, \; u - v - w = 0. \end{cases}$$

Thus $(2, 2, 2)$ is a saddle point.

At $(2, -2, -2)$, we have

$$D_{\mathbf{u}}\big(D_{\mathbf{u}}f\big) = -2(u^2 + v^2 + w^2 + 2uv + 2uw - 2vw)$$
$$= -2[(u + v + w)^2 - 4vw]$$
$$\begin{cases} < 0 & \text{if } v = w = 0, \; u \neq 0 \\ > 0 & \text{if } v = w \neq 0, \; u + v + w = 0. \end{cases}$$

Thus $(2, -2, -2)$ is a saddle point. By symmetry, so are the remaining two critical points.

16. $f(x, y, z) = 4xyz - x^4 - y^4 - z^4$
$D = f(1 + h, 1 + k, 1 + m) - f(1, 1, 1)$
$$= 4(1 + h)(1 + k)(1 + m) - (1 + h)^4 - (1 + k)^4$$
$$- (1 + m)^4 - 1$$
$$= 4(1 + h + k + m + hk + hm + km + hkm)$$
$$- (1 + 4h + 6h^2 + 4h^3 + h^4)$$
$$- (1 + 4k + 6k^2 + 4k^3 + k^4)$$
$$- (1 + 4m + 6m^2 + 4m^3 + m^4) - 1$$
$$= 4(hk + hm + km) - 6(h^2 + k^2 + m^2) + \cdots,$$

where $\cdots$ stands for terms of degree 3 and 4 in the variables h, k, and m. Completing some squares among the quadratic terms we obtain

$$D = -2\big[(h - k)^2 + (k - m)^2 + (h - m)^2 + h^2 + k^2 + m^2\big] + \cdots$$

which is negative if $|h|$, $|k|$ and $|m|$ are small and not all 0. (This is because the terms of degree 3 and 4 are smaller in size than the quadratic terms for small values of the variables.)
Hence f has a local maximum value at $(1, 1, 1)$.

18. $f(x, y) = \dfrac{x}{1 + x^2 + y^2}$
$$f_1(x, y) = \dfrac{1 + y^2 - x^2}{(1 + x^2 + y^2)^2}$$
$$f_2(x, y) = \dfrac{-2xy}{(1 + x^2 + y^2)^2}.$$

For critical points, $x^2 - y^2 = 1$, and $xy = 0$. The critical points are $(\pm 1, 0)$. $f(\pm 1, 0) = \pm\frac{1}{2}$.
Since $f(x, y) \to 0$ as $x^2 + y^2 \to \infty$, the maximum and minimum values of f are $1/2$ and $-1/2$ respectively.

20. $f(x, y) = x + 8y + \dfrac{1}{xy}$, $\quad (x > 0, \quad y > 0)$

$f_1(x, y) = 1 - \dfrac{1}{x^2 y} = 0 \quad \Rightarrow \quad x^2 y = 1$

$f_2(x, y) = 8 - \dfrac{1}{xy^2} = 0 \quad \Rightarrow \quad 8xy^2 = 1.$

The critical points must satisfy

$$\frac{x}{y} = \frac{x^2 y}{xy^2} = 8,$$

that is, $x = 8y$. Also, $x^2 y = 1$, so $64y^3 = 1$. Thus $y = 1/4$, and $x = 2$; the critical point is $\left(2, \frac{1}{4}\right)$. Since $f(x, y) \to \infty$ if $x \to 0+$, $y \to 0+$, or $x^2 + y^2 \to \infty$, the critical point must give a minimum value for f. The minimum value is $f\left(2, \frac{1}{4}\right) = 2 + 2 + 2 = 6$.

22. Let the length, width, and height of the box be x, y, and z, respectively. Then $V = xyz$. If the top and side walls cost $\$k$ per unit area, then the total cost of materials for the box is

$$C = 2kxy + kxy + 2kxz + 2kyz$$
$$= k\left[3xy + 2(x + y)\frac{V}{xy}\right] = k\left[3xy + \frac{2V}{x} + \frac{2V}{y}\right],$$

where $x > 0$ and $y > 0$. Since $C \to \infty$ as $x \to 0+$ or $y \to 0+$ or $x^2 + y^2 \to \infty$, C must have a minimum value at a critical point in the first quadrant. For CP:

$$0 = \frac{\partial C}{\partial x} = k\left(3y - \frac{2V}{x^2}\right)$$
$$0 = \frac{\partial C}{\partial y} = k\left(3x - \frac{2V}{y^2}\right).$$

Thus $3x^2 y = 2V = 3xy^2$, so that $x = y = (2V/3)^{1/3}$ and $z = V/(2V/3)^{2/3} = (9V/4)^{1/3}$.

24. Given that $a > 0$, $b > 0$, $c > 0$, and $a + b + c = 30$, we want to maximize

$$P = ab^2 c^3 = (30 - b - c)b^2 c^3 = 30b^2 c^3 - b^3 c^3 - b^2 c^4.$$

Since $P = 0$ if $b = 0$ or $c = 0$ or $b + c = 30$ (i.e., $a = 0$), the maximum value of P will occur at a critical point (b, c) satisfying $b > 0$, $c > 0$, and $b + c < 30$. For CP:

$$0 = \frac{\partial P}{\partial b} = 60bc^3 - 3b^2 c^3 - 2bc^4 = bc^3(60 - 3b - 2c)$$
$$0 = \frac{\partial P}{\partial c} = 90b^2 c^2 - 3b^3 c^2 - 4b^2 c^3 = b^2 c^2(90 - 3b - 4c).$$

Hence $9b + 6c = 180 = 6b + 8c$, from which we obtain $3b = 2c = 30$. The three numbers are $b = 10$, $c = 15$, and $a = 30 - 10 - 15 = 5$.

26. We will use the second derivative test to classify the two critical points calculated in Exercise 25. To calculate the second partials

$$A = \frac{\partial^2 z}{\partial x^2}, \qquad B = \frac{\partial^2 z}{\partial x \partial y}, \qquad C = \frac{\partial^2 z}{\partial y^2},$$

we differentiate the expressions $(*)$, and $(**)$ obtained in Exercise 25.

Differentiating $(*)$ with respect to x, we obtain

$$e^{2zx-x^2}\left[\left(2x\frac{\partial z}{\partial x} + 2z - 2x\right)^2\right.$$
$$\left. + 4\frac{\partial z}{\partial x} + 2x\frac{\partial^2 z}{\partial x^2} - 2\right]$$
$$- 3e^{2zy+y^2}\left[\left(2y\frac{\partial z}{\partial x}\right)^2 + 2y\frac{\partial^2 z}{\partial x^2}\right] = 0.$$

At a critical point, $\frac{\partial z}{\partial x} = 0$, $z = x$, $z = -y$, and $z^2 = \ln 3$, so

$$3\left(2x\frac{\partial^2 z}{\partial x^2} - 2\right) - \frac{3}{3}\left(2y\frac{\partial^2 z}{\partial x^2}\right) = 0,$$
$$A = \frac{\partial^2 z}{\partial x^2} = \frac{6}{6x - 2y}.$$

Differentiating $(**)$ with respect to y gives

$$e^{2zx-x^2}\left[\left(2x\frac{\partial z}{\partial y}\right)^2 + 2x\frac{\partial^2 z}{\partial y^2}\right]$$
$$- 3e^{2zy+y^2}\left[\left(2y\frac{\partial z}{\partial y} + 2z + 2y\right)^2 + 4\frac{\partial z}{\partial y} + 2y\frac{\partial^2 z}{\partial y^2} + 2\right] = 0,$$

and evaluation at a critical point gives

$$3\left(2x\frac{\partial^2 z}{\partial y^2}\right) - \frac{3}{3}\left(2y\frac{\partial^2 z}{\partial y^2} + 2\right) = 0;$$
$$C = \frac{\partial^2 z}{\partial y^2} = \frac{2}{6x - 2y}.$$

Finally, differentiating $(*)$ with respect to y gives

$$e^{2zx-x^2}\left[\left(2x\frac{\partial z}{\partial x} + 2z - 2x\right)\left(2x\frac{\partial z}{\partial y}\right)\right.$$
$$\left. + 2x\frac{\partial^2 z}{\partial x \partial y} + 2\frac{\partial z}{\partial y}\right]$$
$$- 3e^{2zy+y^2}\left[\left(2y\frac{\partial z}{\partial y} + 2z + 2y\right)\left(2y\frac{\partial z}{\partial x}\right)\right.$$
$$\left. + 2\frac{\partial z}{\partial x} + 2y\frac{\partial^2 z}{\partial x \partial y}\right] = 0,$$

and, evaluating at a critical point,

$$(6x - 2y)\frac{\partial^2 z}{\partial x \partial y} = 0,$$

so that

$$B = \frac{\partial^2 z}{\partial x \partial y} = 0.$$

At the critical point $(\sqrt{\ln 3}, -\sqrt{\ln 3})$ we have

$$A = \frac{6}{8\ln 3}, \qquad B = 0, \qquad C = \frac{2}{8\ln 3},$$

so $B^2 - AC < 0$, and f has a local minimum at that critical point.

At the critical point $(-\sqrt{\ln 3}, \sqrt{\ln 3})$ we have

$$A = -\frac{6}{8\ln 3}, \qquad B = 0, \qquad C = -\frac{2}{8\ln 3},$$

so $B^2 - AC < 0$, and f has a local maximum at that critical point.

28. We have

$$Q(u, v) = Au^2 + 2Buv + Cv^2$$
$$= A\left(u^2 + \frac{2B}{A}uv + \frac{B^2}{A^2}v^2\right) + \left(C - \frac{B^2}{A}\right)v^2$$
$$= A\left(u + \frac{Bv}{A}\right)^2 + \frac{AC - B^2}{A}v^2.$$

If $\begin{vmatrix} A & B \\ B & C \end{vmatrix} = AC - B^2 > 0$, both terms above have the same sign, positive if $A > 0$ and negative if $A < 0$, ensuring that Q is positive definite or negative definite respectively, since the two terms cannot both vanish if $(u, v) \neq (0, 0)$. If $AC - B^2 < 0$, $Q(u, v)$ is a difference of squares, and must be indefinite.

Section 13.2 Extreme Values of Functions Defined on Restricted Domains (page 797)

2. $f(x, y) = xy - 2x$ on
$R = \{(x, y) : -1 \le x \le 1, \ 0 \le y \le 1\}$.
For critical points:

$$0 = f_1(x, y) = y - 2, \qquad 0 = f_2(x, y) = x.$$

The only CP is $(0, 2)$, which lies outside R. Therefore the maximum and minimum values of f on R lie on one of the four boundary segments of R.

On $x = -1$ we have $f(-1, y) = 2 - y$ for $0 \le y \le 1$, which has maximum value 2 and minimum value 1.

On $x = 1$ we have $f(1, y) = y - 2$ for $0 \le y \le 1$, which has maximum value -1 and minimum value -2.

On $y = 0$ we have $f(x, 0) = -2x$ for $-1 \le x \le 1$, which has maximum value 2 and minimum value -2.

On $y = 1$ we have $f(x, 1) = -x$ for $-1 \le x \le 1$, which has maximum value 1 and minimum value -1.

Thus the maximum and minimum values of f on the rectangle R are 2 and -2 respectively.

4. $f(x, y) = x + 2y$ on the closed disk $x^2 + y^2 \leq 1$. Since $f_1 = 1$ and $f_2 = 2$, f has no critical points, and the maximum and minimum values of f, which must exist because f is continuous on a closed, bounded set in the plane, must occur at boundary points of the domain, that is, points of the circle $x^2 + y^2 = 1$. This circle can be parametrized $x = \cos t$, $y = \sin t$, so that

$$f(x, y) = f(\cos t, \sin t) = \cos t + 2 \sin t = g(t), \text{ say.}$$

For critical points of g: $0 = g'(t) = -\sin t + 2 \cos t$. Thus $\tan t = 2$, and $x = \pm 1/\sqrt{5}$, $y = \pm 2/\sqrt{5}$. The critical points are $(-1/\sqrt{5}, -2/\sqrt{5})$, where f has value $-\sqrt{5}$, and $(1/\sqrt{5}, 2/\sqrt{5})$, where f has value $\sqrt{5}$. Thus the maximum and minimum values of $f(x, y)$ on the disk are $\sqrt{5}$ and $-\sqrt{5}$ respectively.

6. $f(x, y) = xy(1 - x - y)$ on the triangle T shown in the figure. Evidently $f(x, y) = 0$ on all three boundary segments of T, and $f(x, y) > 0$ inside T. Thus the minimum value of f on T is 0, and the maximum value must occur at an interior critical point. For critical points:

$$0 = f_1(x, y) = y(1 - 2x - y), \qquad 0 = f_2(x, y) = x(1 - x - 2y).$$

The only critical points are $(0, 0)$, $(1, 0)$ and $(0, 1)$, which are on the boundary of T, and $(1/3, 1/3)$, which is inside T. The maximum value of f over T is $f(1/3, 1/3) = 1/27$.

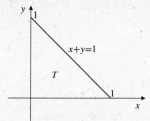

Fig. 13.2.6

8. $f(x, y) = \sin x \sin y \sin(x + y)$ on the triangle T shown in the figure. Evidently $f(x, y) = 0$ on the boundary of T, and $f(x, y) > 0$ at all points inside T. Thus the minimum value of f on T is zero, and the maximum value must occur at an interior critical point. For critical points inside T we must have

$$0 = f_1(x, y) = \cos x \sin y \sin(x + y) + \sin x \sin y \cos(x + y)$$
$$0 = f_2(x, y) = \sin x \cos y \sin(x + y) + \sin x \sin y \cos(x + y).$$

Therefore $\cos x \sin y = \cos y \sin x$, which implies $x = y$ for points inside T, and

$$\cos x \sin x \sin 2x + \sin^2 x \cos 2x = 0$$
$$2 \sin^2 x \cos^2 x + 2 \sin^2 x \cos^2 x - \sin^2 x = 0$$
$$4 \cos^2 x = 1.$$

Thus $\cos x = \pm 1/2$, and $x = \pm \pi/3$. The interior critical point is $(\pi/3, \pi/3)$, where f has the value $3\sqrt{3}/8$. This is the maximum value of f on T.

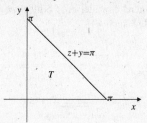

Fig. 13.2.8

10. $f(x, y) = \dfrac{x - y}{1 + x^2 + y^2}$ on the half-plane $y \geq 0$. For critical points:

$$0 = f_1(x, y) = \frac{1 - x^2 + y^2 + 2xy}{(1 + x^2 + y^2)^2}$$
$$0 = f_2(x, y) = \frac{-1 - x^2 + y^2 - 2xy}{(1 + x^2 + y^2)^2}.$$

Any critical points must satisfy $1 - x^2 + y^2 + 2xy = 0$ and $-1 - x^2 + y^2 - 2xy = 0$, and hence $x^2 = y^2$ and $2xy = -1$. Therefore $y = -x = \pm 1/\sqrt{2}$. The only critical point in the region $y \geq 0$ is $(-1/\sqrt{2}, 1/\sqrt{2})$, where f has the value $-1/\sqrt{2}$.

On the boundary $y = 0$ we have

$$f(x, 0) = \frac{x}{1 + x^2} = g(x), \qquad (-\infty < x < \infty).$$

Evidently, $g(x) \to 0$ as $x \to \pm\infty$.
Since $g'(x) = \dfrac{1 - x^2}{(1 + x^2)^2}$, the critical points of g are $x = \pm 1$. We have $g(\pm 1) = \pm \dfrac{1}{2}$.
The maximum and minimum values of f on the upper half-plane $y \geq 0$ are $1/2$ and $-1/\sqrt{2}$ respectively.

12. Let $f(x, y, z) = xz + yz$ on the ball $x^2 + y^2 + z^2 \leq 1$. First look for interior critical points:

$$0 = f_1 = z, \quad 0 = f_2 = z, \quad 0 = f_3 = x + y.$$

All points on the line $z = 0$, $x + y = 0$ are CPs, and $f = 0$ at all such points.

Now consider the boundary sphere $x^2 + y^2 + z^2 = 1$. On it

$$f(x, y, z) = (x + y)z = \pm(x + y)\sqrt{1 - x^2 - y^2} = g(x, y),$$

where g has domain $x^2 + y^2 \le 1$. On the boundary of its domain, g is identically 0, although g takes both positive and negative values at some points inside its domain. Therefore, we need consider only critical points of g in $x^2 + y^2 < 1$. For such CPs:

$$0 = g_1 = \sqrt{1 - x^2 - y^2} + \frac{(x + y)(-2x)}{2\sqrt{1 - x^2 - y^2}}$$

$$= \frac{1 - x^2 - y^2 - x^2 - xy}{\sqrt{1 - x^2 - y^2}}$$

$$0 = g_2 = \frac{1 - x^2 - y^2 - xy - y^2}{\sqrt{1 - x^2 - y^2}}.$$

Therefore $2x^2 + y^2 + xy = 1 = x^2 + 2y^2 + xy$, from which $x^2 = y^2$.
Case I: $x = -y$. Then $g = 0$, so $f = 0$.
Case II: $x = y$. Then $2x^2 + x^2 + x^2 = 1$, so $x^2 = 1/4$ and $x = \pm 1/2$. g (which is really two functions depending on our choice of the "+" or "−" sign) has four CPs, two corresponding to $x = y = 1/2$ and two to $x = y = -1/2$. The values of g at these four points are $\pm 1/\sqrt{2}$.

Since we have considered all points where f can have extreme values, we conclude that the maximum value of f on the ball is $1/\sqrt{2}$ (which occurs at the boundary points $\pm(\frac{1}{2}, \frac{1}{2}, \frac{1}{\sqrt{2}})$) and minimum value $-1/\sqrt{2}$ (which occurs at the boundary points $\pm(\frac{1}{2}, \frac{1}{2}, -\frac{1}{\sqrt{2}})$).

14. $f(x, y) = xy^2 e^{-xy}$ on $Q = \{(x, y) : x \ge 0, \ y \ge 0\}$.
As in Exercise 13, $f(x, 0) = f(0, y) = 0$ and $\lim_{x \to \infty} f(x, kx) = k^2 x^3 e^{-x^2} = 0$.

Also, $f(0, y) = 0$ while $f\left(\frac{1}{y}, y\right) = \frac{y}{e} \to \infty$ as $y \to \infty$, so that f has no limit as $x^2 + y^2 \to \infty$ in Q, and f has no maximum value on Q.

16. Let the dimensions be as shown in the figure. Then $2x + y = 100$, the length of the fence. For maximum area A of the enclosure we will have $x > 0$ and $0 < \theta < \pi/2$. Since $h = x \cos \theta$, the area A is

$$A = xy \cos \theta + 2 \times \frac{1}{2}(x \sin \theta)(x \cos \theta)$$

$$= x(100 - 2x) \cos \theta + x^2 \sin \theta \cos \theta$$

$$= (100x - 2x^2) \cos \theta + \frac{1}{2}x^2 \sin 2\theta.$$

We look for a critical point of A satisfying $x > 0$ and $0 < \theta < \pi/2$.

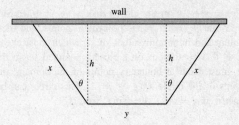

Fig. 13.2.16

$$0 = \frac{\partial A}{\partial x} = (100 - 4x) \cos \theta + x \sin 2\theta$$

$$\Rightarrow \cos \theta (100 - 4x + 2x \sin \theta) = 0$$

$$\Rightarrow 4x - 2x \sin \theta = 100 \Rightarrow x = \frac{50}{2 - \sin \theta}$$

$$0 = \frac{\partial A}{\partial \theta} = -(100x - 2x^2) \sin \theta + x^2 \cos 2\theta$$

$$\Rightarrow x(1 - 2 \sin^2 \theta) + 2x \sin \theta - 100 \sin \theta = 0.$$

Substituting the first equation into the second we obtain

$$\frac{50}{2 - \sin \theta}\left(1 - 2 \sin^2 \theta + 2 \sin \theta\right) - 100 \sin \theta = 0$$

$$50(1 - 2 \sin^2 \theta + 2 \sin \theta) = 100(2 \sin \theta - \sin^2 \theta)$$

$$50 = 100 \sin \theta.$$

Thus $\sin \theta = 1/2$, and $\theta = \pi/6$.
Therefore $x = \dfrac{50}{2 - (1/2)} = \dfrac{100}{3}$, and $y = 100 - 2x = \dfrac{100}{3}$.

The maximum area for the enclosure is

$$A = \left(\frac{100}{3}\right)^2 \frac{\sqrt{3}}{2} + \left(\frac{100}{3}\right)^2 \frac{1}{2}\frac{\sqrt{3}}{2} = \frac{2500}{\sqrt{3}}$$

square units. All three segments of the fence will be the same length, and the bend angles will be $120°$.

18. Minimize $F(x, y, z) = 2x + 3y + 4z$ subject to

$$x \ge 0, \qquad y \ge 0, \qquad z \ge 0,$$
$$x + y \ge 2, \qquad y + z \ge 2, \qquad x + z \ge 2.$$

Here the constraint region has vertices $(1, 1, 1)$, $(2, 2, 0)$, $(2, 0, 2)$, and $(0, 2, 2)$. Since $F(1, 1, 1) = 9$, $F(2, 2, 0) = 10$, $F(2, 0, 2) = 12$, and $F(0, 2, 2) = 14$, the minimum value of F subject to the constraints is 9.

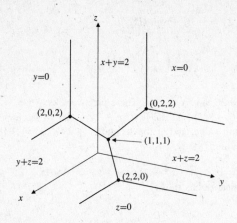

Fig. 13.2.18

20. If the developer builds x houses, y duplex units, and z apartments, his profit will be

$$P = 40,000x + 20,000y + 16,000z.$$

The legal constraints imposed require that

$$\frac{x}{6} + \frac{y}{8} + \frac{z}{12} \le 10, \quad \text{that is } 4x + 3y + 2z \le 240,$$

and also

$$z \ge x + y.$$

Evidently we must also have $x \ge 0$, $y \ge 0$, and $z \ge 0$. The planes $4x + 3y + 2z = 240$ and $z = x + y$ intersect where $6x + 5y = 240$. Thus the constraint region has vertices $(0, 0, 0)$, $(40, 0, 40)$, $(0, 48, 48)$, and $(0, 0, 120)$, which yield revenues of \$0, \$2,240,000, \$1,728,000, and \$1,920,000 respectively.

For maximum profit, the developer should build 40 houses, no duplex units, and 40 apartments.

Section 13.3 Lagrange Multipliers (page 807)

2. a) Let D be the distance from $(3, 0)$ to the point (x, y) on the curve $y = x^2$. Then

$$D^2 = (x - 3)^2 + y^2 = (x - 3)^2 + x^4.$$

For a minimum, $0 = \dfrac{dD^2}{dx} = 2(x - 3) + 4x^3$. Thus $2x^3 + x - 3 = 0$. Clearly $x = 1$ is a root of this cubic equation. Since

$$\frac{2x^3 + x - 3}{x - 1} = 2x^2 + 2x + 3,$$

and $2x^2 + 2x + 3$ has negative discriminant, $x = 1$ is the only critical point. Thus the minimum distance from $(3, 0)$ to $y = x^2$ is $D = \sqrt{(-2)^2 + 1^4} = \sqrt{5}$ units.

b) We want to minimize $D^2 = (x-3)^2 + y^2$ subject to the constraint $y = x^2$. Let $L = (x - 3)^2 + y^2 + \lambda(x^2 - y)$. For critical points of L we want

$$0 = \frac{\partial L}{\partial x} = 2(x - 3) + 2\lambda x$$
$$\Rightarrow \quad (1 + \lambda)x - 3 = 0 \qquad (A)$$
$$0 = \frac{\partial L}{\partial y} = 2y - \lambda \qquad (B)$$
$$0 = \frac{\partial L}{\partial \lambda} = x^2 - y. \qquad (C)$$

Eliminating λ from (A) and (B), we get
$x + 2xy - 3 = 0$.
Substituting (C) then leads to $2x^3 + x - 3 = 0$, or $(x - 1)(2x^2 + 2x + 3) = 0$. The only real solution is $x = 1$, so the point on $y = x^2$ closest to $(3, 0)$ is $(1, 1)$.
Thus the minimum distance from $(3, 0)$ to $y = x^2$ is $D = \sqrt{(1 - 3)^2 + 1^2} = \sqrt{5}$ units.

4. Let $f(x, y, z) = x + y - z$, and define the Lagrangian

$$L = x + y - z + \lambda(x^2 + y^2 + z^2 - 1).$$

Solutions to the constrained problem will be found among the critical points of L. To find these we have

$$0 = \frac{\partial L}{\partial x} = 1 + 2\lambda x,$$
$$0 = \frac{\partial L}{\partial y} = 1 + 2\lambda y,$$
$$0 = \frac{\partial L}{\partial z} = -1 + 2\lambda z,$$
$$0 = \frac{\partial L}{\partial \lambda} = x^2 + y^2 + z^2 - 1.$$

Therefore $2\lambda x = 2\lambda y = -2\lambda z$. Either $\lambda = 0$ or $x = y = -z$. $\lambda = 0$ is not possible. (It implies $0 = 1$ from the first equation.) From $x = y = -z$ we obtain $1 = x^2 + y^2 + z^2 = 3x^2$, so $x = \pm\dfrac{1}{\sqrt{3}}$. L has critical points at $\left(\dfrac{1}{\sqrt{3}}, \dfrac{1}{\sqrt{3}}, -\dfrac{1}{\sqrt{3}} \right)$ and $\left(\dfrac{1}{-\sqrt{3}}, -\dfrac{1}{\sqrt{3}}, \dfrac{1}{\sqrt{3}} \right)$. At the first $f = \sqrt{3}$, which is the maximum value of f on the sphere; at the second $f = -\sqrt{3}$, which is the minimum value.

6. Let $L = x^2 + y^2 + z^2 + \lambda(xyz^2 - 2)$. For critical points:

$$0 = \frac{\partial L}{\partial x} = 2x + \lambda yz^2 \quad \Leftrightarrow \quad -\lambda xyz^2 = 2x^2$$

$$0 = \frac{\partial L}{\partial y} = 2y + \lambda xz^2 \quad \Leftrightarrow \quad -\lambda xyz^2 = 2y^2$$

$$0 = \frac{\partial L}{\partial z} = 2z + 2\lambda xyz \quad \Leftrightarrow \quad -\lambda xyz^2 = z^2$$

$$0 = \frac{\partial L}{\partial \lambda} = xyz^2 - 2.$$

From the first three equations, $x^2 = y^2$ and $z^2 = 2x^2$. The fourth equation then gives $x^2 y^2 4 z^4 = 4$, or $x^8 = 1$. Thus $x^2 = y^2 = 1$ and $z^2 = 2$.
The shortest distance from the origin to the surface $xyz^2 = 2$ is

$$\sqrt{1 + 1 + 2} = 2 \text{ units.}$$

8. Let $L = x^2 + y^2 + \lambda(3x^2 + 2xy + 3y^2 - 16)$. We have

$$0 = \frac{\partial L}{\partial x} = 2x + 6\lambda x + 2\lambda y \qquad (A)$$

$$0 = \frac{\partial L}{\partial y} = 2y + 6\lambda y + 2\lambda x. \qquad (B)$$

Multiplying (A) by y and (B) by x and subtracting we get

$$2\lambda(y^2 - x^2) = 0.$$

Thus, either $\lambda = 0$, or $y = x$, or $y = -x$.
$\lambda = 0$ is not possible, since it implies $x = 0$ and $y = 0$, and the point $(0, 0)$ does not lie on the given ellipse.
If $y = x$, then $8x^2 = 16$, so $x = y = \pm\sqrt{2}$.
If $y = -x$, then $4x^2 = 16$, so $x = -y = \pm 2$.
The points on the ellipse nearest the origin are $(\sqrt{2}, \sqrt{2})$ and $(-\sqrt{2}, -\sqrt{2})$. The points farthest from the origin are $(2, -2)$ and $(-2, 2)$. The major axis of the ellipse lies along $y = -x$ and has length $4\sqrt{2}$. The minor axis lies along $y = x$ and has length 4.

10. Let $L = x + 2y - 3z + \lambda(x^2 + 4y^2 + 9z^2 - 108)$. For CPs of L:

$$0 = \frac{\partial L}{\partial x} = 1 + 2\lambda x \qquad (A)$$

$$0 = \frac{\partial L}{\partial y} = 2 + 8\lambda y \qquad (B)$$

$$0 = \frac{\partial L}{\partial z} = -3 + 18\lambda z \qquad (C)$$

$$0 = \frac{\partial L}{\partial \lambda} = x^2 + 4y^2 + 9z^2 - 108. \qquad (D)$$

From (A), (B), and (C),

$$\lambda = -\frac{1}{2x} = -\frac{2}{8y} = \frac{3}{18z},$$

so $x = 2y = -3z$. From (D):

$$x^2 + 4\left(\frac{x^2}{4}\right) + 9\left(\frac{x^2}{9}\right) = 108,$$

so $x^2 = 36$, and $x = \pm 6$. There are two CPs: $(6, 3, -2)$ and $(-6, -3, 2)$. At the first, $x + 2y - 3z = 18$, the maximum value, and at the second, $x + 2y - 3z = -18$, the minimum value.

12. Let $L = x^2 + y^2 + z^2 + \lambda(x^2 + y^2 - z^2) + \mu(x - 2z - 3)$. For critical points of L:

$$0 = \frac{\partial L}{\partial x} = 2x(1 + \lambda) + \mu \qquad (A)$$

$$0 = \frac{\partial L}{\partial y} = 2y(1 + \lambda) \qquad (B)$$

$$0 = \frac{\partial L}{\partial z} = 2z(1 - \lambda) - 2\mu \qquad (C)$$

$$0 = \frac{\partial L}{\partial \lambda} = x^2 + y^2 - z^2 \qquad (D)$$

$$0 = \frac{\partial L}{\partial \mu} = x - 2z - 3. \qquad (E)$$

From (B), either $y = 0$ or $\lambda = -1$.

CASE I. $y = 0$. Then (D) implies $x = \pm z$.
If $x = z$ then (E) implies $z = -3$, so we get the point $(-3, 0, -3)$.
If $x = -z$ then (E) implies $z = -1$, so we get the point $(1, 0, -1)$.

CASE II. $\lambda = -1$. Then (A) implies $\mu = 0$ and (C) implies $z = 0$. By (D), $x = y = 0$, and this contradicts (E), so this case is not possible.

If $f(x, y, z) = x^2 + y^2 + z^2$, then $f(-3, 0, -3) = 18$ is the maximum value of f on the ellipse $x^2 + y^2 = z^2$, $x - 2z = 3$, and $f(1, 0, -1) = 2$ is the minimum value.

14. The max and min values of $f(x, y, z) = x + y^2 z$ subject to the constraints $y^2 + z^2 = 2$ and $z = x$ will be found among the critical points of

$$L = x + y^2 z + \lambda(y^2 + z^2 - 2) + \mu(z - x).$$

Thus

$$0 = \frac{\partial L}{\partial x} = 1 - \mu = 0,$$

$$0 = \frac{\partial L}{\partial y} = 2yz + 2\lambda y = 0,$$

$$0 = \frac{\partial L}{\partial z} = y^2 + 2\lambda z + \mu = 0,$$

$$0 = \frac{\partial L}{\partial \lambda} = y^2 + z^2 - 2,$$

$$0 = \frac{\partial L}{\partial \mu} = z - x.$$

From the first equation $\mu = 1$. From the second, either $y = 0$ or $z = -\lambda$.
If $y = 0$ then $z^2 = 2$, $z = x$, so critical points are $(\sqrt{2}, 0, \sqrt{2})$ and $(-\sqrt{2}, 0, -\sqrt{2})$. f has the values $\pm\sqrt{2}$ at these points. If $z = -\lambda$ then $y^2 - 2z^2 + 1 = 0$. Thus $2z^2 - 1 = 2 - z^2$, or $z^2 = 1$, $z = \pm 1$. This leads to critical points $(1, \pm 1, 1)$ and $(-1, \pm 1, -1)$ where f has values ± 2. The maximum value of f subject to the constraints is 2; the minimum value is -2.

16. Let $L = x_1 + x_2 + \cdots + x_n + \lambda(x_1^2 + x_2^2 + \cdots + x_n^2 - 1)$.
For critical points of L we have

$$0 = \frac{\partial L}{\partial x_1} = 1 + 2\lambda x_1, \quad \ldots \quad 0 = \frac{\partial L}{\partial x_n} = 1 + 2\lambda x_n$$

$$0 = \frac{\partial L}{\partial \lambda} = x_1^2 + x_2^2 + \cdots + x_n^2 - 1.$$

The first n equations give

$$x_1 = x_2 = \cdots = x_n = -\frac{1}{2\lambda},$$

and the final equation gives

$$\frac{1}{4\lambda^2} + \frac{1}{4\lambda^2} + \cdots + \frac{1}{4\lambda^2} = 1,$$

so that $4\lambda^2 = n$, and $\lambda = \pm\sqrt{n}/2$.
The maximum and minimum values of $x_1 + x_2 + \cdots + x_n$ subject to $x_1^2 + \cdots + x_n^2 = 1$ are $\pm\frac{n}{2\lambda}$, that is, $\sqrt{n}$ and $-\sqrt{n}$ respectively.

18. Let the width, depth, and height of the box be x, y and z respectively. We want to minimize the surface area

$$S = xy + 2xz + 2yz$$

subject to the constraint that $xyz = V$, where V is a given positive volume. Let

$$L = xy + 2xz + 2yz + \lambda(xyz - V).$$

For critical points of L,

$$0 = \frac{\partial L}{\partial x} = y + 2z + \lambda yz \quad \Leftrightarrow \quad -\lambda xyz = xy + 2xz$$

$$0 = \frac{\partial L}{\partial y} = x + 2z + \lambda xz \quad \Leftrightarrow \quad -\lambda xyz = xy + 2yz$$

$$0 = \frac{\partial L}{\partial z} = 2x + 2y + \lambda xy \quad \Leftrightarrow \quad -\lambda xyz = 2xz + 2yz$$

$$0 = \frac{\partial L}{\partial \lambda} = xyz - V.$$

From the first three equations, $xy = 2xz = 2yz$. Since x, y, and z are all necessarily positive, we must therefore have $x = y = 2z$. Thus the most economical box with no top has width and depth equal to twice the height.

20. We want to maximize xyz subject to $xy + 2yz + 3xz = 18$. Let

$$L = xyz + \lambda(xy + 2yz + 3xz - 18).$$

For critical points of L,

$$0 = \frac{\partial L}{\partial x} = yz + \lambda(y + 3z) \quad \Leftrightarrow \quad -xyz = \lambda(xy + 3xz)$$

$$0 = \frac{\partial L}{\partial y} = xz + \lambda(x + 2z) \quad \Leftrightarrow \quad -xyz = \lambda(xy + 2yz)$$

$$0 = \frac{\partial L}{\partial z} = xy + \lambda(2y + 3x) \quad \Leftrightarrow \quad -xyz = \lambda(2yz + 3xz)$$

$$0 = \frac{\partial L}{\partial \lambda} = xy + 2yz + 3xz - 18.$$

From the first three equations $xy = 2yz = 3xz$. From the fourth equation, the sum of these expressions is 18. Thus

$$xy = 2yz = 3xz = 6.$$

Thus the maximum volume of the box is

$$V = xyz = \sqrt{(xy)(yz)(xz)} = \sqrt{6 \times 3 \times 2} = 6 \text{ cubic units.}$$

22. $f(x, y, z) = xy + z^2$ on $B = \{(x, y, z) : x^2 + y^2 + z^2 \leq 1\}$.
For critical points of f,

$$0 = f_1(x, y, z) = y, \qquad 0 = f_2(x, y, z) = x,$$
$$0 = f_3(x, y, z) = 2z.$$

Thus the only critical point is the interior point $(0, 0, 0)$, where f has the value 0, evidently neither a maximum nor a minimum. The maximum and minimum must therefore occur on the boundary of B, that is, on the sphere $x^2 + y^2 + z^2 = 1$. Let

$$L = xy + z^2 + \lambda(x^2 + y^2 + z^2 - 1).$$

For critical points of L,

$$0 = \frac{\partial L}{\partial x} = y + 2\lambda x \qquad (A)$$

$$0 = \frac{\partial L}{\partial y} = x + 2\lambda y \qquad (B)$$

$$0 = \frac{\partial L}{\partial z} = 2z(1 + \lambda) \qquad (C)$$

$$0 = \frac{\partial L}{\partial \lambda} = x^2 + y^2 + z^2 - 1. \qquad (D)$$

From (C) either $z = 0$ or $\lambda = -1$.

CASE I. $z = 0$. (A) and (B) imply that $y^2 = x^2$ and (D) then implies that $x^2 = y^2 = 1/2$. At the four points

$$\left(\frac{1}{\sqrt{2}}, \pm\frac{1}{\sqrt{2}}, 0\right) \quad \text{and} \quad \left(-\frac{1}{\sqrt{2}}, \pm\frac{1}{\sqrt{2}}, 0\right)$$

f takes the values $\frac{1}{2}$ and $-\frac{1}{2}$.

CASE II. $\lambda = -1$. (A) and (B) imply that $x = y = 0$, and so by (D), $z = \pm 1$. f has the value 1 at the points $(0, 0, \pm 1)$.

Thus the maximum and minimum values of f on B are 1 and $-1/2$ respectively.

24. Let $L = \sin \frac{x}{2} \sin \frac{y}{2} \sin \frac{z}{2} + \lambda(x + y + z - \pi)$. Then

$$0 = \frac{\partial L}{\partial x} = \frac{1}{2} \cos \frac{x}{2} \sin \frac{y}{2} \sin \frac{z}{2} + \lambda \qquad (A)$$

$$0 = \frac{\partial L}{\partial y} = \frac{1}{2} \sin \frac{x}{2} \cos \frac{y}{2} \sin \frac{z}{2} + \lambda \qquad (B)$$

$$0 = \frac{\partial L}{\partial z} = \frac{1}{2} \sin \frac{x}{2} \sin \frac{y}{2} \cos \frac{z}{2} + \lambda. \qquad (C)$$

For any triangle we must have $0 \le x \le \pi$, $0 \le y \le \pi$ and $0 \le z \le \pi$. Also

$$P = \sin \frac{x}{2} \sin \frac{y}{2} \sin \frac{z}{2}$$

is 0 if any of x, y or z is 0 or π. Subtracting equations (A) and (B) gives

$$\frac{1}{2} \sin \frac{z}{2} \sin \frac{x - y}{2} = 0.$$

It follows that we must have $x = y$; all other possibilities lead to a zero value for P. Similarly, $y = z$. Thus the triangle for which P is maximum must be equilateral: $x = y = z = \pi/3$. Since $\sin(\pi/3) = 1/2$, the maximum value of P is $1/8$.

26. As can be seen in the figure, the minimum distance from $(0, -1)$ to points of the semicircle $y = \sqrt{1 - x^2}$ is $\sqrt{2}$, the closest points to $(0, -1)$ on the semicircle being $(\pm 1, 0)$. These points will not be found by the method of Lagrange multipliers because the level curve $f(x, y) = 2$ of the function f giving the square of the distance from (x, y) to $(0, -1)$ is not tangent to the semicircle at $(\pm 1, 0)$. This could only have happened because $(\pm 1, 0)$ are *endpoints* of the semicircle.

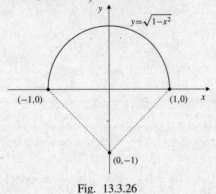

Fig. 13.3.26

Section 13.4 The Method of Least Squares (page 814)

2. We want to minimize $S = \sum_{i=1}^{n}(ax_i^2 - y_i)^2$. Thus

$$0 = \frac{dS}{da} = \sum_{i=1}^{n} 2(ax_i^2 - y_i)x_i^2$$

$$= 2 \sum_{i=1}^{n}(ax_i^4 - x_i^2 y_i),$$

and $a = \left(\sum_{i=1}^{n} x_i^2 y_i\right) \Big/ \left(\sum_{i=1}^{n} x_i^4\right)$.

4. We choose a, b, and c to minimize

$$S = \sum_{i=1}^{n}\left(ax_i + by_i + c - z_i\right)^2.$$

Thus

$$0 = \frac{\partial S}{\partial a} = 2 \sum_{i=1}^{n}(ax_i + by_i + c - z_i)x_i$$

$$0 = \frac{\partial S}{\partial b} = 2 \sum_{i=1}^{n}(ax_i + by_i + c - z_i)y_i$$

$$0 = \frac{\partial S}{\partial c} = 2 \sum_{i=1}^{n}(ax_i + by_i + c - z_i).$$

Let $A = \sum x_i^2$, $B = \sum x_i y_i$, $C = \sum x_i$, $D = \sum y_i^2$, $E = \sum y_i$, $F = \sum x_i z_i$, $G = \sum y_i z_i$, and $H = \sum z_i$. In terms of these quantities the above equations become

$$
\begin{aligned}
Aa &+ Bb &+ Cc &= F \\
Ba &+ Db &+ Ec &= G \\
Ca &+ Eb &+ nc &= H.
\end{aligned}
$$

By Cramer's Rule (Theorem 5 of Section 1.6) the solution is

$$a = \frac{1}{\Delta}\begin{vmatrix} F & B & C \\ G & D & E \\ H & E & n \end{vmatrix}, \qquad b = \frac{1}{\Delta}\begin{vmatrix} A & F & C \\ B & G & E \\ C & H & n \end{vmatrix},$$

$$c = \frac{1}{\Delta}\begin{vmatrix} A & B & F \\ B & D & G \\ C & E & H \end{vmatrix}, \quad \text{where } \Delta = \begin{vmatrix} A & B & C \\ B & D & E \\ C & E & n \end{vmatrix}.$$

6. The relationship $y = p + qx^2$ is linear in p and q, so we choose p and q to minimize

$$S = \sum_{i=1}^{n}(p + qx_i^2 - y_i)^2.$$

Thus

$$0 = \frac{\partial S}{\partial p} = 2\sum_{i=1}^{n}(p + qx_i^2 - y_i)$$

$$0 = \frac{\partial S}{\partial q} = 2\sum_{i=1}^{n}(p + qx_i^2 - y_i)x_i^2,$$

that is,

$$np \quad + \quad \left(\sum x_i^2\right)q \quad = \quad \sum y_i$$
$$\left(\sum x_i^2\right)p \quad + \quad \left(\sum x_i^4\right)q \quad = \quad \sum x_i^2 y_i,$$

so

$$p = \frac{\left(\sum y_i\right)\left(\sum x_i^4\right) - \left(\sum x_i^2 y_i\right)\left(\sum x_i^2\right)}{n\left(\sum x_i^4\right) - \left(\sum x_i^2\right)^2}$$

$$q = \frac{n\left(\sum x_i^2 y_i\right) - \left(\sum y_i\right)\left(\sum x_i^2\right)}{n\left(\sum x_i^4\right) - \left(\sum x_i^2\right)^2}.$$

This is the result obtained by direct linear regression. (No transformation of variables was necessary.)

8. We transform $y = \ln(p + qx)$ into the form $e^y = p + qx$, which is linear in p and q. We let $\eta_i = e^{y_i}$ and use the regression line $\eta = ax + b$ obtained from the data (x_i, η_i), with $a = q$ and $b = p$.
    Using the formulas for a and b obtained in the text, we have

$$q = a = \frac{n\left(\sum x_i e^{y_i}\right) - \left(\sum x_i\right)\left(\sum e^{y_i}\right)}{n\left(\sum x_i^2\right) - \left(\sum x_i\right)^2}$$

$$p = b = \frac{\left(\sum x_i^2\right)\left(\sum e^{y_i}\right) - \left(\sum x_i\right)\left(\sum x_i e^{y_i}\right)}{n\left(\sum x_i^2\right) - \left(\sum x_i\right)^2}.$$

These values of p and q are not the same values that minimize the expression

$$S = \sum_{i=1}^{n}\bigl(\ln(p + qx_i) - y_i\bigr)^2.$$

10. We transform $y = \sqrt{(px + q)}$ into the form $y^2 = px + q$, which is linear in p and q. We let $\eta_i = y_i^2$ and use the regression line $\eta = ax + b$ obtained from the data (x_i, η_i), with $a = p$ and $b = q$.
    Using the formulas for a and b obtained in the text, we have

$$p = a = \frac{n\left(\sum x_i y_i^2\right) - \left(\sum x_i\right)\left(\sum y_i^2\right)}{n\left(\sum x_i^2\right) - \left(\sum x_i\right)^2}$$

$$q = b = \frac{\left(\sum x_i^2\right)\left(\sum y_i^2\right) - \left(\sum x_i\right)\left(\sum x_i y_i^2\right)}{n\left(\sum x_i^2\right) - \left(\sum x_i\right)^2}.$$

These values of p and q are not the same values that minimize the expression

$$S = \sum_{i=1}^{n}\left(\sqrt{px_i + q} - y_i\right)^2.$$

12. We use the result of Exercise 6. We have $n = 6$ and

$$\sum x_i^2 = 115, \qquad \sum x_i^4 = 4051,$$
$$\sum y_i = 55.18, \qquad \sum x_i^2 y_i = 1984.50.$$

Therefore

$$p = \frac{\left(\sum y_i\right)\left(\sum x_i^4\right) - \left(\sum x_i^2 y_i\right)\left(\sum x_i^2\right)}{n\left(\sum x_i^4\right) - \left(\sum x_i^2\right)^2}$$

$$= \frac{55.18 \times 4051 - 1984.50 \times 115}{6 \times 4051 - 115^2} \approx -0.42$$

$$q = \frac{n\left(\sum x_i^2 y_i\right) - \left(\sum y_i\right)\left(\sum x_i^2\right)}{n\left(\sum x_i^4\right) - \left(\sum x_i^2\right)^2}$$

$$= \frac{6 \times 1984.50 - 55.18 \times 115}{6 \times 4051 - 115^2} \approx 0.50.$$

We have (approximately) $y = -0.42 + 0.50x^2$. The predicted value of y at $x = 5$ is $-0.42 + 0.50 \times 25 \approx 12.1$.

14. Since $y = pe^x + q + re^{-x}$ is equivalent to

$$e^x y = p(e^x)^2 + qe^x + r,$$

we let $\xi_i = e^{x_i}$ and $\eta_i = e^{x_i}y_i$ for $i = 1, 2, \ldots, n$. We then have $p = a$, $q = b$, and $r = c$, where a, b, and c are the values calculated by the formulas in Exercise 13, but for the data (ξ_i, η_i) instead of (x_i, y_i).

16. To maximize $I = \int_0^{\pi}\bigl(ax(\pi - x) - \sin x\bigr)^2 dx$, we choose a so that

$$0 = \frac{dI}{da} = \int_0^{\pi} 2\bigl(ax(\pi - x) - \sin x\bigr)x(\pi - x)\,dx$$

$$= 2a\int_0^{\pi} x^2(\pi - x)^2\,dx - 2\int_0^{\pi} x(\pi - x)\sin x\,dx$$

$$= \frac{\pi^5 a}{15} - 8.$$

(We have omitted the details of evaluation of these integrals.) Hence $a = 120/\pi^5$. The minimum value of I is

$$\int_0^{\pi}\left(\frac{120}{\pi^5}x(\pi - x) - \sin x\right)^2 dx = \frac{\pi}{2} - \frac{480}{\pi^5} \approx 0.00227.$$

18. To minimize $\int_0^1 (x^3 - ax^2 - bx - c)^2\, dx$, choose a, b and c so that

$$0 = 2\int_0^1 (x^3 - ax^2 - bx - c)(-x^2)\, dx$$

$$0 = 2\int_0^1 (x^3 - ax^2 - bx - c)(-x)\, dx$$

$$0 = 2\int_0^1 (x^3 - ax^2 - bx - c)(-1)\, dx,$$

that is,

$$\frac{a}{5} + \frac{b}{4} + \frac{c}{3} = \frac{1}{6}$$
$$\frac{a}{4} + \frac{b}{3} + \frac{c}{2} = \frac{1}{5}$$
$$\frac{a}{3} + \frac{b}{2} + c = \frac{1}{4}$$

for which the solution is $a = \dfrac{3}{2}$, $b = -\dfrac{3}{5}$, and $c = \dfrac{1}{20}$.

20. $J = \int_{-1}^1 (x - a\sin\pi x - b\sin 2\pi x - c\sin 3\pi x)^2\, dx.$

To minimize J, choose a, b, and c to satisfy

$$0 = \frac{\partial J}{\partial a}$$
$$= -2\int_{-1}^1 (x - a\sin\pi x - b\sin 2\pi x - c\sin 3\pi x)\sin\pi x\, dx$$
$$= \frac{2}{\pi}(\pi a - 2)$$
$$0 = \frac{\partial J}{\partial b}$$
$$= -2\int_{-1}^1 (x - a\sin\pi x - b\sin 2\pi x - c\sin 3\pi x)\sin 2\pi x\, dx$$
$$= \frac{2}{\pi}(\pi b + 1)$$
$$0 = \frac{\partial J}{\partial c}$$
$$= -2\int_{-1}^1 (x - a\sin\pi x - b\sin 2\pi x - c\sin 3\pi x)\sin 3\pi x\, dx$$
$$= \frac{2}{3\pi}(3\pi c - 2).$$

We have omitted the details of evaluation of these integrals, but note that

$$\int_{-1}^1 \sin m\pi x \sin n\pi x\, dx = 0$$

if m and n are different integers.

The equations above imply that $a = 2/\pi$, $b = -1/\pi$, and $c = 2/(3\pi)$. These are the values that minimize J.

22. The Fourier sine series coefficients for $f(x) = x$ on $(0, \pi)$ are

$$b_n = \frac{2}{\pi}\int_0^\pi x\sin(nx)\, dx = (-1)^{n-1}\frac{2}{n}$$

for $n = 1, 2, \ldots$. Thus the series is

$$\sum_{n=0}^\infty (-1)^{n-1}\frac{2}{n}\sin nx.$$

Since x and the functions $\sin nx$ are all odd functions, we would also expect the series to converge to x on $(-\pi, 0)$.

24. We are given that $x_1 \le x_2 \le x_3 \le \ldots \le x_n$. To motivate the method, look at a special case, $n = 5$ say.

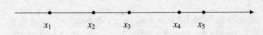

Fig. 13.4.24

If $x = x_3$, then

$$\sum_{i=1}^5 |x - x_i|$$
$$= (x_3 - x_1) + (x_3 - x_2) + 0 + (x_4 - x_3) + (x_5 - x_3)$$
$$= (x_5 - x_1) + (x_4 - x_2).$$

If x moves away from x_3 in either direction, then

$$\sum_{i=1}^5 |x - x_i| = (x_5 - x_1) + (x_4 - x_2) + |x - x_3|.$$

Thus the minimum sum occurs if $x = x_3$.
In general, if n is odd, then $\sum_{i=1}^n |x - x_i|$ is minimum if $x = x_{(n+1)/2}$, the middle point of the set of points $\{x_1, x_2, \ldots, x_n\}$. The value of x is unique in this case. If n is even and x satisfies $x_{n/2} \le x \le x_{(n/2)+1}$, then

$$\sum_{i=1}^n |x - x_i| = \sum_{i=1}^{n/2} |x_{n+1-i} - x_i|,$$

and the sum will increase if x is outside that interval. In this case the value of x which minimizes the sum is not unique unless it happens that $x_{n/2} = x_{(n/2)+1}$.

Section 13.5 Parametric Problems (page 824)

2. $\displaystyle\int_{-\infty}^{\infty} e^{-u^2}\, du = \sqrt{\pi}$ Let $u = xt$

$$du = x\, dt$$

$$\int_{-\infty}^{\infty} e^{-x^2 t^2}\, dt = \frac{\sqrt{\pi}}{x}.$$

Differentiate with respect to x:

$$\int_{-\infty}^{\infty} -2xt^2 e^{-t^2 x^2}\, dt = -\frac{\sqrt{\pi}}{x^2}$$

$$\int_{-\infty}^{\infty} t^2 e^{-x^2 t^2}\, dt = \frac{\sqrt{\pi}}{2x^3}. \qquad (*)$$

If $x = 1$ we get $\displaystyle\int_{-\infty}^{\infty} t^2 e^{-t^2}\, dt = \frac{\sqrt{\pi}}{2}$.

Differentiate $(*)$ with respect to x again:

$$\int_{-\infty}^{\infty} -2xt^4 e^{-x^2 t^2}\, dt = -\frac{3\sqrt{\pi}}{2x^4}.$$

Divide by -2 and let $x = 1$:

$$\int_{-\infty}^{\infty} t^4 e^{-t^2}\, dt = \frac{3\sqrt{\pi}}{4}.$$

4. Let $\displaystyle I(x, y) = \int_0^1 \frac{t^x - t^y}{\ln t}\, dt$, where $x > -1$ and $y > -1$.
Then

$$\frac{\partial I}{\partial x} = \int_0^1 t^x\, dt = \frac{1}{x+1}$$

$$\frac{\partial I}{\partial y} = -\frac{1}{y+1}.$$

Thus

$$I(x, y) = \int \frac{dx}{x+1} = \ln(x+1) + C_1(y)$$

$$\frac{-1}{y+1} = \frac{\partial I}{\partial y} = \frac{\partial C_1}{\partial y} \Rightarrow C_1(y) = -\ln(y+1) + C_2$$

$$I(x, y) = \ln\left(\frac{x+1}{y+1}\right) + C_2.$$

But $I(x, x) = 0$, so $C_2 = 0$. Thus

$$I(x, y) = \int_0^1 \frac{t^x - t^y}{\ln t}\, dt = \ln\left(\frac{x+1}{y+1}\right)$$

for $x > -1$ and $y > -1$.

6. $\displaystyle F(x) = \int_0^{\infty} e^{-xt}\, \frac{\sin t}{t}\, dt$

$$F'(x) = \int_0^{\infty} -e^{-xt} \sin t\, dt = -\frac{1}{1+x^2} \quad (x > 0).$$

Therefore $\displaystyle F(x) = -\int \frac{dx}{1+x^2} = -\tan^{-1} x + C$.
Now, make the change of variable $xt = s$ in the integral defining $F(x)$, and obtain

$$F(x) = \int_0^{\infty} e^{-s}\, \frac{\sin(s/x)}{s/x}\, \frac{ds}{x} = \int_0^{\infty} \frac{e^{-s}}{s} \sin\frac{s}{x}\, ds.$$

Since $|\sin(s/x)| \le s/x$ if $s > 0$, $x > 0$, we have

$$|F(x)| \le \frac{1}{|x|} \int_0^{\infty} e^{-s}\, ds = \frac{1}{|x|} \to 0 \quad \text{as } x \to \infty.$$

Hence $-\dfrac{\pi}{2} + C = 0$, and $C = \dfrac{\pi}{2}$. Therefore

$$F(x) = \int_0^{\infty} e^{-xt}\, \frac{\sin t}{t}\, dt = \frac{\pi}{2} - \tan^{-1} x.$$

In particular, $\int_0^{\infty} \frac{\sin t}{t}\, dt = \lim_{x \to 0} F(x) = \frac{\pi}{2}$.

8. $\displaystyle\int_0^x \frac{dt}{x^2 + t^2} = \frac{1}{x} \tan^{-1} \frac{t}{x}\bigg|_0^x = \frac{\pi}{4x}$ for $x > 0$.
Differentiate with respect to x:

$$\frac{1}{2x^2} + \int_0^x \frac{-2x\, dt}{(x^2 + t^2)^2} = -\frac{\pi}{4x^2}$$

$$\int_0^x \frac{dt}{(x^2 + t^2)^2} = -\frac{1}{2x}\left[-\frac{\pi}{4x^2} - \frac{1}{2x^2}\right]$$

$$= \frac{\pi}{8x^3} + \frac{1}{4x^3}.$$

Differentiate with respect to x again:

$$\frac{1}{4x^4} + \int_0^x \frac{-4x\, dt}{(x^2 + t^2)^3} = -\frac{3}{x^4}\left[\frac{\pi}{8} + \frac{1}{4}\right]$$

$$\int_0^x \frac{dt}{(x^2 + t^2)^3} = -\frac{1}{4x}\left[-\frac{3\pi}{8x^4} - \frac{3}{4x^4} - \frac{1}{4x^4}\right]$$

$$= \frac{3\pi}{32x^5} + \frac{1}{4x^5}.$$

10. $\displaystyle f(x) = Cx + D + \int_0^x (x - t) f(t)\, dt \quad \Rightarrow \quad f(0) = D$

$$f'(x) = C + \int_0^x f(t)\, dt \quad \Rightarrow \quad f'(0) = C$$

$$f''(x) = f(x) \quad \Rightarrow \quad f(x) = A \cosh x + B \sinh x$$

$$D = f(0) = A, \qquad C = f'(0) = B$$

$$\Rightarrow f(x) = D \cosh x + C \sinh x.$$

12.
$$f(x) = 1 + \int_0^1 (x+t) f(t)\, dt$$

$$f'(x) = \int_0^1 f(t)\, dt = C, \quad \text{say,}$$

since the integral giving $f'(x)$ does not depend on x. Thus $f(x) = A + Cx$, where $A = f(0)$. Substituting this expression into the given equation, we obtain

$$A + Cx = 1 + \int_0^1 (x+t)(A + Ct)\, dt$$

$$= 1 + Ax + \frac{A}{2} + \frac{Cx}{2} + \frac{C}{3}.$$

Therefore

$$\frac{A}{2} - 1 - \frac{C}{3} + x\left(\frac{C}{2} - A\right) = 0.$$

This can hold for all x only if

$$\frac{A}{2} - 1 - \frac{C}{3} = 0 \quad \text{and} \quad \frac{C}{2} - A = 0.$$

Thus $C = 2A$ and $\dfrac{A}{2} - \dfrac{2A}{3} = 1$, so that $A = -6$ and $C = -12$. Therefore $f(x) = -6 - 12x$.

14. We eliminate c from the pair of equations

$$f(x, y, c) = y - (x - c)\cos c - \sin c = 0$$

$$\frac{\partial}{\partial c} f(x, y, c) = \cos c + (x - c)\sin c - \cos c = 0.$$

Thus $c = x$ and $y - 0 - \sin x = 0$. The envelope is $y = \sin x$.

16. We eliminate c from the pair of equations

$$f(x, y, c) = \frac{x}{\cos c} + \frac{y}{\sin c} - 1 = 0$$

$$\frac{\partial}{\partial c} f(x, y, c) = \frac{x \sin c}{\cos^2 c} - \frac{y \cos c}{\sin^2 c} = 0.$$

From the second equation, $y = x \tan^3 c$. Thus

$$\frac{x}{\cos c}(1 + \tan^2 c) = 1$$

which implies that $x = \cos^3 c$, and hence $y = \sin^3 c$. The envelope is the astroid $x^{2/3} + y^{2/3} = 1$.

18. We eliminate c from the pair of equations

$$f(x, y, c) = (x - c)^2 + (y - c)^2 - 1 = 0$$

$$\frac{\partial}{\partial c} f(x, y, c) = 2(c - x) + 2(c - y) = 0.$$

Thus $c = (x + y)/2$, and

$$\left(\frac{x-y}{2}\right)^2 + \left(\frac{y-x}{2}\right)^2 = 1$$

or $x - y = \pm\sqrt{2}$. These two parallel lines constitute the envelope of the given family which consists of circles of radius 1 with centres along the line $y = x$.

20. The curve $x^2 + (y - c)^2 = kc^2$ is a circle with centre $(0, c)$ and radius $\sqrt{k}\,c$, provided $k > 0$. Consider the system:

$$f(x, y, c) = x^2 + (y - c)^2 - kc^2 = 0$$

$$\frac{\partial}{\partial c} f(x, y, c) = -2(y - c) - 2kc = 0.$$

The second equation implies that $y - c = -kc$, and the first equation then says that $x^2 = k(1 - k)c^2$. This is only possible if $0 \le k \le 1$.

The cases $k = 0$ and $k = 1$ are degenerate. If $k = 0$ the "curves" are just points on the y-axis. If $k = 1$ the curves are circles, all of which are tangent to the x-axis at the origin. There is no reasonable envelope in either case. If $0 < k < 1$, the envelope is the pair of lines given by $x^2 = \dfrac{k}{1 - k} y^2$, that is, the lines $\sqrt{1 - k}\, x = \pm\sqrt{k}\, y$. These lines make angle $\sin^{-1}\sqrt{k}$ with the y-axis.

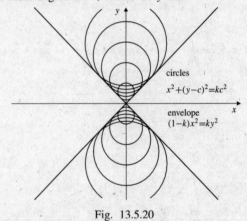

Fig. 13.5.20

22. If the family of surfaces $f(x, y, z, \lambda, \mu) = 0$ has an envelope, that envelope will have parametric equations

$$x = x(\lambda, \mu), \qquad y = y(\lambda, \mu), \qquad z = z(\lambda, \mu),$$

giving the point on the envelope where the envelope is tangent to the particular surface in the family having parameter values λ and μ. Thus

$$f\big(x(\lambda, \mu), y(\lambda, \mu), z(\lambda, \mu), \lambda, \mu\big) = 0.$$

Differentiating with respect to λ, we obtain

$$f_1 \frac{\partial x}{\partial \lambda} + f_2 \frac{\partial y}{\partial \lambda} + f_3 \frac{\partial z}{\partial \lambda} + f_4 = 0.$$

However, since for fixed μ, the parametric curve

$$x = x(t, \mu), \qquad y = y(t, \mu), \qquad z = z(t, \mu)$$

is tangent to the surface $f(x, y, z, \lambda, \mu) = 0$ at $t = \lambda$, its tangent vector there,

$$\mathbf{T} = \frac{\partial x}{\partial \lambda}\mathbf{i} + \frac{\partial y}{\partial \lambda}\mathbf{j} + \frac{\partial z}{\partial \lambda}\mathbf{k},$$

is perpendicular to the normal

$$\mathbf{N} = \boldsymbol{\nabla} f = f_1\mathbf{i} + f_2\mathbf{j} + f_3\mathbf{k},$$

so

$$f_1\frac{\partial x}{\partial \lambda} + f_2\frac{\partial y}{\partial \lambda} + f_3\frac{\partial z}{\partial \lambda} = 0.$$

Hence we must also have $\dfrac{\partial f}{\partial \lambda} = f_4(x, y, z, \lambda, \mu) = 0$.

Similarly, $\dfrac{\partial f}{\partial \mu} = 0$.

The parametric equations of the envelope must therefore satisfy the three equations

$$f(x, y, z, \lambda, \mu) = 0$$

$$\frac{\partial}{\partial \lambda} f(x, y, z, \lambda, \mu) = 0$$

$$\frac{\partial}{\partial \mu} f(x, y, z, \lambda, \mu) = 0.$$

The envelope can be found by eliminating λ and μ from these three equations.

24. $(x - \lambda)^2 + (y - \mu)^2 + z^2 = \dfrac{\lambda^2 + \mu^2}{2}$.

Differentiate with respect to λ and μ:

$$-2(x - \lambda) = \lambda, \qquad -2(y - \mu) = \mu.$$

Thus $\lambda = 2x$, $\mu = 2y$, and

$$x^2 + y^2 + z^2 = 2x^2 + 2y^2.$$

The envelope is the cone $z^2 = x^2 + y^2$.

26. $y^2 + \epsilon e^{-y^2} = 1 + x^2$

$$2yy_\epsilon + e^{-y^2} - 2y\epsilon e^{-y^2} y_\epsilon = 0$$

$$2y\left(1 - \epsilon e^{-y^2}\right) y_\epsilon + e^{-y^2} = 0$$

$$2y_\epsilon\left(1 - \epsilon e^{-y^2}\right)y_\epsilon - 2ye^{-y^2} y_\epsilon + 2y\left(2y\epsilon e^{-y^2} y_\epsilon\right)y_\epsilon$$
$$+ 2y\left(1 - \epsilon e^{-y^2}\right)y_{\epsilon\epsilon} - 2ye^{-y^2} y_\epsilon = 0.$$

At $\epsilon = 0$ we have $y(x, 0) = \sqrt{1 + x^2}$, and

$$2\sqrt{1 + x^2}\, y_\epsilon(x, 0) + e^{-(1+x^2)} = 0$$

$$y_\epsilon(x, 0) = -\frac{1}{2\sqrt{1 + x^2}}e^{-(1+x^2)}$$

$$2y_\epsilon^2 - 4ye^{-y^2} y_\epsilon + 2yy_{\epsilon\epsilon} = 0$$

$$yy_{\epsilon\epsilon} = 2yy_\epsilon e^{-y^2} - y_\epsilon^2$$

$$y_{\epsilon\epsilon}(x, 0) = -\left(\frac{1}{\sqrt{1 + x^2}} + \frac{1}{4(1 + x^2)^{3/2}}\right)e^{-2(1+x^2)}.$$

Thus

$$y = y(x, \epsilon) = y(x, 0) + \epsilon y_\epsilon(x, 0) + \frac{\epsilon^2}{2!}y_{\epsilon\epsilon}(x, 0) + \cdots$$

$$= \sqrt{1 + x^2} - \frac{\epsilon}{2\sqrt{1 + x^2}}e^{-(1+x^2)}$$

$$- \frac{\epsilon^2}{2}\left(\frac{1}{\sqrt{1 + x^2}} + \frac{1}{4(1 + x^2)^{3/2}}\right)e^{-2(1+x^2)} + \cdots.$$

28. Let $y(x, \epsilon)$ be the solution of $y + \epsilon y^5 = \dfrac{1}{2}$. Then we have

$$y_\epsilon\left(1 + 5\epsilon y^4\right) + y^5 = 0$$

$$y_{\epsilon\epsilon}\left(1 + 5\epsilon y^4\right) + 20\epsilon y^3 y_\epsilon^2 + 10y^4 y_\epsilon = 0$$

$$y_{\epsilon\epsilon\epsilon}\left(1 + 5\epsilon y^4\right) + y_{\epsilon\epsilon}\left(60\epsilon y^3 y_\epsilon + 15y^4\right)$$
$$+ 60\epsilon y_\epsilon^3 y^2 + 60y^3 y_\epsilon^2 = 0.$$

At $\epsilon = 0$ we have

$$y(x, 0) = \frac{1}{2}$$

$$y_\epsilon(x, 0) = -\frac{1}{32}$$

$$y_{\epsilon\epsilon}(x, 0) = -\frac{10}{16}\left(-\frac{1}{32}\right) = \frac{5}{16^2}$$

$$y_{\epsilon\epsilon\epsilon}(x, 0) = -\frac{5}{16^2}\left(\frac{15}{16}\right) - \frac{60}{8}\left(-\frac{1}{32}\right)^2 = -\frac{105}{4096}.$$

For $\epsilon = \dfrac{1}{100}$ we have

$$y = \frac{1}{2} - \frac{1}{32} \times \frac{1}{100} + \frac{5}{256} \times \frac{1}{2 \times 100^2}$$
$$- \frac{105}{4096} \times \frac{1}{6 \times 100^3} + \cdots$$
$$\approx 0.49968847$$

with error less than 10^{-8} in magnitude.

Section 13.6   Newton's Method   (page 828)

For each of Exercises 1–6, and 9, we sketch the graphs of the two given equations, $f(x, y) = 0$ and $g(x, y) = 0$, and use their intersections to make initial guesses x_0 and y_0 for the solutions. These guesses are then refined using the formulas

$$x_{n+1} = x_n - \frac{fg_2 - gf_2}{f_1g_2 - g_1f_2}\bigg|_{(x_n,y_n)}, \qquad y_{n+1} = y_n - \frac{f_1g - g_1f}{f_1g_2 - g_1f_2}\bigg|_{(x_n,y_n)}.$$

NOTE: The numerical values in the tables below were obtained by programming a microcomputer to calculate the iterations of the above formulas. In most cases the computer was using more significant digits than appear in the tables, and did not truncate the values obtained at one step before using them to calculate the next step. If you use a calculator, and use the numbers as quoted on one line of a table to calculate the numbers on the next line, your results may differ slightly (in the last one or two decimal places).

2.

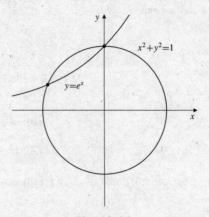

Fig. 13.6.2

$$f(x, y) = x^2 + y^2 - 1 \qquad f_1(x, y) = 2x \qquad g_1(x, y) = -e^x.$$
$$g(x, y) = y - e^x \qquad f_2(x, y) = 2y \qquad g_2(x, y) = 1$$

Evidently one solution is $x = 0$, $y = 1$. The second solution is near $(-1, 0)$. We try $x_0 = -0.9$, $y_0 = 0.2$.

n	x_n	y_n	$f(x_n, y_n)$	$g(x_n, y_n)$
0	−0.9000000	0.2000000	−0.1500000	−0.2065697
1	−0.9411465	0.3898407	0.0377325	−0.0003395
2	−0.9170683	0.3995751	0.0006745	−0.0001140
3	−0.9165628	0.3998911	0.0000004	−0.0000001
4	−0.9165626	0.3998913	0.0000000	0.0000000

The second solution is $x = -0.9165626$, $y = 0.3998913$.

4.

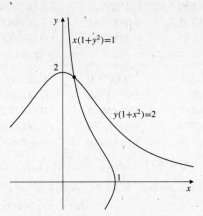

Fig. 13.6.4

$$f(x,y) = x(1+y^2) - 1 \qquad f_1(x,y) = 1 + y^2 \qquad g_1(x,y) = 2xy$$
$$g(x,y) = y(1+x^2) - 2 \qquad f_2(x,y) = 2xy \qquad g_2(x,y) = 1 + x^2$$

The solution appears to be near $x = 0.2$, $y = 1.8$.

n	x_n	y_n	$f(x_n, y_n)$	$g(x_n, y_n)$
0	0.2000000	1.8000000	−0.1520000	−0.1280000
1	0.2169408	1.9113487	0.0094806	0.0013031
2	0.2148268	1.9117785	−0.0000034	0.0000081
3	0.2148292	1.9117688	0.0000000	0.0000000

The solution is $x = 0.2148292$, $y = 1.9117688$.

6.

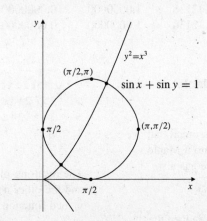

Fig. 13.6.6

$$f(x,y) = \sin x + \sin y - 1 \qquad f_1(x,y) = \cos x \qquad g_1(x,y) = -3x^2.$$
$$g(x,y) = y^2 - x^3 \qquad f_2(x,y) = \cos y \qquad g_2(x,y) = 2y$$

There are infinitely many solutions for the given pair of equations, since the level curve of $f(x,y) = 0$ is repeated periodically throughout the plane. We will find the two solutions closest to the origin in the first quadrant. From the figure, it appears that these solutions are near $(0.6, 0.4)$ and $(2, 3)$.

n	x_n	y_n	$f(x_n, y_n)$	$g(x_n, y_n)$
0	0.6000000	0.4000000	−0.0459392	−0.0560000
1	0.5910405	0.4579047	−0.0007050	0.0032092

n	x_n	y_n	$f(x_n, y_n)$	$g(x_n, y_n)$
2	0.5931130	0.4567721	−0.0000015	−0.0000063
3	0.5931105	0.4567761	0.0000000	0.0000000
4	0.5931105	0.4567761	0.0000000	0.0000000

n	x_n	y_n	$f(x_n, y_n)$	$g(x_n, y_n)$
0	2.0000000	3.0000000	0.0504174	1.0000000
1	2.0899016	3.0131366	−0.0036336	−0.0490479
2	2.0854887	3.0116804	−0.0000086	−0.0001199
3	2.0854779	3.0116770	0.0000000	0.0000000
4	2.0854779	3.0116770	0.0000000	0.0000000

The solutions are $x = 0.5931105$, $y = 0.4567761$, and $x = 2.0854779$, $y = 3.0116770$.

8.

$f(x, y, z) = y^2 + z^2 - 3$	$g(x, y, z) = x^2 + z^2 - 2$	$h(x, y, z) = x^2 - z$
$f_1(x, y, z) = 0$	$g_1(x, y, z) = 2x$	$h_1(x, y, z) = 2x$
$f_2(x, y, z) = 2y$	$g_2(x, y, z) = 0$	$h_2(x, y, z) = 0$
$f_3(x, y, z) = 2z$	$g_3(x, y, z) = 2z$	$h_3(x, y, z) = -1$

It is easily seen that the system

$$f(x, y, z) = 0, \qquad g(x, y, z) = 0, \qquad h(x, y, z) = 0$$

has first-quadrant solution $x = z = 1$, $y = \sqrt{2}$. Let us start at the "guess" $x_0 = y_0 = z_0 = 2$.

n	x_n	y_n	z_n	$f(x_n, y_n, z_n)$	$g(x_n, y_n, z_n)$	$h(x_n, y_n, z_n)$
0	2.0000000	2.0000000	2.0000000	5.0000000	6.0000000	2.0000000
1	1.3000000	1.5500000	1.2000000	0.8425000	1.1300000	0.4900000
2	1.0391403	1.4239564	1.0117647	0.0513195	0.1034803	0.0680478
3	1.0007592	1.4142630	1.0000458	0.0002313	0.0016104	0.0014731
4	1.0000003	1.4142136	1.0000000	0.0000000	0.0000006	0.0000006
5	1.0000000	1.4142136	1.0000000	0.0000000	0.0000000	0.0000000

Section 13.7 Calculations with Maple (page 833)

For the problems in this Section it is assumed that the file `newton.def` has been read into a Maple worksheet, and that the `plots` package has also been loaded.

2. The equation $y = \sin z$ can be used to reduce the given system of three equations in three variables to a system of 2 equations in two variables:

$$x^4 + \sin^2 z + z^2 = 1$$
$$z + z^3 + z^4 = x + \sin z.$$

The first equation can only be satisfied by points (x, z) satisfying $|x| \le 1$ and $|z| \le 1$. We use implicitplot to locate suitable starting points for `newtroot`.

```
> implicitplot(
>   {x^2+(sin(z))^2+z^2-1,
>   z+z^3+z^4=x+sin(z}, x=-1..1, z=-
1..1);
```

The resulting plot shows two roots, one near $(.6, .9)$ and the other near $(-.2, -.7)$. We prepare a vector-valued function whose zeros are the desired roots.

```
> u := (x,y) ->
>   [x^2+(sin(z))^2+z^2-
1,z+z^3+z^4=x+sin(z)];
```

The following command locates the root near $(.6, .9)$.

```
> newtroot(u,[.6,.9],10,.00001);
```

The result is 5 iterations (which we won't reproduce here) and the final returned value

[.5976010418, .6862588847]

The corresponding value of y can be found from $y = \sin(z)$:

```
> y := sin(%[2]);
        y := .6336474142
```

Thus one solution of the given system, rounded to 5 decimal places, is $(.59760, .63365, .68626)$.

The following command locates the root near $(-.2, -.7)$.

```
> newtroot(u,[-.2,-.7],10,.00001);
```

The result is 3 iterations (which we won't reproduce here) and the final returned value
$$[-.1707125527, -.7387417742]$$

The corresponding value of z can be found from $z = xy$:

```
> y := sin(%[2]);
        z := y := -.6733582179
```

Thus the other solution of the given system, rounded to 5 decimal places, is $(-.17071, -.67336, -.73874)$.

4. $f(x, y, z) = 1 - 10x^4 - 8y^4 - 7z^4$

$g(x, y, z) = f(x, y, z) + yz - xyz - x - 2y + z$

Since $g(0, 0, 0) = 1$ and $g \to -\infty$ as $x^2 + y^2 + z^2$ increases, the maximum value of g will be near $(0, 0, 0)$.

```
> g := (x,y,z) -> 1-10*x^4-8*y^4
>      -7*z^4+y*z-x*y*z-x-2*y+z;
```

We can try

```
> newtcp(g,[a,b,c],20,.00001);
```

for various choices of a, b, and c in the interval $[-1, 1]$. Some choices (for instance $a = b = c = 0$) yield errors; many others fail to converge within the allowed 20 iterations. Eventually, you will get ones that converge within 20 iterations. For instance

```
> newtcp(g,[-.25,0,0],20,.00001);
```
converges in 9 iterations to yield
$$[-.282429, -.372953, .265109], 1.91367$$

The procedure also yields three negative eigenvalues for the Hessian, indicating a local (at least) maximum. More trials all either fail to converge or converge to this same point. The absolute maximum value of g is 1.91367 (to five decimal places).

6. First define the function:

```
> f := (x,y,z) ->
>   (x+y-z+.1)/(1+x^2+y^2);
```

Since $f(x, y, z) \to 0$ as $x^2 + y^2 + z^2 \to \infty$ we expect f to have maximum and minimum values in some neighbourhood of the origin. If the ".1" term in the numerator were not present, we would expect the extreme values to occur along the line $x = y = -z$. Accordingly, we use starting points along this line.

```
> newtcp(f,[1,1,-1],10,.00001);
```

fails to converge, but

```
> newtcp(f,[.5,.5,-.5],10,.00001);
```

converges in 4 iterations to yield a maximum value of $.91747$ at $(.54498, .54498, -.54498)$. Similarly,

```
> newtcp(f,[-.5,-.5,.5],10,.00001);
```

converges in 4 iterations to yield a minimum value of $-.81747$ at $(-.61165, -.61165, .61165)$. Other starting points either result in the same critical points or fail to converge.

Review Exercises 13 (page 834)

2. $f(x, y) = x^2y - 2xy^2 + 2xy$

$f_1(x, y) = 2xy - 2y^2 + 2y = 2y(x - y + 1)$

$f_2(x, y) = x^2 - 4xy + 2x = x(x - 4y + 2)$

$A = f_{11} = 2y$

$B = f_{12} = 2x - 4y + 2$

$C = f_{22} = -4x$.

For CP: either $y = 0$ or $x - y + 1 = 0$, and either $x = 0$ or $x - 4y + 2 = 0$. The CPs are $(0, 0)$, $(0, 1)$, $(-2, 0)$, and $(-2/3, 1/3)$.

CP	A	B	C	$AC - B^2$	class
$(0,0)$	0	2	0	-4	saddle
$(0,1)$	2	-2	0	-4	saddle
$(-2,0)$	0	-2	8	-4	saddle
$(-\frac{2}{3}, \frac{1}{3})$	$\frac{2}{3}$	$-\frac{2}{3}$	$\frac{8}{3}$	$\frac{4}{3}$	loc. min

4. $f(x,y) = x^2 y(2 - x - y) = 2x^2 y - x^3 y - x^2 y^2$

$f_1(x,y) = 4xy - 3x^2 y - 2xy^2 = xy(4 - 3x - 2y)$

$f_2(x,y) = 2x^2 - x^3 - 2x^2 y = x^2(2 - x - 2y)$

$A = f_{11} = 4y - 6xy - 2y^2$

$B = f_{12} = 4x - 3x^2 - 4xy$

$C = f_{22} = -2x^2$.

$(0,y)$ is a CP for any y. If $x \neq 0$ but $y = 0$, then $x = 2$ from the second equation. Thus $(2,0)$ is a CP.

If neither x nor y is 0, then $x + 2y = 2$ and $3x + 2y = 4$, so that $x = 1$ and $y = 1/2$. The third CP is $(1, 1/2)$.

CP	A	B	C	$AC - B^2$	class
$(0,y)$	$4y - 2y^2$	0	0	0	?
$(2,0)$	0	-4	-8	-16	saddle
$(1, \frac{1}{2})$	$-\frac{3}{2}$	-1	-2	2	loc. max

The second derivative test is unable to classify the line of critical points along the y-axis. However, direct inspection of $f(x,y)$ shows that these are local minima if $y(2 - y) > 0$ (that is, if $0 < y < 2$) and local maxima if $y(2 - y) < 0$ (that is, if $y < 0$ or $y > 2$). The points $(0,0)$ and $(0,2)$ are neither maxima nor minima, so they are saddle points.

6. $x^2 + y^2 + z^2 - xy - xz - yz$

$= \frac{1}{2}\left[(x^2 - 2xy + y^2) + (x^2 - 2xz + z^2)\right.$

$\left. + (y^2 - 2yz + z^2)\right]$

$= \frac{1}{2}\left[(x - y)^2 + (x - z)^2 + (y - z)^2\right] \geq 0.$

The minimum value, 0, is assumed at the origin and at all points of the line $x = y = z$.

8. $f(x,y) = (4x^2 - y^2)e^{-x^2 + y^2}$

$f_1(x,y) = e^{-x^2 + y^2} 2x(4 - 4x^2 + y^2)$

$f_2(x,y) = e^{-x^2 + y^2}(-2y)(1 - 4x^2 + y^2).$

f has CPs $(0,0)$, $(\pm 1, 0)$. $f(0,0) = 0$.

$f(\pm 1, 0) = 4/e$.

a) Since $f(0,y) = -y^2 e^{y^2} \to -\infty$ as $y \to \pm\infty$, and since $f(x,x) = 3x^2 e^0 = 3x^2 \to \infty$ as $x \to \pm\infty$, f does not have a minimum or a maximum value on the xy-plane.

b) On $y = 3x$, $f(x, 3x) = -5x^2 e^{8x^2} \to -\infty$ as $x \to \infty$. Thus f can have no minimum value on the wedge $0 \leq y \leq 3x$. However, as noted in (a), $f(x,x) \to \infty$ as $x \to \infty$. Since (x,x) is in the wedge for $x > 0$, f cannot have a maximum value on the wedge either.

10.   Let the length, width, and height of the box be x, y, and z in, respectively. Then the girth is $g = 2x + 2y$. We require $g + z \leq 120$ in. The volume $V = xyz$ of the box will be maximized under the constraint $2x + 2y + z = 120$, so we look for CPs of

$$L = xyz + \lambda(2x + 2y + z - 120).$$

For CPs:

$$0 = \frac{\partial L}{\partial x} = yz + 2\lambda \qquad (A)$$

$$0 = \frac{\partial L}{\partial y} = xz + 2\lambda \qquad (B)$$

$$0 = \frac{\partial L}{\partial z} = xy + \lambda \qquad (C)$$

$$0 = \frac{\partial L}{\partial \lambda} = 2x + 2y + z - 120. \qquad (D)$$

Comparing (A), (B), and (C), we see that $x = y = z/2$. Then (D) implies that $3z = 120$, so $z = 40$ and $x = y = 20$ in. The largest box has volume

$$V = (20)(20)(40) = 16,000 \text{ in}^3,$$

or, about 9.26 cubic feet.

12.   The ellipsoid $(x/a)^2 + (y/b)^2 + (z/c)^2 = 1$ contains the rectangle $-1 \leq x \leq 1$, $-2 \leq y \leq 2$, $-3 \leq z \leq 3$, provided $(1/a^2) + (4/b^2) + (9/c^2) = 1$. The volume of the ellipsoid is $V = 4\pi abc/3$. We minimize V by looking for critical points of

$$L = \frac{4\pi}{3}abc + \lambda\left(\frac{1}{a^2} + \frac{4}{b^2} + \frac{9}{c^2} - 1\right).$$

For CPs:

$$0 = \frac{\partial L}{\partial a} = \frac{4\pi}{3}bc - \frac{2\lambda}{a^3} \qquad (A)$$

$$0 = \frac{\partial L}{\partial b} = \frac{4\pi}{3}ac - \frac{8\lambda}{b^3} \qquad (B)$$

$$0 = \frac{\partial L}{\partial c} = \frac{4\pi}{3}ab - \frac{18\lambda}{c^3} \qquad (C)$$

$$0 = \frac{\partial L}{\partial \lambda} = \frac{1}{a^2} + \frac{4}{b^2} + \frac{9}{c^2} - 1. \qquad (D)$$

Multiplying (A) by a, (B) by b, and (C) by c, we obtain $2\lambda/a^2 = 8\lambda/b^2 = 18\lambda/c^2$, so that either $\lambda = 0$ or $b = 2a$, $c = 3a$. Now $\lambda = 0$ implies $bc = 0$, which is inconsistent with (D). If $b = 2a$ and $c = 3a$, then (D) implies that $3/a^2 = 1$, so $a = \sqrt{3}$. The smallest volume of the ellipsoid is

$$V = \frac{4\pi}{3}(\sqrt{3})(2\sqrt{3})(3\sqrt{3}) = 24\sqrt{3}\pi \text{ cubic units.}$$

14.

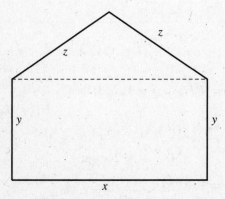

Fig. R-13.14

The area of the window is

$$A = xy + \frac{x}{2}\sqrt{z^2 - \frac{x^2}{4}},$$

or, since $x + 2y + 2z = L$,

$$A = \frac{x}{2}\left(L - x - 2z + \sqrt{z^2 - \frac{x^2}{4}}\right).$$

For maximum A, we look for critical points:

$$0 = \frac{\partial A}{\partial x} = \frac{1}{2}\left(L - x - 2z + \sqrt{z^2 - \frac{x^2}{4}}\right)$$
$$+ \frac{x}{2}\left(-1 - \frac{x}{4\sqrt{z^2 - \frac{x^2}{4}}}\right)$$
$$= \frac{L}{2} - x - z + \frac{2z^2 - x^2}{4\sqrt{z^2 - \frac{x^2}{4}}} \qquad \text{(A)}$$

$$0 = \frac{\partial A}{\partial z} = -x + \frac{xz}{2\sqrt{z^2 - \frac{x^2}{4}}}. \qquad \text{(B)}$$

Now (B) implies that either $x = 0$ or $z = 2\sqrt{z^2 - (x^2/4)}$. But $x = 0$ gives zero area rather than maximum area, so the second alternative must hold, and it implies that $z = x/\sqrt{3}$. Then (A) gives

$$\frac{L}{2} = \left(1 + \frac{1}{\sqrt{3}}\right)x + \frac{x}{2\sqrt{3}},$$

from which we obtain $x = L/(2 + \sqrt{3})$. The maximum area of the window is, therefore,

$$A\bigg|_{x=\frac{L}{2+\sqrt{3}},\ z=\frac{L/\sqrt{3}}{2+\sqrt{3}}} = \frac{1}{4}\frac{L^2}{2 + \sqrt{3}}$$
$$\approx 0.0670L^2 \text{ sq. units.}$$

16. The envelope of $y = (x - c)^3 + 3c$ is found by eliminating c from that equation and

$$0 = \frac{\partial}{\partial c}[(x - c)^3 + 3c] = -3(x - c)^2 + 3.$$

This later equation implies that $(x - c)^2 = 1$, so $x - c = \pm 1$.
The envelope is $y = (\pm 1)^3 + 3(x \mp 1)$, or $y = 3x \pm 2$.

18. a) $G(y) = \int_0^\infty \frac{\tan^{-1}(xy)}{x}\,dx$

$G'(y) = \int_0^\infty \frac{1}{x}\frac{x}{1 + x^2 y^2}\,dx$ Let $u = xy$
$\qquad\qquad\qquad\qquad\qquad\qquad\qquad\quad du = y\,dx$

$\qquad = \frac{1}{y}\int_0^\infty \frac{du}{1 + u^2} = \frac{\pi}{2y}$ for $y > 0$.

b) $\int_0^\infty \frac{\tan^{-1}(\pi x) - \tan^{-1} x}{x}\,dx$

$\qquad = G(\pi) - G(1) = \int_1^\pi G'(y)\,dy = \frac{\pi}{2}\int_1^\pi \frac{dy}{y} = \frac{\pi \ln \pi}{2}.$

Challenging Problems 13   (page 834)

2. If $f(x) = \begin{cases} 0 & \text{for } -\pi \leq x < 0 \\ x & \text{for } 0 \leq x \leq \pi \end{cases}$, then

$$a_0 = \frac{1}{\pi} \int_0^\pi x\, dx = \frac{\pi}{2}$$

$$a_k = \frac{1}{\pi} \int_0^\pi x \cos kx\, dx$$

$$\begin{aligned} U = x \qquad & dV = \cos kx\, dx \\ dU = dx \qquad & V = \frac{1}{k} \sin kx \end{aligned}$$

$$= \frac{1}{\pi k} \left(x \sin kx \Big|_0^\pi - \int_0^\pi \sin kx\, dx \right)$$

$$= \frac{\cos k\pi - 1}{\pi k^2} = \begin{cases} 0 & \text{if } k \text{ is even} \\ -\dfrac{2}{\pi k^2} & \text{if } k \text{ is odd} \end{cases}$$

$$b_k = \frac{1}{\pi} \int_0^\pi x \sin kx\, dx$$

$$\begin{aligned} U = x \qquad & dV = \sin kx\, dx \\ dU = dx \qquad & V = -\frac{1}{k} \cos kx \end{aligned}$$

$$= -\frac{1}{\pi k} \left(x \cos kx \Big|_0^\pi - \int_0^\pi \cos kx\, dx \right)$$

$$= \frac{(-1)^{k+1}}{k}.$$

Because of the properties of trigonometric integrals listed in the solution to Problem 1,

$$\int_{-\pi}^\pi \left(\frac{a_0}{2} + \sum_{k=1}^n (a_k \cos kx + b_k \sin kx) \right)^2 dx$$

$$= \frac{\pi a_0^2}{2} + \pi \sum_{k=0}^n (a_k^2 + b_k^2)$$

$$\int_{-\pi}^\pi f(x) \left(\frac{a_0}{2} + \sum_{k=1}^n (a_k \cos kx + b_k \sin kx) \right) dx$$

$$= \frac{\pi a_0^2}{2} + \pi \sum_{k=0}^n (a_k^2 + b_k^2).$$

Therefore

$$I_n = \int_{-\pi}^\pi \left[f(x) - \left(\frac{a_0}{2} + \sum_{k=1}^n (a_k \cos kx + b_k \sin kx) \right) \right]^2 dx$$

$$= \int_{-\pi}^\pi (f(x))^2\, dx - 2 \left(\frac{\pi a_0^2}{2} + \pi \sum_{k=0}^n (a_k^2 + b_k^2) \right)$$

$$+ \frac{\pi a_0^2}{2} + \pi \sum_{k=0}^n (a_k^2 + b_k^2)$$

$$= \int_{-\pi}^\pi (f(x))^2\, dx - \left(\frac{\pi a_0^2}{2} + \pi \sum_{k=0}^n (a_k^2 + b_k^2) \right).$$

In fact, it can be shown that $I_n \to 0$ as $n \to \infty$.

4.

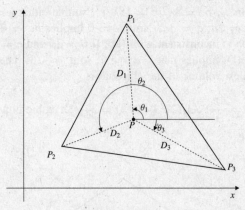

Fig. C-13.4

If $D_i = |PP_i|$ for $i = 1, 2, 3$, then

$$D_i^2 = (x - x_i)^2 + (y - y_i)^2$$

$$2D_i \frac{\partial D_i}{\partial x} = 2(x - x_i)$$

$$\frac{\partial D_i}{\partial x} = \frac{x - x_i}{D_i} = \cos \theta_i$$

where θ_i is the angle between $\overrightarrow{PP_i}$ and **i**. Similarly $\partial D_i / \partial y = \sin \theta_i$. To minimize $S = D_1 + D_2 + D_3$ we look for critical points:

$$0 = \frac{\partial S}{\partial x} = \cos \theta_1 + \cos \theta_2 + \cos \theta_3$$

$$0 = \frac{\partial S}{\partial y} = \sin \theta_1 + \sin \theta_2 + \sin \theta_3.$$

Thus $\cos \theta_1 + \cos \theta_2 = -\cos \theta_3$ and $\sin \theta_1 + \sin \theta_2 = -\sin \theta_3$. Squaring and adding these two equations we get

$$2 + 2(\cos \theta_1 \cos \theta_2 + \sin \theta_1 \sin \theta_2) = 1,$$

or $\cos(\theta_1 - \theta_2) = -1/2$. Thus $\theta_1 - \theta_2 = \pm 2\pi/3$. Similarly $\theta_1 - \theta_3 = \theta_2 - \theta_3 = \pm 2\pi/3$. Thus P should be chosen so that $\overrightarrow{PP_1}$, $\overrightarrow{PP_2}$, and $\overrightarrow{PP_3}$ make $120°$ angles with each other. This is possible only if all three angles of the triangle are less than $120°$. If the triangle has an angle of $120°$ or more (say at P_1), then P should be that point on the side P_2P_3 such that $PP_1 \perp P_2P_3$.

CHAPTER 14. MULTIPLE INTEGRATION

Section 14.1 Double Integrals (page 841)

2. $R = 1 \times \big[f(1,1) + f(1,2) + f(2,1) + f(2,2)$
$\qquad + f(3,1) + f(3,2) \big]$
$\quad = 3 + 2 + 2 + 1 + 1 + 0 = 9$

4. $R = 1 \times \big[f(1,0) + f(1,1) + f(2,0) + f(2,1)$
$\qquad + f(3,0) + f(3,1) \big]$
$\quad = 4 + 3 + 3 + 2 + 2 + 1 = 15$

6. $I = \iint_D (5 - x - y)\, dA$ is the volume of the solid in the figure.

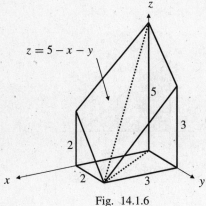

Fig. 14.1.6

The solid is split by the vertical plane through the z-axis and the point $(3, 2, 0)$ into two pyramids, each with a trapezoidal base; one pyramid's base is in the plane $y = 0$ and the other's is in the plane $z = 0$. I is the sum of the volumes of these pyramids:

$$I = \frac{1}{3} \left(\frac{5+2}{2}(3)(2) \right) + \frac{1}{3} \left(\frac{5+3}{2}(2)(3) \right) = 15.$$

8. $R = 4 \times 1 \times [4 + 4 + 4 + 3 + 0] = 60$

10. $J = \text{area of disk} = \pi(5^2) \approx 78.54$

12. $f(x, y) = x^2 + y^2$
$\quad R = 4 \times 1 \times \big[f(\tfrac{1}{2}, \tfrac{1}{2}) + f(\tfrac{3}{2}, \tfrac{1}{2}) + f(\tfrac{5}{2}, \tfrac{1}{2}) + f(\tfrac{7}{2}, \tfrac{1}{2})$
$\qquad + f(\tfrac{9}{2}, \tfrac{1}{2}) + f(\tfrac{1}{2}, \tfrac{3}{2}) + f(\tfrac{3}{2}, \tfrac{3}{2}) + f(\tfrac{5}{2}, \tfrac{3}{2})$
$\qquad + f(\tfrac{7}{2}, \tfrac{3}{2}) + f(\tfrac{9}{2}, \tfrac{3}{2})$
$\qquad + f(\tfrac{1}{2}, \tfrac{5}{2}) + f(\tfrac{3}{2}, \tfrac{5}{2}) + f(\tfrac{5}{2}, \tfrac{5}{2}) + f(\tfrac{7}{2}, \tfrac{5}{2})$
$\qquad + f(\tfrac{1}{2}, \tfrac{7}{2}) + f(\tfrac{3}{2}, \tfrac{7}{2}) + f(\tfrac{5}{2}, \tfrac{7}{2}) + f(\tfrac{1}{2}, \tfrac{9}{2}) + f(\tfrac{3}{2}, \tfrac{9}{2}) \big]$
$\quad = 918$

14. $\displaystyle \iint_D (x + 3)\, dA = \iint_D x\, dA + 3 \iint_D dA$
$\qquad = 0 + 3(\text{area of } D)$
$\qquad = 3 \times \frac{\pi 2^2}{2} = 6\pi.$

The integral of x over D is zero because D is symmetrical about $x = 0$.

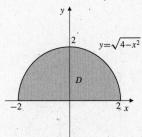

Fig. 14.1.14

16. $\displaystyle \iint_{|x|+|y|\leq 1} \big(x^3 \cos(y^2) + 3\sin y - \pi \big)\, dA$
$\quad = 0 + 0 - \pi \big(\text{area bounded by } |x| + |y| = 1$
$\quad = -\pi \times 4 \times \frac{1}{2}(1)(1) = -2\pi.$

(Each of the first two terms in the integrand is an odd function of one of the variables, and the square is symmetrical about each coordinate axis.)

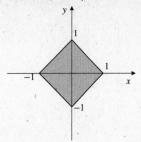

Fig. 14.1.16

18. $\displaystyle \iint_{x^2 + y^2 \leq a^2} \sqrt{a^2 - x^2 - y^2}\, dA$

$= \text{volume of hemisphere shown in the figure}$

$= \frac{1}{2} \left(\frac{4}{3}\pi a^3 \right) = \frac{2}{3}\pi a^3.$

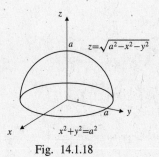

Fig. 14.1.18

271

20. By the symmetry of S with respect to x and y we have

$$\iint_S (x+y)\, dA = 2\iint_S x\, dA$$

$$= 2\times \text{(volume of wedge shown in the figure)}$$

$$= 2\times \frac{1}{2}(a^2)a = a^3.$$

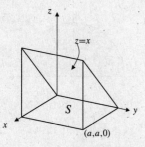

Fig. 14.1.20

22. $\displaystyle \iint_R \sqrt{b^2-y^2}\, dA$

$= $ volume of the quarter cylinder shown in the figure

$$= \frac{1}{4}(\pi b^2)a = \frac{1}{4}\pi ab^2.$$

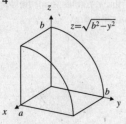

Fig. 14.1.22

Section 14.2 Iteration of Double Integrals in Cartesian Coordinates (page 849)

2. $\displaystyle \int_0^1\int_0^y (xy+y^2)\, dx\, dy$

$$= \int_0^1 \left(\frac{x^2 y}{2}+xy^2\right)\Bigg|_{x=0}^{x=y} dy$$

$$= \frac{3}{2}\int_0^1 y^3\, dy = \frac{3}{8}.$$

4. $\displaystyle \int_0^2 dy \int_0^y y^2 e^{xy}\, dx$

$$= \int_0^2 y^2\, dy \left(\frac{1}{y}e^{xy}\Big|_{x=0}^{x=y}\right).$$

$$= \int_0^2 y(e^{y^2}-1)\, dy = \frac{e^{y^2}-y^2}{2}\Bigg|_0^2 = \frac{e^4-5}{2}.$$

6. $\displaystyle \iint_R x^2 y^2\, dA = \int_0^a x^2\, dx \int_0^b y^2\, dy$

$$= \frac{a^3}{3}\frac{b^3}{3} = \frac{a^3 b^3}{9}.$$

8. $\displaystyle \iint_T (x-3y)\, dA = \int_0^a dx \int_0^{b(1-(x/a))} (x-3y)\, dy$

$$= \int_0^a dx \left(xy-\frac{3}{2}y^2\right)\Bigg|_{y=0}^{y=b(1-(x/a))}$$

$$= \int_0^a \left[b\left(x-\frac{x^2}{a}\right)-\frac{3}{2}b^2\left(1-\frac{2x}{a}+\frac{x^2}{a^2}\right)\right] dx$$

$$= \left(b\frac{x^2}{2}-\frac{b}{a}\frac{x^3}{3}-\frac{3}{2}b^2 x+\frac{3}{2}\frac{b^2 x^2}{a}-\frac{1}{2}\frac{b^2 x^3}{a^2}\right)\Bigg|_0^a$$

$$= \frac{a^2 b}{6}-\frac{ab^2}{2}.$$

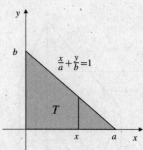

Fig. 14.2.8

10. $\displaystyle \iint_D x\cos y\, dA$

$$= \int_0^1 x\, dx \int_0^{1-x^2} \cos y\, dy$$

$$= \int_0^1 x\, dx\, (\sin y)\Big|_{y=0}^{y=1-x^2}$$

$$= \int_0^1 x\sin(1-x^2)\, dx \quad \text{Let } u = 1-x^2$$

$$\qquad\qquad\qquad\qquad\qquad du = -2x\, dx$$

$$= -\frac{1}{2}\int_1^0 \sin u\, du = \frac{1}{2}\cos u\Big|_1^0 = \frac{1-\cos(1)}{2}.$$

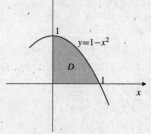

Fig. 14.2.10

12.
$$\iint_T \sqrt{a^2 - y^2}\, dA = \int_0^a \sqrt{a^2 - y^2}\, dy \int_y^a dx$$
$$= \int_0^a (a - y)\sqrt{a^2 - y^2}\, dy$$
$$= a \int_0^a \sqrt{a^2 - y^2}\, dy - \int_0^a y\sqrt{a^2 - y^2}\, dy$$

$$\text{Let } u = a^2 - y^2$$
$$du = -2y\, dy$$

$$= a\,\frac{\pi a^2}{4} + \frac{1}{2}\int_{a^2}^0 u^{1/2}\, du$$
$$= \frac{\pi a^3}{4} - \frac{1}{3}u^{3/2}\Big|_0^{a^2} = \left(\frac{\pi}{4} - \frac{1}{3}\right)a^3.$$

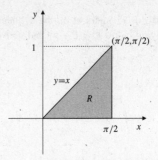

Fig. 14.2.16

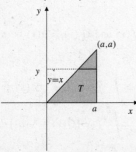

Fig. 14.2.12

14.
$$\iint_T \frac{xy}{1 + x^4}\, dA = \int_0^1 \frac{x}{1 + x^4}\, dx \int_0^x y\, dy$$
$$= \frac{1}{2}\int_0^1 \frac{x^3}{1 + x^4}\, dx$$
$$= \frac{1}{8}\ln(1 + x^4)\Big|_0^1 = \frac{\ln 2}{8}.$$

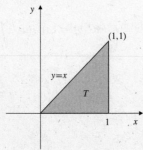

Fig. 14.2.14

18.
$$\int_0^1 dx \int_x^{x^{1/3}} \sqrt{1 - y^4}\, dy$$
$$= \iint_R \sqrt{1 - y^4}\, dA \quad (R \text{ as shown})$$
$$= \int_0^1 y\sqrt{1 - y^4}\, dy - \int_0^1 y^3\sqrt{1 - y^4}\, dy$$

$$\text{Let } u = y^2 \qquad \text{Let } v = 1 - y^4$$
$$du = 2y\, dy \qquad dv = -4y^3\, dy$$

$$= \frac{1}{2}\int_0^1 \sqrt{1 - u^2}\, du + \frac{1}{4}\int_1^0 v^{1/2}\, dv$$
$$= \frac{1}{2}\left(\frac{\pi}{4} \times 1^2\right) + \frac{1}{6}v^{3/2}\Big|_1^0 = \frac{\pi}{8} - \frac{1}{6}.$$

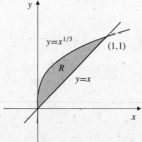

Fig. 14.2.18

16.
$$\int_0^{\pi/2} dy \int_y^{\pi/2} \frac{\sin x}{x}\, dx = \iint_R \frac{\sin x}{x}\, dA \quad (R \text{ as shown})$$
$$= \int_0^{\pi/2} \frac{\sin x}{x}\, dx \int_0^x dy = \int_0^{\pi/2} \sin x\, dx = 1.$$

20.
$$V = \int_0^1 dy \int_0^y (1 - x^2)\, dx$$
$$= \int_0^1 \left(y - \frac{y^3}{3}\right) dy = \frac{1}{2} - \frac{1}{12} = \frac{5}{12} \text{ cu. units.}$$

22. $z = 1 - y^2$ and $z = x^2$ intersect on the cylinder $x^2 + y^2 = 1$. The volume lying below $z = 1 - y^2$ and above $z = x^2$ is

$$V = \iint_{x^2 + y^2 \le 1} (1 - y^2 - x^2)\, dA$$

$$= 4 \int_0^1 dx \int_0^{\sqrt{1-x^2}} (1 - x^2 - y^2)\, dy$$

$$= 4 \int_0^1 dx \left. \left((1 - x^2)y - \frac{y^3}{3} \right) \right|_{y=0}^{y=\sqrt{1-x^2}}$$

$$= \frac{8}{3} \int_0^1 (1 - x^2)^{3/2}\, dx \quad \text{Let } x = \sin u$$
$$\hspace{4cm} dx = \cos u\, du$$

$$= \frac{8}{3} \int_0^{\pi/2} \cos^4 u\, du = \frac{2}{3} \int_0^{\pi/2} (1 + \cos 2u)^2\, du$$

$$= \frac{2}{3} \int_0^{\pi/2} \left(1 + 2\cos 2u + \frac{1 + \cos 4u}{2} \right) du$$

$$= \frac{2}{3} \frac{3}{2} \frac{\pi}{2} = \frac{\pi}{2} \text{ cu. units.}$$

24. $V = \int_0^{\pi^{1/4}} dy \int_0^y x^2 \sin(y^4)\, dx$

$$= \frac{1}{3} \int_0^{\pi^{1/4}} y^3 \sin(y^4)\, dy \quad \text{Let } u = y^4$$
$$\hspace{4cm} du = 4y^3\, dy$$

$$= \frac{1}{12} \int_0^{\pi} \sin u\, du = \frac{1}{6} \text{ cu. units.}$$

26. $\text{Vol} = \iint_T \left(2 - \frac{x}{a} - \frac{y}{b} \right) dA$

$$= \int_0^a dx \int_0^{b(1-(x/a))} \left(2 - \frac{x}{a} - \frac{y}{b} \right) dy$$

$$= \int_0^a \left[\left(2 - \frac{x}{a} \right) b \left(1 - \frac{x}{a} \right) - \frac{1}{2b} b^2 \left(1 - \frac{x}{a} \right)^2 \right] dx$$

$$= \frac{b}{2} \int_0^a \left(3 - \frac{4x}{a} + \frac{x^2}{a^2} \right) dx$$

$$= \frac{b}{2} \left. \left(3x - \frac{2x^2}{a} + \frac{x^3}{3a^2} \right) \right|_0^a = \frac{2}{3} ab \text{ cu. units.}$$

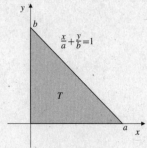

Fig. 14.2.26

28. The part of the plane $z = 8 - x$ lying inside the elliptic cylinder $x^2 = 2y^2 = 8$ lies above $z = 0$. The part of the plane $z = y - 4$ inside the cylinder lies below $z = 0$. Thus the required volume is

$$\text{Vol} = \iint_{x^2 + 2y^2 \le 8} \left(8 - x - (y - 4) \right) dA$$

$$= \iint_{x^2 + 2y^2 \le 8} 12\, dA \quad \text{(by symmetry)}$$

$$= 12 \times \text{area of ellipse } \frac{x^2}{8} + \frac{y^2}{4} = 1$$

$$= 12 \times \pi(2\sqrt{2})(2) = 48\sqrt{2}\pi \text{ cu. units.}$$

30. Since $F'(x) = f(x)$ and $G'(x) = g(x)$ on $a \le x \le b$, we have

$$\iint_T f(x)g(x)\, dA = \int_a^b f(x)\, dx \int_a^x G'(y)\, dy$$

$$= \int_a^b f(x) \left(G(x) - G(a) \right) dx$$

$$= \int_a^b f(x)G(x)\, dx - G(a)F(b) + G(a)F(a)$$

$$\iint_T f(x)g(x)\, dA = \int_a^b g(y)\, dy \int_y^b F'(x)\, dx$$

$$= \int_a^b g(y) \left(F(b) - F(y) \right) dy$$

$$= F(b)G(b) - F(b)G(a) - \int_a^b F(y)g(y)\, dx.$$

Thus

$$\int_a^b f(x)G(x)\, dx = F(b)G(b) - F(a)G(a) - \int_a^b g(y)F(y)\, dy.$$

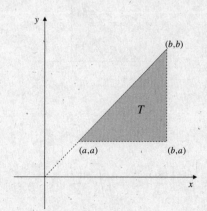

Fig. 14.2.30

Section 14.3 Improper Integrals and a Mean-Value Theorem (page 855)

2.
$$\iint_Q \frac{dA}{(1+x^2)(1+y^2)} = \int_0^\infty \frac{dx}{1+x^2} \int_0^\infty \frac{dy}{1+y^2}$$
$$= \left(\lim_{R\to\infty} (\tan^{-1}x)\Big|_0^R\right)^2 = \frac{\pi^2}{4}$$
(converges)

4.
$$\iint_T \frac{1}{x\sqrt{y}}\, dA = \int_0^1 \frac{dx}{x} \int_x^{2x} \frac{dy}{\sqrt{y}}$$
$$= \int_0^1 \frac{2(\sqrt{2x} - \sqrt{x})}{x}\, dx$$
$$= 2(\sqrt{2}-1)\int_0^1 \frac{dx}{\sqrt{x}} = 4(\sqrt{2}-1) \text{ (converges)}$$

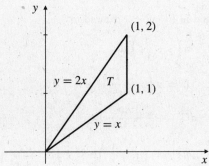

Fig. 14.3.4

6.
$$\iint_H \frac{dA}{1+x+y} = \int_0^\infty dx \int_0^1 \frac{1}{1+x+y}\, dy$$
$$= \int_0^\infty \left(\ln(1+x+y)\Big|_{y=0}^{y=1}\right) dx$$
$$= \int_0^\infty \ln\left(\frac{2+x}{1+x}\right) dx = \int_0^\infty \ln\left(1+\frac{1}{1+x}\right) dx.$$

Since $\displaystyle\lim_{u\to 0+} \frac{\ln(1+u)}{u} = 1$, we have $\ln(1+u) \geq u/2$ on some interval $(0, u_0)$. Therefore

$$\ln\left(1+\frac{1}{1+x}\right) \geq \frac{1}{2(1+x)}$$

on some interval (x_0, ∞), and

$$\int_0^\infty \ln\left(1+\frac{1}{1+x}\right) dx \geq \int_{x_0}^\infty \frac{1}{2(1+x)}\, dx,$$

which diverges to infinity. Thus the given double integral diverges to infinity by comparison.

8. On the strip S between the parallel lines $x + y = 0$ and $x + y = 1$ we have $e^{-|x+y|} = e^{-(x+y)} \geq 1/e$. Since S has infinite area,

$$\iint_S e^{-|x+y|}\, dA = \infty.$$

Since $e^{-|x+y|} > 0$ for all (x, y) in $\mathbb{R}^2$, we have

$$\iint_{\mathbb{R}^2} e^{-|x+y|}\, dA > \iint_S e^{-|x+y|}\, dA,$$

and the given integral diverges to infinity.

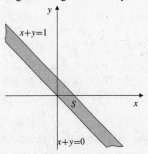

Fig. 14.3.8

10.
$$\iint_T \frac{dA}{x^2+y^2} = \int_1^\infty dx \int_0^x \frac{dy}{x^2+y^2}$$
$$= \int_1^\infty dx \left(\frac{1}{x}\tan^{-1}\frac{y}{x}\Big|_{y=0}^{y=x}\right)$$
$$= \frac{\pi}{4}\int_1^\infty \frac{dx}{x} = \infty$$
(The integral diverges to infinity.)

12.
$$\iint_R \frac{1}{x}\sin\frac{1}{x}\, dA = \int_{2/\pi}^\infty \frac{1}{x}\sin\frac{1}{x}\, dx \int_0^{1/x} dy$$
$$= \int_{2/\pi}^\infty \frac{1}{x^2}\sin\frac{1}{x}\, dx \quad \text{Let } u = 1/x$$
$$\qquad\qquad du = -1/x^2\, dx$$
$$= -\int_{\pi/2}^0 \sin u\, du = \cos u\Big|_{\pi/2}^0 = 1$$
(The integral converges.)

14.
$$\text{Vol} = \iint_S \frac{2xy}{x^2+y^2}\, dA$$
$$= 4\iint_T \frac{2xy}{x^2+y^2}\, dA \quad (T \text{ as in \#9(b)})$$
$$= 4\int_0^1 x\, dx \int_0^x \frac{y\, dy}{x^2+y^2} \quad \text{Let } u = x^2 + y^2$$
$$\qquad\qquad du = 2y\, dy$$
$$= 2\int_0^1 x\, dx \int_{x^2}^{2x^2} \frac{du}{u}$$
$$= 2\ln 2 \int_0^1 x\, dx = \ln 2 \text{ cu. units.}$$

16. $\iint_{D_k} y^b \, dA = \int_0^1 dx \int_0^{x^k} y^b \, dy = \int_0^1 \frac{x^{k(b+1)}}{b+1} \, dx$ if

$b > -1$. This latter integral converges if $k(b + 1) > -1$.
Thus, the given integral converges if $b > -1$ and
$k > -1/(b + 1)$.

18. $\iint_{R_k} \frac{dA}{y^b} = \int_1^\infty dx \int_0^{x^k} \frac{dy}{y^b} = \int_1^\infty \frac{x^{k(1-b)}}{1-b} \, dx$ if $b < 1$.

This latter integral converges if $k(1 - b) < -1$. Thus, the
given integral converges if $b < 1$ and $k < -1/(1 - b)$.

20. $\iint_{R_k} x^a y^b \, dA = \int_1^\infty x^a \, dx \int_0^{x^k} y^b \, dy = \int_1^\infty \frac{x^{a+(b+1)k}}{b+1} \, dx$,

if $b > -1$. This latter integral converges if
$a + (b + 1)k < -1$. Thus, the given integral converges if
$b > -1$ and $k < -(a + 1)/(b + 1)$.

22. The average value of x^2 over the rectangle R is

$$\frac{1}{(b-a)(d-c)} \cdot \iint_R x^2 \, dA$$

$$= \frac{1}{(b-a)(d-c)} \int_a^b x^2 \, dx \int_c^d dy$$

$$= \frac{1}{b-a} \frac{b^3 - a^3}{3} = \frac{a^2 + ab + b^2}{3}.$$

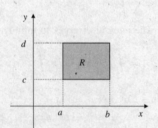

Fig. 14.3.22

24. The area of region R is

$$\int_0^1 (\sqrt{x} - x^2) \, dx = \frac{1}{3} \text{ sq. units.}$$

The average value of $1/x$ over R is

$$3 \iint_R \frac{dA}{x} = 3 \int_0^1 \frac{dx}{x} \int_{x^2}^{\sqrt{x}} dy$$

$$= 3 \int_0^1 \left(x^{-1/2} - x \right) \, dx = \frac{9}{2}.$$

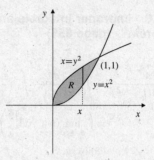

Fig. 14.3.24

26. Let R be the region $0 \le x < \infty$, $0 \le y \le 1/(1 + x^2)$. If
$f(x, y) = x$, then

$$\int_R f(x, y) \, dA = \int_0^\infty x \, dx \int_0^{1/(1+x^2)} dy = \int_0^\infty \frac{x \, dx}{1 + x^2}$$

which diverges to infinity. Thus f has no average value
on R.

28. The integral in Example 2 reduced to

$$\int_1^\infty \ln\left(1 + \frac{1}{x^2}\right) dx$$

$$\begin{aligned} U &= \ln\left(1 + \frac{1}{x^2}\right) & dV &= dx \\ & & V &= x \\ dU &= -\frac{2\,dx}{x(x^2+1)} & & \end{aligned}$$

$$= \lim_{R\to\infty} \left[x \ln\left(1 + \frac{1}{x^2}\right) \Big|_1^R + 2\int_1^R \frac{dx}{1+x^2} \right]$$

$$= 2\left(\frac{\pi}{2} - \frac{\pi}{4}\right) - \ln 2 + \lim_{R\to\infty} \frac{\ln\left(1 + (1/R^2)\right)}{1/R}$$

$$= \frac{\pi}{2} - \ln 2 + \lim_{R\to\infty} \frac{-(2/R^3)}{\left(1 + (1/R^2)\right)(-1/R^2)}$$

$$= \frac{\pi}{2} - \ln 2.$$

30. If $R = \{(x, y) : a \le x \le a + h,\ b \le y \le b + k\}$, then

$$\iint_R f_{12}(x, y) \, dA = \int_a^{a+h} dx \int_b^{b+k} f_{12}(x, y) \, dy$$

$$= \int_a^{a+h} \left[f_1(x, b+k) - f_1(x, b) \right] dx$$

$$= f(a+h, b+k) - f(a, b+k) - f(a+h, b) + f(a, b)$$

$$\iint_R f_{21}(x, y) \, dA = \int_b^{b+k} dy \int_a^{a+h} f_{21}(x, y) \, dx$$

$$= \int_b^{b+k} \left[f_2(a+h, y) - f_2(a, y) \right] dy$$

$$= f(a+h, b+k) - f(a+h, b) - f(a, b+k) + f(a, b).$$

Thus

$$\iint_R f_{12}(x, y) \, dA = \iint_R f_{21}(x, y) \, dA.$$

Divide both sides of this identity by hk and let $(h, k) \to (0, 0)$ to obtain, using the result of Exercise 31,

$$f_{12}(a, b) = f_{21}(a, b).$$

Section 14.4   Double Integrals in Polar Coordinates   (page 865)

2. $\displaystyle\iint_D \sqrt{x^2 + y^2}\, dA = \int_0^{2\pi} d\theta \int_0^a r\, r\, dr = \frac{2\pi a^3}{3}$

4. $\displaystyle\iint_D |x|\, dA = 4\int_0^{\pi/2} d\theta \int_0^a r\cos\theta\, r\, dr$

$$= 4\sin\theta \Big|_0^{\pi/2} \frac{a^3}{3} = \frac{4a^3}{3}$$

6. $\displaystyle\iint_D x^2 y^2\, dA = 4\int_0^{\pi/2} d\theta \int_0^a r^4 \cos^2\theta \sin^2\theta\, r\, dr$

$$= \frac{a^6}{6} \int_0^{\pi/2} \sin^2(2\theta)\, d\theta$$

$$= \frac{a^6}{12} \int_0^{\pi/2} \left(1 - \cos(4\theta)\right) d\theta = \frac{\pi a^6}{24}$$

8. $\displaystyle\iint_Q (x + y)\, dA = \frac{2a^3}{3}$; by symmetry, the value is twice that obtained in the previous exercise.

10. $\displaystyle\iint_Q \frac{2xy}{x^2 + y^2}\, dA = \int_0^{\pi/2} d\theta \int_0^a \frac{2r^2 \sin\theta\cos\theta}{r^2}\, r\, dr$

$$= \frac{a^2}{2} \int_0^{\pi/2} \sin(2\theta)\, d\theta = -\frac{a^2 \cos(2\theta)}{4} \Big|_0^{\pi/2} = \frac{a^2}{2}$$

12. $\displaystyle\iint_S x\, dA = 2\int_0^{\pi/4} d\theta \int_{\sec\theta}^{\sqrt{2}} r\cos\theta\, r\, dr$

$$= \frac{2}{3} \int_0^{\pi/4} \cos\theta \left(2\sqrt{2} - \sec^3\theta\right) d\theta$$

$$= \frac{4\sqrt{2}}{3} \sin\theta \Big|_0^{\pi/4} - \frac{2}{3}\tan\theta \Big|_0^{\pi/4}$$

$$= \frac{4}{3} - \frac{2}{3} = \frac{2}{3}$$

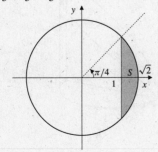

Fig. 14.4.12

14. $\displaystyle\iint_{x^2+y^2 \le 1} \ln(x^2 + y^2)\, dA = \int_0^{2\pi} d\theta \int_0^1 (\ln r^2) r\, dr$

$$= 4\pi \int_0^1 r\ln r\, dr$$

$$U = \ln r \qquad dV = r\, dr$$

$$dU = \frac{dr}{r} \qquad V = \frac{r^2}{2}$$

$$= 4\pi \left[\frac{r^2}{2}\ln r \Big|_0^1 - \frac{1}{2}\int_0^1 r\, dr \right]$$

$$= 4\pi \left[0 - 0 - \frac{1}{4} \right] = -\pi$$

(Note that the integral is improper, but converges since $\lim_{r\to 0+} r^2 \ln r = 0$.)

16. The annular region R: $0 < a \le \sqrt{x^2 + y^2} \le b$ has area $\pi(b^2 - a^2)$. The average value of $e^{-(x^2+y^2)}$ over the region is

$$\frac{1}{\pi(b^2 - a^2)} \iint_R e^{-(x^2+y^2)}\, dA$$

$$= \frac{1}{\pi(b^2 - a^2)} \int_0^{2\pi} d\theta \int_a^b e^{-r^2} r\, dr \quad \text{Let } u = r^2$$
$$\qquad\qquad\qquad\qquad\qquad\qquad\qquad du = 2r\, dr$$

$$= \frac{1}{\pi(b^2 - a^2)} (2\pi) \frac{1}{2} \int_{a^2}^{b^2} e^{-u}\, du$$

$$= \frac{1}{b^2 - a^2} \left(e^{-a^2} - e^{-b^2} \right).$$

18. $\displaystyle\iint_{\mathbb{R}^2} \frac{dA}{(1 + x^2 + y^2)^k}$

$$= \int_0^{2\pi} d\theta \int_0^\infty \frac{r\, dr}{(1 + r^2)^k} \quad \text{Let } u = 1 + r^2$$
$$\qquad\qquad\qquad\qquad\qquad\qquad du = 2r\, dr$$

$$= \pi \int_1^\infty u^{-k}\, du = \frac{-\pi}{1 - k} \text{ if } k > 1.$$

The integral converges to $\dfrac{\pi}{k - 1}$ if $k > 1$.

20. $\displaystyle\iint_C y\, dA = \int_0^\pi d\theta \int_0^{1+\cos\theta} r\sin\theta\, r\, dr$

$$= \frac{1}{3} \int_0^\pi \sin\theta (1 + \cos\theta)^3\, d\theta \quad \text{Let } u = 1 + \cos\theta$$
$$\qquad\qquad\qquad\qquad\qquad\qquad\qquad du = -\sin\theta\, d\theta$$

$$= \frac{1}{3} \int_0^2 u^3\, du = \frac{u^4}{12} \Big|_0^2 = \frac{4}{3}$$

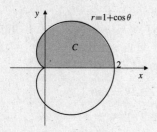

Fig. 14.4.20

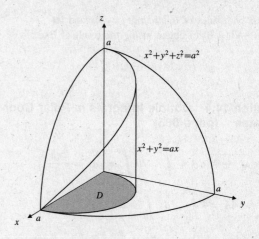

Fig. 14.4.22

24.   Volume $= \displaystyle\int_0^{2\pi} d\theta \int_0^2 (r\cos\theta + r\sin\theta + 4)r\,dr$

$$= \int_0^{2\pi} (\cos\theta + \sin\theta)\,d\theta \int_0^2 r^2\,dr + 8\pi \int_0^2 r\,dr$$

$$= 0 + 4\pi(2^2) = 16\pi \ \text{cu. units.}$$

22.   One quarter of the required volume lies in the first octant. (See the figure.) In polar coordinates the cylinder $x^2 + y^2 = ax$ becomes $r = a\cos\theta$. Thus, the required volume is

26.   One quarter of the required volume V is shown in the figure. We have

$$V = 4 \iint_D \sqrt{y}\,dA$$

$$= 4 \int_0^{\pi/2} d\theta \int_0^{2\sin\theta} \sqrt{r\sin\theta}\,r\,dr$$

$$= 4 \int_0^{\pi/2} \sqrt{\sin\theta}\,d\theta \left(\frac{2}{5}r^{5/2} \Big|_0^{2\sin\theta} \right)$$

$$= \frac{32\sqrt{2}}{5} \int_0^{\pi/2} \sin^3\theta\,d\theta = \frac{64\sqrt{2}}{15} \ \text{cu. units.}$$

$$V = 4 \iint_D \sqrt{a^2 - x^2 - y^2}\,dA$$

$$= 4 \int_0^{\pi/2} d\theta \int_0^{a\cos\theta} \sqrt{a^2 - r^2}\,r\,dr \quad \text{Let } u = a^2 - r^2$$
$$\qquad\qquad\qquad\qquad\qquad\qquad\qquad\quad du = -2r\,dr$$

$$= 2 \int_0^{\pi/2} d\theta \int_{a^2\sin^2\theta}^{a^2} u^{1/2}\,du$$

$$= \frac{4}{3} \int_0^{\pi/2} d\theta \left(u^{3/2} \Big|_{a^2\sin^2\theta}^{a^2} \right)$$

$$= \frac{4}{3}a^3 \int_0^{\pi/2} (1 - \sin^3\theta)\,d\theta$$

$$= \frac{4}{3}a^3 \left(\frac{\pi}{2} - \int_0^{\pi/2} \sin\theta(1 - \cos^2\theta)\,d\theta \right)$$

$$\qquad\qquad \text{Let } v = \cos\theta$$
$$\qquad\qquad dv = -\sin\theta\,d\theta$$

$$= \frac{2\pi a^3}{3} - \frac{4a^3}{3} \int_0^1 (1 - v^2)\,dv$$

$$= \frac{2\pi a^3}{3} - \frac{4a^3}{3} \left(v - \frac{v^3}{3} \right) \Big|_0^1$$

$$= \frac{2\pi a^3}{3} - \frac{8a^3}{9} = \frac{2}{9}a^3(3\pi - 4) \ \text{cu. units.}$$

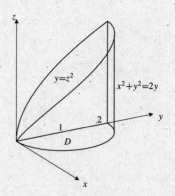

Fig. 14.4.26

28. The area of S is $(4\pi - 3\sqrt{3})/3$ sq. units. Thus

$$\bar{x} = \frac{3}{4\pi - 3\sqrt{3}} \iint_S x\, dA$$

$$= \frac{6}{4\pi - 3\sqrt{3}} \int_0^{\pi/3} d\theta \int_{\sec\theta}^2 r\cos\theta\, r\, dr$$

$$= \frac{2}{4\pi - 3\sqrt{3}} \int_0^{\pi/3} \cos\theta(8 - \sec^3\theta)\, d\theta$$

$$= \frac{2}{4\pi - 3\sqrt{3}} \left(4\sqrt{3} - \tan\theta\Big|_0^{\pi/3}\right) = \frac{6\sqrt{3}}{4\pi - 3\sqrt{3}}.$$

The segment has centroid $\left(\dfrac{6\sqrt{3}}{4\pi - 3\sqrt{3}}, 0\right)$.

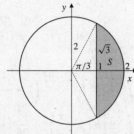

Fig. 14.4.28

30. We use the same regions and change of variables as in the previous exercise. The required volume is

$$V = \iint_E \left(1 - \frac{x^2}{a^2} - \frac{y^2}{b^2}\right) dx\, dy$$

$$= ab \iint_Q (1 - u^2 - v^2)\, du\, dv.$$

Now transform to polar coordinates in the uv-plane: $u = r\cos\theta$, $v = r\sin\theta$.

$$V = ab \int_0^{\pi/2} d\theta \int_0^1 (1 - r^2)r\, dr$$

$$= \frac{\pi ab}{2} \left(\frac{r^2}{2} - \frac{r^4}{4}\right)\Big|_0^1 = \frac{\pi ab}{8} \text{ cu. units.}$$

32. The parallelogram P bounded by $x + y = 1$, $x + y = 2$, $3x + 4y = 5$, and $3x + 4y = 6$ corresponds to the square S bounded by $u = 1$, $u = 2, v = 5$, and $v = 6$ under the transformation

$$u = x + y, \qquad v = 3x + 4y,$$

or, equivalently,

$$x = 4u - v, \qquad y = v - 3u.$$

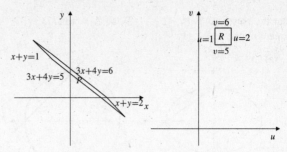

Fig. 14.4.32a   Fig. 14.4.32b

We have

$$\frac{\partial(x, y)}{\partial(u, v)} = \begin{vmatrix} 4 & -1 \\ -3 & 1 \end{vmatrix} = 1,$$

so $dx\, dy = du\, dv$. Also

$$x^2 + y^2 = (4u - v)^2 + (v - 3u)^2 = 25u^2 - 14uv + 2v^2.$$

Thus we have

$$\iint_P (x^2 + y^2)\, dx\, dy = \iint_S (25u^2 - 14uv + 2v^2)\, du\, dv$$

$$= \int_1^2 du \int_5^6 (25u^2 - 14uv + 2v^2)\, dv = \frac{7}{2}.$$

34. Under the transformation $u = x^2 - y^2$, $v = xy$, the region R in the first quadrant of the xy-plane bounded by $y = 0$, $y = x$, $xy = 1$, and $x^2 - y^2 = 1$ corresponds to the square S in the uv-plane bounded by $u = 0$, $u = 1$, $v = 0$, and $v = 1$. Since

$$\frac{\partial(u, v)}{\partial(x, y)} = \begin{vmatrix} 2x & -2y \\ y & x \end{vmatrix} = 2(x^2 + y^2),$$

we therefore have

$$(x^2 + y^2)\, dx\, dy = \frac{1}{2} du\, dv.$$

Hence,

$$\iint_R (x^2 + y^2)\, dx\, dy = \iint_S \frac{1}{2} du\, dv = \frac{1}{2}.$$

36. The region R whose area we must find is shown in part (a) of the figure. The change of variables $x = 3u$, $y = 2v$ maps the ellipse $4x^2 + 9y^2 = 36$ to the circle $u^2 + v^2 = 1$, and the line $2x + 3y = 1$ to the line $u + v = 1$. Thus it maps R to the region S in part (b) of the figure. Since

$$dx\, dy = \left|\begin{vmatrix} 3 & 0 \\ 0 & 2 \end{vmatrix}\right| du\, dv = 6\, du\, dv,$$

the area of R is

$$A = \iint_R dx\, dy = 6 \iint_S du\, dv.$$

But the area of S is $(\pi/4) - (1/2)$, so $A = (3\pi/2) - 3$ square units.

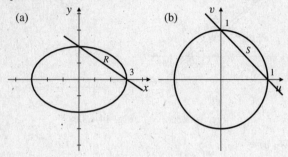

Fig. 14.4.36

38.

a) $\Gamma(x) = \int_0^\infty t^{x-1} e^{-t}\, dt$ Let $t = s^2$
$$dt = 2s\, ds$$
$$= 2 \int_0^\infty s^{2x-1} e^{-s^2}\, ds.$$

b) $\Gamma\left(\frac{1}{2}\right) = 2 \int_0^\infty e^{-s^2}\, ds = 2 \frac{\sqrt{\pi}}{2} = \sqrt{\pi}$

$\Gamma\left(\frac{3}{2}\right) = \frac{1}{2}\Gamma\left(\frac{1}{2}\right) = \frac{1}{2}\sqrt{\pi}.$

c) $B(x, y) = \int_0^1 t^{x-1}(1-t)^{y-1}\, dt$ $(x > 0,\ y > 0)$

let $t = \cos^2\theta$, $dt = -2\sin\theta\cos\theta\, d\theta$

$$= 2 \int_0^{\pi/2} \cos^{2x-1}\theta \sin^{2y-1}\theta\, d\theta.$$

d) If Q is the first quadrant of the st-plane,

$$\Gamma(x)\Gamma(y) = \left(2\int_0^\infty s^{2x-1}e^{-s^2}\, ds\right)\left(2\int_0^\infty t^{2y-1}e^{-t^2}\, dt\right)$$

$$= 4 \iint_Q s^{2x-1} t^{2y-1} e^{-(s^2+t^2)}\, ds\, dt$$

(change to polar coordinates)

$$= 4 \int_0^{\pi/2} d\theta \int_0^\infty r^{2x-1}\cos^{2x-1}\theta\, r^{2y-1}\sin^{2y-1}\theta\, e^{-r^2} r\, dr$$

$$= \left(2\int_0^{\pi/2}\cos^{2x-1}\theta\sin^{2y-1}\theta\, d\theta\right)$$

$$\times \left(2\int_0^\infty r^{2(x+y)-1}e^{-r^2}\, dr\right)$$

$$= B(x, y)\Gamma(x + y)$$ by (a) and (c).

Thus $B(x, y) = \dfrac{\Gamma(x)\Gamma(y)}{\Gamma(x + y)}.$

Section 14.5   Triple Integrals   (page 873)

2.
$$\iiint_B xyz\, dV = \int_0^1 x\, dx \int_{-2}^0 y\, dy \int_1^4 z\, dz$$
$$= \frac{1}{2}\left(-\frac{4}{2}\right)\left(\frac{16-1}{2}\right) = -\frac{15}{2}.$$

4.
$$\iiint_R x\, dV = \int_0^a x\, dx \int_0^{b\left(1-\frac{x}{a}\right)} dy \int_0^{c\left(1-\frac{x}{a}-\frac{y}{b}\right)} dz$$

$$= c \int_0^a x\, dx \int_0^{b\left(1-\frac{x}{a}\right)} \left(1 - \frac{x}{a} - \frac{y}{b}\right) dy$$

$$= c \int_0^a x \left[b\left(1-\frac{x}{a}\right)^2 - \frac{b^2}{2b}\left(1-\frac{x}{a}\right)^2\right] dx$$

$$= \frac{bc}{2} \int_0^a \left(1-\frac{x}{a}\right)^2 x\, dx \quad \text{Let } u = 1 - (x/a)$$
$$du = -(1/a)\, dx$$

$$= \frac{a^2bc}{2} \int_0^1 u^2(1 - u)\, du = \frac{a^2bc}{24}.$$

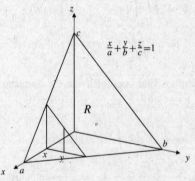

Fig. 14.5.4

6. As in Exercise 5,

$$\iiint_R (x^2 + y^2 + z^2)\, dV = 3 \iiint_R x^2\, dV = \frac{3}{3} = 1.$$

8. R is the cube $0 \le x, y, z \le 1$. We have

$$\iiint_R yz^2 e^{-xyz}\, dV$$

$$= \int_0^1 z\, dz \int_0^1 dy \left(-e^{-xyz}\right)\Big|_{x=0}^{x=1}$$

$$= \int_0^1 z\, dz \int_0^1 (1 - e^{-yz})\, dy$$

$$= \int_0^1 z\left(1 + \frac{1}{z} e^{-yz}\Big|_{y=0}^{y=1}\right) dz$$

$$= \frac{1}{2} + \int_0^1 (e^{-z} - 1)\, dz$$

$$= \frac{1}{2} - 1 - e^{-z}\Big|_0^1 = \frac{1}{2} - \frac{1}{e}.$$

10. $\displaystyle\iiint_R y\, dV = \int_0^1 y\, dy \int_{1-y}^1 dz \int_0^{2-y-z} dx$

$$= \int_0^1 y\, dy \int_{1-y}^1 (2 - y - z)\, dz$$

$$= \int_0^1 y\, dy \left((2 - y)z - \frac{z^2}{2}\right)\Big|_{z=1-y}^{z=1}$$

$$= \int_0^1 y\left((2 - y)y - \frac{1}{2}\left(1 - (1 - y)^2\right)\right) dy$$

$$= \int_0^1 \frac{1}{2}\left(2y^2 - y^3\right) dy = \frac{5}{24}.$$

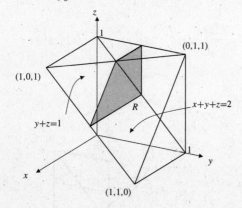

Fig. 14.5.10

12. We have

$$\iiint_R \cos x \cos y \cos z\, dV$$

$$= \int_0^\pi \cos x\, dx \int_0^{\pi - x} \cos y\, dy \int_0^{\pi - x - y} \cos z\, dz$$

$$= \int_0^\pi \cos x\, dx \int_0^{\pi - x} \cos y\, dy\, (\sin z)\Big|_{z=0}^{z=\pi - x - y}$$

$$= \int_0^\pi \cos x\, dx \int_0^{\pi - x} \cos y \sin(x + y)\, dy$$

recall that $\sin a \cos b = \frac{1}{2}\left(\sin(a + b) + \sin(a - b)\right)$

$$= \int_0^\pi \cos x\, dx \int_0^{\pi - x} \frac{1}{2}\left[\sin(x + 2y) + \sin x\right] dy$$

$$= \frac{1}{2} \int_0^\pi \cos x\, dx \left[-\frac{\cos(x + 2y)}{2} + y \sin x\right]\Big|_{y=0}^{y=\pi - x}$$

$$= \frac{1}{2} \int_0^\pi \left(-\frac{\cos x \cos(2\pi - x)}{2} + \frac{\cos^2 x}{2}\right.$$

$$\left. + (\pi - x) \cos x \sin x\right) dx$$

$$= \frac{1}{2} \int_0^\pi \frac{\pi - x}{2} \sin 2x\, dx$$

$$U = \pi - x \quad dV = \sin 2x\, dx$$

$$dU = -dx \quad V = -\frac{\cos 2x}{2}$$

$$= \frac{1}{4}\left[-\frac{\pi - x}{2} \cos 2x\Big|_0^\pi - \frac{1}{2} \int_0^\pi \cos 2x\, dx\right]$$

$$= \frac{1}{8}\left[\pi - \frac{\sin 2x}{2}\Big|_0^\pi\right] = \frac{\pi}{8}.$$

14. Let E be the elliptic disk bounded by $x^2 + 4y^2 = 4$. Then E has area $\pi(2)(1) = 2\pi$ square units. The volume of the region of 3-space lying above E and beneath the plane $z = 2 + x$ is

$$V = \iint_E (2 + x)\, dA = 2 \iint_E dA = 4\pi \text{ cu. units,}$$

since $\iint_E x\, dA = 0$ by symmetry.

16.

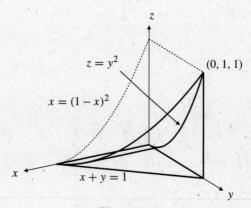

Fig. 14.5.16

$$\iiint_R f(x,y,z)\,dV = \int_0^1 dx \int_0^{1-x} dy \int_0^{y^2} f(x,y,z)\,dz$$

$$= \int_0^1 dy \int_0^{1-y} dx \int_0^{y^2} f(x,y,z)\,dz$$

$$= \int_0^1 dy \int_0^{y^2} dz \int_0^{1-y} f(x,y,z)\,dx$$

$$= \int_0^1 dz \int_{\sqrt{z}}^1 dy \int_0^{1-y} f(x,y,z)\,dx$$

$$= \int_0^1 dx \int_0^{(1-x)^2} dz \int_{\sqrt{z}}^{1-x} f(x,y,z)\,dy$$

$$= \int_0^1 dz \int_0^{1-\sqrt{z}} dx \int_{\sqrt{z}}^{1-x} f(x,y,z)\,dy.$$

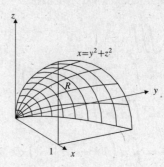

Fig. 14.5.20

22. $I = \int_0^1 dz \int_z^1 dy \int_0^y f(x,y,z)\,dx.$
The given iteration corresponds to

$$0 \le z \le 1, \quad z \le y \le 1, \quad 0 \le x \le y.$$

Thus $0 \le x \le 1$, $x \le y \le 1$, $0 \le z \le y$, and

$$I = \int_0^1 dx \int_x^1 dy \int_0^y f(x,y,z)\,dz.$$

18. $\int_0^1 dz \int_z^1 dy \int_0^y f(x,y,z)\,dx$

$$= \iiint_R f(x,y,z)\,dV \quad (R \text{ is the pyramid in the figure})$$

$$= \int_0^1 dx \int_x^1 dy \int_0^y f(x,y,z)\,dz.$$

24. $I = \int_0^1 dy \int_0^{\sqrt{1-y^2}} dz \int_{y^2+z^2}^1 f(x,y,z)\,dx.$
The given iteration corresponds to

$$0 \le y \le 1, \quad 0 \le z \le \sqrt{1-y^2}, \quad y^2+z^2 \le x \le 1.$$

Thus $0 \le x \le 1$, $0 \le y \le \sqrt{x}$, $0 \le z \le \sqrt{x-y^2}$, and

$$I = \int_0^1 dx \int_0^{\sqrt{x}} dy \int_0^{\sqrt{x-y^2}} f(x,y,z)\,dz.$$

26.

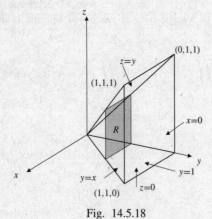

Fig. 14.5.18

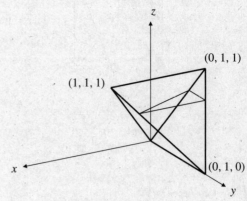

Fig. 14.5.26

20. $\int_0^1 dy \int_0^{\sqrt{1-y^2}} dz \int_{y^2+z^2}^1 f(x,y,z)\,dx$

$$= \iiint_R f(x,y,z)\,dV \quad (R \text{ is the paraboloid in the figure})$$

$$= \int_0^1 dx \int_0^{\sqrt{x}} dy \int_0^{\sqrt{x-y^2}} f(x,y,z)\,dz.$$

$$I = \int_0^1 dx \int_x^1 dy \int_x^y f(x,y,z)\,dz = \iiint_P f(x,y,z)\,dV,$$

where P is the triangular pyramid (see the figure) with vertices at $(0, 0, 0)$, $(0, 1, 0)$, $(0, 1, 1)$, and $(1, 1, 1)$. If we reiterate I to correspond to the horizontal slice shown then

$$\int_0^1 dz \int_z^1 dy \int_0^z f(x, y, z)\, dx.$$

28.
$$\int_0^1 dx \int_0^{1-x} dy \int_y^1 \frac{\sin(\pi z)}{z(2-z)} dz$$

$$= \iiint_R \frac{\sin(\pi z)}{z(2-z)} dV \qquad (R \text{ is the pyramid in the figure})$$

$$= \int_0^1 \frac{\sin(\pi z)}{z(2-z)} dz \int_0^z dy \int_0^{1-y} dx$$

$$= \int_0^1 \frac{\sin(\pi z)}{z(2-z)} dz \int_0^z (1-y)\, dy$$

$$= \int_0^1 \frac{\sin(\pi z)}{z(2-z)} \left(z - \frac{z^2}{2} \right) dz$$

$$= \frac{1}{2} \int_0^1 \sin(\pi z)\, dz = \frac{1}{\pi}.$$

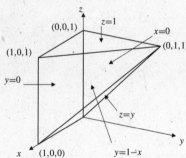

Fig. 14.5.28

30. If the function $f(x, y, z)$ is continuous on a closed, bounded, connected set D in 3-space, then there exists a point (x_0, y_0, z_0) in D such that

$$\iiint_D f(x, y, z)\, dV = f(x_0, y_0, z_0) \times (\text{volume of } D).$$

Apply this with $D = B_\epsilon(a, b, c)$, which has volume $\frac{4}{3}\pi\epsilon^3$, to get

$$\iiint_{B_\epsilon(a,b,c)} f(x, y, z)\, dV = f(x_0, y_0, z_0)\frac{4}{3}\pi\epsilon^3$$

for some (x_0, y_0, z_0) in $B_\epsilon(a, b, c)$. Thus

$$\lim_{\epsilon \to 0} \frac{3}{4\pi\epsilon^3} \iiint_{B_\epsilon(a,b,c)} f(x, y, z)\, dV$$

$$= \lim_{\epsilon \to 0} f(x_0, y_0, z_0) = f(a, b, c)$$

since f is continuous at (a, b, c).

Section 14.6   Change of Variables in Triple Integrals   (page 882)

2. Cartesian: $(2, -2, 1)$;
Cylindrical: $[2\sqrt{2}, -\pi/4, 1]$;
Spherical: $[3, \cos^{-1}(1/3), -\pi/4]$.

4. Spherical: $[1, \phi, \theta]$; Cylindrical: $[r, \pi/4, r]$.

$$x = \sin\phi\cos\theta = r\cos\pi/4 = r/\sqrt{2}$$
$$y = \sin\phi\sin\theta = r\sin\pi/4 = r/\sqrt{2}$$
$$z = \cos\phi = r.$$

Thus $x = y$, $\theta = \pi/4$, and $r = \sin\phi = \cos\phi$. Hence $\phi = \pi/4$, so $r = 1/\sqrt{2}$. Finally: $x = y = 1/2$, $z = 1/\sqrt{2}$. Cartesian: $(1/2, 1/2, 1/\sqrt{2})$.

6. $\phi = 2\pi/3$ represents the lower half of the right-circular cone with vertex at the origin, axis along the z-axis, and semi-vertical angle $\pi/3$. Its Cartesian equation is $z = -\sqrt{(x^2 + y^2)/3}$.

8. $\rho = 4$ represents the sphere of radius 4 centred at the origin.

10. $\rho = z$ represents the positive half of the z-axis.

12. $\rho = 2x$ represents the half-cone with vertex at the origin, axis along the positive x-axis, and semi-vertical angle $\pi/3$. Its Cartesian equation is $x = \sqrt{(y^2 + z^2)/3}$.

14. $r = 2\cos\theta \Rightarrow x^2 + y^2 = r^2 = 2r\cos\theta = 2x$, or $(x - 1)^2 + y^2 = 1$. Thus the given equation represents the circular cylinder of radius 1 with axis along the vertical line $x = 1$, $y = 0$.

16. The surface $z = \sqrt{r}$ intersects the sphere $r^2 + z^2 = 2$ where $r^2 + r - 2 = 0$. This equation has positive root $r = 1$. The required volume is

$$V = \int_0^{2\pi} d\theta \int_0^1 r\, dr \int_{\sqrt{r}}^{\sqrt{2-r^2}} dz$$

$$= \int_0^{2\pi} d\theta \int_0^1 \left(\sqrt{2 - r^2} - \sqrt{r} \right) r\, dr$$

$$= 2\pi \left(\int_0^1 r\sqrt{2 - r^2}\, dr - \frac{2}{5} \right) \quad \begin{array}{l} \text{Let } u = 2 - r^2 \\ du = -2r\, dr \end{array}$$

$$= \pi \int_1^2 u^{1/2}\, du - \frac{4\pi}{5}$$

$$= \frac{2\pi}{3}(2\sqrt{2} - 1) - \frac{4\pi}{5} = \frac{4\sqrt{2}\pi}{3} - \frac{22\pi}{15} \quad \text{cu. units.}$$

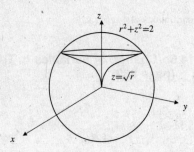

Fig. 14.6.16

18. The paraboloid $z = r^2$ intersects the sphere $r^2 + z^2 = 12$ where $r^4 + r^2 - 12 = 0$, that is, where $r = \sqrt{3}$. The required volume is

$$V = \int_0^{2\pi} d\theta \int_0^{\sqrt{3}} \left(\sqrt{12 - r^2} - r^2\right) r \, dr$$

$$= 2\pi \int_0^{\sqrt{3}} r\sqrt{12 - r^2} \, dr - \frac{9\pi}{2} \quad \begin{array}{l} \text{Let } u = 12 - r^2 \\ du = -2r \, dr \end{array}$$

$$= \pi \int_9^{12} u^{1/2} \, du - \frac{9\pi}{2}$$

$$= \frac{2\pi}{3}\left(12^{3/2} - 27\right) - \frac{9\pi}{2} = 16\sqrt{3}\pi - \frac{45\pi}{2} \text{ cu. units.}$$

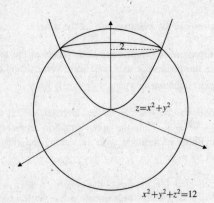

Fig. 14.6.18

20. The required volume V lies above $z = 0$, below $z = 1 - r^2$, and between $\theta = -\pi/4$ and $\theta = \pi/3$. Thus

$$V = \int_{-\pi/4}^{\pi/3} d\theta \int_0^1 (1 - r^2) r \, dr$$

$$= \frac{7\pi}{12}\left(\frac{1}{2} - \frac{1}{4}\right) = \frac{7\pi}{48} \text{ cu. units.}$$

22. One eighth of the required volume V lies in the first octant. Call this region R. Under the transformation

$$x = au, \qquad y = bv, \qquad z = cw,$$

R corresponds to the region S in the first octant of uvw-space bounded by $w = 0$, $w = 1$, and $u^2 + v^2 - w^2 = 1$. Thus

$$V = 8abc \times (\text{volume of } S).$$

The volume of S can be determined by using horizontal slices:

$$V = 8abc \int_0^1 \frac{\pi}{4}(1 + w^2) \, dw = \frac{8}{3}\pi abc \text{ cu. units.}$$

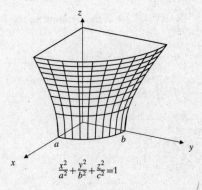

$$\frac{x^2}{a^2} + \frac{y^2}{b^2} + \frac{z^2}{c^2} = 1$$

Fig. 14.6.22

24. $\displaystyle\iiint_R (x^2 + y^2 + z^2) \, dV$

$$= \int_0^{2\pi} d\theta \int_0^a r \, dr \int_0^h (r^2 + z^2) \, dz$$

$$= 2\pi \int_0^a \left(r^3 h + \frac{1}{3} r h^3\right) dr$$

$$= 2\pi \left(\frac{a^4 h}{4} + \frac{a^2 h^3}{6}\right) = \frac{\pi a^4 h}{2} + \frac{\pi a^2 h^3}{3}.$$

26. $\displaystyle\iiint_B (x^2 + y^2 + z^2) \, dV$

$$= \int_0^{2\pi} d\theta \int_0^\pi \sin\phi \, d\phi \int_0^a R^4 \, dR = \frac{4\pi a^5}{5}.$$

28. $\displaystyle\iiint_R (x^2 + y^2) \, dV$

$$= \int_0^{2\pi} d\theta \int_0^{\tan^{-1}(1/c)} \sin^3\phi \, d\phi \int_0^a R^4 \, dR$$

$$= \frac{2\pi a^5}{5} \int_0^{\tan^{-1}(1/c)} \sin\phi(1 - \cos^2\phi) \, d\phi \quad \begin{array}{l} \text{Let } u = \cos\phi \\ du = -\sin\phi \, d\phi \end{array}$$

$$= \frac{2\pi a^5}{5} \int_{c/\sqrt{c^2+1}}^1 (1 - u^2) \, du$$

$$= \frac{2\pi a^5}{5}\left(u - \frac{u^3}{3}\right)\Bigg|_{c/\sqrt{c^2+1}}^1$$

$$= \frac{2\pi a^5}{5}\left(\frac{2}{3} - \frac{c}{\sqrt{c^2+1}} + \frac{c^3}{3(c^2+1)^{3/2}}\right).$$

30. By symmetry, both integrals have the same value:

$$\iiint_R x\,dV = \iiint_R z\,dV$$

$$= \int_0^{\pi/2} d\theta \int_0^{\pi/2} \cos\phi\sin\phi\,d\phi \int_0^a R^3\,dR$$

$$= \frac{\pi}{2}\left(\frac{1}{2}\right)\frac{a^4}{4} = \frac{\pi a^4}{16}.$$

32. If

$$x = au, \qquad y = bv, \qquad z = cw,$$

then the volume of a region R in xyz-space is abc times the volume of the corresponding region S in uvw-space.

If R is the region inside the ellipsoid

$$\frac{x^2}{a^2} + \frac{y^2}{b^2} + \frac{z^2}{c^2} = 1$$

and above the plane $y + z = b$, then the corresponding region S lies inside the sphere

$$u^2 + v^2 + w^2 = 1$$

and above the plane $bv + cw = b$. The distance from the origin to this plane is

$$D = \frac{b}{\sqrt{b^2 + c^2}} \qquad \text{(assuming } b > 0)$$

by Example 7 of Section 1.4. By symmetry, the volume of S is equal to the volume lying inside the sphere $u^2 + v^2 + w^2 = 1$ and above the plane $w = D$. We calculate this latter volume by slicing; it is

$$\pi \int_D^1 (1 - w^2)\,dw = \pi\left(w - \frac{w^3}{3}\right)\Big|_D^1$$

$$= \pi\left(\frac{2}{3} - D + \frac{D^3}{3}\right).$$

Hence, the volume of R is

$$\pi abc\left(\frac{2}{3} - \frac{b}{\sqrt{b^2 + c^2}} + \frac{b^3}{3(b^2 + c^2)^{3/2}}\right) \text{ cu. units.}$$

34. Cylindrical and spherical coordinates are related by

$$z = \rho\cos\phi, \qquad r = \rho\sin\phi.$$

(The θ coordinates are identical in the two systems.) Observe that z, r, ρ, and ϕ play, respectively, the same roles that x, y, r, and θ play in the transformation from Cartesian to polar coordinates in the plane. We can exploit this correspondence to avoid repeating the calculations of partial derivatives of a function u, since the results correspond to calculations made (for a function z) in Example 10 of Section 3.5. Comparing with the calculations in that Example, we have

$$\frac{\partial u}{\partial \rho} = \cos\phi\frac{\partial u}{\partial z} + \sin\phi\frac{\partial u}{\partial r}$$

$$\frac{\partial u}{\partial \phi} = -\rho\sin\phi\frac{\partial u}{\partial z} + \rho\cos\phi\frac{\partial u}{\partial r}$$

$$\frac{\partial^2 u}{\partial \rho^2} = \cos^2\phi\frac{\partial^2 u}{\partial z^2} + 2\cos\phi\sin\phi\frac{\partial^2 u}{\partial z\partial r} + \sin^2\phi\frac{\partial^2 u}{\partial r^2}$$

$$\frac{\partial^2 u}{\partial \phi^2} = -\rho\frac{\partial u}{\partial \rho} + \rho^2\left(\sin^2\phi\frac{\partial^2 u}{\partial z^2}\right.$$

$$\left. - 2\cos\phi\sin\phi\frac{\partial^2 u}{\partial z\partial r} + \cos^2\phi\frac{\partial^2 u}{\partial r^2}\right).$$

Substituting these expressions into the expression for Δu given in the statement of this exercise in terms of spherical coordinates, we obtain the expression in terms of cylindrical coordinates established in the previous exercise:

$$\frac{\partial^2 u}{\partial \rho^2} + \frac{2}{\rho}\frac{\partial u}{\partial \rho} + \frac{\cot\phi}{\rho^2}\frac{\partial u}{\partial \phi} + \frac{1}{\rho^2}\frac{\partial^2 u}{\partial \phi^2} + \frac{1}{\rho^2\sin^2\phi}\frac{\partial^2 u}{\partial \theta^2}$$

$$= \frac{\partial^2 u}{\partial r^2} + \frac{1}{r}\frac{\partial u}{\partial r} + \frac{1}{r^2}\frac{\partial^2 u}{\partial \theta^2} + \frac{\partial^2 u}{\partial z^2}$$

$$= \frac{\partial^2 u}{\partial x^2} + \frac{\partial^2 u}{\partial y^2} = \Delta u$$

by Exercise 33.

Section 14.7 Applications of Multiple Integrals (page 891)

2. $z = (3x - 4y)/5$, $\quad \dfrac{\partial z}{\partial x} = \dfrac{3}{5}$, $\quad \dfrac{\partial z}{\partial y} = \dfrac{4}{5}$

$$dS = \sqrt{1 + \frac{3^2 + 4^2}{5^2}}\,dA = \sqrt{2}\,dA$$

$$S = \iint_{(x/2)^2 + y^2 \le 1} \sqrt{2}\,dA = \sqrt{2}\,\pi(2)(1) = 2\sqrt{2}\pi \text{ sq. units.}$$

4.　$z = 2\sqrt{1 - x^2 - y^2}$

$$\frac{\partial z}{\partial x} = -\frac{2x}{\sqrt{1 - x^2 - y^2}}, \quad \frac{\partial z}{\partial y} = -\frac{2y}{\sqrt{1 - x^2 - y^2}}$$

$$dS = \sqrt{1 + \frac{4(x^2 + y^2)}{1 - x^2 - y^2}}\, dA = \sqrt{\frac{1 + 3(x^2 + y^2)}{1 - x^2 - y^2}}\, dA$$

$$S = \iint_{x^2 + y^2 \le 1} dS$$

$$= \int_0^{2\pi} d\theta \int_0^1 \sqrt{\frac{1 + 3r^2}{1 - r^2}}\, r\, dr \quad \begin{array}{l} \text{Let } u^2 = 1 - r^2 \\ u\, du = -r\, dr \end{array}$$

$$= 2\pi \int_0^1 \sqrt{4 - 3u^2}\, du \quad \begin{array}{l} \text{Let } \sqrt{3}\, u = 2\sin v \\ \sqrt{3}\, du = 2\cos v\, dv \end{array}$$

$$= 2\pi \int_0^{\pi/3} (2\cos^2 v)\frac{2\, dv}{\sqrt{3}}$$

$$= \frac{4\pi}{\sqrt{3}} \int_0^{\pi/3} (1 + \cos 2v)\, dv$$

$$= \frac{4\pi}{\sqrt{3}} \left(v + \frac{\sin 2v}{2} \right)\Big|_0^{\pi/3} = \frac{4\pi^2}{3\sqrt{3}} + \pi \text{ sq. units.}$$

6.　$z = 1 - x^2 - y^2, \quad \dfrac{\partial z}{\partial x} = -2x, \quad \dfrac{\partial z}{\partial y} = -2y$

$$dS = \sqrt{1 + 4x^2 + 4y^2}\, dA$$

$$S = \iint_{x^2 + y^2 \le 1,\ x \ge 0,\ y \ge 0} \sqrt{1 + 4(x^2 + y^2)}\, dA$$

$$= \int_0^{\pi/2} d\theta \int_0^1 \sqrt{1 + 4r^2}\, r\, dr \quad \begin{array}{l} \text{Let } u = 1 + 4r^2 \\ du = 8r\, dr \end{array}$$

$$= \frac{\pi}{16} \int_1^5 u^{1/2}\, du$$

$$= \frac{\pi}{16} \left(\frac{2}{3} u^{3/2} \right)\Big|_1^5 = \frac{\pi(5\sqrt{5} - 1)}{24} \text{ sq. units.}$$

8.　$z = \sqrt{x}, \quad \dfrac{\partial z}{\partial x} = \dfrac{1}{2\sqrt{x}}, \quad dS = \sqrt{1 + \dfrac{1}{4x}}\, dA$

$$S = \int_0^1 dx \int_0^{\sqrt{x}} \sqrt{1 + \frac{1}{4x}}\, dy = \int_0^1 \sqrt{\frac{4x + 1}{4x}}\, \sqrt{x}\, dx$$

$$= \frac{1}{2} \int_0^1 \sqrt{4x + 1}\, dx \quad \begin{array}{l} \text{Let } u = 4x + 1 \\ du = 4\, dx \end{array}$$

$$= \frac{1}{8} \int_1^5 u^{1/2}\, du = \frac{1}{8} \left(\frac{2}{3} u^{3/2} \right)\Big|_1^5 = \frac{5\sqrt{5} - 1}{12} \text{ sq. units.}$$

10.　The area elements on $z = 2xy$ and $z = x^2 + y^2$, respectively, are

$$dS_1 = \sqrt{1 + (2y)^2 + (2x)^2}\, dA = \sqrt{1 + 4x^2 + 4y^2}\, dx\, dy,$$
$$dS_2 = \sqrt{1 + (2x)^2 + (2y)^2}\, dA = \sqrt{1 + 4x^2 + 4y^2}\, dx\, dy.$$

Since these elements are equal, the area of the parts of both surfaces defined over any region of the xy-plane will be equal.

12.　As the figure suggests, the area of the canopy is the area of a hemisphere of radius $\sqrt{2}$ minus four times the area of half of a spherical cap cut off from the sphere $x^2 + y^2 + z^2 = 2$ by a plane at distance 1 from the origin, say the plane $z = 1$. Such a spherical cap, $z = \sqrt{2 - x^2 - y^2}$, lies above the disk $x^2 + y^2 \le 2 - 1 = 1$. Since $\dfrac{\partial z}{\partial x} = -x/z$ and $\dfrac{\partial z}{\partial y} = -y/z$ on it, the area of the spherical cap is

$$\iint_{x^2 + y^2 \le 1} \sqrt{1 + \frac{x^2 + y^2}{z^2}}\, dA$$

$$= 2\sqrt{2}\pi \int_0^1 \frac{r\, dr}{\sqrt{2 - r^2}} \quad \begin{array}{l} \text{Let } u = 2 - r^2 \\ du = -2r\, dr \end{array}$$

$$= \sqrt{2}\pi \int_1^2 u^{-1/2}\, du = 2\sqrt{2}(\sqrt{2} - 1) = 4 - 2\sqrt{2}.$$

Thus the area of the canopy is

$$S = 2\pi(\sqrt{2})^2 - 4 \times \frac{1}{2} \times (4 - 2\sqrt{2}) = 4(\pi + \sqrt{2}) - 8 \text{ sq. units.}$$

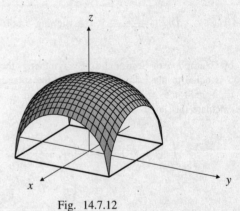

Fig. 14.7.12

14.　A slice of the ball at height z, having thickness dz, is a circular disk of radius $\sqrt{a^2 - z^2}$ and areal density $\delta\, dz$. As calculated in the text, this disk attracts mass m at $(0, 0, b)$ with vertical force

$$dF = 2\pi km\delta dz \left(1 - \frac{b - z}{\sqrt{a^2 - z^2 + (b - z)^2}} \right).$$

Thus the ball attracts m with vertical force

$$F = 2\pi km\delta \int_{-a}^{a} \left(1 - \frac{b-z}{\sqrt{a^2 + b^2 - 2bz}} \right) dz$$

$$\text{let } v = a^2 + b^2 - 2bz, \quad dv = -2b\, dz$$

$$\text{then } b - z = b - \frac{a^2 + b^2 - v}{2b} = \frac{b^2 - a^2 + v}{2b}$$

$$= 2\pi km\delta \left[2a - \frac{1}{4b^2} \int_{(b-a)^2}^{(b+a)^2} \frac{b^2 - a^2 + v}{\sqrt{v}}\, dv \right]$$

$$= 2\pi km\delta \left[2a - \frac{b^2 - a^2}{2b^2} \big(b + a - (b - a) \big) \right.$$

$$\left. - \frac{1}{6b^2} \big((b+a)^3 - (b-a)^3 \big) \right]$$

$$= \frac{4\pi km\delta a^3}{3b^2} = \frac{kmM}{b^2},$$

where $M = (4/3)\pi a^3 \delta$ is the mass of the ball. Thus the ball attracts the external mass m as though the ball were a point mass M located at its centre.

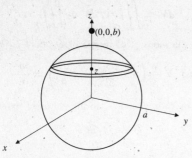

Fig. 14.7.14

16. The force is

$$F = 2\pi km\delta \int_{0}^{b} \left(1 - \frac{b-z}{\sqrt{a^2(b-z)^2 + (b-z)^2}} \right) dz$$

$$= 2\pi km\delta \int_{0}^{b} \left(1 - \frac{1}{\sqrt{a^2 + 1}} \right) dz$$

$$= 2\pi km\delta b \left(1 - \frac{1}{\sqrt{a^2 + 1}} \right).$$

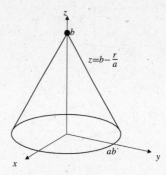

Fig. 14.7.16

18.
$$m = \int_{0}^{a} dx \int_{0}^{a} dy \int_{0}^{a} (x^2 + y^2 + z^2)\, dz$$

$$= 3 \int_{0}^{a} x^2\, dx \int_{0}^{a} dy \int_{0}^{a} dz = a^5$$

$$M_{x=0} = \int_{0}^{a} x\, dx \int_{0}^{a} dy \int_{0}^{a} (x^2 + y^2 + z^2)\, dz$$

$$= \int_{0}^{a} x\, dx \int_{0}^{a} \left(a(x^2 + y^2) + \frac{a^3}{3} \right) dy$$

$$= \int_{0}^{a} \left(\frac{2a^4}{3} + a^2 x^2 \right) x\, dx = \frac{7a^6}{12}.$$

Thus $\bar{x} = M_{x=0}/m = \dfrac{7a}{12}$.

By symmetry, the centre of mass is $\left(\dfrac{7a}{12}, \dfrac{7a}{12}, \dfrac{7a}{12} \right)$.

20. Volume of region $= \displaystyle\int_{0}^{2\pi} d\theta \int_{0}^{\infty} e^{-r^2} r\, dr = \pi$. By symmetry, the moments about $x = 0$ and $y = 0$ are both zero. We have

$$M_{z=0} = \int_{0}^{2\pi} d\theta \int_{0}^{\infty} r\, dr \int_{0}^{e^{-r^2}} z\, dz$$

$$= \pi \int_{0}^{\infty} r e^{-2r^2}\, dr = \frac{\pi}{4}.$$

The centroid is $(0, 0, 1/4)$.

22. The cube has centroid $(1/2, 1/2, 1/2)$. The tetrahedron lying above the plane $x + y + x = 2$ has centroid $(3/4, 3/4, 3/4)$ and volume $1/6$. Therefore the part of the cube lying below the plane has centroid (c, c, c) and volume $5/6$, where

$$\frac{5}{6} c + \frac{3}{4} \times \frac{1}{6} = \frac{1}{2} \times 1.$$

Thus $c = 9/20$; the centroid is $\left(\dfrac{9}{20}, \dfrac{9}{20}, \dfrac{9}{20} \right)$.

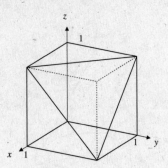

Fig. 14.7.22

24. $I = \delta \int_0^{2\pi} d\theta \int_0^a r^3\, dr \int_0^h dz$

$\quad = 2\pi\delta h \left(\dfrac{a^4}{4}\right) = \dfrac{\pi\delta h a^4}{2}.$

$\quad m = \pi\delta a^2 h, \qquad \bar{D} = \sqrt{I/m} = \dfrac{a}{\sqrt{2}}.$

26. $I = \delta \int_0^{2\pi} d\theta \int_0^a r^3\, dr \int_0^{h(1-(r/a))} dz$

$\quad = 2\pi\delta h \int_0^a r^3 \left(1 - \dfrac{r}{a}\right) dr = \dfrac{\pi\delta a^4 h}{10},$

$\quad m = \dfrac{\pi\delta a^2 h}{3}, \qquad \bar{D} = \sqrt{I/m} = \sqrt{\dfrac{3}{10}}\,a.$

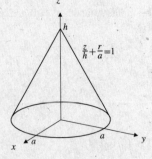

Fig. 14.7.26

28. $I = \delta \iiint_Q (x^2 + y^2)\, dV$

$\quad = 2\delta \int_0^a x^2\, dx \int_0^a dy \int_0^a dz = \dfrac{2\delta a^5}{3},$

$\quad m = \delta a^3, \qquad \bar{D} = \sqrt{I/m} = \sqrt{\dfrac{2}{3}}\,a.$

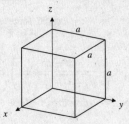

Fig. 14.7.28

30. The line L through the origin parallel to the vector $\mathbf{v} = \mathbf{i} + \mathbf{j} + \mathbf{k}$ is a diagonal of the cube Q. By Example 8 of Section 1.4, the distance from the point with position vector $\mathbf{r} = x\mathbf{i} + y\mathbf{j} + z\mathbf{k}$ to L is $s = |\mathbf{v} \times \mathbf{r}|/|\mathbf{v}|$. Thus, the square of the distance from (x, y, z) to L is

$$s^2 = \frac{(x - y)^2 + (y - z)^2 + (z - x)^2}{3}$$

$$= \frac{2}{3}\left(x^2 + y^2 + z^2 - xy - xz - yz\right).$$

We have

$$\iiint_Q x^2\, dV = \iiint_Q y^2\, dV = \iiint_Q z^2\, dV = \frac{a^5}{3}$$

$$\iiint_Q xy\, dV = \iiint_Q yz\, dV = \iiint_Q xz\, dV = \frac{a^5}{4}.$$

Therefore, the moment of inertia of Q about L is

$$I = \frac{2\delta}{3}\left(3 \times \frac{a^5}{3} - 3 \times \frac{a^5}{4}\right) = \frac{\delta a^5}{6}.$$

The mass of Q is $m = \delta a^3$, so the radius of gyration is

$$\bar{D} = \sqrt{I/m} = \frac{a}{\sqrt{6}}.$$

32. $I = \delta \int_0^{2\pi} d\theta \int_0^c dz \int_a^b r^3\, dr = \dfrac{\pi\delta c(b^4 - a^4)}{2},$

$\quad m = \pi\delta c(b^2 - a^2), \qquad \bar{D} = \sqrt{\dfrac{b^2 + a^2}{2}}.$

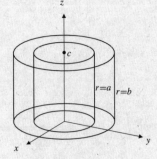

Fig. 14.7.32

34. By Exercise 26, the cylinder has moment of inertia

$$I = \frac{\pi\delta a^4 h}{2} = \frac{ma^2}{2},$$

where m is its mass. Following the method of Example 4(b), the kinetic energy of the cylinder rolling down the inclined plane with speed v is

$$KE = \frac{1}{2}mv^2 + \frac{1}{2}I\Omega^2$$
$$= \frac{1}{2}mv^2 + \frac{1}{4}ma^2\frac{v^2}{a^2} = \frac{3}{4}mv^2.$$

The potential energy of the cylinder when it is at height h is mgh, so, by conservation of energy,

$$\frac{3}{4}mv^2 + mgh = \text{constant}.$$

Differentiating this equation with respect to time t, we obtain

$$0 = \frac{3}{2}mv\frac{dv}{dt} + mg\frac{dh}{dt}$$
$$= \frac{3}{2}mv\frac{dv}{dt} + mgv\sin\alpha.$$

Thus the cylinder rolls down the plane with acceleration

$$-\frac{dv}{dt} = \frac{2}{3}g\sin\alpha.$$

36. The kinetic energy of the oscillating pendulum is

$$KE = \frac{1}{2}I\left(\frac{d\theta}{dt}\right)^2.$$

The potential energy is mgh, where h is the distance of C above A. In this case, $h = -a\cos\theta$. By conservation of energy,

$$\frac{1}{2}I\left(\frac{d\theta}{dt}\right)^2 - mga\cos\theta = \text{constant}.$$

Differentiating with respect to time t, we obtain

$$I\left(\frac{d\theta}{dt}\right)\frac{d^2\theta}{dt^2} + mga\sin\theta\left(\frac{d\theta}{dt}\right) = 0,$$

or

$$\frac{d^2\theta}{dt^2} + \frac{mga}{I}\sin\theta = 0.$$

For small oscillations we have $\sin\theta \approx \theta$, and the above equation is approximated by

$$\frac{d^2\theta}{dt^2} + \omega^2\theta = 0,$$

where $\omega^2 = mga/I$. The period of oscillation is

$$T = \frac{2\pi}{\omega} = 2\pi\sqrt{\frac{I}{mga}}.$$

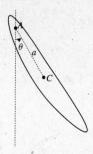

Fig. 14.7.36

38. The moment of inertia of the ball about the point where it contacts the plane is, by Example 4(b) and Exercise 39,

$$I = \frac{8}{15}\pi\delta a^5 + \left(\frac{4}{3}\pi\delta a^3\right)a^2$$
$$= \left(\frac{2}{5} + 1\right)ma^2 = \frac{7}{5}ma^2.$$

The kinetic energy of the ball, regarded as rotating about the point of contact with the plane, is therefore

$$KE = \frac{1}{2}I\Omega^2 = \frac{7}{10}ma^2\frac{v^2}{a^2} = \frac{7}{10}mv^2.$$

Review Exercises 14   (page 892)

2.
$$\iint_P (x^2 + y^2)\,dA = \int_0^1 dy \int_y^{2+y} (x^2 + y^2)\,dx$$
$$= \int_0^1 \left(\frac{x^3}{3} + xy^2\right)\Bigg|_{x=y}^{x=2+y} dy$$
$$= \int_0^1 \left(\frac{(2+y)^3}{3} + y^2(2+y) - \frac{y^3}{3} - y^3\right) dy$$
$$= \int_0^1 \left(\frac{8}{3} + 4y + 4y^2\right) dy = \frac{8}{3} + 2 + \frac{4}{3} = 6$$

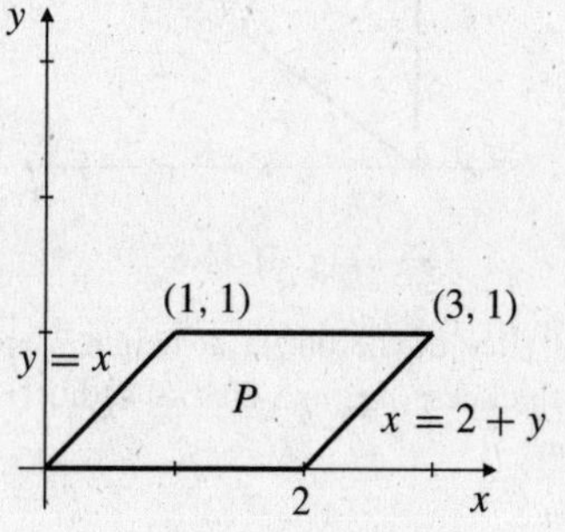

Fig. R-14.2

4.   a) $I = \displaystyle\int_0^{\sqrt{3}} dy \int_{y/\sqrt{3}}^{\sqrt{4-y^2}} e^{-x^2-y^2} \, dx$

$= \displaystyle\iint_R e^{-x^2-y^2} \, dA$

where R is as shown in the figure.

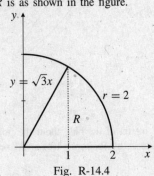

Fig. R-14.4

   b) $I = \displaystyle\int_0^1 dx \int_0^{\sqrt{3}x} e^{-x^2-y^2} \, dy$

$+ \displaystyle\int_1^2 dx \int_0^{\sqrt{4-x^2}} e^{-x^2-y^2} \, dy$

   c) $I = \displaystyle\int_0^{\pi/3} d\theta \int_0^2 e^{-r^2} r \, dr$

   d) $I = \dfrac{\pi}{3} \left(-\dfrac{e^{-r^2}}{2} \right) \Big|_0^2 = \dfrac{\pi(1-e^{-4})}{6}$

6. $I = \displaystyle\int_0^2 dy \int_0^y f(x,y) \, dx + \int_2^6 dy \int_0^{\sqrt{6-y}} f(x,y) \, dx$

$= \displaystyle\iint_R f(x,y) \, dA,$

where R is as shown in the figure. Thus

$I = \displaystyle\int_0^2 dx \int_x^{6-x^2} f(x,y) \, dy.$

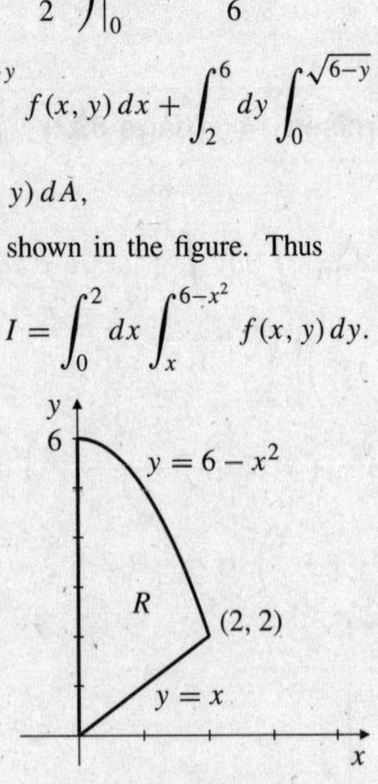

$y = 6 - x^2$

R

$(2, 2)$

$y = x$

Fig. R-14.6

8.   A horizontal slice of the object at height z above the base, and having thickness dz, is a disk of radius $r = \frac{1}{2}(10-z)$ m. Its volume is

$dV = \pi \dfrac{(10-z)^2}{4} \, dz \text{ m}^3.$

The density of the slice is $\delta = kz^2$ kg/m^3. Since $\delta = 3,000$ when $z = 10$, we have $k = 30$.

   a) The mass of the object is

$m = \displaystyle\int_0^{10} 30z^2 \dfrac{\pi}{4} (10-z)^2 \, dz$

$= \dfrac{15\pi}{2} \displaystyle\int_0^{10} (100z^2 - 20z^3 + z^4) \, dz$

$= \dfrac{15\pi}{2} \left(\dfrac{100,000}{3} - 50,000 + 20,000 \right) \approx 78,540 \text{ kg}.$

   b) The moment of inertia (about its central axis) of the disk-shaped slice at height z is

$dI = 30z^2 \, dz \displaystyle\int_0^{2\pi} d\theta \int_0^{(10-z)/2} r^3 \, dr.$

Thus the moment of inertia about the whole solid cone is

$I = \displaystyle\int_0^{10} 30z^2 \, dz \int_0^{2\pi} d\theta \int_0^{(10-z)/2} r^3 \, dr.$

10.   If $f(x,y) = \lfloor x+y \rfloor$, then $f = 0, 1$, or 2, in parts of the quarter disk Q, as shown in the figure.

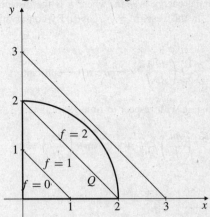

Fig. R-14.10

Thus

$\displaystyle\iint_Q f(x,y) \, dA = 0 \left(\dfrac{1}{2} \right) + 1 \left(\dfrac{3}{2} \right) + 2 \, (\pi - 2) = 2\pi - \dfrac{5}{2},$

and $\bar{f} = \dfrac{1}{\pi} \left(2\pi - \dfrac{5}{2} \right) = 2 - \dfrac{5}{2\pi}.$

12. The solid S lies above the region in the xy-plane bounded by the circle $x^2 + y^2 = 2ay$, which has polar equation $r = 2a \sin \theta$, $(0 \le \theta \le \pi)$. It lies below the cone $z = \sqrt{x^2 + y^2} = r$. The moment of inertia of S about the z-axis is

$$I = \iiint_S (x^2 + y^2)\, dV = \int_0^\pi d\theta \int_0^{2a \sin \theta} r^3\, dr \int_0^r dz$$

$$= \int_0^\pi d\theta \int_0^{2a \sin \theta} r^4\, dr = \frac{32a^5}{5} \int_0^\pi \sin^5 \theta\, d\theta$$

$$= \frac{32a^5}{5} \int_0^\pi (1 - \cos^2 \theta)^2 \sin \theta\, d\theta \quad \text{Let } u = \cos \theta$$
$$\qquad\qquad\qquad\qquad\qquad\qquad\qquad du = -\sin \theta\, d\theta$$

$$= \frac{32a^5}{5} \int_{-1}^1 (1 - 2u^2 + u^4)\, du$$

$$= \frac{64a^5}{5}\left(1 - \frac{2}{3} + \frac{1}{5}\right) = \frac{512a^5}{75}.$$

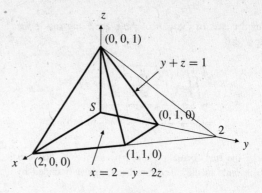

Fig. R-14.14

16. The plane $z = 2x$ intersects the paraboloid $z = x^2 + y^2$ on the circular cylinder $x^2 + y^2 = 2x$, (that is, $(x - 1)^2 + y^2 = 1$), which has radius 1. Since $dS = \sqrt{1 + 2^2}\, dA = \sqrt{5}\, dA$ on the plane, the area of the part of the plane inside the paraboloid (and therefore inside the cylinder) is $\sqrt{5}$ times the area of a circle of radius 1, that is, $\sqrt{5}\pi$ square units.

18. The region R inside the ellipsoid $\frac{x^2}{36} + \frac{y^2}{9} + \frac{z^2}{4} = 1$ and above the plane $x + y + z = 1$ is transformed by the change of variables

$$x = 6u, \quad y = 3v, \quad z = 2w$$

to the region S inside the sphere $u^2 + v^2 + w^2 = 1$ and above the plane $6u + 3v + 2w = 1$. The distance from the origin to this plane is

$$D = \frac{1}{\sqrt{6^2 + 3^2 + 2^2}} = \frac{1}{7},$$

so, by symmetry, the volume of S is equal to the volume inside the sphere and above the plane $w = 1/7$, that is,

$$\int_{1/7}^1 \pi(1 - w^2)\, dw = \pi\left(w - \frac{w^3}{3}\right)\Big|_{1/7}^1 = \frac{180\pi}{343} \text{ units}^3.$$

Since $|\partial(x, y, z)/\partial(u, v, w)| = 6 \cdot 3 \cdot 2 = 18$, the volume of R is $18 \times (180\pi/343) = 3240\pi/343 \approx 29.68$ cu. units.

Challenging Problems 14   (page 893)

2. The plane $(x/a) + (y/b) + (z/c) = 1$ intersects the ellipsoid $(x/a)^2 + (y/b)^2 + (z/c)^2 = 1$ above the region R in the xy-plane bounded by the ellipse

$$\frac{x^2}{a^2} + \frac{y^2}{b^2} + \left(1 - \frac{x}{a} - \frac{y}{b}\right)^2 = 1,$$

or, equivalently,

$$\frac{x^2}{a^2} + \frac{y^2}{b^2} + \frac{xy}{ab} - \frac{x}{a} - \frac{y}{b} = 0.$$

14.
$$V = \iiint_S dV = \int_0^1 dy \int_0^{1-y} dz \int_0^{2-y-2z} dx$$

$$= \int_0^1 dy \int_0^{1-y} (2 - y - 2z)\, dz$$

$$= \int_0^1 [(2 - y)(1 - y) - (1 - y)^2]\, dy$$

$$= \int_0^1 (1 - y)\, dy = \frac{1}{2}$$

$$M_{x=0} = \iiint_S x\, dV = \int_0^1 dy \int_0^{1-y} dz \int_0^{2-y-2z} x\, dx$$

$$= \frac{1}{2} \int_0^1 dy \int_0^{1-y} [(2 - y)^2 - 4(2 - y)z + 4z^2]\, dz$$

$$= \frac{1}{2} \int_0^1 \left[(2 - y)^2(1 - y) - 2(2 - y)(1 - y)^2\right.$$

$$\left. + \frac{4}{3}(1 - y)^3\right] dy \quad \text{Let } u = 1 - y$$
$$\qquad\qquad\qquad\qquad\qquad\qquad du = -dy$$

$$= \frac{1}{2} \int_0^1 \left[(u + 1)^2 u - 2(u + 1)u^2 + \frac{4}{3}u^3\right] du$$

$$= \frac{1}{2} \int_0^1 \left[\frac{1}{3}u^3 + u\right] du = \frac{7}{24}$$

$$\bar{x} = \frac{7}{24} \Big/ \frac{1}{2} = \frac{7}{12}$$

Thus the area of the part of the plane lying inside the ellipsoid is

$$S = \iint_R \sqrt{1 + \frac{c^2}{a^2} + \frac{c^2}{b^2}}\, dx\, dy$$
$$= \frac{\sqrt{a^2 b^2 + a^2 c^2 + b^2 c^2}}{ab}\, (\text{area of } R).$$

Under the transformation $x = a(u + v)$, $y = b(u - v)$, R corresponds to the ellipse in the uv-plane bounded by

$$(u + v)^2 + (u - v)^2 + (u^2 - v^2) - (u + v) - (u - v) = 0$$
$$3u^2 + v^2 - 2u = 0$$
$$3\left(u^2 - \frac{2}{3}u + \frac{1}{9}\right) + v^2 = \frac{1}{3}$$
$$\frac{(u - 1/3)^2}{1/9} + \frac{v^2}{1/3} = 1,$$

an ellipse with area $\pi(1/3)(1/\sqrt{3}) = \pi/(3\sqrt{3})$ sq. units. Since

$$dx\, dy = \left| \begin{vmatrix} a & a \\ b & -b \end{vmatrix} \right|\, du\, dv = 2ab\, du\, dv,$$

we have

$$S = \frac{2\pi}{3\sqrt{3}} \sqrt{a^2 b^2 + a^2 c^2 + b^2 c^2} \text{ sq. units.}$$

4. Under the transformation $u = \mathbf{a} \bullet \mathbf{r}$, $v = \mathbf{b} \bullet \mathbf{r}$, $w = \mathbf{c} \bullet \mathbf{r}$, where $\mathbf{r} = x\mathbf{i} + y\mathbf{j} + z\mathbf{k}$, the parallelepiped P corresponds to the rectangle R specified by $0 \le u \le d_1$, $0 \le v \le d_2$, $0 \le w \le d_3$. If $\mathbf{a} = a_1\mathbf{i} + a_2\mathbf{j} + a_3\mathbf{k}$ and similar expressions hold for $\mathbf{b}$ and $\mathbf{c}$, then

$$\frac{\partial(u, v, w)}{\partial(x, y, z)} = \begin{vmatrix} a_1 & a_2 & a_3 \\ b_1 & b_2 & b_3 \\ c_1 & c_2 & c_3 \end{vmatrix} = \mathbf{a} \bullet (\mathbf{b} \times \mathbf{c}).$$

Therefore

$$dx\, dy\, dz = \left| \frac{\partial(x, y, z)}{\partial(u, v, w)} \right|\, du\, dv\, dw = \frac{du\, dv\, dw}{|\mathbf{a} \bullet (\mathbf{b} \times \mathbf{c})|},$$

and we have

$$\iiint_P (\mathbf{a} \bullet \mathbf{r})(\mathbf{b} \bullet \mathbf{r})(\mathbf{c} \bullet \mathbf{r})\, dx\, dy\, dz$$
$$= \iiint_R \frac{uvw}{|\mathbf{a} \bullet (\mathbf{b} \times \mathbf{c})|}\, du\, dv\, dw$$
$$= \frac{1}{|\mathbf{a} \bullet (\mathbf{b} \times \mathbf{c})|} \int_0^{d_1} u\, du \int_0^{d_2} v\, dv \int_0^{d_3} w\, dw$$
$$= \frac{d_1^2 d_2^2 d_3^2}{8|\mathbf{a} \bullet (\mathbf{b} \times \mathbf{c})|}.$$

6. Under the transformation $x = u^3$, $y = v^3$, $z = w^3$, the region R bounded by the surface $x^{2/3} + y^{2/3} + z^{2/3} = a^{2/3}$ gets mapped to the ball B bounded by $u^2 + v^2 + w^2 = a^{2/3}$. Assume that $a > 0$. Since

$$\frac{\partial(x, y, z)}{\partial(u, v, w)} = 27u^2 v^2 w^2,$$

the volume of R is

$$V = 27 \iiint_B u^2 v^2 w^2\, du\, dv\, dw.$$

Now switch to polar coordinates $[\rho, \phi, \theta]$ in uvw-space. Since

$$uvw = (\rho \sin\phi \cos\theta)(\rho \sin\phi \sin\theta)(\rho \cos\phi),$$

we have

$$V = 27 \int_0^{2\pi} \cos^2\theta \sin^2\theta\, d\theta \int_0^{\pi} \sin^5\phi \cos^2\phi\, d\phi \int_0^{a^{1/3}} \rho^8\, d\rho$$
$$= 3a^3 \int_0^{2\pi} \frac{\sin^2(2\theta)}{4}\, d\theta \int_0^{\pi} (1 - \cos^2\phi)^2 \cos^2\phi \sin\phi\, d\phi$$

Let $t = \cos\phi$, $dt = -\sin\phi\, d\phi$

$$= 3a^3 \int_0^{2\pi} \frac{1 - \cos(4\theta)}{8}\, d\theta \int_{-1}^{1} (1 - t^2)^2 t^2\, dt$$
$$= \frac{3a^3}{8}(2\pi)2 \int_0^1 (t^2 - 2t^4 + t^6)\, dt = \frac{4\pi a^3}{35} \text{ cu. units.}$$

CHAPTER 15. VECTOR FIELDS

Section 15.1 Vector and Scalar Fields (page 900)

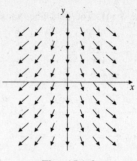

Fig. 15.1.6

2. $\mathbf{F} = x\mathbf{i} + y\mathbf{j}$.

The field lines satisfy $\dfrac{dx}{x} = \dfrac{dy}{y}$.

Thus $\ln y = \ln x + \ln C$, or $y = Cx$. The field lines are straight half-lines emanating from the origin.

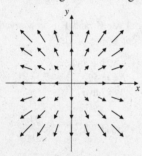

Fig. 15.1.2

8. $\mathbf{F} = \cos y\,\mathbf{i} - \cos x\,\mathbf{j}$.

The field lines satisfy $\dfrac{dx}{\cos y} = -\dfrac{dy}{\cos x}$, that is, $\cos x\,dx + \cos y\,dy = 0$. Thus they are the curves $\sin x + \sin y = C$.

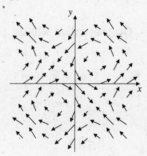

Fig. 15.1.8

4. $\mathbf{F} = \mathbf{i} + \sin x\,\mathbf{j}$.

The field lines satisfy $dx = \dfrac{dy}{\sin x}$.

Thus $\dfrac{dy}{dx} = \sin x$. The field lines are the curves $y = -\cos x + C$.

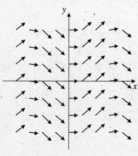

Fig. 15.1.4

10. $\mathbf{v}(x, y, z) = x\mathbf{i} + y\mathbf{j} - x\mathbf{k}$.

The streamlines satisfy $\dfrac{dx}{x} = \dfrac{dy}{y} = -\dfrac{dz}{x}$. Thus $z + x = C_1$, $y = C_2 x$. The streamlines are straight half-lines emanating from the z-axis and perpendicular to the vector $\mathbf{i} + \mathbf{k}$.

12. $\mathbf{v} = \dfrac{x\mathbf{i} + y\mathbf{j}}{(1 + z^2)(x^2 + y^2)}$.

The streamlines satisfy $dz = 0$ and $\dfrac{dx}{x} = \dfrac{dy}{y}$. Thus $z = C_1$ and $y = C_2 x$. The streamlines are horizontal half-lines emanating from the z-axis.

14. $\mathbf{v} = e^{xyz}(x\mathbf{i} + y^2\mathbf{j} + z\mathbf{k})$. The field lines satisfy

$$\frac{dx}{x} = \frac{dy}{y^2} = \frac{dz}{z},$$

so they are given by $z = C_1 x$, $\ln|x| = \ln|C_2| - (1/y)$ (or, equivalently, $x = C_2 e^{-1/y}$).

6. $\mathbf{F} = \boldsymbol{\nabla}(x^2 - y) = 2x\mathbf{i} - \mathbf{j}$.

The field lines satisfy $\dfrac{dx}{2x} = \dfrac{dy}{-1}$. They are the curves $y = -\dfrac{1}{2}\ln x + C$.

16. $\mathbf{v}(x, y) = x\mathbf{i} + (x + y)\mathbf{j}$. The field lines satisfy

$$\frac{dx}{x} = \frac{dy}{x + y}$$

$$\frac{dy}{dx} = \frac{x + y}{x} \qquad \text{Let } y = xv(x)$$

$$\frac{dy}{dx} = v + x\frac{dv}{dx}$$

$$v + x\frac{dv}{dx} = \frac{x(1 + v)}{x} = 1 + v.$$

Thus $dv/dx = 1/x$, and so $v(x) = \ln|x| + C$. The field lines have equations $y = x \ln|x| + Cx$.

18. $\mathbf{F} = \hat{\mathbf{r}} + \theta\hat{\boldsymbol{\theta}}$. The field lines satisfy $dr = r\, d\theta/\theta$, or $dr/r = d\theta/\theta$, so they are the spirals $r = C\theta$.

20. $\mathbf{F} = r\hat{\mathbf{r}} - \hat{\boldsymbol{\theta}}$. The field lines satisfy $dr/r = -r\, d\theta$, or $-dr/r^2 = d\theta$, so they are the spirals $1/r = \theta + C$, or $r = 1/(\theta + C)$.

Section 15.2 Conservative Fields (page 909)

2. $\mathbf{F} = y\mathbf{i} + x\mathbf{j} + z^2\mathbf{k}$, $F_1 = y$, $F_2 = x$, $F_3 = z^2$. We have

$$\frac{\partial F_1}{\partial y} = 1 = \frac{\partial F_2}{\partial x},$$

$$\frac{\partial F_1}{\partial z} = 0 = \frac{\partial F_3}{\partial x},$$

$$\frac{\partial F_2}{\partial z} = 0 = \frac{\partial F_3}{\partial y}.$$

Therefore, $\mathbf{F}$ may be conservative. If $\mathbf{F} = \nabla\phi$, then

$$\frac{\partial\phi}{\partial x} = y, \qquad \frac{\partial\phi}{\partial y} = x, \qquad \frac{\partial\phi}{\partial z} = z^2.$$

Therefore,

$$\phi(x, y, z) = \int y\, dx = xy + C_1(y, z)$$

$$x = \frac{\partial\phi}{\partial y} = x + \frac{\partial C_1}{\partial y} \Rightarrow \frac{\partial C_1}{\partial y} = 0$$

$$C_1(y, z) = C_2(z), \qquad \phi(x, y, z) = xy + C_2(z)$$

$$z^2 = \frac{\partial\phi}{\partial z} = C_2'(z) \Rightarrow C_2(z) = \frac{z^3}{3}.$$

Thus $\phi(x, y, z) = xy + \dfrac{z^3}{3}$ is a potential for $\mathbf{F}$, and $\mathbf{F}$ is conservative on $\mathbb{R}^3$.

4. $\mathbf{F} = \dfrac{x\mathbf{i} + y\mathbf{j}}{x^2 + y^2}$, $F_1 = \dfrac{x}{x^2 + y^2}$, $F_2 = \dfrac{y}{x^2 + y^2}$. We have

$$\frac{\partial F_1}{\partial y} = -\frac{2xy}{(x^2 + y^2)^2} = \frac{\partial F_2}{\partial x}.$$

Therefore, $\mathbf{F}$ may be conservative. If $\mathbf{F} = \nabla\phi$, then

$$\frac{\partial\phi}{\partial x} = \frac{x}{x^2 + y^2}, \qquad \frac{\partial\phi}{\partial y} = \frac{y}{x^2 + y^2}.$$

Therefore,

$$\phi(x, y) = \int \frac{x}{x^2 + y^2}\, dx = \frac{\ln(x^2 + y^2)}{2} + C_1(y)$$

$$\frac{y}{x^2 + y^2} = \frac{\partial\phi}{\partial y} = \frac{y}{x^2 + y^2} + c_1'(y) \Rightarrow c_1'(y) = 0.$$

Thus we can choose $C_1(y) = 0$, and

$$\phi(x, y) = \frac{1}{2}\ln(x^2 + y^2)$$

is a scalar potential for $\mathbf{F}$, and $\mathbf{F}$ is conservative everywhere on $\mathbb{R}^2$ except at the origin.

6. $\mathbf{F} = e^{x^2 + y^2 + z^2}(xz\mathbf{i} + yz\mathbf{j} + xy\mathbf{k})$.
$F_1 = xze^{x^2 + y^2 + z^2}$, $F_2 = yze^{x^2 + y^2 + z^2}$,
$F_3 = xye^{x^2 + y^2 + z^2}$. We have

$$\frac{\partial F_1}{\partial y} = 2xyze^{x^2 + y^2 + z^2} = \frac{\partial F_2}{\partial x},$$

$$\frac{\partial F_1}{\partial z} = (x + 2xz^2)e^{x^2 + y^2 + z^2},$$

$$\frac{\partial F_3}{\partial x} = (y + 2x^2y)e^{x^2 + y^2 + z^2} \neq \frac{\partial F_1}{\partial z}.$$

Thus $\mathbf{F}$ cannot be conservative.

8. $\dfrac{\partial}{\partial x}\ln|\mathbf{r}| = \dfrac{1}{|\mathbf{r}|}\dfrac{\mathbf{r} \bullet \dfrac{\partial\mathbf{r}}{\partial x}}{|\mathbf{r}|} = \dfrac{x}{|\mathbf{r}|^2}$

$\nabla\ln|\mathbf{r}| = \dfrac{x\mathbf{i} + y\mathbf{j} + z\mathbf{k}}{|\mathbf{r}|^2} = \dfrac{\mathbf{r}}{|\mathbf{r}|^2}.$

10. $\mathbf{F} = \dfrac{2x}{z}\mathbf{i} + \dfrac{2y}{z}\mathbf{j} - \dfrac{x^2 + y^2}{z^2}\mathbf{k} = \mathbf{G} + \mathbf{k}$,
where $\mathbf{G}$ is the vector field $\mathbf{F}$ of Exercise 9. Since $\mathbf{G}$ is conservative (except on the plane $z = 0$), so is $\mathbf{F}$, which has scalar potential

$$\phi(x, y, z) = \frac{x^2 + y^2}{z} + z = \frac{x^2 + y^2 + z^2}{z},$$

since $\dfrac{x^2 + y^2}{z}$ is a potential for $\mathbf{G}$ and z is a potential for the vector $\mathbf{k}$.

The equipotential surfaces of $\mathbf{F}$ are $\phi(x, y, z) = C$, or

$$x^2 + y^2 + z^2 = Cz$$

which are spheres tangent to the xy-plane having centres on the z-axis.

The field lines of **F** satisfy

$$\frac{dx}{\frac{2x}{z}} = \frac{dy}{\frac{2y}{z}} = \frac{dz}{1 - \frac{x^2 + y^2}{z^2}}.$$

As in Exercise 9, the first equation has solutions $y = Ax$, representing vertical planes containing the z-axis. The remaining equations can then be written in the form

$$\frac{dz}{dx} = \frac{z^2 - x^2 - y^2}{2xz} = \frac{z^2 - (1 + A^2)x^2}{2zx}.$$

This first order DE is of homogeneous type (see Section 9.2), and can be solved by a change of dependent variable: $z = xv(x)$. We have

$$v + x\frac{dv}{dx} = \frac{dz}{dx} = \frac{x^2v^2 - (1 + A^2)x^2}{2x^2v}$$

$$x\frac{dv}{dx} = \frac{v^2 - (1 + A^2)}{2v} - v = -\frac{v^2 + (1 + A^2)}{2v}$$

$$\frac{2v\,dv}{v^2 + (1 + A^2)} = -\frac{dx}{x}$$

$$\ln\left(v^2 + (1 + A^2)\right) = -\ln x + \ln B$$

$$v^2 + 1 + A^2 = \frac{B}{x}$$

$$\frac{z^2}{x^2} + 1 + A^2 = \frac{B}{x}$$

$$z^2 + x^2 + y^2 = Bx.$$

These are spheres centred on the x-axis and passing through the origin. The field lines are the intersections of the planes $y = Ax$ with these spheres, so they are vertical circles passing through the origin and having centres in the xy-plane. (The technique used to find these circles excludes those circles with centres on the y-axis, but they are also field lines of **F**.)

Note: In two dimensions, circles passing through the origin and having centres on the x-axis intersect perpendicularly circles passing through the origin and having centres on the y-axis. Thus the nature of the field lines of **F** can be determined geometrically from the nature of the equipotential surfaces.

12. The scalar potential for the source-sink system is

$$\phi(x, y, z) = \phi(\mathbf{r}) = -\frac{2}{|\mathbf{r}|} + \frac{1}{|\mathbf{r} - \mathbf{k}|}.$$

Thus, the velocity field is

$$\mathbf{v} = \boldsymbol{\nabla}\phi = \frac{2\mathbf{r}}{|\mathbf{r}|^3} - \frac{\mathbf{r} - \mathbf{k}}{|\mathbf{r} - \mathbf{k}|^3}$$

$$= \frac{2(x\mathbf{i} + y\mathbf{j} + z\mathbf{k})}{(x^2 + y^2 + z^2)^{3/2}} - \frac{x\mathbf{i} + y\mathbf{j} + (z - 1)\mathbf{k}}{(x^2 + y^2 + (z - 1)^2)^{3/2}}.$$

For vertical velocity we require

$$\frac{2x}{(x^2 + y^2 + z^2)^{3/2}} = \frac{x}{(x^2 + y^2 + (z - 1)^2)^{3/2}},$$

and a similar equation for y. Both equations will be satisfied at all points of the z-axis, and also wherever

$$2\left(x^2 + y^2 + (z - 1)^2\right)^{3/2} = \left(x^2 + y^2 + z^2\right)^{3/2}$$

$$2^{2/3}\left(x^2 + y^2 + (z - 1)^2\right) = x^2 + y^2 + z^2$$

$$x^2 + y^2 + (z - K)^2 = K^2 - K,$$

where $K = 2^{2/3}/(2^{2/3} - 1)$. This latter equation represents a sphere, S, since $K^2 - K > 0$. The velocity is vertical at all points on S, as well as at all points on the z-axis.

Since the source at the origin is twice as strong as the sink at $(0, 0, 1)$, only half the fluid it emits will be sucked into the sink. By symmetry, this half will the half emitted into the half-space $z > 0$. The rest of the fluid emitted at the origin will flow outward to infinity. There is one point where $\mathbf{v} = \mathbf{0}$. This point (which is easily calculated to be $(0, 0, 2 + \sqrt{2})$) lies inside S. Streamlines emerging from the origin parallel to the xy-plane lead to this point. Streamlines emerging into $z > 0$ cross S and approach the sink. Streamlines emerging into $z < 0$ flow to infinity. Some of these cross S twice, some others are tangent to S, some do not intersect S anywhere.

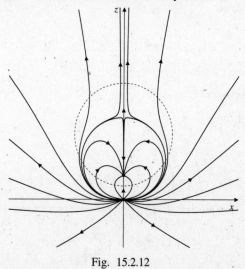

Fig. 15.2.12

14. For $\mathbf{v}(x, y) = \dfrac{m(x\mathbf{i} + y\mathbf{j})}{x^2 + y^2}$, we have

$$\frac{\partial v_1}{\partial y} = -\frac{2mxy}{(x^2 + y^2)^2} = \frac{\partial v_2}{\partial x},$$

so **v** may be conservative, except at $(0, 0)$. We have

$$\phi(x, y) = m \int \frac{x\, dx}{x^2 + y^2} = \frac{m}{2} \ln(x^2 + y^2) + C_1(y)$$

$$\frac{my}{x^2 + y^2} = \frac{\partial \phi}{\partial y} = \frac{my}{x^2 + y^2} + \frac{dC_1}{dy}.$$

Thus we may take $C_1(y) = 0$, and obtain

$$\phi(x, y) = \frac{m}{2} \ln(x^2 + y^2) = m \ln |\mathbf{r}|,$$

as a scalar potential for the velocity field **v** of a line source of strength of m.

16. The equipotential curves for the two-dimensional dipole have equations $y = 0$ or

$$-\frac{\mu y}{x^2 + y^2} = \frac{1}{C}$$
$$x^2 + y^2 + \mu C y = 0$$
$$x^2 + \left(y + \frac{\mu C}{2}\right)^2 = \frac{\mu^2 C^2}{4}.$$

These equipotentials are circles tangent to the x-axis at the origin.

18. The velocity field for a point source of strength $m\, dt$ at $(0, 0, t)$ is

$$\mathbf{v}_t(x, y, z) = \frac{m\left(x\mathbf{i} + y\mathbf{j} + (z - t)\mathbf{k}\right)}{\left(x^2 + y^2 + (z - t)^2\right)^{3/2}}.$$

Hence we have

$$\int_{-\infty}^{\infty} \mathbf{v}_t(x, y, z)\, dt$$

$$= m \int_{-\infty}^{\infty} \frac{x\mathbf{i} + y\mathbf{j} + (z - t)\mathbf{k}}{\left(x^2 + y^2 + (z - t)^2\right)^{3/2}}\, dt$$

$$= m(x\mathbf{i} + y\mathbf{j}) \int_{-\infty}^{\infty} \frac{dt}{\left(x^2 + y^2 + (z - t)^2\right)^{3/2}}$$

Let $z - t = \sqrt{x^2 + y^2} \tan\theta$

$-dt = \sqrt{x^2 + y^2} \sec^2\theta\, d\theta$

$$= \frac{m(x\mathbf{i} + y\mathbf{j})}{x^2 + y^2} \int_{-\pi/2}^{\pi/2} \cos\theta\, d\theta$$

$$= \frac{2m(x\mathbf{i} + y\mathbf{j})}{x^2 + y^2},$$

which is the velocity field of a line source of strength $2m$ along the z-axis.

The definition of strength of a point source in 3-space was made to ensure that the velocity field of a source of strength 1 had speed 1 at distance 1 from the source. This corresponds to fluid being emitted from the source at a volume rate of 4π. Similarly, the definition of strength of a line source guaranteed that a source of strength 1 gives rise to fluid speed of 1 at unit distance 1 from the line source. This corresponds to a fluid emission at a volume rate 2π per unit length along the line. Thus, the integral of a 3-dimensional source gives twice the volume rate of a 2-dimensional source, per unit length along the line.

The potential of a point source $m\, dt$ at $(0, 0, t)$ is

$$\phi(x, y, z) = -\frac{m}{\sqrt{x^2 + y^2 + (x - t)^2}}.$$

This potential cannot be integrated to give the potential for a line source along the z-axis because the integral

$$-m \int_{-\infty}^{\infty} \frac{dt}{\sqrt{x^2 + y^2 + (z - t)^2}}$$

does not converge, in the usual sense in which convergence of improper integrals was defined.

20. If $\mathbf{F} = F_r(r, \theta)\hat{\mathbf{r}} + F_\theta(r, \theta)\hat{\boldsymbol{\theta}}$ is conservative, then $\mathbf{F} = \nabla\phi$ for some scalar field $\phi(r, \theta)$, and by Exercise 19,

$$\frac{\partial \phi}{\partial r} = F_r, \quad \frac{1}{r}\frac{\partial \phi}{\partial \theta} = F_\theta.$$

For the equality of the mixed second partial derivatives of ϕ, we require that

$$\frac{\partial F_r}{\partial \theta} = \frac{\partial}{\partial r}(r F_\theta) = F_\theta + r\frac{\partial F_\theta}{\partial r},$$

that is, $\dfrac{\partial F_r}{\partial \theta} - r\dfrac{\partial F_\theta}{\partial r} = F_\theta.$

22. If $\mathbf{F} = r^2 \cos\theta\,\hat{\mathbf{r}} + \alpha r^\beta \sin\theta\,\hat{\boldsymbol{\theta}} = \nabla\phi(r, \theta)$, then we must have

$$\frac{\partial \phi}{\partial r} = r^2 \cos\theta, \quad \frac{1}{r}\frac{\partial \phi}{\partial \theta} = \alpha r^\beta \sin\theta.$$

From the first equation

$$\phi(r, \theta) = \frac{r^3}{3}\cos\theta + C(\theta).$$

The second equation then gives

$$C'(\theta) - \frac{r^3}{3}\sin\theta = \frac{\partial \phi}{\partial \theta} = \alpha r^{\beta+1}\sin\theta.$$

This equation can be solved for a function $C(\theta)$ independent of r only if $\alpha = -1/3$ and $\beta = 2$. In this case, $C(\theta) = C$ (a constant). $\mathbf{F}$ is conservative if α and β have these values, and a potential for it is $\phi = \frac{1}{3}r^3\cos\theta + C$.

Section 15.3   Line Integrals   (page 915)

2. $\mathcal{C}: x = t\cos t,\ y = t\sin t,\ z = t,\ (0 \le t \le 2\pi)$. We have

$$ds = \sqrt{(\cos t - t\sin t)^2 + (\sin t + t\cos t)^2 + 1}\, dt$$
$$= \sqrt{2 + t^2}\, dt.$$

Thus

$$\int_{\mathcal{C}} z\, ds = \int_0^{2\pi} t\sqrt{2 + t^2}\, dt \quad \text{Let } u = 2 + t^2$$
$$\phantom{\int_{\mathcal{C}} z\, ds =} du = 2t\, dt$$
$$= \frac{1}{2}\int_2^{2+4\pi^2} u^{1/2}\, du$$
$$= \frac{1}{3}u^{3/2}\Big|_2^{2+4\pi^2} = \frac{(2 + 4\pi^2)^{3/2} - 2^{3/2}}{3}.$$

4. The wire of Example 3 lies in the first octant on the surfaces $z = x^2$ and $z = 2 - x^2 - 2y^2$, and, therefore, also on the surface $x^2 = 2 - x^2 - 2y^2$, or $x^2 + y^2 = 1$, a circular cylinder. Since it goes from $(1, 0, 1)$ to $(0, 1, 0)$ it can be parametrized

$$\mathbf{r} = \cos t\mathbf{i} + \sin t\mathbf{j} + \cos^2 t\,\mathbf{k}, \quad (0 \le t \le \pi/2)$$
$$\mathbf{v} = -\sin t\mathbf{i} + \cos t\mathbf{j} - 2\cos t\sin t\mathbf{k}$$
$$v = \sqrt{1 + \sin^2(2t)} = \sqrt{2 - \cos^2(2t)}.$$

Since the wire has density $\delta = xy = \sin t\cos t = \frac{1}{2}\sin(2t)$, its mass is

$$m = \frac{1}{2}\int_0^{\pi/2} \sqrt{2 - \cos^2(2t)}\,\sin(2t)\, dt \quad \text{Let } v = \cos(2t)$$
$$ dv = -2\sin(2t)\, dt$$
$$= \frac{1}{4}\int_{-1}^1 \sqrt{2 - v^2}\, dv = \frac{1}{2}\int_0^1 \sqrt{2 - v^2}\, dv,$$

which is the same integral obtained in Example 3, and has value $(\pi + 2)/8$.

6. $\mathcal{C}$ is the same curve as in Exercise 5. We have

$$\int_{\mathcal{C}} e^z\, ds = \int_0^{2\pi} e^t\sqrt{1 + 2e^{2t}}\, dt \quad \text{Let } \sqrt{2}e^t = \tan\theta$$
$$\phantom{\int_{\mathcal{C}} e^z\, ds =} \sqrt{2}e^t\, dt = \sec^2\theta\, d\theta$$
$$= \frac{1}{\sqrt{2}}\int_{t=0}^{t=2\pi} \sec^3\theta\, d\theta$$
$$= \frac{1}{2\sqrt{2}}\Big[\sec\theta\tan\theta + \ln|\sec\theta + \tan\theta|\Big]\Big|_{t=0}^{t=2\pi}$$
$$= \frac{\sqrt{2}e^t\sqrt{1 + 2e^{2t}} + \ln(\sqrt{2}e^t + \sqrt{1 + 2e^{2t}})}{2\sqrt{2}}\Big|_0^{2\pi}$$
$$= \frac{e^{2\pi}\sqrt{1 + 2e^{4\pi}} - \sqrt{3}}{2}$$
$$\quad + \frac{1}{2\sqrt{2}}\ln\frac{\sqrt{2}e^{2\pi} + \sqrt{1 + 2e^{4\pi}}}{\sqrt{2} + \sqrt{3}}.$$

8. The curve $\mathcal{C}$ of intersection of $x^2 + z^2 = 1$ and $y = x^2$ can be parametrized

$$\mathbf{r} = \cos t\mathbf{i} + \cos^2 t\mathbf{j} + \sin t\mathbf{k}, \quad (0 \le t \le 2\pi).$$

Thus

$$ds = \sqrt{\sin^2 t + 4\sin^2 t\cos^2 t + \cos^2 t}\, dt = \sqrt{1 + \sin^2 2t}\, dt.$$

We have

$$\int_{\mathcal{C}} \sqrt{1 + 4x^2z^2}\, ds$$
$$= \int_0^{2\pi} \sqrt{1 + 4\cos^2 t\sin^2 t}\sqrt{1 + \sin^2 2t}\, dt$$
$$= \int_0^{2\pi} (1 + \sin^2 2t)\, dt$$
$$= \int_0^{2\pi} \left(1 + \frac{1 - \cos 4t}{2}\right) dt$$
$$= \frac{3}{2}(2\pi) = 3\pi.$$

10. Here the wire of Exercise 9 extends only from $t = 0$ to $t = \pi$:

$$m = \sqrt{2}\int_0^\pi t\, dt = \frac{\pi^2\sqrt{2}}{2}$$
$$M_{x=0} = \sqrt{2}\int_0^\pi t\cos t\, dt = -2\sqrt{2}$$
$$M_{y=0} = \sqrt{2}\int_0^\pi t\sin t\, dt = \pi\sqrt{2}$$
$$M_{z=0} = \sqrt{2}\int_0^\pi t^2\, dt = \frac{\pi^3\sqrt{2}}{3}.$$

The centre of mass is $\left(-\dfrac{4}{\pi^2}, \dfrac{2}{\pi}, \dfrac{2\pi}{3}\right)$.

12.
$$m = \int_0^1 (e^t + e^{-t})\, dt = \frac{e^2 - 1}{e}$$
$$M_{x=0} = \int_0^1 e^t(e^t + e^{-t})\, dt = \frac{e^2 + 1}{2}$$
$$M_{y=0} = \int_0^1 \sqrt{2}t(e^t + e^{-t})\, dt = \frac{2\sqrt{2}(e - 1)}{e}$$
$$M_{z=0} = \int_0^1 e^{-t}(e^t + e^{-t})\, dt = \frac{3e^2 - 1}{2e^2}$$

The centroid is $\left(\dfrac{e^3 + e}{2e^2 - 2}, \dfrac{2\sqrt{2}}{e + 1}, \dfrac{3e^2 - 1}{2e^3 - 2e}\right)$.

14. On $\mathcal{C}$, we have

$$z = \sqrt{1 - x^2 - y^2} = \sqrt{1 - x^2 - (1-x)^2} = \sqrt{2(x - x^2)}.$$

Thus $\mathcal{C}$ can be parametrized

$$\mathbf{r} = t\mathbf{i} + (1-t)\mathbf{j} + \sqrt{2(t - t^2)}\mathbf{k}, \quad (0 \le t \le 1).$$

Hence

$$ds = \sqrt{1 + 1 + \frac{(1 - 2t)^2}{2(t - t^2)}}\, dt = \frac{dt}{\sqrt{2(t - t^2)}}.$$

We have

$$\int_{\mathcal{C}} z\, ds = \int_0^1 \sqrt{2(t - t^2)} \frac{dt}{\sqrt{2(t - t^2)}} = 1.$$

16. $\mathcal{C}: y = x^2,\ z = y^2$, from $(0,0,0)$ to $(2,4,16)$. Parametrize $\mathcal{C}$ by

$$\mathbf{r} = t\mathbf{i} + t^2\mathbf{j} + t^4\mathbf{k}, \quad (0 \le t \le 2).$$

Since $ds = \sqrt{1 + 4t^2 + 16t^6}\, dt$, we have

$$\int_{\mathcal{C}} xyz\, ds = \int_0^2 t^7 \sqrt{1 + 4t^2 + 16t^6}\, dt.$$

18. The straight line $\mathcal{L}$ with equation $Ax + By = C,\ (C \ne 0)$, lies at distance $D = \sqrt{|C|}/\sqrt{A^2 + B^2}$ from the origin. So does the line $\mathcal{L}_1$ with equation $y = D$. Since $x^2 + y^2$ depends only on distance from the origin, we have, by symmetry,

$$\int_{\mathcal{L}} \frac{ds}{x^2 + y^2} = \int_{\mathcal{L}_1} \frac{ds}{x^2 + y^2}$$

$$= \int_{-\infty}^{\infty} \frac{dx}{x^2 + D^2}$$

$$= \frac{2}{D} \tan^{-1}\frac{x}{D} \Big|_0^{\infty} = \frac{2}{D}\left(\frac{\pi}{2} - 0\right)$$

$$= \frac{\pi}{D} = \frac{\pi\sqrt{A^2 + B^2}}{|C|}.$$

Section 15.4 Line Integrals of Vector Fields (page 923)

2. $\mathbf{F} = \cos x\, \mathbf{i} - y\mathbf{j} = \nabla\left(\sin x - \frac{y^2}{2}\right)$.

$\mathcal{C}: \quad y = \sin x$ from $(0,0)$ to $(\pi, 0)$.

$$\int_{\mathcal{C}} \mathbf{F} \bullet d\mathbf{r} = \left(\sin x - \frac{y^2}{2}\right)\Big|_{(0,0)}^{(\pi,0)} = 0.$$

4. $\mathbf{F} = z\mathbf{i} - y\mathbf{j} + 2x\mathbf{k}$.
$\mathcal{C}: \mathbf{r} = t\mathbf{i} + t^2\mathbf{j} + t^3\mathbf{k},\ (0 \le t \le 1)$.

$$\int_{\mathcal{C}} \mathbf{F} \bullet d\mathbf{r} = \int_0^1 [t^3 - t^2(2t) + 2t(3t^2)]\, dt$$

$$= \int_0^1 5t^3\, dt = \frac{5t^4}{4}\Big|_0^1 = \frac{5}{4}.$$

6. $\mathbf{F} = (x - z)\mathbf{i} + (y - z)\mathbf{j} - (x + y)\mathbf{k}$
$$= \nabla\left(\frac{x^2 + y^2}{2} - (x + y)z\right).$$
$\mathcal{C}$ is a given polygonal path from $(0,0,0)$ to $(1,1,1)$ (but any other piecewise smooth path from the first point to the second would do as well).

$$\int_{\mathcal{C}} \mathbf{F} \bullet d\mathbf{r} = \left(\frac{x^2 + y^2}{2} - (x + y)z\right)\Big|_{(0,0,0)}^{(1,1,1)} = 1 - 2 = -1.$$

8. $\mathcal{C}$ is made up of four segments as shown in the figure.
On $\mathcal{C}_1$, $y = 0$, $dy = 0$, and x goes from 0 to 1.
On $\mathcal{C}_2$, $x = 1$, $dx = 0$, and y goes from 0 to 1.
On $\mathcal{C}_3$, $y = 1$, $dy = 0$, and x goes from 1 to 0.
On $\mathcal{C}_4$, $x = 0$, $dx = 0$, and y goes from 1 to 0.
Thus

$$\int_{\mathcal{C}_1} x^2 y^2\, dx + x^3 y\, dy = 0$$

$$\int_{\mathcal{C}_2} x^2 y^2\, dx + x^3 y\, dy = \int_0^1 y\, dy = \frac{1}{2}$$

$$\int_{\mathcal{C}_3} x^2 y^2\, dx + x^3 y\, dy = \int_1^0 x^2\, dx = -\frac{1}{3}$$

$$\int_{\mathcal{C}_4} x^2 y^2\, dx + x^3 y\, dy = 0.$$

Finally, therefore,

$$\int_{\mathcal{C}} x^2 y^2\, dx + x^3 y\, dy = 0 + \frac{1}{2} - \frac{1}{3} + 0 = \frac{1}{6}.$$

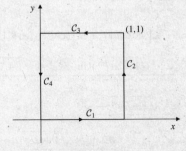

Fig. 15.4.8

10. $\mathbf{F} = (axy + z)\mathbf{i} + x^2\mathbf{j} + (bx + 2z)\mathbf{k}$ is conservative if

$$\frac{\partial F_1}{\partial y} = \frac{\partial F_2}{\partial x} \quad \Leftrightarrow \quad a = 2$$

$$\frac{\partial F_1}{\partial z} = \frac{\partial F_3}{\partial x} \quad \Leftrightarrow \quad b = 1$$

$$\frac{\partial F_2}{\partial z} = \frac{\partial F_3}{\partial y} \quad \Leftrightarrow \quad 0 = 0.$$

If $a = 2$ and $b = 1$, then $\mathbf{F} = \nabla\phi$ where

$$\phi = \int (2xy + z)\,dx = x^2 y + xz + C_2(y, z)$$

$$\frac{\partial C_1}{\partial y} + x^2 = F_2 = x^2 \quad \Rightarrow \quad C_1(y, z) = C_2(z)$$

$$\frac{dC_2}{dz} + x = F_3 = x + 2z \quad \Rightarrow \quad C_2(z) = z^2 + C.$$

Thus $\phi = x^2 y + xz + z^2 + C$ is a potential for $\mathbf{F}$.

12. $\mathbf{F} = (y^2 \cos x + z^3)\mathbf{i} + (2y \sin x - 4)\mathbf{j} + (3xz^2 + 2)\mathbf{k}$
$= \nabla(y^2 \sin x + xz^3 - 4y + 2z).$
The curve $\mathcal{C}$: $x = \sin^{-1} t$, $y = 1 - 2t$, $z = 3t - 1$,
$(0 \le t \le 1)$, goes from $(0, 1, -1)$ to $(\pi/2, -1, 2)$. The
work done by $\mathbf{F}$ in moving a particle along $\mathcal{C}$ is

$$W = \int_{\mathcal{C}} \mathbf{F} \bullet d\mathbf{r}$$

$$= (y^2 \sin x + xz^3 - 4y + 2z)\Big|_{(0,1,-1)}^{(\pi/2,-1,2)}$$

$$= 1 + 4\pi + 4 + 4 - 0 - 0 + 4 + 2 = 15 + 4\pi.$$

14. a) $S = \{(x, y) : x > 0, y \ge 0\}$ is a simply connected domain.

b) $S = \{(x, y) : x = 0, y \ge 0\}$ is not a domain. (It has empty interior.)

c) $S = \{(x, y) : x \ne 0, y > 0\}$ is a domain but is not connected. There is no path in S from $(-1, 1)$ to $(1, 1)$.

d) $S = \{(x, y, z) : x^2 > 1\}$ is a domain but is not connected. There is no path in S from $(-2, 0, 0)$ to $(2, 0, 0)$.

e) $S = \{(x, y, z) : x^2 + y^2 > 1\}$ is a connected domain but is not simply connected. The circle $x^2 + y^2 = 2$, $z = 0$ lies in S, but cannot be shrunk through S to a point since it surrounds the cylinder $x^2 + y^2 \le 1$ which is outside S.

f) $S = \{(x, y, z) : x^2 + y^2 + z^2 > 1\}$ is a simply connected domain even though it has a ball-shaped "hole" in it.

16. $\mathcal{C}$ is the curve $\mathbf{r} = a \cos t\mathbf{i} + b \sin t\mathbf{j}$, $(0 \le t \le 2\pi)$.

$$\oint_{\mathcal{C}} x\,dy = \int_0^{2\pi} a \cos t\, b \cos t\,dt = \pi ab$$

$$\oint_{\mathcal{C}} y\,dx = \int_0^{2\pi} b \sin t(-a \sin t)\,dt = -\pi ab.$$

18. $\mathcal{C}$ is made up of four segments as shown in the figure.
On $\mathcal{C}_1$, $y = 0$, $dy = 0$, and x goes from 0 to 1.
On $\mathcal{C}_2$, $x = 1$, $dx = 0$, and y goes from 0 to 1.
On $\mathcal{C}_3$, $y = 1$, $dy = 0$, and x goes from 1 to 0.
On $\mathcal{C}_4$, $x = 0$, $dx = 0$, and y goes from 1 to 0.

$$\oint_{\mathcal{C}} x\,dy = \int_{\mathcal{C}_1} + \int_{\mathcal{C}_2} + \int_{\mathcal{C}_3} + \int_{\mathcal{C}_4}$$

$$= 0 + \int_0^1 dy + 0 + 0 = 1$$

$$\oint_{\mathcal{C}} y\,dx = \int_{\mathcal{C}_1} + \int_{\mathcal{C}_2} + \int_{\mathcal{C}_3} + \int_{\mathcal{C}_4}$$

$$= 0 + 0 + \int_1^0 dx + 0 = -1.$$

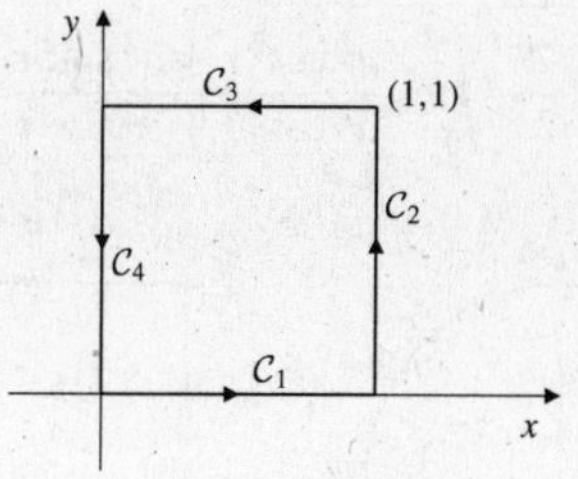

Fig. 15.4.18

20. Conjecture: If D is a domain in $\mathbb{R}^2$ whose boundary is a closed, non-self-intersecting curve $\mathcal{C}$, oriented counter-clockwise, then

$$\oint_{\mathcal{C}} x\,dy = \text{area of } D,$$

$$\oint_{\mathcal{C}} y\,dx = -\text{area of } D.$$

Proof for a domain D that is x-simple and y-simple:
Since D is x-simple, it can be specified by the inequalities

$$c \le y \le d, \qquad f(y) \le x \le g(y).$$

Let $\mathcal{C}$ consist of the four parts shown in the figure. On $\mathcal{C}_1$ and $\mathcal{C}_3$, $dy = 0$.
On $\mathcal{C}_2$, $x = g(y)$, where y goes from c to d.
On $\mathcal{C}_2$, $x = f(y)$, where y goes from d to c. Thus

$$\oint_{\mathcal{C}} x\,dy = \int_{\mathcal{C}_1} + \int_{\mathcal{C}_2} + \int_{\mathcal{C}_3} + \int_{\mathcal{C}_4}$$

$$= 0 + \int_c^d g(y)\,dy + 0 + \int_d^c f(y)\,dy$$

$$= \big(g(y) - f(y)\big)\,dy = \text{area of } D.$$

The proof that $\displaystyle\oint_{\mathcal{C}} y\,dx = -(\text{area of } D)$ is similar, and uses the fact that D is y-simple.

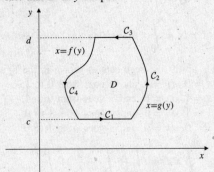

Fig. 15.4.20

22. a) $\mathcal{C}:\ x = a\cos t,\ x = a\sin t,\ 0 \le t \le 2\pi.$

$$\frac{1}{2\pi}\oint_{\mathcal{C}}\frac{x\,dy - y\,dx}{x^2 + y^2}$$
$$= \frac{1}{2\pi}\int_0^{2\pi}\frac{a^2\cos^2 t + a^2\sin^2 t}{a^2\cos^2 t + a^2\sin^2 t}\,dt = 1.$$

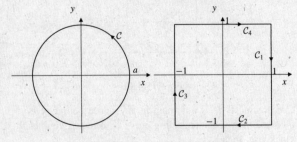

Fig. 15.4.22(a)   Fig. 15.4.22(b)

b) See the figure. $\mathcal{C}$ has four parts.
On $\mathcal{C}_1$, $x = 1$, $dx = 0$, y goes from 1 to -1.
On $\mathcal{C}_2$, $y = -1$, $dy = 0$, x goes from 1 to -1.
On $\mathcal{C}_3$, $x = -1$, $dx = 0$, y goes from -1 to 1.
On $\mathcal{C}_4$, $x = 1$, $dx = 0$, y goes from 1 to -1.

$$\frac{1}{2\pi}\oint_{\mathcal{C}}\frac{x\,dy - y\,dx}{x^2 + y^2}$$
$$= \frac{1}{2\pi}\left[\int_1^{-1}\frac{dy}{1 + y^2} + \int_1^{-1}\frac{dx}{x^2 + 1}\right.$$
$$\left.\int_{-1}^{1}\frac{-dy}{1 + y^2} + \int_{-1}^{1}\frac{-dx}{x^2 + 1}\right]$$
$$= -\frac{2}{\pi}\int_{-1}^{1}\frac{dt}{1 + t^2}$$
$$= -\frac{2}{\pi}\tan^{-1}t\,\Big|_{-1}^{1} = -\frac{2}{\pi}\left(\frac{\pi}{4} + \frac{\pi}{4}\right) = -1.$$

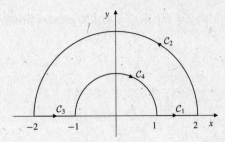

Fig. 15.4.22

c) See the figure. $\mathcal{C}$ has four parts.
On $\mathcal{C}_1$, $y = 0$, $dy = 0$, x goes from 1 to 2.
On $\mathcal{C}_2$, $x = 2\cos t$, $y = 2\sin t$, t goes from 0 to π.
On $\mathcal{C}_3$, $y = 0$, $dy = 0$, x goes from -2 to -1.
On $\mathcal{C}_4$, $x = \cos t$, $y = \sin t$, t goes from π to 0.

$$\frac{1}{2\pi}\oint_{\mathcal{C}}\frac{x\,dy - y\,dx}{x^2 + y^2}$$
$$= \frac{1}{2\pi}\left[0 + \int_0^{\pi}\frac{4\cos^2 t + 4\sin^2 t}{4\cos^2 t + 4\sin^2 t}\,dt\right.$$
$$\left. + 0 + \int_{\pi}^{0}\frac{\cos^2 t + \sin^2 t}{\cos^2 t + \sin^2 t}\,dt\right]$$
$$= \frac{1}{2\pi}(\pi - \pi) = 0.$$

24. If $\mathcal{C}$ is a closed, piecewise smooth curve in $\mathbb{R}^2$ having equation $\mathbf{r} = \mathbf{r}(t)$, $a \le t \le b$, and if $\mathcal{C}$ does not pass through the origin, then the polar angle function $\theta = \theta\big(x(t), y(t)\big) = \theta(t)$ can be defined so as to vary continuously on $\mathcal{C}$. Therefore,

$$\theta(x, y)\,\Big|_{t=a}^{t=b} = 2\pi \times w(\mathcal{C}),$$

where $w(\mathcal{C})$ is the number of times $\mathcal{C}$ winds around the origin in a counterclockwise direction. For example, $w(\mathcal{C})$ equals 1, -1 and 0 respectively, for the curves $\mathcal{C}$ in parts (a), (b) and (c) of Exercise 22. Since

$$\nabla\theta = \frac{\partial\theta}{\partial x}\mathbf{i} + \frac{\partial\theta}{\partial y}\mathbf{j}$$
$$= \frac{-y\mathbf{i} + x\mathbf{j}}{x^2 + y^2},$$

we have

$$\frac{1}{2\pi}\oint_{\mathcal{C}}\frac{x\,dy - y\,dx}{x^2 + y^2} = \frac{1}{2\pi}\oint_{\mathcal{C}}\nabla\theta\bullet d\mathbf{r}$$
$$= \frac{1}{2\pi}\theta(x, y)\,\Big|_{t=a}^{t=b} = w(\mathcal{C}).$$

Section 15.5   Surfaces and Surface Integrals (page 935)

2. The area element dS is bounded by the curves in which the coordinate planes at θ and $\theta + d\theta$ and the coordinate cones at ϕ and $\phi + d\phi$ intersect the sphere $R = a$. (See the figure.) The element is rectangular with sides $a\,d\phi$ and $a\sin\phi\,d\theta$. Thus

$$dS = a^2 \sin\phi\,d\phi\,d\theta.$$

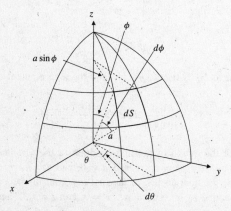

Fig. 15.5.2

4. One-quarter of the required area is shown in the figure. It lies above the semicircular disk R bounded by $x^2 + y^2 = 2ay$, or, in terms of polar coordinates, $r = 2a\sin\theta$. On the sphere $x^2 + y^2 + z^2 = 4a^2$, we have

$$2z\frac{\partial z}{\partial x} = -2x, \qquad \text{or} \qquad \frac{\partial z}{\partial x} = -\frac{x}{z}.$$

Similarly, $\dfrac{\partial z}{\partial y} = -\dfrac{y}{z}$, so the surface area element on the sphere can be written

$$dS = \sqrt{1 + \frac{x^2 + y^2}{z^2}}\,dx\,dy = \frac{2a\,dx\,dy}{\sqrt{4a^2 - x^2 - y^2}}.$$

The required area is

$$S = 4\iint_R \frac{2a}{\sqrt{4a^2 - x^2 - y^2}}\,dx\,dy$$

$$= 8a\int_0^{\pi/2} d\theta \int_0^{2a\sin\theta} \frac{r\,dr}{\sqrt{4a^2 - r^2}} \quad \begin{array}{l}\text{Let } u = 4a^2 - r^2 \\ du = -2r\,dr\end{array}$$

$$= 4a\int_0^{\pi/2} d\theta \int_{4a^2\cos^2\theta}^{4a^2} u^{-1/2}\,du$$

$$= 8a\int_0^{\pi/2} (2a - 2a\cos\theta)\,d\theta$$

$$= 16a^2(\theta - \sin\theta)\Big|_0^{\pi/2} = 8a^2(\pi - 2) \text{ sq. units.}$$

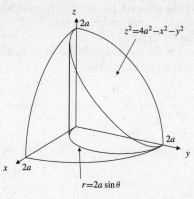

Fig. 15.5.4

6. The cylinder $x^2 + y^2 = 2ay$ intersects the sphere $x^2 + y^2 + z^2 = 4a^2$ on the parabolic cylinder $2ay + z^2 = 4a^2$. By Exercise 5, the area element on $x^2 + y^2 - 2ay = 0$ is

$$dS = \left|\frac{2x\mathbf{i} + (2y - 2a)\mathbf{j}}{2x}\right|\,dy\,dz$$

$$= \sqrt{1 + \frac{(y - a)^2}{2ay - y^2}}\,dy\,dz$$

$$= \sqrt{\frac{2ay - y^2 + y^2 - 2ay + a^2}{2ay - y^2}}\,dy\,dz = \frac{a}{\sqrt{2ay - y^2}}\,dy\,dz.$$

The area of the part of the cylinder inside the sphere is 4 times the part shown in Figure 15.23 in the text, that is, 4 times the double integral of dS over the region $0 \le y \le 2a$, $0 \le z \le \sqrt{4a^2 - 2ay}$, or

$$S = 4\int_0^{2a} \frac{a\,dy}{\sqrt{2ay - y^2}} \int_0^{\sqrt{4a^2 - 2ay}} dz$$

$$= 4a\int_0^{2a} \frac{\sqrt{2a(2a - y)}}{\sqrt{y(2a - y)}}\,dy = 4\sqrt{2}a^{3/2}\int_0^{2a} \frac{dy}{\sqrt{y}}$$

$$= 4\sqrt{2}a^{3/2}(2\sqrt{y})\Big|_0^{2a} = 16a^2 \text{ sq. units.}$$

8. The normal to the cone $z^2 = x^2 + y^2$ makes a $45°$ angle with the vertical, so $dS = \sqrt{2}\,dx\,dy$ is a surface area element for the cone. Both *nappes* (halves) of the cone pass through the interior of the cylinder $x^2 + y^2 = 2ay$, so the area of that part of the cone inside the cylinder is $2\sqrt{2}\pi a^2$ square units, since the cylinder has a circular cross-section of radius a.

10. One-eighth of the required area lies in the first octant, above the triangle T with vertices $(0, 0, 0)$, $(a, 0, 0)$ and $(a, a, 0)$. (See the figure.)

The surface $x^2 + z^2 = a^2$ has normal $\mathbf{n} = x\mathbf{i} + z\mathbf{k}$, so an area element on it can be written

$$dS = \frac{|\mathbf{n}|}{|\mathbf{n} \bullet \mathbf{k}|} \, dx\, dy = \frac{a}{z} \, dx\, dy = \frac{a\, dx\, dy}{\sqrt{a^2 - x^2}}.$$

The area of the part of that cylinder lying inside the cylinder $y^2 + z^2 = a^2$ is

$$S = 8 \iint_T \frac{a\, dx\, dy}{\sqrt{a^2 - x^2}} = 8a \int_0^a \frac{dx}{\sqrt{a^2 - x^2}} \int_0^x dy$$

$$= 8a \int_0^a \frac{x\, dx}{\sqrt{a^2 - x^2}}$$

$$= -8a\sqrt{a^2 - x^2}\,\Big|_0^a = 8a^2 \text{ sq. units.}$$

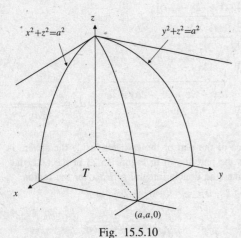

Fig. 15.5.10

12. We want to find A_1, the area of that part of the cylinder $x^2 + z^2 = a^2$ inside the cylinder $y^2 + z^2 = b^2$, and A_2, the area of that part of $y^2 + z^2 = b^2$ inside $x^2 + z^2 = a^2$. We have

$$A_1 = 8 \times (\text{area of } \mathcal{S}_1),$$
$$A_2 = 8 \times (\text{area of } \mathcal{S}_2),$$

where $\mathcal{S}_1$ and $\mathcal{S}_2$ are the parts of these surfaces lying in the first octant, as shown in the figure.

A normal to $\mathcal{S}_1$ is $\mathbf{n}_1 = x\mathbf{i} + z\mathbf{k}$, and the area element on $\mathcal{S}_1$ is

$$dS_1 = \frac{|\mathbf{n}_1|}{|\mathbf{n}_1 \bullet \mathbf{i}|} \, dy\, dz = \frac{a\, dy\, dz}{\sqrt{a^2 - z^2}}.$$

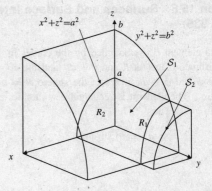

Fig. 15.5.12

A normal to $\mathcal{S}_2$ is $\mathbf{n}_2 = x\mathbf{j} + z\mathbf{k}$, and the area element on $\mathcal{S}_2$ is

$$dS_2 = \frac{|\mathbf{n}_2|}{|\mathbf{n}_2 \bullet \mathbf{j}|} \, dx\, dz = \frac{b\, dx\, dz}{\sqrt{b^2 - z^2}}.$$

Let R_1 be the region of the first quadrant of the yz-plane bounded by $y^2 + z^2 = b^2$, $y = 0$, $z = 0$, and $z = a$.

Let R_2 be the quarter-disk in the first quadrant of the xz-plane bounded by $x^2 + z^2 = a^2$, $x = 0$, and $z = 0$. Then

$$A_1 = 8 \iint_{R_1} dS_1 = 8a \int_0^a \frac{dz}{\sqrt{a^2 - z^2}} \int_0^{\sqrt{b^2 - z^2}} dy$$

$$= 8a \int_0^a \frac{\sqrt{b^2 - z^2}}{\sqrt{a^2 - z^2}} \, dz \quad \begin{array}{l} \text{Let } z = a\sin t \\ dz = a\cos t\, dt \end{array}$$

$$= 8a \int_0^{\pi/2} \sqrt{b^2 - a^2 \sin^2 t} \, dt$$

$$= 8ab \int_0^{\pi/2} \sqrt{1 - \frac{a^2}{b^2} \sin^2 t} \, dt$$

$$= 8ab\, E\left(\frac{a}{b}\right) \text{ sq. units.}$$

$$A_2 = 8 \iint_{R_2} dS_2 = 8b \int_0^a \frac{dz}{\sqrt{b^2 - z^2}} \int_0^{\sqrt{a^2 - z^2}} dx$$

$$= 8b \int_0^a \frac{\sqrt{a^2 - z^2}}{\sqrt{b^2 - z^2}} \, dz \quad \begin{array}{l} \text{Let } z = b\sin t \\ dz = b\cos t\, dt \end{array}$$

$$= 8b \int_0^{\sin^{-1}(a/b)} \sqrt{a^2 - b^2 \sin^2 t} \, dt$$

$$= 8ab \int_0^{\sin^{-1}(a/b)} \sqrt{1 - \frac{b^2}{a^2} \sin^2 t} \, dt$$

$$= 8ab\, E\left(\frac{b}{a}, \sin^{-1}\frac{a}{b}\right) \text{ sq. units.}$$

14. Continuing the above solution, the cone $z = \sqrt{2(x^2 + y^2)}$ has area element

$$dS = \sqrt{1 + \left(\frac{\partial z}{\partial x}\right)^2 + \left(\frac{\partial z}{\partial y}\right)^2}\, dx\, dy$$

$$= \sqrt{1 + \frac{4(x^2 + y^2)}{z^2}}\, dx\, dy = \sqrt{3}\, dx\, dy.$$

If $\mathcal{S}$ is the part of the cone lying below the plane $z = 1 + y$, then

$$\iint_{\mathcal{S}} y\, dS = \sqrt{3} \iint_E y\, dx\, dy = \sqrt{3}\, A\bar{y} = \sqrt{6}\pi.$$

16. The surface $z = \sqrt{2xy}$ has area element

$$dS = \sqrt{1 + \frac{y}{2x} + \frac{x}{2y}}\, dx\, dy$$

$$= \sqrt{\frac{2xy + y^2 + x^2}{2xy}}\, dx\, dy = \frac{|x + y|}{\sqrt{2xy}}\, dx\, dy.$$

If its density is kz, the mass of the specified part of the surface is

$$m = \int_0^5 dx \int_0^2 k\sqrt{2xy}\, \frac{x + y}{\sqrt{2xy}}\, dy$$

$$= k \int_0^5 dx \int_0^2 (x + y)\, dy$$

$$= k \int_0^5 (2x + 2)\, dx = 35k \text{ units.}$$

18. The upper half of the spheroid $\frac{x^2}{a^2} + \frac{y^2}{a^2} + \frac{z^2}{c^2} = 1$ has a circular disk of radius a as projection onto the xy-plane. Since

$$\frac{2x}{a^2} + \frac{2z}{c^2}\frac{\partial z}{\partial x} = 0 \quad \Rightarrow \quad \frac{\partial z}{\partial x} = -\frac{c^2 x}{a^2 z},$$

and, similarly, $\frac{\partial z}{\partial y} = -\frac{c^2 y}{a^2 z}$, the area element on the spheroid is

$$dS = \sqrt{1 + \frac{c^4}{a^4}\frac{x^2 + y^2}{z^2}}\, dx\, dy$$

$$= \sqrt{1 + \frac{c^2}{a^2}\frac{x^2 + y^2}{a^2 - x^2 - y^2}}\, dx\, dy$$

$$= \sqrt{\frac{a^4 + (c^2 - a^2)r^2}{a^2(a^2 - r^2)}}\, r\, dr\, d\theta$$

in polar coordinates. Thus the area of the spheroid is

$$S = \frac{2}{a} \int_0^{2\pi} d\theta \int_0^a \sqrt{\frac{a^4 + (c^2 - a^2)r^2}{a^2 - r^2}}\, r\, dr$$

$$\text{Let } u^2 = a^2 - r^2$$

$$u\, du = -r\, dr$$

$$= \frac{4\pi}{a} \int_0^a \sqrt{a^4 + (c^2 - a^2)(a^2 - u^2)}\, du$$

$$= \frac{4\pi}{a} \int_0^a \sqrt{a^2 c^2 - (c^2 - a^2)u^2}\, du$$

$$= 4\pi c \int_0^a \sqrt{1 - \frac{c^2 - a^2}{a^2 c^2}\, u^2}\, du.$$

For the case of a prolate spheroid $0 < a < c$, let $k^2 = \frac{c^2 - a^2}{a^2 c^2}$. Then

$$S = 4\pi c \int_0^a \sqrt{1 - k^2 u^2}\, du \quad \text{Let } ku = \sin v$$

$$k\, du = \cos v\, dv$$

$$= \frac{4\pi c}{k} \int_0^{\sin^{-1}(ka)} \cos^2 v\, dv$$

$$= \frac{2\pi c}{k} (v + \sin v \cos v)\Big|_0^{\sin^{-1}(ka)}$$

$$= \frac{2\pi a c^2}{\sqrt{c^2 - a^2}} \sin^{-1}\frac{\sqrt{c^2 - a^2}}{c} + 2\pi a^2 \text{ sq. units.}$$

20. $x = au \cos v$, $y = au \sin v$, $z = bv$, $(0 \le u \le 1, \ 0 \le v \le 2\pi)$. This surface is a spiral (helical) ramp of radius a and height $2\pi b$, wound around the z-axis. (It's like a circular staircase with a ramp instead of stairs.) We have

$$\frac{\partial(x, y)}{\partial(u, v)} = \begin{vmatrix} a\cos v & -au\sin v \\ a\sin v & au\cos v \end{vmatrix} = a^2 u$$

$$\frac{\partial(y, z)}{\partial(u, v)} = \begin{vmatrix} a\sin v & au\cos v \\ 0 & b \end{vmatrix} = ab\sin v$$

$$\frac{\partial(z, x)}{\partial(u, v)} = \begin{vmatrix} 0 & b \\ a\cos v & -au\sin v \end{vmatrix} = -ab\cos v$$

$$dS = \sqrt{a^4 u^2 + a^2 b^2 \sin^2 v + a^2 b^2 \cos^2 v}\, du\, dv$$

$$= a\sqrt{a^2 u^2 + b^2}\, du\, dv.$$

The area of the ramp is

$$A = a \int_0^1 \sqrt{a^2 u^2 + b^2}\, du \int_0^{2\pi} dv$$

$$= 2\pi a \int_0^1 \sqrt{a^2 u^2 + b^2}\, du \quad \text{Let } au = b\tan\theta$$
$$\qquad\qquad\qquad\qquad\qquad a\, du = b\sec^2\theta\, d\theta$$

$$= 2\pi b^2 \int_{u=0}^{u=1} \sec^3\theta\, d\theta$$

$$= \pi b^2 \left(\sec\theta\tan\theta + \ln|\sec\theta + \tan\theta| \right)\Big|_{u=0}^{u=1}$$

$$= \pi b^2 \left(\frac{au\sqrt{a^2 u^2 + b^2}}{b^2} + \ln\left|\frac{au + \sqrt{a^2 u^2 + b^2}}{b}\right| \right)\Big|_0^1$$

$$= \pi a \sqrt{a^2 + b^2} + \pi b^2 \ln\left(\frac{a + \sqrt{a^2 + b^2}}{b} \right) \text{ sq. units.}$$

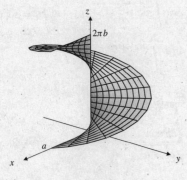

Fig. 15.5.20

22. Use spherical coordinates. The area of the eighth-sphere $\mathcal{S}$ is

$$A = \frac{1}{8}(4\pi a^2) = \frac{\pi a^2}{2} \text{ sq. units.}$$

The moment about $z = 0$ is

$$M_{z=0} = \iint_{\mathcal{S}} z\, dS$$

$$= \int_0^{\pi/2} d\theta \int_0^{\pi/2} a\cos\phi\, a^2 \sin\phi\, d\phi$$

$$= \frac{\pi a^3}{2} \int_0^{\pi/2} \frac{\sin 2\phi}{2}\, d\phi = \frac{\pi a^3}{4}.$$

Thus $\bar{z} = \dfrac{M_{z=0}}{A} = \dfrac{a}{2}$. By symmetry, $\bar{x} = \bar{y} = \bar{z}$, so the centroid of that part of the surface of the sphere $x^2 + y^2 + z^2 = a^2$ lying in the first octant is $\left(\dfrac{a}{2}, \dfrac{a}{2}, \dfrac{a}{2} \right)$.

24. By symmetry, the force of attraction of the hemisphere shown in the figure on the mass m at the origin is vertical. The vertical component of the force exerted by area element $dS = a^2 \sin\phi\, d\phi\, d\theta$ at the position with spherical coordinates (a, ϕ, θ) is

$$dF = \frac{km\sigma\, dS}{a^2}\cos\phi = km\sigma \sin\phi\cos\phi\, d\phi\, d\theta.$$

Thus, the total force on m is

$$F = km\sigma \int_0^{2\pi} d\theta \int_0^{\pi/2} \sin\phi\cos\phi\, d\phi = \pi km\sigma \text{ units.}$$

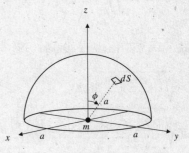

Fig. 15.5.24

26. $\mathcal{S}$ is the cylindrical surface $x^2 + y^2 = a^2$, $0 \le z \le h$, with areal density σ. Its mass is $m = 2\pi ah\sigma$. Since all surface elements are at distance a from the z-axis, the radius of gyration of the cylindrical surface about the z-axis is $\bar{D} = a$. Therefore the moment of inertia about that axis is

$$I = m\bar{D}^2 = ma^2 = 2\pi\sigma a^3 h.$$

28. The surface area element for a conical surface $\mathcal{S}$,

$$z = h\left(1 - \frac{\sqrt{x^2 + y^2}}{a} \right),$$

having base radius a and height h, was determined in the solution to Exercise 23 to be

$$dS = \frac{\sqrt{a^2 + h^2}}{a}\, dx\, dy.$$

The mass of $\mathcal{S}$, which has areal density σ, was also determined in that exercise: $m = \pi\sigma a\sqrt{a^2 + h^2}$. The moment of inertia of $\mathcal{S}$ about the z-axis is

$$I = \sigma \iint_{\mathcal{S}} (x^2 + y^2)\, dS$$

$$= \frac{\sigma\sqrt{a^2 + h^2}}{a} \int_0^{2\pi} d\theta \int_0^a r^2\, r\, dr$$

$$= \frac{2\pi\sigma\sqrt{a^2 + h^2}}{a}\frac{a^4}{4} = \frac{\pi\sigma a^3 \sqrt{a^2 + h^2}}{2}.$$

The radius of gyration is $\bar{D} = \sqrt{I/m} = \dfrac{a}{\sqrt{2}}$.

Section 15.6 Oriented Surfaces and Flux Integrals (page 943)

2. On the sphere S with equation $x^2 + y^2 + z^2 = a^2$ we have

$$\hat{\mathbf{N}} = \frac{x\mathbf{i} + y\mathbf{j} + z\mathbf{k}}{a}.$$

If $\mathbf{F} = x\mathbf{i} + y\mathbf{j} + z\mathbf{k}$, then $\mathbf{F} \bullet \hat{\mathbf{N}} = a$ on S. Thus the flux of $\mathbf{F}$ out of S is

$$\iint_S \mathbf{F} \bullet \hat{\mathbf{N}}\, dS = a \times 4\pi a^2 = 4\pi a^3.$$

4. $\mathbf{F} = y\mathbf{i} + z\mathbf{k}$. Let S_1 be the conical surface and S_2 be the base disk. The flux of $\mathbf{F}$ outward through the surface of the cone is

$$\iint_S \mathbf{F} \bullet \hat{\mathbf{N}} = \iint_{S_1} + \iint_{S_2}.$$

On S_1: $\hat{\mathbf{N}} = \dfrac{1}{\sqrt{2}}\left(\dfrac{x\mathbf{i} + y\mathbf{j}}{\sqrt{x^2 + y^2}} + \mathbf{k} \right)$, $dS = \sqrt{2}\,dx\,dy$. Thus

$$\iint_{S_1} \mathbf{F} \bullet \hat{\mathbf{N}}\, dS$$
$$= \iint_{x^2 + y^2 \leq 1} \left(\frac{xy}{\sqrt{x^2 + y^2}} + 1 - \sqrt{x^2 + y^2} \right) dx\, dy$$
$$= 0 + \pi \times 1^2 - \int_0^{2\pi} d\theta \int_0^1 r^2\, dr$$
$$= \pi - \frac{2\pi}{3} = \frac{\pi}{3}.$$

On S_2: $\hat{\mathbf{N}} = -\mathbf{k}$ and $z = 0$, so $\mathbf{F} \bullet \hat{\mathbf{N}} = 0$. Thus, the total flux of $\mathbf{F}$ out of the cone is $\pi/3$.

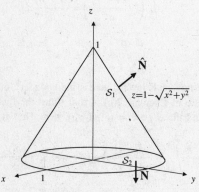

Fig. 15.6.4

6. For $z = x^2 - y^2$ the upward surface element is

$$\hat{\mathbf{N}}\, dS = \frac{-2x\mathbf{i} + 2y\mathbf{j} + \mathbf{k}}{1}\, dx\, dy.$$

The flux of $\mathbf{F} = x\mathbf{i} + x\mathbf{j} + \mathbf{k}$ upward through S, the part of $z = x^2 - y^2$ inside $x^2 + y^2 = a^2$ is

$$\iint_S \mathbf{F} \bullet \hat{\mathbf{N}}\, dS = \iint_{x^2 + y^2 \leq a^2} (-2x^2 + 2xy + 1)\, dx\, dy$$
$$= -2 \int_0^{2\pi} \cos^2 \theta\, d\theta \int_0^a r^3\, dr + 0 + \pi a^2$$
$$= \pi a^2 - 2(\pi)\frac{a^4}{4} = \frac{\pi}{2} a^2 (2 - a^2).$$

8. The upward vector surface element on the top half of $x^2 + y^2 + z^2 = a^2$ is

$$\hat{\mathbf{N}}\, dS = \frac{2x\mathbf{i} + 2y\mathbf{j} + 2z\mathbf{k}}{2z}\, dx\, dy = \left(\frac{x\mathbf{i} + y\mathbf{j}}{z} + \mathbf{k} \right) dx\, dy.$$

The flux of $\mathbf{F} = z^2\mathbf{k}$ upward through the first octant part S of the sphere is

$$\iint_S \mathbf{F} \bullet \hat{\mathbf{N}}\, dS = \int_0^{\pi/2} d\theta \int_0^a (a^2 - r^2) r\, dr = \frac{\pi a^4}{8}.$$

10. S: $\mathbf{r} = u^2 v\mathbf{i} + uv^2\mathbf{j} + v^3\mathbf{k}$, ($0 \leq u \leq 1$, $0 \leq v \leq 1$), has upward surface element

$$\hat{\mathbf{N}}\, dS = \frac{\partial \mathbf{r}}{\partial u} \times \frac{\partial \mathbf{r}}{\partial v}\, du\, dv$$
$$= (2uv\mathbf{i} + v^2\mathbf{j}) \times (u^2\mathbf{i} + 2uv\mathbf{j} + 3v^2\mathbf{k})\, du\, dv$$
$$= (3v^4\mathbf{i} - 6uv^3\mathbf{j} + 3u^2 v^2\mathbf{k})\, du\, dv.$$

The flux of $\mathbf{F} = 2x\mathbf{i} + y\mathbf{j} + z\mathbf{k}$ upward through S is

$$\iint_S \mathbf{F} \bullet \hat{\mathbf{N}}\, dS$$
$$= \int_0^1 du \int_0^1 (6u^2 v^5 - 6u^2 v^5 + 3u^2 v^5)\, dv$$
$$= \frac{1}{2} \int_0^1 u^2\, du = \frac{1}{6}.$$

12. S: $\mathbf{r} = e^u \cos v\mathbf{i} + e^u \sin v\mathbf{j} + u\mathbf{k}$, ($0 \leq u \leq 1$, $0 \leq v \leq \pi$), has upward surface element

$$\hat{\mathbf{N}}\, dS = \frac{\partial \mathbf{r}}{\partial u} \times \frac{\partial \mathbf{r}}{\partial v}\, du\, dv$$
$$= (-e^u \cos v\mathbf{i} - e^u \sin v\mathbf{j} + e^{2u}\mathbf{k})\, du\, dv.$$

The flux of $\mathbf{F} = yz\mathbf{i} - xz\mathbf{j} + (x^2 + y^2)\mathbf{k}$ upward through $\mathcal{S}$ is

$$\iint_{\mathcal{S}} \mathbf{F} \bullet \hat{\mathbf{N}} \, dS$$

$$= \int_0^1 du \int_0^{\pi} (-ue^{2u} \sin v \cos v + ue^{2u} \sin v \cos v + e^{4u}) \, dv$$

$$= \int_0^1 e^{4u} \, du \int_0^{\pi} dv = \pi \frac{(e^4 - 1)}{4}.$$

14. The flux of $\mathbf{F} = \dfrac{m\mathbf{r}}{|\mathbf{r}|^3}$ out of the cube $1 \le x, y, z \le 2$ is equal to three times the total flux out of the pair of opposite faces $z = 1$ and $z = 2$, which have outward normals $-\mathbf{k}$ and $\mathbf{k}$ respectively. This latter flux is $2mI_2 - mI_1$, where

$$I_k = \int_1^2 dx \int_1^2 \frac{dy}{(x^2 + y^2 + k^2)^{3/2}}$$

$$\text{Let } y = \sqrt{x^2 + k^2} \tan u$$

$$dy = \sqrt{x^2 + k^2} \sec^2 u \, du$$

$$= \int_1^2 \frac{dx}{x^2 + k^2} \int_{y=1}^{y=2} \cos u \, du$$

$$= \int_1^2 \frac{dx}{x^2 + k^2} \left(\sin u \right) \Big|_{y=1}^{y=2}$$

$$= \int_1^2 \frac{dx}{x^2 + k^2} \left(\frac{y}{\sqrt{x^2 + y^2 + k^2}} \Big|_1^2 \right) = J_{k2} - J_{k1},$$

where

$$J_{kn} = n \int_1^2 \frac{dx}{(x^2 + k^2)\sqrt{x^2 + n^2 + k^2}}$$

$$\text{Let } x = \sqrt{n^2 + k^2} \tan v$$

$$dx = \sqrt{n^2 + k^2} \sec^2 v \, dv$$

$$= n \int_{x=1}^{x=2} \frac{\sec^2 v \, dv}{\left[(n^2 + k^2) \tan^2 v + k^2 \right] \sec v}$$

$$= n \int_{x=1}^{x=2} \frac{\cos v \, dv}{(n^2 + k^2) \sin^2 v + k^2 \cos^2 v}$$

$$= n \int_{x=1}^{x=2} \frac{\cos v \, dv}{k^2 + n^2 \sin^2 v} \quad \text{Let } w = n \sin v$$

$$\quad\quad\quad\quad\quad\quad\quad\quad\quad\quad\quad\quad dw = n \cos v \, dv$$

$$= \int_{x=1}^{x=2} \frac{dw}{k^2 + w^2} = \frac{1}{k} \tan^{-1} \frac{w}{k} \Big|_{x=1}^{x=2}$$

$$= \frac{1}{k} \tan^{-1} \frac{n \sin v}{k} \Big|_{x=1}^{x=2}$$

$$= \frac{1}{k} \tan^{-1} \frac{nx}{k\sqrt{x^2 + n^2 + k^2}} \Big|_1^2$$

$$= \frac{1}{k} \left(\tan^{-1} \frac{2n}{k\sqrt{4 + n^2 + k^2}} - \tan^{-1} \frac{n}{k\sqrt{1 + n^2 + k^2}} \right).$$

Thus

$$I_k = \frac{1}{k} \left[\tan^{-1} \frac{4}{k\sqrt{8 + k^2}} - 2\tan^{-1} \frac{2}{k\sqrt{5 + k^2}} \right.$$

$$\left. + \tan^{-1} \frac{1}{k\sqrt{2 + k^2}} \right].$$

The contribution to the total flux from the pair of surfaces $z = 1$ and $z = 2$ of the cube is

$$2mI_2 - mI_1$$

$$= m \left[\tan^{-1} \frac{1}{\sqrt{3}} - 2\tan^{-1} \frac{1}{3} + \tan^{-1} \frac{1}{2\sqrt{6}} \right.$$

$$\left. - \tan^{-1} \frac{4}{3} + 2\tan^{-1} \frac{2}{\sqrt{6}} - \tan^{-1} \frac{1}{\sqrt{3}} \right].$$

Using the identities

$$2\tan^{-1} a = \tan^{-1} \frac{2a}{1 - a^2}, \quad \text{and}$$

$$\tan^{-1} a = \frac{\pi}{2} - \tan^{-1} \frac{1}{a},$$

we calculate

$$-2\tan^{-1} \frac{1}{3} = -\tan^{-1} \frac{3}{4} = -\frac{\pi}{2} + \tan^{-1} \frac{4}{3}$$

$$2\tan^{-1} \frac{2}{\sqrt{6}} = \tan^{-1} \frac{12}{\sqrt{6}} = \frac{\pi}{2} - \tan^{-1} \frac{1}{2\sqrt{6}}.$$

Thus the net flux out of the pair of opposite faces is 0. By symmetry this holds for each pair, and the total flux out of the cube is 0. (You were warned this would be a difficult calculation!)

16. $\mathbf{F} = -\dfrac{x\mathbf{i} + y\mathbf{j}}{x^2 + y^2}.$

a) The flux of $\mathbf{F}$ inward across the circle of Exercise 7(a) is

$$-\oint_C \left(-\frac{x\mathbf{i} + y\mathbf{j}}{a^2} \right) \bullet \frac{x\mathbf{i} + y\mathbf{j}}{a} \, ds$$

$$= \oint_C \frac{a^2}{a^3} \, ds = \frac{1}{a} \times 2\pi a = 2\pi.$$

b) The flux of $\mathbf{F}$ inward across the boundary of the square of Exercise 7(b) is four times the flux inward across the edge $x = 1$, $-1 \le y \le 1$. Thus it is

$$-4 \int_{-1}^1 \left(-\frac{\mathbf{i} + y\mathbf{j}}{1 + y^2} \right) \bullet \mathbf{i} \, dy = 4 \int_{-1}^1 \frac{dy}{1 + y^2}$$

$$= 4\tan^{-1} y \Big|_{-1}^1 = 2\pi.$$

18. Let $\mathbf{F} = F_1\mathbf{i} + F_2\mathbf{j} + F_3\mathbf{k}$ be a constant vector field.

a) If R is a rectangular box, we can choose the origin and coordinate axes in such a way that the box is $0 \leq x \leq a$, $0 \leq y \leq b$, $0 \leq z \leq c$. On the faces $x = 0$ and $x = a$ we have $\hat{\mathbf{N}} = -\mathbf{i}$ and $\hat{\mathbf{N}} = \mathbf{i}$ respectively. Since F_1 is constant, the total flux out of the box through these two faces is

$$\iint_{\substack{0 \leq y \leq b \\ 0 \leq z \leq c}} (F_1 - F_1)\, dy\, dz = 0.$$

The flux out of the other two pairs of opposite faces is also 0. Thus the total flux of $\mathbf{F}$ out of the box is 0.

b) If $\mathcal{S}$ is a sphere of radius a we can choose the origin so that $\mathcal{S}$ has equation $x^2 + y^2 + z^2 = a^2$, and so its outward normal is

$$\hat{\mathbf{N}} = \frac{x\mathbf{i} + y\mathbf{j} + z\mathbf{k}}{a}.$$

Thus the flux out of $\mathcal{S}$ is

$$\frac{1}{a} \iint_{\mathcal{S}} (F_1 x + F_2 y + F_3 z)\, ds = 0,$$

since the sphere $\mathcal{S}$ is symmetric about the origin.

Review Exercises 15   (page 944)

2. $\mathcal{C}$ can be parametrized $x = t$, $y = 2t$, $z = t + 4t^2$, $(0 \leq t \leq 2)$. Thus

$$\int_{\mathcal{C}} 2y\, dx + x\, dy + 2\, dz$$

$$= \int_0^2 [4t(1) + t(2) + 2(1 + 8t)]\, dt$$

$$= \int_0^2 (22t + 2)\, dt = 48.$$

4. The plane $x + y + z = 1$ has area element $dS = \sqrt{3}\, dx\, dy$. If $\mathcal{S}$ is the part of the plane in the first octant, then the projection of $\mathcal{S}$ on the xy-plane is the triangle $0 \leq x \leq 1$, $0 \leq y \leq 1 - x$. Thus

$$\iint_{\mathcal{S}} xyz\, dS = \sqrt{3} \int_0^1 x\, dx \int_0^{1-x} y(1 - x - y)\, dy$$

$$= \sqrt{3} \int_0^1 \frac{x(1-x)^3}{6}\, dx \quad \text{Let } u = 1 - x$$

$$\qquad\qquad\qquad\qquad\qquad du = -dx$$

$$= \frac{\sqrt{3}}{6} \int_0^1 u^3(1 - u)\, du = \frac{\sqrt{3}}{6}\left(\frac{1}{4} - \frac{1}{5}\right) = \frac{\sqrt{3}}{120}.$$

6. The plane $x + 2y + 3z = 6$ has downward vector surface element

$$\hat{\mathbf{N}}\, dS = \frac{-\mathbf{i} - 2\mathbf{j} - 3\mathbf{k}}{3}\, dx\, dy.$$

If $\mathcal{S}$ is the part of the plane in the first octant, then the projection of $\mathcal{S}$ on the xy-plane is the triangle $0 \leq y \leq 3$, $0 \leq x \leq 6 - 2y$. Thus

$$\iint_{\mathcal{S}} (x\mathbf{i} + y\mathbf{j} + z\mathbf{k}) \bullet \hat{\mathbf{N}}\, dS$$

$$= -\frac{1}{3} \int_0^3 dy \int_0^{6-2y} (x + 2y + 6 - x - 2y)\, dx$$

$$= -2 \int_0^3 (6 - 2y) = -36 + 18 = -18.$$

8. $\int_{\mathcal{C}} \mathbf{F} \bullet d\mathbf{r}$ can be determined using only the endpoints of $\mathcal{C}$, provided

$$\mathbf{F} = (axy + 3yz)\mathbf{i} + (x^2 + 3xz + by^2z)\mathbf{j} + (bxy + cy^3)\mathbf{k}$$

is conservative, that is, if

$$ax + 3z = \frac{\partial F_1}{\partial y} = \frac{\partial F_2}{\partial x} = 2x + 3z$$

$$3y = \frac{\partial F_1}{\partial z} = \frac{\partial F_3}{\partial x} = by$$

$$3x + by^2 = \frac{\partial F_2}{\partial z} = \frac{\partial F_3}{\partial y} = bx + 3cy^2.$$

Thus we need $a = 2$, $b = 3$, and $c = 1$. With these values, $\mathbf{F} = \boldsymbol{\nabla}(x^2y + 3xyz + y^3z)$. Thus

$$\int_{\mathcal{C}} \mathbf{F} \bullet d\mathbf{r} = (x^2y + 3xyz + y^3z)\Big|_{(0,1,-1)}^{(2,1,1,)} = 11 - (-1) = 12.$$

10. a) $\mathbf{F} = (1 + x)e^{x+y}\mathbf{i} + (xe^{x+y} + 2y)\mathbf{j} - 2z\mathbf{k}$
$$= \boldsymbol{\nabla}(xe^{x+y} + y^2 - z^2).$$
Thus $\mathbf{F}$ is conservative.

b) $\mathbf{G} = (1 + x)e^{x+y}\mathbf{i} + (xe^{x+y} + 2z)\mathbf{j} - 2y\mathbf{k}$
$$= \mathbf{F} + 2(z - y)(\mathbf{j} + \mathbf{k}).$$
$\mathcal{C} : \mathbf{r} = (1 - t)e^t\mathbf{i} + t\mathbf{j} + 2t\mathbf{k}$, $(0 \leq t \leq 1)$.
$\mathbf{r}(0) = (1, 0, 0)$, $\mathbf{r}(1) = (0, 1, 2)$. Thus

$$\int_{\mathcal{C}} \mathbf{G} \bullet d\mathbf{r} = \int_{\mathcal{C}} \mathbf{F} \bullet d\mathbf{r} + \int_{\mathcal{C}} 2(z - y)(\mathbf{j} + \mathbf{k}) \bullet d\mathbf{r}$$

$$= (xe^{x+y} + y^2 - z^2)\Big|_{(1,0,0)}^{(0,1,2)}$$

$$\qquad + 2 \int_0^1 (2t - t)(1 + 2)\, dt$$

$$= -3 - e + 3t^2\Big|_0^1 = -e.$$

12. The first octant part of the cylinder $y^2 + z^2 = 16$ has outward vector surface element

$$\hat{\mathbf{N}}\,dS = \frac{2y\mathbf{j} + 2z\mathbf{k}}{2z}\,dx\,dy = \left(\frac{y}{\sqrt{16 - y^2}}\mathbf{j} + \mathbf{k}\right) dx\,dy.$$

The flux of $3z^2 x\mathbf{i} - x\mathbf{j} - y\mathbf{k}$ outward through the specified surface $\mathcal{S}$ is

$$\mathbf{F}\bullet\hat{\mathbf{N}}\,dS = \int_0^5 dx \int_0^4 \left(0 - \frac{xy}{\sqrt{16 - y^2}} - y\right) dy$$

$$= \int_0^5 \left(x\sqrt{16 - y^2} - \frac{y^2}{2}\right)\Bigg|_{y=0}^{y=4} dx$$

$$= -\int_0^5 (4x + 8)\,dx = -90.$$

Challenging Problems 15   (page 944)

2. This is a trick question. Observe that the given parametrization $\mathbf{r}(u, v)$ satisfies

$$\mathbf{r}(u + \pi, v) = \mathbf{r}(u, -v).$$

Therefore the surface $\mathcal{S}$ is traced out twice as u goes from 0 to 2π. (It is a Möbius band. See Figure 15.28 in the text.) If $\mathcal{S}_1$ is the part of the surface corresponding to $0 \le u \le \pi$, and $\mathcal{S}_2$ is the part corresponding to $\pi \le u \le 2\pi$, then $\mathcal{S}_1$ and $\mathcal{S}_2$ coincide as point sets, but their normals are oppositely oriented: $\hat{\mathbf{N}}_2 = -\hat{\mathbf{N}}_1$ at corresponding points on the two surfaces. Hence

$$\iint_{\mathcal{S}_1} \mathbf{F}\bullet\hat{\mathbf{N}}_1\,dS = -\iint_{\mathcal{S}_2} \mathbf{F}\bullet\hat{\mathbf{N}}_2\,dS,$$

for any smooth vector field, and

$$\iint_{\mathcal{S}} \mathbf{F}\bullet\hat{\mathbf{N}}\,dS = \iint_{\mathcal{S}_1} \mathbf{F}\bullet\hat{\mathbf{N}}_1\,dS + \iint_{\mathcal{S}_2} \mathbf{F}\bullet\hat{\mathbf{N}}_2\,dS = 0.$$

CHAPTER 16. VECTOR CALCULUS

Section 16.1 Gradient, Divergence, and Curl (page 953)

2. $\mathbf{F} = y\mathbf{i} + x\mathbf{j}$

$\operatorname{\mathbf{div}} \mathbf{F} = \dfrac{\partial}{\partial x}(y) + \dfrac{\partial}{\partial y}(x) + \dfrac{\partial}{\partial z}(0) = 0 + 0 = 0$

$\operatorname{\mathbf{curl}} \mathbf{F} = \begin{vmatrix} \mathbf{i} & \mathbf{j} & \mathbf{k} \\ \dfrac{\partial}{\partial x} & \dfrac{\partial}{\partial y} & \dfrac{\partial}{\partial z} \\ y & x & 0 \end{vmatrix} = (1-1)\mathbf{k} = \mathbf{0}$

4. $\mathbf{F} = yz\mathbf{i} + xz\mathbf{j} + xy\mathbf{k}$

$\operatorname{\mathbf{div}} \mathbf{F} = \dfrac{\partial}{\partial x}(yz) + \dfrac{\partial}{\partial y}(xz) + \dfrac{\partial}{\partial z}(xy) = 0$

$\operatorname{\mathbf{curl}} \mathbf{F} = \begin{vmatrix} \mathbf{i} & \mathbf{j} & \mathbf{k} \\ \dfrac{\partial}{\partial x} & \dfrac{\partial}{\partial y} & \dfrac{\partial}{\partial z} \\ yz & xz & xy \end{vmatrix}$

$\qquad = (x-x)\mathbf{i} + (y-y)\mathbf{j} + (z-z)\mathbf{k} = \mathbf{0}$

6. $\mathbf{F} = xy^2\mathbf{i} - yz^2\mathbf{j} + zx^2\mathbf{k}$

$\operatorname{\mathbf{div}} \mathbf{F} = \dfrac{\partial}{\partial x}\left(xy^2\right) + \dfrac{\partial}{\partial y}\left(-yz^2\right) + \dfrac{\partial}{\partial z}\left(zx^2\right)$

$\qquad = y^2 - z^2 + x^2$

$\operatorname{\mathbf{curl}} \mathbf{F} = \begin{vmatrix} \mathbf{i} & \mathbf{j} & \mathbf{k} \\ \dfrac{\partial}{\partial x} & \dfrac{\partial}{\partial y} & \dfrac{\partial}{\partial z} \\ xy^2 & -yz^2 & zx^2 \end{vmatrix}$

$\qquad = 2yz\mathbf{i} - 2xz\mathbf{j} - 2xy\mathbf{k}$

8. $\mathbf{F} = f(z)\mathbf{i} - f(z)\mathbf{j}$

$\operatorname{\mathbf{div}} \mathbf{F} = \dfrac{\partial}{\partial x} f(z) + \dfrac{\partial}{\partial y}\left(-f(z)\right) = 0$

$\operatorname{\mathbf{curl}} \mathbf{F} = \begin{vmatrix} \mathbf{i} & \mathbf{j} & \mathbf{k} \\ \dfrac{\partial}{\partial x} & \dfrac{\partial}{\partial y} & \dfrac{\partial}{\partial z} \\ f(z) & -f(z) & 0 \end{vmatrix} = f'(z)(\mathbf{i} + \mathbf{j})$

10. $\mathbf{F} = \hat{\mathbf{r}} = \cos\theta\mathbf{i} + \sin\theta\mathbf{j}$

$\operatorname{\mathbf{div}} \mathbf{F} = \dfrac{\sin^2\theta}{r} + \dfrac{\cos^2\theta}{r} = \dfrac{1}{r} = \dfrac{1}{\sqrt{x^2+y^2}}$

$\operatorname{\mathbf{curl}} \mathbf{F} = \begin{vmatrix} \mathbf{i} & \mathbf{j} & \mathbf{k} \\ \dfrac{\partial}{\partial x} & \dfrac{\partial}{\partial y} & \dfrac{\partial}{\partial z} \\ \cos\theta & \sin\theta & 0 \end{vmatrix}$

$\qquad = -\left(\dfrac{\cos\theta\sin\theta}{r} - \dfrac{\cos\theta\sin\theta}{r}\right)\mathbf{k} = \mathbf{0}$

12. We use the Maclaurin expansion of $\mathbf{F}$, as presented in the proof of Theorem 1:

$$\mathbf{F} = \mathbf{F}_0 + \mathbf{F}_1 x + \mathbf{F}_2 y + \mathbf{F}_3 z + \cdots,$$

where

$\mathbf{F}_0 = \mathbf{F}(0,0,0)$

$\mathbf{F}_1 = \dfrac{\partial}{\partial x}\mathbf{F}(x,y,z)\bigg|_{(0,0,0)} = \left(\dfrac{\partial F_1}{\partial x}\mathbf{i} + \dfrac{\partial F_2}{\partial x}\mathbf{j} + \dfrac{\partial F_3}{\partial x}\mathbf{k}\right)\bigg|_{(0,0,0)}$

$\mathbf{F}_2 = \dfrac{\partial}{\partial y}\mathbf{F}(x,y,z)\bigg|_{(0,0,0)} = \left(\dfrac{\partial F_1}{\partial y}\mathbf{i} + \dfrac{\partial F_2}{\partial y}\mathbf{j} + \dfrac{\partial F_3}{\partial y}\mathbf{k}\right)\bigg|_{(0,0,0)}$

$\mathbf{F}_3 = \dfrac{\partial}{\partial z}\mathbf{F}(x,y,z)\bigg|_{(0,0,0)} = \left(\dfrac{\partial F_1}{\partial z}\mathbf{i} + \dfrac{\partial F_2}{\partial z}\mathbf{j} + \dfrac{\partial F_3}{\partial z}\mathbf{k}\right)\bigg|_{(0,0,0)}$

and where $\cdots$ represents terms of degree 2 and higher in x, y, and z.

On the top of the box $B_{a,b,c}$, we have $z = c$ and $\hat{\mathbf{N}} = \mathbf{k}$.
On the bottom of the box, we have $z = -c$ and $\hat{\mathbf{N}} = -\mathbf{k}$.
On both surfaces $dS = dx\,dy$. Thus

$$\left(\iint_{\text{top}} + \iint_{\text{bottom}}\right)\mathbf{F}\bullet\hat{\mathbf{N}}\,dS$$

$$= \int_{-a}^{a} dx \int_{-b}^{b} dy\big(c\mathbf{F}_3\bullet\mathbf{k} - c\mathbf{F}_3\bullet(-\mathbf{k})\big) + \cdots$$

$$= 8abc\mathbf{F}_3\bullet\mathbf{k} + \cdots = 8abc\dfrac{\partial}{\partial z}F_3(x,y,z)\bigg|_{(0,0,0)} + \cdots,$$

where $\cdots$ represents terms of degree 4 and higher in a, b, and c.

Similar formulas obtain for the two other pairs of faces, and the three formulas combine into

$$\oiint_{B_{a,b,c}} \mathbf{F}\bullet\hat{\mathbf{N}}\,dS = 8abc\operatorname{\mathbf{div}}\mathbf{F}(0,0,0) + \cdots.$$

It follows that

$$\lim_{a,b,c\to 0+} \dfrac{1}{8abc}\oiint_{B_{a,b,c}} \mathbf{F}\bullet\hat{\mathbf{N}}\,dS = \operatorname{\mathbf{div}}\mathbf{F}(0,0,0).$$

14. We use the same Maclaurin expansion for $\mathbf{F}$ as in Exercises 12 and 13. On $\mathcal{C}_\epsilon$ we have

$\mathbf{r} = \epsilon\cos\theta\mathbf{i} + \epsilon\sin\theta\mathbf{j}, \quad (0 \le \theta \le 2\pi)$

$d\mathbf{r} = -\epsilon\sin\theta\mathbf{i} + \epsilon\cos\theta\mathbf{j}$

$\mathbf{F}\bullet d\mathbf{r} = \big(-\epsilon\sin\theta\mathbf{F}_0\bullet\mathbf{i} + \epsilon\cos\theta\mathbf{F}_0\bullet\mathbf{j}$

$\qquad - \epsilon^2\sin\theta\cos\theta\mathbf{F}_1\bullet\mathbf{i} + \epsilon^2\cos^2\theta\mathbf{F}_1\bullet\mathbf{j}$

$\qquad - \epsilon^2\sin^2\theta\mathbf{F}_2\bullet\mathbf{i} + \epsilon^2\sin\theta\cos\theta\mathbf{F}_2\bullet\mathbf{j} + \cdots\big)\,ds,$

where $\cdots$ represents terms of degree 3 or higher in ϵ. Since

$$\int_0^{2\pi} \sin\theta\,d\theta = \int_0^{2\pi} \cos\theta\,d\theta = \int_0^{2\pi} \sin\theta\cos\theta\,d\theta = 0$$

$$\int_0^{2\pi} \cos^2\theta\,d\theta = \int_0^{2\pi} \sin^2\theta\,d\theta = \pi,$$

we have

$$\frac{1}{\pi\epsilon^2}\oint_{C_\epsilon}\mathbf{F}\bullet d\mathbf{r}=\mathbf{F}_1\bullet\mathbf{j}-\mathbf{F}_2\bullet\mathbf{i}+\cdots,$$

where $\cdots$ represents terms of degree at least 1 in ϵ.
Hence

$$\lim_{\epsilon\to0+}\frac{1}{\pi\epsilon^2}\oint_{C_\epsilon}\mathbf{F}\bullet d\mathbf{r}=\mathbf{F}_1\bullet\mathbf{j}-\mathbf{F}_2\bullet\mathbf{i}$$
$$=\frac{\partial F_2}{\partial x}-\frac{\partial F_1}{\partial y}$$
$$=\operatorname{curl}\mathbf{F}\bullet\mathbf{k}=\operatorname{curl}\mathbf{F}\bullet\hat{\mathbf{N}}.$$

Section 16.2   Some Identities Involving Grad, Div, and Curl   (page 961)

2. Theorem 3(b):

$$\boldsymbol{\nabla}\bullet(\phi\mathbf{F})=\frac{\partial}{\partial x}(\phi F_1)+\frac{\partial}{\partial y}(\phi F_2)+\frac{\partial}{\partial z}(\phi F_3)$$
$$=\frac{\partial\phi}{\partial x}F_1+\phi\frac{\partial F_1}{\partial x}+\cdots+\frac{\partial\phi}{\partial z}F_3+\phi\frac{\partial F_3}{\partial z}+\cdots$$
$$=\boldsymbol{\nabla}\phi\bullet\mathbf{F}+\phi\boldsymbol{\nabla}\bullet\mathbf{F}.$$

4. Theorem 3(f). The first component of $\boldsymbol{\nabla}(\mathbf{F}\bullet\mathbf{G})$ is

$$\frac{\partial F_1}{\partial x}G_1+F_1\frac{\partial G_1}{\partial x}+\frac{\partial F_2}{\partial x}G_2+F_2\frac{\partial G_2}{\partial x}+\frac{\partial F_3}{\partial x}G_3+F_3\frac{\partial G_3}{\partial x}.$$

We calculate the first components of the four terms on the right side of the identity to be proved.
The first component of $\mathbf{F}\times(\boldsymbol{\nabla}\times\mathbf{G})$ is

$$F_2\left(\frac{\partial G_2}{\partial x}-\frac{\partial G_1}{\partial y}\right)-F_3\left(\frac{\partial G_1}{\partial z}-\frac{\partial G_3}{\partial x}\right).$$

The first component of $\mathbf{G}\times(\boldsymbol{\nabla}\times\mathbf{F})$ is

$$G_2\left(\frac{\partial F_2}{\partial x}-\frac{\partial F_1}{\partial y}\right)-G_3\left(\frac{\partial F_1}{\partial z}-\frac{\partial F_3}{\partial x}\right).$$

The first component of $(\mathbf{F}\bullet\boldsymbol{\nabla})\mathbf{G}$ is

$$F_1\frac{\partial G_1}{\partial x}+F_2\frac{\partial G_1}{\partial y}+F_3\frac{\partial G_1}{\partial z}.$$

The first component of $(\mathbf{G}\bullet\boldsymbol{\nabla})\mathbf{F}$ is

$$G_1\frac{\partial F_1}{\partial x}+G_2\frac{\partial F_1}{\partial y}+G_3\frac{\partial F_1}{\partial z}.$$

When we add these four first components, eight of the fourteen terms cancel out and the six remaining terms are the six terms of the first component of $\boldsymbol{\nabla}(\mathbf{F}\bullet\mathbf{G})$, as calculated above. Similar calculations show that the second and third components of both sides of the identity agree. Thus

$$\boldsymbol{\nabla}(\mathbf{F}\bullet\mathbf{G})=\mathbf{F}\times(\boldsymbol{\nabla}\times\mathbf{G})+\mathbf{G}\times(\boldsymbol{\nabla}\times\mathbf{F})+(\mathbf{F}\bullet\boldsymbol{\nabla})\mathbf{G}+(\mathbf{G}\bullet\boldsymbol{\nabla})\mathbf{F}.$$

6. Theorem 3(i). We examine the first components of the terms on both sides of the identity

$$\boldsymbol{\nabla}\times(\boldsymbol{\nabla}\times\mathbf{F})=\boldsymbol{\nabla}(\boldsymbol{\nabla}\bullet\mathbf{F})-\boldsymbol{\nabla}^2\mathbf{F}.$$

The first component of $\boldsymbol{\nabla}\times(\boldsymbol{\nabla}\times\mathbf{F})$ is

$$\frac{\partial}{\partial y}\left(\frac{\partial F_2}{\partial x}-\frac{\partial F_1}{\partial y}\right)-\frac{\partial}{\partial z}\left(\frac{\partial F_1}{\partial z}-\frac{\partial F_3}{\partial x}\right)$$
$$=\frac{\partial^2 F_2}{\partial y\partial x}-\frac{\partial^2 F_1}{\partial y^2}-\frac{\partial^2 F_1}{\partial z^2}+\frac{\partial^2 F_3}{\partial z\partial x}.$$

The first component of $\boldsymbol{\nabla}(\boldsymbol{\nabla}\bullet\mathbf{F})$ is

$$\frac{\partial}{\partial x}\boldsymbol{\nabla}\bullet\mathbf{F}=\frac{\partial^2 F_1}{\partial x^2}+\frac{\partial^2 F_2}{\partial x\partial y}+\frac{\partial^2 F_3}{\partial x\partial z}.$$

The first component of $-\boldsymbol{\nabla}^2\mathbf{F}$ is

$$-\boldsymbol{\nabla}^2 F_1=-\frac{\partial^2 F_1}{\partial x^2}-\frac{\partial^2 F_1}{\partial y^2}-\frac{\partial^2 F_1}{\partial z^2}.$$

Evidently the first components of both sides of the given identity agree. By symmetry, so do the other components.

8. If $\mathbf{r}=x\mathbf{i}+y\mathbf{j}+z\mathbf{k}$ and $r=|\mathbf{r}|$, then

$$\boldsymbol{\nabla}\bullet\mathbf{r}=3,\qquad\boldsymbol{\nabla}\times\mathbf{r}=\mathbf{0},\qquad\boldsymbol{\nabla}r=\frac{\mathbf{r}}{r}.$$

If $\mathbf{c}$ is a constant vector, then its divergence and curl are both zero. By Theorem 3(d), (e), and (f) we have

$$\boldsymbol{\nabla}\bullet(\mathbf{c}\times\mathbf{r})=(\boldsymbol{\nabla}\times\mathbf{c})\bullet\mathbf{r}-\mathbf{c}\bullet(\boldsymbol{\nabla}\times\mathbf{r})=\mathbf{0}$$
$$\boldsymbol{\nabla}\times(\mathbf{c}\times\mathbf{r})=(\boldsymbol{\nabla}\bullet\mathbf{r})\mathbf{c}+(\mathbf{r}\bullet\boldsymbol{\nabla})\mathbf{c}-(\boldsymbol{\nabla}\bullet\mathbf{c})\mathbf{r}-(\mathbf{c}\bullet\boldsymbol{\nabla})\mathbf{r}$$
$$=3\mathbf{c}+\mathbf{0}-\mathbf{0}-\mathbf{c}=2\mathbf{c}$$

$$\boldsymbol{\nabla}(\mathbf{c}\bullet\mathbf{r})=\mathbf{c}\times(\boldsymbol{\nabla}\times\mathbf{r})+\mathbf{r}\times(\boldsymbol{\nabla}\times\mathbf{c})+(\mathbf{c}\bullet\boldsymbol{\nabla})\mathbf{r}+(\mathbf{r}\bullet\boldsymbol{\nabla})\mathbf{c}$$
$$=\mathbf{0}+\mathbf{0}+\mathbf{c}+\mathbf{0}=\mathbf{c}.$$

10. Given that $\operatorname{div}\mathbf{F}=0$ and $\operatorname{curl}\mathbf{F}=\mathbf{0}$, Theorem 3(i) implies that $\boldsymbol{\nabla}^2\mathbf{F}=\mathbf{0}$ too. Hence the components of $\mathbf{F}$ are harmonic functions.
If $\mathbf{F}=\boldsymbol{\nabla}\phi$, then

$$\boldsymbol{\nabla}^2\phi=\boldsymbol{\nabla}\bullet\boldsymbol{\nabla}\phi=\boldsymbol{\nabla}\bullet\mathbf{F}=0,$$

so ϕ is also harmonic.

12. If $\nabla^2 \phi = 0$ and $\nabla^2 \psi = 0$, then

$$\nabla \bullet (\phi \nabla \psi - \psi \nabla \phi)$$
$$= \nabla \phi \bullet \nabla \psi + \phi \nabla^2 \psi - \nabla \psi \bullet \nabla \phi - \psi \nabla^2 \phi = 0,$$

so $\phi \nabla \psi - \psi \nabla \phi$ is solenoidal.

14. By Theorem 3(b), (d), and (h), we have

$$\nabla \bullet \left(f(\nabla g \times \nabla h) \right)$$
$$= \nabla f \bullet (\nabla g \times \nabla h) + f \nabla \bullet (\nabla g \times \nabla h)$$
$$= \nabla f \bullet (\nabla g \times \nabla h) + f \left((\nabla \times \nabla g) \bullet \nabla h - \nabla g \bullet (\nabla \times \nabla h) \right)$$
$$= \nabla f \bullet (\nabla g \times \nabla h) + \mathbf{0} - \mathbf{0} = \nabla f \bullet (\nabla g \times \nabla h).$$

16. If $\nabla \times \mathbf{G} = \mathbf{F} = -y\mathbf{i} + x\mathbf{j}$, then

$$\frac{\partial G_3}{\partial y} - \frac{\partial G_2}{\partial z} = -y$$
$$\frac{\partial G_1}{\partial z} - \frac{\partial G_3}{\partial x} = x$$
$$\frac{\partial G_2}{\partial x} - \frac{\partial G_1}{\partial y} = 0.$$

As in Example 1, we try to find a solution with $G_2 = 0$. Then

$$G_3 = -\int y \, dy = -\frac{y^2}{2} + M(x, z).$$

Again we try $M(x, z) = 0$, so $G_3 = -\dfrac{y^2}{2}$. Thus $\dfrac{\partial G_3}{\partial x} = 0$ and

$$G_1 = \int x \, dz = xz + N(x, y).$$

Since $\dfrac{\partial G_1}{\partial y} = 0$ we may take $N(x, y) = 0$.

$\mathbf{G} = xz\mathbf{i} - \dfrac{1}{2}y^2\mathbf{k}$ is a vector potential for $\mathbf{F}$. (Of course, this answer is not unique.)

18. For (x, y, z) in D let $\mathbf{v} = x\mathbf{i} + y\mathbf{j} + z\mathbf{k}$. The line segment $\mathbf{r}(t) = t\mathbf{v}$, $(0 \le t \le 1)$, lies in D, so $\mathbf{div}\,\mathbf{F} = 0$ on the path. We have

$$\mathbf{G}(x, y, z) = \int_0^1 t \mathbf{F}(\mathbf{r}(t)) \times \mathbf{v} \, dt$$
$$= \int_0^1 t \mathbf{F}(\xi(t), \eta(t), \zeta(t)) \times \mathbf{v} \, dt$$

where $\xi = tx$, $\eta = ty$, $\zeta = tz$. The first component of **curl G** is

$$(\mathbf{curl}\,\mathbf{G})_1$$
$$= \int_0^1 t \left(\mathbf{curl}\,(\mathbf{F} \times \mathbf{v}) \right)_1 dt$$
$$= \int_0^1 t \left(\frac{\partial}{\partial y}(\mathbf{F} \times \mathbf{v})_3 - \frac{\partial}{\partial z}(\mathbf{F} \times \mathbf{v})_2 \right) dt$$
$$= \int_0^1 t \left(\frac{\partial}{\partial y}(F_1 y - F_2 x) - \frac{\partial}{\partial z}(F_3 x - F_1 z) \right) dt$$
$$= \int_0^1 \left(t F_1 + t^2 y \frac{\partial F_1}{\partial \eta} - t^2 x \frac{\partial F_2}{\partial \eta} - t^2 x \frac{\partial F_3}{\partial \zeta} \right.$$
$$\left. + t F_1 + t^2 z \frac{\partial F_1}{\partial \zeta} \right) dt$$
$$= \int_0^1 \left(2t F_1 + t^2 x \frac{\partial F_1}{\partial \xi} + t^2 y \frac{\partial F_1}{\partial \eta} + t^2 z \frac{\partial F_1}{\partial \zeta} \right) dt.$$

To get the last line we used the fact that $div\,\mathbf{F} = 0$ to replace $-t^2 x \dfrac{\partial F_2}{\partial \eta} - t^2 x \dfrac{\partial F_3}{\partial \zeta}$ with $t^2 x \dfrac{\partial F_1}{\partial \xi}$. Continuing the calculation, we have

$$(\mathbf{curl}\,\mathbf{G})_1 = \int_0^1 \frac{d}{dt}\left(t^2 F_1(\xi, \eta, \zeta) \right) dt$$
$$= t^2 F_1(tx, ty, tz)\Big|_0^1 = F_1(x, y, z).$$

Similarly, $(\mathbf{curl}\,\mathbf{G})_2 = F_2$ and $(\mathbf{curl}\,\mathbf{G})_3 = F_3$. Thus **curl G = F**, as required.

Section 16.3 Green's Theorem in the Plane (page 965)

2.
$$\oint_C (x^2 - xy)\, dx + (xy - y^2)\, dy$$
$$= -\iint_T \left[\frac{\partial}{\partial x}(xy - y^2) - \frac{\partial}{\partial y}(x^2 - xy) \right] dA$$
$$= -\iint_T (y + x)\, dA$$
$$= -(\bar{y} + \bar{x}) \times (\text{area of } T) = -\left(\frac{1}{3} + 1 \right) \times 1 = -\frac{4}{3}.$$

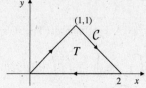

Fig. 16.3.2

4. Let D be the region $x^2 + y^2 \le 9$, $y \ge 0$. Since C is the clockwise boundary of D,

$$\oint_C x^2 y \, dx - xy^2 \, dy$$

$$= -\iint_D \left[\frac{\partial}{\partial x}(-xy^2) - \frac{\partial}{\partial y}(x^2 y) \right] dx \, dy$$

$$= \iint_D (y^2 + x^2) \, dA = \int_0^\pi d\theta \int_0^3 r^3 \, dr = \frac{81\pi}{4}.$$

6. Let R, C, and $\mathbf{F}$ be as in the statement of Green's Theorem. As noted in the proof of Theorem 7, the unit tangent $\hat{\mathbf{T}}$ to C and the unit exterior normal $\hat{\mathbf{N}}$ satisfy $\hat{\mathbf{N}} = \hat{\mathbf{T}} \times \mathbf{k}$. Let

$$\mathbf{G} = F_2(x, y)\mathbf{i} - F_1(x, y)\mathbf{j}.$$

Then $\mathbf{F} \bullet \hat{\mathbf{T}} = \mathbf{G} \bullet \hat{\mathbf{N}}$. Applying the 2-dimensional Divergence Theorem to $\mathbf{G}$, we obtain

$$\int_C F_1 \, dx + F_2 \, dy = \int_C \mathbf{F} \bullet \hat{\mathbf{T}} \, ds = \int_C \mathbf{G} \bullet \hat{\mathbf{N}} \, ds$$

$$= \iint_R \operatorname{div} \mathbf{G} \, dA$$

$$= \iint_R \left(\frac{\partial F_2}{\partial x} - \frac{\partial F_1}{\partial y} \right) dA$$

as required

8. a) $\mathbf{F} = x^2 \mathbf{j}$

$$\oint_C \mathbf{F} \bullet d\mathbf{r} = \oint_C x^2 \, dy = \iint_R 2x \, dA = 2A\bar{x}.$$

b) $\mathbf{F} = xy\mathbf{i}$

$$\oint_C \mathbf{F} \bullet d\mathbf{r} = \oint_C xy \, dx = -\iint_R x \, dA = -A\bar{x}.$$

c) $\mathbf{F} = y^2 \mathbf{i} + 3xy\mathbf{j}$

$$\oint_C \mathbf{F} \bullet d\mathbf{r} = \oint_C y^2 \, dx + 3xy \, dy$$

$$= \iint_R (3y - 2y) \, dA = A\bar{y}.$$

Section 16.4   The Divergence Theorem in 3-Space   (page 971)

2. If $\mathbf{F} = ye^z \mathbf{i} + x^2 e^z \mathbf{j} + xy\mathbf{k}$, then $\operatorname{div} \mathbf{F} = 0$, and

$$\oiint_S \mathbf{F} \bullet \hat{\mathbf{N}} \, dS = \iiint_B 0 \, dV = 0.$$

4. If $\mathbf{F} = x^3 \mathbf{i} + 3yz^2 \mathbf{j} + (3y^2 z + x^2)\mathbf{k}$, then $\operatorname{div} \mathbf{F} = 3x^2 + 3z^2 + 3y^2$, and

$$\oiint_S \mathbf{F} \bullet \hat{\mathbf{N}} \, dS = 3 \iiint_B (x^2 + y^2 + z^2) \, dV$$

$$= 3 \int_0^{2\pi} d\theta \int_0^\pi \sin\phi \, d\phi \int_0^a \rho^4 \, d\rho$$

$$= \frac{12}{5} \pi a^5.$$

6. If $\mathbf{F} = x^2 \mathbf{i} + y^2 \mathbf{j} + z^2 \mathbf{k}$, then $\operatorname{div} \mathbf{F} = 2(x + y + z)$. Therefore the flux of $\mathbf{F}$ out of any solid region R is

$$\text{Flux} = \iiint_R \operatorname{div} \mathbf{F} \, dV$$

$$= 2 \iiint_R (x + y + z) \, dV = 2(\bar{x} + \bar{y} + \bar{z})V$$

where $(\bar{x}, \bar{y}, \bar{z})$ is the centroid of R and V is the volume of R.

If R is the ellipsoid $x^2 + y^2 + 4(z - 1)^2 \le 4$, then $\bar{x} = 0$, $\bar{y} = 0$, $\bar{z} = 1$, and $V = (4\pi/3)(2)(2)(1) = 16\pi/3$. The flux of $\mathbf{F}$ out of R is $2(0 + 0 + 1)(16\pi/3) = 32\pi/3$.

8. If $\mathbf{F} = x^2 \mathbf{i} + y^2 \mathbf{j} + z^2 \mathbf{k}$, then $\operatorname{div} \mathbf{F} = 2(x + y + z)$. Therefore the flux of $\mathbf{F}$ out of any solid region R is

$$\text{Flux} = \iiint_R \operatorname{div} \mathbf{F} \, dV$$

$$= 2 \iiint_R (x + y + z) \, dV = 2(\bar{x} + \bar{y} + \bar{z})V$$

where $(\bar{x}, \bar{y}, \bar{z})$ is the centroid of R and V is the volume of R.

If R is the cylinder $x^2 + y^2 \le 2y$ (or, equivalently, $x^2 + (y - 1)^2 \le 1$), $0 \le z \le 4$, then $\bar{x} = 0$, $\bar{y} = 1$, $\bar{z} = 2$, and $V = (\pi 1^2)(4) = 4\pi$. The flux of $\mathbf{F}$ out of R is $2(0 + 1 + 2)(4\pi) = 24\pi$.

10. The required surface integral,

$$I = \iint_S \nabla\phi \bullet \hat{\mathbf{N}} \, dS,$$

can be calculated directly by the methods of Section 6.6. We will do it here by using the Divergence Theorem instead. S is one face of a tetrahedral domain D whose other faces are in the coordinate planes, as shown in the figure. Since $\phi = xy + z^2$, we have

$$\nabla\phi = y\mathbf{i} + x\mathbf{j} + 2z\mathbf{k}, \qquad \nabla \bullet \nabla\phi = \nabla^2 \phi = 2.$$

Thus

$$\iiint_D \nabla \bullet \nabla\phi \, dV = 2 \times \frac{abc}{6} = \frac{abc}{3},$$

the volume of the tetrahedron D being $abc/6$ cubic units.

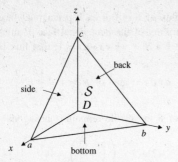

Fig. 16.4.10

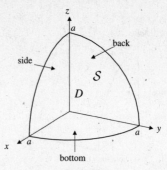

Fig. 16.4.12

The flux of $\nabla\phi$ out of D is the sum of its fluxes out of the four faces of the tetrahedron.

On the bottom, $\hat{\mathbf{N}} = -\mathbf{k}$ and $z = 0$, so $\nabla\phi \bullet \hat{\mathbf{N}} = 0$, and the flux out of the bottom face is 0.

On the side, $y = 0$ and $\hat{\mathbf{N}} = -\mathbf{j}$, so $\nabla\phi \bullet \hat{\mathbf{N}} = -x$. The flux out of the side face is

$$\iint_{\text{side}} \nabla\phi \bullet \hat{\mathbf{N}}\, dS = -\iint_{\text{side}} x\, dx\, dz = -\frac{ac}{2} \times \frac{a}{3} = -\frac{a^2 c}{6}.$$

(We used the fact that $M_{x=0} = \text{area} \times \bar{x}$ and $\bar{x} = a/3$ for that face.)

On the back face, $x = 0$ and $\hat{\mathbf{N}} = -\mathbf{i}$, so the flux out of that face is

$$\iint_{\text{back}} \nabla\phi \bullet \hat{\mathbf{N}}\, dS = -\iint_{\text{back}} y\, dy\, dz = -\frac{bc}{2} \times \frac{b}{3} = -\frac{b^2 c}{6}.$$

Therefore, by the Divergence Theorem

$$I - \frac{a^2 c}{6} - \frac{b^2 c}{6} + 0 = \frac{abc}{3},$$

so $\displaystyle\iint_{S} \nabla\phi \bullet \hat{\mathbf{N}}\, dS = I = \frac{abc}{3} + \frac{c(a^2 + b^2)}{6}.$

12. $\mathbf{F} = (y + xz)\mathbf{i} + (y + yz)\mathbf{j} - (2x + z^2)\mathbf{k}$
$\text{div}\,\mathbf{F} = z + (1 + z) - 2z = 1.$ Thus

$$\iiint_{D} \text{div}\,\mathbf{F}\, dV = \text{volume of } D = \frac{\pi a^3}{6},$$

where D is the region in the first octant bounded by the sphere and the coordinate planes. The boundary of D consists of the spherical part S and the four planar parts, called the bottom, side, and back in the figure.

On the side, $y = 0$, $\hat{\mathbf{N}} = -\mathbf{j}$, $\mathbf{F} \bullet \hat{\mathbf{N}} = 0$, so

$$\iint_{\text{side}} \mathbf{F} \bullet \hat{\mathbf{N}}\, dS = 0.$$

On the back, $x = 0$, $\hat{\mathbf{N}} = -\mathbf{i}$, $\mathbf{F} \bullet \hat{\mathbf{N}} = -y$, so

$$\iint_{\text{back}} \mathbf{F} \bullet \hat{\mathbf{N}}\, dS = -\int_0^{\pi/2} d\theta \int_0^a r\cos\theta\, r\, dr$$
$$= -\sin\theta \Big|_0^{\pi/2} \times \frac{a^3}{3} = -\frac{a^3}{3}.$$

On the bottom, $z = 0$, $\hat{\mathbf{N}} = -\mathbf{k}$, $\mathbf{F} \bullet \hat{\mathbf{N}} = 2x$, so

$$\iint_{\text{bottom}} \mathbf{F} \bullet \hat{\mathbf{N}}\, dS = 2\int_0^{\pi/2} d\theta \int_0^a r\cos\theta\, r\, dr = \frac{2a^3}{3}.$$

By the Divergence Theorem

$$\iint_{S} \mathbf{F} \bullet \hat{\mathbf{N}}\, dS + 0 - \frac{a^3}{3} + \frac{2a^3}{3} = \frac{\pi a^3}{6}.$$

Hence the flux of $\mathbf{F}$ upward through S is

$$\iint_{S} \mathbf{F} \bullet \hat{\mathbf{N}}\, dS = \frac{\pi a^3}{6} - \frac{a^3}{3}.$$

14. Let D be the domain bounded by S, the coordinate planes, and the plane $x = 1$. If

$$\mathbf{F} = 3xz^2\mathbf{i} - x\mathbf{j} - y\mathbf{k},$$

then $\text{div}\,\mathbf{F} = 3z^2$, so the total flux of $\mathbf{F}$ out of D is

$$\oiint_{\text{bdry of } D} \mathbf{F} \bullet \hat{\mathbf{N}}\, dS = \iiint_{D} 3z^2\, dV$$
$$= 3\int_0^1 dx \int_0^{\pi/2} d\theta \int_0^1 r^2 \cos^2\theta\, r\, dr$$
$$= 3 \times \frac{1}{4} \times \frac{\pi}{4} = \frac{3\pi}{16}.$$

The boundary of D consists of the cylindrical surface S and four planar surfaces, the side, bottom, back, and front.

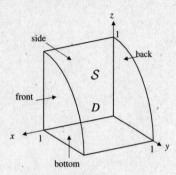

Fig. 16.4.14

On the side, $y = 0$, $\hat{\mathbf{N}} = -\mathbf{j}$, $\mathbf{F} \bullet \hat{\mathbf{N}} = x$, so

$$\iint_{\text{side}} \mathbf{F} \bullet \hat{\mathbf{N}} \, dS = \int_0^1 x \, dx \int_0^1 dz = \frac{1}{2}.$$

On the bottom, $z = 0$, $\hat{\mathbf{N}} = -\mathbf{k}$, $\mathbf{F} \bullet \hat{\mathbf{N}} = y$, so

$$\iint_{\text{bottom}} \mathbf{F} \bullet \hat{\mathbf{N}} \, dS = \int_0^1 y \, dy \int_0^1 dx = \frac{1}{2}.$$

On the back, $x = 0$, $\hat{\mathbf{N}} = -\mathbf{i}$, $\mathbf{F} \bullet \hat{\mathbf{N}} = 0$, so

$$\iint_{\text{back}} \mathbf{F} \bullet \hat{\mathbf{N}} \, dS = 0.$$

On the front, $x = 1$, $\hat{\mathbf{N}} = \mathbf{i}$, $\mathbf{F} \bullet \hat{\mathbf{N}} = 3z^2$, so

$$\iint_{\text{front}} \mathbf{F} \bullet \hat{\mathbf{N}} \, dS = 3 \int_0^{\pi/2} d\theta \int_0^1 r^2 \cos^2 \theta \, r \, dr = \frac{3\pi}{16}.$$

Hence,

$$\iint_S (3xz^2 \mathbf{i} - x\mathbf{j} - y\mathbf{k}) \bullet \hat{\mathbf{N}} \, dS = \frac{3\pi}{16} - \frac{1}{2} - \frac{1}{2} - 0 - \frac{3\pi}{16} = -1.$$

16. $\mathbf{F} = x\mathbf{i} + y\mathbf{j} + z\mathbf{k}$ implies that $\mathbf{div}\,\mathbf{F} = 3$. The total flux of $\mathbf{F}$ out of D is

$$\oiint_{\text{bdry of } D} \mathbf{F} \bullet \hat{\mathbf{N}} \, dS = 3 \iiint_D dV = 12,$$

since the volume of D is half that of a cube of side 2, that is, 4 square units.
D has three triangular faces, three pentagonal faces, and a hexagonal face. By symmetry, the flux of $\mathbf{F}$ out of each triangular face is equal to that out of the triangular face T in the plane $z = 1$. Since $\mathbf{F} \bullet \hat{\mathbf{N}} = \mathbf{k} \bullet \mathbf{k} = 1$ on that face, these fluxes are

$$\iint_T dx \, dy = \text{area of } T = \frac{1}{2}.$$

Similarly, the flux of $\mathbf{F}$ out of each pentagonal face is equal to the flux out of the pentagonal face P in the plane $z = -1$, where $\mathbf{F} \bullet \hat{\mathbf{N}} = -\mathbf{k} \bullet (-\mathbf{k}) = 1$; that flux is

$$\iint_P dx \, dy = \text{area of } P = 4 - \frac{1}{2} = \frac{7}{2}.$$

Thus the flux of $\mathbf{F}$ out of the remaining hexagonal face H is

$$12 - 3 \times \left(\frac{1}{2} + \frac{7}{2} \right) = 0.$$

(This can also be seen directly, since $\mathbf{F}$ radiates from the origin, so is everywhere tangent to the plane of the hexagonal face, the plane $x + y + z = 0$.)

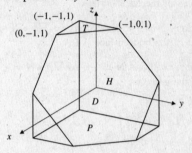

Fig. 16.4.16

18. $\phi = x^2 - y^2 + z^2$, $\mathbf{G} = \frac{1}{3}(-y^3 \mathbf{i} + x^3 \mathbf{j} + z^3 \mathbf{k})$.
$\mathbf{F} = \nabla\phi + \mu\,\mathbf{curl}\,\mathbf{G}$.

Let R be the region of 3-space occupied by the sandpile. Then R is bounded by the upper surface S of the sandpile and by the disk D: $x^2 + y^2 \leq 1$ in the plane $z = 0$. The outward (from R) normal on D is $-\mathbf{k}$. The flux of $\mathbf{F}$ out of R is given by

$$\iint_S \mathbf{F} \bullet \hat{\mathbf{N}} \, dS + \iint_D \mathbf{F} \bullet (-\mathbf{k}) \, dA = \iiint_R \mathbf{div}\,\mathbf{F} \, dV.$$

Now $\mathbf{div}\,\mathbf{curl}\,\mathbf{G} = 0$ by Theorem 3(g). Also $\mathbf{div}\,\nabla\phi = \mathbf{div}\,(2x\mathbf{i} - 2y\mathbf{j} + 2z\mathbf{k}) = 2 - 2 + 2 = 2$. Therefore

$$\iiint_R \mathbf{div}\,\mathbf{F} \, dV = \iiint_R (2 + \mu \times 0) \, dV = 2(5\pi) = 10\pi.$$

In addition,

$$\mathbf{curl}\,\mathbf{G} = \frac{1}{3} \begin{vmatrix} \mathbf{i} & \mathbf{j} & \mathbf{k} \\ \dfrac{\partial}{\partial x} & \dfrac{\partial}{\partial y} & \dfrac{\partial}{\partial z} \\ -y^3 & x^3 & z^3 \end{vmatrix} = 3(x^2 + y^2)\mathbf{k},$$

and $\nabla\phi \bullet \mathbf{k} = 2z = 0$ on D, so

$$\iint_D \mathbf{F} \bullet \mathbf{k} \, dA = 3\mu \int_0^{2\pi} d\theta \int_0^1 r^3 \, dr = \frac{3\pi\mu}{2}.$$

The flux of $\mathbf{F}$ out of $\mathcal{S}$ is $10\pi + (3\pi\mu)/2$.

20. If $\mathbf{r} = x\mathbf{i} + y\mathbf{j} + z\mathbf{k}$, then $\operatorname{div}\mathbf{r} = 3$ and

$$\frac{1}{3} \oiint_{\mathcal{S}} \mathbf{r} \bullet \hat{\mathbf{N}}\, dS = \frac{1}{3} \iiint_{D} 3\, dV = V.$$

22. Taking $\mathbf{F} = \nabla\phi$ in the first identity in Theorem 7(a), we have

$$\oiint_{\mathcal{S}} \nabla\phi \times \hat{\mathbf{N}}\, dS = -\iiint_{D} \operatorname{curl} \nabla\phi\, dV = 0,$$

since $\nabla \times \nabla\phi = 0$ by Theorem 3(h).

24. If $\mathbf{F} = \nabla\phi$ in the previous exercise, then $\operatorname{div}\mathbf{F} = \nabla^2\phi$ and

$$\iiint_{D} \phi\nabla^2\phi\, dV + \iiint_{D} |\nabla\phi|^2\, dV = \oiint_{\mathcal{S}} \phi\nabla\phi \bullet \hat{\mathbf{N}}\, dS.$$

If $\nabla^2\phi = 0$ in D and $\phi = 0$ on $\mathcal{S}$, then

$$\iiint_{D} |\nabla\phi|^2\, dV = 0.$$

Since ϕ is assumed to be smooth, $\nabla\phi = 0$ throughout D, and therefore ϕ is constant on each connected component of D. Since $\phi = 0$ on $\mathcal{S}$, these constants must all be 0, and $\phi = 0$ on D.

26. Re-examine the solution to Exercise 24 above. If $\nabla^2\phi = 0$ in D and $\partial\phi/\partial n = \nabla\phi \bullet \hat{\mathbf{N}} = 0$ on $\mathcal{S}$, then we can again conclude that

$$\iiint_{D} |\nabla\phi|\, dV = 0$$

and $\nabla\phi = 0$ throughout D. Thus ϕ is constant on the connected components of D. (We can't conclude the constant is 0 because we don't know the value of ϕ on $\mathcal{S}$.) If u and v are solutions of the given Neumann problem, then $\phi = u - v$ satisfies

$$\nabla^2\phi = \nabla^2 u - \nabla^2 v = f - f = 0 \text{ on } D$$
$$\frac{\partial\phi}{\partial n} = \frac{\partial u}{\partial n} - \frac{\partial v}{\partial n} = g - g = 0 \text{ on } \mathcal{S},$$

so ϕ is constant on any connected component of $\mathcal{S}$, and u and v can only differ by a constant on $\mathcal{S}$.

28. By Theorem 3(b),

$$\operatorname{div}(\phi\nabla\psi - \psi\nabla\phi)$$
$$= \nabla\phi \bullet \nabla\psi + \phi\nabla^2\psi - \nabla\psi \bullet \nabla\phi - \psi\nabla^2\phi$$
$$= \phi\nabla^2\psi - \psi\nabla^2\phi.$$

Hence, by the Divergence Theorem,

$$\iiint_{D} (\phi\nabla^2\psi - \psi\nabla^2\phi)\, dV = \iiint_{D} \operatorname{div}(\phi\nabla\psi - \psi\nabla\phi)\, dV$$
$$= \oiint_{\mathcal{S}} (\phi\nabla\psi - \psi\nabla\phi) \bullet \hat{\mathbf{N}}\, dS$$
$$= \oiint_{\mathcal{S}} \left(\phi\frac{\partial\psi}{\partial n} - \psi\frac{\partial\phi}{\partial n}\right) dS.$$

30.
$$\frac{1}{\operatorname{vol}(D_\epsilon)} \oiint_{\mathcal{S}_\epsilon} \mathbf{F} \bullet \hat{\mathbf{N}}\, dS = \frac{1}{\operatorname{vol}(D_\epsilon)} \iiint_{D_\epsilon} \operatorname{div}\mathbf{F}\, dV$$
$$= \frac{1}{\operatorname{vol}(D_\epsilon)}\left[\iiint_{D_\epsilon} \operatorname{div}\mathbf{F}(P_0)\, dV\right.$$
$$\left.+ \iiint_{D_\epsilon} \big(\operatorname{div}\mathbf{F} - \operatorname{div}\mathbf{F}(P_0)\big)\, dV\right]$$
$$= \operatorname{div}\mathbf{F}(P_0) + \frac{1}{\operatorname{vol}(D_\epsilon)} \iiint_{D_\epsilon} \big(\operatorname{div}\mathbf{F} - \operatorname{div}\mathbf{F}(P_0)\big)\, dV.$$

Thus

$$\left|\frac{1}{\operatorname{vol}(D_\epsilon)} \oiint_{\mathcal{S}_\epsilon} \mathbf{F} \bullet \hat{\mathbf{N}}\, dS - \operatorname{div}\mathbf{F}(P_0)\right|$$
$$\leq \frac{1}{\operatorname{vol}(D_\epsilon)} \iiint_{D_\epsilon} |\operatorname{div}\mathbf{F} - \operatorname{div}\mathbf{F}(P_0)|\, dV$$
$$\leq \max_{P \text{ in } D_\epsilon} |\operatorname{div}\mathbf{F} - \operatorname{div}\mathbf{F}(P_0)|$$

$\to 0$ as $\epsilon \to 0+$ assuming $\operatorname{div}\mathbf{F}$ is continuous.

$$\lim_{\epsilon\to 0+} \frac{1}{\operatorname{vol}(D_\epsilon)} \oiint_{\mathcal{S}_\epsilon} \mathbf{F} \bullet \hat{\mathbf{N}}\, dS = \operatorname{div}\mathbf{F}(P_0).$$

Section 16.5   Stokes's Theorem   (page 977)

2. Let $\mathcal{S}$ be the part of the surface $z = y^2$ lying inside the cylinder $x^2 + y^2 = 4$, and having upward normal $\hat{\mathbf{N}}$. Then $\mathcal{C}$ is the oriented boundary of $\mathcal{S}$. Let D be the disk $x^2 + y^2 \leq 4$, $z = 0$, that is, the projection of $\mathcal{S}$ onto the xy-plane.

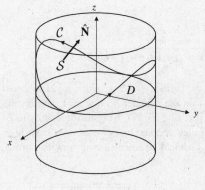

Fig. 16.5.2

If $\mathbf{F} = y\mathbf{i} - x\mathbf{j} + z^2\mathbf{k}$, then

$$\text{curl}\,\mathbf{F} = \begin{vmatrix} \mathbf{i} & \mathbf{j} & \mathbf{k} \\ \dfrac{\partial}{\partial x} & \dfrac{\partial}{\partial y} & \dfrac{\partial}{\partial z} \\ y & -x & z^2 \end{vmatrix} = -2\mathbf{k}.$$

Since $dS = \dfrac{dx\,dy}{\mathbf{k}\bullet\hat{\mathbf{N}}}$ on $\mathcal{S}$, we have

$$\oint_{\mathcal{C}} y\,dx - x\,dy + z^2\,dz = \oint_{\mathcal{C}} \mathbf{F}\bullet d\mathbf{r} = \iint_{\mathcal{S}} \text{curl}\,\mathbf{F}\bullet\hat{\mathbf{N}}\,dS$$
$$= \iint_{D} -2\mathbf{k}\bullet\hat{\mathbf{N}}\,\frac{dx\,dy}{\mathbf{k}\bullet\hat{\mathbf{N}}} = -8\pi.$$

4. The surface $\mathcal{S}$ with equation

$$x^2 + y^2 + 2(z-1)^2 = 6, \qquad z \geq 0,$$

with outward normal $\hat{\mathbf{N}}$, is that part of an ellipsoid of revolution about the z-axis, centred at $(0,0,1)$, and lying above the xy-plane. The boundary of $\mathcal{S}$ is the circle $\mathcal{C}$: $x^2 + y^2 = 4$, $z = 0$, oriented counterclockwise as seen from the positive z-axis. $\mathcal{C}$ is also the oriented boundary of the disk $x^2 + y^2 \leq 4$, $z = 0$, with normal $\hat{\mathbf{N}} = \mathbf{k}$.
If $\mathbf{F} = (xz - y^3\cos z)\mathbf{i} + x^3 e^z\mathbf{j} + xyze^{x^2+y^2+z^2}\mathbf{k}$, then, on $z = 0$, we have

$$\text{curl}\,\mathbf{F}\bullet\mathbf{k} = \left(\frac{\partial}{\partial x} x^3 e^z - \frac{\partial}{\partial y}(xz - y^3\cos z)\right)\Bigg|_{z=0}$$
$$= \left(3x^2 e^z + 3y^2\cos z\right)\Big|_{z=0} = 3(x^2 + y^2).$$

Thus

$$\iint_{\mathcal{S}} \text{curl}\,\mathbf{F}\bullet\hat{\mathbf{N}}\,dS = \oint_{\mathcal{C}} \mathbf{F}\bullet d\mathbf{r} = \iint_{D} \text{curl}\,\mathbf{F}\bullet\mathbf{k}\,dA$$
$$= \int_0^{2\pi} d\theta \int_0^2 3r^2\,r\,dr = 24\pi.$$

6. The curve $\mathcal{C}$:

$$\mathbf{r} = \cos t\,\mathbf{i} + \sin t\,\mathbf{j} + \sin 2t\,\mathbf{k}, \qquad 0 \leq t \leq 2\pi,$$

lies on the surface $z = 2xy$, since $\sin 2t = 2\cos t\sin t$. It also lies on the cylinder $x^2 + y^2 = 1$, so it is the boundary of that part of $z = 2xy$ lying inside that cylinder. Since $\mathcal{C}$ is oriented counterclockwise as seen from high on the z-axis, $\mathcal{S}$ should be oriented with upward normal,

$$\hat{\mathbf{N}} = \frac{-2y\mathbf{i} - 2x\mathbf{j} + \mathbf{k}}{\sqrt{1 + 4(x^2 + y^2)}},$$

and has area element

$$dS = \sqrt{1 + 4(x^2 + y^2)}\,dx\,dy.$$

If $\mathbf{F} = (e^x - y^3)\mathbf{i} + (e^y + x^3)\mathbf{j} + e^z\mathbf{k}$, then

$$\text{curl}\,\mathbf{F} = \begin{vmatrix} \mathbf{i} & \mathbf{j} & \mathbf{k} \\ \dfrac{\partial}{\partial x} & \dfrac{\partial}{\partial y} & \dfrac{\partial}{\partial z} \\ e^x - y^3 & e^y + x^3 & e^z \end{vmatrix} = 3(x^2 + y^2)\mathbf{k}.$$

If D is the disk $x^2 + y^2 \leq 1$ in the xy-plane, then

$$\oint_{\mathcal{C}} \mathbf{F}\bullet d\mathbf{r} = \iint_{\mathcal{S}} \text{curl}\,\mathbf{F}\bullet\hat{\mathbf{N}}\,dS = \iint_{D} 3(x^2 + y^2)\,dx\,dy$$
$$= 3\int_0^{2\pi} d\theta \int_0^1 r^2\,r\,dr = \frac{3\pi}{2}.$$

8. The closed curve

$$\mathbf{r} = (1 + \cos t)\mathbf{i} + (1 + \sin t)\mathbf{j} + (1 - \cos t - \sin t)\mathbf{k},$$

$(0 \leq t \leq 2\pi)$, lies in the plane $x + y + z = 3$ and is oriented counterclockwise as seen from above. Therefore it is the boundary of a region $\mathcal{S}$ in that plane with normal field $\hat{\mathbf{N}} = (\mathbf{i} + \mathbf{j} + \mathbf{k})/\sqrt{3}$. The projection of $\mathcal{S}$ onto the xy-plane is the circular disk D of radius 1 with centre at $(1,1)$.

If $\mathbf{F} = ye^x\mathbf{i} + (x^2 + e^x)\mathbf{j} + z^2 e^z\mathbf{k}$, then

$$\text{curl}\,\mathbf{F} = \begin{vmatrix} \mathbf{i} & \mathbf{j} & \mathbf{k} \\ \dfrac{\partial}{\partial x} & \dfrac{\partial}{\partial y} & \dfrac{\partial}{\partial z} \\ ye^x & x^2 + e^x & z^2 + e^z \end{vmatrix} = 2x\mathbf{k}.$$

By Stokes's Theorem,

$$\oint_{\mathcal{C}} \mathbf{F}\bullet d\mathbf{r} = \iint_{\mathcal{S}} \text{curl}\,\mathbf{F}\bullet\hat{\mathbf{N}}\,dS$$
$$= \iint_{\mathcal{S}} \frac{2x}{\sqrt{3}}\,dS = \iint_{D} \frac{2x}{\sqrt{3}}(\sqrt{3})\,dx\,dy$$
$$= 2\bar{x}A = 2\pi,$$

where $\bar{x} = 1$ is the x-coordinate of the centre of D, and $A = \pi 1^2 = \pi$ is the area of D.

10. The curve $\mathcal{C}$: $(x-1)^2 + 4y^2 = 16$, $2x + y + z = 3$, oriented counterclockwise as seen from above, bounds an elliptic disk $\mathcal{S}$ on the plane $2x + y + z = 3$. $\mathcal{S}$ has normal $\hat{\mathbf{N}} = (2\mathbf{i} + \mathbf{j} + \mathbf{k})/\sqrt{6}$. Since its projection onto the xy-plane is an elliptic disk with centre at $(1,0,0)$ and area $\pi(4)(2) = 8\pi$, therefore $\mathcal{S}$ has area $8\sqrt{6}\pi$ and centroid $(1,0,1)$. If

$$\mathbf{F} = (z^2 + y^2 + \sin x^2)\mathbf{i} + (2xy + z)\mathbf{j} + (xz + 2yz)\mathbf{k},$$

then

$$\text{curl}\,\mathbf{F} = \begin{vmatrix} \mathbf{i} & \mathbf{j} & \mathbf{k} \\ \dfrac{\partial}{\partial x} & \dfrac{\partial}{\partial y} & \dfrac{\partial}{\partial z} \\ z^2 + y^2 + \sin x^2 & 2xy + z & xz + 2yz \end{vmatrix}$$
$$= (2z - 1)\mathbf{i} + z\mathbf{j}.$$

By Stokes's Theorem,

$$\oint_C \mathbf{F} \bullet d\mathbf{r} = \iint_S \operatorname{curl} \mathbf{F} \bullet \hat{\mathbf{N}} \, dS$$

$$= \frac{1}{\sqrt{6}} \iint_S (2(2z - 1) + z) \, dS$$

$$= \frac{5\bar{z} - 2}{\sqrt{6}} (8\sqrt{6}\pi) = 24\pi.$$

12. We are given that $\mathcal{C}$ bounds a region R in a plane P with unit normal $\hat{\mathbf{N}} = a\mathbf{i} + b\mathbf{j} + c\mathbf{k}$. Therefore, $a^2 + b^2 + c^2 = 1$. If $\mathbf{F} = (bz - cy)\mathbf{i} + (cx - az)\mathbf{j} + (ay - bx)\mathbf{k}$, then

$$\operatorname{curl} \mathbf{F} = \begin{vmatrix} \mathbf{i} & \mathbf{j} & \mathbf{k} \\ \dfrac{\partial}{\partial x} & \dfrac{\partial}{\partial y} & \dfrac{\partial}{\partial z} \\ bz - cy & cx - az & ay - bx \end{vmatrix}$$

$$= 2a\mathbf{i} + 2b\mathbf{j} + 2c\mathbf{k}.$$

Hence $\operatorname{curl} \mathbf{F} \bullet \hat{\mathbf{N}} = 2(a^2 + b^2 + c^2) = 2$. We have

$$\frac{1}{2} \oint_C (bz - cy) \, dx + (cx - az) \, dy + (ay - bx) \, dz$$

$$= \frac{1}{2} \oint_C \mathbf{F} \bullet d\mathbf{r} = \frac{1}{2} \iint_R \operatorname{curl} \mathbf{F} \bullet \hat{\mathbf{N}} \, dS$$

$$= \frac{1}{2} \iint_R 2 \, dS = \text{area of } R.$$

Section 16.6 Some Physical Applications of Vector Calculus (page 985)

2. The first component of $\mathbf{F}(\mathbf{G} \bullet \hat{\mathbf{N}})$ is $(F_1 \mathbf{G}) \bullet \hat{\mathbf{N}}$. Applying the Divergence Theorem and Theorem 3(b), we obtain

$$\iint_S (F_1 \mathbf{G}) \bullet \hat{\mathbf{N}} \, dS = \iiint_D \operatorname{div}(F_1 \mathbf{G}) \, dV$$

$$= \iiint_D \left(\boldsymbol{\nabla} F_1 \bullet \mathbf{G} + F_1 \boldsymbol{\nabla} \bullet \mathbf{G} \right) dS.$$

But $\boldsymbol{\nabla} F_1 \bullet \mathbf{G}$ is the first component of $(\mathbf{G} \bullet \boldsymbol{\nabla})\mathbf{F}$, and $F_1 \boldsymbol{\nabla} \bullet \mathbf{G}$ is the first component of $\mathbf{F} \operatorname{div} \mathbf{G}$. Similar results obtain for the other components, so

$$\iint_S \mathbf{F}(\mathbf{G} \bullet \hat{\mathbf{N}}) \, dS = \iiint_D \left(\mathbf{F} \operatorname{div} \mathbf{G} + (\mathbf{G} \bullet \boldsymbol{\nabla})\mathbf{F} \right) dV.$$

4. If f is continuous and vanishes outside a bounded region (say the ball of radius R centred at $\mathbf{r}$), then $|f(\xi, \eta, \zeta)| \le K$, and, if (ρ, ϕ, θ) denote spherical co-ordinates centred at $\mathbf{r}$, then

$$\iiint_{\mathbb{R}^3} \frac{|f(\mathbf{s})|}{|\mathbf{r} - \mathbf{s}|} \, dV_s \le K \int_0^{2\pi} d\theta \int_0^\pi \sin\phi \, d\phi \int_0^R \frac{\rho^2}{\rho} \, d\rho$$

$$= 2\pi K R^2 \quad \text{a constant.}$$

6. Since $\mathbf{r} = x\mathbf{i} + y\mathbf{j} + z\mathbf{k}$ and $\mathbf{b} = b_1\mathbf{i} + b_2\mathbf{j} + b_3\mathbf{k}$, we have

$$|\mathbf{r} - \mathbf{b}|^2 = (x - b_1)^2 + (y - b_2)^2 + (z - b_3)^2$$

$$2|\mathbf{r} - \mathbf{b}| \frac{\partial}{\partial x} |\mathbf{r} - \mathbf{b}| = 2(x - b_1)$$

$$\frac{\partial}{\partial x} |\mathbf{r} - \mathbf{b}| = \frac{x - b_1}{|\mathbf{r} - \mathbf{b}|}.$$

Similar formulas hold for the other first partials of $|\mathbf{r} - \mathbf{b}|$, so

$$\boldsymbol{\nabla} \left(\frac{1}{|\mathbf{r} - \mathbf{b}|} \right)$$

$$= \frac{-1}{|\mathbf{r} - \mathbf{b}|^2} \left(\frac{\partial}{\partial x} |\mathbf{r} - \mathbf{b}|\mathbf{i} + \cdots + \frac{\partial}{\partial z} |\mathbf{r} - \mathbf{b}|\mathbf{k} \right)$$

$$= \frac{-1}{|\mathbf{r} - \mathbf{b}|^2} \frac{(x - b_1)\mathbf{i} + (y - b_2)\mathbf{j} + (z - b_3)\mathbf{k}}{|\mathbf{r} - \mathbf{b}|}$$

$$= -\frac{\mathbf{r} - \mathbf{b}}{|\mathbf{r} - \mathbf{b}|^3}.$$

8. For any element $d\mathbf{s}$ on the filament $\mathcal{F}$, we have

$$\operatorname{div} \left(d\mathbf{s} \times \frac{\mathbf{r} - \mathbf{s}}{|\mathbf{r} - \mathbf{s}|^3} \right) = 0$$

by Exercise 5, since the divergence is taken with respect to $\mathbf{r}$, and so $\mathbf{s}$ and $d\mathbf{s}$ can be regarded as constant. Hence

$$\operatorname{div} \oint_{\mathcal{F}} \frac{d\mathbf{s} \times (\mathbf{r} - \mathbf{s})}{|\mathbf{r} - \mathbf{s}|^3} = \oint_{\mathcal{F}} \operatorname{div} \left(d\mathbf{s} \times \frac{\mathbf{r} - \mathbf{s}}{|\mathbf{r} - \mathbf{s}|^3} \right) = 0.$$

10. The first component of $(d\mathbf{s} \bullet \boldsymbol{\nabla})\mathbf{F}(s)$ is $\boldsymbol{\nabla} F_1(s) \bullet d\mathbf{s}$. Since $\mathcal{F}$ is closed and $\boldsymbol{\nabla} F_1$ is conservative,

$$\mathbf{i} \bullet \oint_{\mathcal{F}} (d\mathbf{s} \bullet \boldsymbol{\nabla})\mathbf{F}(s) = \oint_{\mathcal{F}} \boldsymbol{\nabla} F_1(s) \bullet d\mathbf{s} = 0.$$

Similarly, the other components have zero line integrals, so

$$\oint_{\mathcal{F}} (d\mathbf{s} \bullet \boldsymbol{\nabla})\mathbf{F}(s) = \mathbf{0}.$$

12. By analogy with the filament case, the current in volume element dV at position $\mathbf{s}$ is $\mathbf{J}(\mathbf{s}) \, dV$, which gives rise at position $\mathbf{r}$ to a magnetic field

$$d\mathbf{H}(\mathbf{r}) = \frac{1}{4\pi} \frac{\mathbf{J}(\mathbf{s}) \times (\mathbf{r} - \mathbf{s})}{|\mathbf{r} - \mathbf{s}|^3} \, dV.$$

If R is a region of 3-space outside which $\mathbf{J}$ is identically zero, then at any point $\mathbf{r}$ in 3-space, the total magnetic field is

$$\mathbf{H}(\mathbf{r}) = \frac{1}{4\pi} \iiint_R \frac{\mathbf{J}(\mathbf{s}) \times (\mathbf{r} - \mathbf{s})}{|\mathbf{r} - \mathbf{s}|^3} \, dV.$$

Now $\mathbf{A}(\mathbf{r})$ was defined to be

$$\mathbf{A}(\mathbf{r}) = \frac{1}{4\pi} \iiint_R \frac{\mathbf{J}(\mathbf{s})}{|\mathbf{r} - \mathbf{s}|} \, dV.$$

We have

$$\mathbf{curl}\,\mathbf{A}(\mathbf{r}) = \frac{1}{4\pi} \iiint_R \nabla_{\mathbf{r}} \times \left(\frac{1}{|\mathbf{r} - \mathbf{s}|} \mathbf{J}(\mathbf{s}) \right) dV$$

$$= \frac{1}{4\pi} \iiint_R \nabla_{\mathbf{r}} \frac{1}{|\mathbf{r} - \mathbf{s}|} \times \mathbf{J}(\mathbf{s}) \, dV$$

(by Theorem 3(c))

$$= -\frac{1}{4\pi} \iiint_R \frac{(\mathbf{r} - \mathbf{s}) \times \mathbf{J}(\mathbf{s})}{|\mathbf{r} - \mathbf{s}|^3} \, dV$$

(by Exercise 4)

$$= \mathbf{H}(\mathbf{r}).$$

14. $\mathbf{A}(\mathbf{r}) = \dfrac{1}{4\pi} \iiint_R \dfrac{\mathbf{J}(\mathbf{s}) \, dV}{|\mathbf{r} - \mathbf{s}|}$, where R is a region of 3-space such that $\mathbf{J}(\mathbf{s}) = \mathbf{0}$ outside R. We assume that $\mathbf{J}(\mathbf{s})$ is continuous, so $\mathbf{J}(\mathbf{s}) = \mathbf{0}$ on the surface $\mathcal{S}$ of R.
In the following calculations we use subscripts $\mathbf{s}$ and $\mathbf{r}$ to denote the variables with respect to which derivatives are taken. By Theorem 3(b),

$$\mathbf{div}_{\mathbf{s}} \frac{\mathbf{J}(\mathbf{s})}{|\mathbf{r} - \mathbf{s}|} = \left(\nabla_{\mathbf{s}} \frac{1}{|\mathbf{r} - \mathbf{s}|} \right) \bullet \mathbf{J}(\mathbf{s}) + \frac{1}{|\mathbf{r} - \mathbf{s}|} \nabla_{\mathbf{s}} \bullet \mathbf{J}(\mathbf{s})$$

$$= -\nabla_{\mathbf{r}} \left(\frac{1}{|\mathbf{r} - \mathbf{s}|} \right) \bullet \mathbf{J}(\mathbf{s}) + 0$$

because $\nabla_{\mathbf{r}} |\mathbf{r} - \mathbf{s}| = -\nabla_{\mathbf{s}} |\mathbf{r} - \mathbf{s}|$, and because $\nabla \bullet \mathbf{J} = \nabla \bullet (\nabla \times \mathbf{H}) = 0$ by Theorem 3(g). Hence

$$\mathbf{div}\,\mathbf{A}(\mathbf{r}) = \frac{1}{4\pi} \iiint_R \left(\nabla_{\mathbf{r}} \frac{1}{|\mathbf{r} - \mathbf{s}|} \right) \bullet \mathbf{J}(\mathbf{s}) \, dV$$

$$= -\frac{1}{4\pi} \iiint_R \nabla_{\mathbf{s}} \bullet \frac{\mathbf{J}(\mathbf{s})}{|\mathbf{r} - \mathbf{s}|} \, dV$$

$$= -\frac{1}{4\pi} \iint_{\mathcal{S}} \frac{\mathbf{J}(\mathbf{s})}{|\mathbf{r} - \mathbf{s}|} \bullet \hat{\mathbf{N}} \, dS = 0$$

since $\mathbf{J}(\mathbf{s}) = \mathbf{0}$ on $\mathcal{S}$.
By Theorem 3(i),

$$\mathbf{J} = \nabla \times \mathbf{H} = \nabla \times (\nabla \times \mathbf{A}) = \nabla(\nabla \bullet \mathbf{A}) - \nabla^2 \mathbf{A} = -\nabla^2 \mathbf{A}.$$

16. The heat content of an arbitrary region R (with surface $\mathcal{S}$) at time t is

$$H(t) = \delta c \iiint_R T(x, y, z, t) \, dV.$$

This heat content increases at (time) rate

$$\frac{dH}{dt} = \delta c \iiint_R \frac{\partial T}{\partial t} \, dV.$$

If heat is not "created" or "destroyed" (by chemical or other means) within R, then the increase in heat content must be due to heat flowing into R across $\mathcal{S}$.
The rate of flow of heat into R across surface element dS with outward normal $\hat{\mathbf{N}}$ is

$$-k \nabla T \bullet \hat{\mathbf{N}} \, dS.$$

Therefore, the rate at which heat enters R through $\mathcal{S}$ is

$$k \iint_{\mathcal{S}} \nabla T \bullet \hat{\mathbf{N}} \, dS.$$

By conservation of energy and the Divergence Theorem we have

$$\delta c \iiint_R \frac{\partial T}{\partial t} \, dV = k \iint_{\mathcal{S}} \nabla T \bullet \hat{\mathbf{N}} \, dS$$

$$= k \iiint_R \nabla \bullet \nabla T \, dV$$

$$= k \iiint_R \nabla^2 T \, dV.$$

Thus, $\displaystyle\iiint_R \left(\frac{\partial T}{\partial t} - \frac{k}{\delta c} \nabla^2 T \right) dV = 0.$

Since R is arbitrary, and the temperature T is assumed to be smooth, the integrand must vanish everywhere. Thus

$$\frac{\partial T}{\partial t} = \frac{k}{\delta c} \nabla^2 T = \frac{k}{\delta c} \left[\frac{\partial^2 T}{\partial x^2} + \frac{\partial^2 T}{\partial y^2} + \frac{\partial^2 T}{\partial z^2} \right].$$

Section 16.7 Orthogonal Curvilinear Coordinates (page 996)

2. $f(\rho, \phi, \theta) = \rho \phi \theta$ (spherical coordinates). By Example 10,

$$\nabla f = \frac{\partial f}{\partial \rho} \hat{\boldsymbol{\rho}} + \frac{1}{\rho} \frac{\partial f}{\partial \phi} \hat{\boldsymbol{\phi}} + \frac{1}{\rho \sin \phi} \frac{\partial f}{\partial \theta} \hat{\boldsymbol{\theta}}$$

$$= \phi \theta \, \hat{\boldsymbol{\rho}} + \theta \, \hat{\boldsymbol{\phi}} + \frac{\phi}{\sin \phi} \, \hat{\boldsymbol{\theta}}.$$

4. $\mathbf{F}(r, \theta, z) = r\hat{\boldsymbol{\theta}}$

$$\operatorname{div}\mathbf{F} = \frac{1}{r}\left[\frac{\partial}{\partial\theta}(r)\right] = 0$$

$$\operatorname{curl}\mathbf{F} = \frac{1}{r}\begin{vmatrix} \hat{\mathbf{r}} & r\hat{\boldsymbol{\theta}} & \mathbf{k} \\ \dfrac{\partial}{\partial r} & \dfrac{\partial}{\partial\theta} & \dfrac{\partial}{\partial z} \\ 0 & r^2 & 0 \end{vmatrix} = 2\mathbf{k}.$$

6. $\mathbf{F}(\rho, \phi, \theta) = \rho\,\hat{\boldsymbol{\phi}}$

$$\operatorname{div}\mathbf{F} = \frac{1}{\rho^2\sin\phi}\left[\frac{\partial}{\partial\phi}\left(\rho^2\sin\phi\right)\right] = \cot\phi$$

$$\operatorname{curl}\mathbf{F} = \frac{1}{\rho^2\sin\phi}\begin{vmatrix} \hat{\boldsymbol{\rho}} & \rho\,\hat{\boldsymbol{\phi}} & \rho\sin\phi\,\hat{\boldsymbol{\theta}} \\ \dfrac{\partial}{\partial\rho} & \dfrac{\partial}{\partial\phi} & \dfrac{\partial}{\partial\theta} \\ 0 & \rho^2 & 0 \end{vmatrix} = 2\,\hat{\boldsymbol{\theta}}.$$

8. $\mathbf{F}(\rho, \phi, \theta) = \rho^2\,\hat{\boldsymbol{\rho}}$

$$\operatorname{div}\mathbf{F} = \frac{1}{\rho^2\sin\phi}\left[\frac{\partial}{\partial\rho}\left(\rho^4\sin\phi\right)\right] = 4\rho$$

$$\operatorname{curl}\mathbf{F} = \frac{1}{\rho^2\sin\phi}\begin{vmatrix} \hat{\boldsymbol{\rho}} & \rho\,\hat{\boldsymbol{\phi}} & \rho\sin\phi\,\hat{\boldsymbol{\theta}} \\ \dfrac{\partial}{\partial\rho} & \dfrac{\partial}{\partial\phi} & \dfrac{\partial}{\partial\theta} \\ \rho^2 & 0 & 0 \end{vmatrix} = \mathbf{0}.$$

10. Since (u, v, z) constitute orthogonal curvilinear coordinates in $\mathbb{R}^3$, with scale factors h_u, h_v and $h_z = 1$, we have, for a function $f(u, v)$ independent of z,

$$\begin{aligned}\nabla f(u, v) &= \frac{1}{h_u}\frac{\partial f}{\partial u}\hat{\mathbf{u}} + \frac{1}{h_v}\frac{\partial f}{\partial v}\hat{\mathbf{v}} + \frac{1}{1}\frac{\partial f}{\partial z}\mathbf{k} \\ &= \frac{1}{h_u}\frac{\partial f}{\partial u}\hat{\mathbf{u}} + \frac{1}{h_v}\frac{\partial f}{\partial v}\hat{\mathbf{v}}.\end{aligned}$$

For $\mathbf{F}(u, v) = F_u(u, v)\hat{\mathbf{u}} + F_v(u, v)\hat{\mathbf{v}}$ (independent of z and having no $\mathbf{k}$ component), we have

$$\operatorname{div}\mathbf{F}(u, v) = \frac{1}{h_u h_v}\left[\frac{\partial}{\partial u}(h_u F_u) + \frac{\partial}{\partial v}(h_v F_v)\right]$$

$$\begin{aligned}\operatorname{curl}\mathbf{F}(u, v) &= \frac{1}{h_u h_v}\begin{vmatrix} h_u\hat{\mathbf{u}} & h_v\hat{\mathbf{v}} & \mathbf{k} \\ \dfrac{\partial}{\partial u} & \dfrac{\partial}{\partial v} & \dfrac{\partial}{\partial z} \\ h_u F_u & h_v F_v & 0 \end{vmatrix} \\ &= \frac{1}{h_u h_v}\left[\frac{\partial}{\partial u}(h_v F_v) - \frac{\partial}{\partial v}(h_u F_u)\right]\mathbf{k}.\end{aligned}$$

12. $x = a\cosh u\cos v, \quad y = a\sinh u\sin v.$

a) u-curves: If $A = a\cosh u$ and $B = a\sinh u$, then

$$\frac{x^2}{A^2} + \frac{y^2}{B^2} = \cos^2 v + \sin^2 v = 1.$$

Since $A^2 - B^2 = a^2(\cosh^2 u - \sinh^2 u) = a^2$, the u-curves are ellipses with foci at $(\pm a, 0)$.

b) v-curves: If $A = a\cos v$ and $B = a\sin v$, then

$$\frac{x^2}{A^2} - \frac{y^2}{B^2} = \cosh^2 u - \sinh^2 u = 1.$$

Since $A^2 + B^2 = a^2(\cos^2 v + \sin^2 v) = a^2$, the v-curves are hyperbolas with foci at $(\pm a, 0)$.

c) The u-curve $u = u_0$ has parametric equations

$$x = a\cosh u_0\cos v, \qquad y = a\sinh u_0\sin v,$$

and therefore has slope at (u_0, v_0) given by

$$m_u = \frac{dy}{dx} = \frac{dy}{dv}\bigg/\frac{dx}{dv}\bigg|_{(u_0, v_0)} = \frac{a\sinh u_0\cos v_0}{-a\cosh u_0\sin v_0}.$$

The v-curve $v = v_0$ has parametric equations

$$x = a\cosh u\cos v_0, \qquad y = a\sinh u\sin v_0,$$

and therefore has slope at (u_0, v_0) given by

$$m_v = \frac{dy}{dx} = \frac{dy}{du}\bigg/\frac{dx}{du}\bigg|_{(u_0, v_0)} = \frac{a\cosh u_0\sin v_0}{a\sinh u_0\cos v_0}.$$

Since the product of these slopes is $m_u m_v = -1$, the curves $u = u_0$ and $v = v_0$ intersect at right angles.

d) $\quad\mathbf{r} = a\cosh u\cos v\,\mathbf{i} + a\sinh u\sin v\,\mathbf{j}$

$$\frac{\partial\mathbf{r}}{\partial u} = a\sinh u\cos v\,\mathbf{i} + a\cosh u\sin v\,\mathbf{j}$$

$$\frac{\partial\mathbf{r}}{\partial v} = -a\cosh u\sin v\,\mathbf{i} + a\sinh u\cos v\,\mathbf{j}.$$

The scale factors are

$$h_u = \left|\frac{\partial\mathbf{r}}{\partial u}\right| = a\sqrt{\sinh^2 u\cos^2 v + \cosh^2 u\sin^2 v}$$

$$h_v = \left|\frac{\partial\mathbf{r}}{\partial v}\right| = a\sqrt{\sinh^2 u\cos^2 v + \cosh^2 u\sin^2 v} = h_u.$$

The area element is

$$\begin{aligned}dA &= h_u h_v\,du\,dv \\ &= a^2\left(\sinh^2 u\cos^2 v + \cosh^2 u\sin^2 v\right)du\,dv.\end{aligned}$$

14. $\quad\nabla f(r, \theta, z) = \dfrac{\partial f}{\partial r}\hat{\mathbf{r}} + \dfrac{1}{r}\dfrac{\partial f}{\partial\theta}\hat{\boldsymbol{\theta}} + \dfrac{\partial f}{\partial z}\mathbf{k}$

$$\nabla^2 f(r, \theta, z) = \operatorname{div}\left(\nabla f(r, \theta, z)\right)$$

$$= \frac{1}{r}\left[\frac{\partial}{\partial r}\left(r\frac{\partial f}{\partial r}\right) + \frac{\partial}{\partial\theta}\left(\frac{1}{r}\frac{\partial f}{\partial\theta}\right) + \frac{\partial}{\partial z}\left(r\frac{\partial f}{\partial z}\right)\right]$$

$$= \frac{\partial^2 f}{\partial r^2} + \frac{1}{r}\frac{\partial f}{\partial r} + \frac{1}{r^2}\frac{\partial^2 f}{\partial\theta^2} + \frac{\partial^2 f}{\partial z^2}.$$

16. $\nabla f(u, v, w) = \dfrac{1}{h_u}\dfrac{\partial f}{\partial u}\hat{\mathbf{u}} + \dfrac{1}{h_v}\dfrac{\partial f}{\partial v}\hat{\mathbf{v}} + \dfrac{1}{h_w}\dfrac{\partial f}{\partial w}\hat{\mathbf{w}}$

$\nabla^2 f(u, v, w) = \mathbf{div}\left(\nabla f(u, v, w)\right)$

$= \dfrac{1}{h_u h_v h_w}\left[\dfrac{\partial}{\partial u}\left(\dfrac{h_v h_w}{h_u}\dfrac{\partial f}{\partial u}\right) + \dfrac{\partial}{\partial v}\left(\dfrac{h_u h_w}{h_v}\dfrac{\partial f}{\partial v}\right)\right.$

$\left. + \dfrac{\partial}{\partial w}\left(\dfrac{h_u h_v}{h_w}\dfrac{\partial f}{\partial w}\right)\right]$

$= \dfrac{1}{h_u^2}\left[\dfrac{\partial^2 f}{\partial u^2} + \left(\dfrac{1}{h_v}\dfrac{\partial h_v}{\partial u} + \dfrac{1}{h_w}\dfrac{\partial h_w}{\partial u} - \dfrac{1}{h_u}\dfrac{\partial h_u}{\partial u}\right)\dfrac{\partial f}{\partial u}\right]$

$+ \dfrac{1}{h_v^2}\left[\dfrac{\partial^2 f}{\partial v^2} + \left(\dfrac{1}{h_u}\dfrac{\partial h_u}{\partial v} + \dfrac{1}{h_w}\dfrac{\partial h_w}{\partial v} - \dfrac{1}{h_v}\dfrac{\partial h_v}{\partial v}\right)\dfrac{\partial f}{\partial v}\right]$

$+ \dfrac{1}{h_w^2}\left[\dfrac{\partial^2 f}{\partial w^2} + \left(\dfrac{1}{h_u}\dfrac{\partial h_u}{\partial w} + \dfrac{1}{h_v}\dfrac{\partial h_v}{\partial w} - \dfrac{1}{h_w}\dfrac{\partial h_w}{\partial w}\right)\dfrac{\partial f}{\partial w}\right].$

Review Exercises 16   (page 997)

2. Let R be the region inside the cylinder S and between the planes $z = 0$ and $z = b$. The oriented boundary of R consists of S and the disks D_1 with normal $\hat{\mathbf{N}}_1 = \mathbf{k}$ and D_2 with normal $\hat{\mathbf{N}}_2 = -\mathbf{k}$ as shown in the figure. For $\mathbf{F} = x\mathbf{i} + \cos(z^2)\mathbf{j} + e^z\mathbf{k}$ we have $\mathbf{div}\,\mathbf{F} = 1 + e^z$ and

$$\iiint_R \mathbf{div}\,\mathbf{F}\,dV = \iint_{D_2}dx\,dy\int_0^b(1 + e^z)\,dz$$

$$= \iint_{D_2}[b + (e^b - 1)]\,dx\,dy$$

$$= \pi a^2 b + \pi a^2(e^b - 1).$$

Also $\displaystyle\iint_{D_2}\mathbf{F}\bullet(-\mathbf{k})\,dA = -\iint_{D_2}e^0\,dA = -\pi a^2$

$\displaystyle\iint_{D_1}\mathbf{F}\bullet\mathbf{k}\,dA = \iint_{D_1}e^b\,dA = \pi a^2 e^b.$

By the Divergence Theorem

$$\iint_S \mathbf{F}\bullet\hat{\mathbf{N}}\,dS + \iint_{D_1}\mathbf{F}\bullet\mathbf{k}\,dA + \iint_{D_2}\mathbf{F}\bullet(-\mathbf{k})\,dA$$

$$= \iiint_R \mathbf{div}\,\mathbf{F}\,dV = \pi a^2 b + \pi a^2(e^b - 1).$$

Therefore, $\displaystyle\iint_S \mathbf{F}\bullet\hat{\mathbf{N}}\,dS = \pi a^2 b.$

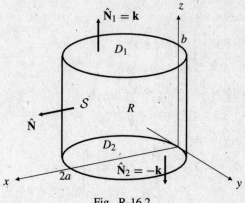

Fig. R-16.2

4. If $\mathbf{F} = -z\mathbf{i} + x\mathbf{j} + y\mathbf{k}$, then

$$\mathbf{curl}\,\mathbf{F} = \begin{vmatrix} \mathbf{i} & \mathbf{j} & \mathbf{k} \\ \dfrac{\partial}{\partial x} & \dfrac{\partial}{\partial y} & \dfrac{\partial}{\partial z} \\ -z & x & y \end{vmatrix} = \mathbf{i} - \mathbf{j} + \mathbf{k}.$$

The unit normal $\hat{\mathbf{N}}$ to a region in the plane $2x + y + 2z = 7$ is

$$\hat{\mathbf{N}} = \pm\frac{2\mathbf{i} + \mathbf{j} + 2\mathbf{k}}{3}.$$

If C is the boundary of a disk D of radius a in that plane, then

$$\oint_C \mathbf{F}\bullet d\mathbf{r} = \iint_D \mathbf{curl}\,\mathbf{F}\bullet\hat{\mathbf{N}}\,dS$$

$$= \pm\iint_D \frac{2 - 1 + 2}{3}\,dS = \pm\pi a^2.$$

6. If S is any surface with upward normal $\hat{\mathbf{N}}$ and boundary the curve C: $x^2 + y^2 = 1$, $z = 2$, then C is oriented counterclockwise as seen from above, and it has parametrization

$$\mathbf{r} = \cos t\mathbf{i} + \sin t\mathbf{j} + 2\mathbf{k}\quad(0 \le 2 \le 2\pi).$$

Thus $d\mathbf{r} = (-\sin t\mathbf{i} + \cos t\mathbf{j})\,dt$, and if $\mathbf{F} = -y\mathbf{i} + x\cos(1 - x^2 - y^2)\mathbf{j} + yz\mathbf{k}$, then the flux of $\mathbf{curl}\,\mathbf{F}$ upward through S is

$$\iint_S \mathbf{curl}\,\mathbf{F}\bullet\hat{\mathbf{N}}\,dS = \oint_C \mathbf{F}\bullet d\mathbf{r}$$

$$= \int_0^{2\pi}(\sin^2 t + \cos^2 t + 0)\,dt = 2\pi.$$

8. If $\mathbf{curl}\,\mathbf{F} = \mu\mathbf{F}$ on $\mathbb{R}^3$, where $\mu \ne 0$ is a constant, then

$$\mathbf{div}\,\mathbf{F} = \frac{1}{\mu}\mathbf{div}\,\mathbf{curl}\,\mathbf{F} = 0$$

by Theorem 3(g) of Section 7.2. By part (i) of the same theorem,

$$\nabla^2 \mathbf{F} = \nabla(\text{div }\mathbf{F}) - \text{curl curl }\mathbf{F}$$
$$= 0 - \mu \text{curl }\mathbf{F} = -\mu^2 \mathbf{F}.$$

Thus $\nabla^2 \mathbf{F} + \mu^2 \mathbf{F} = \mathbf{0}$.

10. Let $\mathcal{C}$ be a simple, closed curve in the xy-plane bounding a region R. If

$$\mathbf{F} = (2y^3 - 3y + xy^2)\mathbf{i} + (x - x^3 + x^2 y)\mathbf{j},$$

then by Green's Theorem, the circulation of $\mathbf{F}$ around $\mathcal{C}$ is

$$\oint_{\mathcal{C}} \mathbf{F} \bullet d\mathbf{r}$$
$$= \iint_R \left[\frac{\partial}{\partial x}(x - x^3 + x^2 y) - \frac{\partial}{\partial y}(2y^3 - 3y + xy^2) \right] dA$$
$$= \iint_R (1 - 3x^2 + 2xy - 6y^2 + 3 - 2xy)\, dA$$
$$= \iint_R (4 - 3x^2 - 6y^2)\, dx\, dy.$$

The last integral has a maximum value when the region R is bounded by the ellipse $3x^2 + 6y^2 = 4$, oriented counterclockwise; this is the largest region in the xy-plane where the integrand is nonnegative.

12. Let $\mathcal{C}$ be a simple, closed curve on the plane $x + y + z = 1$, oriented counterclockwise as seen from above, and bounding a plane region $\mathcal{S}$ on $x + y + z = 1$. Then $\mathcal{S}$ has normal $\hat{\mathbf{N}} = (\mathbf{i} + \mathbf{j} + \mathbf{k})/\sqrt{3}$. If $\mathbf{F} = xy^2\mathbf{i} + (3z - xy^2)\mathbf{j} + (4y - x^2 y)\mathbf{k}$, then

$$\text{curl }\mathbf{F} = \begin{vmatrix} \mathbf{i} & \mathbf{j} & \mathbf{k} \\ \dfrac{\partial}{\partial x} & \dfrac{\partial}{\partial y} & \dfrac{\partial}{\partial z} \\ xy^2 & 3z - xy^2 & 4y - x^2 y \end{vmatrix}$$
$$= (1 - x^2)\mathbf{i} + 2xy\mathbf{j} - (y^2 + 2xy)\mathbf{k}.$$

By Stokes's Theorem we have

$$\oint_{\mathcal{C}} \mathbf{F} \bullet d\mathbf{r} = \iint_{\mathcal{S}} \text{curl }\mathbf{F} \bullet \hat{\mathbf{N}}\, dS = \iint_{\mathcal{S}} \frac{1 - x^2 - y^2}{\sqrt{3}}\, dS.$$

The last integral will be maximum if the projection of $\mathcal{S}$ onto the xy-plane is the disk $x^2 + y^2 \le 1$. This maximum value is

$$\iint_{x^2 + y^2 \le 1} \frac{1 - x^2 - y^2}{\sqrt{3}}\, \sqrt{3}\, dx\, dy$$
$$= \int_0^{2\pi} d\theta \int_0^1 (1 - r^2)r\, dr = 2\pi \left(\frac{1}{2} - \frac{1}{4} \right) = \frac{\pi}{2}.$$

Challenging Problems 16   (page 998)

2. a) The steradian measure of a half-cone of semi-vertical angle α is

$$\int_0^{2\pi} d\theta \int_0^\alpha \sin\phi\, d\phi = 2\pi(1 - \cos\alpha).$$

b) If $\mathcal{S}$ is the intersection of a smooth surface with the general half-cone K, and is oriented with normal field $\hat{\mathbf{N}}$ pointing away from the vertex P of K, and if $\mathcal{S}_a$ is the intersection with K of a sphere of radius a centred at P, with a chosen so that $\mathcal{S}$ and $\mathcal{S}_a$ do not intersect in K, then $\mathcal{S}$, $\mathcal{S}_a$, and the walls of K bound a solid region R that does not contain the origin. If $\mathbf{F} = \mathbf{r}/|\mathbf{r}|^3$, then $\text{div }\mathbf{F} = 0$ in R (see Example 3 in Section 7.1), and $\mathbf{F} \bullet \hat{\mathbf{N}} = 0$ on the walls of K. It follows from the Divergence Theorem applied to $\mathbf{F}$ over R that

$$\iint_{\mathcal{S}} \mathbf{F} \bullet \hat{\mathbf{N}}\, dS = \iint_{\mathcal{S}_a} \mathbf{F} \bullet \frac{\mathbf{r}}{|\mathbf{r}|}\, dS$$
$$= \frac{a^2}{a^4} \iint_{\mathcal{S}_a} dS = \frac{1}{a^2}(\text{area of }\mathcal{S}_a)$$
$$= \text{area of }\mathcal{S}_1.$$

The area of $\mathcal{S}_1$ (the part of the sphere of radius 1 in K) is the measure (in steradians) of the solid angle subtended by K at its vertex P. Hence this measure is given by

$$\iint_{\mathcal{S}} \frac{\mathbf{r}}{|\mathbf{r}|^3} \bullet \hat{\mathbf{N}}\, dS.$$

4. a) Verification of the identity

$$\frac{\partial}{\partial t}\left(\mathbf{G} \bullet \left[\frac{\partial \mathbf{r}}{\partial u} \times \frac{\partial \mathbf{r}}{\partial v} \right] \right) - \frac{\partial}{\partial u}\left(\mathbf{G} \bullet \left[\frac{\partial \mathbf{r}}{\partial t} \times \frac{\partial \mathbf{r}}{\partial v} \right] \right)$$
$$- \frac{\partial}{\partial v}\left(\mathbf{G} \bullet \left[\frac{\partial \mathbf{r}}{\partial u} \times \frac{\partial \mathbf{r}}{\partial t} \right] \right)$$
$$= \frac{\partial \mathbf{F}}{\partial t} \bullet \left[\frac{\partial \mathbf{r}}{\partial u} \times \frac{\partial \mathbf{r}}{\partial v} \right] + (\nabla \bullet \mathbf{F})\frac{\partial \mathbf{r}}{\partial t} \bullet \left[\frac{\partial \mathbf{r}}{\partial u} \times \frac{\partial \mathbf{r}}{\partial v} \right].$$

can be carried out using the following MapleV commands:

321

```
> with(linalg):
> F:=(x,y,z,t)->[F1(x,y,z,t),
>   F2(x,y,z,t),F3(x,y,z,t)];
> r:=(u,v,t)->[x(u,v,t),y(u,v,t),
>   z(u,v,t)];
> ru:=(u,v,t)->diff(r(u,v,t),u);
> rv:=(u,v,t)->diff(r(u,v,t),v);
> rt:=(u,v,t)->diff(r(u,v,t),t);
> G:=(u,v,t)->F(x(u,v,t),
>   y(u,v,t),z(u,v,t),t);
> ruxv:=(u,v,t)->crossprod(ru(u,v,t),
>   rv(u,v,t));
> rtxv:=(u,v,t)->crossprod(rt(u,v,t),
>   rv(u,v,t));
> ruxt:=(u,v,t)->crossprod(ru(u,v,t),
>   rt(u,v,t));
> LH1:=diff(dotprod(G(u,v,t),
>   ruxv(u,v,t)),t);
> LH2:=diff(dotprod(G(u,v,t),
>   rtxv(u,v,t)),u);
> LH3:=diff(dotprod(G(u,v,t),
>   ruxt(u,v,t)),v);
> LHS:=simplify(LH1-LH2-LH3);
> RH1:=dotprod(subs(x=x(u,v,t),
>   y=y(u,v,t),z=z(u,v,t),
>   diff(F(x,y,z,t),t)),ruxv(u,v,t));
> RH2:=(divf(u,v,t))*
>   (dotprod(rt(u,v,t),ruxv(u,v,t)));
> RHS:=simplify(RH1+RH2);
> simplify(LHS-RHS);
```

Again the final output is 0, indicating that the identity is valid.

b) If $\mathcal{C}_t$ is the oriented boundary of $\mathcal{S}_t$ and L_t is the corresponding counterclockwise boundary of the parameter region R in the uv-plane, then

$$\oint_{\mathcal{C}_t} \left(\mathbf{F} \times \frac{\partial \mathbf{r}}{\partial t} \right) \bullet d\mathbf{r}$$

$$= \oint_{L_t} \left(\mathbf{G} \times \frac{\partial \mathbf{r}}{\partial t} \right) \bullet \left(\frac{\partial \mathbf{r}}{\partial u} du + \frac{\partial \mathbf{r}}{\partial v} dv \right)$$

$$= \oint_{L_t} \left[-\mathbf{G} \bullet \left(\frac{\partial \mathbf{r}}{\partial u} \times \frac{\partial \mathbf{r}}{\partial t} \right) + \mathbf{G} \bullet \left(\frac{\partial \mathbf{r}}{\partial t} \times \frac{\partial \mathbf{r}}{\partial v} \right) \right] dt$$

$$= \iint_R \left[\frac{\partial}{\partial u} \left(\mathbf{G} \bullet \left(\frac{\partial \mathbf{r}}{\partial t} \times \frac{\partial \mathbf{r}}{\partial v} \right) \right) \right.$$

$$\left. + \frac{\partial}{\partial v} \left(\mathbf{G} \bullet \left(\frac{\partial \mathbf{r}}{\partial u} \times \frac{\partial \mathbf{r}}{\partial t} \right) \right) \right] du\, dv,$$

by Green's Theorem.

c) Using the results of (a) and (b), we calculate

$$\frac{d}{dt} \iint_{\mathcal{S}_t} \mathbf{F} \bullet \hat{\mathbf{N}} dS = \iint_R \frac{\partial}{\partial t} \left[\mathbf{G} \bullet \left(\frac{\partial \mathbf{r}}{\partial u} \times \frac{\partial \mathbf{r}}{\partial v} \right) \right] du\, dv$$

$$= \iint_R \frac{\partial \mathbf{F}}{\partial t} \bullet \left(\frac{\partial \mathbf{r}}{\partial u} \times \frac{\partial \mathbf{r}}{\partial v} \right) du\, dv$$

$$+ \iint_R (\operatorname{div} \mathbf{F}) \frac{\partial \mathbf{r}}{\partial t} \bullet \left(\frac{\partial \mathbf{r}}{\partial u} \times \frac{\partial \mathbf{r}}{\partial v} \right) du\, dv$$

$$+ \iint_R \left[\frac{\partial}{\partial u} \left(\mathbf{G} \bullet \left(\frac{\partial \mathbf{r}}{\partial t} \times \frac{\partial \mathbf{r}}{\partial v} \right) \right) \right.$$

$$\left. + \frac{\partial}{\partial v} \left(\mathbf{G} \bullet \left(\frac{\partial \mathbf{r}}{\partial u} \times \frac{\partial \mathbf{r}}{\partial t} \right) \right) \right] du\, dv$$

$$= \iint_{\mathcal{S}_t} \frac{\partial \mathbf{F}}{\partial t} \bullet \hat{\mathbf{N}} dS + \iint_{\mathcal{S}_t} (\operatorname{div} \mathbf{F}) \mathbf{v}_S \bullet \hat{\mathbf{N}} dS$$

$$+ \oiint_{\mathcal{C}_t} (\mathbf{F} \times \mathbf{v}_C) \bullet d\mathbf{r}.$$

APPENDICES

Appendix I. Complex Numbers (page A-11)

2. $z = 4 - i$, $\text{Re}(z) = 4$, $\text{Im}(z) = -1$

4. $z = -6$, $\text{Re}(z) = -6$, $\text{Im}(z) = 0$

6. $z = -2$, $|z| = 2$, $\text{Arg}(z) = \pi$
$z = 2(\cos \pi + i \sin \pi)$

8. $z = -5i$, $|z| = 5$, $\text{Arg}(z) = -\pi/2$
$z = 5(\cos(-\pi/2) + i \sin(-\pi/2))$

10. $z = -2 + i$, $|z| = \sqrt{5}$, $\theta = \text{Arg}(z) = \pi - \tan^{-1}(1/2)$
$z = \sqrt{5}(\cos \theta + i \sin \theta)$

12. $z = 3 - 4i$, $|z| = 5$, $\theta = \text{Arg}(z) = -\tan^{-1}(4/3)$
$z = 5(\cos \theta + i \sin \theta)$

14. $z = -\sqrt{3} - 3i$, $|z| = 2\sqrt{3}$, $\text{Arg}(z) = -2\pi/3$
$z = 2\sqrt{3}(\cos(-2\pi/3) + i \sin(-2\pi/3))$

16. If $\text{Arg}(z) = \dfrac{3\pi}{4}$ and $\text{Arg}(w) = \dfrac{\pi}{2}$, then
$\arg(zw) = \dfrac{3\pi}{4} + \dfrac{\pi}{2} = \dfrac{5\pi}{4}$, so
$\text{Arg}(zw) = \dfrac{5\pi}{4} - 2\pi = \dfrac{-3\pi}{4}$.

18. $|z| = 2$, $\arg(z) = \pi \Rightarrow z = 2(\cos \pi + i \sin \pi) = -2$

20. $|z| = 1$, $\arg(z) = \dfrac{3\pi}{4} \Rightarrow z = \left(\cos \dfrac{3\pi}{4} + i \sin \dfrac{3\pi}{4}\right)$
$\Rightarrow z = -\dfrac{1}{\sqrt{2}} + \dfrac{1}{\sqrt{2}}i$

22. $|z| = 0 \Rightarrow z = 0$ for any value of $\arg(z)$

24. $\overline{5 + 3i} = 5 - 3i$

26. $\overline{4i} = -4i$

28. $|z| = 2$ represents all points on the circle of radius 2 centred at the origin.

30. $|z - 2i| \le 3$ represents all points in the closed disk of radius 3 centred at the point $2i$.

32. $\arg(z) = \pi/3$ represents all points on the ray from the origin in the first quadrant, making angle $60°$ with the positive direction of the real axis.

34. $(2 + 5i) + (3 - i) = 5 + 4i$

36. $(4 + i)(4 - i) = 16 - i^2 = 17$

38. $(a + bi)(\overline{2a - bi}) = (a + bi)(2a + bi) = 2a^2 - b^2 + 3abi$

40. $\dfrac{2 - i}{2 + i} = \dfrac{(2 - i)^2}{4 - i^2} = \dfrac{3 - 4i}{5}$

42. $\dfrac{1 + i}{i(2 + 3i)} = \dfrac{1 + i}{-3 + 2i} = \dfrac{(1 + i)(-3 - 2i)}{9 + 4} = \dfrac{-1 - 5i}{13}$

44. If $z = x + yi$ and $w = u + vi$, where x, y, u, and v are real, then
$$\overline{z + w} = \overline{x + u + (y + v)i}$$
$$= x + u - (y + v)i = x - yi + u - vi = \overline{z} + \overline{w}.$$

46. $z = 3 + i\sqrt{3} = 2\sqrt{3}\left(\cos \dfrac{\pi}{6} + i \sin \dfrac{\pi}{6}\right)$
$w = -1 + i\sqrt{3} = 2\left(\cos \dfrac{2\pi}{3} + i \sin \dfrac{2\pi}{3}\right)$
$zw = 4\sqrt{3}\left(\cos \dfrac{5\pi}{6} + i \sin \dfrac{5\pi}{6}\right)$
$\dfrac{z}{w} = \sqrt{3}\left(\cos \dfrac{-\pi}{2} + i \sin \dfrac{-\pi}{2}\right) = -i\sqrt{3}$

48. $\cos(3\theta) + i \sin(3\theta) = (\cos \theta + i \sin \theta)^3$
$= \cos^3 \theta + 3i \cos^2 \theta \sin \theta - 3 \cos \theta \sin^2 \theta - i \sin^3 \theta$
Thus
$$\cos(3\theta) = \cos^3 \theta - 3 \cos \theta \sin^2 \theta = 4 \cos^3 \theta - 3 \cos \theta$$
$$\sin(3\theta) = 3 \cos^2 \theta \sin \theta - \sin^3 \theta = 3 \sin \theta - 4 \sin^3 \theta.$$

50. If $z = w = -1$, then $zw = 1$, so $\sqrt{zw} = 1$. But if we use $\sqrt{z} = \sqrt{-1} = i$ and the same value for $\sqrt{w}$, then $\sqrt{z}\sqrt{w} = i^2 = -1 \ne \sqrt{zw}$.

52. The three cube roots of $-8i = 8\left(\cos \dfrac{3\pi}{2} + i \sin \dfrac{3\pi}{2}\right)$ are of the form $2(\cos \theta + i \sin \theta)$ where $\theta = \pi/2$, $\theta = 7\pi/6$, and $\theta = 11\pi/6$. Thus they are
$$2i, \quad -\sqrt{3} - i, \quad \sqrt{3} - i.$$

54. The four fourth roots of $4 = 4(\cos 0 + i \sin 0)$ are of the form $\sqrt{2}(\cos \theta + i \sin \theta)$ where $\theta = 0$, $\theta = \pi/2$, π, and $\theta = 3\pi/2$. Thus they are $\sqrt{2}$, $i\sqrt{2}$, $-\sqrt{2}$, and $-i\sqrt{2}$.

56. The equation $z^5 + a^5 = 0$ ($a > 0$) has solutions that are the five fifth roots of $-a^5 = a(\cos \pi + i \sin \pi)$; they are of the form $a(\cos \theta + i \sin \theta)$, where $\theta = \pi/5$, $3\pi/5$, π, $7\pi/5$, and $9\pi/5$.

Appendix II.   Continuous Functions (page A-17)

2. To be proved: If $f(x) \leq K$ on $[a, b)$ and $(b, c]$, and if $\lim_{x \to b} f(x) = L$, then $L \leq K$.

Proof: If $L > K$, then let $\epsilon = (L - K)/2$; thus $\epsilon > 0$. There exists $\delta > 0$ such that $\delta < b - a$ and $\delta < c - b$, and such that if $0 < |x - b| < \delta$, then $|f(x) - L| < \epsilon$. In this case

$$f(x) > L - \epsilon = L - \frac{L - K}{2} > K,$$

which contradicts the fact that $f(x) \leq K$ on $[a, b)$ and $(b, c]$. Therefore $L \leq K$.

4.   a) Let $f(x) = C$, $g(x) = x$. Let $\epsilon > 0$ be given and let $\delta = \epsilon$. For any real number x, if $|x - a| < \delta$, then

$$|f(x) - f(a)| = |C - C| = 0 < \epsilon,$$
$$|g(x) - g(a)| = |x - a| < \delta = \epsilon.$$

Thus $\lim_{x \to a} f(x) = f(a)$ and $\lim_{x \to a} g(x) = g(a)$, and f and g are both continuous at every real number a.

6. If P and Q are polynomials, they are continuous everywhere by Exercise 5. If $Q(a) \neq 0$, then $\lim_{x \to a} \frac{P(x)}{Q(x)} = \frac{P(a)}{Q(a)}$ by Theorem 1(a). Hence P/Q is continuous everywhere except at the zeros of Q.

8. By Exercise 5, x^m is continuous everywhere. By Exercise 7, $x^{1/n}$ is continuous at each $a > 0$. Thus for $a > 0$ we have

$$\lim_{x \to a} x^{m/n} = \lim_{x \to a} \left(x^{1/n} \right)^m = \left(\lim_{x \to a} x^{1/n} \right)^m$$
$$= (a^{1/n})^m = a^{m/n},$$

and $x^{m/n}$ is continuous at each positive number.

10. Let $\epsilon > 0$ be given. Let $\delta = \epsilon$. If a is any real number then

$$\big||x| - |a|\big| \leq |x - a| < \epsilon \quad \text{if} \quad |x - a| < \delta.$$

Thus $\lim_{x \to a} |x| = |a|$, and the absolute value function is continuous at every real number.

12. The proof that cos is continuous everywhere is almost identical to that for sin in Exercise 11.

14. Let a be any real number, and let $\epsilon > 0$ be given. Assume (making ϵ smaller if necessary) that $\epsilon < e^a$. Since

$$\ln \left(1 - \frac{\epsilon}{e^a} \right) + \ln \left(1 + \frac{\epsilon}{e^a} \right) = \ln \left(1 - \frac{\epsilon^2}{e^{2a}} \right) < 0,$$

we have $\ln \left(1 + \frac{\epsilon}{e^a} \right) < - \ln \left(1 - \frac{\epsilon}{e^a} \right)$.

Let $\delta = \ln \left(1 + \frac{\epsilon}{e^a} \right)$. If $|x - a| < \delta$, then

$$\ln \left(1 - \frac{\epsilon}{e^a} \right) < x - a < \ln \left(1 + \frac{\epsilon}{e^a} \right)$$
$$1 - \frac{\epsilon}{e^a} < e^{x-a} < 1 + \frac{\epsilon}{e^a}$$
$$\left| e^{x-a} - 1 \right| < \frac{\epsilon}{e^a}$$
$$|e^x - e^a| = e^a |e^{x-a} - 1| < \epsilon.$$

Thus $\lim_{x \to a} e^x = e^a$ and e^x is continuous at every point a in its domain.

16. Let $g(t) = \dfrac{t}{1 + |t|}$. For $t \neq 0$ we have

$$g'(t) = \frac{1 + |t| - t \operatorname{sgn} t}{(1 + |t|)^2} = \frac{1 + |t| - |t|}{(1 + |t|)^2} = \frac{1}{(1 + |t|)^2} > 0.$$

If $t = 0$, g is also differentiable, and has derivative 1:

$$g'(0) = \lim_{h \to 0} \frac{g(h) - g(0)}{h} = \lim_{h \to 0} \frac{1}{1 + |h|} = 1.$$

Thus g is continuous and increasing on $\mathbb{R}$. If f is continuous on $[a, b]$, then

$$h(x) = g\big(f(x) \big) = \frac{f(x)}{1 + |f(x)|}$$

is also continuous there, being the composition of continuous functions. Also, $h(x)$ is bounded on $[a, b]$, since

$$\left| g\big(f(x) \big) \right| \leq \frac{|f(x)|}{1 + |f(x)|} \leq 1.$$

By assumption in this problem, $h(x)$ must assume maximum and minimum values; there exist c and d in $[a, b]$ such that

$$g\big(f(c) \big) \leq g\big(f(x) \big) \leq g\big(f(d) \big)$$

for all x in $[a, b]$. Since g is increasing, so is its inverse g^{-1}. Therefore

$$f(c) \leq f(x) \leq f(d)$$

for all x in $[a, b]$, and f is bounded on that interval.

Appendix III. The Riemann Integral
(page A-23)

2. $f(x) = \begin{cases} 1 & \text{if } x = 1/n \quad (n = 1, 2, 3, \ldots) \\ 0 & \text{otherwise} \end{cases}$

If P is any partition of $[0, 1]$ then $L(f, P) = 0$. Let $0 < \epsilon \le 2$. Let N be an integer such that $N + 1 > \dfrac{2}{\epsilon} \ge N$. A partition P of $[0, 1]$ can be constructed so that the first two points of P are 0 and $\dfrac{\epsilon}{2}$, and such that each of the N points $\dfrac{1}{n}$ $(n = 1, 2, 3, \ldots, n)$ lies in a subinterval of P having length at most $\dfrac{\epsilon}{2N}$. Since every number $\dfrac{1}{n}$ with n a positive integer lies either in $\left[0, \dfrac{\epsilon}{2}\right]$ or one of these other N subintervals of P, and since $\max f(x) = 1$ for these subintervals and $\max f(x) = 0$ for all other subintervals of P, therefore $U(f, P) \le \dfrac{\epsilon}{2} + N\dfrac{\epsilon}{2N} = \epsilon$. By Theorem 3, f is integrable on $[0, 1]$. Evidently

$$\int_0^1 f(x)\, dx = \text{least upper bound } L(f, P) = 0.$$

4. Suppose, to the contrary, that $I_* > I^*$. Let $\epsilon = \dfrac{I_* - I^*}{3}$, so $\epsilon > 0$. By the definition of I_* and I^*, there exist partitions P_1 and P_2 of $[a, b]$, such that $L(f, P_1) \ge I_* - \epsilon$ and $U(f, P_2) \le I^* + \epsilon$. By Theorem 2, $L(f, P_1) \le U(f, P_2)$, so

$$3\epsilon = I_* - I^* \le L(f, P_1) + \epsilon - U(f, P_2) + \epsilon \le 2\epsilon.$$

Since $\epsilon > 0$, it follows that $3 \le 2$. This contradiction shows that we must have $I_* \le I^*$.

6. Let $\epsilon > 0$ be given. Let $\delta = \epsilon^2/2$. Let $0 \le x \le 1$ and $0 \le y \le 1$. If $x < \epsilon^2/4$ and $y < \epsilon^2/4$ then $|\sqrt{x} - \sqrt{y}| \le \sqrt{x} + \sqrt{y} < \epsilon$.
If $|x - y| < \delta$ and either $x \ge \epsilon^2/4$ or $y \ge \epsilon^2/4$ then

$$|\sqrt{x} - \sqrt{y}| = \frac{|x - y|}{\sqrt{x} + \sqrt{y}} < \frac{2}{\epsilon} \times \frac{\epsilon^2}{2} = \epsilon.$$

Thus $f(x) = \sqrt{x}$ is uniformly continuous on $[0, 1]$.

8. Suppose that $|f(x)| \le K$ on $[a, b]$ (where $K > 0$), and that f is integrable on $[a, b]$. Let $\epsilon > 0$ be given, and let $\delta = \epsilon/K$. If x and y belong to $[a, b]$ and $|x - y| < \delta$, then

$$|F(x) - F(y)| = \left| \int_a^x f(t)\, dt - \int_a^y f(t)\, dt \right|$$

$$= \left| \int_y^x f(t)\, dt \right| \le K|x - y| < K\frac{\epsilon}{K} = \epsilon.$$

(See Theorem 3(f) of Section 6.4.) Thus F is uniformly continuous on $[a, b]$.

Appendix IV. Differential Equations
(page A-38)

2. $\dfrac{d^2 y}{dx^2} + x = y$: 2nd order, linear, nonhomogeneous.

4. $y''' + xy' = x \sin x$: 3rd order, linear, nonhomogeneous.

6. $y'' + 4y' - 3y = 2y^2$: 2nd order, nonlinear.

8. $\cos x \dfrac{dx}{dt} + x \sin t = 0$: 1st order, nonlinear, homogeneous.

10. $x^2 y'' + e^x y' = \dfrac{1}{y}$: 2nd order, nonlinear.

12. If $y = e^x$, then $y'' - y = e^x - e^x = 0$; if $y = e^{-x}$, then $y'' - y = e^{-x} - e^{-x} = 0$. Thus e^x and e^{-x} are both solutions of $y'' - y = 0$. Since $y'' - y = 0$ is linear and homogeneous, any function of the form

$$y = Ae^x + Be^{-x}$$

is also a solution. Thus $\cosh x = \frac{1}{2}(e^x + e^{-x})$ is a solution, but neither $\cos x$ nor x^e is a solution.

14. Given that $y_1 = e^{kx}$ is a solution of $y'' - k^2 y = 0$, we suspect that $y_2 = e^{-kx}$ is also a solution. This is easily verified since

$$y_2'' - k^2 y_2 = k^2 e^{-kx} - k^2 e^{-kx} = 0.$$

Since the DE is linear and homogeneous,

$$y = Ay_1 + By_2 = Ae^{kx} + Be^{-kx}$$

is a solution for any constants A and B. It will satisfy

$$0 = y(1) = Ae^k + Be^{-k}$$
$$2 = y'(1) = Ake^k - Bke^{-k},$$

provided $A = e^{-k}/k$ and $B = -e^k/k$. The required solution is

$$y = \frac{1}{k}e^{k(x-1)} - \frac{1}{k}e^{-k(x-1)}.$$

16. $y = e^{rx}$ is a solution of the equation $y'' - y' - 2y = 0$ if $r^2 e^{rx} - re^{rx} - 2e^{rx} = 0$, that is, if $r^2 - r - 2 = 0$. This quadratic has two roots, $r = 2$, and $r = -1$. Since the DE is linear and homogeneous, the function $y = Ae^{2x} + Be^{-x}$ is a solution for any constants A and B. This solution satisfies

$$1 = y(0) = A + B, \quad 2 = y'(0) = 2A - B,$$

provided $A = 1$ and $B = 0$. Thus, the required solution is $y = e^{2x}$.

18. If $y = y_1(x) = -e$, then $y_1' = 0$ and $y_1'' = 0$. Thus $y_1'' - y_1 = 0 + e = e$. By Exercise 12 we know that $y_2 = Ae^x + Be^{-x}$ satisfies the homogeneous DE $y'' - y = 0$. Therefore, by Theorem 2,

$$y = y_1(x) + y_2(x) = -e + Ae^x + Be^{-x}$$

is a solution of $y'' - y = e$. This solution satisfies

$$0 = y(1) = Ae + \frac{B}{e} - e, \quad 1 = y'(1) = Ae - \frac{B}{e},$$

provided $A = (e+1)/(2e)$ and $B = e(e-1)/2$. Thus the required solution is $y = -e + \frac{1}{2}(e+1)e^{x-1} + \frac{1}{2}(e-1)e^{1-x}$.

20. $\dfrac{dy}{dx} = \dfrac{xy}{x^2 + 2y^2}$ Let $y = vx$

$$v + x\frac{dv}{dx} = \frac{vx^2}{(1 + 2v^2)x^2}$$

$$x\frac{dv}{dx} = \frac{v}{1 + 2v^2} - v = -\frac{2v^3}{1 + 2v^2}$$

$$\int \frac{1 + 2v^2}{v^3}\,dv = -2\int \frac{dx}{x}$$

$$-\frac{1}{2v^2} + 2\ln|v| = -2\ln|x| + C_1$$

$$-\frac{x^2}{2y^2} + 2\ln|y| = C_1$$

$$x^2 - 4y^2\ln|y| = Cy^2.$$

22. $\dfrac{dy}{dx} = \dfrac{x^3 + 3xy^2}{3x^2y + y^3}$ Let $y = vx$

$$v + x\frac{dv}{dx} = \frac{x^3(1 + 3v^2)}{x^3(3v + v^3)}$$

$$x\frac{dv}{dx} = \frac{1 + 3v^2}{3v + v^3} - v = \frac{1 - v^4}{v(3 + v^2)}$$

$$\int \frac{(3 + v^2)v\,dv}{1 - v^4} = \int \frac{dx}{x} \qquad \begin{aligned} &\text{Let } u = v^2 \\ &du = 2v\,dv \end{aligned}$$

$$\frac{1}{2}\int \frac{3 + u}{1 - u^2}\,du = \ln|x| + C_1$$

$$\frac{3}{4}\ln\left|\frac{u+1}{u-1}\right| - \frac{1}{4}\ln|1 - u^2| = \ln|x| + C_1$$

$$3\ln\left|\frac{y^2 + x^2}{y^2 - x^2}\right| - \ln\left|\frac{x^4 - y^4}{x^4}\right| = 4\ln|x| + C_2$$

$$\ln\left|\left(\frac{x^2 + y^2}{x^2 - y^2}\right)^3 \frac{1}{x^4 - y^4}\right| = C_2$$

$$\ln\left|\frac{(x^2 + y^2)^2}{(x^2 - y^2)^4}\right| = C_2$$

$$x^2 + y^2 = C(x^2 - y^2)^2.$$

24. $\dfrac{dy}{dx} = \dfrac{y}{x} - e^{-y/x}$ (let y=vx)

$$v + x\frac{dv}{dx} = v - e^{-v}$$

$$e^v\,dv = -\frac{dx}{x}$$

$$e^v = -\ln|x| + \ln|C|$$

$$e^{y/x} = \ln\left|\frac{C}{x}\right|$$

$$y = x\ln\ln\left|\frac{C}{x}\right|.$$

26. If $\xi = x - x_0$, $\eta = y - y_0$, and

$$\frac{dy}{dx} = \frac{ax + by + c}{ex + fy + g},$$

then

$$\begin{aligned} \frac{d\eta}{d\xi} = \frac{dy}{dx} &= \frac{a(\xi + x_0) + b(\eta + y_0) + c}{e(\xi + x_0) + f(\eta + y_0) + g} \\ &= \frac{a\xi + b\eta + (ax_0 + by_0 + c)}{e\xi + f\eta + (ex_0 + fy_0 + g)} \\ &= \frac{a\xi + b\eta}{e\xi + f\eta} \end{aligned}$$

provided x_0 and y_0 are chosen such that

$$ax_0 + by_0 + c = 0, \quad \text{and} \quad ex_0 + fy_0 + g = 0.$$

28. $(xy^2 + y)\,dx + (x^2y + x)\,dy = 0$

$$d\left(\frac{1}{2}x^2y^2 + xy\right) = 0$$

$$x^2y^2 + 2xy = C.$$

30. $e^{xy}(1 + xy)\,dx + x^2e^{xy}\,dy = 0$

$$d(xe^{xy}) = 0 \quad \Rightarrow \quad xe^{xy} = C.$$

32. $(x^2 + 2y)\,dx - x\,dy = 0$

$$M = x^2 + 2y, \qquad N = -x$$

$$\frac{1}{N}\left(\frac{\partial M}{\partial y} - \frac{\partial N}{\partial x}\right) = -\frac{3}{x} \quad \text{(indep. of } y)$$

$$\frac{d\mu}{\mu} = -\frac{3}{x}\,dx \quad \Rightarrow \quad \mu = \frac{1}{x^3}$$

$$\left(\frac{1}{x} + \frac{2y}{x^3}\right)dx - \frac{1}{x^2}\,dy = 0$$

$$d\left(\ln|x| - \frac{y}{x^2}\right) = 0$$

$$\ln|x| - \frac{y}{x^2} = C_1$$

$$y = x^2\ln|x| + Cx^2.$$

34. If $\mu(y)M(x, y)\,dx + \mu(y)N(x, y)\,dy$ is exact, then

$$\frac{\partial}{\partial y}\big(\mu(y)M(x, y)\big) = \frac{\partial}{\partial x}\big(\mu(y)N(x, y)\big)$$

$$\mu'(y)M + \mu\frac{\partial M}{\partial y} = \mu\frac{\partial N}{\partial x}$$

$$\frac{\mu'}{\mu} = \frac{1}{M}\left(\frac{\partial N}{\partial x} - \frac{\partial M}{\partial y}\right).$$

Thus M and N must be such that

$$\frac{1}{M}\left(\frac{\partial N}{\partial x} - \frac{\partial M}{\partial y}\right)$$

depends only on y.

36. Consider $y\,dx - (2x + y^3e^y)\,dy = 0$.
Here $M = y$, $N = -2x - y^3e^y$, $\dfrac{\partial M}{\partial y} = 1$, and $\dfrac{\partial N}{\partial x} = -2$.
Thus

$$\frac{\mu'}{\mu} = -\frac{3}{y} \quad\Rightarrow\quad \mu = \frac{1}{y^3}$$

$$\frac{1}{y^2}\,dx - \left(\frac{2x}{y^3} + e^y\right)\,dy = 0$$

$$d\left(\frac{x}{y^2} - e^y\right) = 0$$

$$\frac{x}{y^2} - e^y = C, \quad\text{or}\quad x - y^2e^y = Cy^2.$$

38. For $\left(x\cos x + \dfrac{y^2}{x}\right)dx - \left(\dfrac{x\sin x}{y} + y\right)dy$ we have

$$M = x\cos x + \frac{y^2}{x}, \qquad N = -\frac{x\sin x}{y} - y$$

$$\frac{\partial M}{\partial y} = \frac{2y}{x}, \qquad \frac{\partial N}{\partial x} = -\frac{\sin x}{y} - \frac{x\cos x}{y}$$

$$\frac{\partial N}{\partial x} - \frac{\partial M}{\partial y} = -\left(\frac{\sin x}{y} + \frac{x\cos x}{y} + \frac{2y}{x}\right)$$

$$xM - yN = x^2\cos x + y^2 + x\sin x + y^2$$

$$\frac{1}{xM - yN}\left(\frac{\partial N}{\partial x} - \frac{\partial M}{\partial y}\right) = -\frac{1}{xy}.$$

Thus, an integrating factor is given by

$$\frac{\mu'(t)}{\mu(t)} = -\frac{1}{t} \quad\Rightarrow\quad \mu(t) = \frac{1}{t}.$$

We multiply the original equation by $1/(xy)$ to make it exact:

$$\left(\frac{\cos x}{y} + \frac{y}{x^2}\right)dx - \left(\frac{\sin x}{y^2} + \frac{1}{x}\right)dy = 0$$

$$d\left(\frac{\sin x}{y} - \frac{y}{x}\right) = 0$$

$$\frac{\sin x}{y} - \frac{y}{x} = C.$$

The solution is $x\sin x - y^2 = Cxy$.

A computer spreadsheet was used in Exercises 39–45. The intermediate results appearing in the spreadsheet are not shown in these solutions.

40. We start with $x_0 = 1$, $y_0 = 0$, and calculate

$$x_{n+1} = x_n + h, \qquad u_{n+1} = y_n + h(x_n + y_n)$$

$$y_{n+1} = y_n + \frac{h}{2}(x_n + y_n + x_{n+1} + u_{n+1}).$$

a) For $h = 0.2$ we get $x_5 = 2$, $y_5 = 2.405416$.

b) For $h = 0.1$ we get $x_{10} = 2$, $y_{10} = 2.428162$.

c) For $h = 0.05$ we get $x_{20} = 2$, $y_{20} = 2.434382$.

42. We start with $x_0 = 0$, $y_0 = 0$, and calculate

$$x_{n+1} = x_n + h, \qquad y_{n+1} = hx_n e^{-y_n}.$$

a) For $h = 0.2$ we get $x_{10} = 2$, $y_{10} = 1.074160$.

b) For $h = 0.1$ we get $x_{20} = 2$, $y_{20} = 1.086635$.

44. We start with $x_0 = 0$, $y_0 = 0$, and calculate

$$x_{n+1} = x_n + h$$

$$p_n = x_n e^{-y_n}$$

$$q_n = \left(x_n + \frac{h}{2}\right)e^{-(y_n + (h/2)p_n)}$$

$$r_n = \left(x_n + \frac{h}{2}\right)e^{-(y_n + (h/2)q_n)}$$

$$s_n = (x_n + h)e^{-(y_n + hr_n)}$$

$$y_{n+1} = y_n + \frac{h}{6}(p_n + 2q_n + 2r_n + s_n).$$

a) For $h = 0.2$ we get $x_{10} = 2$, $y_{10} = 1.098614$.

b) For $h = 0.1$ we get $x_{20} = 2$, $y_{20} = 1.098612$.

46. $u(x) = 1 + 3\displaystyle\int_2^x t^2 u(t)\,dt$

$$\frac{du}{dx} = 3x^2u(x), \qquad u(2) = 1 + 0 = 1$$

$$\frac{du}{u} = 3x^2\,dx \quad\Rightarrow\quad \ln u = x^3 + C$$

$$0 = \ln 1 = \ln u(2) = 2^3 + C \quad\Rightarrow\quad C = -8$$

$$u = e^{x^3 - 8}.$$

48. If $\phi(0) = A \geq 0$ and $\phi'(x) \geq k\phi(x)$ on an interval $[0, X]$, where $k > 0$ and $X > 0$, then

$$\frac{d}{dx}\left(\frac{\phi(x)}{e^{kx}}\right) = \frac{e^{kx}\phi'(x) - ke^{kx}\phi(x)}{e^{2kx}} \geq 0.$$

Thus $\phi(x)/e^{kx}$ is increasing on $[0, X]$. Since its value at $x = 0$ is $\phi(0) = A \geq 0$, therefore $\phi(x)/e^{kx} \geq A$ on $[0, X]$, and $\phi(x) \geq Ae^{kx}$ there.